INTRODUCTION TO DYNAMIC SYSTEMS ANALYSIS

McGraw-Hill Series in Mechanical Engineering

Consulting Editors

Jack P. Holman, *Southern Methodist University*
John R. Lloyd, *Michigan State University*

Anderson: *Modern Compressible Flow: With Historical Perspective*
Arora: *Introduction to Optimum Design*
Bray and Stanley: *Nondestructive Evaluation: A Tool for Design, Manufacturing, and Service*
Burton: *Introduction to Dynamic Systems Analysis*
Culp: *Principles of Energy Conversion*
Dally: *Packaging of Electronic Systems: A Mechanical Engineering Approach*
Dieter: *Engineering Design: A Materials and Processing Approach*
Eckert and Drake: *Analysis of Heat and Mass Transfer*
Edwards and McKee: *Fundamentals of Mechanical Component Design*
Gebhart: *Heat Conduction and Mass Diffusion*
Gibson: *Principles of Composite Material Mechanics*
Hamrock: *Fundamentals of Fluid Film Lubrication*
Heywood: *Internal Combustion Engine Fundamentals*
Hinze: *Turbulence*
Holman: *Experimental Methods for Engineers*
Howell and Buckius: *Fundamentals of Engineering Thermodynamics*
Hutton: *Applied Mechanical Vibrations*
Juvinall: *Engineering Considerations of Stress, Strain, and Strength*
Kane and Levinson *Dynamics: Theory and Applications*
Kays and Crawford: *Convective Heat and Mass Transfer*
Kelly: *Fundamentals of Mechanical Vibrations*
Kimbrell: *Kinematics Analysis and Synthesis*
Martin: *Kinematics and Dynamics of Machines*
Modest: *Radiative Heat Transfer*
Norton: *Design of Machinery*
Phelan: *Fundamentals of Mechanical Design*
Raven: *Automatic Control Engineering*
Reddy: *An Introduction to the Finite Element Method*
Rosenberg and Karnopp: *Introduction to Physical Systems Dynamics*
Schlichting: *Boundary-Layer Theory*
Shames: *Mechanics of Fluids*
Sherman: *Viscous Flow*
Shigley: *Kinematic Analysis of Mechanisms*
Shigley and Mischke: *Mechanical Engineering Design*
Shigley and Uicker: *Theory of Machines and Mechanisms*
Stiffler: *Design with Microprocessors for Mechanical Engineers*
Stoecker and Jones: *Refrigeration and Air Conditioning*
Ullman: *The Mechanical Design Process*
Vanderplaats: *Numerical Optimization: Techniques for Engineering Design, with Applications*
White: *Viscous Fluid Flow*
Zeid: *CAD/CAM Theory and Practice:*

McGraw-Hill Series in Aeronautical and Aerospace Engineering

Consulting Editor

***John D. Anderson, Jr.**, University of Maryland*

Anderson: *Fundamentals of Aerodynamics*
Anderson: *Hypersonic and High Temperature Gas Dynamics*
Anderson: *Introduction to Flight*
Anderson: *Modern Compressible Flow: With Historical Perspective*
Burton: *Introduction to Dynamic Systems Analysis*
D'Azzo and Houpis: *Linear Control System Analysis and Design*
Donaldson: *Analysis of Aircraft Structures: An Introduction*
Gibson: *Principles of Composite Material Mechanics*
Kane, Likins, and Levinson: *Spacecraft Dynamics*
Katz and Plotkin: *Low-Speed Aerodynamics: From Wing Theory to Panel Methods*
Nelson: *Flight Stability and Automatic Control*
Peery and Azar: *Aircraft Structures*
Rivello: *Theory and Analysis of Flight Structures*
Schlichting: *Boundary Layer Theory*
White: *Viscous Fluid Flow*
Wiesel: *Spaceflight Dynamics*

Also Available from McGraw-Hill

Schaum's Outline Series in Mechanical Engineering

Most outlines include basic theory, definitions, hundreds of example problems solved in step-by-step detail, and supplementary problems with answers.

Related titles on the current list include:

Acoustics
Continuum Mechanics
Elementary Statics & Strength of Materials
Engineering Economics
Engineering Mechanics
Fluid Dynamics
Fluid Mechanics & Hydraulics
Heat Transfer
Langrangian Dynamics
Machine Design
Mathematical Handbook of Formulas & Tables
Mechanical Vibrations
Operations Research
Statics & Mechanics of Materials
Strength of Materials
Theoretical Mechanics
Thermodynamics for Engineers
Thermodynamics with Chemical Applications

Schaum's Solved Problems Books

Each title in this series is a complete and expert source of solved problems with solutions worked out in step-by-step detail.

Related titles on the current list include:

3000 Solved Problems in Calculus
2500 Solved Problems in Differential Equations
2500 Solved Problems in Fluid Mechanics & Hydraulics
1000 Solved Problems in Heat Transfer
3000 Solved Problems in Linear Algebra
2000 Solved Problems in Mechanical Engineering Thermodynamics
2000 Solved Problems in Numerical Analysis
700 Solved Problems in Vector Mechanics for Engineers: Dynamics
800 Solved Problems in Vector Mechanics for Engineers: Statics

Available at most college bookstores, or for a complete list of titles and prices, write to: Schaum Division
McGraw-Hill, Inc.
Princeton Road, S-1
Hightstown, NJ 08520

INTRODUCTION TO DYNAMIC SYSTEMS ANALYSIS

T. D. Burton

Washington State University

McGraw-Hill, Inc.

New York St. Louis San Francisco Auckland Bogotá
Caracas Lisbon London Madrid Mexico City Milan
Montreal New Delhi San Juan Singapore Sydney Tokyo Toronto

This book was set in Times Roman by Science Typographers, Inc.
The editors were John J. Corrigan and Eleanor Castellano;
the production supervisor was Louise Karam.
The cover was designed by Carla Bauer.
R. R. Donnelley & Sons Company was printer and binder.

INTRODUCTION TO DYNAMIC SYSTEMS ANALYSIS

This book is printed on recycled, acid-free paper containing a minimum of
50% total recycled fiber with 10% postconsumer de-inked fiber.

1 2 3 4 5 6 7 8 9 0 D O C D O C 9 0 9 8 7 6 5 4

ISBN 0-07-009290-7

Library of Congress Cataloging-in-Publication Data

Burton, T. D. (Thomas D.)
Introduction to dynamic systems analysis / T. D. Burton
p. cm.—(McGraw-Hill series in mechanical engineering)
Includes bibliographical references and index.
ISBN 0-07-009290-7
1. Systems engineering. 2. Dynamics. 3. Linear Systems.
I. Title. II. Series.
TA168.B87 1994
620'.001'1—dc20 93-32113

ABOUT THE AUTHOR

T. D. Burton is Professor of Mechanical Engineering at Washington State University, Pullman, WA. He received his B.S. in Engineering from Caltech and his Ph.D. degree in Mechanical Engineering and Applied Mechanics from the University of Pennsylvania. Professor Burton was an aerospace engineer with the General Electric Company, Missile and Space Division, from 1969–1977. Since 1977 he has been at Washington State University, where his teaching and research have been in the areas of dynamics, vibrations, finite elements, and nonlinear systems.

TO SUE

CONTENTS

PREFACE

This book is intended for junior- or senior-level students in mechanical engineering, aerospace engineering, and engineering science/mechanics. The book is intended to form the basis of a one-quarter or one-semester course in dynamic systems analysis or a two-quarter or two-semester sequence in dynamic systems, vibrations, and controls.

The content of courses in "dynamic systems," as currently taught at the undergraduate level in the United States varies quite a bit from university to university and from instructor to instructor. This appears to be because such courses are relatively new, and standard course content is not as well established as in more classical areas such as strength of materials, fluid mechanics, and thermodynamics. Viewpoints on the orientation of the subject also differ, depending, for example, on whether one prefers to emphasize control systems, use of bond graphs in modeling, basic dynamic systems theory, and so on. It is my view that mechanically oriented engineers well trained in "dynamic systems" should have a working knowledge of eight basic areas: (1) modeling, (2) dynamics, (3) basic systems concepts such as equilibrium, stability, and linearization, (4) linear system theory, (5) mechanical vibrations, (6) control systems, (7) nonlinear system behavior, and (8) computer usage in dynamic systems analysis. These topics constitute the material of this book. Not all of this material can be covered in a single quarter or semester course. A one-quarter or one-semester course, therefore, would depend on the instructor's own view of what is important, on the preparation of the entering students, and on which topics are covered elsewhere in the curriculum.

In viewing this book as a prospective text, the following four comments need to be taken into account:

1. First, I have attempted to approach the subject from a fairly general viewpoint. This reflects the modern trend in dynamic systems analysis, wherein one tries to understand certain common features exhibited by different dynamic systems arising from a variety of physical phenomena. This trend toward generalization may become more pronounced in the future because of the increasing complexity and diversity of the types of dynamic systems with which engineers must deal.
2. The characterization and solution of linear dynamic system models is based on eigenvalue/eigenvector methods and motions in the state space, rather

than on Laplace transformation. Laplace transformation is introduced and used heavily in Chap. 8 on controls but is not otherwise viewed as having special status in relation to the basic concepts of the subject.

3. Although the emphasis in the book is properly on linear systems, linear system models are viewed, in general, as approximations to some nonlinear model that actually governs the behavior of the system. This viewpoint is taken because nonlinearities are a fact of life with which practicing engineers must deal. Furthermore, in engineering practice, the rather complex nonlinear models that often arise are analyzed almost exclusively via computer simulation. Interpretation of numerically obtained results requires an understanding of basic nonlinear phenomena such as limit cycles, nonlinear resonances, and the possibility of multiple steady-state responses.
4. I have incorporated a number of topics which, 30 years ago, might have been considered a bit exotic for undergraduates but which today are routine in industrial applications: (a) an emphasis on the state or phase space approach to describing system behavior, (b) the energy methods of dynamics (Lagrange's equations), (c) spectrum analysis, (d) modal analysis of vibratory systems, (e) finite element applications in structural dynamics, and (f) nonlinear system behavior. Space limitations have precluded more than an introduction of the basic ideas, which has been attempted in a straightforward way.

The organization of the book is described below, chapter by chapter. Suggested coverages for courses or course sequences of various lengths and with various emphases are then discussed.

Chapter 1 provides a short introduction to basic modeling methodology and some review and classification of ordinary differential equations. The objective here is to introduce the student to the modeling approach used in this book: application of simplifying assumptions, physical laws, and constitutive relations to develop ordinary differential equation models of dynamic systems.

Chapter 2 is concerned with the construction of ODE models of systems arising out of newtonian mechanics. Basic dynamics of particles and rigid bodies are reviewed, along with work and energy. Constitutive models used to describe forces and moments that act on mechanical systems are then discussed. Finally, the modeling process is illustrated, mainly by example, for newtonian systems of varying complexity.

Chapter 3 considers modeling of mechanically oriented systems for which some or all of the loadings arise due to hydraulic, pneumatic, electronic, or electromagnetic effects. The treatment here is not exhaustive and is intended mainly to illustrate how such effects can be used to exert forces and moments to move mechanical objects. At the close of this chapter examples of nonengineering systems are considered to illustrate how ODE models may arise in such systems.

Chapter 4 covers the essential ideas of dynamic systems analysis and is perhaps the most important chapter of the book. The basic concepts of visualization of motion in the phase space, equilibrium, stability of equilibrium and linearization of the governing ODE model are first discussed (Sec. 4.1). This is followed by use of eigenvalue/eigenvector methods to solve the autonomous linear models that describe the near-equilibrium behavior of dynamic systems (Sec. 4.2). The chapter concludes with some additional topics that fill out the material covered in the first two sections.

Chapter 5 considers the response of linear systems to external excitations or inputs. Basic ideas such as transient and steady-state response are first described. Then the response of first- and second-order systems to step and harmonic excitations is analyzed. The chapter closes with introduction of frequency response methods and Bode plots as a way to characterize and analyze systems of the first, second, and higher orders.

The first five chapters described in the preceding paragraphs are intended to provide coverage of the essential topics in a dynamic systems course. The material in the remaining five chapters is intended for use selectively, at the discretion of the instructor, to cover additional relevant topics.

Chapter 6 covers two topics not normally included in a book of this type: particle dynamics in three dimensions using rotating coordinate systems and an introduction to lagrangian mechanics. These topics have been included because of their practical utility and because students exposed only to the traditional introductory course in dynamics often do not have a sufficient grasp of the fundamentals of the subject. In particular, the focus of introductory courses on two-dimensional motions causes many students to have difficulty when confronted with three-dimensional systems undergoing rotational motions.

Chapter 7 presents the basics of single and multi-degree of freedom free and forced vibration of simple linear systems. The forced multi-degree of freedom problem is analyzed by introducing modal analysis, which nowadays is indispensable to the practicing structural dynamicist. The chapter closes with a section on spectrum analysis, another area important to engineers who do analysis of experimental data.

Chapter 8 presents an introduction to classical linear control theory. After some introductory material on the basic elements of the controls problem (measurement, feedback, error determination, and actuation), a summary of Laplace transformation is provided. Then transfer functions and block diagrams are introduced. This is followed by a discussion of performance measures and simple control strategies. The root locus and frequency domain (Bode plot) methods are then presented as a way to understand the effects of various control strategies on system response and as a way to design simple controllers in the PID family. The chapter closes with some discussion of disturbance rejection and pole placement from the state variable point of view. While this is the longest chapter in the book, the intent is not to provide a comprehensive treatment of controls but rather to introduce some of the basic ideas.

Chapter 9 covers some aspects of the numerical analysis procedures involved in the study of dynamic systems. Techniques for system identification and solution of the algebraic eigenvalue problem are considered briefly. Then numerical integration of ODEs, emphasizing the fourth-order Runge-Kutta method, is presented. Finally, an introduction to the finite element method in structural dynamics is given, based on the simple problem of the vibrating beam.

Chapter 10 considers the dynamic behavior of the single degree of freedom nonlinear oscillator, as a way to introduce some of the more common nonlinear behaviors. These include the following: (1) nonlinear damping and limit cycles, (2) the frequency-amplitude dependence due to nonlinear stiffness, (3) nonlinear resonances in the presence of a harmonic excitation, and (4) chaotic motion. The harmonic balance method is introduced as a straightforward way to analyze periodic nonlinear phenomena. The intent of this chapter is to provide some flavor as to the types of attractors that can occur in a nonlinear system.

Suggested below are possible strategies for using this book as a text in undergraduate courses of varying durations and with varying emphases:

1. One-quarter course: Chaps. 1, 2, and 3 in whatever detail is deemed necessary, with additional/alternate examples based on the instructors' own preference; Secs. 4.1 and 4.2 of Chap. 4, with topics from Sec. 4.3 as desired; Secs. 5.1, 5.2, and 5.3 of Chap. 5, with additional material from Sec. 5.4 if desired.
2. One-semester course with an emphasis on basic dynamic systems concepts: essentially all of Chaps. 1 through 5, with selections from Secs. 9.2, 9.3, 9.4, and 9.5 as time permits.
3. One-semester course emphasizing basic dynamic systems and vibrations: first 10 weeks, same as (1) above; then Chap. 7 and Sec. 9.6 as time allows.
4. One-semester course with emphasis on basic dynamic systems and controls: first 10 weeks as in (1) above; then Sec. 5.4 and selected topics from Chap. 8.
5. One-semester course with emphasis on basic dynamic systems and dynamics: first 10 weeks as in (1) above; then Chap. 6 and selected material from Chaps. 7 and 9.
6. Two-quarter sequence: After covering the basic material in Chaps. 1 through 5 in the first quarter, any two of Chaps. 6, 7, and 8 could be covered reasonably well in the second quarter, with additional material from other chapters as deemed necessary.

The preceding possibilities are intended only as guidelines since the specific curricula and orientation of the instructor will dictate the details.

ACKNOWLEDGMENTS

Many individuals contributed to the production of this text. I am particularly indebted to Jo Ann Rattey-Hicks and to Danielle Bishop for their excellent typing of the manuscript and to the following people who allowed me to use their results: Carl Baker, Maia Genaux, Zahid Rahman, John Massenburg, Pete Miles, and Mark Young. I would also like to acknowledge the following reviewers, who provided valuable suggestions during the review process: Aldo Ferri, George W. Woodruff School of Mechanical Engineering; R. Rees Fullmer, Iowa State University; Neyram Hemati, Drexel University; Ping Hsu, University of Illinois, Urbana-Champaign; R. Gordon Kirk, Virginia Polytechnic Institute and State University; K. Krishnamurthy, University of Missouri, Rolla; E. Harry Law, Clemson University; Roger W. Mayne, SUNY, Buffalo; Kevin P. Meade, Illinois Institute of Technology; Sanford G. Meek, University of Utah; Robert G. Melton, Penn State University; Simone Mola, GMI & EMI; Ali Seireg, University of Wisconsin, Madison; Andres Soom, SUNY, Buffalo; and T. M. Wu, University of Kentucky. I am grateful to John J. Corrigan and Eleanor Castellano, editors at McGraw-Hill, whose patience and encouragement were indispensable. Finally, I am grateful to my mentors Jake Abel of the University of Pennsylvania and Alan Whitman of Villanova University, who nurtured whatever clarity of thought I possess.

T. D. Burton

CHAPTER 1

INTRODUCTION TO MODELING AND ANALYSIS

1.1 INTRODUCTION

This book is intended to convey the fundamentals of modeling, analysis, and control of dynamic systems of interest to mechanically oriented engineers. By a *system* we mean any collection of interacting objects or components having certain cause-and-effect relations among them. Examples of systems of interest to us include space vehicles, aircraft, automobiles, structures, machinery, and manufacturing systems. Further, we will usually be able to identify components or subsystems that may be considered dynamic systems in their own right. In the automobile, for instance, we can identify the engine, the suspension system, the steering system, and the overall structure as systems in which dynamic behavior is important.

We want to distinguish at the outset a static versus a dynamic system. In *static systems* the descriptive properties that we use (or measure) to describe the behavior of the system, such as displacements, velocities, stresses, or temperatures, do not change with time, although they may vary spatially. Static structural analysis of the type you have seen in your strength-of-materials or statics course fits into this category. For instance, the shear Q, bending moment M, displacement v, and stress σ_x in a statically loaded beam are all functions of the location x along the neutral axis, but the time is not involved.

In a *dynamic system* the descriptive properties of the system will in general change with time, and it is the manner in which such changes occur that is of primary interest to us. For instance, if a beam or structure is subjected to loading that is time-varying, then all of the descriptors—shear, moment, stress, and displacement—will also be time-varying, and an understanding of the dynamic behavior is essential in determining whether the system will successfully perform its intended function.

We also want to note that by "dynamic systems" we do not mean just those systems which operate in accordance with newtonian mechanics. Although such systems, in which objects move in accordance with the forces and moments acting upon them, will certainly be of interest to us, we want to take a broader view, recognizing that there are many important dynamic systems which have nothing to do with newtonian mechanics. Some examples of such systems are considered in Chap. 3. One thing we will want to do eventually is to identify certain common features of dynamic behavior that are relevant for essentially all dynamic systems, regardless of their origin.

We should also be aware that most dynamic systems are capable of exhibiting static behavior, which we view as merely a special case of the more general dynamics problem. In fact, in many systems, the desired operating condition is static, and our job is to eliminate or minimize dynamic effects. Examples of this type include vibrations of machinery, engines and turbines; vibration of structures due to winds or earthquake ground motions; shimmy of automobile front ends or airplane nose gear, and so on.

1.2 BASIC STEPS IN MODELING AND ANALYSIS OF DYNAMIC SYSTEMS

Let us now summarize the basic steps we need to follow in the modeling and analysis of dynamic systems (see Fig. 1.1). First, we suppose that we have identified the particular dynamic system which is to be studied (block 1, Fig. 1.1). This may be a phenomenon that occurs naturally, such as the flow of air over a hill on or near which a structure is to be built; it may be a manufactured system that already exists and whose behavior we need to understand, modify, or control; or it may be a system that is yet to be fabricated, in which case the modeling and analysis are intended to provide guidance in its design.

Next, we have to establish some objectives (block 2). These may vary from simply understanding why the system behaves as it does (always necessary!) to ensuring that the system exhibits certain specific types of behavior. Included in this phase of the effort will be an assessment of the level of sophistication required in our model. For instance, in designing the *Apollo* spacecraft that was used to transport the astronauts to the moon and back, the overall system model had to be very sophisticated. On the other hand, if we need to design a frame structure to support a fan, with the intent of keeping the vibration due to any unbalance of the rotating parts to a minimum, then we may be able to model the system as a simple mechanical oscillator.

Taking 1 and 2 into account, we next attempt to determine which physical effects have an important influence on the system dynamics and which effects are small enough to ignore (block 3, Fig. 1.1). What we try to do, at least to

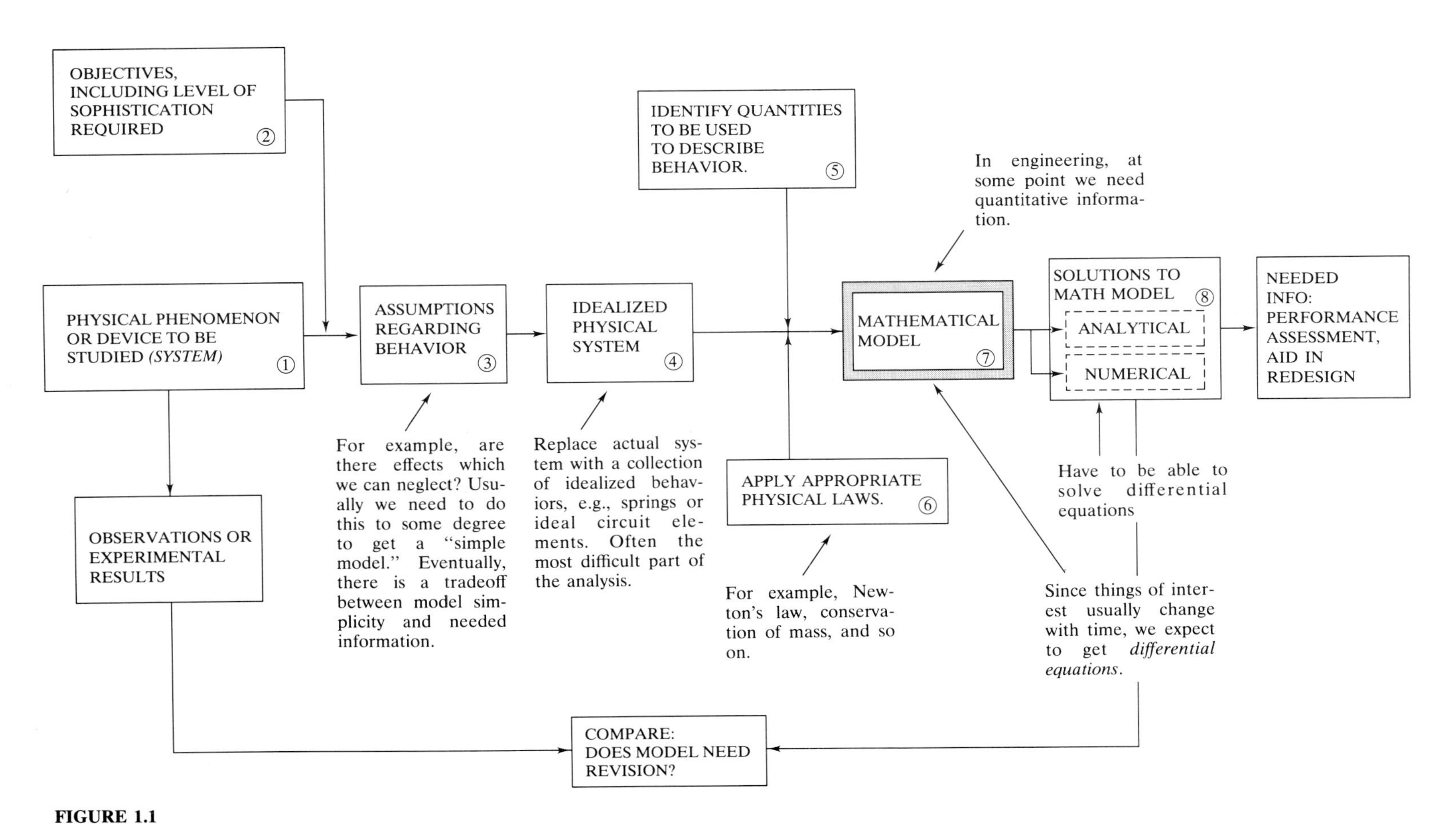

FIGURE 1.1
Some basic steps in modeling and analysis.

start, is to develop a model which is as simple as possible but which retains those effects that are of importance. Of course, there is a tradeoff between model simplicity and level of accuracy. The advantage of a simple model is that it is easier to analyze and to understand the resulting behavior. The potential disadvantage is that we may ignore something that is actually important. Often, if the system under investigation is at all complicated, we will need to construct several models of increasing complexity and then try to establish, by comparison, which is best for our purposes. With 1, 2, and 3 in mind we can now construct an idealized physical system that approximates the actual onc (block 4). This involves the use of idealized system components such as lumped masses, linear springs, and ideal circuit elements. Examples of this process are given in Chaps. 2 and 3.

At this point we are ready to bring some mathematics into the picture. We first have to identify those quantities that are needed to describe quantitatively the behavior of our idealized system (block 5). These will generally be variables such as displacements and velocities, as well as *system constants* such as masses, stiffnesses, and resistances. We will now want to begin construction of a mathematical model that will enable us to calculate how the model variables change with time in specific situations.

The actual mathematical model (block 7) is obtained through appropriate use of physical laws and constitutive relations (block 6). Laws that we often employ include the following:

1. Newton's second law of motion
2. Newton's law of gravitation
3. Conservation of mass
4. Conservation of energy (first law of thermodynamics)
5. Laws of electrodynamics (Maxwell's equations)

We will also need to employ constitutive relations. A *constitutive relation* is a statement of cause and effect connecting two or more of our descriptive variables, often algebraically. A constitutive relation is usually an approximation based on experience. Examples of constitutive relations which we typically use are the perfect gas relation ($p = \rho RT$); Ohm's "law" ($V = RI$); Hooke's "law" ($\sigma = E\varepsilon$); and the relation connecting the force F and displacement x in a linear spring, $F = -kx$ (this will be recognized as a special case of Hooke's law). Note that some of these constitutive relations enjoy the status of laws, but they are really only approximations of behavior. For instance, the perfect gas relation breaks down at very high pressure, and the remaining three relations, which are linear, may not be valid if I, ε, or x is large. In any event we now recognize the two basic types of mathematical models: (1) true laws, which we may consider to be exact and always true, and (2) constitutive relations, which are approximations, often linear, which are based on experience and which relate two or more variables that appear in the model.

Let us assume that we have applied relevant physical laws and constitutive relations and have obtained a mathematical model of the idealized physical system. This mathematical model will describe the manner in which the variables used to describe system behavior change with time. For this reason we should expect the mathematical model to consist of differential equations in which time is an independent variable and in which the descriptive variables are dependent variables (unknowns). An obvious example of this type of situation arises in the newtonian mechanics of a particle of mass m, for which Newton's second law states $\mathbf{F} = m\mathbf{a}$, where $\mathbf{F}$ is the applied force vector and $\mathbf{a}$ the resulting acceleration vector. Now, in formulating such problems, we normally use the position components of the particle (defined, for example, by the cartesian coordinates x, y, and z) as the descriptive variables. Because acceleration is the second derivative of position, Newton's law then automatically takes the form of a second-order ordinary differential equation. This fact actually was a prime motivator of the study of differential equations following Newton's discovery of the calculus, the laws of particle motion, and the law of gravitation.

Before proceeding, we need to say something about the various ways in which mathematical models are used. First, it is apparent that we are going to use the mathematical model as a conceptual substitute for the actual system. Finding solutions to the mathematical model is considered to be equivalent to measuring the response of the actual system during operation, at least to the extent that the model is accurate. We may use our mathematical models in the following ways:

1. To achieve *understanding* of the underlying behavior of the system. This is always necessary.
2. To *predict* the behavior of the actual system in a variety of circumstances (for instance, if we change a system component such as a spring).
3. As a *substitute for experimentation*. If we have confidence in our model, then we may be able to use it in lieu of often costly and time-consuming experimentation. Computer simulation to perform this function is common nowadays.
4. For *system identification*. The mathematical model will generally contain system constants such as stiffnesses, masses, geometric parameters, and other coefficients. Sometimes these quantities will not be known exactly ahead of time. What we can do is to conduct experiments and then adjust the values of the system constants so that the solutions to the math model agree with the experimental results. In this way we identify proper values for the system constants.

We are now, at least temporarily, finished with the modeling of the system. The next step in our sequence is the *analysis* of the mathematical model. This involves finding solutions to the math model in order to determine the response of the system and to understand how it behaves (by *response* we mean the manner in which our descriptive variables or unknowns change with time). It should be clear that this step will involve finding solutions to differential equations, a subject which is essential in the study of dynamic systems.

There are two main lines of attack in the analysis phase. The first involves finding exact (or approximate) solutions *analytically*: We try to find closed-form solutions to the differential equations that comprise the math model. This is often a practical approach if the system is fairly simple—for instance, if the model consists of just a few ordinary differential equations. The second approach is computer simulation; here we use some type of numerical integration scheme to integrate the governing differential equations. This may be the only practical approach if the model is at all complex. Often we attempt to use a combination of these approaches. For example, we may be able to take a fairly complex model, for which computer simulation is required, and devise a simple version of this model that can be solved analytically. From this analytical solution we may be able to infer certain important features of the response which we may then seek to define more accurately via computer simulation.

Analytical solutions are definitely to be preferred if they can be found relatively easily. Such solutions display directly, in a few algebraic relationships, not only the manner in which the descriptive variables change with time for any initial conditions but also how this response will be affected if we change a system property such as a mass, stiffness, or geometrical feature. The disadvantage of a solution obtained via numerical integration on the computer is that we merely obtain graphs and printouts of the time dependence of the dependent variables for a single specific case. Since we have no mathematical relations to describe this response, we may not be able to predict what will happen if we change one or more of the system properties; we have to do more computer simulations. Nevertheless, many, if not most, systems that have to be analyzed in the real world of engineering are sufficiently complex that it is impractical to attempt to find analytical solutions, and, for these types of systems, computer simulation is the only realistic approach. For this reason, it is essential that a thorough understanding of the basic behavior of dynamic systems be achieved. In this book we will often make use of analytical solutions for simple systems as a way to gain such understanding. What we will see is that even complicated systems usually exhibit responses that have direct counterparts in simple systems. We will look for certain underlying properties of solutions that are common to a diverse range of both simple and complex systems.

Suppose we have now modeled and analyzed our dynamic system according to the aforementioned comments. We now have to be concerned with the *validity* of the model. It must always be kept in mind that what we are really doing is using the model as a conceptual substitute for the actual physical system under study. Since virtually all system mathematical models are *approximations*, we have to compare the results of the analysis to the observed behavior of the actual system; this process may require that we revise our model. This type of model verification is often done in parallel with the modeling effort, especially to characterize the behavior (i.e., to develop constitutive relations) of the various individual components of the system. The model, once validated, will then tell us how the actual system will behave in various circumstances of operation. The model enables us to design or redesign components if the original system behaves in a way that is unacceptable or does not perform its intended function.

Portions of the preceding commentary may appear a bit hazy to the student, who has likely not gone through all of the processes described. It is suggested, therefore, that this section be reread occasionally as the student progresses through the book.

1.3 CONTROL OF DYNAMIC SYSTEMS

It is frequently necessary that we wish the dynamic system to exhibit certain specific types of motions. Furthermore, we would like the system to be able to execute these motions automatically, without outside intervention. An example of this type of situation is the "cruise control" in an automobile, wherein a predetermined forward speed V_0 set by the vehicle operator is supposed to be maintained, regardless of hills, winds, road conditions, and so on. A *control system* is used to accomplish this task. Thus, if the automobile is traveling at the desired speed V_0 on a level road and then encounters a steep hill, the forward speed will tend to drop. When this occurs, the cruise control must cause increased fuel flow in order to accelerate the vehicle back to the desired speed V_0. We observe that this control system must incorporate two features not present in uncontrolled dynamic systems:

1. A means to determine, at any time t, the *error*, that is, the difference between the desired forward speed V_0 and the actual forward speed $V(t)$. This error determination necessitates both *measurement* of the actual forward speed and *feedback* to an electronic device that compares the actual to the desired speed to establish the error. Thus, an essential feature of control systems is the inclusion of instruments to measure the dynamic variable(s), here speed $V(t)$, which is (are) to be controlled.
2. *Actuation devices* that, based on the error, cause corrective action to be taken automatically so as to drive the error to zero. In the cruise-control system the actuator exerts forces on the mechanical elements regulating rate of fuel flow, causing increased fuel flow if $V(t) < V_0$ and decreased fuel flow if $V(t) > V_0$.

In summary, measurement, feedback, error detection, and actuation are the essential elements of control systems. From a dynamic systems perspective we may view the controls problem in the following way: We start with an uncontrolled system, and we assume that we can develop a mathematical model describing the dynamics of this uncontrolled system. When we add the control system, we create a new dynamic system, characterized by a new and more complicated mathematical model, which can be used to study the effectiveness of the proposed control system. It is important to understand that the same dynamic systems concepts are needed to understand system behavior, whether or not a control system is involved.

1.4 COMMENTS ON ODE MODELS

The math models that we will develop and study will usually consist of one or more ordinary differential equations (ODEs). Time t will be the independent variable, and the dependent variables will be those quantities that describe the dynamic behavior of the system and that we eventually find by solving the governing differential equations. In this section we discuss certain important definitions, features, and classifications of ODE models that will be important throughout our study of dynamic systems.

1.4.1 System Order and Representation of ODE Models

Recall that the *order* of a differential equation is the order of the highest derivative appearing in the equation. Depending on the nature of the system under study, our ODE models may arise or be cast in several forms: (1) a single ODE of, say, nth order, (2) a set of n, first-order ODEs, and (3) a collection of several ODEs of varying orders. The *order of the system* is defined as the sum of the orders of the individual ODEs comprising the math model.

It is important to realize that the form of a given ODE math model is not unique. For example, a single nth-order ODE may be reexpressed as a set of n, first-order ODEs. It is worthwhile at this point to illustrate this for a simple example. Suppose that, for a given system, the dependent variables for which we need to develop a model are x and y and that the governing ODEs turn out to be as follows:

$$\ddot{x} + \alpha_1 \dot{x} + \alpha_2 x = \alpha_3 y \qquad (1.1a)$$

$$\dot{y} + \beta_1 y = F(t) - \beta_2 x \qquad (1.1b)$$

where a dot denotes a time derivative, the α's and β's are constants, and $F(t)$ is a specified function of time. Our math model consists of one second-order ODE and one first-order ODE, so that the system is of third order. Notice that the equations are *coupled* and have to be solved simultaneously. Each equation contains both x and y, so that what x does will be influenced by what y is doing, and vice versa.

As we'll see later, it is often desirable to express the math model as a set of first-order ODEs. For the example system (1.1) this is done by first defining a new variable $v \equiv \dot{x}$. Then (1.1a) may be rewritten as

$$\dot{v} + \alpha_1 v + \alpha_2 x = \alpha_3 y$$

We are now able to write out the system math model as a set of three first-order

ODEs involving the new set of three dependent variables x, y, and v:

$$\dot{v} + \alpha_1 v + \alpha_2 x = \alpha_3 y \tag{1.2a}$$

$$\dot{y} + \beta_1 y = F(t) - \beta_2 x \tag{1.2b}$$

$$\dot{x} = v \tag{1.2c}$$

Note that the first two equations are just the original system equations (1.1), rewritten in terms of a different set of dependent variables, while the third equation is merely a definition, which we need to include to be able to express the system model as a set of three first-order ODEs. The order of the system must stay the same. Usually, when we take a given mathematical model and convert it to a set of first-order ODEs, we solve for the first derivatives of our variables and write the result as

$$\dot{v} = -\alpha_1 v - \alpha_2 x + \alpha_3 y \tag{1.3a}$$

$$\dot{y} = -\beta_1 y - \beta_2 x + F(t) \tag{1.3b}$$

$$\dot{x} = v \tag{1.3c}$$

Thus (1.3) is simply an alternative way to express the original mathematical model given by (1.1). The reason for wanting to redefine the model in this way may not be clear at this point, but, as we'll see in Chap. 4, the first-order version is most effective when studying general properties of solutions of ODE models. We can generalize the results of this example as follows: Any model of nth order can be written as a system of n first-order ODEs through definition of appropriate new variables and rewriting of the original model in terms of these and the original variables.

Now let's look at yet another way to express the system model. We have just seen that it is possible to express a given nth-order model as a set of n first-order ODEs. It is also possible to express a given nth-order model as a *single* nth-order ODE. To see how this would be done, consider again the example model (1.1). If we take (1.1a) and differentiate it, we obtain

$$\dddot{x} + \alpha_1 \ddot{x} + \alpha_2 \dot{x} = \alpha_3 \dot{y} \tag{1.4}$$

Next, solve (1.1b) for $\dot{y}$:

$$\dot{y} = -\beta_1 y - \beta_2 x + F(t)$$

and substitute this into (1.4) to obtain

$$\dddot{x} + \alpha_1 \ddot{x} + \alpha_2 \dot{x} = \alpha_3[-\beta_1 y - \beta_2 x + F(t)] \tag{1.5}$$

Next, solve (1.1a) for y and use the result to eliminate y on the right-hand side of (1.5):

$$\dddot{x} + \alpha_1 \ddot{x} + \alpha_2 \dot{x} = \alpha_3\left[-\beta_2 x + F(t) - \frac{\beta_1}{\alpha_3}(\ddot{x} + \alpha_1 \dot{x} + \alpha_2 x)\right]$$

Finally, transpose all of the terms involving the dependent variable x to the left-hand side to obtain the following third-order ODE for x:

$$\dddot{x} + (\alpha_1 + \beta_1)\ddot{x} + (\alpha_2 + \beta_1\alpha_1)\dot{x} + (\alpha_2\beta_1 + \alpha_3\beta_2)x = \alpha_3 F(t) \qquad (1.6)$$

Note that the variable y does not appear in the model; if we are going to express the original system model as a single nth-order ODE, then we have to pick one of the original dependent variables to work with. Here we've used x; we could also have developed a third-order equation in terms of y.

To summarize the above comments: We have discussed three ways to express the ODEs comprising an nth-order system model: (1) however they arise naturally (this will of course depend on the particular system being studied), (2) as a set of n first-order ODEs, each defining the rate of change of one of the n variables, and (3) as a single nth-order ODE (here the single dependent variable appearing in the equation may be any one of those appearing in the original formulation).

1.4.2 Initial Conditions and States

Now we have to say something about the *initial conditions* (ICs). If we wish to solve the ODEs in our math model for specific cases, we have to specify certain information (the ICs) at $t = 0$. The number of initial conditions specified must equal the order of the system. To see what quantities need to be specified as ICs, we need to look at the individual equations comprising the model. Recall that for a single nth-order ODE we need to specify the initial value of the dependent variable and all of its time derivatives up to and including the $(n - 1)$th. If the system model consists of multiple ODEs, we apply this rule to each equation. For instance, for the third-order model (1.1), the initial conditions would be $x(0)$, $\dot{x}(0)$, and $y(0)$.

Those quantities that must be specified as initial conditions are referred to as the *states of the dynamical system*. Another way to view this definition is that the states of a dynamic system are those quantities which, if specified at some time, determine completely the evolution of the system for all future times. Clearly, the number of states will equal the order of the system. Inspection of the math model (1.3) for our example system reveals that the variables (x, v, y) are the states of the system, and that the redefined form (1.3) of the model consists of a first-order ODE for each state. For this reason, we call the set (1.3) of first-order ODEs the *state variable form* of the system math model. We have to realize that, regardless of the way in which the system model is presented, the same information will be available to us if we can solve the differential equation(s). Both the state variable form and the single nth-order ODE form are particularly useful.

1.5 CLASSIFICATION OF ODE MODELS

There are several important general properties of ODEs that are used to classify system mathematical models, and these are now summarized.

1.5.1 Linear versus Nonlinear Systems

This is perhaps the most important of the various classifications of system mathematical models. In terms of the cause-and-effect phenomena that govern dynamic behavior, a given phenomenon is linear if the effect is *proportional* to the cause that produces it. For example, if a cantilever beam is statically loaded, the beam will undergo a certain deflection. If a doubling of the applied load causes a doubling of the deflection, then the behavior is linear. Furthermore, if two different loads that produce two different deflections are applied together and result in a deflection which is the sum of the deflections occurring when the loads were separately applied, then the behavior is linear. The preceding statement is an example of the *principle of superposition*, which defines mathematically the property of linearity. This principle is stated as

$$f(\alpha x_1 + \beta x_2) = \alpha f(x_1) + \beta f(x_2) \tag{1.7}$$

where α and β are constants and where x_1 and x_2 are viewed as "causes" and $f(x_1)$ and $f(x_2)$ the resulting "effects." Thus, (1.7) says that, if x represents the applied static load, and if $f(x)$ represents the resulting deflection, then the deflection for any linear combination of different static loading conditions is obtained by summing the individual deflections that occur when the loads are applied separately. In essence "the whole is equal to the sum of its parts." Hooke's law, Ohm's law, and the force-displacement relation for an ideal spring are examples of linear constitutive behavior.

As an example of a constitutive relation that is *nonlinear*, consider the aerodynamic drag force exerted on an object moving through a fluid medium. Experience shows that this drag force F is proportional to the *square* of the object's speed V, $F = \alpha V^2$, where α is a constant (see also Example 2.7 in Chap. 2). Thus, if the drag force generated at a given speed V_1 is $F_1 = \alpha V_1^2$, and if the drag force at a second speed V_2 is $F_2 = \alpha V_2^2$, then the drag force F resulting from a speed $V = V_1 + V_2$ is $F = \alpha(V_1 + V_2)^2 = \alpha(V_1^2 + V_2^2 + 2V_1V_2)$. This violates the superposition principle (1.7); the phenomenon is nonlinear. For example, a doubling of the speed will cause a quadrupling of the resultant force.

In the preceding discussion the mathematical statements describing cause and effect consist of *algebraic*, rather than differential, equations. The deflection of the ideal cantilever is a linear function of the applied load, and the aerodynamic drag force F is a nonlinear function of speed V. The principle of superposition (1.7) applies also to linear systems of ordinary differential equations. If an ODE is linear, and if $x_1(t)$ and $x_2(t)$ are two linearly independent solutions for the dependent variable x, then any linear combination $\alpha x_1(t) + \beta x_2(t)$ is also a solution. This property is what enables us to construct the most general solution to homogeneous linear ODEs of arbitrary order.

In the mathematical ODE models that are used to approximate the behavior of dynamic systems, the system linearity or nonlinearity is determined by the manner in which the system states appear in the mathematical model.

Terms involving the system states will appear as a result of constitutive relations and as a result of application of appropriate physical laws such as Newton's second law of motion.

Let's assume that our dynamic model consists of a set of ODEs in state variable form. If the mathematical model contains any nonlinear function of any state or any combination of states, then the system is nonlinear (otherwise it is linear). For example, if x and y are two states in a math model, then functions like x^2, $\sin x$, $e^{\alpha x}$, $1/(1+x)$, and xy are nonlinear. Thus, the following differential equations would bc nonlinear (the nonlinear term is underlined in each):

$$\ddot{x} + \underline{\sin x} = 0$$

$$\ddot{x} + \dot{x}(\underline{x^2} - 1) + x = p \cos \Omega t$$

$$\dot{x} = \alpha x - \underline{\beta x^3}$$

$$\ddot{x} + \underline{\alpha \dot{x}^3} + x = 0$$

The following set of two first-order ODEs represents a nonlinear system, even though only one of the ODEs is nonlinear:

$$\dot{x} = \alpha x + \beta y$$

$$\dot{y} = -\gamma y + \underline{\delta xy}$$

Note that the presence of the independent variable t has no effect on whether the system is nonlinear. Thus, the following differential equations are linear:

$$\ddot{x} + (1 - \cos 2t)x = 0$$

$$\ddot{x} + x = te^{-\alpha t}$$

$$\ddot{x} + t^3 x = 0$$

Note also that a system is nonlinear even if only a single nonlinear term appears in one of the ODEs, even though the remaining ODEs are linear.

In our work here we are going to spend a lot of time analyzing linear models of dynamic systems. There are two reasons for this: (1) linear models turn out to provide an adequate description of the system behavior in many cases, and (2) the theory of linear ODEs is well developed and pretty easy to follow and apply.

The reader should be aware, however, that for many engineering phenomena, the inclusion of nonlinearities is absolutely essential if the math model is to describe correctly the behavior of the system. Nonlinear systems can exhibit a wealth of interesting responses that cannot occur in the linear world (we'll see some examples of these in Chap. 10). In terms of the "richness" of phenomena exhibited, the difference between linear and nonlinear systems is akin to the difference between Pullman and Paris.

Prior to the widespread use of the digital computer, the tendency of engineers was to "force" system models to be linear so that analytical solutions could be easily obtained, there being no general techniques for finding exact, analytical solutions for nonlinear ODEs.* Nowadays, however, computer simulation of both linear and nonlinear systems is so common that, in order to interpret simulation results, a competent analyst has to have some idea of the manner in which nonlinear systems exhibit behavior that is fundamentally different from that which is possible in a linear system.

1.5.2 Autonomous and Nonautonomous Systems: The System and the Input

If the governing differential equations contain the independent variable t explicitly, then the system model is said to be *nonautonomous*. If t does not appear, then the system model is *autonomous*. For example, the example system model (1.1) is nonautonomous because of the term $F(t)$. This classification is important for the following reason: We want to distinguish between (1) physical effects that are inherently part of the system and (2) external effects that arise outside the system itself but nevertheless influence its behavior. For our purposes any physical effect that is modeled mathematically by functions of the states only is considered to be "part of the system." Any effect that depends only on the time t, and hence arises independently of what the system is doing, is not considered to be a part of the system. We refer to such external, time-dependent effects as *inputs*, or *external excitations*. Inputs, then, will influence the behavior of the system but are not considered to be a part of the system. In (1.1), for example, the input is $F(t)$ and all of the other terms define the system itself. As another example, when we push a child on a swing, the *system* consists of the child, the swing, and the support to which the swing is attached. The *input* is the time-dependent force exerted by us on the child. The motion of the system can be described by the rotation $\theta(t)$ of the swing.

From this discussion we observe that the system is defined by the autonomous version of the mathematical model, that is, with the input removed. It should be kept in mind that our defining the input to be separate from the system itself is based on mathematical, rather than physical, considerations. In many situations the mechanism that produces the input is actually a part of the physical system.

---

*The high-water mark of "forced linearization" was probably achieved by the physicists in the development of quantum mechanics, a basic postulate of which was linearity.

1.5.3 Time Variant versus Time Invariant Systems

With reference to (1.1) the constants α_1, α_2, α_3, β_1, and β_2 are the *system parameters*. If one or more of these parameters is actually a function of time, then the system is said to be time-varying. If such is the case, then the math model is nonautonomous and any term in the model that involves a time-varying property will be a function of both time and one or more states. In this book we will confine our attention to time invariant systems.

1.5.4 Closing Remarks on ODE Models

At this point we can identify the three types of quantities that will appear in our math models. The first quantities are *system states*, that is, the "unknowns" to be found. We have to solve the governing ODEs to determine how these quantities vary with time. The second are *system parameters*, that is, physical constants such as masses, stiffnesses, and geometrical features that are often proportionality constants in some constitutive relation. The third are *inputs*, or external excitations. These are specified functions of time which "excite," or "drive," the system but which are not considered to be part of the system.

In Chaps. 2 and 3 ODE math models are developed for a number of mechanical and other types of systems. For each of these models the classifications, according to the material presented in this section, are discussed. It is suggested that this section (that is, Sec. 1.5) be reread as necessary in order to have a clear understanding of the associated discussions in Chaps. 2 and 3.

1.6 SUMMARY OF WHAT IS TO COME

The following paragraphs provide a summary of the remaining chapters of the book, with comments directed to the student.

In *Chaps. 2 and 3* we consider basic ideas and procedures related to the construction of mathematical models of dynamic systems. In *Chap. 2* we study mechanical systems governed by the laws of newtonian mechanics. In *Chap. 3* we study hydraulic, pneumatic, electrical, and nonengineering systems. For each of these types of systems, useful physical laws and constitutive relations are first discussed and are then applied, through consideration of numerous examples, to develop ODE models of the systems considered. The main thrust is to gain insight into the modeling process through example. Keep in mind that in these chapters the intent is only *to develop* governing ODE models, not to solve them.

In *Chap. 4* the central concepts of dynamic systems theory are introduced: equilibrium, stability, linearization, and visualization of motion in the phase space. These ideas are then applied to the study of motions of linear autonomous systems. This is the first chapter in which our main concern is with the system response, that is, with properties of the *solutions* to ODE models.

In *Chap. 5* we consider the response of linear systems that are driven by external excitations (inputs). The focus here will be on the relation between the input and the resulting response for systems of first, second, and higher orders. Several methods of calculating the response will be presented.

Chapters 2 through 5 contain fundamental material that is essential in understanding the basics of dynamic systems modeling and analysis. The topics in Chaps. 6 through 10 are directed to more specific applications.

Chapter 6 considers additional aspects of classical dynamics that are requisite for a more advanced view of modeling of mechanical systems: three-dimensional rotating coordinate systems and Lagrange's equations. As we shall see, Lagrange's equations provide an alternative to the newtonian formulation of mechanics. Lagrange's equations often allow a mathematical model of a mechanical system to be developed in a much simpler way than does the newtonian formulation.

Chapter 7 presents a description of single and multi-degree-of-freedom linear vibrations. This chapter considers those mechanical systems that inherently tend to exhibit oscillatory behavior. Two topics that are especially important from a practical standpoint are introduced: modal analysis and spectrum analysis.

Chapter 8 introduces the basic ideas of classical linear control theory. Here we consider those situations in which we wish the dynamic system to move in a certain way (for example, to some target), to perform some type of maneuver or to be insensitive to effects of outside disturbances. Thus, we add "control actions" to the system, thereby creating a new system.

Chapter 9 discusses some of the basic methods of computer simulation that are widely used in the real world of engineering. The emphasis is on numerical integration of ODEs and on the introduction of finite element methods in structural dynamics.

Finally, *Chapter 10* presents a study of some basic aspects of behavior of nonlinear systems. Here we consider those phenomena, in both autonomous and nonautonomous nonlinear systems, which have no counterpart in the linear world. We close this chapter with a brief description of "chaotic motion," which was discovered fairly recently and which has resulted in a renaissance in the study of dynamic systems.

REFERENCES

1. Boyce, W. E., and R. C. DiPrima: *Elementary Differential Equations and Boundary Value Problems*, Wiley, New York (1977). This is a good reference to consult for review of ordinary differential equations.
2. Apostol, T. M.: *Calculus*, vol. II, Blaisdell, New York (1965). Chapter 7 contains relevant material on linear ordinary differential equations.

EXERCISES

For each of the following seven ODE systems, determine the order of the system, and identify the states and whether the system is linear or nonlinear (if nonlinear, identify the nonlinearities), and autonomous or nonautonomous. Assume all Greek letters to be known constants.

1.

$$\dot{x} = \sigma x + \beta y$$
$$\dot{y} = \gamma y + \alpha z x$$
$$\dot{z} = \delta z + \varepsilon x y$$

2.

$$\ddot{z} + \dot{z}(1 - z^2) + z = y$$
$$\dot{y} = \alpha y + \delta \cos 3t$$

3.

$$\frac{d^4 x}{dt^4} + 6xte^{-t} = 0$$

4.

$$\ddot{x} + x(1 + \cos 2t) = 0$$

5.

$$\dot{x}_1 = x_1 - x_1^3 + \alpha x_1 x_2$$
$$\dot{x}_2 = -x_2 + \beta x_1 x_2$$

6.

$$\dot{y} = \alpha - \beta y^{1/2} \qquad (y \geq 0)$$

7.

$$\dot{x} + \dot{y} = -3x + xz + te^{-t}$$
$$\dot{y} = x - 2y$$
$$\ddot{z} = -4z + 2xy$$

8. The mathematical model of a certain newtonian system consists of the two second-order ODEs given below:

$$m\ddot{x} - ma\ddot{\theta} = -k_1 x - mg$$

$$I\ddot{\theta} - ma\ddot{x} = -k_2\theta + mga$$

Identify the states of the system, and put the model into state variable form (i.e., as a set of four first-order ODEs, one for each state). Note that a bit more work is required here than was needed in converting (1.1) to (1.3) in the text.

9. Consider the second-order system of Exercise 5. The system equations are in state variable form. Reexpress the mathematical model as a single second-order ODE for the variable x_2, by differentiating the $\dot{x}_2$ equation and eliminating x_1 in the result. This exercise illustrates that conversion of a given system model to a single higher-order ODE in one of the dependent variables can be messy if the system is nonlinear.

10. An important property of linear systems is that the principle of *superposition* applies. Thus, consider the linear, second-order ODE:

$$\ddot{x} + \dot{x} + x = 0 \qquad (P1.1)$$

Suppose that we can find two linearly independent solutions, denoted as $x_1(t)$ and $x_2(t)$, each of which solves $(P1.1)$. Then the principle of superposition says that any linear combination $c_1 x_1(t) + c_2 x_2(t)$ is also a solution, where c_1 and c_2 are any constants. Verify directly that superposition holds for the linear system $(P1.1)$, but not for the nonlinear system $(P1.2)$ below:

$$\ddot{x} + \dot{x} + x^3 = 0 \qquad (P1.2)$$

11. Consider the second-order, linear, inhomogeneous ODE:

$$a_1\ddot{x} + a_2\dot{x} + a_3x = f(t)$$

where a_1, a_2, and a_3 are constants and the input $f(t)$ is a function of time. Suppose that when $f(t) = f_1(t)$, the solution is $x_1(t)$. Suppose that when $f(t) = f_2(t)$, the solution is $x_2(t)$. Show that when $f(t) = \alpha f_1(t) + \beta f_2(t)$, the solution is $x(t) = \alpha x_1(t) + \beta x_2(t)$, that is, that the principle of superposition holds. The intent of this exercise is to illustrate that arbitrary functions of the independent variable t do not affect linearity.

12. Consider the *nonlinear* second-order, inhomogeneous ODE:

$$a_1\ddot{x} + a_2\dot{x} + a_3x^3 = f(t)$$

wherein the nonlinear function x^3 of the state x appears. Let $x_1(t)$ be the solution when $f(t) = f_1(t)$ and let $x_2(t)$ be the solution when $f(t) = f_2(t)$. In this case show that when $f(t) = \alpha f_1(t) + \beta f_2(t)$, the function $\alpha x_1(t) + \beta x_2(t)$ does not satisfy the ODE, so that superposition does not hold.

13. A jogger runs for 1 km, at the rate of 20 km/h, into a 20 km/h headwind. She then turns around and runs 1 km at 20 km/h with the wind at her back. Assume the aerodynamic drag force to be proportional to the square of the speed of the jogger relative to the wind. Determine the percentage difference in the average drag force for the situation described as compared to the case of no wind. What would this percentage difference be if the drag force were a *linear* function of the relative speed?

CHAPTER 2

MODELING OF MECHANICAL SYSTEMS

The objective of this chapter is to present the techniques needed to construct mathematical models of engineering systems (blocks 1 through 7 of Fig. 1.1). In this chapter we will study mechanical systems, for which we will want to model the manner in which "particles" and rigid bodies move as a result of the forces and moments exerted upon them. In Chap. 3 we will then consider hydraulic, pneumatic, electromagnetic, and nonengineering systems. The reader should keep in mind that the objective in this chapter is to develop ODE models and to classify them according to the discussion of Sec. 1.5. We will at this point make no attempt to analyze the ODE models to determine the response of the system. Thus, in this chapter we will not be too concerned with trying to determine how the systems being modeled are going to behave. The system response will be considered in some detail in Chaps. 4 and 5.

2.1 BASICS OF NEWTONIAN MECHANICS

In this section we review the basic laws of newtonian mechanics for a "particle" translating and rotating in three dimensions. In Sec. 2.2 we consider the laws of motion for a rigid body translating and rotating in the plane. Then, in Secs. 2.3 and 2.4 we will see how to apply these laws to construct mathematical models. Since newtonian mechanics is a *vector* formulation, we will need to be conversant with vector operations, which should be reviewed as necessary. Most of the material in Secs. 2.1 and 2.2 should be familiar to the student. At the end of this

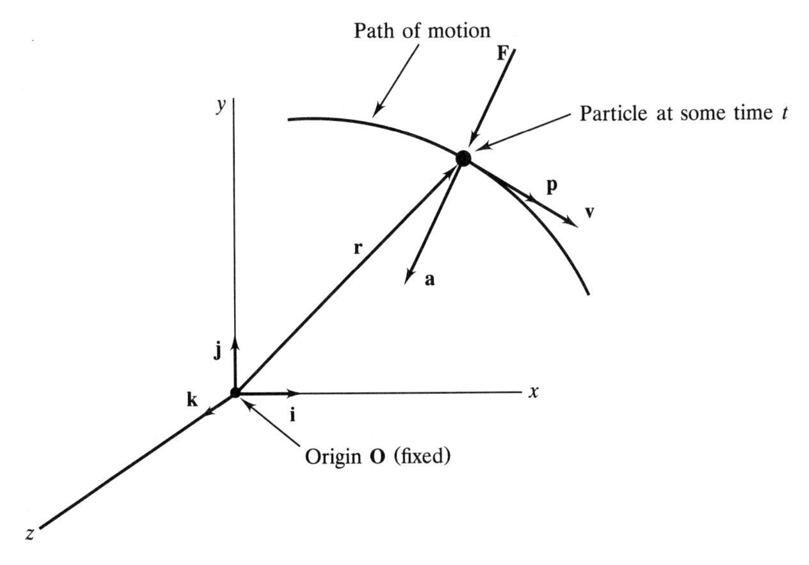

FIGURE 2.1
3-D particle kinematics in cartesian coordinates.

chapter are listed three references that contain basic material on vector analysis and newtonian mechanics.

2.1.1 The Translation of a Particle

Let us begin by considering the motion of a *particle*, a hypothetical object of fixed mass m occupying zero volume. The motion of the particle is described with reference to the cartesian (x, y, z) coordinate system shown in Fig. 2.1.

The position vector **r**, velocity vector **v**, acceleration vector **a**, and linear momentum **p** of the particle are defined in the cartesian system in component form as

$$\mathbf{r} = x\mathbf{i} + y\mathbf{j} + z\mathbf{k} \tag{2.1}$$

$$\mathbf{v} = \frac{d\mathbf{r}}{dt} = \dot{x}\mathbf{i} + \dot{y}\mathbf{j} + \dot{z}\mathbf{k} \tag{2.2}$$

$$\mathbf{a} = \frac{d\mathbf{v}}{dt} = \ddot{x}\mathbf{i} + \ddot{y}\mathbf{j} + \ddot{z}\mathbf{k} \tag{2.3}$$

$$\mathbf{p} = m\mathbf{v} \tag{2.4}$$

where dots denote time derivatives and **i**, **j**, **k** are fixed unit vectors defining the x, y, and z directions, respectively. Recall that a vector (such as **r** or **v**) may have a nonzero rate of change if either its magnitude or its direction is changing.

Newton's second law relates the resultant applied force $\mathbf{F}$, the vector sum of all of the individual applied forces, which we view as a *cause*, to the *effect* it produces, a change in linear momentum:

$$\mathbf{F} = \frac{d\mathbf{p}}{dt} = m\mathbf{a} \tag{2.5}$$

Clearly, if the resultant force $\mathbf{F} = 0$, then linear momentum is conserved and the particle moves in a straight line at constant speed or is at rest. An important aspect of (2.5) to keep in mind is that the acceleration $\mathbf{a}$ in (2.5) must be calculated relative to an inertial frame of reference, which Newton defined as one fixed relative to the distant stars. Thus, technically, we should in computing $\mathbf{a}$ account for the rotation of the earth about its axis, the rotation of the earth about the sun, and so on. In practice, however, we often find that such effects can be ignored, and we can use a frame of reference fixed relative to the earth's surface.

It is often useful to express (2.5) in component form. First the applied force $\mathbf{F}$ is resolved into its cartesian components:

$$\mathbf{F} = F_x\mathbf{i} + F_y\mathbf{j} + F_z\mathbf{k} \tag{2.6}$$

Then, using (2.3) and (2.5), we obtain the three scalar equations:

$$\begin{aligned} m\ddot{x} &= F_x \\ m\ddot{y} &= F_y \\ m\ddot{z} &= F_z \end{aligned} \tag{2.7}$$

Thus, the most general motion of the particle is governed by a set of three second-order ordinary differential equations in which the position components x, y, and z are the dependent variables and in which time is the independent variable. The particle has 3 degrees of freedom; three coordinates are needed to specify the position of the particle. The system order is 6, and the six states are x, y, z, $\dot{x}$, $\dot{y}$, and $\dot{z}$.

Let us now consider the special case in which the motion is rectilinear, so that only one direction, say, x, is involved. Then (2.7) reduces to

$$m\ddot{x} = F_x \tag{2.8}$$

Now an important aspect in the application of (2.8) to actual problems of particle motion is the determination of the functional dependence of the force F_x, which may actually consist of several individual forces that arise independently of each other. In general, F_x may depend on the position x, the velocity $\dot{x}$, and the time t, so to note this functional dependence we write

$$m\ddot{x} = F_x(x, \dot{x}, t) \tag{2.9}$$

We can now identify two basic types of forces: (1) those that depend only on position and/or velocity, $F_x = F_x(x, \dot{x})$ (such forces are *generated* as a result of the position or motion of the particle) and (2) those forces that depend only on time, $F_x = F_x(t)$ (such forces, since they do not depend on the position or motion of the particle, must arise from effects that are independent of the motion). Examples of typical forces include the gravitational force (a function of position), aerodynamic drag forces (a function of velocity), and the torque generated by a motor (a function of time). Sometimes forces arise that are functions of both position (and/or velocity) and time, but we won't consider these for now.

The discussion of the functional dependence of forces is important for the following reason: The construction of mathematical models of the forces that act on the particle (or on any other mechanical system) is often difficult, and one of the first questions to be answered is "What is the force a function of?"* Once we know the functional dependence, we can attempt, through testing and/or analysis, to develop an equation or graph (constitutive model) that describes this dependence. Examples in Secs. 2.4, et seq., will address this point.

In the general case of three-dimensional motion, (2.9) is extended to the following:

$$\begin{aligned} m\ddot{x} &= F_x(x, y, z, \dot{x}, \dot{y}, \dot{z}, t) \\ m\ddot{y} &= F_y(x, y, z, \dot{x}, \dot{y}, \dot{z}, t) \\ m\ddot{z} &= F_z(x, y, z, \dot{z}, \dot{y}, \dot{z}, t) \end{aligned} \tag{2.10}$$

Note that, in general, each force component F_x, F_y, or F_z may depend on all three position and velocity components, so that the three equations are coupled and we cannot solve them independently of each other. Note further that (2.10) does not imply that in all situations the functional dependence has to be of the type indicated; for instance, we may encounter in one-dimensional (1-D) problems a situation in which there exists a single force F_x that is a function only of position x, so that (2.9) would reduce to $m\ddot{x} = F_x(x)$. We use (2.9) and (2.10) here in order to indicate the most general possibility.

Sometimes we find it convenient to describe planar particle motion in terms of polar (r, θ) rather than cartesian (x, y) coordinates. In such cases we describe the vectors **r**, **v**, **a**, and **F** in terms of their radial $\mathbf{e}_r$ and tangential $\mathbf{e}_\theta$ components, as shown in Fig. 2.2. Here the orthogonal radial $\mathbf{e}_r$ and tangential $\mathbf{e}_\theta$ unit vectors rotate, so that $\mathbf{e}_r$ is always directed toward the mass m, with the scalar r measuring the distance to the particle. The relations analogous to (2.1) through (2.3) are as follows:

$$\mathbf{r} = r\mathbf{e}_r \tag{2.11a}$$

$$\mathbf{v} = \dot{r}\mathbf{e}_r + r\dot{\theta}\mathbf{e}_\theta \tag{2.11b}$$

$$\mathbf{a} = (\ddot{r} - r\dot{\theta}^2)\mathbf{e}_r + (r\ddot{\theta} + 2\dot{r}\dot{\theta})\mathbf{e}_\theta \tag{2.11c}$$

*The reader will kindly allow the author the liberty of ending sentences with a preposition.

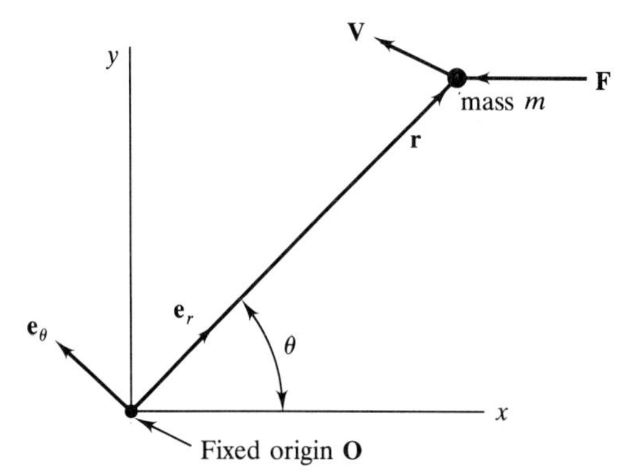

FIGURE 2.2
Plane particle kinematics in polar coordinates.

The resultant force **F** is resolved along the radial and tangential directions, $\mathbf{F} = F_r\mathbf{e}_r + F_\theta\mathbf{e}_\theta$, and the equations of motion analogous to (2.10) are

$$\begin{aligned} m(\ddot{r} - r\dot{\theta}^2) &= F_r(r, \theta, \dot{r}, \dot{\theta}, t) \\ m(r\ddot{\theta} + 2\dot{r}\dot{\theta}) &= F_\theta(r, \theta, \dot{r}, \dot{\theta}, t) \end{aligned} \tag{2.12}$$

The system math model (2.12) is of fourth order. The states are r, θ, $\dot{r}$, and $\dot{\theta}$. The model is intrinsically nonlinear because of the following terms that involve nonlinear functions of the states: $r\dot{\theta}^2$, $r\ddot{\theta}$, and $\dot{r}\dot{\theta}$. These nonlinearities arise directly from the acceleration (2.11c) expressed in polar coordinates. Note that the formulation of particle motion in polar coordinates involves a rotating coordinate system; the unit vectors $\mathbf{e}_r, \mathbf{e}_\theta$ rotate at the angular rate $\dot{\theta}$. This means that when we differentiate the position **r** to obtain the velocity **v**, account must be taken of the nonzero rate of change of the unit vectors. The use of general rotating coordinate systems for three-dimensional motion is presented in detail in Chap. 6. In particular, see Sec. 6.2.2 for details of the polar coordinate formulation of planar particle motion.

2.1.2 Angular Momentum of a Particle

So far we have formulated the equations of motion of a particle by relating the applied force **F** to the rate of change of linear momentum $\mathbf{p} = m\mathbf{v}$. It is also useful to relate the rate of change of angular momentum to the applied moment due to the force **F**. In computing angular momentum and applied moment, we must always define a reference point with respect to which the calculation is to be made.

This reference point need not be fixed but may be moving in an arbitrary manner. With reference to Fig. 2.3 we define a point S as our moment reference. The position of this point relative to the fixed origin **O** is denoted by $\mathbf{R}_s$; the velocity and acceleration of the reference point S are then $\dot{\mathbf{R}}_s$ and $\ddot{\mathbf{R}}_s$. The position of the particle m relative to the moment reference is denoted by **s**.

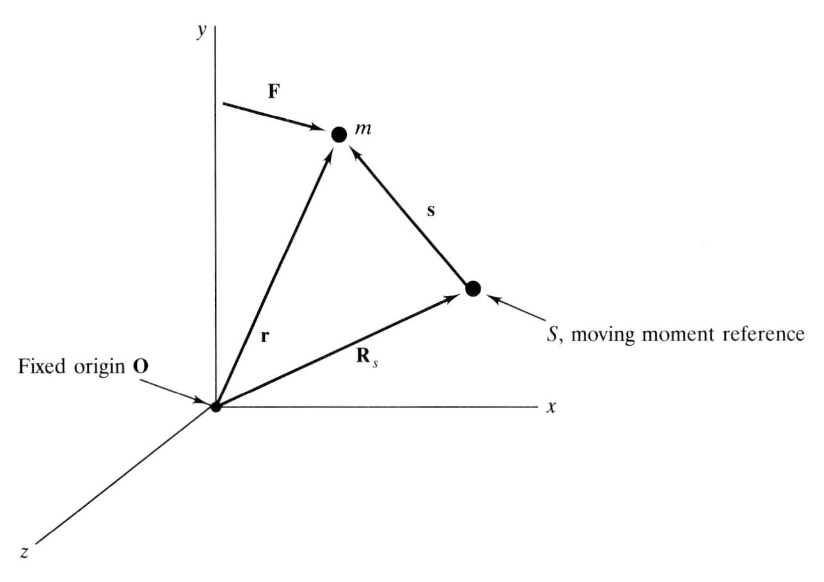

FIGURE 2.3
Notation used for angular momentum and moment calculation.

The angular momentum $\mathbf{H}_s$ with respect to the point S is defined as

$$\mathbf{H}_s \equiv \mathbf{s} \times \mathbf{p} = \mathbf{s} \times m\dot{\mathbf{r}} \tag{2.13}$$

and the moment $\mathbf{M}_s$ with respect to S is defined by

$$\mathbf{M}_s \equiv \mathbf{s} \times \mathbf{F} = \mathbf{s} \times m\ddot{\mathbf{r}} \tag{2.14}$$

We can relate the moment $\mathbf{M}_s$ to the rate of change of angular momentum $\dot{\mathbf{H}}_s$ by first computing $\dot{\mathbf{H}}_s$:

$$\dot{\mathbf{H}}_s = \frac{d}{dt}(\mathbf{s} \times \mathbf{m}\dot{\mathbf{r}}) = \dot{\mathbf{s}} \times m\dot{\mathbf{r}} + \mathbf{s} \times m\ddot{\mathbf{r}}$$

Because $\mathbf{s} \times m\ddot{\mathbf{r}} = \mathbf{s} \times \mathbf{F} = \mathbf{M}_s$ and $\mathbf{r} = \mathbf{R}_s + \mathbf{s}$, we can rewrite this relation as

$$\dot{\mathbf{H}}_s = \dot{\mathbf{s}} \times m\left(\dot{\mathbf{R}}_s + \dot{\mathbf{s}}\right) + \mathbf{M}_s$$

Now the first term on the right-hand side is expanded to $\dot{\mathbf{s}} \times m\dot{\mathbf{R}}_s + \dot{\mathbf{s}} \times m\dot{\mathbf{s}}$, and because $\dot{\mathbf{s}} \times m\dot{\mathbf{s}} = m(\dot{\mathbf{s}} \times \dot{\mathbf{s}}) = 0$, we arrive at the result:

$$\dot{\mathbf{H}}_s = \mathbf{M}_s + \dot{\mathbf{s}} \times m\dot{\mathbf{R}}_s \tag{2.15}$$

It is important to note that if the reference point S is actually *fixed*, so that $\dot{\mathbf{R}}_s = 0$, then (2.15) reduces to the familiar relation for rotation about a fixed point:

$$\dot{\mathbf{H}}_s = \mathbf{M}_s \qquad (S \text{ fixed}) \tag{2.16}$$

This relation is analogous to Newton's second law for particle translation $\mathbf{F} = \dot{\mathbf{p}}$, in which a force induces a rate of change in linear momentum.

We observe that, for rotation about a fixed point, the angular momentum is conserved if the applied moment $\mathbf{M}_s = 0$. This can occur if $\mathbf{F} = 0$, if $\mathbf{r} = 0$, or if $\mathbf{r}$ and $\mathbf{F}$ are colinear. An example of this latter condition is the motion of bodies in a central force field. Thus, the planets move about the sun, satellites move about the earth, and so on in such a way that angular momentum is conserved.* The relations (2.13) through (2.15) will prove useful in Sec. 2.2, in which we summarize the laws of planar motion of a rigid body.

2.1.3 Relative Angular Momentum

In the preceding subsection, the angular momentum $\mathbf{H}_s$ was defined in terms of the "moment arm" $\mathbf{s}$ from the reference point S to the particle and in terms of the actual (i.e., inertial) linear momentum $\mathbf{p} = m\dot{\mathbf{r}}$. It is also useful, however, to define the angular momentum in terms of the linear momentum *relative* to the reference point S, that is, $\mathbf{p}_r = m\dot{\mathbf{s}}$, where the r subscript indicates a quantity measured relative to the reference point S. Thus, we define the relative angular momentum $\mathbf{H}_{s_r}$ as

$$\mathbf{H}_{s_r} \equiv \mathbf{s} \times m\dot{\mathbf{s}} \tag{2.17}$$

Next we establish the relation between the moment $\mathbf{M}_s$ and the rate of change of relative angular momentum. Differentiating (2.17), we obtain

$$\dot{\mathbf{H}}_{s_r} = \cancel{\dot{\mathbf{s}} \times m\dot{\mathbf{s}}} + \mathbf{s} \times m\ddot{\mathbf{s}}$$

Noting that the moment $\mathbf{M}_s = \mathbf{s} \times m\ddot{\mathbf{r}}$ (not $\mathbf{s} \times m\ddot{\mathbf{s}}$!), we substitute $\ddot{\mathbf{s}} = \ddot{\mathbf{r}} - \ddot{\mathbf{R}}_s$ to rewrite $\dot{\mathbf{H}}_{s_r}$ as

$$\dot{\mathbf{H}}_{s_r} = \mathbf{s} \times m(\ddot{\mathbf{r}} - \ddot{\mathbf{R}}_s) = \mathbf{s} \times m\ddot{\mathbf{r}} - \mathbf{s} \times m\ddot{\mathbf{R}}_s$$

or

$$\dot{\mathbf{H}}_{s_r} = \mathbf{M}_s - \mathbf{s} \times m\ddot{\mathbf{R}}_s \tag{2.18}$$

Note that our use of relative linear momentum to define $\mathbf{H}_{s_r}$ has no influence on the moment $\mathbf{M}_s$, which is the same, whether we use $\mathbf{H}_{s_r}$ or $\mathbf{H}_s$.

At this point the reader may wonder why it is useful to define the relative angular momentum (2.17) and the resulting moment law (2.18). One reason is that $\mathbf{H}_{s_r}$ is often a very convenient quantity to calculate, so the formulation (2.18) can be quite useful. More importantly, when we extend our results for a single particle to develop similar relations for rigid bodies, we will see that the quantity used to characterize rotation of rigid bodies is the relative angular momentum $\mathbf{H}_{s_r}$, not $\mathbf{H}_s$. Thus, we must have a clear understanding of the distinction between the two.

---

*This is not quite true, as it ignores the influence of gravitational forces due to the other planets, the moon, and so on, but it is often an adequate approximation.

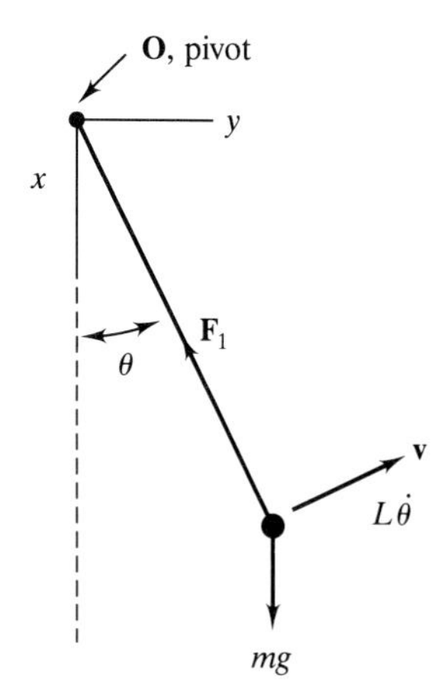

FIGURE 2.4
The simple pendulum.

Now let us consider a couple of examples that illustrate the analysis we have done so far. The first example, Example 2.1 below, illustrates that, for a given system, there may be several ways to set up a mathematical model and that some ways turn out to be very inconvenient. The second example (Example 2.2) illustrates both the use of (2.15) to set up a mathematical model using a moving reference point S and the use of the relative angular momentum $\mathbf{H}_{s_r}$ and associated moment law (2.18).

Example 2.1 The Simple Pendulum. The equations of motion of a particle can *always* be developed using Newton's second law for translation, (2.10) or (2.12). In some situations, however, the moment/angular momentum relations (2.15) and (2.18) are more convenient. This is nicely illustrated by the simple pendulum (Fig. 2.4). A particle of mass m, attached to a massless string of length L, rotates in the plane without friction about the fixed point **O**. We can identify the position **r**, velocity **v**, and acceleration **a** in terms of the cartesian or polar coordinates shown. Hence, either (2.10), (2.12), or (2.16) can be used to develop the mathematical model of the system.

Note that there are two forces acting on the particle: the gravitational force mg (in the plus x direction), and the force F_1 exerted by the string on the particle and directed through the pivot **O**.

Let us first employ the rotational law (2.16) to develop an ODE mathematical model of the system. The angular momentum of the particle m with respect to the fixed point **O** is $\mathbf{H}_O = \mathbf{r} \times m\mathbf{v}$. Using the polar coordinate relations (2.11a and b) and noting that $\dot{r} = 0$ because the particle remains a fixed distance from the origin, the position and velocity vectors are $\mathbf{r} = L\mathbf{e}_r$, $\mathbf{v} = L\dot{\theta}\mathbf{e}_\theta$. Thus, $\mathbf{H}_O = mL^2\dot{\theta}\mathbf{k}$, where $\mathbf{k}$ is the unit vector out of the paper ($\mathbf{k} = \mathbf{e}_r \times \mathbf{e}_\theta$), about which the rotation occurs. This result for $\mathbf{H}_O$ is the familiar result $H = I\omega$ from beginning dynamics, where $I = mL^2$ is the moment of inertia of the particle m with respect to **O** and where $\omega = \dot{\theta}$ is the angular velocity. The rate of change of angular momentum is obtained by differentiating $\mathbf{H}_O$, resulting in $\dot{\mathbf{H}}_O = mL^2\ddot{\theta}\mathbf{k}$. The applied moment $\mathbf{M}_O$ is due only to the gravitational force because the string force F_1 is always directed through the moment reference **O**. Noting that the position vector **r** may be written in cartesians as $\mathbf{r} = L\cos\theta\mathbf{i} + L\sin\theta\mathbf{j}$ and that the force due to gravity is $\mathbf{F} = mg\mathbf{i}$, the moment $\mathbf{M}_O = -mgL\sin\theta\mathbf{k}$.* Thus, the system ODE, or "equation of motion," is obtained by equating components of the vectors $\dot{\mathbf{H}}_O$ and $\mathbf{M}_O$ and is

$$mL^2\ddot{\theta} = -mgL\sin\theta \tag{2.19}$$

This is a second-order autonomous ODE which is nonlinear because of the $\sin\theta$ term. The system order is 2, and the states are θ and $\dot{\theta}$.

Now as a second line of attack, let us use the translational formulation (2.12) in polar coordinates, in which the forces and accelerations are to be resolved into radial and tangential components. Noting that the force components are $F_r = -F_1 + mg\cos\theta$, $F_\theta = -mg\sin\theta$, and that $\dot{r} = \ddot{r} = 0$, the distance from m to **O** being fixed, (2.12) yields the two equations;

$$-mL\dot{\theta}^2 = -F_1 + mg\cos\theta \tag{2.20a}$$

$$mL\ddot{\theta} = -mg\sin\theta \tag{2.20b}$$

Observe that the tangential (second) equation of motion is merely (2.19) divided by L. The radial (first) equation will enable us to solve for the force F_1 (which is initially unknown), provided that we first solve the second equation to find θ and $\dot{\theta}$. This polar coordinate formulation is useful if we need to know the tension F_1 in the string; otherwise, (2.20b) and (2.19) are essentially the same.

It is important to note in this problem that the system possesses a single degree of freedom: only one coordinate θ is needed in order to completely specify the position of the particle.

Now let us try the third formulation, (2.10), in cartesians. With $F_x = mg - F_1\cos\theta$, $F_y = -F_1\sin\theta$, (2.10) yields

$$m\ddot{x} = mg - F_1\cos\theta$$

$$m\ddot{y} = -F_1\sin\theta.$$

We can now eliminate the θ terms from geometry: $\sin\theta = y/L$, $\cos\theta = x/L$, so that

$$m\ddot{x} = mg - \frac{F_1 x}{L} \tag{2.21a}$$

$$m\ddot{y} = \frac{-F_1 y}{L} \tag{2.21b}$$

At this point we encounter a difficulty because there are three unknowns (x, y, F_1) and only two equations. We recognize, however, that x and y are not independent but are related to each other by geometry, $x^2 + y^2 = L^2$, so we now have three equations (two ODEs and one algebraic equation) in three unknowns.

In order to proceed further, we do the following: Solve (2.21a) for F_1 and use the result to eliminate F_1 in (2.21b); this revised version of (2.21b) will then

---

*This can be seen easily by inspection, but we have to be careful about the sign: If θ is positive, then the moment is about an axis *into* the paper, i.e., in the $-\mathbf{k}$ direction (and vice versa).

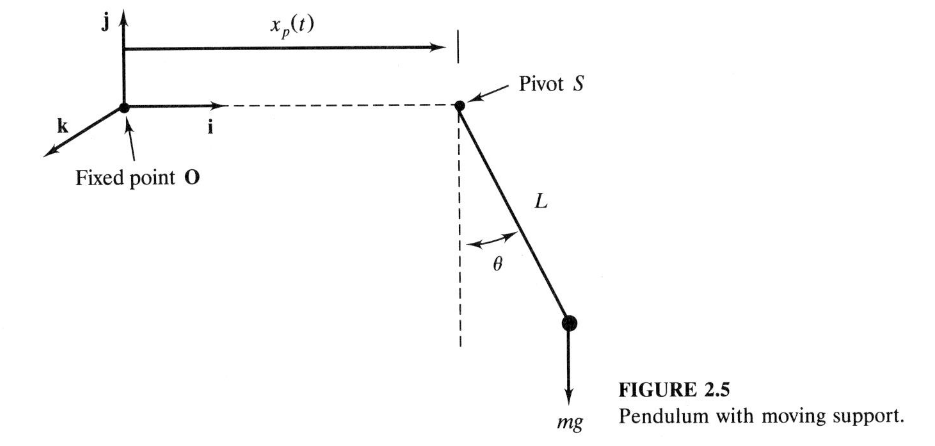

FIGURE 2.5
Pendulum with moving support.

contain terms involving both x and y. The next step is to use the geometric relation $x^2 + y^2 = L^2$ to eliminate all x terms in (2.21b), finally obtaining a single equation involving only the coordinate y. The courageous reader is invited to execute these steps as an exercise. It is clear that this procedure is a difficult way to set up the mathematical model.

The points that this example illustrates are the following: (1) There may be several ways to define coordinates and to set up the equations of motion for a given system, and (2) some ways may be more direct than others.

Example 2.2 The Pendulum with Moving Support. We now take our pendulum of Example 2.1 and assume that the pivot S translates in the horizontal direction in some specified manner; the position of the pivot S from the origin **O** of a fixed cartesian system is denoted by $x_p(t)$, where the input $x_p(t)$ is assumed to be a known function of time (see Fig. 2.5, which is a concrete example of the general situation shown in Fig. 2.3). Otherwise, the system is the same as in Example 2.1.

The mathematical model can be conveniently developed using (2.15), where the moving pivot S is the reference point. First we tabulate the vector quantities needed in the calculation:

$$
\begin{aligned}
\mathbf{r} &= \left[x_p(t) + L\sin\theta\right]\mathbf{i} - L\cos\theta\,\mathbf{j} \\
\dot{\mathbf{r}} &= \left[\dot{x}_p(t) + L\dot{\theta}\cos\theta\right]\mathbf{i} + L\dot{\theta}\sin\theta\,\mathbf{j} \\
\mathbf{s} &= L\sin\theta\,\mathbf{i} - L\cos\theta\,\mathbf{j} \\
\mathbf{R}_s &= x_p(t)\mathbf{i}
\end{aligned}
\tag{2.22}
$$

The angular momentum $\mathbf{H}_s$ with respect to the moving reference S is then

$$\mathbf{H}_s = \mathbf{s} \times m\dot{\mathbf{r}} = \left(mL^2\dot{\theta} + mL\dot{x}_p\cos\theta\right)\mathbf{k}$$

The moment $\mathbf{M}_s$ is the same as in Example 2.1:

$$\mathbf{M}_s = -mgL\sin\theta\,\mathbf{k}$$

We thus obtain from (2.15) the equation of motion:

$$\dot{\mathbf{H}}_s = \left(mL^2\ddot{\theta} + mL\ddot{x}_p \cos\theta - mL\dot{x}_p\dot{\theta}\sin\theta\right)\mathbf{k}$$

$$= -mgL\sin\theta\mathbf{k} + \left(L\dot{\theta}\cos\theta\mathbf{i} + L\dot{\theta}\sin\theta\mathbf{j}\right) \times m\left(\dot{x}_p\mathbf{i}\right)$$

Working out the algebra and noting that the resulting vectors are all directed along **k**, we obtain the second-order differential equation of motion:

$$mL^2\ddot{\theta} = -mgL\sin\theta - mL\ddot{x}_p\cos\theta \qquad (2.23)$$

Observe that, in comparison to (2.19), the support motion exerts a direct influence on the motion of the pendulum. Note also that we *could* have obtained a mathematical model by using (2.16) and taking the fixed point **O** as the moment reference. This would have had the disadvantage of introducing the force F_1 into the formulation, since this force would no longer be applied through the moment reference. The real advantage of (2.15) is the automatic elimination of F_1.

We can also obtain the equation of motion (2.23) by setting up the relative angular momentum (2.17) and then using the moment law (2.18). First we calculate the relative angular momentum $\mathbf{H}_{s_r} = \mathbf{s} \times m\dot{\mathbf{s}}$; using the relation for **s** from (2.22), we find

$$\mathbf{H}_{s_r} = mL^2\dot{\theta}\mathbf{k} \qquad \left(\Rightarrow \dot{\mathbf{H}}_{s_r} = mL^2\ddot{\theta}\mathbf{k}\right)$$

The moment $\mathbf{M}_s = -mgL\sin\theta\mathbf{k}$ as before, and calculation of the extra term in (2.18) gives $-\mathbf{s} \times m\ddot{\mathbf{R}}_s = -mL\ddot{x}_p\cos\theta\mathbf{k}$. Thus, the equation of motion is

$$mL^2\ddot{\theta} = -mgL\sin\theta - mL\ddot{x}_p\cos\theta$$

which is the same as (2.23). The mathematical model (2.23) is of second order, and the system states are θ and $\dot{\theta}$. The model is nonlinear due to the presence of the terms $\sin\theta$ and $\cos\theta$, which are nonlinear functions of the state θ. The model is nonautonomous due to the given function of time $\ddot{x}_p(t)$. The model is also time-varying because the term $\ddot{x}_p\cos\theta$ involves a known function of time multiplying a function of a state.

Example 2.2 is intended to illustrate a situation that commonly occurs in mechanical systems: A body rotates about a pivot that is itself in motion. The motion of the pivot influences the motion of the body, and the basic laws (2.15) and (2.18) through which this effect is modeled need to be understood.

2.2 TRANSLATION AND PLANE ROTATION OF RIGID BODIES

In this section we will apply Newton's second law for particle motion to develop laws of motion for rigid bodies. A *rigid body* is a material object occupying a fixed volume (deformations such as we study in strength of materials are not allowed) and has mass which is continuously distributed over this volume. We will analyze the rigid body by considering it to be a collection of a fixed number

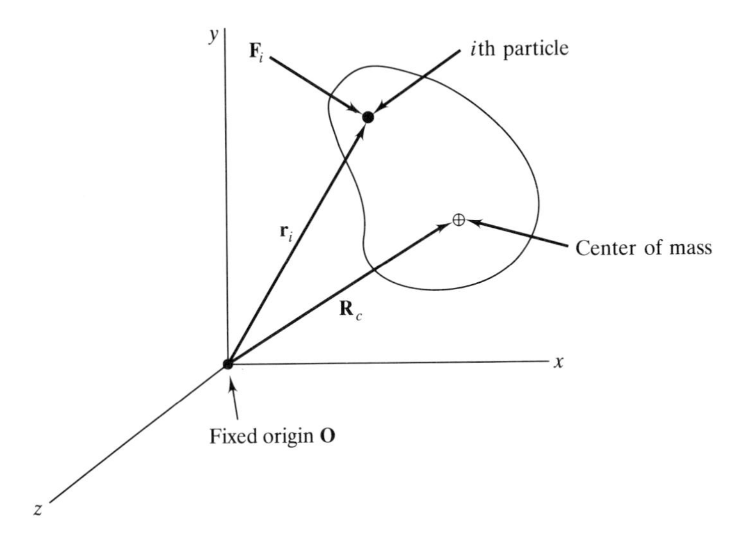

FIGURE 2.6
The rigid body as an assemblage of particles.

N of particles, such that each particle remains a fixed distance from all others. Our collection of N particles will then become a rigid body in the limit $N \to \infty$.

In general, the motion of a rigid body will consist of both translation and rotation; the rotational motion is what distinguishes a rigid body from a particle. To start, let's consider the translation of the N particles that comprise the rigid body. In Fig. 2.6 a representative particle (the ith) has mass m_i and is located by a position vector $\mathbf{r}_i$ from the origin **O**, which is fixed in an inertial frame. The particle m_i is subjected generally to several externally applied forces that have a resultant $\mathbf{F}_i$.*

The center of mass of the rigid body, located by the vector $\mathbf{R}_c$, is defined by

$$\mathbf{R}_c = \frac{1}{M} \sum_{i=1}^{N} m_i \mathbf{r}_i \tag{2.24}$$

where $M = \sum_{i=1}^{N} m_i$ is the mass of the rigid body. Newton's law for a representative particle states that

$$\mathbf{F}_i = m_i \mathbf{a}_i = m_i \ddot{\mathbf{r}}_i$$

where $\mathbf{a}_i = \ddot{\mathbf{r}}_i$ is the (inertial) acceleration of the ith particle. A summation over

---

*Internal forces, exerted by the particles on each other, also exist, but these forces will cancel out in our analysis and are not considered here.

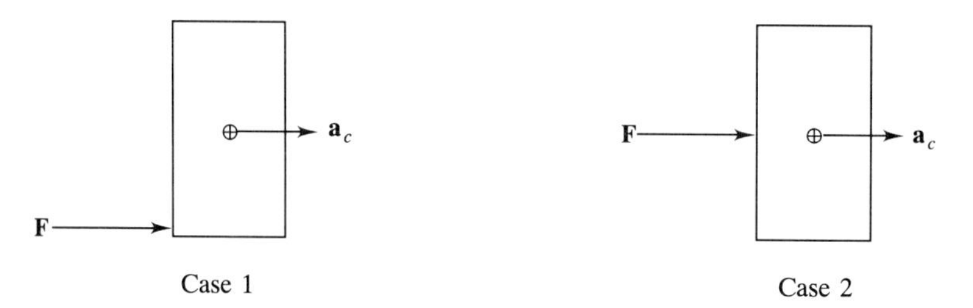

FIGURE 2.7
The instantaneous acceleration of the center of mass is the same for the two cases shown.

the N particles then yields

$$\sum_{i=1}^{N} \mathbf{F}_i = \sum_{i=1}^{N} m_i \ddot{\mathbf{r}}_i$$

Now we define the overall resultant force as $\mathbf{F} = \Sigma \mathbf{F}_i$, and we note that, because each particle has fixed mass, the right-hand side of the preceding equation may be rewritten as follows:

$$\sum m_i \ddot{\mathbf{r}}_i = \frac{d^2}{dt^2}\left(\sum m_i \mathbf{r}_i\right) = \frac{d^2}{dt^2}(M\mathbf{R}_c) = M\ddot{\mathbf{R}}_c$$

Thus, we obtain the important result:

$$\mathbf{F} = M\ddot{\mathbf{R}}_c = M\mathbf{a}_c \tag{2.25}$$

where the subscript $(\ )_c$ refers to the motion of the center of mass. Equation (2.25) shows that the center of mass of a rigid body, under the action of the resultant force $\mathbf{F}$, will translate in exactly the same way as a single particle of mass M. It is this important result that justifies our detailed study of the single particle.

Note that it does not matter whether the resultant force $\mathbf{F}$ is actually applied through the center of mass. Thus, the instantaneous acceleration of the center of mass will be exactly the same in the two situations shown in Fig. 2.7, in which identical forces are applied at different locations on the body. In case 1 of Fig. 2.7 we will also have some rotational acceleration.

Next, let us consider the *planar* rotational mechanics of a rigid body. In Fig. 2.8 we define the coordinates necessary to specify the location of every point on the rigid body. We need two displacements to locate any point on the body. The point selected in Fig. 2.8 is the center of mass, with cartesian coordinates (x_c, y_c). We also need an angle θ to measure the orientation of the body. This angle may be defined as the angle between an arbitrarily specified reference line in the plane and a line painted on the rigid body. Observe that the angular velocity vector ($\boldsymbol{\omega} = \dot{\theta}\mathbf{k}$) and angular acceleration ($\boldsymbol{\alpha} = \ddot{\theta}\mathbf{k}$) are each directed along the z axis normal to the plane in which the motion takes place. This unidirectional nature of the vectors involved in plane rotation (angular

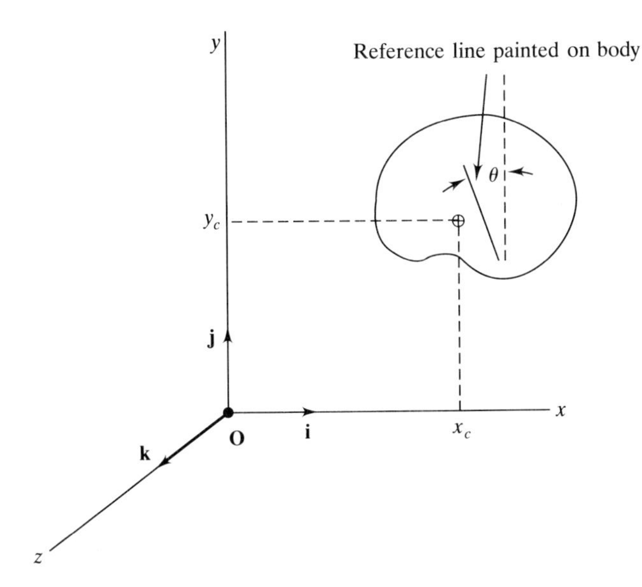

FIGURE 2.8
Coordinates defining the planar motion of a rigid body.

momentum and torque will also be directed along the z axis) is what makes plane rotation such a simple type of motion. We see that a rigid body confined to a planar motion has 3 degrees of freedom. Thus, we will in general have three second order-ordinary differential equations to describe the plane motion of the rigid body: Two of these will be the x and y (or r and θ) components of (2.10) or (2.12), which define the translational motion of the center of mass. The third will be the equation of rotational motion, to which we now turn.

Our objective now is to present the basic laws for planar rotation. We will couch the formulation in terms of the relative angular momentum $\mathbf{H}_{s_r}$ of the rigid body since this is the quantity most often used in calculations. Thus, we will need to understand two things: (1) how to determine $\mathbf{H}_{s_r}$ in specific cases and (2) how to relate the rate of change of the relative angular momentum $\mathbf{H}_{s_r}$ to the applied moment $\mathbf{M}_s$. We again consider the rigid body to be a collection of N particles, each with a resultant force acting upon it. The vector quantities needed in the formulation are defined in Fig. 2.9.

Here the cartesian coordinate system is fixed in an inertial frame, and the (inertial) position of the ith particle is denoted by $\mathbf{r}_i$. The inertial position of the center of mass is $\mathbf{R}_c$, while $\mathbf{R}_s$ locates the position of the moment reference, which may be moving in an arbitrary fashion. The vector $\boldsymbol{\rho}_i$ locates the ith particle relative to the center of mass, while the vector $\mathbf{s}_i$ locates the ith particle relative to the moment reference. Finally, $\mathbf{r}_{c/s}$ locates the center of mass relative to the moment reference. Thus, the relative angular momentum $\mathbf{H}_{s_r}$ for the system of particles is defined by summing (2.17) over the collection of particles:

$$\mathbf{H}_{s_r} \equiv \sum_{i=1}^{N} \mathbf{s}_i \times m_i \dot{\mathbf{s}}_i \tag{2.26}$$

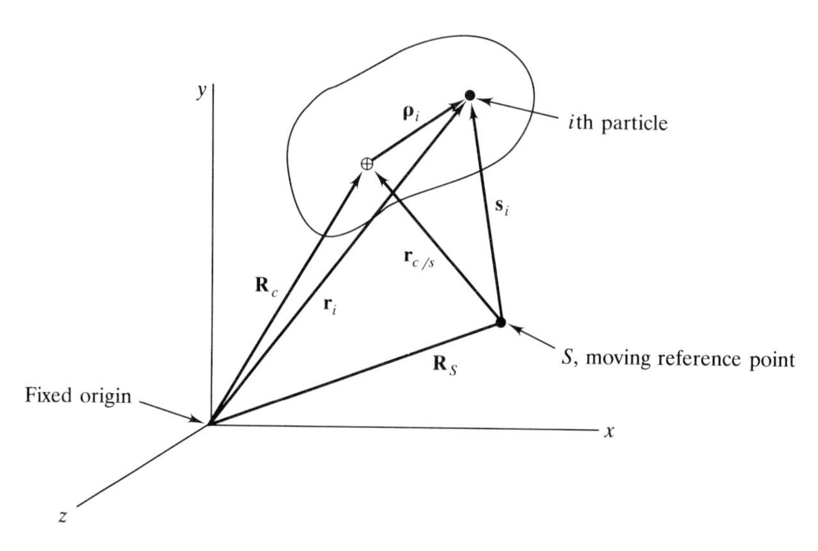

FIGURE 2.9
Definition of vector quantities for analysis of rigid body plane rotation.

To begin, we will develop the relation between the rate of change of angular momentum $\dot{\mathbf{H}}_{s_r}$ and applied moment $\mathbf{M}_s$, which is defined for the collection of particles by

$$\mathbf{M}_s = \sum_{i=1}^{N} \mathbf{s}_i \times \mathbf{F}_i = \sum_{i=1}^{M} \mathbf{s}_i \times m_i \ddot{\mathbf{r}}_i$$

Then we will discuss the calculation of $\mathbf{H}_{s_r}$.

Now from (2.26) we find $\dot{\mathbf{H}}_{s_r}$ for the system of particles to be

$$\dot{\mathbf{H}}_{s_r} = \sum_{i=1}^{N} \left(\cancel{\dot{\mathbf{s}}_i \times m_i \dot{\mathbf{s}}_i} + \mathbf{s}_i \times m_i \ddot{\mathbf{s}}_i \right)$$

In order to bring the applied moment $\mathbf{M}_s$ into the formulation, we use the vector identity $\mathbf{s}_i = \mathbf{r}_i - \mathbf{R}_s$ to rewrite the $\ddot{\mathbf{s}}_i$ term in $\dot{\mathbf{H}}_{s_r}$, so that

$$\dot{\mathbf{H}}_{s_r} = \sum_{i=1}^{N} \mathbf{s}_i \times m_i \ddot{\mathbf{r}}_i - \sum_{i=1}^{N} \mathbf{s}_i \times m_i \ddot{\mathbf{R}}_s$$

Next, note that $\sum \mathbf{s}_i \times m_i \ddot{\mathbf{R}}_s = (\sum m_i \mathbf{s}_i) \times \ddot{\mathbf{R}}_s$, since $\ddot{\mathbf{R}}_s$ does not participate in the summation; further, $\sum m_i \mathbf{s}_i = M\mathbf{r}_{c/s}$, so that we arrive at the result:

$$\dot{\mathbf{H}}_{s_r} = \mathbf{M}_s - M\mathbf{r}_{c/s} \times \ddot{\mathbf{R}}_s \tag{2.27}$$

This is the basic relation that we will use in applications; it is very close to the single particle relation (2.18).

Now that the basic rigid body law (2.27) has been developed,* let us see how to determine the relative angular momentum $\mathbf{H}_{s_r}$ in specific cases. Going back to the definition (2.26), we first use the vector relation $\mathbf{s}_i = \mathbf{r}_{c/s} + \boldsymbol{\rho}_i$ to rewrite (2.26) as

$$\mathbf{H}_{s_r} = \sum_{i=1}^{N} \left[(\mathbf{r}_{c/s} + \boldsymbol{\rho}_i) \times m_i (\dot{\mathbf{r}}_{c/s} + \dot{\boldsymbol{\rho}}_i) \right]$$

Expanding this relation, we obtain

$$\mathbf{H}_{s_r} = \sum_{i=1}^{N} \left(\mathbf{r}_{c/s} \times m_i \dot{\mathbf{r}}_{c/s} + \boldsymbol{\rho}_i \times m_i \dot{\boldsymbol{\rho}}_i + \mathbf{r}_{c/s} \times m_i \dot{\boldsymbol{\rho}}_i + \boldsymbol{\rho}_i \times m_i \dot{\mathbf{r}}_{c/s} \right)$$

The last two terms vanish because of the definition of the center of mass, and we are left with

$$\mathbf{H}_{s_r} = \sum_{i=1}^{N} \boldsymbol{\rho}_i \times m_i \dot{\boldsymbol{\rho}}_i + \mathbf{r}_{c/s} \times M \dot{\mathbf{r}}_{c/s} \tag{2.28}$$

The result (2.28) applies generally, whether the rotational motion is planar or three-dimensional. Let us now specialize this result for the case of plane rotation. First we note that the velocity $\dot{\boldsymbol{\rho}}_i$ of the ith particle relative to the center of mass will be due to rotation only, so that $\dot{\boldsymbol{\rho}}_i = \boldsymbol{\omega} \times \boldsymbol{\rho}_i$, where the angular velocity $\boldsymbol{\omega} = \dot{\theta}\mathbf{k}$. If we let $\boldsymbol{\rho}_i = x_i\mathbf{i} + y_i\mathbf{j}$, where $\mathbf{i}$ and $\mathbf{j}$ are the x and y direction unit vectors originating at the center of mass, so that (x_i, y_i) locate the ith particle relative to the center of mass, then the first term in $\mathbf{H}_{s_r}$ works out to be

$$\sum_{i=1}^{N} \boldsymbol{\rho}_i \times m_i \dot{\boldsymbol{\rho}}_i = \dot{\theta} \left[\sum_{i=1}^{N} m_i (x_i^2 + y_i^2) \right] \mathbf{k}$$

Now as we pass to the limit $N \to \infty$, so that the mass is distributed continuously, we replace the summation $\Sigma_{i=1}^{N}$ by an integral over the area occupied by the rigid body, and replace m_i by μdA, where μ is the mass per unit area and $dA = dx\,dy$ is an incremental area. Thus, $\mathbf{H}_{s_r}$ becomes

$$\mathbf{H}_{s_r} = \left[\int_A \mu (x^2 + y^2)\, dA \right] \dot{\theta}\mathbf{k} + \mathbf{r}_{c/s} \times M \dot{\mathbf{r}}_{c/s}$$

Now we recognize that the quantity $\int_A \mu (x^2 + y^2)\, dA$ is the moment of inertia, with respect to the center of mass, for rotation about an axis passing through the center of mass and aligned with the $z(\mathbf{k})$ direction. Defining

$$I_c = \int_A \mu (x^2 + y^2)\, dA$$

---

*This law (2.27) actually applies generally whether the rotational motion is planar or three-dimensional.

we obtain the desired result for $\mathbf{H}_{s_r}$ in the form (planar motion!):

$$\mathbf{H}_{s_r} = I_c\dot{\theta}\mathbf{k} + \mathbf{r}_{c/s} \times M\dot{\mathbf{r}}_{c/s} \tag{2.29}$$

This basic result shows that the relative angular momentum $\mathbf{H}_{s_r}$ of a rigid body consists of the sum of two parts: (1) the angular momentum relative to the center of mass, plus (2) the relative angular momentum of the center of mass (acting like a single particle) with respect to the moment reference.

Now the basic results that we need to construct mathematical models for plane rotation are (2.27) and (2.29). Let us see how these results simplify in certain special cases.

Case 1. The center of mass is selected as the moment reference. In this case, because the vector $\mathbf{r}_{c/s} = 0$, the relations (2.27) and (2.29) for $\mathbf{H}_{s_r}$ and $\dot{\mathbf{H}}_{s_r}$ reduce to

$$\mathbf{H}_{c_r} = I_c\dot{\theta}\mathbf{k}$$

$$\dot{\mathbf{H}}_{c_r} = \mathbf{M}_c$$

where *a* $(\ )_c$ subscript is used to signify that the center of mass is the moment reference. The equation of rotational motion is then simply

$$I_c\ddot{\theta} = M_c \tag{2.30}$$

where $\mathbf{M}_c \equiv M_c\mathbf{k}$; that is, M_c is the scalar moment (about the $\mathbf{k}$ axis) with respect to the center of mass. The simple relation (2.30) is what renders the center of mass such a useful moment reference. The result applies even if the center of mass is translating in an arbitrary manner.

Case 2. The moment reference S is a *fixed* point, other than the center of mass, on the rigid body; we now have "plane rotation about a fixed point." In this case $\ddot{\mathbf{R}}_s = 0$ in (2.27); furthermore, (2.29) simplifies to

$$\mathbf{H}_{s_r} = \left(I_c + Md^2\right)\dot{\theta}\mathbf{k} \tag{2.31}$$

where d is the distance from the reference S to the center of mass. This result reflects the familiar parallel axis theorem. Thus, for plane rotation about a fixed point S, we obtain the following equation of rotational motion:

$$\left(I_c + Md^2\right)\ddot{\theta} = M_s \tag{2.32}$$

Case 3. The moment reference S is a point on the rigid body, other than the center of mass, and the point S translates in an arbitrary manner. In this case $\mathbf{H}_{s_r}$ is still given by (2.31), but we need to retain the extra term in (2.27), as neither $\mathbf{r}_{c/s}$ nor $\ddot{\mathbf{R}}_s$ is zero. In this case the equation of motion reads

$$\left(I_c + md^2\right)\ddot{\theta} = M_s - \left(M\mathbf{r}_{c/s} \times \ddot{\mathbf{R}}_s\right)_{\mathbf{k}} \tag{2.33}$$

where the notation indicates that the **k** component of $M\mathbf{r}_{c/s} \times \ddot{\mathbf{R}}_s$ is to be used. Here we have to set up the vectors $\mathbf{r}_{c/s}$ and $\mathbf{R}_s$ and do the indicated calculation.

The basic rotational laws for the plane rotation of a rigid body, Eqs. (2.30), (2.32), and (2.33), will be applied to several example problems in Sec. 2.5.

2.3 WORK AND KINETIC AND POTENTIAL ENERGIES

In the preceding Secs. 2.1 and 2.2, we have been concerned with the essentials of the *vector* formulation of the laws of motion. There are, however, certain *scalar* quantities that we may also use to model the behavior of mechanical systems. Foremost among these are kinetic energy, work, and potential energy, which we now discuss.

We will start by recalling the definition of the kinetic energy T enjoyed by a single particle:

$$T = \tfrac{1}{2}m\mathbf{v} \cdot \mathbf{v} \tag{2.34}$$

In cartesian coordinates, use of (2.2) leads to

$$T = \tfrac{1}{2}m\left(\dot{x}^2 + \dot{y}^2 + \dot{z}^2\right)$$

while in polar coordinates, (2.11*b*) leads to

$$T = \tfrac{1}{2}m(\dot{r}^2 + r^2\dot{\theta}^2)$$

Next, we define the work W done by the applied force **F** as the particle moves along its path of motion. With reference to Fig. 2.10, we can pick any two points 1 and 2 (defined by position vectors $\mathbf{r}_1$ and $\mathbf{r}_2$), and the work W_{12} is then

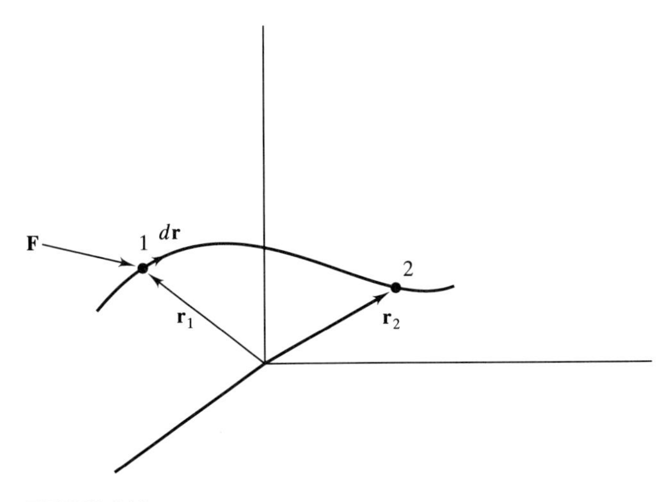

FIGURE 2.10
Work done by the applied force **F**.

defined by the following line integral:

$$W_{12} = \int_{\mathbf{r}_1}^{\mathbf{r}_2} \mathbf{F} \cdot d\mathbf{r} \tag{2.35}$$

According to this definition, only the component of the force in the direction of motion, which is defined by the infinitesimal vector $d\mathbf{r}$, does any work. For instance, the string force $\mathbf{F}_1$ in the simple pendulum (Example 2.1, Fig. 2.4) does no work, as it is always directed normal to the direction of motion.

Equation (2.35) can be used to relate the work done to the kinetic energy by converting (2.35) to a time, rather than line, integral. Since $d\mathbf{r} = \mathbf{v}\,dt$, (2.35) can be rewritten as

$$W_{12} = \int_{t_1}^{t_2} \mathbf{F} \cdot \mathbf{v}\,dt \tag{2.36}$$

The quantity $\mathbf{F} \cdot \mathbf{v}$ is recognized as the power, the rate at which work is done by the force $\mathbf{F}$. Next, we can replace $\mathbf{F}$ in (2.36) by $m\mathbf{a} = m(d\mathbf{v}/dt)$ to obtain

$$W_{12} = \int_{t_1}^{t_2} m\frac{d\mathbf{v}}{dt} \cdot \mathbf{v}\,dt$$

Next, we note that $(d\mathbf{v}/dt) \cdot \mathbf{v} = \frac{1}{2}(d/dt)(\mathbf{v} \cdot \mathbf{v})$, the dot product being commutative, so that

$$W_{12} = \int_{t_1}^{t_2} \frac{1}{2}m\frac{d}{dt}(\mathbf{v} \cdot \mathbf{v})\,dt$$

or

$$W_{12} = \frac{1}{2}mv^2\Big|_{t_1}^{t_2} = \frac{1}{2}mv_2^2 - \frac{1}{2}mv_1^2$$

where $v^2 \equiv (\mathbf{v} \cdot \mathbf{v})$ is the square of the particle speed. In other words,

$$W_{12} = T_2 - T_1 \tag{2.37}$$

that is, the *work done is equal to the change in the kinetic energy*. This is a general result, regardless of the nature of the forces that act on the particle.

There is a very nice interpretation of *certain types* of forces which depend only on the *position* of the particle; such forces are denoted notationally as $\mathbf{F} = \mathbf{F}(\mathbf{r})$. These forces are characterized by what we call static *fields of force*, or simply *force fields*: At each point (x, y, z) in space we can place an arrow to model the magnitude and direction of the force field at that point. Examples include gravitational forces, electromagnetic forces, forces due to mechanical stiffness, and so on. Note that forces which depend on velocity, such as damping, or aerodynamic drag forces, are not of this type. Thus, the particle has a force exerted on it merely as a result of its particular position in space. Force fields are examples of *vector fields*; to each point in space is assigned a vector, and the whole collection of such vectors forms the *vector field*.

Now let's picture the particle moving through a force field from one point $\mathbf{r}_1$ to another $\mathbf{r}_2$. It turns out that the work done by the force $\mathbf{F}(\mathbf{r})$ may be

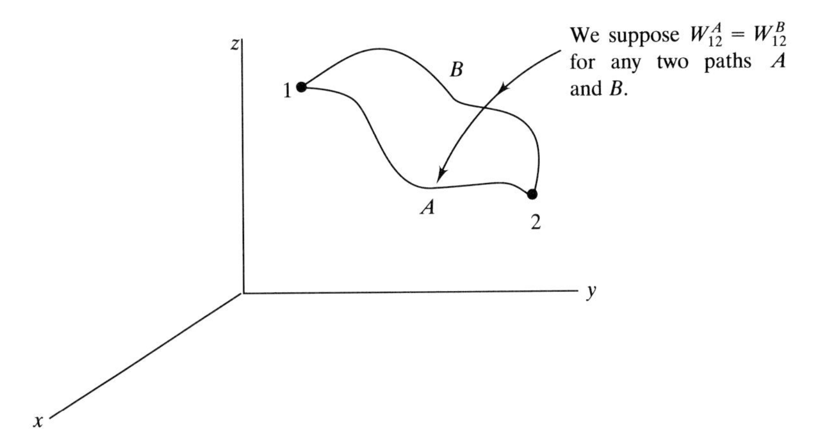

FIGURE 2.11
Path independence of work for conservative forces.

independent of the *path* but dependent only on the end points $\mathbf{r}_1$ and $\mathbf{r}_2$. See Fig. 2.11.

Let's see what this condition of path independence, if it exists, says mathematically. First, we note that the only way to get a line integral defined between two points to depend only on these end points is if the integral is of the form $\int d\phi$, that is, we have to be integrating the *perfect differential* of some *scalar* function (call it ϕ) since, by definition, $\int d\phi = \phi(\mathbf{r}_2) - \phi(\mathbf{r}_1)$. Here $\phi = \phi(\mathbf{r})$ or $\phi(x, y, z)$ is a *scalar field*: To each point (x, y, z) in space we assign a scalar value ϕ, rather than a vector value, as in the case of the force field. The temperature in a three-dimensional solid is an example of a scalar field. Thus, we have to conclude that if the work done is to be independent of the path, then we must be able to write $\mathbf{F} \cdot d\mathbf{r} = d\phi$, where ϕ at this point is just some scalar field.

To see why this stuff is useful, let's continue the mathematical development. If we write the condition $\mathbf{F} \cdot d\mathbf{r} = d\phi$ out the long way by first noting that

$$\mathbf{F} = \mathbf{F}_x\mathbf{i} + F_y\mathbf{j} + F_z\mathbf{k}$$

$$d\mathbf{r} = dx\,\mathbf{i} + dy\,\mathbf{j} + dz\,\mathbf{k}$$

so that

$$\mathbf{F} \cdot d\mathbf{r} = F_x\,dx + F_y\,dy + F_z\,dz \tag{2.38}$$

and if we then write $d\phi$ out using the chain rule with $\phi = \phi(x, y, z)$,

$$d\phi = \frac{\partial \phi}{\partial x}\,dx + \frac{\partial \phi}{\partial y}\,dy + \frac{\partial \phi}{\partial z}\,dz \tag{2.39}$$

we conclude that the respective coefficients of dx, dy, and dz in (2.38) and (2.39) have to be equal since (2.38) and (2.39) must hold for *any* combination of the infinitesimals dx, dy, and dz. Thus, we conclude that the three scalar components of $\mathbf{F}$ must be equal to the three partial derivatives of ϕ as follows:

$$F_x = \frac{\partial \phi}{\partial x}, \qquad F_y = \frac{\partial \phi}{\partial y}, \qquad F_z = \frac{\partial \phi}{\partial z}$$

A more compact way to write this result is

$$\mathbf{F} = \nabla \phi$$

where the operator ∇ stands for the "gradient," defined by

$$\nabla \phi = \frac{\partial \phi}{\partial x}\mathbf{i} + \frac{\partial \phi}{\partial y}\mathbf{j} + \frac{\partial \phi}{\partial z}\mathbf{k}$$

The gradient of a *scalar* function is a *vector*. At any point (x, y, z) the gradient points in the direction of the *largest rate of change* of the scalar function, and it is numerically equal to the rate of change of the scalar function along that direction. The idea of the gradient is perhaps easy to visualize if you picture the temperature gradients in a solid.

Now, then, how is all of this related to the potential energy? Well, the potential energy, which we'll call V, a scalar function of position, $V = V(x, y, z)$, is just $-\phi$. The minus sign is used for reasons that will emerge shortly. The result is this: If a particular force that depends only on position has the additional property that the work it does depends only on the end points but not on the path connecting these end points, then it is possible to express such a force as the negative gradient of the "potential energy":

$$\mathbf{F} = -\nabla V \qquad F_x = -\frac{\partial V}{\partial x} \qquad F_y = -\frac{\partial V}{\partial y} \qquad F_z = -\frac{\partial V}{\partial z} \tag{2.40}$$

Thus, we see that the potential energy is just an alternate way to formulate mathematically a particular type of force. Assuming we know the force $\mathbf{F}(\mathbf{r})$, then we can use (2.40) to find the corresponding potential energy. Simple examples that we commonly encounter are the following 1-D forces:

Gravity: $F_z = -mg$, $V = mgz$, where z is the height of the mass m

The linear spring: $F_x = -kx$, $V = \frac{1}{2}kx^2$, where x is the displacement of the spring

The simple pendulum: $V = mgz = mgL(1 - \cos\theta)$

To see a primary reason for the utility of the potential energy, let's suppose that *all* of the forces involved in the given particle problem are position dependent only and produce path-independent work. Then we can define an overall potential energy V for the system and compute the work done in terms of it:

$$\int_{\mathbf{r}_1}^{\mathbf{r}_2} \mathbf{F} \cdot d\mathbf{r} = \phi_2 - \phi_1 = V_1 - V_2$$

Thus, as with the kinetic energy, the work done by such forces is equal to the (negative) change in potential energy. Recalling that

$$\int_{\mathbf{r}_1}^{\mathbf{r}_2} \mathbf{F} \cdot d\mathbf{r} = T_2 - T_1$$

we can combine the preceding two equations to arrive at the law of *conservation of total energy* E (the total energy E is here defined as $E \equiv T + V$):

$$E = T_1 + V_1 = T_2 + V_2$$

Inasmuch as this result applies for *any* two points $\mathbf{r}_1$ and $\mathbf{r}_2$, it must hold *in general*, that is, for all time, so that

$$E = T + V = \text{constant} \tag{2.41}$$

In practice we would evaluate T and V at $t = 0$ using the initial conditions; thereafter T and V will generally both change as the particle moves around, but such changes in T and V will occur so as to conserve their sum. For this reason forces of the type we are now considering are called *conservative*. Forces that depend on velocity (such as damping) and/or time explicitly (such as an external, time-dependent driving force) do not qualify as conservative forces. In addition, not all forces that depend only on position are conservative, as we can easily devise force fields $\mathbf{F} = \mathbf{F}(\mathbf{r})$ for which the work done is path dependent.

It is important to keep in mind the following distinction between the kinetic and potential energies: The fact that the work done is equal to the change in kinetic energy is a general result, valid for *any* type of force. On the other hand, the fact that the work done may also equal the (negative) change in the potential energy is *not* a general result; it is true only for a certain type of force (a conservative one).

So far we have been concerned mainly with the fundamentals underlying the development of ODE models of particle motion. If we can solve these ODEs, then we are able to determine the position and velocity of the particle at every instant of time. It sometimes occurs, however, that such a detailed description of the motion is not necessary; we may only need to look at the state of the system at certain times or at certain locations. In such cases, conservation of energy can be usefully applied, assuming, of course, that we can ignore nonconservative forces. For example, consider again the simple pendulum (Example 2.1), with the problem stated as follows: For the system initially at rest, $\theta(0) = 0$, what initial angular velocity $\dot{\theta}_0$ is required in order that the pendulum execute a whirling rotation rather than an oscillatory swinging motion? To solve this problem, we can calculate the initial angular velocity $\dot{\theta}_0$ needed to make the pendulum just come to rest at $\theta = 180°$, as follows:

$$\tfrac{1}{2}mL^2\dot{\theta}_0^2 = 2mgL$$

where the term on the left is the initial energy (all kinetic) and where $2mgL$ is the energy (all potential) at $\theta = \pi$. Thus, for the system to execute whirling motion, we need $\dot{\theta}_0 > (4g/L)^{1/2}$. In Example 2.11 and in later chapters we will see other useful ways to utilize work and energy in the description of the motion of dynamic systems.

2.4 CONSTITUTIVE MODELS FOR MECHANICAL SYSTEMS

In Secs. 2.1 and 2.2 we studied basic laws of motion based on newtonian mechanics. These laws relate applied loadings (forces and moments) to the motions which they produce, as measured by the rate of change of linear or angular momentum. In the construction of mathematical models of mechanical systems using these laws, much of the work will center on the development of constitutive relations that describe the various individual forces/moments acting on the system. In this section we consider certain basic types of forces/moments and state simple versions of the constitutive relations for each. Then, in Sec. 2.5 the basic laws and constitutive relations are combined for a variety of examples to develop mathematical models of typical mechanical systems.

Recalling the discussion of Sec. 2.1, we note that, in general, the forces and moments will be functions of position, velocity, and/or time. Our first task in the modeling of specific loadings will thus be to establish their functional dependence. Then we can attempt to devise a mathematical relation that states this dependence, at least approximately. We now consider several generic types of forces/moments, classified according to their functional dependence.

2.4.1 The Force / Moment Is a Function of Position

To fix ideas, suppose that we have a system with only 1 degree of freedom (a translational coordinate x or a rotational coordinate θ) and that the force or moment is a function of position only, $F_x = F_x(x)$ or $M = M(\theta)$. Thus, the loading is a *static* function, depending only on position (not on velocity nor explicitly on time).

An important force of this type is the force due to gravity. This force, exerted mutually by any two interacting masses, is proportional to the product of the masses and inversely proportional to the square of the distance between them. For bodies moving over the earth's surface, a convenient mathematical representation of this force is as follows: Let R be the radius of an assumed spherical earth ($R \cong 6371$ km), and let z be the distance of an object of mass m above the earth's surface, measured positive upward. Then the gravitational force F_G exerted by the earth on the mass m can be expressed as

$$F_G(z) = -mg\left(\frac{R}{R+z}\right)^2 \tag{2.42}$$

where $g = 9.8\ \text{m/s}^2$ is the gravitational acceleration at the surface of the earth.

We observe that this force is a nonlinear function of the coordinate z. In many applications z is small compared to R, so that $R^2/(R+z)^2$ is essentially unity, and we may use $F \cong -mg$; the force has reduced to a constant. Notice that a negative sign has been included in (2.42) in order to signify that the force is directed opposite the direction of increasing z (it is crucial to keep track of signs in force/moment modeling!).

A second generic type of position-dependent loading may be classified as "stiffness." *Stiffness* may be defined as the property whereby, if one attempts to displace a body from its rest position, an opposing force is *generated* as a result of this displacement, and the resulting generated force tries to push the body back to its rest position. Such loadings are often referred to as *restoring forces* (they attempt to restore the body to a rest position). Stiffness effects can arise from a variety of sources, including structural stiffness, gravity, and magnetic and pneumatic stiffness. Structural elements such as beams, plates, trusses, frames, and airplane wings, all have the property that if one displaces them from rest, a restoring force is generated which is a function of the amount of displacement. Thus, structural systems inherently possess stiffness. Oftentimes, components are designed specifically to produce restoring forces, such as coil or leaf springs in an automobile suspension system. The second stiffness effect is gravity. The static moment produced by gravity in the simple pendulum (Example 2.1) and the self-righting tendency of boats in rolling motion are examples of *gravitational stiffness*, that is, restoring forces or moments produced by gravity. Other stiffness effects include magnetic and pneumatic stiffness. In these cases stiffness forces may be generated by the position of an object in a magnetic field or through compression of a gas (these types of stiffening will be considered in Chap. 3).

The idealized "lumped" device that we use to represent structural stiffness is the coil spring system, shown schematically in Fig. 2.12. We define a coordinate x which measures the amount of displacement from rest (at rest, $x = 0$; the spring generates no force). Then, if we were to slowly displace the spring end by various amounts and measure the force generated, i.e., the force exerted *by* the spring *on* the agent that is moving the spring end, we might obtain the "force-deflection plot" shown in Fig. 2.13. Note that when $x = 0$, $F_x = 0$ and when $x > 0$, $F_x < 0$, indicating the restoring nature of the force; i.e., when x is positive, the force exerted by the spring is in the negative x direction.

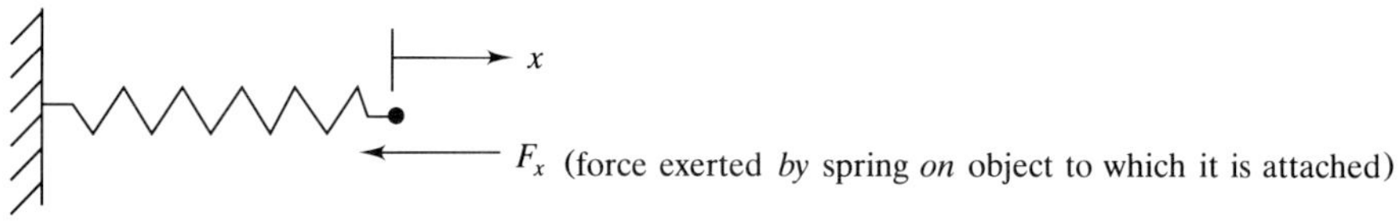

x

F_x (force exerted *by* spring *on* object to which it is attached)

FIGURE 2.12
The idealized linear stiffening element: The coil spring.

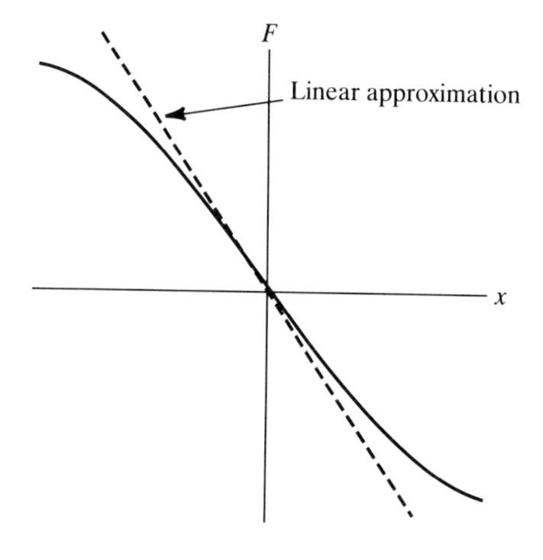

FIGURE 2.13
Force/deflection plot to characterize system stiffness.

Now assuming the force to be a smooth function of position, we can formally expand $F(x)$ in a Taylor series about the rest position $x = 0$:

$$F(x) = F(x = 0) + \left.\frac{dF}{dx}\right|_{x=0} x + \frac{1}{2}\left.\left(\frac{d^2F}{dx^2}\right)\right|_{x=0} x^2 + \cdots \tag{2.43}$$

where the derivatives $[(dF/dx)]_{x=0}, [(d^2F/dx^2)]_{x=0}$ are to be evaluated at the rest position $x = 0$ and hence are assumed to be known constants. To simplify the notation, we can rewrite (2.43) as

$$F(x) = a_1 x + a_2 x^2 + a_3 x^3 + \cdots \tag{2.44}$$

where $a_1 = [(dF/dx)]_{x=0}$ and so on.

For a restoring force, $a_1 < 0$. The purpose in developing (2.44) is to show that, if we know the force to be a smooth function of position only, then we can express the dependence as the Taylor series (2.44). What we usually do next, in order to obtain a simple model, is to retain only the linear term in (2.44) and to define the *linear spring* by

$$F(x) \cong -kx \tag{2.45}$$

where $k \equiv -(dF/dx)|_{x=0}$ is the spring constant, in units of force per unit displacement (N/m or lb/ft). This linear approximation would be as shown in Fig. 2.13. What we observe is that this linear spring model should work well for "small displacements" x. For a given set of measured data which Fig. 2.13 represents, we can identify a "linear range of operation." We have to keep in mind, however, that the linear model will generally break down if the displacements become large enough.

For rotational motion the analogous physical effect is *torsional stiffness*, in which a rotation of a body from its rest position results in a restoring torque which is a function of rotation angle θ. We denote such an effect schematically with the torsional spring, as shown in Fig. 2.14 and described by the following

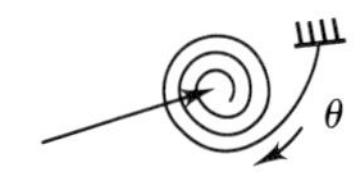

FIGURE 2.14
The torsional spring model.

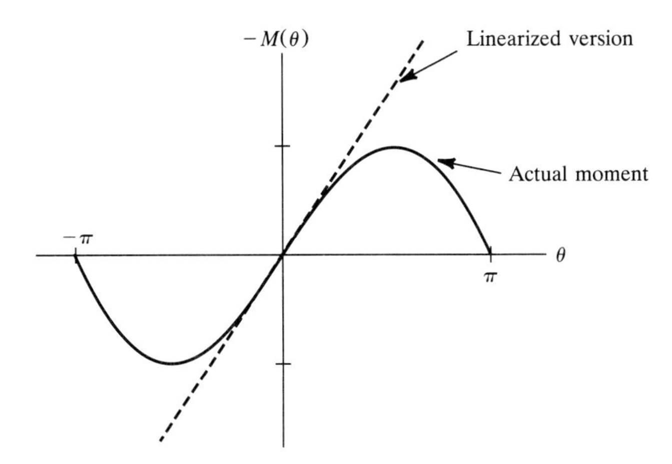

FIGURE 2.15
The actual and linearized moments for the simple pendulum.

relation which is analogous to (2.45):

$$M(\theta) = -k_T\theta \tag{2.46}$$

Here the restoring moment, according to a linear model of the effect, is $M(\theta) = -k_T\theta$. The torsional stiffness k_T has units of torque per unit rotation: N-m/rad or ft · lb/rad. Note that in (2.45) and (2.46) the constants k and k_T are positive for a restoring force, and the negative signs indicate this explicitly. Note further that we expect the linear model (2.46) to be valid only in some limited range of operation, that is, for relatively small rotations.

For example, in the simple pendulum system Example 2.1, the moment due to gravity is a static function of the angle θ and is given by

$$M(\theta) = -mgL\sin\theta$$

This moment is a nonlinear function of rotation angle θ and is shown in Fig. 2.15.

If we expand $\sin\theta$ in a Taylor series, $\sin\theta = \theta - \frac{1}{3!}\theta^3 + \frac{1}{5!}\theta^5 + \cdots$, we obtain the moment $M(\theta)$ as

$$M(\theta) = -mgL\left(\theta - \tfrac{1}{6}\theta^3 + \tfrac{1}{120}\theta^5 + \cdots\right)$$

which is analogous to (2.44) for the force $F(x)$. Retaining only the linear term,

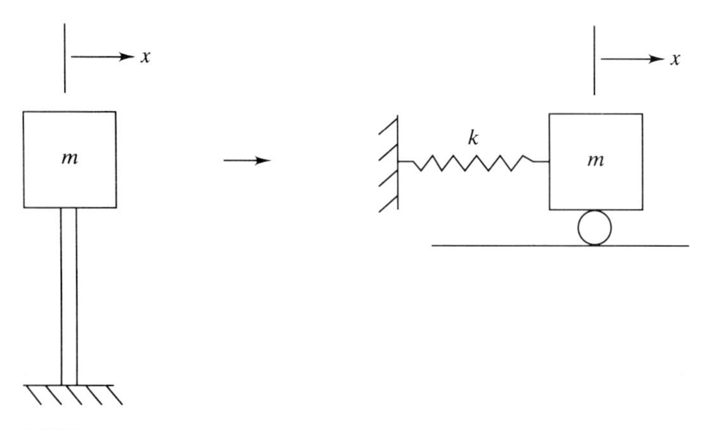

FIGURE 2.16
Idealization of a translational mass/structure system as a mass/linear spring system.

we have the approximate moment model:

$$M(\theta) \cong -mgL\theta \tag{2.47}$$

The torsional stiffness k_T in (2.46) is then $k_T = mgL$. We can see from Fig. 2.15 that (2.47) is a reasonable approximation in the approximate range of $-30° < \theta < 30°$

Frequently in the construction of mathematical models of actual systems, we idealize the actual mechanism producing stiffness by using the linear spring. Thus, a mass m supported on a simple structural element may be modeled as a mass m restrained by a linear spring, as shown in Fig. 2.16. The same idea applies to torsional effects produced by structural members (Fig. 2.17).

Of course, in using idealized components such as linear springs, it is necessary to establish appropriate values for the spring constant k or k_T. This may be done through testing, as in the force-deflection plot, or sometimes, by

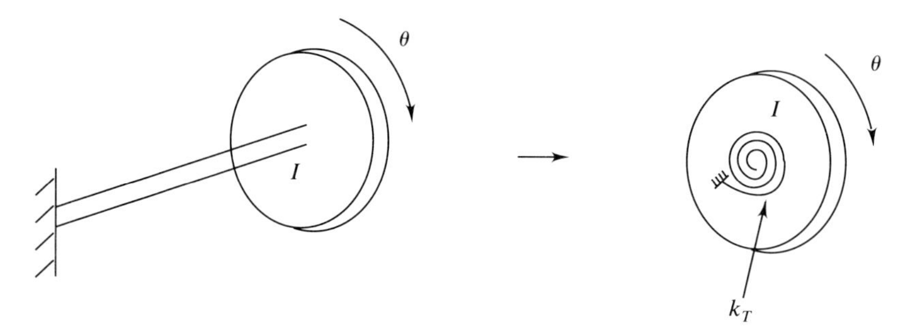

FIGURE 2.17
Idealization of a rotational mass/structure system as an inertia I restrained by a torsional spring.

calculation; from strength of materials we may be able to determine the relation between an applied static load and the resulting deflection of the system. Because stiffness forces are *static*, we can often utilize the techniques of static structural analysis to determine analytically the stiffness properties of a given system.

An important characteristic of the idealized stiffness-producing elements discussed here is that they are inherently conservative. The work done to displace the ideal spring through a certain distance is completely recoverable. We can determine the potential energy stored in an idealized spring to be $V = \frac{1}{2}kx^2$ for a translational spring and $V = \frac{1}{2}k_T\theta^2$ for a torsional spring. Thus, if bodies move under the action of only these types of forces or moments, the total energy will be conserved.

In Sec. 2.5 we will see examples of the use of idealized stiffness elements to construct fairly realistic models of mechanical systems, in which several springs may be used to represent the different stiffnesses that restrain the various masses comprising the system.

2.4.2 The Force / Moment Is a Function of Velocity

Velocity-dependent forces and moments often arise as a result of fluid motion. Examples include the aerodynamic forces and moments exerted on vehicles in flight, the effect of air resistance on the motion of a vibrating structure, and the forces generated in a shock absorber due to motion of a piston in a surrounding fluid. In order to characterize such forces and moments, let us again assume for illustration that the force/moment of interest is a function of a single coordinate, say, $\dot{x}$, so that a graph of force versus velocity might look like Fig. 2.18. If the force is assumed to be a smooth function of velocity only, but not of position or time, then we can use the Taylor series representation to express the functional dependence of F on $\dot{x}$ as

$$F(\dot{x}) = b_1\dot{x} + b_2\dot{x}^2 + b_3\dot{x}^3 + \cdots \tag{2.48}$$

where b_n are the coefficients in the Taylor series expansion of $F(\dot{x})$,

$$b_n = \frac{1}{n!}\left(\frac{d^nF}{d\dot{x}^n}\right)\bigg|_{\dot{x}=0}$$

and where we have assumed that $F(0) = 0$. Let us now start with the linear version of (2.48) and define the idealized "linear viscous damper" as a device or effect having the constitutive relation

$$F(\dot{x}) = -c\dot{x} \tag{2.49}$$

where the constant c (assumed positive) is the so-called damping constant, having units of force per unit velocity (N-s/m or lb-s/ft). According to this model, if $c > 0$, then a force is exerted opposite the direction of motion, with $-c$ being the proportionality constant in the Taylor series expansion of $F(\dot{x})$.

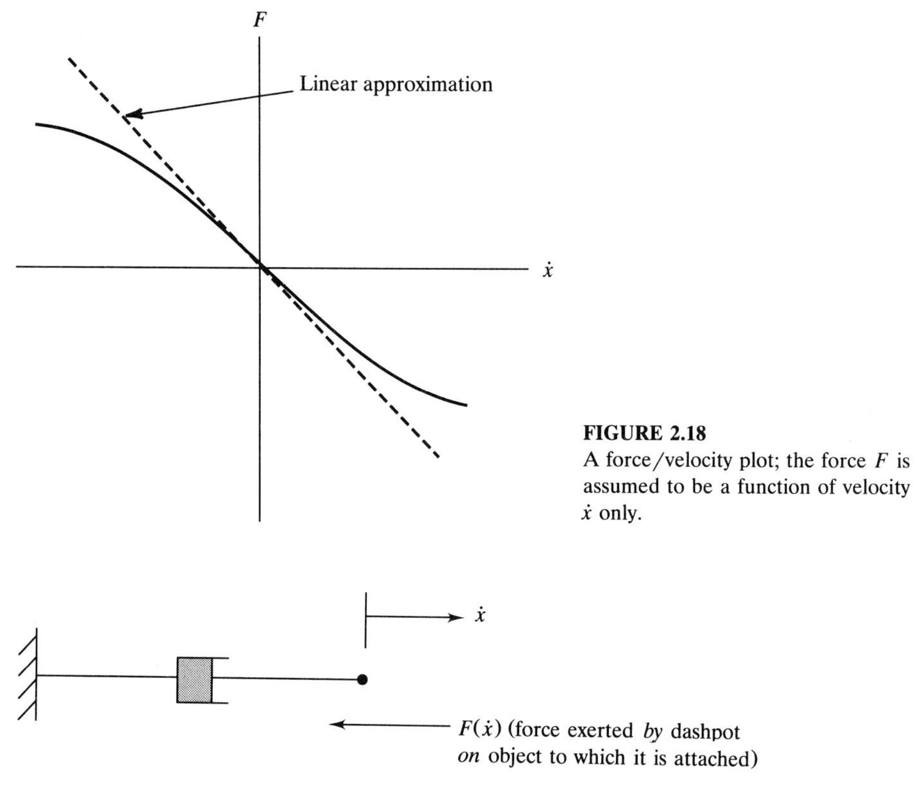

FIGURE 2.18
A force/velocity plot; the force F is assumed to be a function of velocity $\dot{x}$ only.

FIGURE 2.19
The linear viscous damper (dashpot).

The idealized physical model of this effect is the *dashpot*, shown schematically in Fig. 2.19.

Here a piston moves in a cylinder containing a viscous fluid. The viscous resistance to piston motion produces a force that is modeled as being proportional to the instantaneous velocity and in the opposite direction. The device is called a *damper* because the work done is not recoverable; hence the damper is a nonconservative device. For instance, the work done by the damping force between two arbitrary times t_1 and t_2 during which the system is in motion will be [from (2.36)]

$$W_{12} = \int_{t_1}^{t_2} (-c\dot{x})\dot{x}\,dt = \int_{t_1}^{t_2} -c\dot{x}^2\,dt$$

Because $-c\dot{x}^2$ is always negative, the overall system mechanical energy (kinetic plus potential) is continually being decreased. What happens is that the damper continually converts mechanical (kinetic, potential) energy into other forms, such as heat or motion of the surrounding fluid, which cannot be recovered.

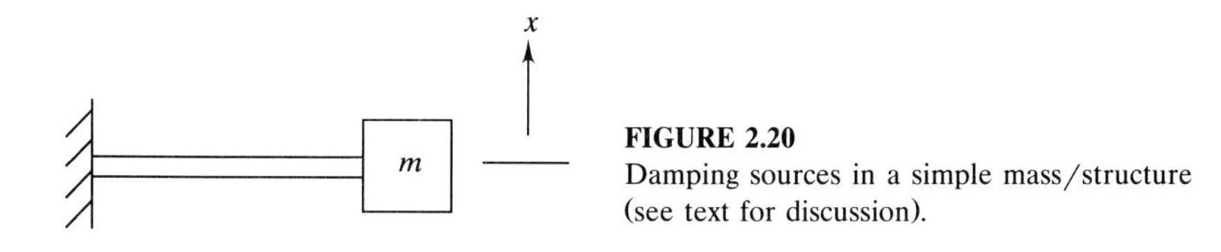

FIGURE 2.20
Damping sources in a simple mass/structure (see text for discussion).

People often speak of "energy dissipation due to viscous damping"; what is really meant by this phrase is that some energy is converted to another form, so that the mechanical energy associated with the motion of the system is reduced.

The linear damping device is often used to represent any effect that tends to "dissipate" energy. As an example, consider a bar clamped at one end and supporting a mass m at the other, Fig. 2.20. As the mass m moves, we can identify three obvious sources of energy dissipation: (1) damping due to the surrounding air (the moving system is subjected to viscous resistance and also induces motion of, i.e., imparts kinetic energy to, the surrounding fluid); (2) dissipation in the supporting member; small amounts of heat are generated when a solid is stressed, so that a small amount of the work done to deflect the stiffness element is actually converted to heat; and (3) rubbing and dry friction at the point at which the beam is clamped. These three types of energy dissipation mechanisms are referred to as *fluid damping*, *material damping*, and *dry friction damping*, respectively. Any or all of these mechanisms are usually present in realistic mechanical systems.

From the preceding discussion it seems clear that it would be exceedingly difficult to attempt to calculate a priori the value of an overall damping constant c to represent all of the damping mechanisms that may be present. One usually has to resort to testing, as discussed, for example, in Sec. 7.2.1, in order to obtain a reasonable estimate for the system damping constants, and this represents a case in which "system identification" is especially important. In any case, we can now represent the system of Fig. 2.20 by a model consisting of a mass and two ideal force-producing elements to represent the stiffness and damping, as shown in Fig. 2.21.

Regarding velocity-dependent forces, we close with two remarks:

1. The generation of velocity-dependent forces that result from fluid motion, as, for example, in an actual automotive shock absorber, is a very complicated phenomenon that often results in a functional dependence $F(\dot{x})$, which is nonlinear. For example, Fig. 2.22 shows force versus velocity data measured for several standard automotive shock absorbers (after Ref. 2.4).

   One observes significant nonlinearity. In addition, there is a *deadband*, a region of velocities near $\dot{x} = 0$ for which $F(\dot{x}) = 0$. The functions $F(\dot{x})$ shown are, therefore, not smooth functions of $\dot{x}$ and cannot be expanded in Taylor series. Thus, in order to find a damping constant c that represents one of these devices, one needs to find a straight line which in some sense best represents the data, at least for small to moderate velocities. Clearly, for

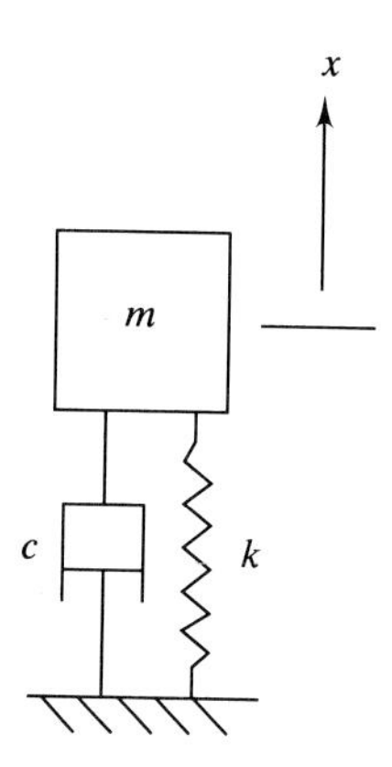

FIGURE 2.21
Idealization of the system of Fig. 2.20 by a mass and linear damping and stiffness elements.

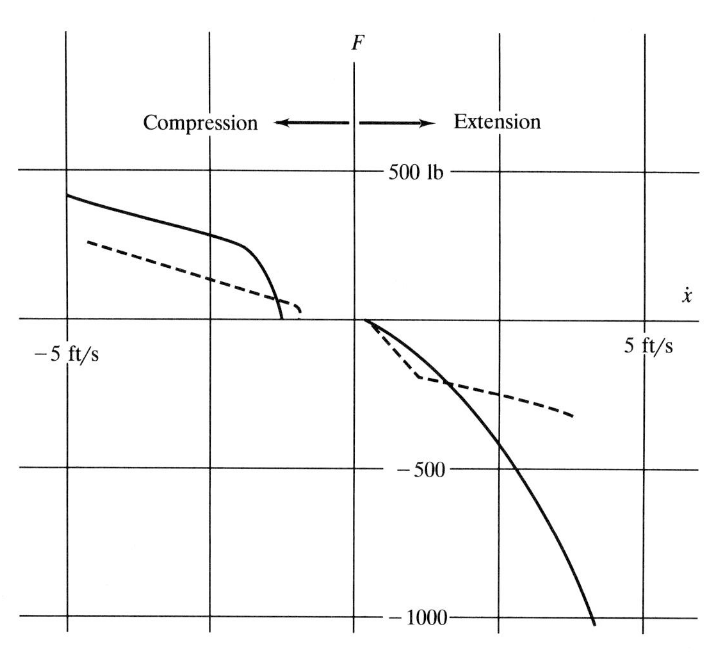

FIGURE 2.22
Force/velocity plots for shock absorbers (Ref. 2.4).

the data shown, a linear damping model may be suspect if one desires quantitative accuracy in the model.

2. In some situations velocity-dependent forces are inherently nonlinear and simply cannot be represented by a linear model. For example, consider the aerodynamic drag force exerted on a flight vehicle or on a car at moderate to high speed. This force turns out to be proportional to the *square* of the velocity

$$F(\dot{x}) = -\alpha \dot{x}^2 \tag{2.50}$$

where α depends on the size and shape of the object and on the density of the fluid medium.* In the context of (2.48), the linear term in the Taylor series has dropped out. A linear model would not be appropriate in this situation.

In the preceding discussion we have presented idealized linear force/moment models for loadings that depend only on position or only on velocity. It is also possible that the loadings will depend simultaneously on both position and velocity (an example of this situation is given in Example 2.7 in Sec. 2.5). In any case any force $F(x, \dot{x})$ that depends on position and/or velocity, but not explicitly on time, is *generated* by the motion of the system and is, therefore, considered to be a part of the system, from the standpoint of mathematical modeling.

2.4.3 The Force / Moment Depends Explicitly on Time

Loadings that depend only on time, and hence arise independently of the motion of the system, are *inputs* that are (mathematically, at least) not considered to be part of the system itself. Examples of such loadings include the thrust forces produced by rocket engines and prescribed motions of the supports by which a system is grounded. For example, the motion of the ground during an earthquake acts as an input that drives the resulting vibratory motions of buildings. Forces and moments generated by rotating or reciprocating parts in an engine or machine are other examples of explicitly time-dependent external excitations (inputs).

2.4.4 Concluding Remarks

In the next section we will illustrate the modeling ideas discussed here with several examples of simple mechanical systems. In doing so, we have to keep several procedural aspects in mind. Our first step will be to identify the various motions that the system can exhibit, or which we want to consider, and to then define a set of coordinates, one for each degree of freedom, which can be used to describe these motions. Then we identify the various loadings present and devise a constitutive model for each. Finally, we apply an appropriate physical law from newtonian mechanics. The resulting mathematical models will then serve as conceptual substitutes for the actual systems under study.

---

*Actually, we should express this model as $F = -\alpha\dot{x}|\dot{x}|$ in order that the force always be opposite the direction of motion.

2.5 MECHANICAL SYSTEM MODELING: EXAMPLES

In this section the modeling ideas presented in Chap. 1 and in the preceding sections of Chap. 2 are illustrated by example. Keep in mind that the intent is only to develop governing ODE models, not to solve them. For each example, the classifications according to Sec. 1.5 are discussed, as are other features relevant to the modeling process.

Example 2.3 The Harmonic Oscillator. This is an idealized model in which the forces are due to two sources: (1) *stiffness*, a static force, depending only on position, and (2) *damping*, a velocity-dependent force that opposes the motion. Pictorially we represent the system as in Fig. 2.23.

The stiffness is represented by a linear spring, so that the force F_s exerted on the mass m by the spring is defined by $F_s = -kx$; the spring force is assumed to be proportional to the deflection x; k is the *spring constant*, in units of lb/ft or N/m. The negative sign in F_s is important and states that the force opposes the displacement; if we move the mass m to the right, $x > 0$, the spring force is to the left, $F_s < 0$, and vice versa.

The damping is represented by the dashpot, which exerts a force F_d on the mass m, given by $F_d = -c\dot{x}$. The *damping constant* c has units of lb-s/ft or N-s/m. Note again that the negative sign indicates that if $\dot{x} > 0$, then F_d is in the negative x direction, and vice versa.

Putting the two forces into (2.8), we obtain the "equation of motion" of the system:

$$m\ddot{x} = -c\dot{x} - kx \tag{2.51}$$

Notice that the two forces appearing in (2.51) arise as a result of the motion of the mass m. Sometimes we encounter external, time-dependent forces $F(t)$ that exist independently of the motion. The addition of such a force would produce the equation of motion:

$$m\ddot{x} + c\dot{x} + kx = F(t) \tag{2.52}$$

We usually leave explicit time-dependent forces (inputs) on the right-hand side of the equation of motion and move all terms involving the dependent variable x and its derivatives to the left-hand side; thus all terms that define the *system* appear on

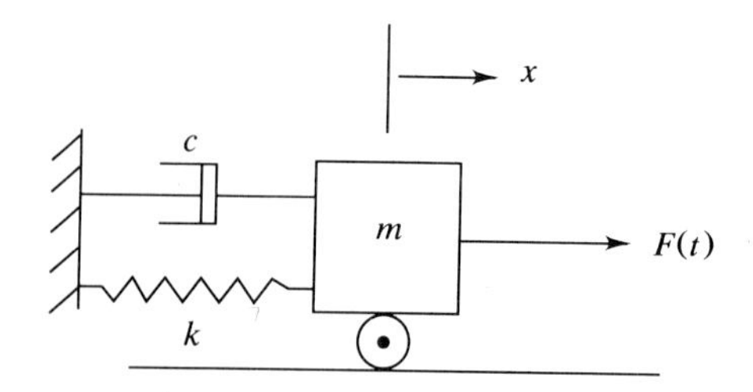

FIGURE 2.23
The linear harmonic oscillator model.

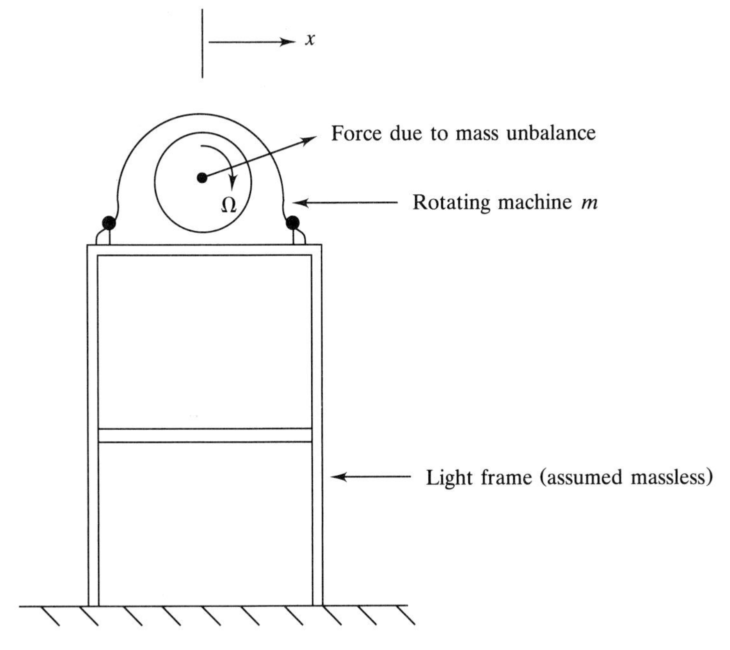

FIGURE 2.24
A rotating machine supported on a light frame structure.

the left-hand side, while the *input* $F(t)$ appears on the right-hand side. The system model (2.52) is second order, linear (by construction), and nonautonomous. The states are x and $\dot{x}$, and the input is $F(t)$.

The harmonic oscillator is a useful idealized model because it mimics many phenomena, which, through simplifying assumptions, can often be modeled by (2.52). An example follows.

Consider a rotating machine (a large fan, for example) mounted on a light frame structure as shown in Fig. 2.24. Suppose the rotating parts are out of balance, so that a centrifugal "force" is exerted on the mass m. This force rotates at the operating speed Ω, as shown. If we first consider only the horizontal component F_H of this force, we see that it will be sinusoidal, $F_H(t) = F_0 \cos \Omega t$, or $F_0 \sin \Omega t$, depending on how one defines the starting time $t = 0$. Thus, the machine will move in the horizontal direction in response to this horizontal input force. This motion, however, will be restrained by the frame, which provides stiffness and may be modeled as a linear spring, at least to start with. Likewise, there will be some damping forces due to air resistance and energy dissipation in the frame as it deflects, and a simple model of these effects would be the dashpot. Note that, on physical grounds, the damping forces produced by a dashpot are much different from those that really occur in the above system. Nevertheless, mathematically, the resulting forces may be assumed to have the same form, at least to start with. Thus, the simplest model for the horizontal motion of the rotating machine is just (2.52), with $F(t) = F_0 \cos \Omega t$, a harmonic load. Once we arrive at this model, we must then determine appropriate constants m, c, and k to put into the model. For the present system some testing would be required in order to establish appropriate values for the system constants.

Now, to be complete, we should also consider motion of the fan in the vertical direction. Usually, however, the frame is so stiff in this direction that such motions will be small, and we will ignore them here. In the present system violent horizontal vibration of the machine will occur if the mass unbalance is large enough and if the rotational operating speed Ω is in a certain range. This situation is discussed in Chap. 7. Other examples of vibrations due to unbalanced rotating parts would be in steam turbines, jet engines, automobile tires, or a washing machine with several pairs of dungarees on the same side of the barrel (I have, inadvertently, done experiments on this phenomenon in my basement).

Example 2.4 Automobile Suspension System (Simple). The purpose of an automotive suspension system is to isolate the car body from road irregularities. Thus, while the axles may undergo fairly violent motions in response to bumps, the car body is not supposed to be affected by them. We'll look at two models that might be used to study this problem and to design a suspension system.

We'll start off by making the following, obviously idealized assumptions: (1) Only vertical translations of the axles and car body are to be allowed, (2) the axles move up and down together as a single mass (this is a significant simplifying assumption), (3) front and rear suspensions (coil springs, leaf springs, shock

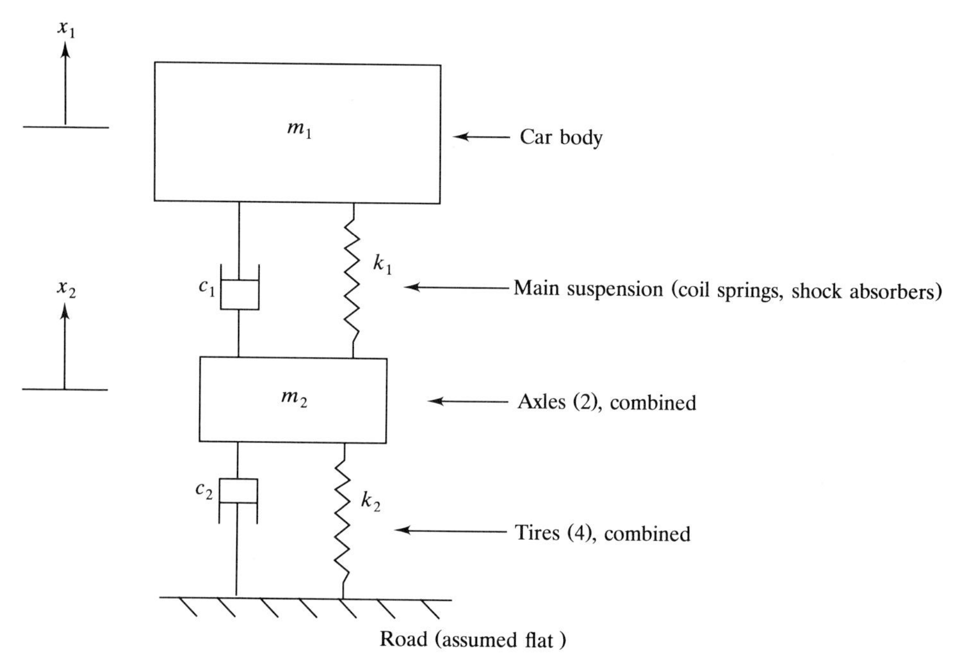

FIGURE 2.25
Idealized model of a 2 degree of freedom automobile-suspension system.

absorbers) are lumped together; this is a consequence of 2, (4) ignore gravity, and (5) the road over which the vehicle travels is smooth. Thus, our idealized model of the automobile-suspension system is as shown in Fig. 2.25.

Some notes on this model:

1. Since the car body and axles can move independently, we need two coordinates (x_1, x_2) to describe these two positions. We assume that at rest $x_1 = x_2 = 0$.
2. The main suspension elements (combined coil and leaf springs, shock absorbers) are modeled by a linear spring and a linear viscous damper; the linear damper is not always a good model of the shock absorber, as we have seen.
3. We have to account for the stiffness k_2 of the tires (tires are actually stiffer than the main suspension elements), and we might as well add some tire damping c_2.

Now we can proceed to develop the mathematical model. We do this by applying Newton's law for each of the two masses in the system. First, for the car body, the only forces acting are those due to the main suspension (k_1, c_1). First consider the spring force on m_1. The important thing to note is that this force will depend on *both* x_1 and x_2: The force is proportional to the distance through which the spring has stretched (or compressed), which is $x_1 - x_2$; for example, if $x_1 = x_2$, then there is no spring force. Thus, the spring force $F_{s_1} = -k_1(x_1 - x_2)$; if $x_1 - x_2 > 0$, then F_{s_1} is in the negative x_1 direction, so we need the minus sign. By similar reasoning, the damping force is given by $F_{D_1} = -c_1(\dot{x}_1 - \dot{x}_2)$; for example, if m_2 and m_1 are each moving upward at the same velocity, then the damping force F_{D_1} is zero. Thus, we can write down the car body equation of motion as

$$m_1\ddot{x}_1 = -c_1(\dot{x}_1 - \dot{x}_2) - k_1(x_1 - x_2) \tag{2.53}$$

Now the equation of motion for the axles m_2 is a bit more complicated, as both the tires and the main suspension exert forces on the axles. First, note that the forces exerted by the main suspension (k_1, c_1) will be equal and opposite those given in (2.53), a consequence of Newton's law of action and reaction. The tire forces depend only on the motion of the axles and hence have a form identical to those derived in Example 2.3 for the harmonic oscillator. Thus, we can write down the equation of motion of the axles as follows:

$$m_2\ddot{x}_2 = -c_1(\dot{x}_2 - \dot{x}_1) - k_1(x_2 - x_1) - c_2\dot{x}_2 - k_2x_2 \tag{2.54}$$

Rewriting (2.53) and (2.54) in the usual form in which all terms involving the dependent variables (x_1, x_2) and their derivatives appear on the left-hand side, we obtain the mathematical model for the idealized system as

$$m_1\ddot{x}_1 + c_1(\dot{x}_1 - \dot{x}_2) + k_1(x_1 - x_2) = 0 \tag{2.55a}$$

$$m_2\ddot{x}_2 + (c_1 + c_2)\dot{x}_2 - c_1\dot{x}_1 + (k_1 + k_2)x_2 - k_1x_1 = 0 \tag{2.55b}$$

This model consists of two, second-order linear, autonomous ODEs. Note that the linearity of this system has been imposed by us at the outset by assuming that all damping and stiffness effects can be modeled using linear constitutive relations. The system order is 4, and the states are x_1, $\dot{x}_1$, x_2, and $\dot{x}_2$. There are no inputs; the input for this type of system is provided by a bumpy road, as in the following, more realistic version of the same problem.

Example 2.5 Automobile Suspension System (More Realistic). One of the most limiting assumptions made in Example 2.4 is the lumping of the two axles into a single mass. We now relax this restriction and separate the axles. In doing so, we must also allow the car body to rotate, as well as to translate vertically. The idealized model is shown in Fig. 2.26. We now have a system with 4 degrees of freedom: vertical translation and plane rotation of the car body and vertical translation of each axle.

Nomenclature:

m_1, m_2, m_3	Car body, front axle, rear axle masses, respectively
I_c	Car body moment of inertia for plane rotation about center of mass
x_1, θ	Car body vertical translation of center of mass (x_1) and rotation (θ); θ assumed small
x_2, x_3	Vertical translation of front and rear axles
$k's, c's$	Main suspension and tire spring and damping constants
a_2, a_3	Distance from car body center of mass to point of attachment of front suspension (a_2) and rear suspension (a_3)
$x_f(t)$	Instantaneous height of road, relative to a "smooth road," under front wheels
$x_r(t)$	Same as $x_f(t)$ for rear wheels; note that $x_r(t)$ is just $x_f(t)$, delayed by a certain time τ, which depends on the wheelbase and speed of the car

Now we can derive the equations of motion. Starting with the vertical translation of the car body center of mass, the forces that must be modeled are those exerted by the front and rear suspensions. Let's start with the front suspension and determine the spring force. First, we can see that the amount of stretch of the front suspension springs will be given for small θ by $(x_1 + a_2\theta - x_2)$, where $a_2\theta$ is the vertical translation of the point of attachment of the front suspension relative to the center of mass. Thus, the front suspension spring force on the car body will be $-k_1(x_1 + a_2\theta - x_2)$. In a like manner, the rear suspension spring force will be $-k_2(x_1 - a_3\theta - x_3)$. Note here that a positive θ *decreases* the rear spring stretch. The damping forces are similar, with $c's$ replacing $k's$ and velocities replacing positions. Thus, the equation of vertical translation of the car body center of mass is as follows:

$$m_1\ddot{x}_1 + c_1\left(\dot{x}_1 + a_2\dot{\theta} - \dot{x}_2\right) + c_2\left(\dot{x}_1 - a_3\dot{\theta} - \dot{x}_3\right) + k_1(x_1 + a_2\theta - x_2)$$

$$+k_2(x_1 - a_3\theta - x_3) = 0 \quad (2.56)$$

Now let's move on to plane rotation of the car body about its center of mass. Choosing the center of mass as the moment reference enables us to use the law (2.30), and we must calculate the torques exerted about the center of mass. First, the front suspension force produces the following torque M_{s_1} about the center of

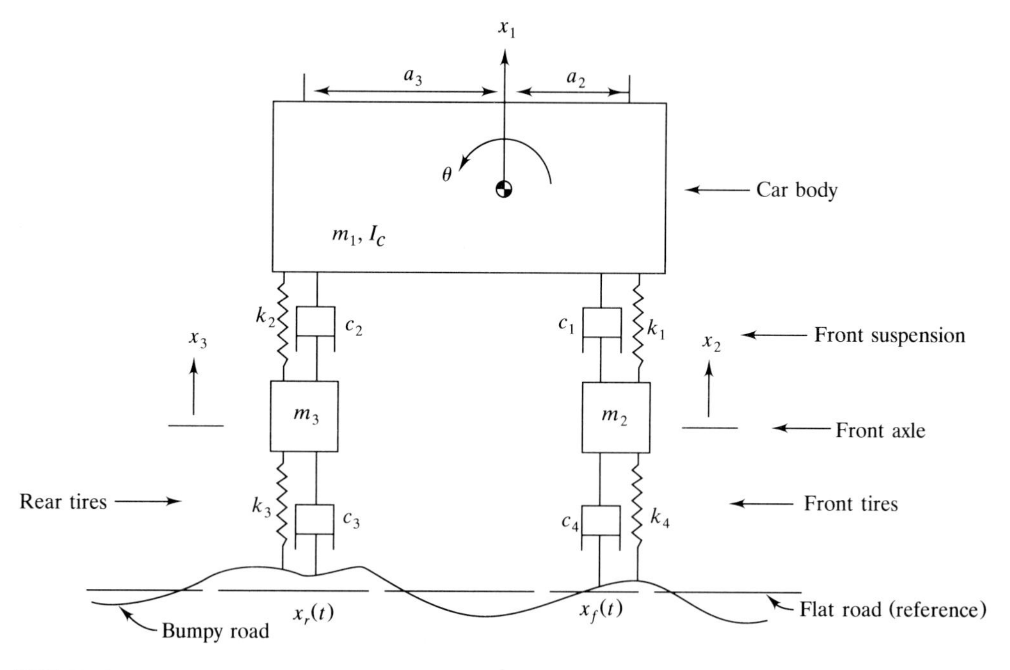

FIGURE 2.26
Idealized model of a 4 degree of freedom automobile-suspension system.

mass:

$$M_{s_1} = -a_2 k_1 \underbrace{(x_1 + a_2\theta - x_2)}_{\text{Spring stretch}}$$

$\uparrow$ moment arm

If $x_1 + a_2\theta - x_2 > 0$, then the torque is in the clockwise direction, i.e., negative θ, so we need a minus sign.

Likewise, the rear suspension spring torque M_{s_2} will be

$$M_{s_2} = +a_3 k_2 (x_1 - a_3\theta - x_3)$$

$\uparrow$

If $x_1 - a_3\theta - x_3 > 0$, then the torque is in the counterclockwise direction of positive θ, so we need a plus sign.

We can calculate the damping torques in an analogous fashion, so the equation of motion for car body rotation turns out to be

$$I_c\ddot{\theta} + c_1 a_2\left(\dot{x}_1 + a_2\dot{\theta} - \dot{x}_2\right) - c_2 a_3\left(\dot{x}_1 - a_3\dot{\theta} - \dot{x}_3\right) + k_1 a_2(x_1 + a_2\theta - x_2) - k_2 a_3(x_1 - a_3\theta - x_3) = 0 \qquad (2.57)$$

Now we move on to the front and rear axles. The equations of motion for the axles will be similar to equation (2.55*b*) in Example 2.4, with the following differences:

(1) the effect of car body rotation has to be included in the main suspension forces, and (2) the effect of road irregularity on the tire forces has to be included. It is assumed that the road irregularity, that is, the actual road height measured relative to a smooth datum, is a known function of position along the road. This may then be converted to the known functions of time $x_f(t)$ and $x_r(t)$ if we assume a constant vehicle speed V. We visualize the car to be contacting the "road" at two points that are moving up and down in the prescribed manners $x_f(t)$ and $x_r(t)$. Then the tire spring force exerted on the front axle will be $-k_4[x_2 - x_f(t)]$; we see that the effect of the road irregularity enters through its influence on the tire forces. A similar result obtains for the rear axle. We thus arrive at the following equations of motion for the front and rear axles:

$$m_2\ddot{x}_2 + c_1(\dot{x}_2 - \dot{x}_1 - a_2\dot{\theta}) + c_4\dot{x}_2 + k_1(x_2 - x_1 - a_2\theta) + k_4x_2$$
$$= k_4x_f(t) + c_4\dot{x}_f(t) \tag{2.58}$$

$$m_3\ddot{x}_3 + c_2(\dot{x}_3 - \dot{x}_1 + a_3\dot{\theta}) + c_3\dot{x}_3 + k_2(x_3 - x_1 + a_3\theta) + k_3x_3$$
$$= k_3x_r(t) + c_3\dot{x}_r(t) \tag{2.59}$$

Our model consists of four second-order linear ODEs that form a model with a fair degree of complexity. The order of the system is 8, and the states are x_1, x_2, x_3, θ, $\dot{x}_1$, $\dot{x}_2$, $\dot{x}_3$, and $\dot{\theta}$. The system is nonautonomous; the inputs are $x_r(t)$ and $x_f(t)$, which excite a vibratory motion of the system. If we were to use this model to design a suspension system, we might proceed as follows. First, assume we can develop a computer program that will do numerical integrations of equations (2.56) through (2.59). We might feed in a number of "typical" road irregularity functions $[x_f(t), x_r(t)]$, integrate (2.56) through (2.59) numerically, and so determine the response functions $x_1(t)$, $\theta(t)$, $x_2(t)$, and $x_3(t)$. The objective would be to design the main suspension system elements so that the motions of the car body (x_1, θ) remain acceptably small, even if the axle motions are fairly violent, for the set of road irregularity functions likely to be encountered in practice.

This example illustrates a couple of points: (1) In fairly complex systems we can often develop a hierarchy of mathematical models of increasing complexity; we try to use the simplest model that retains all of the effects of importance, and (2) in fairly complex mechanical systems the most difficult part of the modeling process is often to establish adequate mathematical models for the forces and moments that act on the system. In this regard another comment is required. Inspection of (2.56) through (2.59) shows that a fairly large number of system constants $(m_1, I_c, m_2, m_3, k_1, k_2, k_3, k_4, c_1, c_2, c_3, c_4, a_2, a_3)$ is required in order to fully specify the model. If a given auto-suspension system were to be analyzed, then values of these constants would have to be determined, and we consider this to be part of the modeling process. Often some of the system constants will be found through testing.

On the other hand, in a design process we may need first to determine values for some of the constants (say, k_1, c_1, k_2, c_2, which define the primary suspension) which will result in acceptable performance of the system; then we have to design components such as shocks and coil springs that will realize these values of the design constants.

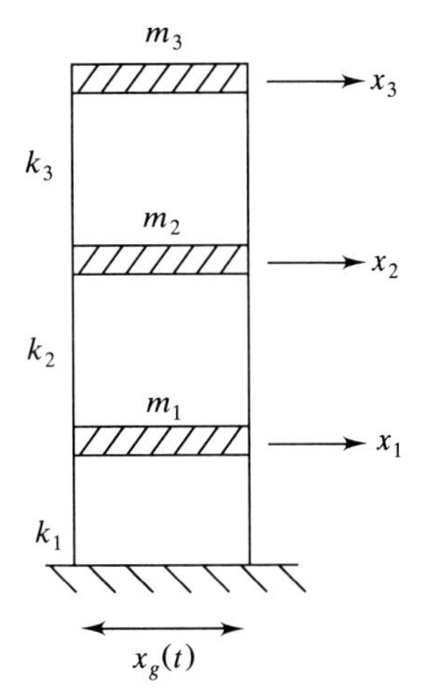

FIGURE 2.27
Idealized three-story building model for analysis of earthquake-induced building motions.

Example 2.6 Earthquake/Building Model. During earthquakes the violent lateral motions of the ground can induce significant motions of buildings, with attendant large stresses that can cause damage. Let's now consider a simple model of this phenomenon for a hypothetical three-story building, which is shown in Fig. 2.27. We will make the following assumptions in order to obtain a reasonably simple model:

1. All of the mass is concentrated in the three floors (m_1, m_2, m_3).
2. The walls provide stiffness which restrains relative horizontal motion of any two adjacent floors (k_1, k_2, k_3).
3. The motion takes place in the plane of the paper; each floor moves only horizontally and does not rotate.
4. The horizontal ground motion $x_g(t)$ is a specified function of time.

With these assumptions we have a system of 3 degrees of freedom, defined by the coordinates x_1, x_2, and x_3, and we can see that the system is equivalent to the configuration of masses and springs shown in Fig. 2.28.

The equations of motion are derived by application of Newton's second law to each mass. The various spring forces are determined in a manner analogous to

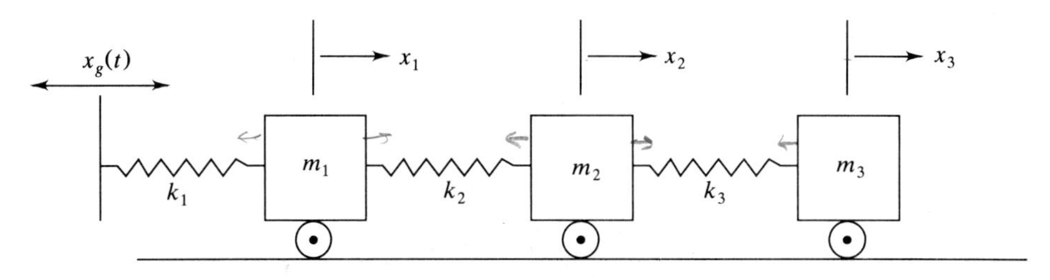

FIGURE 2.28
Equivalent spring/mass system for building model of Fig. 2.27.

Examples 2.4 and 2.5, and the equations of motion turn out to be as follows (check it out!):

$$m_1\ddot{x}_1 + k_1x_1 + k_2(x_1 - x_2) = k_1x_g(t)$$
$$m_2\ddot{x}_2 + k_2(x_2 - x_1) + k_3(x_2 - x_3) = 0 \tag{2.60}$$
$$m_3\ddot{x}_3 + k_3(x_3 - x_2) = 0$$

Here we have a linear system of sixth order; the states are x_1, x_2, x_3, $\dot{x}_1$, $\dot{x}_2$, and $\dot{x}_3$. The model is nonautonomous, and the input is the horizontal ground motion $x_g(t)$. Note that the mathematical model bears many similarities to the models of Examples 2.4 and 2.5, even though the physical system is much different.

Several comments need to be made regarding this example: (1) We should have included some damping effects. Can you see what the damping terms in (2.60) would be if we were to assume linear viscous damping? (2) Our lumping of all the mass at the floors is clearly an approximation. (3) The ground motion $x_g(t)$ acts in essentially the same way as does the rough road in the auto-suspension model. These are examples of what is known as a *support excitation*, in which the support to which a structure is attached moves in some fashion, providing an external excitation. (4) The ground motion $x_g(t)$ is highly irregular and impossible to predict ahead of time. What analysts may do in building design is to subject a given building model to previously measured earthquake ground motions and determine the resulting response, stresses, and so on. (5) The determination of building damping properties is very difficult and has to be done through testing. Years ago in southern California, numerous existing buildings were instrumented and their motions were monitored during the San Fernando earthquake of 1971 (Ref. 2.5). The damping properties were then determined by comparing the calculated responses to those measured during the earthquake. These results enable typical modern building damping properties to be estimated.

Example 2.7 Vertical Descent through the Atmosphere. In this example, we'll consider an object descending vertically through the earth's atmosphere and acted upon by forces due to gravity (F_G) and aerodynamic drag (F_A), as shown in Fig. 2.29. The coordinate z measures the height of the vehicle above the earth's surface. In order to model the gravity force, first note that at the earth's surface ($z = 0$), the force will be $F_G(z = 0) = -mg$; the minus sign indicates that F_G is in the negative z direction. Above the earth's surface the gravitational force decreases in an inverse square fashion, measured from the earth's center, and, to account for this, we can use the constitutive relation (2.42) to write

$$F_G = -mg\left(\frac{R_e}{R_e + z}\right)^2$$

where R_e is the radius of a "spherical earth" ($R_e \approx 6371$ km). Clearly, for many applications we can simply use $F_G \cong -mg$.

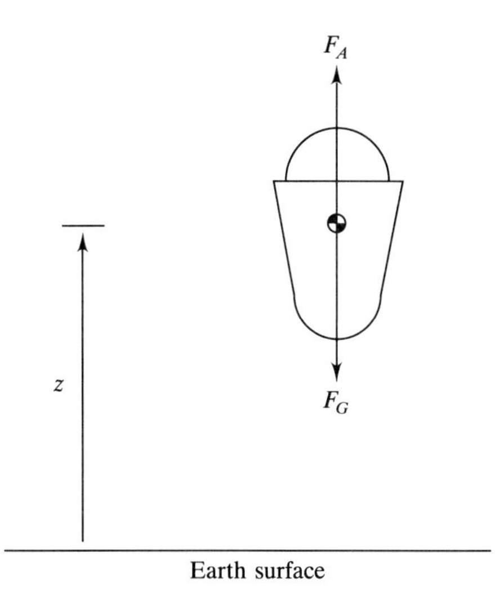

FIGURE 2.29
Flight vehicle descending vertically through the atmosphere.

The aerodynamic force F_A is modeled in terms of the dimensionless drag coefficient C_D, such that

$$F_A = \tfrac{1}{2}\rho V^2 C_D A$$

where ρ is the free stream air density, A the frontal area of the object, and V the speed (here $\dot{z}$). Thus, we can now write out Newton's law for the object as follows:

$$m\ddot{z} = -mg\left(\frac{R_e}{R_e + z}\right)^2 + \left(\frac{1}{2}C_D A\rho\right)\dot{z}^2 \tag{2.61}$$

At this point we also need to realize that the free stream air density ρ is a function of the altitude z, varying approximately exponentially:

$$\rho(z) \cong \rho_0 e^{-z/h}$$

where ρ_0 is the surface density and h a "scale height"; h is anywhere from about 22,000 to 30,000 ft, depending on the atmospheric model being used. The formula states that at $z = h$, the density has attenuated by a factor of e. Thus, if we use the exponential atmosphere model, we have, after dividing by m, the equation of motion:

$$\ddot{z} = -g\left(\frac{R_e}{R_e + z}\right)^2 + \left(\frac{\rho_0 C_D A}{2m}\right)e^{-z/h}\dot{z}^2 \tag{2.62}$$

For descents very near the earth's surface, for which the density and gravitational forces may be assumed constant, this model reduces to

$$\ddot{z} \cong -g + \left(\frac{\rho_0 C_D A}{2m}\right)\dot{z}^2 \tag{2.63}$$

This example once again illustrates that for mechanical systems (loosely speaking, those governed by Newton's law), it is essential to be able to come up with a

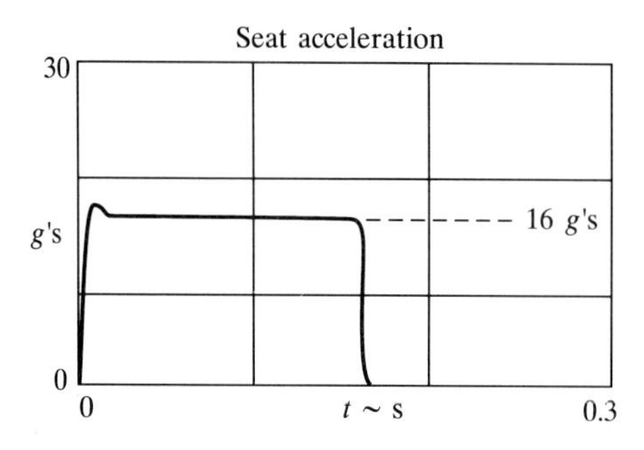

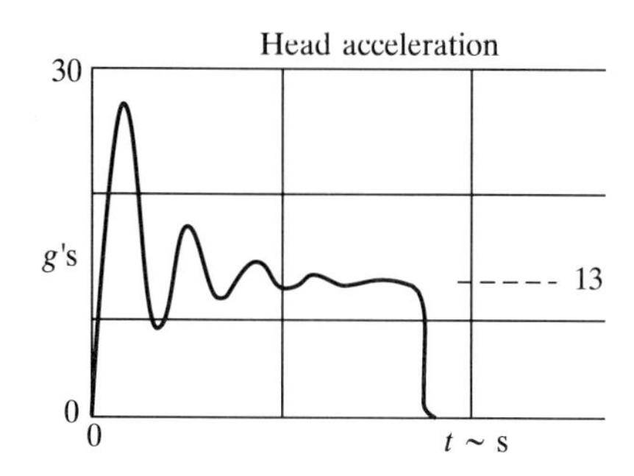

FIGURE 2.30
Hypothetical seat and pilot head accelerations during pilot ejection.

proper model for the forces. Even after writing down (2.62), we have to determine a value of the drag coefficient C_D, which depends on the shape of the vehicle and on the Reynolds and Mach numbers of the flow.

The system model (2.62) is of second order and is autonomous and nonlinear (because of both the drag term $\dot{z}^2 e^{-z/h}$ and the gravity term $R_e^2/(R_e + z)^2$). This system is a good example of a case in which a nonlinearity (here $\dot{z}^2 e^{-z/h}$) has to be retained if the model is to have any validity. Note also that the aerodynamic drag force is a function of both position z and velocity $\dot{z}$, and for descent from high altitude, the density variation has to be taken into account.

Example 2.8 Pilot Ejection from Aircraft (After Steidel, Ref. 7.4). During ejection of pilots from high-performance jet aircraft, the seat to which the pilot is strapped is accelerated rapidly for a short period of time (say, 0.15 s). During this period the maximum acceleration of the pilot's head may be substantially higher than that of the seat, potentially resulting in pilot blackout. See Fig. 2.30, which shows a hypothetical history of seat and pilot head accelerations, typical of actual maneuvers.

In order to construct a simple model that would allow one to describe this phenomenon, and to effect redesign if necessary, we start by assuming that the spinal column/seat behaves as a combined spring/damper and the head as a simple mass. Thus, our idealized model would be as shown in Fig. 2.31. The input here is the seat acceleration $\ddot{y}$, and we need eventually to calculate the pilot head acceleration $\ddot{x}$. Application of Newton's second law yields the mathematical model for the system as

$$m\ddot{x} + c(\dot{x} - \dot{y}) + k(x - y) = 0$$

A more useful form of the model can be obtained by defining the pilot head motion in terms of the *relative* displacement $z \equiv x - y$, so that, with $\ddot{x} = \ddot{z} + \ddot{y}$, the mathematical model in terms of the new dependent variable z would be

$$m\ddot{z} + c\dot{z} + kz = -m\ddot{y} \tag{2.64}$$

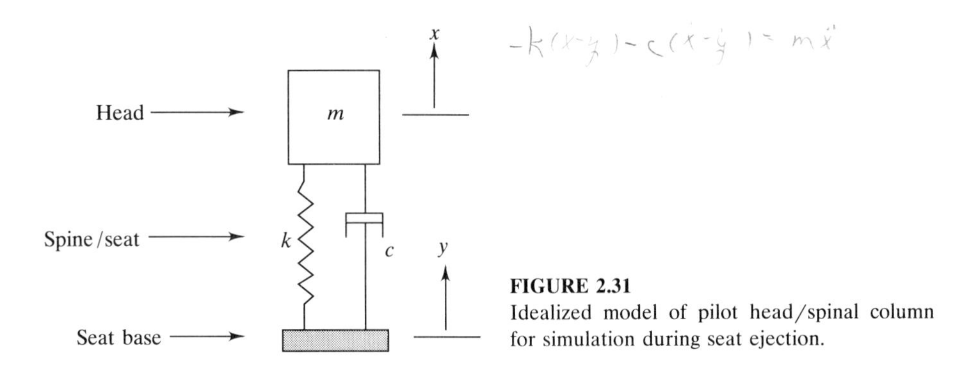

FIGURE 2.31
Idealized model of pilot head/spinal column for simulation during seat ejection.

This formulation has the advantage that the input seat acceleration $\ddot{y}$ appears directly. Thus, the model is of second order, linear and nonautonomous; we also see another example of an input that is a support excitation. In order to see whether our model is valid, we have to determine the response $z(t)$ to an input seat acceleration of the type shown, then calculate the history of pilot head acceleration $\ddot{x} = \ddot{z} + \ddot{y}$ and compare this to the measured head acceleration shown. The model turns out not to be too bad, provided that we can select appropriate values for m, c, and k. It can then be used, for example, to redefine the seat acceleration so that the maneuver can be performed successfully with lower peak accelerations of the pilot's head; we will return to this example in Chap. 5.

Example 2.9 Airfoil in Wind Tunnel. An airfoil of mass m and moment of inertia I_c with respect to the center of mass is mounted in a wind tunnel as shown in Fig. 2.32. The airfoil is suspended on a linear spring of stiffness k, and rotation is restrained by a torsional spring k_T. The center of mass C is located at a distance d ahead of the point of support S. The objective is to derive the equations of motion for the case of no aerodynamic forces/moments. If we assume the motion to be

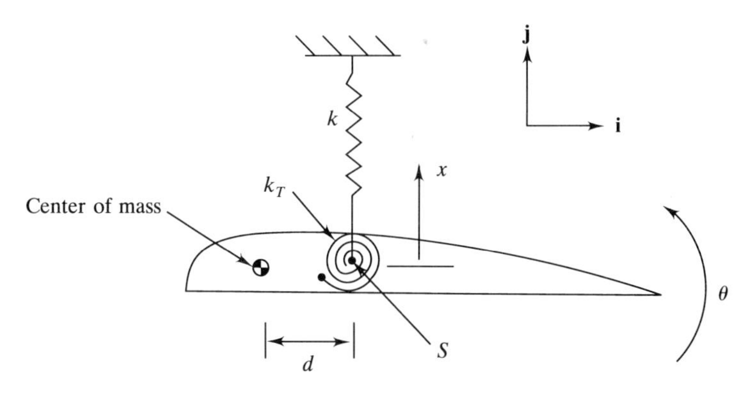

FIGURE 2.32
Airfoil suspended in a wind tunnel.

planar and that the point of support S moves only in the vertical direction, then there are 2 degrees of freedom, and we define the coordinates x and θ shown; x is the vertical position of the point S (defined to be zero when the spring is unstretched), and the rotation θ is zero when the line segment SC is horizontal.

To develop the equations of motion, we use the support point S as moment reference, as the torsional spring torque is exerted directly about this point. Thus, for the rotational motion we will need (2.33) for rotation about a moving reference point located on the rigid body but not at the center of mass. The equation of vertical translation of the center of mass is obtained by treating the center of mass as a particle of mass m, according to (2.25). The equation of vertical translation is then obtained for small angles θ as

$$m(\ddot{x} - d\ddot{\theta}) = -kx - mg \tag{2.65}$$

It is important to note that in applying (2.25), we must use the vertical acceleration *of the center of mass*, which for small θ is $\ddot{x} - d\ddot{\theta}$, rather than the acceleration $\ddot{x}$ of the attachment point S.

The rotational equation of motion is obtained from (2.33). The moment $\mathbf{M}_s$ ($\mathbf{k}$ component understood) is $M_s = mgd - k_T\theta$ (make sure the signs are understood!). The term $\mathbf{r}_{c/s} \times \ddot{\mathbf{R}}_s$ is determined by noting that, for small θ, $\mathbf{r}_{c/s} = -d\mathbf{i}$ and $\ddot{\mathbf{R}}_s = \ddot{x}\mathbf{j}$, so that $\mathbf{r}_{c/s} \times \ddot{\mathbf{R}}_s = -d\ddot{x}\mathbf{k}$. Combining the various terms according to (2.33), we obtain the equation of rotational motion as

$$\left(I_c + md^2\right)\ddot{\theta} = -k_T\theta + mgd + md\ddot{x} \tag{2.66}$$

The mathematical model is of fourth order, linear and autonomous; the constant terms $-mg$ and mgd are considered to be part of the system, rather than inputs. In this example we observe that, because the physical system was defined at the outset in terms of idealized components, the only modeling "difficulty" was in the correct application of the laws of dynamics presented in Sec. 2.2.

Example 2.10 Two-Link Planar Manipulator. Let's now consider another example in which the correct application of the moment laws of Sec. 2.2 is essential in the derivation of a system mathematical model. The system to be considered is a two-link planar manipulator, as shown in Fig. 2.33. Each link is driven by an actuator, located at S_1 and at S_2; we assume these actuators exert time-dependent moments $M_1(t)$ and $M_2(t)$ which maneuver the manipulator links. For simplicity we'll assume that all of the mass is concentrated at the massless link ends (m_1, m_2). The system has 2 degrees of freedom, and the coordinates chosen are θ_1 and θ_2 as shown. Note that θ_2 measures the orientation of link 2 *relative* to link 1, a common practice in robotics. Our job is to derive the equations of motion of this system. In order to do this, the natural thing to try is use of the moment laws with S_1 and S_2 as the moment references since the actuator torques are applied directly about these points. Let us start by applying the law (2.33) for rotation about the moving reference point S_2. Note that the angular momentum $\mathbf{H}_{s_r}$ appearing in (2.33) is the total *system* angular momentum. With S_2 as moment reference, however, only m_2 will contribute, since m_1 is located at S_2. Noting that the net angular velocity of link 2 is $(\dot{\theta}_1 + \dot{\theta}_2)$, the relative angular momentum of the system with respect to S_2 is

$$\mathbf{H}_{s_{2r}} = m_2 L_2^2\left(\dot{\theta}_1 + \dot{\theta}_2\right)\mathbf{k} \tag{2.67}$$

Next we calculate the term $\mathbf{r}_{c/s_2} \times \ddot{\mathbf{R}}_{s_2}$. These vectors can conveniently be written

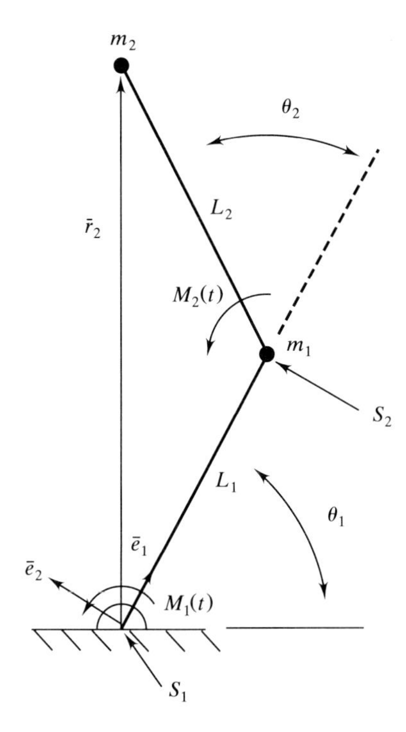

FIGURE 2.33
Simplified model of a two-link planar manipulator.

in terms of a (polar) coordinate system that rotates with link 1: The unit vector $\mathbf{e}_1$ is always directed toward S_2, and $\mathbf{e}_2$ is normal to $\mathbf{e}_1$ and points in the direction of increasing θ_1. In terms of these unit vectors we have

$$\mathbf{r}_{c/s_2} = L_2 \cos\theta_2 \mathbf{e}_1 + L_2 \sin\theta_2 \mathbf{e}_2$$

$$\ddot{\mathbf{R}}_{s_2} = -L_1\dot{\theta}_1^2 \mathbf{e}_1 + L_1\ddot{\theta}_1 \mathbf{e}_2$$

where the latter relation was obtained by setting $\dot{r} = \ddot{r} = 0$ in the polar coordinate relation (2.11c). We next compute $m_2\mathbf{r}_{c/s} \times \ddot{\mathbf{R}}_s$, which turns out to be

$$m_2\mathbf{r}_{c/s} \times \ddot{\mathbf{R}}_s = m_2\left(L_1L_2\cos\theta_2\ddot{\theta}_1 + L_1L_2\sin\theta_2\dot{\theta}_1^2\right)\mathbf{k}$$

The moment exerted about S_2, including effects of gravity, is

$$\mathbf{M}_{s_2} = \left[M_2(t) - m_2L_2 g\cos(\theta_1 + \theta_2)\right]\mathbf{k}$$

Finally, by differentiating (2.67) and combining the preceding terms according to (2.27) or (2.33), we arrive at the first equation of motion:

$$m_2L_2^2\left(\ddot{\theta}_1 + \ddot{\theta}_2\right) = M_2(t) - m_2L_2g\cos(\theta_1 + \theta_2) - m_2L_1L_2\left(\ddot{\theta}_1\cos\theta_2 + \dot{\theta}_1^2\sin\theta_2\right) \qquad (2.68)$$

Now, to develop the second equation of motion, we select the fixed point S_1 as moment reference, so the law (2.32), $\dot{\mathbf{H}}_{s_1} = \mathbf{M}_{s_1}$ applies. The total system angular momentum $\mathbf{H}_{s_1}$ consists of the sum of the angular momenta of the two particles m_1 and m_2. The angular momentum of m_1 is simply $\mathbf{H}_{s_1}^{(1)} = m_1L_1^2\dot{\theta}_1\mathbf{k}$. To find the angular momentum of particle m_2, we have to proceed directly from the definition

$\mathbf{H}_{s_2}^{(2)} = m_2\mathbf{r}_2 \times \dot{\mathbf{r}}_2$, where $\mathbf{r}_2$ is the vector from S_1 to m_2, as shown. Working out the geometry, we have, in terms of the rotating $\mathbf{e}_1, \mathbf{e}_2$ unit vectors,

$$\mathbf{r}_2 = (L_1 + L_2\cos\theta_2)\mathbf{e}_1 + L_2\sin\theta_2\mathbf{e}_2$$

The velocity $\mathbf{r}_2$ is obtained by differentiation, and we have to be careful to note that $\dot{\mathbf{e}}_1 = \dot{\theta}_1\mathbf{e}_2$ and $\dot{\mathbf{e}}_2 = -\dot{\theta}_1\mathbf{e}_1$, as the unit vectors are changing direction.* The resulting velocity $\dot{\mathbf{r}}_2$ is

$$\dot{\mathbf{r}}_2 = -L_2\sin\theta_2(\dot{\theta}_1 + \dot{\theta}_2)\mathbf{e}_1 + \left[L_2\dot{\theta}_2\cos\theta_2 + (L_1 + L_2\cos\theta_2)\dot{\theta}_1\right]\mathbf{e}_2$$

The angular momentum of m_2 with respect to S_1 is then found, after some algebra, to be

$$\mathbf{H}_{s_1}^{(2)} = m_2\left[(L_1^2 + L_2^2)\dot{\theta}_1 + L_2^2\dot{\theta}_2 + 2L_1L_2\dot{\theta}_1\cos\theta_2 + L_1L_2\dot{\theta}_2\cos\theta_2\right]\mathbf{k}$$

Differentiating the sum $\mathbf{H}_{s_1}^{(1)} + \mathbf{H}_{s_1}^{(2)}$ to obtain the rate of change of system angular momentum, and then equating the result to the applied torque yields the second equation of motion as follows:

$$\begin{aligned} m_1L_1^2\ddot{\theta}_1 + m_2\Big[(L_1^2 + L_2^2)\ddot{\theta}_1 &+ L_2^2\ddot{\theta}_2 + 2L_1L_2\ddot{\theta}_1\cos\theta_2 - 2L_1L_2\dot{\theta}_1\dot{\theta}_2\sin\theta_2 \\ &+ L_1L_2\ddot{\theta}_2\cos\theta_2 - L_1L_2\dot{\theta}_2^2\sin\theta_2\Big] \\ &= M_1(t) - m_2L_2g\cos(\theta_1 + \theta_2) - (m_1 + m_2)L_1g\cos\theta_1 \end{aligned} \tag{2.69}$$

The system model, consisting of (2.68) and (2.69), is of fourth order, nonlinear, and nonautonomous. The inputs are $M_1(t)$ and $M_2(t)$ and the states are θ_1, θ_2, $\dot{\theta}_1$, and $\dot{\theta}_2$. This model may appear to be very complex, and it is true that with the myriad nonlinear terms present, solutions to these equations would have to be obtained via computer simulation. The student should note, however, that this example represents a very simple manipulator. Thus, the study of more general manipulators would require one to have a clear understanding of the basics of kinematics and dynamics.

Example 2.11 A Gear/Track Assembly. In the preceding examples, ODE models of mechanical systems were obtained by application of the vector laws of newtonian mechanics. This type of approach can be used for any mechanical system. Sometimes, however, the configuration of the system is such that other approaches may be more direct and insightful. The gear/track assembly shown in Fig. 2.34 is a simple example of a general class of such systems having the following property: There are interconnected moving parts that are connected in such a way that the system has only a single degree of freedom. For such systems the kinetic and potential energies (Sec. 2.3) can be used to formulate the mathematical model. We now show how this would be done for the system of Fig. 2.34.

In the gear/track assembly shown, we assume that a motor provides a known, possibly time-varying input torque $M(t)$. This torque drives the shaft, on which a gear of radius a is mounted. The combined shaft/gear moment of inertia about the axis of rotation is I, and the shaft/gear rotation is defined by an angle θ.

*The rate of change of rotating unit vectors is discussed in some detail in Chap. 6.

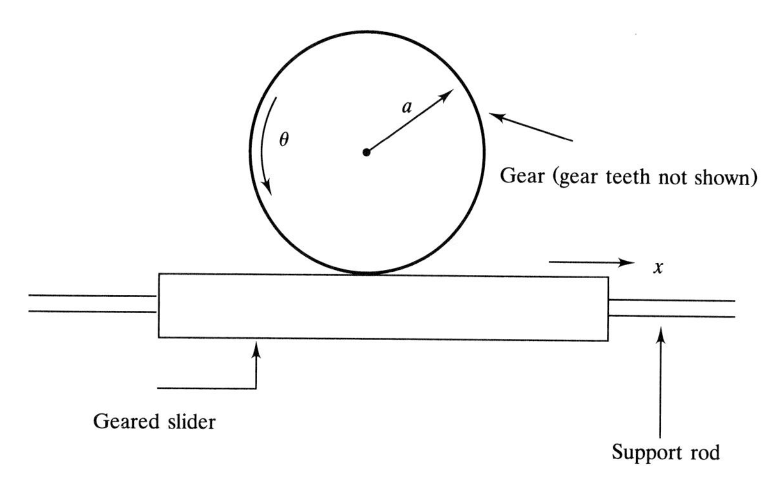

FIGURE 2.34
A gear/track assembly. The gear is mounted on a shaft, normal to the paper, about which a moment $M(t)$ is exerted.

The gear in turn drives a geared track of mass m, causing the track to translate horizontally, as measured by the coordinate x.

The two coordinates x and θ are related by a *geometric constraint relation*, $x = a\theta$. Because of this constraint the system possesses a single degree of freedom: Specification of either x or θ is sufficient to fully determine the configuration of the system. The goal of this analysis will be to determine the ODE math model in terms of the single coordinate θ.

To develop the ODE math model, we first note that, if the track m were removed from the system, the equation of shaft/gear rotational motion would be simply

$$I\ddot{\theta} = M(t) \tag{2.70}$$

In physical terms the presence of the track m will have the effect of increasing the system mass or inertia. In order to take this into account, we can calculate an "equivalent" moment of inertia I_e from work-energy principles. In Sec. 2.3 it was shown that the work W done on the system will always equal the change in the kinetic energy T. The equivalent moment of inertia I_e must be defined so that the quantity $\frac{1}{2}I_e\dot{\theta}^2$ is at all times equal to the actual kinetic energy T of the system. This kinetic energy is defined as

$$T = \tfrac{1}{2}I\dot{\theta}^2 + \tfrac{1}{2}m\dot{x}^2$$

Substitution into the above relation of the constraint $\dot{x} = a\dot{\theta}$ enables T to be written as

$$T = \tfrac{1}{2}(I + ma^2)\dot{\theta}^2$$

From this result we conclude that the equivalent moment of inertia $I_e = I + ma^2$. The equation of motion of the gear/track system of Figure 2.34 is thus

$$(I + ma^2)\ddot{\theta} = M(t) \tag{2.71}$$

The same procedure can be used no matter how many interconnected parts comprise the single degree of freedom system under consideration. One writes down the system kinetic energy and then uses the constraint relations to eliminate all but the one coordinate in terms of which the analysis is to be conducted.

Notice that the kinetic energy equivalence has allowed development of the ODE math model without any consideration of the forces of interaction between the gear and the track, which would have to be included if the vector newtonian mechanics were applied (see Exercise 23).

2.6 CONCLUDING REMARKS ON MODELING METHODOLOGY

Examples 2.1 through 2.11 have been intended to illustrate the process of construction of ODE models of systems governed by newtonian mechanics. Table 2.1 is a summary of these ODE models, and it lists, for each example, the math model equations; the system order and states; classification (linear or nonlinear, autonomous or nonautonomous); the source of any inputs; and the laws, constitutive relations, and assumptions employed in the model development. The methodology followed in the ODE model construction is that outlined in Fig. 1.1 and discussed in Sec. 1.2, and it is recommended that Sec. 1.2 (as well as Secs. 1.4 and 1.5) be reread at this point.

Based on the modeling examples presented in this chapter and on the discussion of modeling methodology in Sec. 1.2, we can identify four critical areas that one must bear in mind when constructing ODE models of newtonian systems:

1. Assumptions or simplifications regarding the types of motion to be considered or allowed in the model system. For instance, the restriction of the building model (Example 2.6) and the auto-suspension models (Examples 2.4 and 2.5) to planar motions, and the descending space vehicle (Example 2.7) to a vertical descent, are restrictions that do not necessarily occur in the actual systems being modeled. In addition, the two models used in the auto suspension problem (Examples 2.4 and 2.5) differed in the motions which were assumed to be allowed. Generally, simplifications of the type discussed here allow the development of simple models which will retain at least some of the important physical effects which the model is supposed to describe. In real-life engineering applications, however, it is necessary to make decisions as to the motions that are to be considered. Often, one would develop several models of increasing complexity and then use the simplest one which retains all of the physical effects of importance.
2. The representation of the actual physical system by an idealized one, often involving collections of lumped masses, springs, and dampers to represent the physical effects of inertia, stiffness, and energy dissipation.
3. The development of constitutive relations that describe all of the forces and moments which act on the system. In many of the examples involving the

TABLE 2.1
Summary of ODE models

Example	Math model	Order (states)	Classification	Input (if nonautonomous)	Laws	Constitutive Relations	Assumptions
2.1 Simple pendulum	(2.19)	$2(\theta, \dot{\theta})$	Nonlinear, autonomous		Newton: rotation, gravity	None	Planar, no friction
2.2 Pendulum with moving support	(2.23)	$2(\theta, \dot{\theta})$	Nonlinear, nonautonomous	Support motion	Newton: rotation, gravity	None	Planar, no friction
2.3 Mechanical oscillator	(2.52)	$2(x, \dot{x})$	Linear, nonautonomous	External force	Newton: 2nd law	Spring, damper	1-D, ideal elements
2.4 2-DOF auto/suspension	(2.55)	$4(x_1, x_2, \dot{x}_1, \dot{x}_2)$	Linear, autonomous		Newton: 2nd law	Springs, dampers	2-D, ideal elements
2.5 4-DOF auto/suspension	(2.56)–(2.59)	$8(x_1, x_2, x_3, \theta, \dot{x}_1, \dot{x}_2, \dot{x}_3, \dot{\theta})$	Linear, nonautonomous	Road irregularity	Newton: RB dyn, particles	Springs, dampers	4-D, ideal elements
2.6 Earthquake/building	(2.60)	$6(x_1, x_2, x_3, \dot{x}_1, \dot{x}_2, \dot{x}_3)$	Linear, nonautonomous	Ground motion	Newton: particles	Stiffness, damping	Planar, ideal elements
2.7 Vertical descent	(2.62)	$2(z, \dot{z})$	Nonlinear, autonomous		Newton, gravity	Density, drag	1-D, idealized constitutive relation
2.8 Pilot ejection	(2.64)	$2(x, \dot{x})$	Linear, nonautonomous	Seat acceleration	Newton: 2nd law	Head/spine model	Idealization of head/spine
2.9 Airfoil/wind tunnel	(2.65), (2.66)	$4(x, \theta, \dot{x}, \dot{\theta})$	Linear, autonomous		Newton: RB rot, trans	Linear/tors stiffnesses	Planar, no friction, no aero
2.10 2 Link manipulator	(2.68), (2.69)	$4(\theta_1, \theta_2, \dot{\theta}_1, \dot{\theta}_2)$	Nonlinear, nonautonomous	Applied torques	Newton: particle rotation	None	Planar, mass concentration at tips
2.11 Gear/track assembly	(2.71)	$2(\theta, \dot{\theta})$	Linear, nonautonomous	Applied moment	Newton: rot about fixed point	Equivalence of kinetic energies	No friction

physical effects of stiffness and damping, linear springs and dampers were used at the outset. These idealized linear devices often provide a good description of behavior. As we have seen for actual shock absorbers, however, nonlinearities may be important; one must always be aware that linear constitutive models are approximations that break down if displacements and/or velocities become large. We also have to keep in mind that the development of constitutive relations really involves two steps. The first is the development of the mathematical *form* of the model (for example, is a force a linear function of velocity, as for the ideal damper, or is the force a quadratic function of velocity, as in the case of aerodynamic drag?). The second is determination of values for the constants appearing in the model (system identification).

4. Use of the laws of dynamics. The importance of correct application of the basic force and moment laws should be clear. Failure in this area will obviously result in an incorrect mathematical model. Furthermore, many systems of practical importance to mechanically oriented engineers involve three-dimensional rotation (some coverage of this topic is contained in Chap. 6), and for such systems an understanding of basic dynamics beyond what we have presented in this section is essential.

For actual engineering systems adequate modeling can be the most challenging part of the overall systems analysis effort. The student should realize that this modeling effort may take a substantial amount of time; it is usually not something that can be done in a couple of hours (as they say in the National Football League, Denver wasn't built in a day). One usually becomes proficient at modeling only with experience; the types of examples considered here and in the exercises are intended to provide the beginner with a start in this direction.

REFERENCES

2.1. Meriam, J. L., and L. G. Kraige: *Engineering Mechanics*, vol. II, *Dynamics*, 3rd ed., Wiley, New York (1992).
2.2. Beer, F. P., and E. R. Johnston, *Vector Mechanics for Engineers: Dynamics*, 5th ed., McGraw-Hill, New York (1977).
2.3. Shames, I. H., *Engineering Mechanics—Statics and Dynamics*, 3rd ed., Prentice-Hall, Englewood Cliffs, N.J. (1980).
The preceding three references contain a comprehensive treatment of dynamics at the introductory level:
2.4. *Car and Driver Magazine*, September 1981.
2.5. Blevins, R. D., *Flow Induced Vibration*, Van Nostrand Rinehold, New York (1977). See Chapter 8, which contains a nice description of damping properties of real-world engineering structures.

EXERCISES

1. Consider a mass m, constrained to move in the horizontal direction without friction, which is restrained by a linear spring of stiffness k, as shown in Fig. $P2.1$.

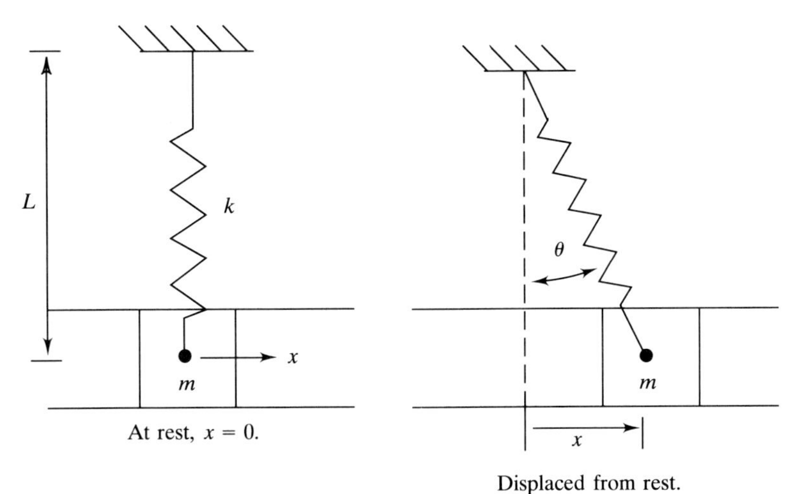

FIGURE ***P*2.1**

When $x = 0$ the spring is vertical and is unstretched (the spring force is zero). Derive the equation of motion of the mass m in terms of the dependent variable x. Is the system linear or nonlinear?

2. Consider a simple pendulum for which the support O is moved vertically in a specified manner $y(t)$, as shown in Fig. $P2.2$. Derive the equation of motion for the θ coordinate (θ need not be small).

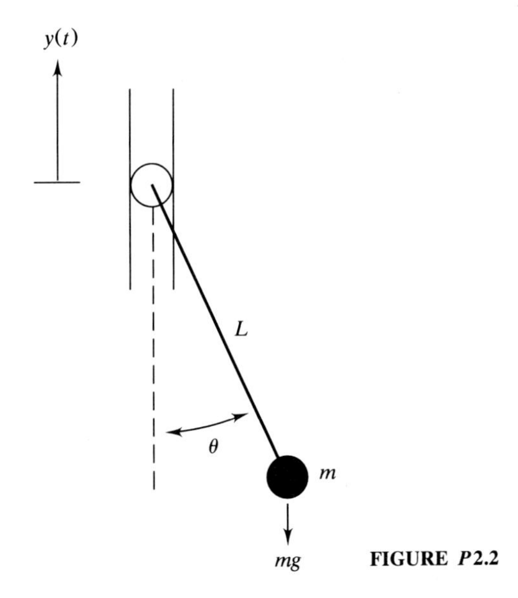

FIGURE ***P*2.2**

Determine whether the system is linear or nonlinear and autonomous or nonautonomous, and express the mathematical model in state variable form.

3. A satellite of mass m is to orbit the earth in a circular orbit a distance of 500 km above the earth's surface. Find the necessary satellite velocity.

4. Shown below in Fig. $P2.4$ is a plot of the static force F_s generated as a function of displacement x for a certain stiffening device.

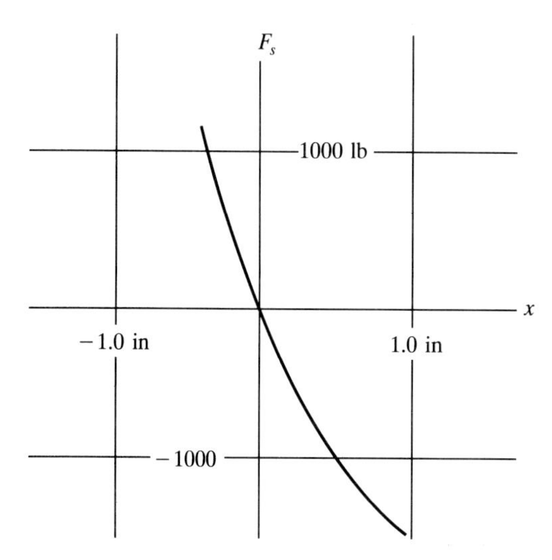

FIGURE $P2.4$

Estimate the linear spring constant k which models the constitutive behavior of this device for small displacements x.

5. The gravitational force F_g exerted on a body of mass m a height z above the surface of the earth was given as

$$F_g = -mg\left(\frac{R}{R+z}\right)^2 \qquad (P2.1)$$

Use the Taylor series expansion to obtain an approximation to $(P2.1)$ of the following form for the case $z \ll R$.

$$F_g \cong -mg + \alpha z \qquad (P2.2)$$

i.e., find α. Find from $(P2.1)$, the heights z at which F_g is 98 and 90% of its value at the surface of the earth. Then determine how close the approximation $(P2.2)$ is for the two cases.

6. For the shock absorber data shown as the solid line in Fig. 2.22, obtain an approximate, linear constitutive model for this device. The model should work reasonably well for small to moderate velocities. Does it?

7. Consider a mass m restrained by two linear springs in series, as shown in Fig. $P2.7(a)$.

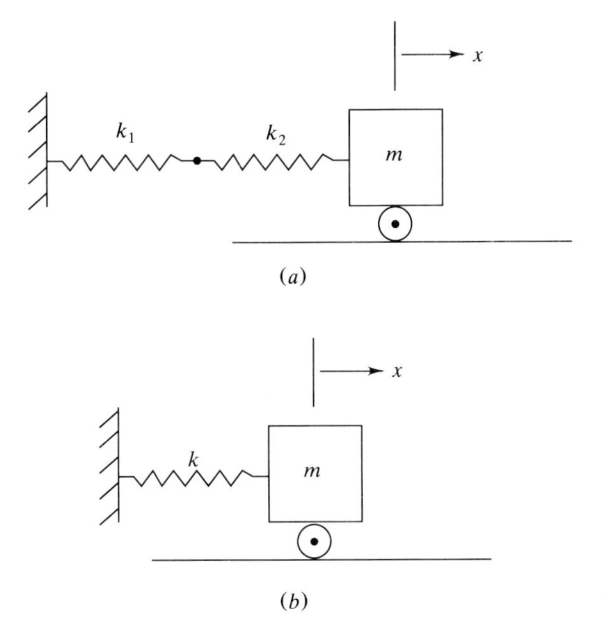

FIGURE $P2.7$

Determine the equation of motion of the mass m and deduce that, if a single overall spring constant k is correctly defined, then the system of Fig. $P2.7(a)$ is equivalent to that of Fig. $P2.7(b)$.

8. Consider the spring-mass-damper system shown in Fig. $P2.8$. The system is excited by moving the support with a prescribed motion $z(t)$.

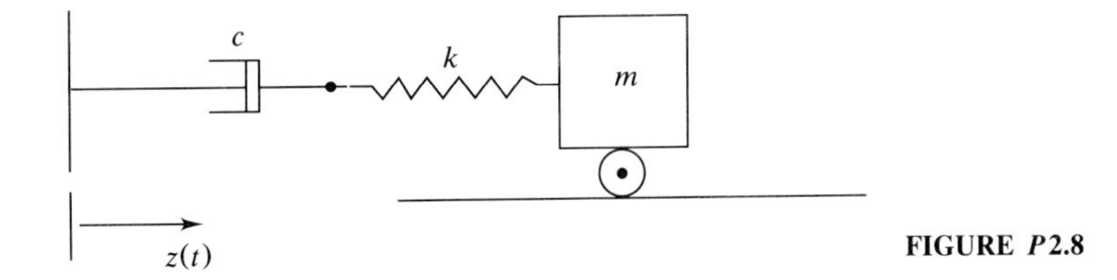

FIGURE $P2.8$

Define a set of coordinates sufficient to describe the system motion, and determine the equations of motion in terms of these coordinates. Determine the order of the system and express the model in state variable form.

9. A uniform bar of mass m and length L is free to rotate about the fixed pivot O, as shown in Fig. $P2.9$. The motion of the bar is restrained by a linear spring and a linear damper. A specified time-dependent force $F(t)$ acts in the vertical direction at a distance $L/3$ from the pivot O.

Derive the equation of motion in terms of the coordinate θ, assuming θ is small, so that the points at which the spring and damper are attached to the bar move essentially in the vertical direction.

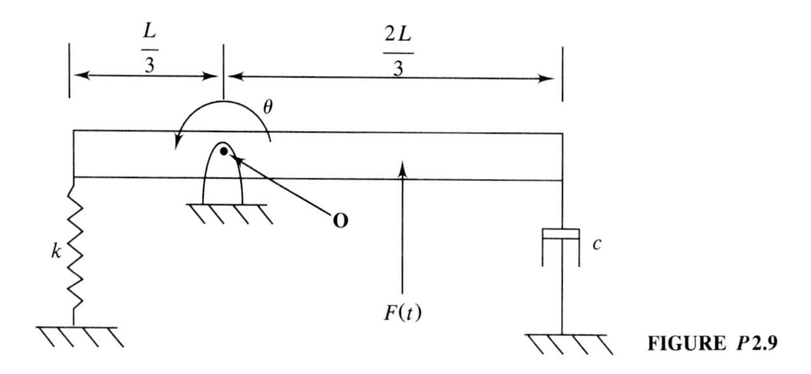

FIGURE $P2.9$

10. A mass m moves horizontally under the influence of two equal linear springs of stiffness k. There is a "deadband" of length $2d(-d < x < d)$, inside which the springs exert no force on m (Fig. $P2.10$); a spring force is exerted only if $x > d$ or if $x < -d$. The coordinate x is zero when the mass m is centered exactly in the deadband. Graph the spring force F_s exerted on m as a function of the displacement x. Is the system linear? Can a Taylor series expansion be used to represent $F_s(x)$?

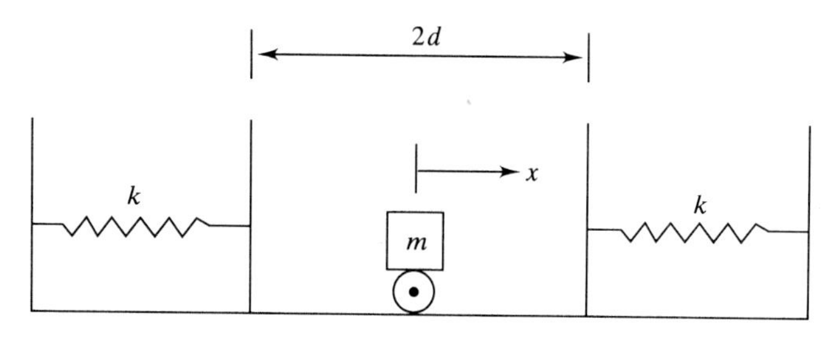

FIGURE $P2.10$

11. Consider the 4 degree of freedom automobile model of Example 2.5. Suppose we wish to take account of the fact that the driver's seat possesses stiffness and damping, in order to determine the motion of the auto driver during operation over a rough road. Determine an idealized model, analogous to Fig. 2.26, which could be used, and develop any additional equations of motion which would be needed. Is any change required in the existing equations (2.56) through (2.59)?

12. A uniform cantilever beam of length L and flexural stiffness EI (E = Young's modulus, I = area moment of inertia) is loaded statically by a load P exerted at the beam end, as in Fig. $P2.12(a)$.

The resulting static deflection, with the end restrained from rotating, is

$$\Delta = \frac{PL^3}{12EI}$$

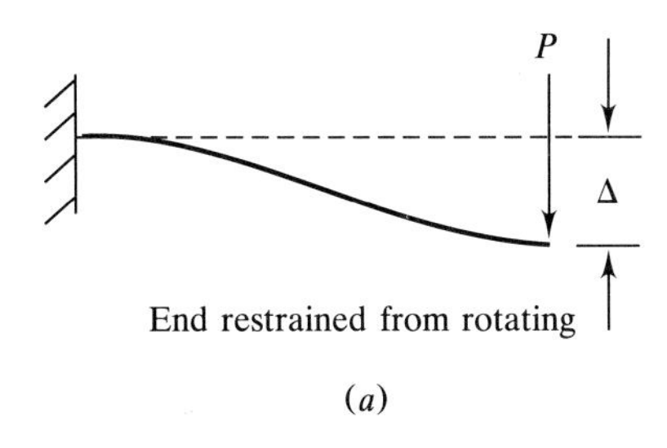

(a)

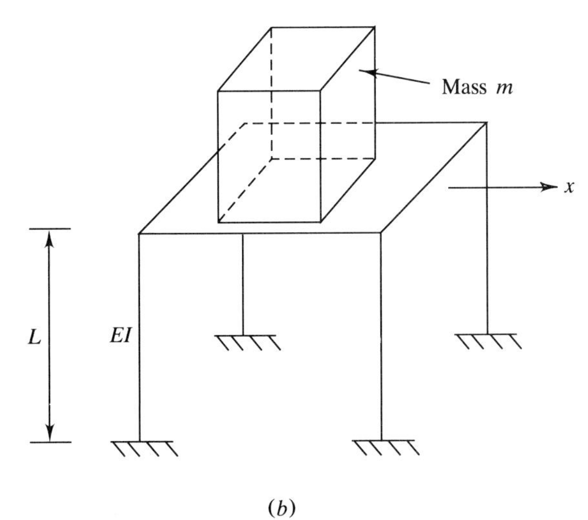

(b)

FIGURE P2.12

Use this result to estimate the lateral stiffness of the frame structure, shown in Fig. 2.12(*b*), on which a mass m is to be supported. Assume the only motion to be a horizontal motion x of the mass m. Each frame member is of length L and flexural stiffness EI.

13. Consider a rigid body, of mass m and moment of inertia I_c with respect to the center of mass, undergoing plane rotation and vertical translation as shown in Fig. $P2.13$.

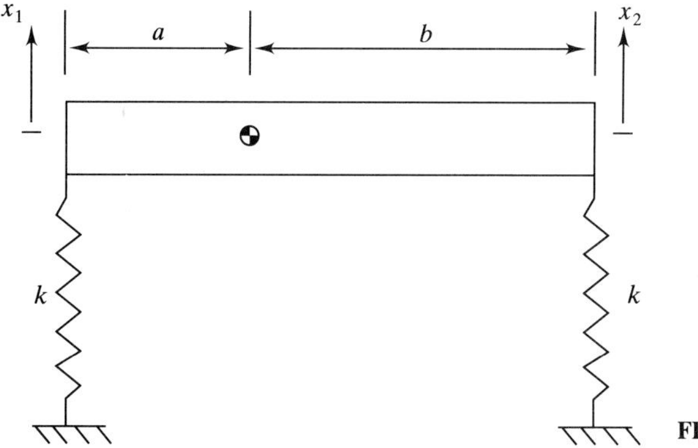

FIGURE P2.13

The motion is restrained by two linear springs, and effects of gravity can be ignored. Derive the equations of motion in terms of the coordinates x_1 and x_2 shown. Assume the ends of the bar move essentially in the vertical direction (the bar rotations are small).

Note: The coordinates x_1 and x_2 selected might be used in order to obtain simple relations for the spring forces (but be careful of the accelerations!).

14. A machine of weight $W = mg$ is to be mounted on four rubber pads in order to isolate the machine from the surroundings. When the machine is placed on the pads, they are observed to deflect (compress) an amount Δ (Fig. $P2.14$).

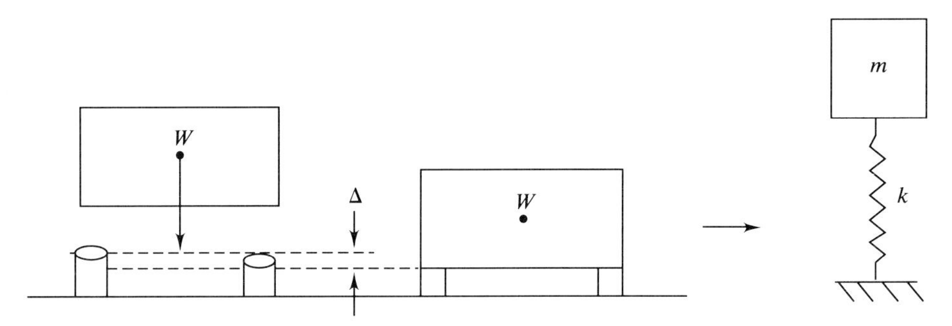

FIGURE P**2.14**

Determine the stiffness k, in the vertical direction, provided by the four pads, in terms of Δ and W.

15. A mass m is restrained by a nonlinear spring for which the dependence of spring force F_s on displacement x is

$$F_s = -k_1 x - k_3 x^3 \qquad (k_1, k_3 > 0)$$

Determine the potential energy stored when the spring is deflected. Find the maximum displacement of the mass if it is set into motion with initial conditions $x(0) = 0$, $\dot{x}(0) = \dot{x}_0$. Will this maximum displacement be larger or smaller than that occurring for the linear spring force $F_s = -k_1 x$?

16. Consider an elastic string which is stretched under a uniform, known tension T. A mass m is located at the center of the string, Fig. $P2.16$.

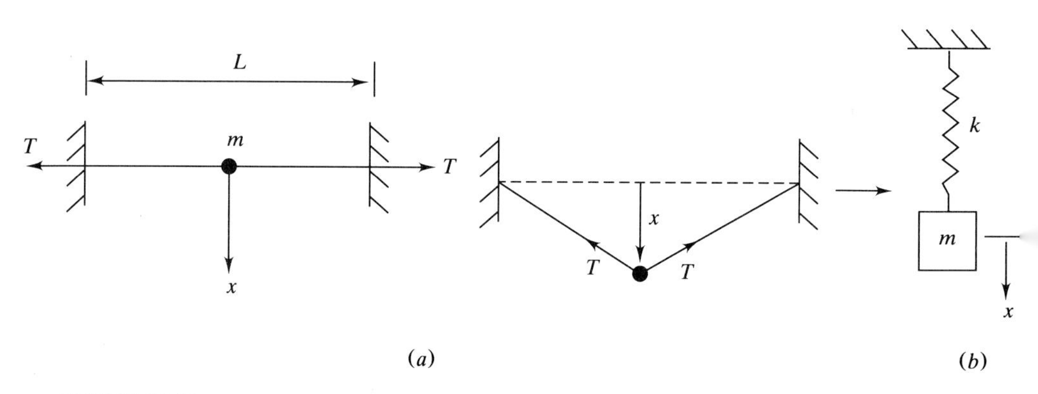

FIGURE P**2.16**

For small motions x find the spring constant k which allows one to model the actual system Fig. 2.16(a) as the idealized system Fig. $P2.16(b)$.

17. To simulate the motions of buildings during earthquakes, one sometimes assumes the building to be a rigid body (unlike in Example 2.6). The ground is assumed to provide linear and torsional stiffnesses which restrain lateral translation and rotation (Fig. $P2.17$).

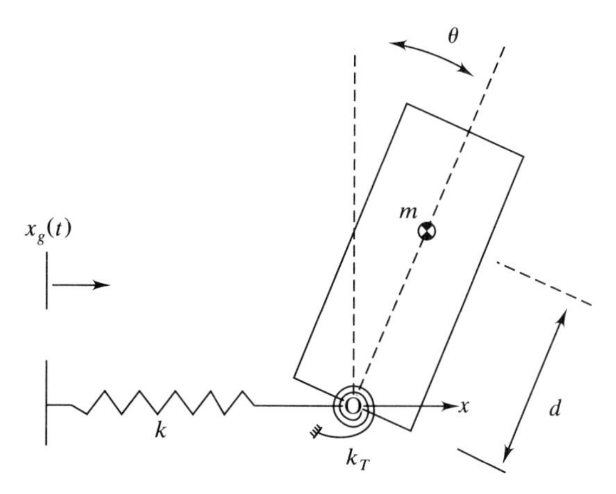

FIGURE $P2.17$

The building has mass m and moment of inertia I_c with respect to the center of mass, which is located a distance d from the point O at which the building is "attached" to the ground. The ground restraint is modeled by the linear spring k and torsional spring k_T. The earthquake is assumed to cause a horizontal translation of the ground, modeled by the specified ground motion $x_g(t)$. Assuming the motion to be planar and that the angle θ is small, derive the equations of motion for the x and θ coordinates. Be sure to consider which of the moment laws is applicable for the rotational equation of motion.

18. Consider the two-dimensional translation of a mass m through the atmosphere close to the surface of the earth. The position of the mass is measured by the cartesian coordinates x and y (Fig. $P2.18$) such that $y = 0$ at the earth's surface.

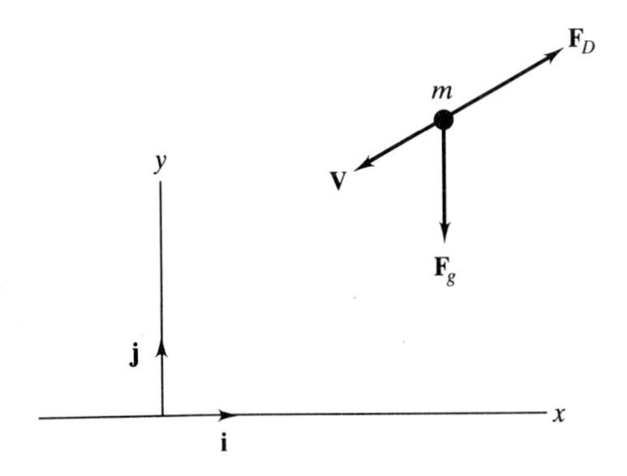

FIGURE $P2.18$

The forces on m are gravity $\mathbf{F}_g$, which may be assumed constant, and aerodynamic drag $\mathbf{F}_D$, which is always directed opposite the velocity $\mathbf{V}$, and which is proportional to the square of the speed. Derive the equations of motion in the form

$$m\ddot{x} = F_x(x, y, \dot{x}, \dot{y})$$
$$m\ddot{y} = F_y(x, y, \dot{x}, \dot{y})$$

19. Consider a geared system in which the input gear is driven by a known torque $M(t)$. Fig. $P2.19(a)$.

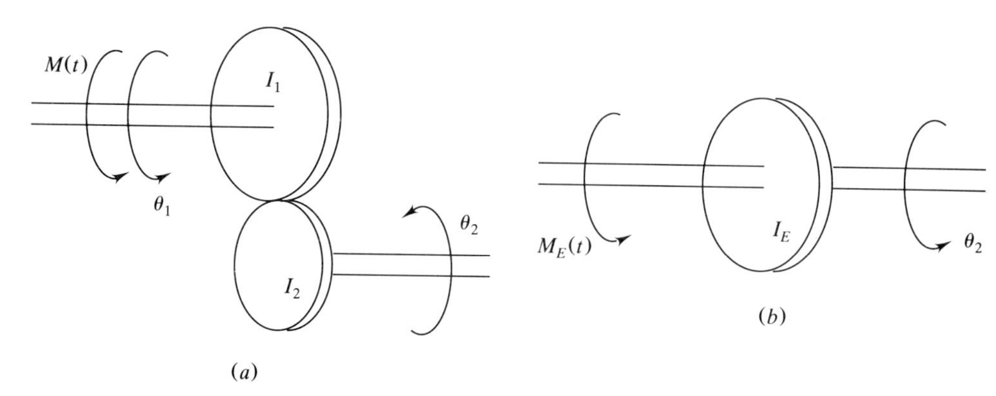

FIGURE $P2.19$

Use the system kinetic energy, as in Example 2.11, to determine the equation of motion in terms of the coordinate θ_2, by defining an equivalent moment of inertia I_E and an equivalent moment $M_E(t)$, as in Fig. $P2.19(b)$.

20. In Example 2.9 the equations of motion of the suspended airfoil were obtained as (2.65) and (2.66). Identify the system states, and put the model equations into state variable form.

21. For the manipulator model equations (2.68) and (2.69), identify all of the nonlinear terms.

22. For the gear/track assembly of Example 2.11, suppose the motion of the track m is restrained by a linear spring of stiffness k, as shown in Fig. $P2.22$.

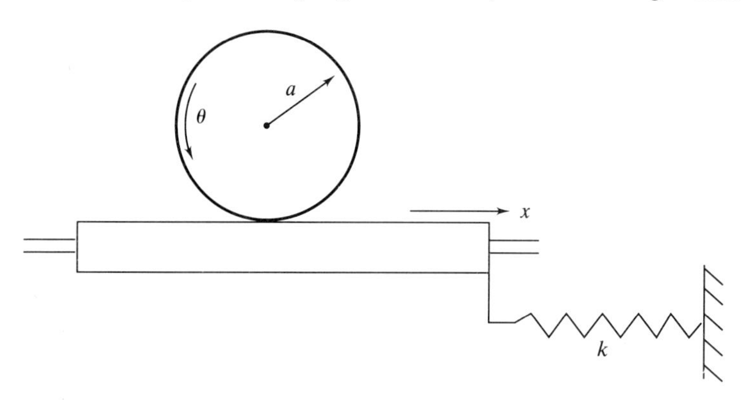

FIGURE $P2.22$

Determine the equation of motion for the coordinate θ. *Hint*: Equivalence of potential energy can be utilized. Thus, define an equivalent torsional stiffness K_T so that the potential energy $V = \frac{1}{2}K_T\theta^2$. Then develop the equation of motion for an equivalent system having a torsional spring.

23. Derive the equation of motion for the gear/track system of Example 2.11 directly from newtonian mechanics by including a force F exerted by the gear on the track and an equal and opposite force exerted by the track on the gear.

24. Derive the equation of motion for the θ coordinate for the system of Fig. *P*2.24, which is a variation of the gear/track assembly of Example 2.11. As the shaft spins, the added mass m_2 is raised (or lowered) on an inextensible wire that wraps (or unwraps) around the shaft, which is of radius b. The moment of inertia of the gear is I, and the geared track m_1 translates normal to the paper, as in Example 2.11.

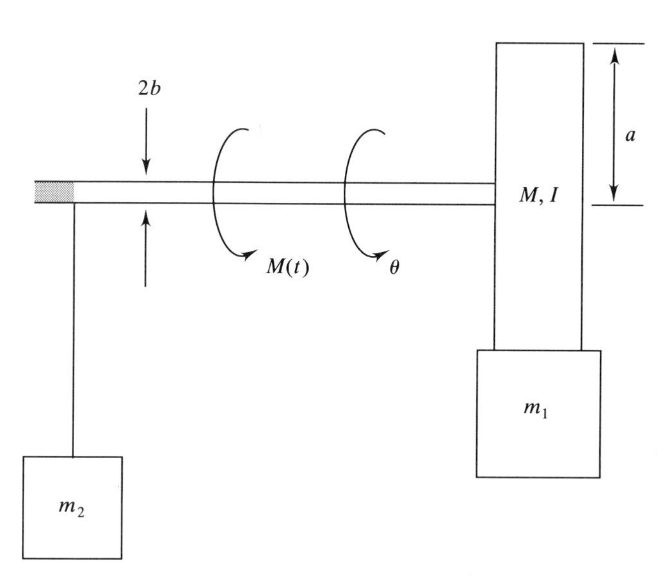

FIGURE *P*2.24

25. Derive the equations of motion (for x_1 and x_2) for the system shown in Fig. *P*2.25.

Note: x_2 and x_1 are defined with respect to the same reference line (that is, x_2 is *not* a *relative* displacement).

26. The following are measured values of damping force F_D versus velocity $\dot{x}$ for a certain damping device. Determine an approximate linear constitutive model (valid for small $\dot{x}$) for the damping force F_D:

$\dot{x}$ (m/s)	0.4	0.3	0.2	− 0.1	0	− 0.1	− 0.2	− 0.3	− 0.4
$F_D(N)$	−550	−375	−220	−100	0	100	180	250	310

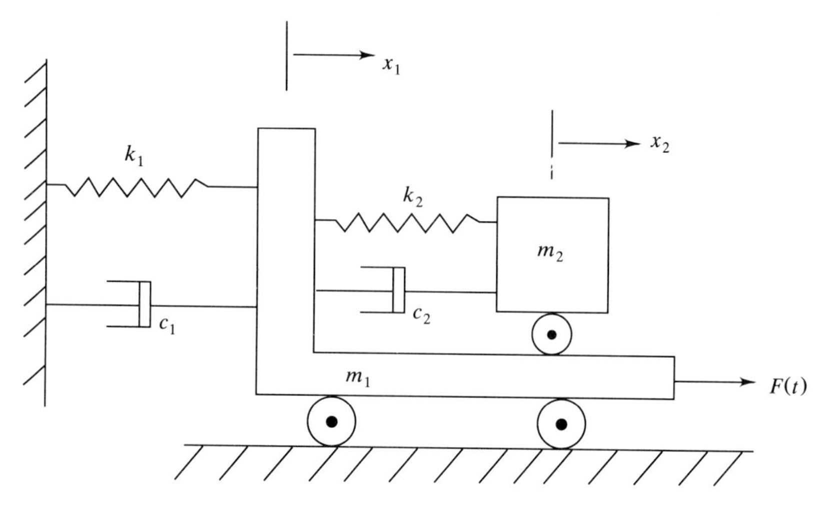

FIGURE ***P*2.25**

27. A mass m attached to a massless rod of length a rotates in the plane under the action of a moment M_0, which is applied about the point S. The point S moves around a circular path of radius R at the fixed angular rate ω (Fig. P2.27).

a. Determine the *relative* angular momentum H_{S_r} of the mass m with respect to the reference point S.

b. Is it true for this system that $\dot{\mathbf{H}}_{S_r} = \mathbf{M}_S$, where $\mathbf{M}_S = M_0\mathbf{k}$? Explain.

28. A rigid body of mass m_1 and moment of inertia with respect to the center of mass I_c rotates about the fixed frictionless pivot O. The body is connected via a linear spring to a mass m_2 which moves horizontally as shown (Fig. P2.28). Determine the equations of motion of the system, assuming that θ is always a small angle.

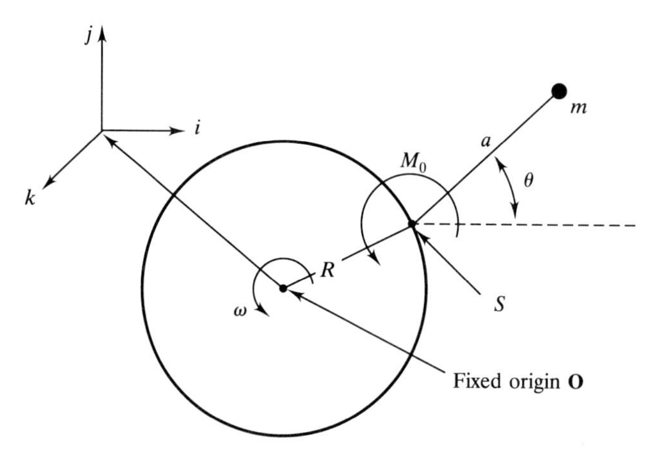

FIGURE ***P*2.27**

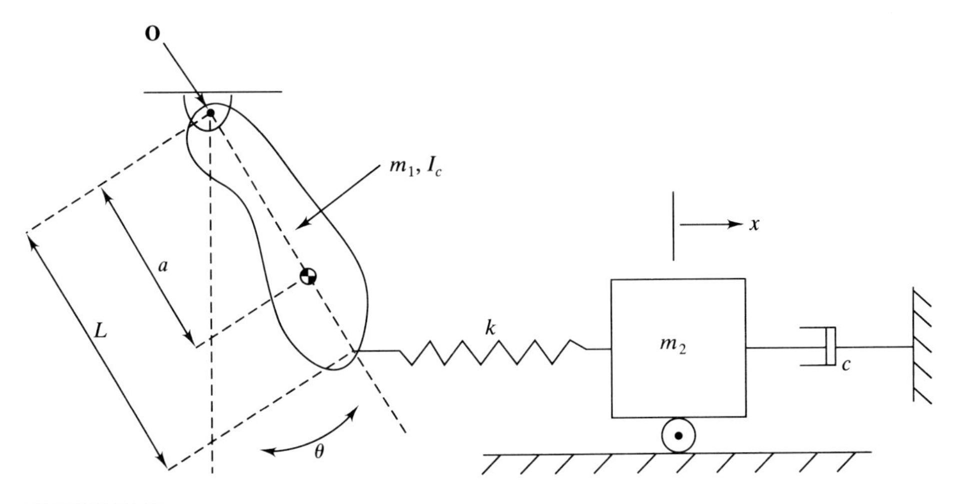

FIGURE *P*2.28

29. A mass $m = 20$ kg is mounted on a flexible beam (Fig. *P*2.29). Test results are obtained as follows:

a. A *static* load of 10^3 N, when applied in the x direction, results in a static deflection of the mass m of 5 mm.

b. If the mass m is held fixed and the damper attachment moved at constant speed of 1.0 m/s, a force of 10^2 N is exerted on the mass m.

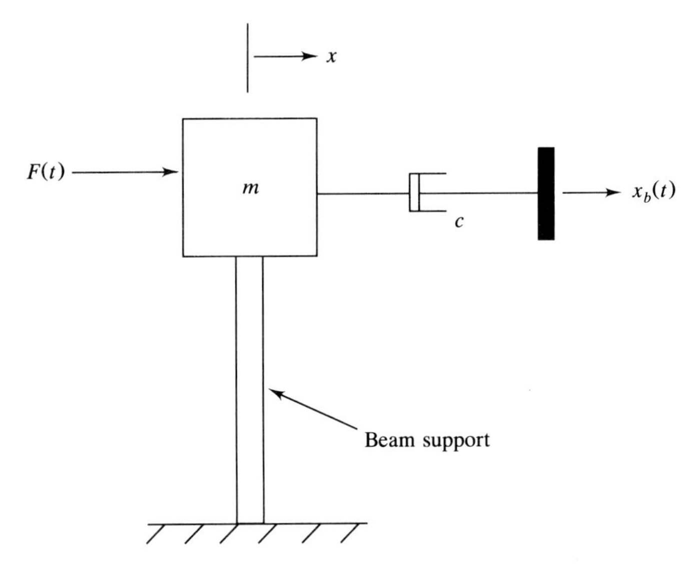

FIGURE *P*2.29

For the case of specified applied force $F(t)$ and specified damper attachment motion $x_b(t)$, determine the equation of motion for the system. Also determine numerical values for the system constants which appear in the model.

30. A mass m is restrained by linear springs, each of stiffness $k/2$, and slides without friction in a cylinder. The cylinder is attached to a shaft which rotates at the constant rate ω as shown in Fig. $P2.30$. A coordinate x measures the movement of the mass m along the cylinder; when $x = 0$, the springs exert no force and the mass m is a distance L from the axis of rotation. Determine the equation of motion of the mass m. Is this equation of motion linear?

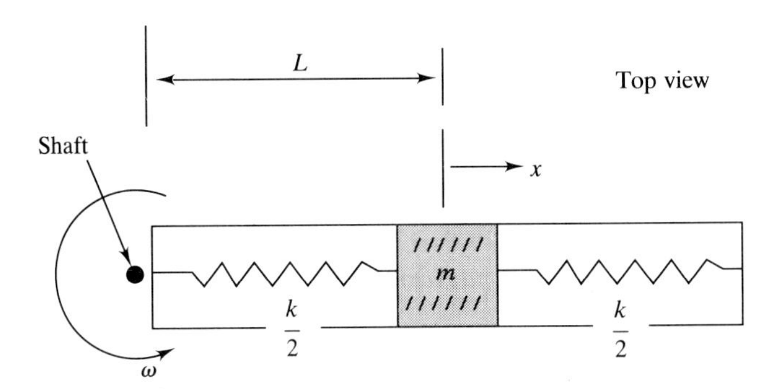

FIGURE $P2.30$

CHAPTER 3

MODELING OF HYDRAULIC, PNEUMATIC, ELECTROMAGNETIC, AND NONENGINEERING SYSTEMS

In the preceding chapter we considered mathematical models of "lumped systems" (all of the masses involved were modeled as particles or rigid bodies) in which the loadings were generated by "mechanical" means. In this chapter we focus attention on hydraulic, pneumatic, and electromagnetic (HPE for short) systems. These types of systems are of interest in their own right, but our main interest here will be fairly narrowly directed: We will be concerned mainly with the ways in which HPE effects can be used to generate loadings that move or restrain mechanical objects. Hence, in most of the systems we will consider, we will still be concerned with the motions of mechanical objects, but now the loadings will be generated by combined mechanical and HPE means, and we will often require ODEs to model not only the motions of our mechanical objects but also what is happening in the HPE part of the system.

Common examples of HPE systems of interest to us include hydraulic actuators which maneuver robotic manipulators, which deflect aerodynamic surfaces on an airplane, which rotate wheels and landing gear in auto and airplane steering and landing gear systems, pneumatic devices such as air shocks and air springs, and electromagnetic devices such as motors, loudspeakers, and instrumentation. We will not consider HPE systems in the same depth as we did newtonian systems in Chap. 2. The objective here is only to cover some of the fundamental aspects of HPE systems and to show how simple HPE systems are modeled and how they may interact with the newtonian systems of Chap. 2.

At the end of this chapter we also explore briefly some common types of ODE models used to describe dynamic phenomena not arising in engineering. Such models are prevalent nowadays in population biology and ecology, chemical kinetics, economics, and other areas. The purpose of this last section (Sec. 3.6) will be to broaden our view of the subject, so that we begin to think of dynamic systems in more general terms.

3.1 HYDRAULIC SYSTEMS

3.1.1 Introduction

There are many applications in which a liquid under pressure is used to perform work. Hydraulic actuators are used for this purpose in many types of vehicles and machines. Advantages of such systems include the ability to generate large loadings with relatively small components and the ease with which a loading may be exerted at a distance through use of geometrically complex fluid transfer networks.

Hydraulic actuator systems generally possess four basic components: a pump that makes available high-pressure fluid, a reservoir that contains excess fluid, valves that are used to control the flow of fluid in the system, and an actuator, often a piston moving in a cylinder with both sides of the piston exposed to hydraulic fluid, which moves a mechanical load. Our main concern here will be with the latter two of these components. For reference, the book *System Dynamics* by Ogata (Ref. 3.1) contains additional information on hydraulic systems.

To begin our study of hydraulic systems, we first define relevant fluid physical properties and descriptive variables. By *descriptive variables* we mean the fluid pressure and velocity, which have the same status in fluid flow as the coordinates (or states) in a newtonian system. The pressure and velocity are the "unknowns" that we will need to construct mathematical models for, in order to define what is happening in the hydraulic part of the system. The physical properties of concern to us for hydraulic fluids are the density, the viscosity, and the bulk modulus. These properties have status analogous to the system constants in a newtonian model.

The *density* ρ measures the amount of mass per unit volume. Its units are kg/m^3 or $slugs/ft^3$.* Because hydraulic liquids are nearly incompressible, we will sometimes assume the density to be uniform in the liquid system to be considered. The *viscosity* μ, measured in units of $N\text{-}s/m^2$ or $lb\text{-}s/ft^2$, is the frictional property by which energy dissipation occurs in a flowing fluid. This

---

*In the English system the natural unit of mass is the slug, *not* the so-called pound mass.

friction causes drag of objects moving through a fluid and also results in pressure losses as a fluid flows through a tube, for example. The *bulk modulus* K (in units of N/m^2 or lb/ft^2, the same units as pressure) measures the ability of a liquid to be compressed and is defined by the differential relation

$$\frac{d\rho}{\rho} = \frac{dp}{K} \tag{3.1}$$

where $d\rho$ is the incremental change in the density of a liquid which starts at a density ρ and is then subjected to an incremental pressure change dp. For most hydraulic liquids the bulk modulus is very large, of the order of 10^5 psi, indicating that very large pressure changes are necessary in order to change the density by even a small amount; this often justifies our assumption that liquids are essentially incompressible. Equation (3.1) may be viewed as a constitutive relation that connects the pressure and density of a liquid.

The state of a flowing fluid is defined by its velocity and static pressure. The pressure p, a force per unit area, is measured in units of N/m^2 (pascals) or lb/ft^2 (the lb/in^2 or psi is also commonly used). Static pressures in typical hydraulic systems may vary from ambient (i.e., atmospheric) to several thousand pounds per square inch. The pressure p in a fluid system is analogous to the potential energy V in a newtonian system. From Sec. 2.3 we recall that the force exerted by a conservative force on a mechanical object is the negative gradient of the potential energy. In a like manner, the pressure force exerted on a small "fluid particle" is the negative gradient of the pressure; fluid is accelerated along directions of decreasing pressure.

In order to define the velocity of a flowing fluid, we need to specify three velocity components, usually denoted as u, v, and w in the x, y, and z directions, respectively, just as for a newtonian particle. So the state of a flowing liquid is determined by a total of four dependent variables: u, v, w, and p. If we consider a fluid flow in some region of space, then we have to note that these four variables will generally be changing not only with time but also with location in the region of interest. Thus, each of the four dependent variables will be functions of time t and location (say, x, y, z in a cartesian system). We have a system with four dependent variables (u, v, w, p) and four independent variables (x, y, z, t). The general fluid problem will be governed by a set of four partial differential equations (we need as many equations as we have dependent variables), not ODEs. These partial differential equations are obtained through use of conservation of mass, yielding the so-called continuity equation, and by application of Newton's second law per unit volume, in each of the three directions, yielding the three so-called momentum equations.

For our purposes we will not attempt to study the general equations of fluid motion. The student should consult one of the fluid mechanics references at the end of this chapter for additional information. We will make use of a simplified version that describes approximately the specific situations of interest to us. Thus, the basic relations we will employ to model fluid problems are the following two:

1. *Conservation of mass*: If we consider a certain volume of fluid, then the rate of change of mass $\dot{m}$ within the volume must equal the rate of mass flow q across the volume boundaries and into the volume, $\dot{m} = q$, where we define the mass flow rate q (in kg/s or slugs/s) to be positive if mass is flowing into the region.
2. *Conservation of energy*: If the effects of viscosity are ignored, so that there is no energy dissipation, and if the liquid flow is assumed incompressible, steady (the flow velocities and pressure do not change with time at any given location in the flow), and irrotational, then the sum of kinetic and potential energies, per unit volume, is conserved:

$$\tfrac{1}{2}\rho V^2 + \rho g z + p = \text{constant} \tag{3.2}$$

This relation is known as *Bernoulli's equation*. The kinetic energy per unit volume is $\frac{1}{2}\rho V^2$, where V is the velocity magnitude, and we have two sources of potential energy (per unit volume): that due to gravity, $\rho g z$, where z is the height above an arbitrarily established reference level, and that due to the pressure p. We often apply (3.2) by equating the total energy at any two locations, or "stations" 1 and 2, in the fluid, so that

$$\tfrac{1}{2}\rho V_1^2 + \rho g z_1 + p_1 = \tfrac{1}{2}\rho V_2^2 + \rho g z_2 + p_2 \tag{3.3}$$

where $(\ )_1$ and $(\ )_2$ refer to locations 1 and 2. Thus, (3.3) shows that the velocity will increase between stations 1 and 2 if station 2 is below station 1 or if the pressure p_2 is less than p_1.

It should be noted that (3.3) for a liquid is analogous to the energy conservation relation (2.41) for a particle acted upon by conservative forces. It is safe to say that the conditions under which (3.3) is exact are never realized in practice, so that (3.3) is an approximation that works well for nearly steady, incompressible flows for which pressure losses due to viscosity can be assumed small. Let us now consider an example illustrating the application of conservation of mass and energy.

Example 3.1 A Liquid Level System. A tank of cross-section A contains water that is allowed to flow through a small opening of area A_0 at the bottom (Fig. 3.1).

Meanwhile, water is poured in at the top at the fixed mass flow rate q_0. We assume the water to be incompressible. The objective is to develop a mathematical model that describes the variation with time of the water level $z(t)$ in the tank.

We can first apply conservation of mass, as follows: The total mass m of water in the tank at any instant is $m = \rho A z$. The rate of change of this quantity must equal the net rate at which mass enters the tank, which is $q_0 - \rho A_0 V$. Thus, conservation of mass yields the first-order differential equation

$$\rho A \dot{z} = q_0 - \rho A_0 V \tag{3.4}$$

Next, we realize that the exit velocity V will be a function of z which can be

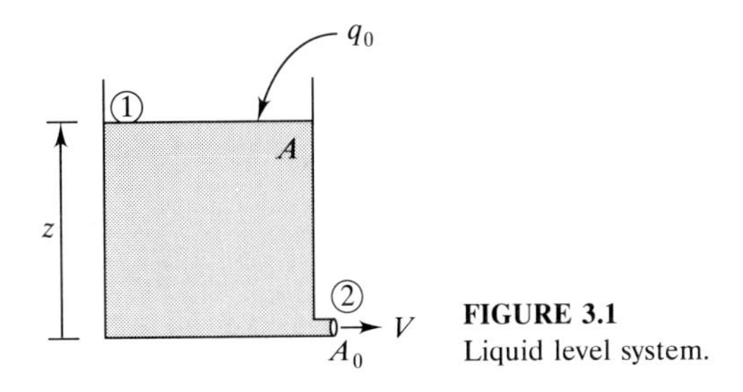

FIGURE 3.1
Liquid level system.

calculated approximately from Bernoulli's equation. Defining stations 1 and 2 as shown in Fig. 3.1, and noting that $p_1 = p_2$ (atmospheric pressure), (3.3) reduces to

$$\tfrac{1}{2}\rho V_1^2 + \rho g z_1 = \tfrac{1}{2}\rho V_2^2 + \rho g z_2$$

Next, let $z_1 = z$, $z_2 = 0$, and, assuming that $A_0 \ll A$, ignore the velocity V_1, so that $\rho g z \cong \frac{1}{2}\rho V^2$. Solving this relation for the exit velocity V, we obtain

$$V \cong (2gz)^{1/2} \tag{3.5}$$

Notice that this result is the same as the velocity of a particle dropped from rest at a height z in the gravitational field, with no air resistance. Having expressed the exit velocity V as a function of the dependent variable of interest z, we now obtain an ODE for z by combining (3.4) and (3.5) to obtain

$$\dot{z} = \frac{q_0}{\rho A} - \frac{A_0}{A}(2gz)^{1/2} \tag{3.6}$$

Our model which describes variations in water level z is of first order, is nonlinear because of the $z^{1/2}$, and is autonomous. Note that, in devising this model, we are not overly concerned with the details of the fluid motion in the tank. Furthermore, the ODE model differs from those for newtonian systems obtained in Chap. 2 in that it does not consist of second-order differential equations. This is fairly typical in the applications we will consider; conservation of mass yields a first-order relation usually involving two or more unknowns. Bernoulli's equation then functions as a constitutive relation that enables one of the unknowns to be eliminated.

3.1.2 Flow through an Orifice

In many practical applications of hydraulic devices, fluid is driven through small valve openings ("orifices") across which a relatively large pressure differential exists. We now develop an approximate model to describe the relation between the mass flow through, and the pressure drop across, an orifice. The idealized situation is shown in Fig. 3.2.

Fluid in a chamber at high pressure p_1 is driven through a small orifice of area A_0 (A_0 is assumed much smaller than the cross-sectional area of the chambers), into a second chamber at lower pressure p_2. In order to pass through the orifice, the fluid must accelerate to a relatively high velocity, and this occurs at the expense of a pressure drop. We'll assume the fluid pressure

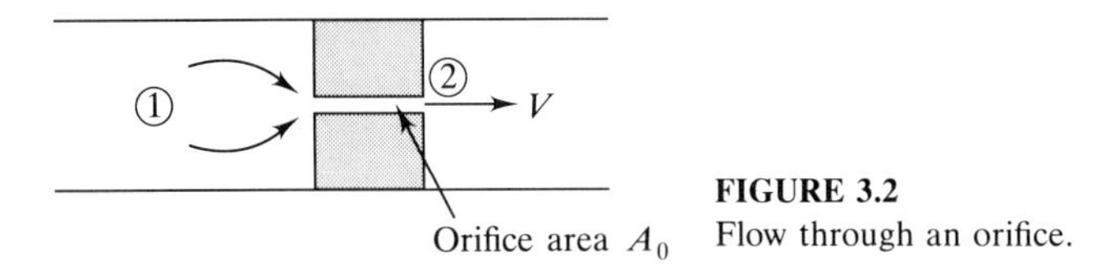

FIGURE 3.2
Flow through an orifice.

and velocity to be related by Bernoulli's equation (3.3). We'll take point 1 to be upstream of the orifice, where the velocity V_1 will be assumed small enough to ignore. Point 2 will be at the orifice exit. Thus, we have from (3.3)

$$p_2 + \tfrac{1}{2}\rho V_2^2 \cong p_1$$

Solving this relation for the velocity V_2 of flow through the orifice, we obtain

$$V_2 \approx \sqrt{\frac{2}{\rho}(p_1 - p_2)}$$

Next we can obtain the mass flow rate q by multiplying this relation by ρA_0, so the mass flow rate through the orifice is

$$q = A_0\sqrt{2\rho(p_1 - p_2)} \tag{3.7}$$

For actual orifice flows, some frictional losses occur, and the mass flow rate q is less than that given by (3.7). In order to account for such losses, we can introduce a multiplicative "loss factor" $c_0 < 1$, which might be determined experimentally, so that the basic orifice flow relation is

$$q = c_0 A_0\sqrt{2\rho(p_1 - p_2)} \tag{3.8}$$

It is important to note the functional dependence in (3.8). The mass flow is linear in area but varies as the *square root* of the pressure difference. Equation (3.8) is the constitutive relation describing the flow of liquid through an orifice. This constitutive relation connects the variables involved, mass flow rate, and pressure difference, in an inherently nonlinear way. Now let us apply the results we have developed so far to another example.

Example 3.2 A "V^2 Damper." We consider an idealized damping device consisting of a piston/cylinder arrangement, in which the cylinder is occupied by a hydraulic fluid that is allowed to pass through small orifices of total area A_0 in the piston (Fig. 3.3). For illustration we suppose the damper piston to be rigidly connected to a mass m, as shown. In order to develop a mathematical model, let us first suppose that the mass m is moving with a velocity $\dot{x} > 0$. Then fluid will be driven from chamber 2 to chamber 1 through the orifices A_0. This flow from one chamber to the other must be accompanied by a pressure drop (that is, $p_2 > p_1$). The pressure differential $(p_2 - p_1)$ exerts a force on the piston, which is communicated directly to the mass m. Thus, the equation of motion of the mass m is simply

$$m\ddot{x} = -(p_2 - p_1)A \tag{3.9}$$

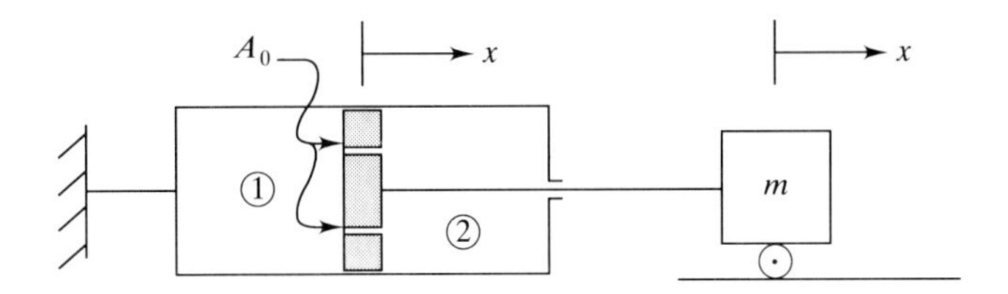

FIGURE 3.3
Schematic of the "V^2 damper."

where A is the cross-sectional area of the piston. Next, we need to reformulate (3.9) so that it involves the single dependent variable x. Thus we look for a connection between the pressure drop $(p_2 - p_1)$ and the piston velocity $\dot{x}$. This can be done using conservation of mass, along with the constitutive relation for orifice flow (3.8). First, assuming the liquid to be incompressible, conservation of mass yields for the orifice mass flow rate q:

$$q = \rho A \dot{x} \tag{3.10}$$

where, if $\dot{x} > 0$, q is understood to be from chamber 2 to chamber 1. The constitutive relation (3.8) also provides an equation for q:

$$q = c_0 A_0 [2\rho(p_2 - p_1)]^{1/2} \qquad \text{for } p_2 > p_1 \tag{3.11}$$

Combining (3.10) and (3.11), we obtain the pressure difference $p_2 - p_1$ as a function of velocity $\dot{x}$ as

$$p_2 - p_1 = \frac{\rho A^2 \dot{x}^2}{2 A_0^2 c_0^2} \tag{3.12}$$

At this point we need to be careful about the signs involved; the preceding relation applies only when $\dot{x} > 0$, in which case $p_2 > p_1$. If, however, $\dot{x} < 0$, then $p_2 < p_1$ and (3.12) would read

$$p_2 - p_1 = -\frac{\rho A^2 \dot{x}^2}{2 A_0^2 c_0^2} \qquad (\dot{x} < 0) \tag{3.13}$$

The signs can be accounted for correctly in a single relation by writing

$$p_2 - p_1 = \frac{\rho A^2}{2 A_0^2 c_0^2} \dot{x}|\dot{x}| \tag{3.14}$$

The sign of $\dot{x}$ now provides the correct overall sign for the relation. Thus, the mathematical model for a mass restrained by the idealized V^2 damper now reads

$$m\ddot{x} + \left(\frac{\rho A^3}{2 A_0^2 c_0^2} \right) \dot{x}|\dot{x}| = 0 \tag{3.15}$$

Because the damping force is proportional to the square of the velocity, the device is often called a "V^2 damper"; it provides a damping force that is inherently nonlinear. Note that the force is not a function of position x. The piston may be located statically in any position x and, provided $\dot{x} = 0$, there is no force generated. In comparison to the linear viscous damper, the V^2 damper will dissipate relatively more energy if the velocities are high and relatively less energy if the velocities are low (see Sec. 10.2.1 for some results for the motion of an oscillator with a V^2 damper).

Next, let us consider another example of great practical importance, the hydraulic servovalve, in which the fluid flow through small orifices is controlled by a sliding valve which opens or closes the orifices, in turn controlling the forces exerted on a mechanical load that is to be moved.

Example 3.3 A Hydraulic Servovalve. This example is a bit more complex than preceding ones and represents a fairly realistic analysis of a heavily used engineering device. We'll consider here a system in which a fluid under pressure is used as a mechanism of force transmission. The example is that of a hydraulic servovalve, with the setup shown in Fig. 3.4.

We note that hydraulic systems are often used to transmit forces in situations which are essentially static, so that dynamic behavior is not of interest. Examples of this type include braking systems and hydraulically actuated clutches in automobiles. In other situations, however, the dynamic behavior may be important, as in the nose landing gear steering system of an airplane or in the use of hydraulic

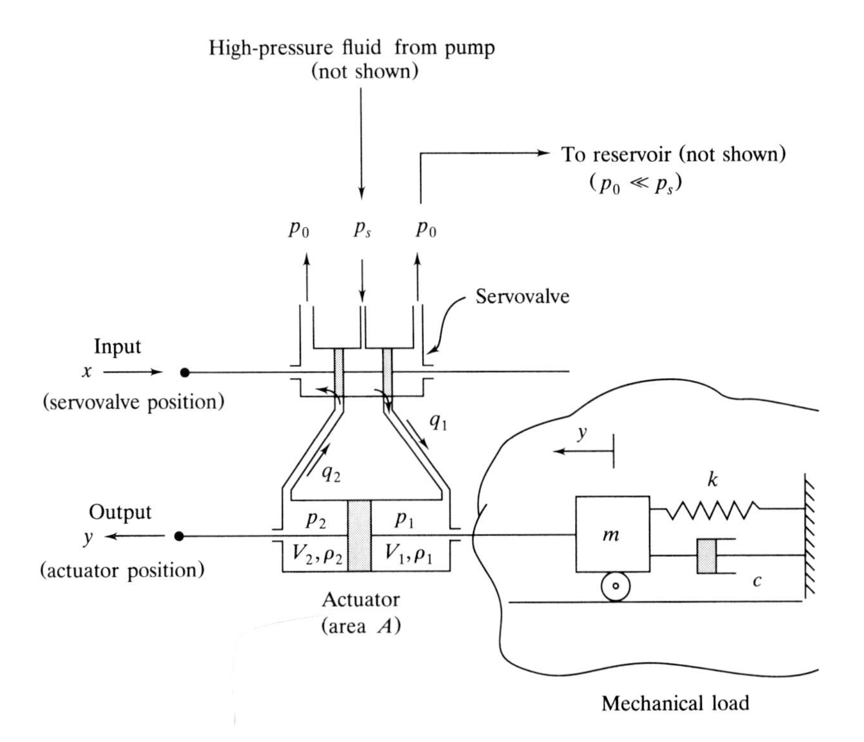

FIGURE 3.4
The hydraulic servovalve. Flow directions shown are for $x > 0$.

actuators to control the motion of mechanical objects such as large robotic manipulators. The present example falls into this latter category. We assume that the objective is to apply a force on the mass m (Fig. 3.4) in order to cause the mass m to move in a certain way. In developing the mathematical model, we have to take into account dynamic behavior in both the mechanical and the fluid parts of the system.

The operation of the hydraulic system of Fig. 3.4 is as follows: A reservoir of hydraulic fluid, such as brake fluid, is maintained at a constant, high "supply" pressure p_s (assumed here to be maintained by a pump, not shown). Depending on the position of the sliding servovalve, chamber 1 or chamber 2 of the actuator is exposed to the high-pressure fluid ($x > 0$ and $x < 0$, respectively). Suppose $x > 0$; then fluid will tend to be forced into chamber 1, moving the actuator piston in the $+y$ direction, and vice versa if $x < 0$. This tendency, however, is resisted by the mechanical load: What actually happens is that there is a two-way interaction of the motion of the mass m and the pressures, flows, and so on in the hydraulic system.

Note that here the input $x(t)$ is assumed to be a *known* function; for example, x might be proportional to the orientation of a steering wheel, which is under the control of the vehicle operator. What we need to do is to develop a mathematical model that connects this known input $x(t)$ to the motion $y(t)$ of the mass m. In order to do this, we will need to model certain aspects of the fluid flows and pressures in the hydraulic portion of the system.

Initially, however, we can write down the equation of motion for the mass m as follows:

$$m\ddot{y} + c\dot{y} + ky = (p_1 - p_2)A \tag{3.16}$$

Here A is the cross-sectional area of the actuator piston, and $(p_1 - p_2)A$ is the pressure force exerted on this piston, and hence on the mass m. We see that, in order to complete the model, we will need to develop equations to tell us what the pressures p_1 and p_2 are doing; we also have to make sure that the input $x(t)$ gets into the model somewhere.

In order to continue the model development, let's take a look at *conservation of mass* applied to the fluid in chamber 1; this will tell us something about the pressure p_1. First, note that the total mass of fluid m_1 in chamber 1 at any time t is given by $m_1 = \rho_1 V_1$. Conservation of mass states that the rate $\dot{m}_1$ at which the mass in chamber 1 increases must equal the rate q_1 at which mass flows into the chamber. Thus, we have

$$q_1 = \dot{m}_1 = \dot{\rho}_1 V_1 + \rho_1 \dot{V}_1 \tag{3.17}$$

The right-hand side of this equation shows that a rate of mass increase may be due to an increasing density $\dot{\rho}_1$, or to an increasing volume $\dot{V}_1$ (due to piston motion).

In the examples up to now we have assumed the hydraulic fluid to be incompressible. In realistic versions of the present example, however, it is common during operation to have pressure fluctuations of several thousand pounds per square inch. The resulting density fluctuations, although still small, may not be ignorable and have to be taken into account.

The next step in our formulation is to rewrite the $\dot{\rho}_1$ term in (3.17) in terms of the rate of change of pressure $\dot{p}_1$. This can be done using the differential constitutive relation (3.1); divide each side of (3.1) by a time increment dt to obtain $\dot{\rho} = (\rho/K)\dot{p}$. Then (3.17) becomes

$$q_1 = \frac{\rho_1 V_1}{K}\dot{p}_1 + \rho_1 \dot{V}_1 \tag{3.18}$$

Let's now look at the volume V_1 terms in (3.18); then we'll tackle the flow q_1. First, when $y = 0$, define the chamber volume V_1 as $V_1 = AL$; here L is the "length" of the chamber (the same for chambers 1 and 2) at rest. Then when $y \neq 0$, the chamber volume V_1 will be $V_1 = A(L + y)$; likewise we will get $V_2 = A(L - y)$. We also find the rate of volume increase $\dot{V}_1$ as $\dot{V}_1 = A\dot{y}$. So we are now able to replace the volume terms in (3.18) with terms involving the dependent variable y, and we obtain the following:

$$q_1 = \frac{\rho_1 A(L + y)}{K}\dot{p}_1 + \rho_1 A\dot{y} \tag{3.19}$$

The final step in the manipulation of the mass conservation relation is to utilize the constitutive relation (3.8) for flow through an orifice, which in the present notation is

$$q_1 = c_0 A_0 \sqrt{2\rho_1(p_s - p_1)}$$

If we assume that the orifice area A_0 is linearly proportional to the valve opening x, then this relation may be rewritten as

$$q_1 = cx\sqrt{p_s - p_1} \tag{3.20}$$

where the $(2\rho)^{1/2}$ and the loss coefficient c_0 have been absorbed into the constant c. Using (3.20) in (3.19), we obtain

$$cx\sqrt{p_s - p_1} = \frac{\rho_1 A(L + y)}{K}\dot{p}_1 + \rho_1 A\dot{y}$$

Next we solve the preceding equation for $\dot{p}_1$ to obtain an equation defining the rate of change of pressure $\dot{p}_1$ as

$$\dot{p}_1 = \frac{K}{\rho_1 A(L + y)}\left[cx(p_s - p_1)^{1/2} - \rho_1 A\dot{y}\right] \tag{3.21}$$

Note that the input servovalve position $x(t)$ appears in this equation.

We can apply the same type of analysis to determine what's going on in chamber 2. The reader may verify that the result will be as follows:

$$\dot{p}_2 = \frac{K}{\rho_2 A(L - y)}\left[-cx(p_2 - p_0)^{1/2} + \rho_2 A\dot{y}\right] \tag{3.22}$$

We are not quite done, however, because it is clear that (3.21) and (3.22) will apply only if $x > 0$. If $x < 0$, then the flow q_1 will depend on $(p_1 - p_0)$ rather than on $(p_s - p_1)$, and the flow q_2 will depend on $(p_s - p_2)$ rather than on $(p_2 - p_0)$. We can take this into account by changing the $(p_s - p_1)$ in (3.21) to $(p_1 - p_0)$ whenever $x < 0$ and by changing the $(p_2 - p_0)$ in (3.22) to $(p_s - p_2)$ when $x < 0$.

As a final comment on the model, the density ρ in (3.21) and (3.22) is assumed to be a known function of pressure. We'll assume that we have a graph of measurements of ρ versus p that apply for the hydraulic fluid being used. So then, let's summarize the model and then say a few words about it:

$$
\begin{aligned}
&m\ddot{y} + c\dot{y} + ky = (p_1 - p_2)A \\
&\dot{p}_1 = \frac{K}{\rho_1 A(L+y)}\left\{cx\begin{bmatrix}(p_s - p_1)^{1/2} & \text{if } x > 0 \\ (p_1 - p_0)^{1/2} & \text{if } x < 0\end{bmatrix} - \rho_1 A\dot{y}\right\} \qquad (3.23) \\
&\dot{p}_2 = \frac{K}{\rho_2 A(L-y)}\left\{-cx\begin{bmatrix}(p_2 - p_0)^{1/2} & \text{if } x > 0 \\ (p_s - p_2)^{1/2} & \text{if } x < 0\end{bmatrix} + \rho_2 A\dot{y}\right\} \\
&\rho = \rho(p)
\end{aligned}
$$

The system is of fourth order, and the states are y, $\dot{y}$, p_1, and p_2. The input variable is $x(t)$, and we can picture putting in some function $x(t)$ and integrating the equations to find the resulting responses y, p_1, p_2. There are several nonlinearities in the model: (1) the square root terms involving the pressures p_1 and p_2, (2) the terms $1/(L+y)$ and $1/(L-y)$ in the pressure equations, and (3) the terms $1/\rho_1$ and $1/\rho_2$ in the pressure equations (since ρ is a function of p). Generally, we would have to solve Eq. (3.23) on a computer using numerical integration methods. In some circumstances it may be possible to "linearize" the system equations; this type of procedure will be discussed in Chap. 4. But we are never going to get the full story by using a linear model of this system.

A final comment on this example involves the modeling methodology whereby the system ODE model (3.23) was eventually obtained. Inspection of Fig. 3.4 and visualization of the movement of the valve, mass, and system flows show that there is a total of 10 "unknown" quantities that will generally be changing with time during system operation: p_2, V_2, ρ_2, p_1, V_1, ρ_1, q_1, q_2, y, and $\dot{y}$ [we do not consider the valve position $x(t)$ in this category because it is assumed to be a specified function of time]. Only 4 of these 10 quantities are states, however (y, $\dot{y}$, p_1, and p_2). This has occurred because the other 6 quantities ($\rho_1, V_1, q_1, \rho_2, V_2, q_2$) may be expressed as functions of the four states and of the input valve position $x(t)$. Thus, of the 10 time-dependent measures of behavior, only 4 are considered independent, and the other 6 are expressed as functions of these 4 through constitutive relations. The process of developing constitutive relations to reduce the number of unknowns in the mathematical model to the minimum number is typical in the modeling process and is an important part of the analysis, especially for systems of moderate complexity.

3.1.3 Comments on Hydraulic Systems

1. The examples of this section illustrate that the mathematical models involving hydraulic systems tend to be inherently nonlinear because of the nonlinear relation between velocity (and mass flow rate) and pressure (or altitude) in Bernoulli's equation. In Chap. 4 we will see that it is possible in many cases to replace the actual nonlinear relations with linear ones that work well for certain operating ranges of the systems. In general, however, it should be borne in mind that nonlinearities tend to be the rule, rather than the exception, in hydraulic systems.

2. From the standpoint of basic fluid mechanics, our modeling has been very simple: We have used only conservation of mass and conservation of energy (Bernoulli's equation) in developing our mathematical models. While conservation of mass has the status of a law (it is always exact), Bernoulli's equation is only an approximation for the interaction among velocity, pressure, and height in a flowing fluid. There are many phenomena of importance to engineers in which Bernoulli's equation provides an incorrect statement of these interactions. Our view here is that these situations lie outside the province of the introductory level material presented here.
3. It should also be realized that we have considered a very limited number of specific types of hydraulic devices. For example, pumps of all types and servovalve/actuator systems of many different configurations fall under the aegis of hydraulics. The interested student may wish to consult some of the references listed at the end of this chapter for further information. Reference 3.1 is a good place to start.

3.2 PNEUMATIC SYSTEMS

In this section we'll consider pneumatic effects in which gases, often air, under pressure are used to perform work on mechanical loads. Pneumatic systems typically operate at much lower pressures and generate lower power in comparison to hydraulic devices. One advantage of pneumatic systems is that air can often be employed as the working fluid, so that used fluid can be exhausted to the atmosphere, obviating the need for the return lines necessary in a hydraulic system, which must operate as a closed system.

In comparing pneumatic and hydraulic systems from the standpoint of mathematical modeling, we have to be careful to keep track of the differences in the two types of systems. There are two main differences:

1. In hydraulic systems we usually assume the working fluid to be incompressible; even if we include the effects of compressibility (as, for example, in Example 3.3, the hydraulic servovalve), we assume the density ρ to be a known function of the pressure, $\rho = \rho(p)$, and the actual percentage changes in density are small. For a pneumatic system, on the other hand, large changes in density nearly always occur, and this compressibility is an essential aspect of gas behavior that has to be taken into account. One consequence of gas compressibility is that pneumatic systems respond more slowly than hydraulic systems to changes in operating state.
2. In hydraulic systems the temperature T and the associated thermal energy are generally not involved in the modeling process.* In pneumatic systems,

*The physical properties of hydraulic fluids, especially viscosity, are temperature dependent, so to this extent temperature is involved. However, we usually do not need to use temperature as a variable in the formulation of the mathematical model.

on the other hand, substantial temperature changes can occur as mechanical energy is converted into thermal energy (temperature increase) and vice versa, according to the laws of thermodynamics.

Thus, in analyzing a hydraulic system, we need to be aware of the basic laws of fluid mechanics, but thermodynamics is not involved. But in pneumatic systems, we need to use both fluid mechanics and thermodynamics in order to be able to analyze the dynamic behavior of the system: The interaction of all three of the basic thermodynamic variables (pressure p, density ρ, and temperature T) must be taken into account, so that the analysis of pneumatic systems is inherently more involved than that of hydraulic systems.

For our purposes the main similarity between hydraulic and pneumatic systems is that the pressure p is the main fluid/thermodynamic variable of interest because the pressure is what results in loads on mechanical objects. Hence, in analyzing pneumatic systems, one of our objectives will be to eliminate the density and temperature where they appear and to express all thermodynamic effects in terms of pressure only, if we can. In the ensuing discussion, we first state some of the basic laws and constitutive relations that we will use to model pneumatic systems. Then several examples are presented. The student should bear in mind that the treatment here is not detailed; the purpose, as with hydraulic systems, is only to introduce some of the important features that need to be taken into account in the mathematical modeling of pneumatic systems. No attempt is made to describe the various types of pneumatic devices used in engineering, such as compressors, valves, and fluidic devices. Several references are listed at the end of the chapter for those interested in further information.

3.2.1 Basic Laws and Constitutive Relations

Since the law of *conservation of mass* applies generally (it is a true law), we will make use of it in exactly the same way as for a liquid.

The conservation of mechanical energy (kinetic plus potential), Bernoulli's equation, does not apply for pneumatic systems, as it is based on the assumption of incompressibility. It is possible, however, to state an analogous result in differential form, as follows:

$$V\,dV + \frac{dp}{\rho} + g\,dz = 0 \tag{3.24}$$

which applies for steady, irrotational, inviscid flow. In this relation dV, dp, and dz are the incremental changes in the variables V, p, and z that occur along a streamline segment of incremental length ds (a streamline is the path followed by a small packet of fluid in steady flow; the fluid velocity is tangent to a streamline). One observes that Bernoulli's equation (3.3) is just an integrated version of (3.24): If we take the differential of (3.3), with $d\rho = 0$, then we obtain (3.24). For a compressible fluid, however, (3.24) cannot be integrated to Bernoulli's equation because the density ρ is not constant. Thus, for a gas, we have to use (3.24) instead of (3.3). In our work here conservation of mass and the differential relation (3.24) will be used to describe the fluid mechanics aspects of pneumatic systems, and it should be borne in mind that, for many phenomena associated with compressible gas flow, a more sophisticated approach is necessary.

In order to account for changes in the thermodynamic variables, we will make use of two constitutive relations. The first is the perfect gas law (Boyle's law) which for a gas in thermodynamic equilibrium is

$$p = \rho R T \tag{3.25}$$

where R is the gas constant, which for air is $R = 1717\ \mathrm{ft}^2/(s^2 R) = 287\ m^2/(s^2 K)$. This constitutive relation is an approximation that works well for the pressure levels encountered in most pneumatic systems.

The second constitutive relation we will use will employ the assumption that the thermodynamic processes involved in our pneumatic systems are *isentropic*. This means that they are both *reversible* (frictionless) and *adiabatic* (no heat is added from or transferred to the surroundings; the system is perfectly insulated). For isentropic processes [which never actually occur, just as the assumptions underlying (3.3) or (3.24) never are actually realized], the pressure and density are related by the relation

$$p\rho^{-\gamma} = \text{constant} \tag{3.26}$$

where $\gamma = c_p/c_v$ is the ratio of specific heats, which for air is $\gamma = 1.4$.

For a pneumatic system the constitutive relations (3.25) and (3.26) give us two equations connecting the three thermodynamic variables p, ρ, and T. We can, therefore, utilize (3.25) and (3.26) to eliminate ρ and T and so express all thermodynamic effects in terms of pressure only [provided, of course, that (3.25) and (3.26) are reasonably close to reality]. Let us now consider an example of a simple pneumatic system.

Example 3.4 Airspring. Pneumatic effects can be used to provide stiffness (restoring forces) that result from the differential pressures generated by compression/expansion of gases on either side of a piston. A schematic of such a device, used to restrain a mass m, is shown in Fig. 3.5. A (massless) piston of cross-section A slides without friction in a cylinder containing a gas. At rest ($x = 0$) the piston is centered, and the thermodynamic variables p, ρ, and T have the known values p_0, ρ_0, and T_0 in each of the two chambers. The "length" of each chamber at rest is d.

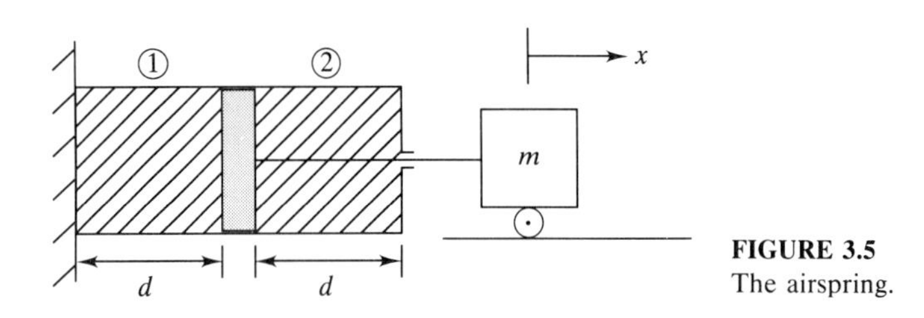

FIGURE 3.5
The airspring.

In physical terms we can see how the airspring produces a restoring force as follows: If $x > 0$, then the gas in chamber 2 is compressed so that $p_2 > p_0$; the gas in chamber 1 has expanded so that $p_1 < p_0$. The result is a pressure force in the negative x direction, that is, a restoring force, which will be a function only of position x. Let us now construct a mathematical model of the system. First, the equation of motion of the mass m is

$$m\ddot{x} = (p_1 - p_2)A \tag{3.27}$$

Now let us attempt to express the pressures p_1 and p_2 as functions of the displacement x, so that the single unknown x appears in (3.27). We can start with conservation of mass. Since the total masses m_1 and m_2 of gas in chambers 1 and 2 remain fixed, we have from geometry

$$m_1 = \rho_1 A(d + x) = \rho_0 Ad$$

$$m_2 = \rho_2 A(d - x) = \rho_0 Ad$$

Thus,

$$\rho_1 = \frac{\rho_0 d}{d + x} = \rho_0\left(\frac{1}{1 + x/d}\right)$$

$$\rho_2 = \rho_0\left(\frac{1}{1 - x/d}\right) \tag{3.28}$$

Next we can use the isentropic gas relation to reexpress (3.28) in terms of pressures rather than densities. From (3.26) we have

$$p_1\rho_1^{-\gamma} = C = p_0\rho_0^{-\gamma}$$

$$p_2\rho_2^{-\gamma} = C = p_0\rho_0^{-\gamma} \tag{3.29}$$

so that

$$\rho_{1,2}^{\gamma} = \frac{p_{1,2}}{C} \tag{3.30}$$

Combining (3.29) and (3.30) and using the definition of the constant $C = p_0\rho_0^{-\gamma}$, we obtain the chamber pressures as functions of the displacement x as

$$p_1 = p_0\left(\frac{1}{1 + x/d}\right)^{\gamma}$$

$$p_2 = p_0\left(\frac{1}{1 - x/d}\right)^{\gamma} \tag{3.31}$$

The equation of motion, therefore, can be expressed in terms of x alone as

$$m\ddot{x} + Ap_0\left[\left(\frac{1}{1 - x/d}\right)^{\gamma} - \left(\frac{1}{1 + x/d}\right)^{\gamma}\right] = 0 \tag{3.32}$$

We observe that the system is of the second order, is autonomous, and is nonlinear; the restoring force produced by gas compression/expansion is a nonlinear function of displacement x because of the isentropic relation connecting p and ρ. A graph of the bracketed term in (3.32), which is the negative of the gas pressure force divided by Ap_0, is shown in Fig. 3.6 for air, $\gamma = 1.4$. Note that if x/d is small

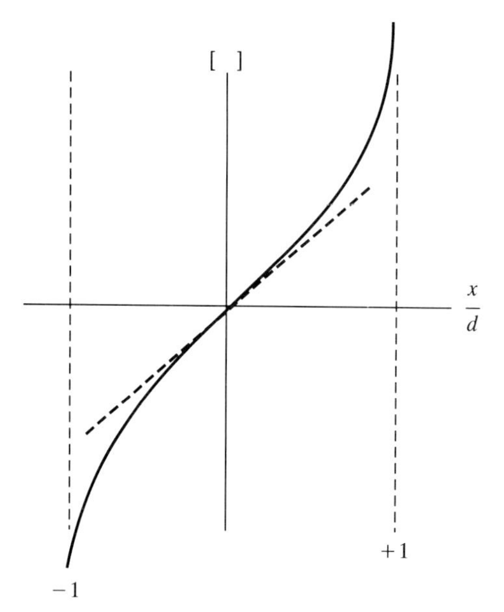

FIGURE 3.6
The bracketed term in (3.32); this is the negative of the gas pressure force, divided by Ap_0.

compared to unity, then the stiffness behavior appears linear, so that, for small x/d, the airspring behaves as an idealized linear spring. We will return to the question of linear operation of this system in Chap. 4.

From a tactical standpoint our use of the mass conservation law in the preceding example is worth a comment. In our application of this law to the hydraulic systems considered in Examples 3.2 and 3.3, we had to account for mass flows into and out of given regions of the system. Thus, we used the "dynamic" form of mass conservation, $\dot{m} = q$. In the present example, however, there are no mass flows ($q = 0$) into either chamber, so the mass conservation law is simply $m_1 = m_2 =$ constant; that is, we use the "static" form of mass conservation.

As a practical example combining both the pneumatic aspects of Example 3.4 and the hydraulic aspects of Example 3.2 (V^2 damper), consider the "gas pressure shock absorber" shown schematically in Fig. 3.7.

There are three fluid chambers in this device. Chambers 1 and 2 contain hydraulic fluid that is allowed to move between chambers 1 and 2 through small holes in the piston, in the manner of Example 3.2, the V^2 damper. A free-floating piston m separates the hydraulic chamber 2 from chamber 3, which contains gas (typically nitrogen) at several hundred pounds per square inch. This arrangement allows for smoother operation of the device. In order to mathematically model the gas pressure shock, one needs to combine the basic ideas of hydraulics and pneumatics. In addition, the motion of the free piston m has to be modeled (for the stout of heart, see Exercise 3.8).

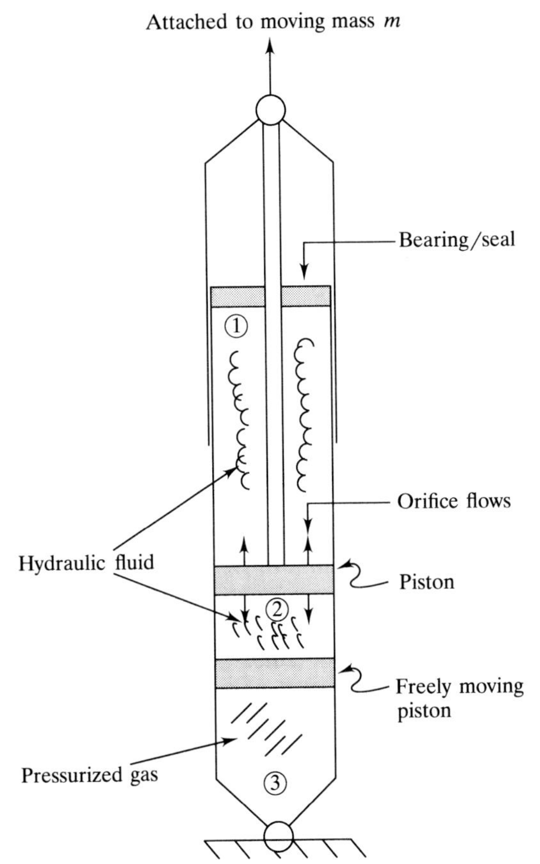

FIGURE 3.7
The gas pressure shock absorber.

3.2.2 Flow of Gas through an Orifice

In pneumatic, as well as hydraulic, systems it is often necessary to model the flow through orifices. For hydraulic systems we obtained the result (3.8), which is based solely on Bernoulli's equation. For gases, however, compressibility makes things more complicated. To start the analysis, we consider the same flow situation as depicted in Fig. 3.2, except we now consider a compressible gas rather than a liquid. We also assume the properties on the high-pressure side (p_1, ρ_1, T_1) to be known. Our goal is to derive an expression, as a function of the pressures p_1 and p_2, for the mass flow rate $\dot{m} = \rho_2 A_0 V_2$ through the orifice. If we neglect elevation differences in (3.24), then we have

$$V\,dV + \frac{dp}{\rho} = 0 \tag{3.33}$$

We would still like to use this relation to determine the velocity V_2 through the orifice. If we assume the flow to be isentropic, then (3.33) can be manipulated so that we can integrate it to obtain a relation analogous to (3.3). First, we determine the differential of the isentropic gas relation ($p\rho^{-\gamma}$ = constant) to obtain

$$d\rho = \frac{\rho\, dp}{\gamma p} \tag{3.34}$$

Next, let us examine the differential $d(p/\rho)$. This is seen to be

$$d\left(\frac{p}{\rho}\right) = \frac{dp}{\rho} - \frac{p}{\rho^2}\, d\rho \tag{3.35}$$

In view of (3.34) we can rewrite (3.35) as

$$d\left(\frac{p}{\rho}\right) = \left(\frac{\gamma - 1}{\gamma}\right)\frac{dp}{\rho}$$

Thus the term dp/ρ in (3.33) may be expressed as the perfect differential

$$\frac{dp}{\rho} = \frac{\gamma}{\gamma - 1} d\left(\frac{p}{\rho}\right)$$

and (3.33) can be expressed as

$$V\,dV + \frac{\gamma}{\gamma - 1} d\left(\frac{p}{\rho}\right) = 0$$

or, moving the differential operator out front,

$$d\left(\frac{V^2}{2} + \frac{\gamma}{\gamma - 1}\frac{p}{\rho}\right) = 0$$

This result implies that the quantity in brackets is a constant, so we obtain a relation analogous to Bernoulli's equation but one that applies for isentropic flow of a compressible gas,

$$\frac{1}{2}V^2 + \frac{\gamma}{\gamma - 1}\frac{p}{\rho} = \text{constant} \tag{3.36}$$

This result can now be applied to stations 1 and 2, as shown in Fig. 3.2. If we assume the velocity V_1 on the high-pressure side to be negligible in comparison to V_2, then we can solve for V_2 as

$$V_2 = \left[\frac{2\gamma}{\gamma - 1}\left(\frac{p_1}{\rho_1} - \frac{p_2}{\rho_2}\right)\right]^{1/2} \tag{3.37}$$

Next, rewrite ρ_2 in this relation using (3.26) to obtain $\rho_2 = (p_2/p_1)^{1/\gamma}\rho_1$ so that, with a little manipulation, (3.37) becomes

$$V_2 = \left\{\frac{2\gamma}{\gamma - 1}\frac{p_1}{\rho_1}\left[1 - \left(\frac{p_2}{p_1}\right)^{(\gamma-1)/\gamma}\right]\right\}^{1/2}$$

The mass flow rate $q = A_0 V_2 \rho_2$, so we find

$$q = \rho_2 A_0 \left\{ \frac{2\gamma}{\gamma - 1} \frac{p_1}{\rho_1} \left[1 - \left(\frac{p_2}{p_1} \right)^{(\gamma-1)/\gamma} \right] \right\}^{1/2}$$

We next eliminate the ρ_2 appearing in this relation using $\rho_2 = (p_2/p_1)^{1/\gamma} \rho_1$, so that the mass flow rate is

$$q = A_0 (p_1 \rho_1)^{1/2} \left\{ \frac{2\gamma}{\gamma - 1} \left[\left(\frac{p_2}{p_1} \right)^{2/\gamma} - \left(\frac{p_2}{p_1} \right)^{(\gamma+1)/\gamma} \right] \right\} \tag{3.38}$$

We have now expressed q in terms of the (assumed known) properties at station 1 and the pressure ratio p_2/p_1. This relation is often expressed in terms of p_1 and T_1, rather than p_1 and ρ_1, by using the perfect gas law to write $\rho_1 = p_1/RT_1$. Furthermore, let us include an orifice loss coefficient c_0 ($c_0 < 1$), which accounts for any losses not included in our idealized model (this coefficient is normally determined experimentally), so that the mass flow rate q is

$$q = \frac{c_0 p_1 A_0}{\sqrt{RT_1}} \left\{ \frac{2\gamma}{\gamma - 1} \left[\left(\frac{p_2}{p_1} \right)^{2/\gamma} - \left(\frac{p_2}{p_1} \right)^{(\gamma+1)/\gamma} \right] \right\}^{1/2} \tag{3.39}$$

Equation (3.39) is the result of the analysis. For given p_1 and T_1, q is a linear function of the orifice area A_0 and a complicated nonlinear function of the pressure ratio p_2/p_1 [note that by definition $p_2/p_1 < 1$ so that, because $2/\gamma < (\gamma + 1)/\gamma$, then $(p_2/p_1)^{2/\gamma} \geq (p_2/p_1)^{(\gamma+1)/\gamma}$, as must occur for us to be able to take the square root in (3.39)].

The basic relation (3.39) is considerably more complicated than the relation (3.8) for a hydraulic fluid. Furthermore, inspection of (3.39) reveals that the mass flow rate will be zero when $p_2/p_1 = 1$ and when $p_2/p_1 = 0$. Thus, q must attain some maximum value between $(p_2/p_1) = 0$ and $(p_2/p_1) = 1$. If we were to make a graph of q versus p_2/p_1 for given p_1 and T_1, we would find a curve of the form shown in Fig. 3.8.

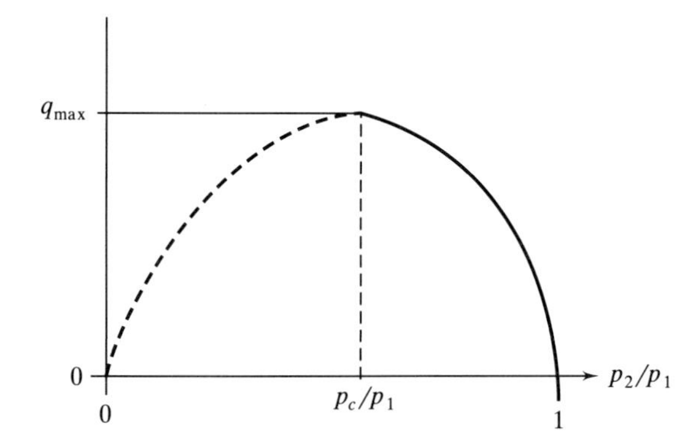

FIGURE 3.8
Mass flow rate of a gas through an orifice as a function of p_2/p_1.

It can be shown that the maximum mass flow rate q_{max} occurs at a critical value p_c of the pressure p_2 given by $p_c/p_1 = (2/(\gamma + 1))^{\gamma/\gamma - 1}$. For air $p_c/p_1 = 0.528$, and the maximum mass flow rate is

$$q_{max} = \frac{c_0 p_1 A_0}{\sqrt{RT_1}} \left[\frac{2\gamma}{\gamma + 1} \left(\frac{2}{\gamma + 1} \right)^{2/(\gamma - 1)} \right]^{1/2} \tag{3.40}$$

Now when $p_2 = p_c$, the flow through the orifice is sonic (that is, V_2 is the speed of sound). Furthermore, for all $p_2 < p_c$, the actual mass flow rate q is maintained at the level q_{max}, so that, for $(p_2/p_1) < 0.528$ (for air) the orifice mass flow is constant. Thus, the actual variation of q with p_2/p_1 is given by the solid line in Fig. 3.8. For subsonic flow conditions $(p_2/p_1 > 0.528)$, q is given by (3.39), while for $(p_2/p_1) < 0.528$, (3.40) must be used. In the analysis of systems involving gas flow through orifices, care must be taken to utilize the correct mass flow relations. Let us now consider an example in which orifice mass flow is an essential element of the system.

Example 3.5 Air Cushion Landing System (ACLS). We'll consider a simple model of an air cushion landing system (ACLS) (Fig. 3.9). Here pressurized air is the working fluid, providing a cushion that may be used to replace conventional landing gear. Such devices are used for landings on rough terrain and also form the basis of support for sea-going hovercraft. We'll consider a very simplified model in which the objective will be to see how the air cushion operation affects the dynamics of the airplane.

The "skirt" shown consists of a set of inflated tubes (picture an upside down river raft) which form an "air cushion" under the vehicle. A fan provides a

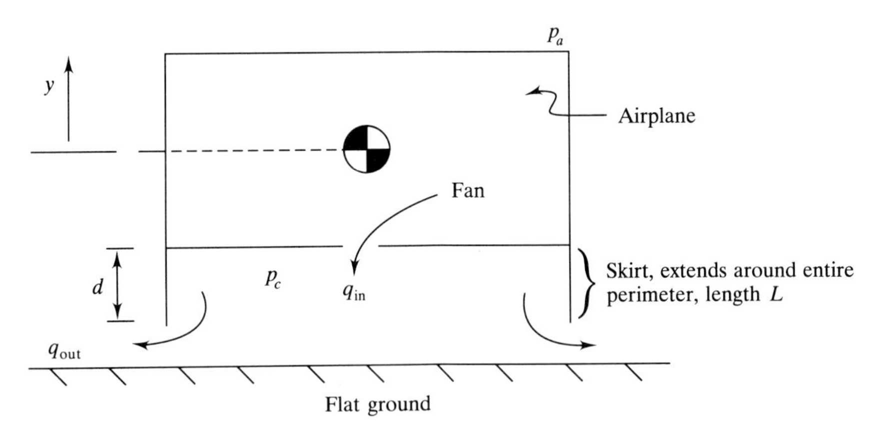

FIGURE 3.9
Schematic of air cushion landing system (ACLS).

constant airflow q_{in} into this cushion, maintaining a cushion pressure $p_c > p_a$* (p_a is atmospheric pressure), which is supposed to support the vehicle. Vehicles equipped with these ACLSs have sometimes been observed to experience a violent up and down (y) oscillation, often coupled with a pitching motion (not considered here), resulting from the interaction of the mechanical motion and the fluid behavior. Thus, we need a mathematical model that will enable us to understand this phenomenon and which can be used to redesign the system so that the violent oscillatory motion can be eliminated.

We'll assume that all airflows are subsonic and that thermodynamic equilibrium exists in the cushion volume. Now we can immediately write down the equation of vertical motion, which is

$$m\ddot{y} = (p_c - p_a)A - mg \tag{3.41}$$

where A is the cushion area. If there is no motion ($\ddot{y} = 0$), (3.41) tells us the cushion pressure p_c needed to support the weight of the vehicle. However, if y happens to fluctuate, then so will the cushion pressure p_c: Picture our suddenly increasing y by a certain amount; this will be accompanied by a drop in p_c as more air is able to flow out of the cushion. Then as p_c drops, the force on the airplane falls and it falls; as it falls, the cushion pressure rises, and so on. A possibility, then, might be some type of vertical oscillation of the airplane, accompanied by an oscillatory cushion pressure p_c. Thus we need an equation telling us how p_c will change with time. We can find the required relation by first considering mass conservation, applied to the mass m_c of air in the cushion. First, the total amount of mass in the cushion is

$$m_c = \rho_c V_c$$

where the cushion volume $V_c = A(d + y)$ (when $y = 0$ the vehicle is sitting on the ground and $V_c = Ad$, Fig. 3.9). The rate of change of cushion mass $\dot{m}_c$ is determined by the flows into and out of the cushion:

$$\dot{m}_c = q_{\text{in}} - q_{\text{out}}$$

where q_{in} is the fixed inflow and q_{out} depends on y (that is, the flow area) and on the pressures p_c and p_a. We also have

$$\dot{m}_c = \dot{\rho}_c V_c + \rho_c \dot{V}_c \tag{3.42}$$

What we need to do now is to formulate $\dot{m}_c$ in terms of the dependent variables y and p_c.

Now to get the cushion pressure p_c into the model, we'll assume that all thermodynamic processes are *isentropic*, so that (3.26) applies. Taking the time

*Typically p_c is only 1 to 2 psi greater than p_a.

derivative of the isentropic relation (3.26), we obtain

$$\dot{p}\rho^{-\gamma} - p\gamma\dot{\rho}\rho^{-\gamma-1} = 0$$

or

$$\dot{p} = p\gamma\frac{\dot{\rho}}{\rho}$$

or

$$\dot{\rho} = \frac{\rho\dot{p}}{\gamma p}$$

Thus mass conservation and the isentropic relation allow (3.42) to be written as follows (using c subscripts to denote cushion variables) where $\dot{V}_c = A\dot{y}$:

$$\frac{\rho_c V_c \dot{p}_c}{\gamma p_c} + \rho_c \dot{V}_c = q_{\text{in}} - q_{\text{out}}$$

From the perfect gas law we also see that $\rho_c = p_c/RT_c$, so that

$$\frac{V_c \dot{p}_c}{\gamma R T_c} + \frac{p_c A\dot{y}}{RT_c} = q_{\text{in}} - q_{\text{out}}$$

Now we can solve this equation for $\dot{p}_c$:

$$\dot{p}_c = \frac{\gamma R T_c}{V_c}\left[q_{\text{in}} - q_{\text{out}} - \frac{p_c A\dot{y}}{RT_c}\right]$$

Note that if p_c is known, then ρ_c is known ($p\rho^{-\gamma}$ = constant) and then so is T_c.

The next step is to express the "orifice flow" q_{out} from the cushion to the surroundings as a function of the cushion and ambient pressures p_c and p_a. For realistic ACLS systems the sonic flow condition does not occur (p_a/p_c is nearly always greater than about 0.8), and the flow relation (3.39) applies, with the orifice area A_0 given by $A_0 = Ly$. We can, therefore, summarize the mathematical model as follows:

$$m\ddot{y} = (p_c - p_a)A - mg \tag{3.43}$$

$$\dot{p}_c = \frac{\gamma R T_c}{A(d+y)}\left[q_{\text{in}} - q_{\text{out}} - \frac{p_c A\dot{y}}{RT_c}\right] \tag{3.44}$$

$$q_{\text{out}} = \frac{c_0 p_c L y}{\sqrt{RT_c}}\left\{\frac{2\gamma}{\gamma - 1}\left[\left(\frac{p_a}{p_c}\right)^{2/\gamma} - \left(\frac{p_a}{p_c}\right)^{(\gamma+1)/\gamma}\right]\right\}^{1/2} \tag{3.45}$$

We observe that the model is of third order; the states are y, $\dot{y}$, and p_c. The model is autonomous, (highly) nonlinear (can you identify the nonlinearities?), and fairly complex. The possibility of developing a linearized version of this model which can be utilized if the system does not move too far from rest is considered in Exercise 4.29 of Chap. 4, in which the process of linearization of nonlinear models is discussed (Sec. 4.1.4).

3.2.3 Concluding Remarks

1. Based on the analysis and examples presented in this section, we observe that, as with hydraulic systems, pneumatic systems tend naturally to be nonlinear because of the basic relations which connect the fluid and thermal variables involved. While it may be possible in many situations to replace the actual nonlinear models with linear versions that correctly describe the dynamics, at least for some conditions, it is also true that for real engineering systems utilizing hydraulic and/or pneumatic components, it is often necessary to study the full, nonlinear models in order to understand system behavior.
2. A comment on the status of the pressure variables involved in the modeling of pneumatic systems is worthwhile. In Example 3.5 (ACLS) we ended up with the *differential equation* (3.44) which defined the cushion pressure p_c. On the other hand, for the airspring (Example 3.4) the pressures p_1 and p_2 were *static* functions of the displacement x; we did not need any ODEs to model the pressures. The distinction is that, for Example 3.5, we had to account for the mass flows ($q_{\text{in}}, q_{\text{out}}$) into and out of the cushion, so that the differential form of mass conservation, $\dot{m} = q$, was used. For a compressible fluid this automatically introduces $\dot{p}$ terms into the math model ($\dot{m}$ is proportional to $\dot{\rho}$, which is proportional to $\dot{p}$). For Example 3.4, though, the mass in each chamber is conserved, so that mass conservation is simply $m_1 = m_2 =$ constant. Here the pressures will only appear *statically*. Thus, in general we conclude the following: If there are nonzero mass flows into and out of the regions of interest in a pneumatic system, then we have to expect to obtain differential equations to describe the associated pressure variations. But if the system involves pneumatic chambers of fixed gas mass, then the pressures will appear statically in the mathematical formulation.
3. The disclaimer (comment 3) at the end of Sec. 3.1.3 applies here as well.

3.3 ELECTRICAL SYSTEMS

Electronic circuits and electromagnetic devices are used extensively by mechanically oriented engineers in instrumentation and for conversion of electromagnetic energy into mechanical energy. We have seen that the production of loadings on mechanical objects by hydraulic and pneumatic devices is newtonian in nature, as work is done by fluid pressure forces, which arise from newtonian effects. Thus, in the broad view (except for the involvement of thermodynamics), hydraulic and pneumatic systems may be classed under the aegis of newtonian mechanics. Electrical and electromagnetic effects are of course fundamentally different in physical terms, but there are many similarities in the underlying mathematics.

In this section we first define some basic descriptive properties and constitutive relations for electric circuits and their components. This is followed by a summary of Kirchhoff's circuit laws, which form the basis of the mathematical modeling of circuits. In this phase of modeling work, we will be concerned only with the development of mathematical models for electric circuits.

Then in Sec. 3.4 our concern will be with electromechanical systems, in which electromagnetic effects are used to produce or sense loads on mechanical objects (examples include motors, generators, loudspeakers, and microphones). As for hydraulic and pneumatic systems, the modeling work here will involve mathematical descriptions of both the mechanical and nonmechanical portions of the system, with particular attention to the interactions between the two. Central to this work will be a description of the laws of magnetic induction, the phenomenon of most practical importance to us.

3.3.1 Electric Circuits

The basic system variables that describe electric circuits are charge q, current I, and voltage V. The unit of charge is the coulomb (C); an electron or proton has charge $\pm 1.60206 \times 10^{-19}$ C. The current I is the rate at which charge flows through a given cross section, so that $I = \dot{q}$. The unit of I is the ampere A ($1\text{ A} = IC/\text{s}$). The unit of voltage is the volt V (1 V = 1 joule/C); the change in voltage $V_a - V_b$ between any two points a and b in space is the work done to move a charge q from a to b in an electric field $\mathbf{E}$. The voltage is merely a scalar description of the electric field vector $\mathbf{E}$; $\mathbf{E}$ and V are connected by $\mathbf{E} = -\nabla V$, which is exactly analogous to the connection (2.40) between force $\mathbf{F}$ and potential energy V for a conservative force. Voltage sources, such as batteries, provide the energy that drives current in an electric circuit: Current flows in the direction of decreasing voltage, just as fluid flows in the direction of decreasing pressure (we have seen that fluid pressure is a form of potential energy; thus the analogy between fluid pressure/mass flow and voltage/current is fairly direct).

3.3.2 Circuit Elements and Constitutive Laws

In Sec. 2.4 ("Constitutive Laws for Mechanical Systems"), the development of constitutive relations was focused on modeling the relation of forces generated to the position and velocity of a mechanical object. For electrical systems our job will be to relate the voltage change which occurs to the current (or charge) in the circuit. Three classical circuit elements will be considered: resistors, capacitors, and inductors.

RESISTORS. It is observed that, if a given potential difference $V_a - V_b \equiv V$ is applied across different conducting materials of the same geometry, the current I which flows through the material will be proportional to the voltage difference for a fairly wide range of operating conditions, but the constant of proportional-

R

a —/\/\/\/— b $\quad$ $\longrightarrow I$

$$V_a - V_b = RI$$

FIGURE 3.10
Schematic of the linear resistor.

ity will vary from one material to another. The mathematical statement which describes this behavior is Ohm's "law":

$$V = RI \tag{3.46}$$

where the resistance R (measured in ohms (Ω); $1\ \Omega = 1V/A$) is the proportionality constant. We represent a resistor schematically as shown in Fig. 3.10.

Resistors dissipate energy (that is, they convert electrical energy into heat) at a rate $P = I^2R$ (watts). This effect forms the basis of standard household devices such as toasters, electric stoves, and electric heaters. Thus, resistance in an electrical system is analogous to damping in a mechanical system. Note that the constitutive relation (3.46) is only an approximation which works well provided that the current I is sufficiently small (just as our newtonian constitutive relations usually work well if displacements or velocities are small). Furthermore, the resistance R usually increases with the temperature of the resistive element, especially if the material is metallic.

CAPACITORS. If a potential difference is applied across two parallel conducting plates separated by a small distance d, then a charge q will accumulate on the plates (positive on one plate and negative on the other). For a linear capacitor the connection between the voltage difference V and the charge q is defined by

$$q = CV \tag{3.47}$$

where the capacitance C, measured in units of farads F (1 F = 1 coulomb per volt), depends on the geometry and on the material between the plates. We represent the capacitor as shown in Fig. 3.11.

The capacitor is an energy-conserving element that stores electrical energy in the amount $\frac{1}{2}CV^2$. This is the amount of work required to charge the capacitor, and it can be recovered by allowing the capacitor to discharge.

INDUCTORS. A consequence of Faraday's law of magnetic induction (Sec. 3.4.1) is that a coil of wire will have a potential difference induced across it if the current I flowing through the coil is changing with time. If $\dot{I} > 0$, then there is a voltage drop V defined for a linear inductor by the relation

$$V = L\dot{I} \tag{3.48}$$

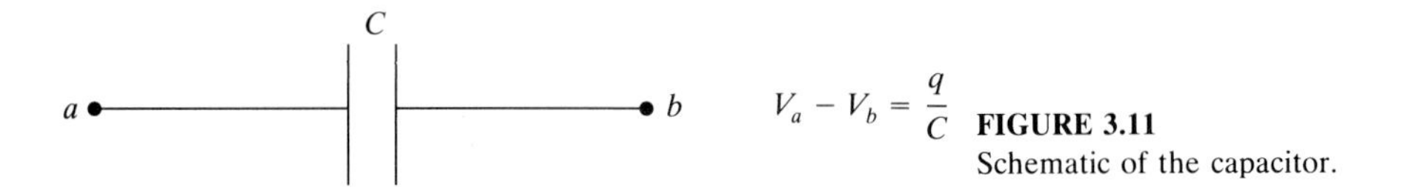

FIGURE 3.11
Schematic of the capacitor.

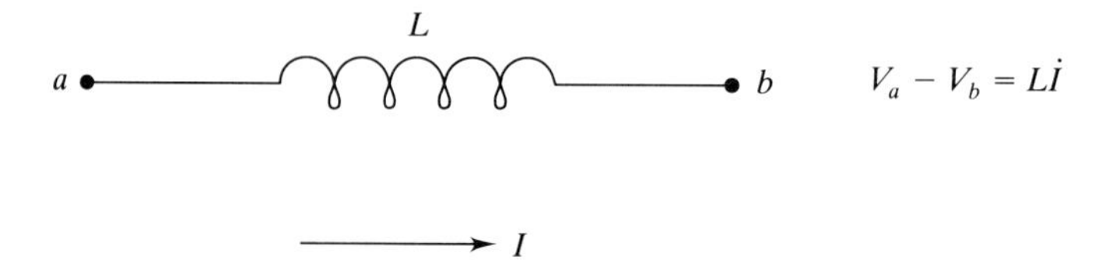

FIGURE 3.12
Schematic of the inductor.

where L is the inductance, measured in units of henries (H; 1 H = 1 V-s/A). We represent the inductor as shown in Fig. 3.12.

Keep in mind that, in the constitutive relations (3.46), (3.47), and (3.48), the voltage V represents the voltage *difference* across the element and it is actually negative if I and q are positive (i.e., as shown in the diagrams); by convention the negative signs do not appear in the constitutive relations, and a voltage drop is understood.

3.3.3 Basic Circuit Laws

In formulating mathematical models for electric circuits, we will utilize the two basic circuit laws of Kirchhoff, which are stated below:

1. Kirchhoff's first law says that, at any node in an electrical circuit, the algebraic sum of the currents entering and leaving the node must be zero. This law is based on the conservation of charge, which cannot accumulate at a node.
2. Kirchhoff's second law states that, for any closed loop in an electric circuit, the sum of the voltages around the loop must be zero. Thus, if we have a circuit consisting of several circuit elements and a voltage source, then we add the voltage drops and increases around the circuit, and the sum must vanish.

We'll now consider an example of the mathematical modeling of an electric circuit.

Example 3.6 Linear *LRC* Circuit. We consider a single-loop circuit consisting of a resistor, a capacitor, and an inductor in series, along with a voltage source $V(t)$ which is assumed to be time dependent in some known manner (Fig. 3.13).

Applying Kirchhoff's voltage law, we sum the voltages around the circuit to obtain the circuit equation in the form

$$V(t) - RI - \frac{q}{C} - L\dot{I} = 0$$

Let us use the charge q rather than the current I, as the dependent variable, in order to avoid an integrodifferential equation. Then the mathematical model can

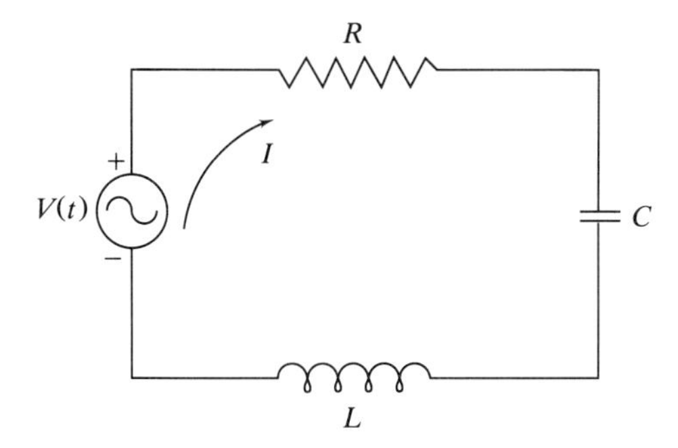

FIGURE 3.13
The linear *LRC* circuit.

be written

$$L\ddot{q} + R\dot{q} + \frac{1}{C}q = V(t) \tag{3.49}$$

The model is linear, nonautonomous, and of second order. The input is the applied voltage $V(t)$, and the system parameters are L, R, and C, while the states are q and $\dot{q}$. We notice the analogy with the mechanical oscillator (2.52): The inductance L is analogous to the mass m, the resistance R is analogous to the viscous damping c, and the capacitance C is analogous to the reciprocal of the spring constant k ($1/k$ is sometimes called the *compliance* in a mechanical system). The voltage $V(t)$ is analogous to the input force. Circuits of the type shown may be used to simulate the behavior of mechanical oscillators through appropriate selection of the system parameters.

Note that (3.49) can be rewritten using the current I, rather than the charge q, as the dependent variable, by differentiating (3.49) and utilizing the relation $I = \dot{q}$ to obtain

$$L\ddot{I} + R\dot{I} + \frac{1}{C}I = \dot{V}(t) \tag{3.50}$$

The system model for the current I is the same as for the charge q except that the input is now the rate of change of voltage $\dot{V}(t)$. The relation (3.50) is generally preferable, as current I is easier to measure than the charge q. Note also that if there were no capacitor, then the mathematical model is of *first* order in the current I:

$$L\dot{I} + RI = V(t) \tag{3.51}$$

3.4 ELECTROMECHANICAL SYSTEMS

We now focus attention on those electromagnetic effects that form the basis of devices used to generate loadings on mechanical objects. The systems of interest to us will consist of mechanical and electromagnetic subsystems. The mechanical subsystem will consist of one or more mass elements, with damping and/or stiffness effects generally present. The purpose of the electromagnetic part of the system will be to generate, via the laws of magnetic induction, a loading that is exerted on the mechanical object, thus converting electromagnetic energy into mechanical energy. In constructing mathematical models of electromechanical systems, we need to be able to do the following: (1) apply the laws of newtonian mechanics to develop ODEs describing the motions of the mechanical objects in the system, (2) apply Kirchhoff's laws and the circuit element constitutive relations to develop ODEs describing the behavior of the electrical part of the system, and (3) apply the laws of magnetic induction to model the interactions of the mechanical and electrical subsystems.

3.4.1 Laws of Magnetic Induction

Let's first consider a magnetic field $\mathbf{B}$, such as exists between the poles of a permanent magnet constructed of a ferromagnetic material such as iron. Now the electric field $\mathbf{E}$ generates voltages in an electric circuit ($\mathbf{E} = -\nabla V$) as a result of the *static* force $\mathbf{F}$ it exerts on a charged particle, $\mathbf{F} = q\mathbf{E}$. A magnetic field $\mathbf{B}$, on the other hand, generates a force on a charged particle only if the charged particle is in motion:

$$\mathbf{F} = q\mathbf{V} \times \mathbf{B} \tag{3.52}$$

where $\mathbf{V}$ is the charged particle velocity and $\mathbf{B}$ is the magnetic field vector, measured in units of newtons per amp-meter [1 gauss (G) = 10^{-4} N/A-m; the magnetic field in the vicinity of the earth's surface is about 0.3 G]. Note that the force exerted on a moving charged particle by a magnetic field does no work (the force is normal to the velocity) and hence will not alter the particle's kinetic energy.

If we now consider a segment of wire carrying a current I and positioned in a magnetic field $\mathbf{B}$, the wire will have a force exerted on it as a result of (3.52), with $q\mathbf{V}$ replaced by the product of current I and length of wire. For an incremental wire segment of length $d\ell$, as shown in Fig. 3.14, the incremental force $d\mathbf{F}$ generated will be given by

$$d\mathbf{F} = I\,\mathbf{d\ell} \times \mathbf{B} \tag{3.53}$$

where $\mathbf{d\ell}$ is an incremental vector of length $d\ell$ which points in the direction of current flow.

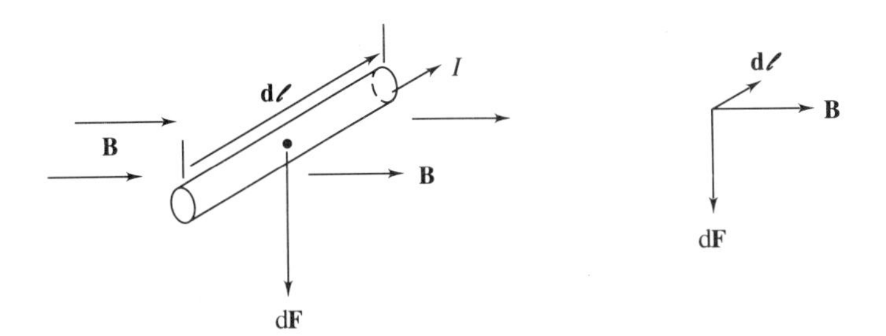

FIGURE 3.14
The incremental force exerted on a current-carrying wire in a magnetic field.

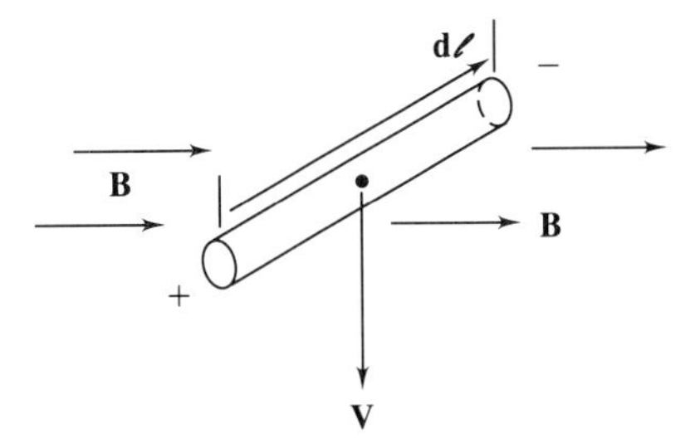

FIGURE 3.15
The induced voltage along a wire moving in a magnetic field.

For a finite length of wire the net force **F** exerted on the wire is obtained by computing the line integral of (3.53) along the wire:

$$\mathbf{F} = \int I\, \mathbf{d}\boldsymbol{\ell} \times \mathbf{B} \tag{3.54}$$

This line integral allows us to account for spatial variations in magnetic field **B**, current I, and the direction of current flow. The electromagnetic effect defined by (3.54) is the basis of actuation devices such as motors and solenoids.

The second electromagnetic effect of interest to us is based on Faraday's law of magnetic induction: If a conducting wire is in motion in a magnetic field, then a potential (voltage) gradient is generated along the wire (Fig. 3.15). For an incremental segment of wire of length $d\ell$, moving with velocity **V** in a magnetic field **B**, the potential difference dV generated is given by

$$dV = \mathbf{V} \times \mathbf{B} \cdot \mathbf{d}\boldsymbol{\ell} \tag{3.55}$$

the induced voltage dV is such that V is increasing in the direction of $\mathbf{V} \times \mathbf{B}$.

In order to obtain the voltage V induced in a finite length of wire, we compute the line integral of (3.55) to obtain

$$V = \int \mathbf{V} \times \mathbf{B} \cdot \mathbf{d}\boldsymbol{\ell} \tag{3.56}$$

The voltage generated according to (3.56) is the basis of nearly all conventional power generation systems: For example, the energy of motion of mechanical objects as produced by steam turbines is converted to electrical energy by rotating a wire coil in a magnetic field.

In the preceding discussion the two electromagnetic effects of most interest to us have been briefly described: the force exerted on a current-carrying wire in a magnetic field and the voltage induced in a moving wire in a magnetic field. If we consider the simultaneous occurrence of these two effects, we can see clearly that the mechanical and electrical parts of an electromechanical system will interact with each other. Suppose that a current-carrying wire is attached rigidly to a mechanical object which is to be moved. If this object is placed in a magnetic field, the resulting force will cause the object to accelerate (either linearly or rotationally, depending on the configuration). As the object moves, its velocity will result in a voltage being generated. This, in turn, will affect the current flow in the electrical part of the system, which, in turn, will affect the force exerted on the mechanical object, and so on. Taking proper account of this electromechanical coupling is essential in the modeling process. Let's now look at a few examples of electromechanical systems.

Example 3.7 The Loudspeaker (after Close and Frederick, Ref. 3.2). In this example we consider an electromechanical system in which the electromagnetic effect is used to generate motion of a mechanical object (a speaker cone) which in turn generates sound, which is transmitted to the ear through the surrounding air.

The essential features of loudspeaker operation are as follows: An electrical signal (voltage from the tape, record, or compact disc unit) is amplified (power amp) and passed through a coil that surrounds a permanent magnet. This causes a force to be exerted on the coil, which is attached to the speaker cone, so the speaker cone moves in response to this force. The system is supposed to be designed so that the speaker cone motion, which produces sound, is proportional to the original input signal, regardless of the frequency of the incoming signal; this gives so-called flat frequency response.

An idealized physical model of the loudspeaker appears in Fig. 3.16. The speaker cone is modeled essentially as an oscillator having an effective mass m, stiffness k, and damping c. The electrical circuit model contains the input voltage $V(t)$, an inductor L and resistor R in series, and the coil which surrounds the permanent magnet. The dynamic model will involve the operation of the speaker cone and the equation governing the electrical circuit.

Let's first consider the equation of motion of the speaker cone m. The damping and stiffness forces are simply modeled with linear elements, so we focus attention on the force generated by the interaction of coil current I and the magnetic field due to the permanent magnets. The coil geometry (Fig. 3.17) is defined by the coil radius a and number of turns n of wire, so that the total length

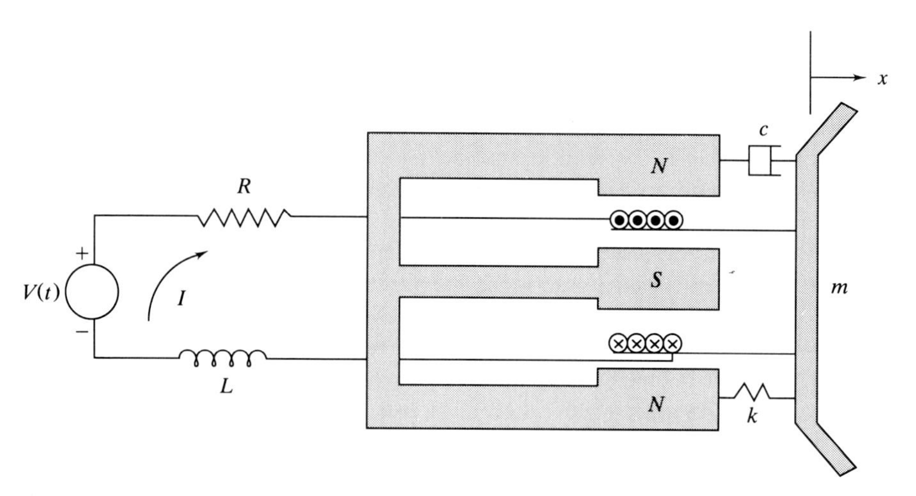

FIGURE 3.16
Idealized model of a typical loudspeaker.

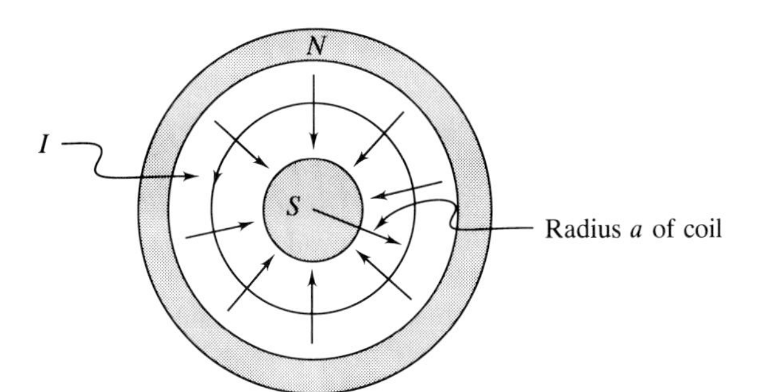

FIGURE 3.17
Loudspeaker magnet/coil geometry (forward looking aft).

of wire is $2\pi na$. We will assume the magnetic field **B** to be radial throughout the region occupied by the coil; this assumption is really valid only for concentric magnets of length large compared to that of the coil. We denote the magnitude of the magnetic field at a radial distance a by the scalar B. As seen in Fig. 3.17, a positive current flow is in the counterclockwise direction (when forward looking aft). Thus, we observe that, at any point along the coil wire, the line segment $\boldsymbol{d\ell}$ and magnetic field **B** are perpendicular, with the vector product $I\,\boldsymbol{d\ell} \times \mathbf{B}$ directed out of the paper (that is, in the $+x$ direction) and of magnitude $IB\,d\ell$.

The force F_x exerted on the coil and transmitted directly to the mass m is then obtained easily from (3.54) as

$$F_x = 2\pi anBI \equiv \alpha I \tag{3.57}$$

where the constant $\alpha \equiv 2\pi anB$ depends on the coil geometry and magnetic field strength. The calculation of the integral (3.54) is particularly simple because of the angular symmetry of the setup. Having calculated the force F_x exerted on the speaker cone, we can now write down the equation of motion for the speaker cone m as

$$m\ddot{x} + c\dot{x} + kx = \alpha I \tag{3.58}$$

Thus, the speaker cone acts, according to our idealized model, like a harmonic oscillator driven by a force proportional to the current flowing through the speaker coil. Since this current is unknown, we have to find a mathematical model from which it may be determined. In order to do this, we have to account for the fact that the speaker cone velocity $\dot{x}$ will have an effect on the electrical part of the system because of the voltage induced by the motion of the wire in the magnetic field.

In order to quantify this induced voltage effect, we note that, when $\dot{x} > 0$, the vector product $\mathbf{V} \times \mathbf{B}$ at any point along the coil wire will be in the clockwise direction (when forward looking aft). Since this is the direction of negative current flow, the induced voltage V_i will be negative for $\dot{x} > 0$ and positive for $\dot{x} < 0$. From (3.56) we then find the induced voltage V_i to be

$$V_i = -2\pi anB\dot{x} = -\alpha\dot{x}$$

We can now write down the circuit equation by summing voltage drops around the circuit according to Kirchhoff's second law. This procedure gives us an equation for the current I as

$$V(t) - IR - L\dot{I} - \alpha\dot{x} = 0$$

or

$$L\dot{I} + RI = V(t) - \alpha\dot{x} \tag{3.59}$$

Here $V(t)$ is the input voltage signal that drives the electrical circuit (and the speaker cone). Note that the system equations are coupled since the motion of the speaker is a response to the current I, but the speaker motion also has an effect on what this current I is going to be. In practice the design of loudspeakers is challenging because acceptable performance has to be established over a wide frequency range (say, 40 to several thousand hertz). We have developed a third-order linear model consisting of one second-order and one first-order ODE. The states are I, x, and $\dot{x}$, and the model is nonautonomous.

In order to gain some physical insight into the effects involved in this system, let's consider the special case for which the inductor L is removed from the circuit. Then (3.59) becomes

$$RI = V(t) - \alpha\dot{x} \tag{3.60}$$

Thus the relation for the current I is now static. If we solve (3.60) for the current I,

$$I = \frac{1}{R}[V(t) - \alpha\dot{x}]$$

and use this result to eliminate I on the right-hand side of (3.58), we obtain the following relation which connects the speaker cone motion directly to the input voltage $V(t)$:

$$m\ddot{x} + \left(c + \frac{\alpha^2}{R}\right)\dot{x} + kx = \frac{\alpha}{R}V(t) \tag{3.61}$$

An important result is that the voltage induced by the motion of the coil in the magnetic field has increased the overall damping of the system. The effective viscous damping constant is now $c + (\alpha^2/R)$. The model is now of the second, rather than of the third, order.

Example 3.8 The Armature-Controlled dc Motor. The basic laws used to develop the ODE model of the loudspeaker can also be applied to obtain the ODE model for a direct current motor. The geometry and motion of moving parts is, however, significantly different. The motor converts electrical energy to mechanical energy by utilizing magnetic effects to produce a torque to do mechanical work. Common examples include the starter motor in an auto engine, servomotors that deflect aerodynamic surfaces on commercial airliners, control rod drives in nuclear power plants, motors that drive fans and pumps, and motors used as positioning devices, such as in an *x-y* plotter. Generally, motors are utilized in situations for which power requirements are light to moderate. Extreme power demands are generally met with hydraulic devices (e.g., cranes or large manufacturing operations).

A schematic cross section of an idealized dc motor is shown in Fig. 3.18. Rather than using permanent magnets, a magnetic field is produced by a "field coil," which is wrapped around an iron core and which has a constant current I_F passing through it. This field current I_F produces an approximately radial magnetic

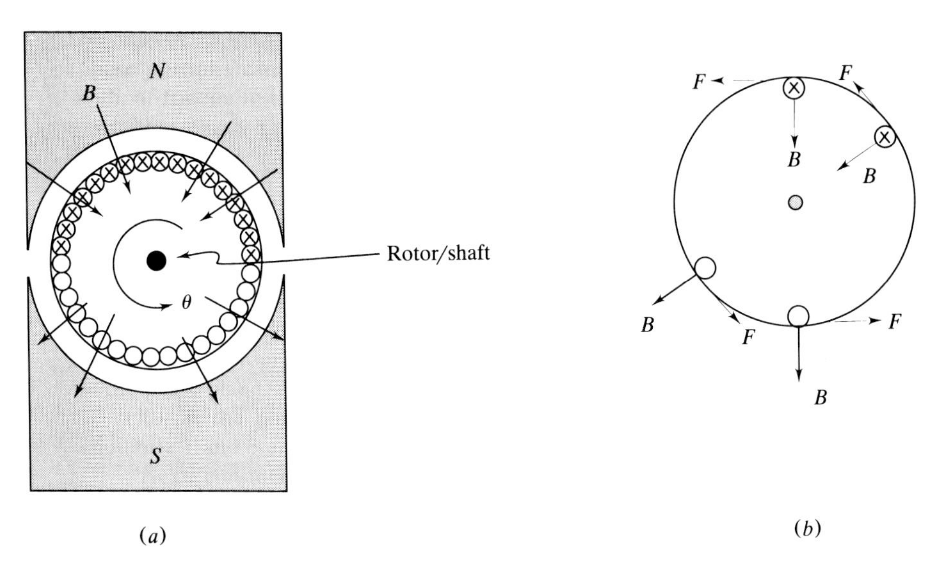

FIGURE 3.18
Magnet and rotor geometry of the armature-controlled dc motor. (*a*) The magnetic field and direction of current flow in the rotor windings; (*b*) the forces generated on individual wires.

field in the airgap between the magnet faces. A rotor, free to rotate about a shaft (denoted by the point in the center of Fig. 3.18), is positioned centrally inside the airgap. N turns of wire are wound on the rotor periphery. A current I_A flows through this coil, which is wound so that the current flow is out of the paper on the bottom half of the coil and into the paper on the top half. This rotor/coil assembly is the armature. The armature circuit in which the armature current I_A flows is shown in Fig. 3.19 and is discussed in succeeding paragraphs.

The basic physical effect that generates torque is the current flow in a wire in a magnetic field, according to (3.53). Figure 3.18(b) shows the directions of the forces generated on the individual wires of the armature coil at various locations around the periphery. The configuration of positive and negative current flows results in a moment exerted on the armature. If the magnetic field is assumed to be radial, then the net torque M exerted about the armature shaft can be calculated from (3.57) to be

$$M = LaBnI_A \equiv \beta I_A \tag{3.62}$$

where a is the armature radius, L the length of the armature, B the magnetic field strength at the radial location of the armature coil, and where the torque M is directed out of the paper.

The torque produced will tend to rotate the armature. Calling the rotation angle θ, with $\theta = 0$ in the configuration shown in Fig. 3.19, we observe that if θ increases to 90°, the torque will disappear, while if θ were 180°, the torque would be equal and opposite (into paper) that given by (3.62). Actual motors employ commutators to maintain the current flow in a "stationary" configuration, even as the armature rotates. Schematically, a pair of fixed (nonrotating) conducting

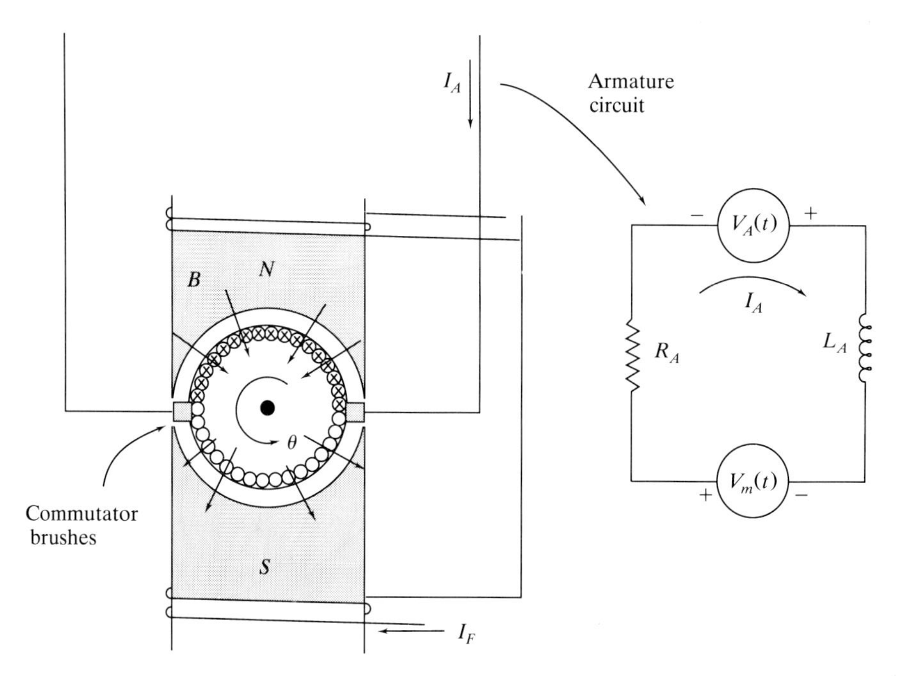

FIGURE 3.19
Armature-controlled dc motor schematic.

brushes that are in contact with the terminals of the armature coil switch the direction of current flow as the wires pass by. The current flows thus may be assumed for modeling purposes to be static as in Figs. 3.18(a) and 3.19.

The complete armature circuit is chematically shown in Fig. 3.19. This circuit consists of the armature voltage $V_A(t)$, a resistance element R_A, an inductance element L_A, and the voltage $V_m(t)$, which is the induced voltage in the armature circuit due to movement of the armature coil wires in the magnetic field (this is similar to the induced voltage in the loudspeaker circuit).

Based on the preceding discussion, the equation of rotational motion of the rotor is obtained as

$$I_e\ddot{\theta} + c\dot{\theta} = \beta I_A + M_L(\theta, \dot{\theta}, t) \tag{3.63}$$

where
θ = Rotor/shaft rotation angle
βI_A = Torque generated according to (3.62)
I_e = Effective moment of inertia of all masses being moved or rotated; consists of the moment of inertia I of the rotor/shaft plus the effective moment of inertia of all other masses connected to the motor shaft
c = Rotational damping constant describing the assumed viscous damping due to air resistance and friction in the rotor/shaft bearings.
$M_L(\theta, \dot{\theta}, t)$ = *Load torque*: any other torques produced by agents external to the motor. These torques may be dependent, in general, on θ, $\dot{\theta}$ and/or time t.

The math model is completed by writing the circuit equation for the armature circuit:

$$L_A \dot{I}_A + R_A I_A = V_A(t) - \beta\dot{\theta} \tag{3.64}$$

where $-\beta\dot{\theta}$ is the (negative) induced voltage due to motion in the magnetic field of the current-carrying armature wires.

The system ODE model, consisting of (3.63) and (3.64), is of third order, is nonautonomous, and is linear (provided that the load torque M_L is a linear function of θ and $\dot{\theta}$). The model is very similar to that developed in Example 3.7 for the loudspeaker. The input is the armature voltage $V_A(t)$, and the output is shaft rotation angle $\theta(t)$. The term *armature controlled* means that the rotation θ is controlled by supplying the appropriate armature voltage $V_A(t)$. Thus, in theory, if we wish the motor to execute some specified "maneuver" $\theta(t)$, then the armature voltage $V_A(t)$ must be such that the desired maneuver is realized. Control systems (Chap. 8) are often utilized to ensure proper execution in such situations.

The model (3.63) and (3.64) is of third order. If, however, the load torque M_L is not a function of rotation angle θ, then we may recast (3.63) as a *first-order* ODE in the state $\omega = \dot{\theta}$, and the model is reduced to one of second order, as below:

$$\begin{aligned} I_e\dot{\omega} + c\omega &= \beta I_A + M_L(\omega, t) \\ L_A \dot{I}_A + R_A I_A &= V_A(t) - \beta\omega \end{aligned} \tag{3.65}$$

Thus, for given input $V_A(t)$, (3.65) can be solved for $\omega(t)$. Then $\theta(t)$ can be found directly from $\theta(t) = \theta(t = 0) + \int_0^t \omega(t)\,dt$. The type of order reduction possible here is not uncommon and occurs in mechanical systems whenever the displacement or rotation angle does not appear in the ODE model. Another simple example is the oscillator model of Example 2.3 with the spring removed, for which

$$m\ddot{x} + c\dot{x} = F(t)$$

or, with $v \equiv \dot{x}$,

$$m\dot{v} + cv = F(t)$$

This model is of first order in the single state v.

The reader is cautioned that the analysis presented in the preceding example is idealized. Possibly significant effects, such as end effects and distortion of the magnetic field due to the rotating armature coil, have been ignored. Furthermore, there exist many different motor types, with differing designs possible for each type, including those which are driven by alternating voltage. The intent here is only to provide some idea of operation and modeling of such devices.

3.5 A THERMAL SYSTEM

In the preceding Secs. 3.1. to 3.4, the motivation in studying HPE systems was the practical utility of HPE effects as mechanisms to generate loads to move mechanical objects. Many engineering systems, however, exhibit dynamic

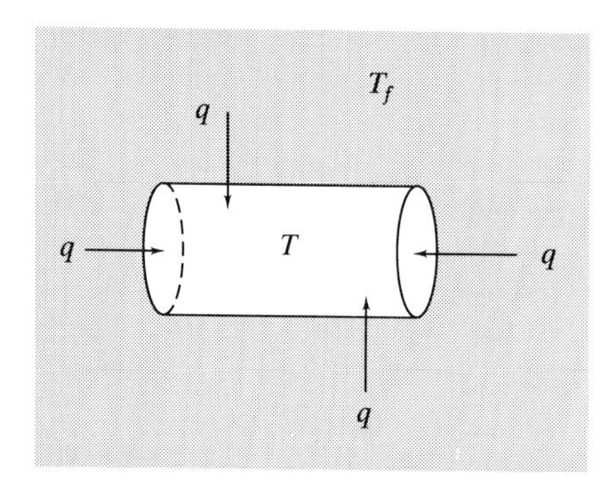

FIGURE 3.20
A billet, shown as a cylinder, is immersed in a hot fluid of temperature T_f.

behavior that is physically unrelated to newtonian mechanics, and the dynamics of such systems may be of interest in their own right. We now consider one such example, a thermal system, for which the temperature $T(t)$ in a solid is the dynamic variable to be modeled.

Here we consider the following physical problem: a solid object such as a metal "billet" is to be heated or cooled by immersion in a fluid, such as a water bath (Fig. 3.20). In general, as the billet is heated or cooled, the temperature T within the billet will be a function of both time and location within the billet. Thus if the cartesians (x, y, z) locate points in the billet, then $T = T(x, y, z, t)$. The temperature T is thus a time-dependent scalar field which is a function of four independent variables. In order to simplify the problem here, we consider only those materials that possess thermal conductivities high enough that, as the billet is heated (cooled), its temperature does not vary spatially. The assumption is that, as heat is transferred from the surrounding fluid to the billet, this additional heat is immediately distributed evenly throughout the billet, so that the billet temperature is a function of time but not location, $T = T(t)$. Based on this simplifying assumption, the mathematical model is derived below for the case of a heated billet, but the results will apply equally well if the billet is being cooled.

With reference to Fig. 3.20, the billet mass is m, its total surface area is A, and its specific heat is c_p. The temperature T_f of the surrounding fluid is assumed to be fixed, and the temperature of the billet is $T(t)$. The basic law by which a mathematical model for the temperature T is derived is the conservation of thermal energy, stated here as follows: The thermal energy (heat) which flows from the surroundings into the billet must equal the increase in the thermal energy contained within the billet. Notice that this statement is essentially the same as the law of conservation of mass (with *thermal energy* replacing *mass*) of Sec. 3.1. If the energy conservation principle is applied over a small time increment Δt, the following results:

$$mc_p \, \Delta T = hA(T_f - T) \, \Delta t \tag{3.66}$$

In (3.66) ΔT is the temperature change occurring in the time interval Δt. The quantity $mc_p \, \Delta T$ is the incremental change in thermal energy stored in the billet. On the right-hand side, the quantity $hA(T_f - T)$ is a mathematical model for the rate at which heat is transferred to the billet from the surrounding fluid. This heat transfer rate is assumed to be proportional to the temperature difference $T_f - T$ and to the total surface area A over which energy transfer occurs. The constant of proportionality is h. Thus the quantity $hA(T_f - T)\,\Delta t$ is the amount of heat transferred to the billet during the time interval Δt. Division of (3.66) by Δt and taking the limit as $\Delta t \to 0$ then yields the following differential equation model for the temperature T within the billet:

$$\dot{T} = \left(\frac{hA}{mc_p}\right)(T_f - T) \tag{3.67}$$

This model is linear and is of first order; (3.67) says that the rate at which the billet temperature increases is proportional to the temperature difference between the billet and surroundings. Keep in mind that such a simple first-order model is based on the assumption that the billet temperature T does not vary spatially, which never actually is satisfied. Thus, this model is recognized from the start to be approximate, rather than exact.

3.6 MODELS OF NONENGINEERING SYSTEMS

In the mathematical modeling work done so far, we have applied two basic types of mathematical statements: (1) true physical laws (such as Newton's second law, conservation of mass, and Kirchhoff's laws), which are *exact* and (2) constitutive "laws," which define cause-and-effect relations among the dynamic variables and which are usually *approximations*, often linear, based on experience (such as Ohm's "law," the perfect gas "law," or Hooke's "law"). In the engineering systems we have studied, the applicable laws and constitutive relations have evolved over the past several hundred years and hence are well established.

Nowadays, however, the ideas of dynamic systems analysis are being applied in nonengineering fields such as ecology, economics, social science, and biology. As with our engineering work, investigations in these fields are concerned with systems whose descriptive variables change with time and for which one seeks an ODE model to describe the behavior. If such an ODE model can be formulated, then it can be analyzed in the same way as would be done for an engineering system (from the standpoint of analysis, once we have the ODEs, it doesn't really matter where they came from). The main difference between engineering and nonengineering systems is in the mathematical modeling. For many, if not most, nonengineering systems of current interest, true "laws" of the type we apply in engineering do not exist. This means that the mathematical modeling of such systems has to be done by applying common sense and intuition to devise mathematical relations that approximate reality. We now consider two examples of this type.

Example 3.9 The Single-Species Population Model. In this example we consider the mathematical model of the population growth of a single species in certain idealized circumstances. Let $x(t)$ denote the population of the species at time t. Let us assume that the population $x(t)$ is small relative to the space, food, and resources available to its members. Further, we will ignore interactions with other species, disease, natural disasters, and social and economic restraints. Under these conditions it is reasonable to assume that, at any time t, the rate at which the population grows, $\dot{x}(t)$, is proportional to the current population $x(t)$. This is stated mathematically as

$$\dot{x} = \alpha x \tag{3.68}$$

where $\alpha > 0$. This linear first-order ODE comprises the "exponential growth" model, which is actually observable when growing cultures in a petri dish.

The problem with this linear model, however, is that it leads ultimately to a population approaching infinity. Realistically, at some point the population will become sufficiently large that factors such as overcrowding and finite food supply have to be taken into account. These factors will tend to slow and ultimately limit further population growth. To account for these population-limiting factors, we can add to the right-hand side of (3.68) a negative (growth-retarding) term which becomes important if the population $x(t)$ is large but which has little effect if the population is small. A term having this property is the nonlinear term $-\beta x^2$, where $\beta > 0$. Addition of this nonlinear term would produce the nonlinear first-order model

$$\dot{x} = \alpha x - \beta x^2 \tag{3.69}$$

This model, while clearly simplistic, exhibits two important features which are qualitatively reasonable: (1) When the population is sufficiently small, the nonlinear term is essentially negligible and the population growth is essentially exponential, (2) as the population becomes large, the nonlinearity provides a *limiting* tendency; in fact the population will grow until $\alpha x = \beta x^2$, at which time the growth and limiting tendencies are in balance, so that $\dot{x} = 0$, and the population remains fixed.

Example 3.10 An Ecological System: The Prey-Predator Problem. Our second example is the predator-prey problem, which goes something like this: A certain ecosystem is populated by foxes and rabbits; the foxes eat the rabbits, and the rabbits eat other food, which is plentiful. The objective is to develop a simple model which describes how the fox and rabbit populations change with time and, in particular, how they interact with each other.

If left to themselves, the rabbits will tend to multiply at a rate (we assume) proportional to their current population. The foxes, however, if left to themselves, would tend to die out (we assume) at a rate proportional to their current population. Thus, simple growth models for the rabbits and fox populations *by themselves* would look like this:

$$\dot{r} = \alpha r$$

$$\dot{f} = -\beta f$$

where $r(t)$ is the number of rabbits at time t, $f(t)$ is the number of foxes, the constants α, $\beta > 0$, and r, $f \geq 0$.

Now let's allow the rabbits and foxes to interact. First, let's assume that the number of "interactions" per unit time between the two populations is proportional to the product of the populations rf. Then we can assume that the rate at which rabbits get eaten is also proportional to rf, as is the rate at which the fox population is sustained. Thus, we arrive at the following simple mathematical model:

$$\dot{r} = \alpha r - \gamma r f$$

$$\dot{f} = -\beta f + \delta r f \tag{3.70}$$

Two new proportionality constants, γ, $\delta > 0$ have appeared.

The model consists of two first-order nonlinear differential equations. The nonlinear terms describe the interaction of the two populations and hence are essential if such interactions are to be modeled. Note that in arriving at this model, we have not invoked any "laws" but rather have simply made some "reasonable" suppositions and written down the mathematical statements which follow. The result is an autonomous dynamical system of the second order.

Of course, this idealized model ignores factors such as finite rabbit food supply, interactions with other animals (hawks, humans), environmental factors such as weather and disease, aging, and so on. The biologists and ecologists use fairly sophisticated models of the present type to try to understand the dynamics of various ecosystems. Chemically reacting mixtures sometimes also are modeled similarly: The concentrations of various chemical species grow or decline at the expense of the concentration of other species.

3.7 CONCLUDING REMARKS ON MODELING METHODOLOGY

Examples 3.1 through 3.10, as well as the thermal system of Sec. 3.5, have been presented to illustrate the process of ODE model construction for systems in which HPE, thermal, and nonengineering phenomena are involved. A summary of the resulting ODE models is contained in Table 3.1.

It is to be kept in mind that for the HPE systems considered in Secs. 3.1 through 3.4, the viewpoint was limited to consideration of HPE effects as ways to generate loads on mechanical objects. For these situations the modeling strategy has been very similar to that employed in Chap. 2, and the comments of Sec. 2.6 also apply to the systems considered in Secs. 3.1 through 3.4. In addition, however, we need to make use of appropriate laws and constitutive relations which describe HPE phenomena.

Let us summarize the physical laws and constitutive relations used so far in Chaps. 2 and 3 (see below). Recall the important distinction between the two: laws are considered to be exact and always true; constitutive relations are generally approximations which relate two or more dynamic variables. Constitutive relations may be essentially exact in some situations, but may break down in others.

TABLE 3.1
Summary of ODE Model Examples of Chap. 3

Example	Math model	Order (states)	Classification	Input (if nonauton)	Laws	Constitutive relations	Assumptions
E3.1 Water tank	(3.6)	$1(z)$	Nonlinear, autonomous		Conservation of mass	Bernoulli	Simple fluid mechanics
E3.2 V^2 damper	(3.15)	$2(x, \dot{x})$	Nonlinear, autonomous		Newton, conservation of mass	Orifice flow	Simple fluid mechanics
E3.3 Hydraulic servo-valve	(3.23)	$4(x, \dot{x}, p_1, p_2)$	Nonlinear, nonautonomous	Servovalve motion	Newton, conservation of mass	Orifice flow, spring/damper	Ideal fluid mechanics, mechanical elements
E3.4 Airspring	(3.32)	$2(x, \dot{x})$	Nonlinear, autonomous		Newton, conservation of mass	Isentropic gas	Ideal thermodynamics
E3.5 Air cushion landing system	(3.43)–(3.45)	$3(y, \dot{y}, p_c)$	Nonlinear, autonomous		Newton, conservation of mass	Orifice flow	1-D, ideal fluid/thermodynamics
E3.6 LRC circuit	(3.49)	$2(q, \dot{q})$	Linear, nonautonomous	Applied voltage	Kirchhoff's law	Circuit elements	Ideal circuit elements
E3.7 Loudspeaker	(3.58), (3.59)	$3(x, \dot{x}, I)$	Linear, nonautonomous	Applied voltage	Newton, Kirchhoff, magnetic induction	Circuit, mechanical elements	1-D, lumped mechanical effects
E3.8 DC motor	(3.63), (3.64)	$3(\theta, \dot{\theta}, I_A)$	Linear, nonautonomous	Applied voltage	Newton, Kirchhoff, magnetic inductions	Circuit, mechanical elements	Ideal circuit elements, geometry
S3.5 Thermal system	(3.67)	$1(T)$	Linear, autonomous		Conservation of energy	Convective heat transfer	Uniform temperature distribution
E3.9 One species population model	(3.69)	$1(x)$	Nonlinear, autonomous		None		Growth, limiting tendencies
E3.10 Foxes and rabbits	(3.70)	$2(r, f)$	Nonlinear, autonomous		None		Frequency of interaction

Laws

1. Newton's second law of motion
2. Newton's law of gravitation
3. Laws derived from Newton's second law such as (2.30), (2.32), and (2.33)
4. Conservation of mass
5. Conservation of energy (for conservative mechanical systems; generally applicable for thermal systems)
6. Kirchhoff's circuit laws
7. Laws of magnetic induction

Constitutive relations

1. Hooke's law (ideal spring, $F_s = -kx$)
2. Ideal linear damper ($F_d = -c\dot{x}$)
3. Ideal circuit elements
4. Boyle's law ($p = \rho RT$)
5. Isentropic gas relation ($p\rho^{-\gamma} = \text{constant}$)
6. Bernoulli's equation (3.2)
7. Drag force model (Example 2.7)
8. Convective heat transfer model (Sec. 3.5)
9. Orifice flow relations (3.8) and (3.39)

The laws and constitutive relations listed above are not inclusive but are typical in the modeling of engineering systems.

The nonengineering Examples 3.9 and 3.10 aside, an important and fairly general property exhibited by the systems modeled in the examples of Chaps. 2 and 3 is the following: in each example in which an ODE model was derived, the differential equations comprising the model all resulted from application of a physical law. Thus, the ODE's in the example models all come directly from physical laws. The constitutive relations, on the other hand, were generally used to eliminate excess time dependent variables, so that the eventual ODEs contained the proper set of dependent variables (see Examples 3.3 and 3.5 for illustration of this).

REFERENCES

The following references contain additional information on modeling of the types of systems considered in this chapter:

3.1. Ogata, K.: *System Dynamics*, 2d ed., Prentice-Hall, Englewood Cliffs, N.J. (1992). This book contains an extensive treatment of modeling of hydraulic and pneumatic systems and devices.
3.2. Close, C. M., and D. K. Frederick: *Modeling and Analysis of Dynamic Systems*, 2d ed., Houghton Mifflin, Princeton, N.J. (1993). See especially the chapters on modeling of electrical and electromechanical systems, which are done in a fair amount of detail.

3.3. Cochin, I.: *Analysis and Design of Dynamic Systems*, Harper & Row, New York (1980). Many interesting real-world modeling examples are contained throughout this book.

For references on basic fluid mechanics (liquids and gases), the following are recommended:

3.4. Roberson, J. A., and C. T. Crowe: *Engineering Fluid Mechanics*, 5th ed., Houghton Mifflin, Princeton, N.J. (1993).

3.5. Fox, R. W., and A. T. McDonald: *Introduction to Fluid Mechanics*, 4th ed., Wiley, New York (1992).

3.6. White, F. M.: *Fluid Mechanics*, McGraw-Hill, New York (1979).

3.7. Sabersky, R. H., and A. J. Acosta: *Fluid Flow, A First Course in Fluid Mechanics*, Macmillan, New York (1964).

For references on basic electric circuits and electromagnetic phenomena, the following are recommended:

3.8. Johnson, D. E., Johnson, J. R., and J. L. Hilburn: *Electric Circuit Analysis*, 2nd ed., Prentice-Hall, Englewood Cliffs, N.J. (1992).

3.9. Smith, R. J., and R. C. Dorf: *Circuits, Systems, and Devices*, 5th ed., Wiley, New York (1992).

EXERCISES

1. Water is poured at the fixed rate q_0 (kg/s) into a tank of variable cross section (Fig. *P*3.1). At any height z the tank cross section is circular, with the radius $r(z)$ given as $r(z) = r_0 + az$. Water exits through a small opening of area A_0 at the bottom of the tank. Determine the ODE model for the height z of water in the tank.

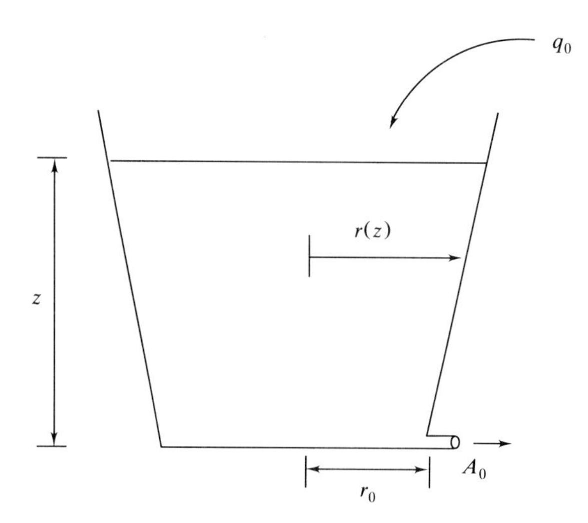

FIGURE *P*3.1

2. Water exits a tank through an orifice of area $A_0 = 1\ \text{cm}^2$. When the height of water in the tank is $z = 3$ m, the exit mass flow rate is observed to be $q = 0.7$ kg/s. From this result estimate the orifice loss coefficient c_0, which applies for flow through the orifice.

3. Shown in Fig. *P*3.3 is a system of two water tanks. Mass flows into the top tank at the fixed rate q_0 and exits this tank through an orifice of area A_1. This exit water

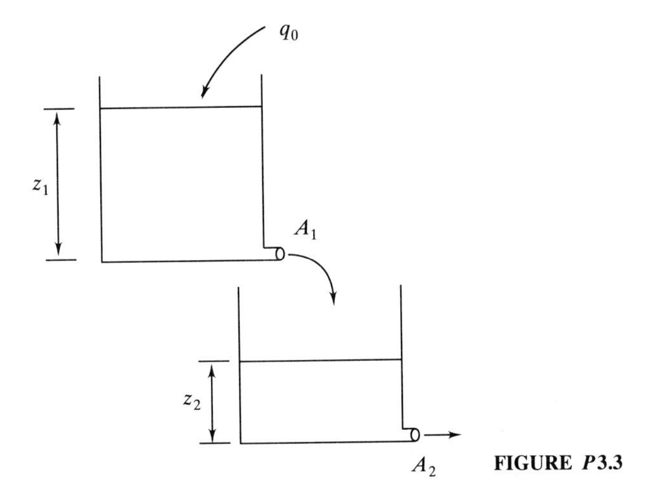

FIGURE *P*3.3

then flows into a second tank which has an orifice of area A_2 at its bottom. Each tank has constant cross section of area A, with $A \gg A_1, A_2$. Determine the ODE model which governs the heights z_1 and z_2 of water in the system.

4. An orifice, through which hydraulic fluid flows, has a specially shaped inlet and exit in an attempt to minimize losses. During tests the measured variations of mass flow rate q with pressure drop $(p_1 - p_2)$ across the orifice yield the data shown in Fig. *P*3.4. From these data estimate the orifice loss coefficient c_0. (Hydraulic fluid density $\rho = 800 \text{ kg/m}^3$, orifice area $A_0 = 1 \text{ mm}^2$, and exhaust pressure p_2 is atmospheric.)

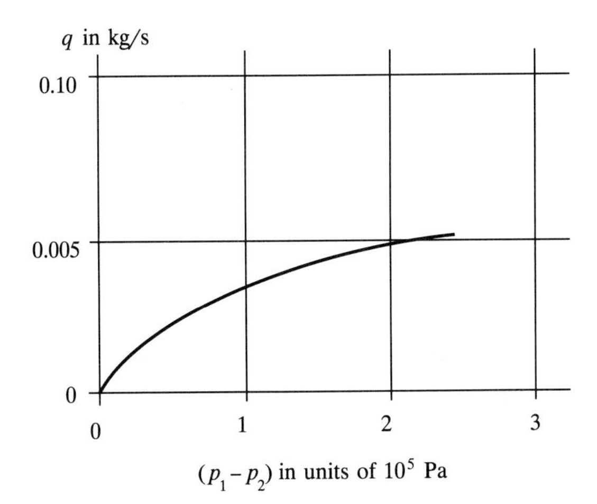

FIGURE *P*3.4

5. In Fig. *P*3.5 we have a schematic diagram of a loaded hydraulic actuator. Derive the mathematical model which relates the input $x(t)$ to the output $y(t)$. Note that the servovalve position $x_v(t)$ is related by geometry to the input x and to the coordinate z. Assume that the rigid, massless bar AB rotates from the vertical by only small

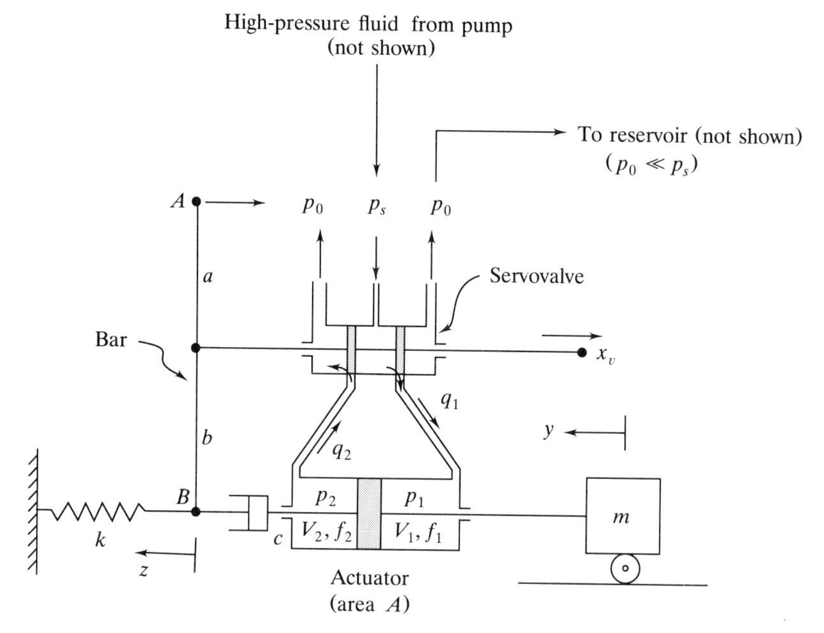

FIGURE ***P*3.5**
(Floor directions indicated are for $x_v > 0$.)

amounts, so that points A and B move essentially in the horizontal direction. In order to solve the differential equations in your math model, which quantities must be specified as initial conditions?

6. For the hydraulic servosystem of Example 3.3, reformulate the mathematical model for the simplified case in which the hydraulic fluid is assumed to be incompressible. Note that in this case (3.8) and (3.17) can be used to establish algebraic equations relating the pressures p_1 and p_2 directly to the velocity $\dot{y}$. Thus, it is possible to express the mathematical model as a single second-order ODE in y.

7. Shown in Fig. 3.7 is a schematic of a gas pressure shock absorber, with the various components identified. Determine the following:

a. The geometric, mass, and fluid parameters that need to be specified in order to construct the mathematical model

b. Those quantities (minimum number) that will be the states in the system ODE model

8. If you are sufficiently stouthearted, develop the ODE model for the gas pressure shock absorber of Fig. 3.7. Classify the model according to the various categories discussed in Sec. 1.5 and identify all nonlinearities.

For the electrical circuits of Exercises 9 through 13, derive the ODE model and construct the equivalent spring-mass-damper system. State the order of the system and the system states (use charge q as the dependent variable, analogous to position x in the oscillator system).

9. (Figure $P3.9$)

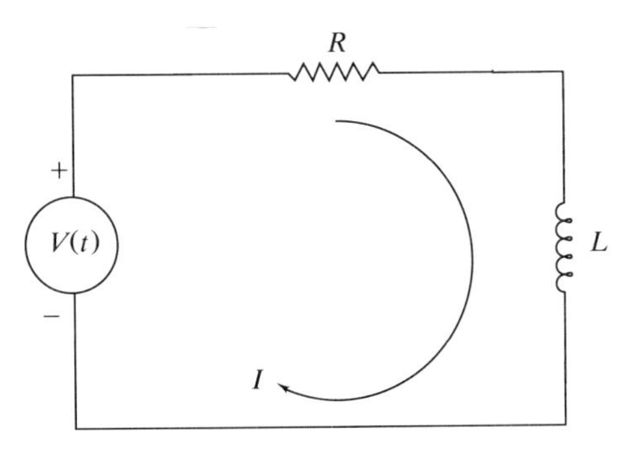

FIGURE $P3.9$

10. (Figure $P3.10$)

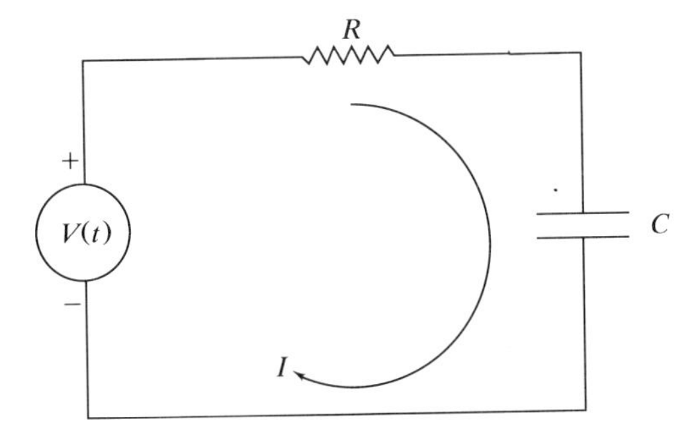

FIGURE $P3.10$

11. (Figure P3.11) Comment as to whether you think this to be a realistic model of an actual circuit.

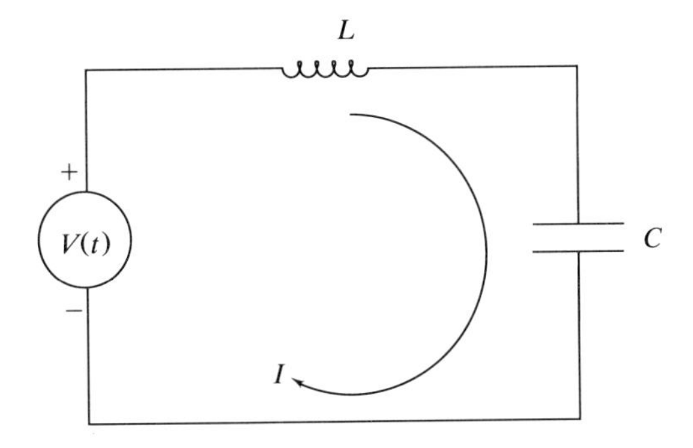

FIGURE $P3.11$

12. (Figure $P3.12$)

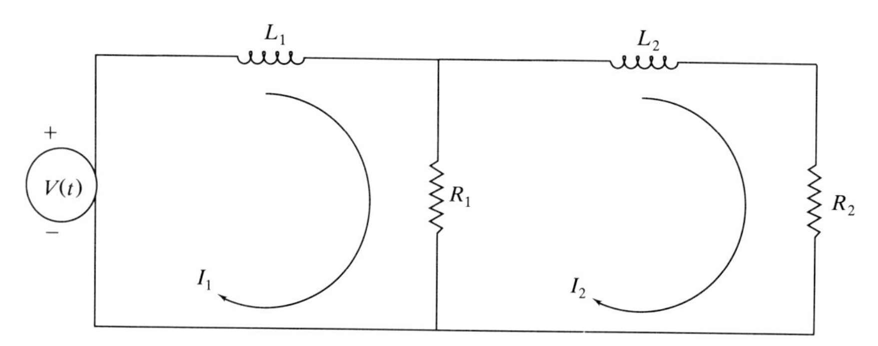

FIGURE ***P*3.12**

13. (Figure *P*3.13)

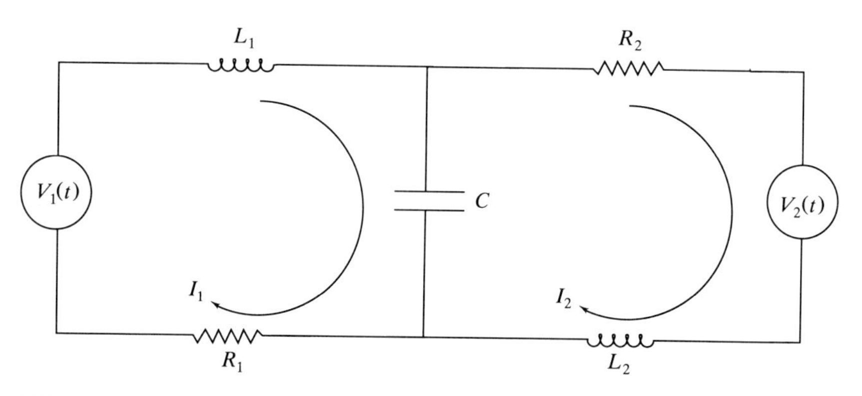

FIGURE ***P*3.13**

14. A mass m, restrained by a linear spring of stiffness k, moves under the influence of a pair of permanent magnets of the same polarity (Fig. *P*3.14). One of the magnets is attached to m, while the other is fixed, as shown. Because the magnets have the same polarity, magnet 2 exerts a repulsive force on magnet 1. This force is inversely

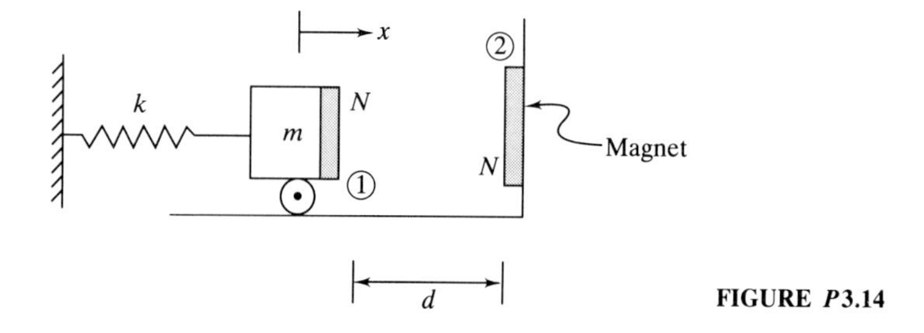

FIGURE ***P*3.14**

proportional to the square of the distance separating the magnets. When $x = 0$ the magnet separation is d. Derive the equation of motion, introducing any constants of proportionality which are needed. Is the system linear or nonlinear?

15. An ecologist postulates a single-species population model for a certain group of animals in a particular ecosystem. The model proposed is

$$\dot{x} = \alpha x - \beta x^2$$

where $x(t)$ is the population, $\alpha = 10^{-2}$ (1/s) and $\beta = 10^{-6}$ (1/s-animal). By direct calculation of the terms αx and $-\beta x^2$ for populations x equal to 10,100, 1000, and 10,000, estimate the approximate population level at which the self-limiting tendency of the nonlinear term becomes appreciable (note that there is no one "answer" to this problem).

16. For the single-species population model (3.69), it is desired to include a term to model the spread of a fatal, communicable disease among the population. Candidate terms to appear on the right-hand side of (3.69) are $-\gamma x$ and $-\epsilon x^2$, where γ and ϵ are positive constants. Discuss which, if either, of these terms might more accurately model the effect of the disease on the population.

17. Consider the math model (3.65) for the armature-controlled dc motor with $M_L(\dot{\theta}, t) = 0$. Rewrite this math model as a single second-order ODE for the angular velocity ω and, from the result, deduce an overall equivalent damping constant which includes mechanical and electromagnetic effects.

CHAPTER 4

FUNDAMENTALS OF DYNAMIC SYSTEMS ANALYSIS

In the preceding two chapters our objective was to see how to develop ODE models for various types of dynamic systems. Such things as idealization of the system through simplifying assumptions, use of physical laws, and development of constitutive relations were of primary interest. Now we turn our attention to the behavior of the system as implied by the dynamic model which we use to represent it. Our interest, therefore, now shifts to the study of the properties of the solutions to the ODEs comprising our mathematical model. This chapter is divided into three sections. In Sec. 4.1 we introduce the concepts which are central to the study of all dynamic systems, regardless of their origin: the phase space, equilibrium, stability of equilibrium, and linearization. In this section our approach is a general one: We start by assuming that we have a system mathematical model consisting of a set of ODEs, but we don't worry about the specific type of physical system from which the ODEs came. In Sec. 4.2 we study the autonomous linear system, which describes the behavior of linear systems with no inputs and of nonlinear systems in the vicinity of their equilibrium states. In Sec. 4.3 are presented several topics that extend or qualify the material covered in Secs. 4.1 and 4.2.

In what follows, the reader should be aware of the question of linear versus nonlinear system behavior. Basic concepts such as motion in phase space, equilibrium, and the stability of equilibrium (Sec. 4.1) are defined and understood without finding solutions to the system ODE model. The basic ideas are the same whether the system is linear or nonlinear. In this sense Sec. 4.1 is fairly general.

On the other hand, Sec. 4.2 and essentially all of Sec. 4.3 are devoted to a quantitative study of the properties of motion of systems governed by *linear* ODE models. Thus, the present chapter does not address the question of how nonlinear systems actually behave, except in situations for which this behavior can be approximated by a linear ODE model. It is important to bear this in mind because nonlinear systems may exhibit a host of interesting and practically important behaviors for which there is no analogy in the linear world. A discussion of some of these nonlinear phenomena appears in Chap. 10. In particular, see Sec. 10.2, which is similar in spirit to this chapter in that the behavior of nonlinear autonomous systems is considered.

4.1 FUNDAMENTAL IDEAS

As we have seen in Chaps. 1 through 3, ODE mathematical models may arise as sets of ODEs of various orders. In analyzing these ODEs, we may leave them as they naturally arise, we may convert them to state variable form (a set of first-order ODEs, one for each state), or we may convert them to a single ODE of the same order as the system. The latter approach (single ODE model), while useful for linear systems, is usually impractical if the system is nonlinear. Our approach, therefore, will be to use the state variable form. Most system models, whether linear or nonlinear, can be converted easily to this form. Furthermore, numerical integration methods usually assume the mathematical model to be in state variable form. Finally, the state variable form is the most useful form for studying basic concepts and *general* properties of system behavior.

Let's assume, then, that we have developed a mathematical model, consisting of a set of ordinary differential equations, for a dynamic system. We identify the *states* of the system, those quantities which, if specified at some time, uniquely determine the future of the system, and we express the mathematical model in state variable form as a set of n, first-order ODEs, one for each state:

$$\begin{aligned}\dot{x}_1 &= f_1(x_1, x_2, \ldots, x_n, t)\\ \dot{x}_2 &= f_2(x_1, x_2, \ldots, x_n, t)\\ &\vdots\\ \dot{x}_n &= f_n(x_1, x_2, \ldots, x_n, t)\end{aligned} \tag{4.1}$$

We can express this general form more compactly as

$$\dot{x}_i = f_i(x_1, x_2, \ldots, x_n, t) \qquad i = 1, \ldots, n \tag{4.2}$$

Here the time t is the independent variable and the n states $x_1, \ldots, x_n$ are the dependent variables. The functional notation indicates that generally, the rate of change of a given state will depend on the instantaneous values of all of the states (the states are "coupled") and, if the system is nonautonomous, also on the time explicitly. We will assume that any explicit time dependence in (4.1) is due to some specified external effect, that is, an input, which arises from outside sources unrelated to the system itself. If we were to remove all inputs, therefore, the mathematical model (4.1) would become autonomous. Generally the functions $f_i(x_1, \ldots, x_n, t)$ may be either linear or nonlinear functions of the states.

If we are to solve (4.1) in specific cases, we need to specify the initial conditions, values $x_1(0), x_2(0), \ldots, x_n(0)$ for each state at some starting time which we pick, taken here to be zero.

4.1.1 The Phase Space

Now let's suppose that for a given set of initial conditions, we can solve (4.1), either numerically or analytically, to obtain a set of solutions $x_i(t)$, $i = 1, \ldots, n$. These solutions tell us how each state evolves in time from its starting value $x_i(0)$. The solutions $x_i(t)$, taken together, constitute the *response* of the system. We would like to represent this response pictorially so that we can more easily deduce how the system behaves. One way to do this is to plot *graphs*, each graph consisting of a state $x_i(t)$ versus time. While this is a useful procedure that is almost always employed in practice, we can present a more comprehensive picture by simultaneously displaying all of the states in what is called the *phase space*, or *state space*. This construction is made as follows: For a system with n states, define a cartesian n space, an example of which is shown in Fig. 4.1 for a three-state system. Each of the n orthogonal axes measures a different state of the system.

At any given time t, the n values of the states $x_1(t), x_2(t), \ldots, x_n(t)$ will define a point in the phase space in the same way that the cartesian coordinates x, y, and z define the location of a point mass m in the ordinary newtonian mechanics problem. We can also view the "system point" in phase space as defining an n-dimensional *state vector* $x(t)$, which is measured from the origin

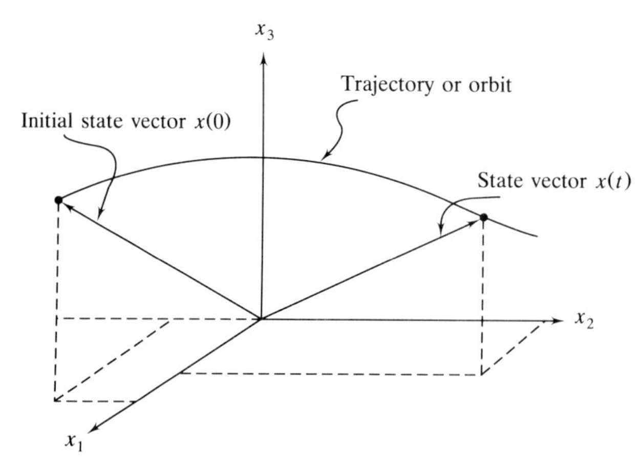

FIGURE 4.1
The "system point," or state vector, traces out an orbit in the phase space.

of the phase space, $x(t) \equiv (x_1(t), x_2(t), \ldots, x_n(t))$. Thus, for a given set of initial conditions, we visualize the solution to (4.1) as follows: The initial state vector, $x(0) = [x_1(0), \ldots, x_n(0)]$ defines a starting point in the phase space. Then for $t > 0$ the states will change in accordance with (4.1). In the phase space the continuously changing states generate a continuous curve in phase space, emanating from the initial point $x(0)$. Each point on this curve defines the values of all the states at some time t. This solution curve in phase space is referred to as an *orbit* or *trajectory*. We draw arrows along the orbit to indicate the direction the orbit takes as time increases. The advantage of the phase space construction is that we obtain an understanding of what all of the states are doing for a given set of initial conditions, by looking at a single curve in the phase space. The disadvantage is that we cannot actually construct the "phase portraits" for $n > 3$; we can, however, "visualize" them, just as we visualize vectors in Euclidean spaces of dimension higher than 3. It should also be noted that the phase portrait does not tell us the time t associated with a given point along an evolving orbit. For our purposes this will not be a problem.

In visualizing motions in the phase space, let us next picture the following situation: Pick a large number of points in the phase space, each of which is to represent a different set of initial conditions. Then simultaneously watch all of the orbits which emanate from these starting points. In the limit in which the number of initial points extends to the entire phase space, the resulting collection of orbits is referred to as the *phase flow*, or simply the *flow*, of the dynamical system. The analogy with fluid flow is apparent; the individual orbits in our phase space representation are analogous to the streamlines along which individual fluid particles move in a fluid flow. We conceive this representation as a way to simultaneously monitor what all of the states are doing for all possible initial conditions. In practice, for two- and three-state systems we represent the flow by drawing a number of different orbits sufficient to indicate the overall structure of the flow.

The orbits in the phase space are determined by the functions $f_1(x_1, x_2, \ldots, x_n, t), \ldots, f_n(x_1, \ldots, x_n, t)$ in (4.1). For an n-state system the n functions $f_1, f_2, \ldots, f_n$ form a *vector field*. This term means that at each point $(x_1, x_2, \ldots, x_n)$ in phase space is associated an n-dimensional vector $(f_1, f_2, \ldots, f_n)$. This is analogous to the three-dimensional flow of a fluid, for which the velocity is a vector field; at each point in physical space is associated a vector which defines the velocity of a fluid particle at that point. The path of motion of fluid particles is everywhere tangent to the velocity vector field. Similarly, in phase space a given orbit must be tangent to the vector field $(f_1, \ldots, f_n)$ at each point. If each of the functions $f_1, \ldots, f_n$ is independent of time, then the vector field is time invariant, and this situation is analogous to the steady flow of a fluid. By constructing a picture of the vector field directly from the state equations (4.1), it is possible to obtain some insight into the nature of the phase space orbits. This is illustrated for a two-state system in the following example.

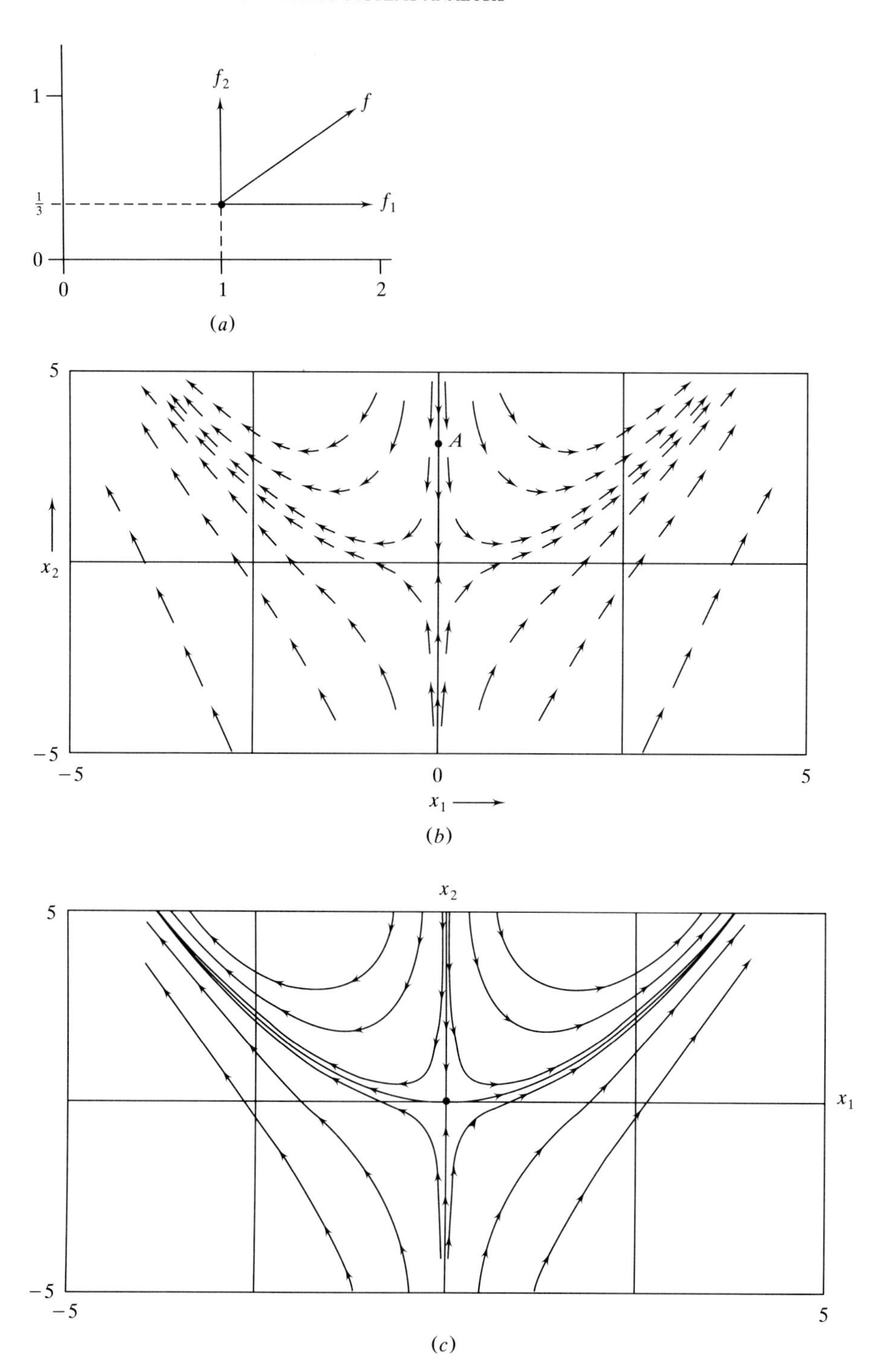

FIGURE 4.2
(*a*) Construction of the vector field f for the system (4.4) at the point $x = (1, \frac{1}{3})$. (*b*) The vector field f for the system (4.4). (*c*) Orbits of the system (4.4).

Example 4.1 Example of a Vector Field. Consider the two-state ODE model:

$$\dot{x}_1 = x_1$$

$$\dot{x}_2 = -x_2 + x_1^2 \tag{4.3}$$

This system is nonlinear, due to the x_1^2, and autonomous. The functions $f_1(x_1, x_2)$ and $f_2(x_1, x_2)$ comprising the system vector field f are

$$f_1(x_1, x_2) = x_1$$

$$f_2(x_1, x_2) = -x_2 + x_1^2 \tag{4.4}$$

The resulting vector field is shown in Fig. 4.2. Figure 4.2(a) depicts the construction of the vector f at the point $x_1 = 1, x_2 = \frac{1}{3}$, at which $f_1 = 1$ and $f_2 = \frac{2}{3}$. An orbit passing through the point $(1, \frac{1}{3})$ must be tangent to the vector f at that point. In Fig. 4.2(b) is shown the vector field obtained by calculating f at a large number of points in the two-dimensional phase space. This picture allows certain features of the phase space orbits to be deduced. For instance, an orbit initiated on the positive x_2 axis at some point A will move along a vertical line toward the origin. An orbit initiated at some point B slightly to the right of A will move initially down toward the origin. Then, as the origin is approached, the orbit will swing outward and move away from the origin along a curved path. Some orbits of this system are shown in Fig. 4.2(c).

If the system is autonomous, so that t does not appear explicitly in the state variable equations (4.1), then the vector field is time independent. In this case an important characteristic of phase space orbits is that they may not cross each other transversally. To do so would violate the condition that orbits must be everywhere tangent to the vectors f. If the vector field f is time dependent, however, orbits may cross transversally. In this case, for which the vectors $(f_1, \ldots, f_n)$ change with time, an orbit may return to a phase space point previously visited; at the later time the direction of f may have changed, allowing a transversal crossing of orbits.

For autonomous systems, note that it is not implied that a given point $(x_1, \ldots, x_n)$ in phase space can be visited only once by a given orbit. For example, orbits consisting of closed curves are possible. In this case an orbit moves around the closed curve endlessly, repeating itself periodically. Each point on the orbit is repeatedly visited. But the direction of travel is the same (tangent to the vector f) each time the point is visited.

In our study of dynamical systems we will use both the phase space and graphical [that is, $x_i(t)$ versus t] representations of system response. As we will see, the phase space representation is especially valuable in achieving an understanding of the fundamental concepts of equilibrium and stability.

4.1.2 Equilibrium

We are familiar with the idea of equilibrium in newtonian systems. The net force exerted on a body vanishes, so that the body is at rest relative to some reference location or moves with constant linear momentum. We speak of *static equilibrium* to denote the condition of rest. For a dynamic system defined by the general ODE model (4.1), *equilibrium* is defined as the condition for which none of the states is changing with time, so that $\dot{x}_i = 0$, $i = 1, \ldots, n$. Thus, in the phase space an equilibrium solution will be represented by any point having the property that an orbit initiated exactly at that point will remain at that point forever.

The particular equilibrium configurations that a given system may admit are considered to be basic properties of the system, unrelated to any inputs. Recall that our definition of an input is any externally imposed, time-dependent effect, which may influence the behavior of the system but which is not considered to be part of the system itself. Thus, when we study equilibria in dynamic systems, we disallow inputs and consider only the autonomous version of the system model,

$$\dot{x}_i = f_i(x_1, \ldots, x_n) \qquad i = 1, \ldots, n \tag{4.5}$$

as this autonomous form defines all of the basic properties of the *system*.

Mathematically, the equilibrium solutions are found by setting to zero the state derivatives $\dot{x}_i$ in the autonomous version (4.5) of the mathematical model. Thus, equilibrium configurations of the system are found by finding solutions to the set of equations

$$\begin{aligned} f_1(x_1, \ldots, x_n) &= 0 \\ f_2(x_1, \ldots, x_n) &= 0 \\ &\vdots \\ f_n(x_1, \ldots, x_n) &= 0 \end{aligned} \tag{4.6}$$

This set consists of n *algebraic* equations, and a solution consists of the values of the n states which, taken together, will satisfy (4.6). We will denote equilibrium values of the states with an overbar (not to be confused with the vector symbol). Thus, an equilibrium solution is denoted as $\bar{x} = (\bar{x}_1, \bar{x}_2, \ldots, \bar{x}_n)$.

We make the following observations: (1) If the system model (4.5) is *linear*, then (4.6) are a set of linear, algebraic equations which will have a single equilibrium solution, provided the equations (4.6) are linearly independent. Frequently the states in a linear system model are defined in such a way that, at equilibrium, all of the states are zero, so the origin of the phase space represents the equilibrium configuration of the system. (2) If the algebraic equations (4.6) are nonlinear, then multiple equilibrium configurations may exist, as a set of simultaneous nonlinear algebraic equations may admit more than one solution. These points will be explored in the examples.

The equilibrium solutions are important for at least two reasons: (1) one of the equilibrium solutions is often the desired operating state of the system; that is, if the system at any time is not at equilibrium, we want it to move to equilibrium or to drive it to equilibrium if we can, (2) in many systems, the response $x_i(t)$ will approach $\bar{x}_i$ at long times; that is, $x_i(t) \rightarrow \bar{x}_i$ as $t \rightarrow \infty$, so equilibrium is relevant if we are interested in long-term behavior, as we almost always are.

The process of finding equilibrium solutions is now illustrated by example for several dynamic systems.

Example 4.2 The Linear Oscillator. We first consider the simple spring-mass system, with no damping and no forcing, depicted in Fig. 2.23. The equation of motion is

$$m\ddot{x} + kx = 0 \tag{4.7a}$$

where the displacement x is defined to be zero when the spring is unstretched. Defining the states as $x_1 = x$ and $x_2 = \dot{x}$, the state variable form of the system model is

$$\dot{x}_1 = x_2 \tag{4.7b}$$

$$\dot{x}_2 = -\frac{k}{m}x_1 \tag{4.7c}$$

At equilibrium, $\dot{x}_1 = \dot{x}_2 = 0$, so (4.7$b$ and c) then tell us that $\bar{x}_1 = \bar{x}_2 = 0$; at equilibrium the mass m is at rest ($\dot{x} = 0$) with the spring undeflected ($x = 0$), as we expect on physical grounds.

Now let us consider the same mass-spring system suspended vertically in the gravitational field, as shown in Fig. 4.3.

We arbitrarily define x to be positive upward, and, again defining x to be zero when the spring is unstretched, the equation of motion, including the gravitational force $-mg$, is

$$m\ddot{x} + kx = -mg \tag{4.8}$$

The state variable form of the system model is

$$\dot{x}_1 = x_2$$

$$\dot{x}_2 = -\frac{k}{m}x_1 - g$$

At equilibrium we then find $\bar{x}_2 = 0$, $\bar{x}_1 = -mg/k$. At equilibrium the mass m is at rest ($\bar{x}_2 = \bar{\dot{x}} = 0$), but it is displaced an amount $-mg/k$ downward, such that the spring force balances the force due to gravity. This example illustrates that for a linear system, an equilibrium for which the states are not all zero requires the

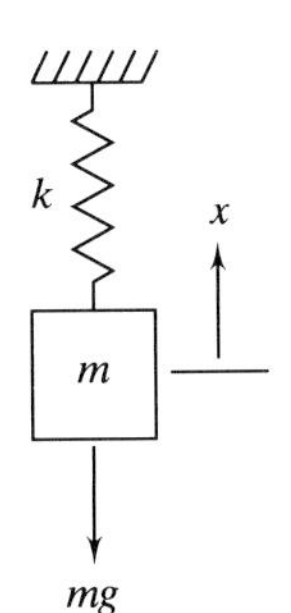

FIGURE 4.3
The undamped oscillator suspended vertically in the gravitational field.

presence of a constant, here $-mg$, in at least one of the state equations. We should also point out that, in order to find the equilibrium condition(s), it is not actually necessary that the mathematical model be in state variable form. In this example, or in any other, we can simply eliminate all time derivatives in the mathematical model and solve the resulting algebraic equation(s). Doing this in (4.8) results in $\bar{x} = -mg/k$.

Example 4.3 The Simple Pendulum. As a second example we consider the simple undamped pendulum (Example 2.1) for which the equation of motion was found to be

$$mL^2\ddot{\theta} + mgL\sin\theta = 0 \tag{4.9}$$

This system is nonlinear because of the $\sin\theta$ term. The states of the system are $x_1 = \theta$, $x_2 = \dot{\theta}$, and the mathematical model in state variable form is then

$$\begin{aligned} \dot{x}_1 &= x_2 \\ \dot{x}_2 &= -\frac{g}{L}\sin x_1 \end{aligned} \tag{4.10}$$

At equilibrium $\dot{x}_1 = \dot{x}_2 = 0$, so we conclude that $\bar{x}_2 = 0$ and $\sin\bar{x}_1 = 0$. The second of these conditions yields two equilibrium solutions for the angular displacement θ: $\bar{x}_1 = 0$ and $\bar{x}_1 = \pi$. The pendulum hangs vertically downward or vertically upward, and in either position the moment due to gravity vanishes. Thus, the two equilibrium solutions are $\bar{x} = (0, 0)$ and $\bar{x} = (\pi, 0)$. The existence of more than one equilibrium configuration is a result of the system's being nonlinear.

Example 4.4 The Single-Species Population Model. The single-species, nonlinear population model of Example 3.9 is given by

$$\dot{x} = \alpha x - \beta x^2 \qquad \alpha, \beta > 0 \tag{4.11}$$

Because this is a single-state system, the equilibrium solutions are obtained directly by setting $\dot{x} = 0$ in (4.11) to obtain

$$\alpha\bar{x} - \beta\bar{x}^2 = \bar{x}(\alpha - \beta\bar{x}) = 0$$

The equilibrium solutions are seen to be

$$\begin{aligned} \bar{x} &= 0 \\ \bar{x} &= \frac{\alpha}{\beta} \end{aligned}$$

The nonzero equilibrium condition $\bar{x} = \alpha/\beta$ reflects a balance between the tendency for population growth (αx) and the self-limiting tendency $(-\beta x^2)$, which becomes important when the population becomes sufficiently large. Again we observe that the nonlinearity results in the existence of multiple equilibria.

Example 4.5 The Water Tank. For the water tank system of Example 3.1 the differential equation for the height z of water was derived as

$$\dot{z} = \frac{q_0}{\rho A} - \frac{A_0}{A}(2gz)^{1/2} \tag{4.12}$$

This model is nonlinear due to the $z^{1/2}$ term. At equilibrium $\dot{z} = 0$, so the equilibrium condition is defined by

$$\bar{z} = \left(\frac{q_0}{\rho A_0}\right)^2 \frac{1}{2g} \tag{4.13}$$

At this water level the water inflow q_0 is exactly balanced by the water flowing out the small hole A_0.

Two comments are relevant for this example: (1) Although the system is nonlinear, there is only one equilibrium solution. This illustrates the fact that, while multiple equilibria can occur in a nonlinear system, it is also possible that nonlinear systems will exhibit a single equilibrium state. (2) In *physical* terms the water inflow q_0 would normally be viewed as an input. This inflow is, however, *constant*, and for this reason we consider it to be, mathematically, a part of the system. The situation is mathematically similar to the effect of gravity on the spring-mass system of Example 4.2. Suppose we were to consider an input flow q consisting of a steady component q_0, with a small fluctuating component $q_1(t)$ superposed on it, so that $q(t) = q_0 + q_1(t)$. In this type of situation the steady component q_0 is still considered to be part of the system, while $q_1(t)$ is considered to be the input, because it is an explicitly time-dependent function.

Example 4.6 The Rolling Ball in a Double Well. In this example we consider a ball or bead of mass m which rolls without friction on a surface with two "wells" and one "hump" as shown in Fig. 4.4.

The shape of the surface along which the ball rolls is assumed to be given by

$$y(x) = -\tfrac{1}{2}x^2 + \tfrac{1}{4}x^4 \tag{4.14}$$

which defines the height y at a given horizontal position x. The positions x at which $dy/dx = 0$, for which the tangents to the surface are horizontal, are $x = 0, \pm 1$, as indicated by the small crosses. We assume the ball always to be in contact with the surface and that the motion takes place in the plane of the paper. This is a newtonian system with a single degree of freedom, which we take to be the horizontal position x. If this horizontal position x is known, then so is the vertical position y, through (4.14). The mathematical model of this system,

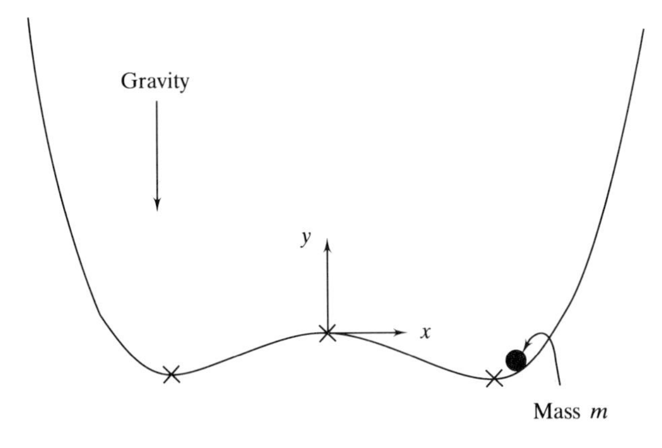

FIGURE 4.4
The rolling ball in a frictionless double well.

obtained in Chap. 6 (Example 6.12) using Lagrange's equation,* is highly nonlinear and turns out to be as follows:

$$\ddot{x}\left[1 + (-x + x^3)^2\right] + x\dot{x}^2(-1 + x^2)(-1 + 3x^2) + gx(-1 + x^2) = 0 \quad (4.15)$$

The states are x and $\dot{x}$; at equilibrium, then, $\dot{x}$ and $\ddot{x}$ vanish, and we have the following static condition defining the equilibrium configuration:

$$g\bar{x}(-1 + \bar{x}^2) = 0$$

Thus, we find three equilibrium positions $\bar{x} = 0$, $\bar{x} = -1$, and $\bar{x} = 1$. The three equilibrium solutions are then $\bar{x} = (0, 0)$, $\bar{x} = (-1, 0)$, and $\bar{x} = (1, 0)$. At equilibrium the mass is at rest ($\dot{x} = 0$) at one of the three positions at which $dy/dx = 0$. In physical terms at these locations the weight of the object is balanced by the normal force of contact exerted on the mass m by the surface, and there are no forces in the x direction.

Example 4.7 Foxes and Rabbits. The nonlinear, two species population model describing the evolution of interacting fox and rabbit populations was (Example 3.10):

$$\dot{r} = \alpha r - \gamma rf$$

$$\dot{f} = -\beta f + \delta rf \quad (4.16)$$

The system model arises naturally in state variable form, and the equilibrium solutions occur when $\dot{r} = \dot{f} = 0$, so that

$$\alpha\bar{r} - \gamma\bar{r}\bar{f} = 0$$

$$-\beta\bar{f} + \delta\bar{r}\bar{f} = 0$$

Factoring an $\bar{r}$ in the first relation and an $\bar{f}$ in the second, we have

$$\bar{r}(\alpha - \gamma\bar{f}) = 0$$

$$\bar{f}(-\beta + \delta\bar{r}) = 0$$

We must find $\bar{r}$ and $\bar{f}$ such that both of these conditions are satisfied simultaneously. We observe that the first relation is satisfied if either $\bar{r} = 0$ or $\bar{f} = \alpha/\gamma$, while the second relation is satisfied if either $\bar{f} = 0$ or $\bar{r} = \beta/\delta$. The only combinations of $\bar{r}$ and $\bar{f}$ which will simultaneously satisfy both relations are the two solutions $\bar{x} = (\bar{r}, \bar{f}) = (0, 0)$ and $(\beta/\delta, \alpha/\gamma)$. Thus, there exists a nonzero equilibrium solution at which both populations remain at fixed levels. The rate of rabbit multiplication equals the rate at which the foxes eat the rabbits, and the foxes eat just enough rabbits to compensate for the rate at which they die off.

*Lagrange's equation, discussed in Chap. 6, is an alternate way to derive the equations of motion for newtonian systems. It is based on the *scalar* system quantities kinetic energy, potential energy due to conservative forces, and work done by nonconservative forces, rather than on the *vector* quantities, force and momentum, that we used in Chap. 2. For some systems, including the present example, the Lagrangian approach is much easier to apply than the newtonian methods used in Chap. 2.

Example 4.8 Air Cushion Landing System. The mathematical model of the idealized ACLS (Example 3.5) is a three-state, nonlinear model with states y, $\dot{y}$, and p_c as given by (3.43) through (3.45). At equilibrium the derivatives of the states, here $\ddot{y}$, $\dot{y}$, and $\dot{p}_c$, will all be zero, so the model reduces to the following algebraic equations which apply at equilibrium:

$$
\begin{aligned}
(p_c - p_a)A - mg &= 0 \\
q_{\text{in}} &= q_{\text{out}} \\
q_{\text{out}} &= q_{\text{out}}(y, p_c)
\end{aligned}
\tag{4.17}
$$

The parameters p_a, A, mg, and q_{in} are assumed to be specified constants, and our job is to find the equilibrium cushion pressure $\bar{p}_c$ and height $\bar{y}$ of the vehicle (we already know that at equilibrium the third state $\bar{\dot{y}} = 0$).

In physical terms let us see what the conditions (4.17) tell us. From (4.17*a*) we obtain the equilibrium cushion pressure $\bar{p}_c$ as

$$\bar{p}_c = p_a + \frac{mg}{A} \tag{4.18}$$

This is the cushion pressure just sufficient to support the weight mg of the vehicle. From the second relation (4.17*b*) we deduce that the airflow from the cushion to the surrounding atmosphere must equal the inflow to the cushion provided by the fan, so that the mass of air in the cushion remains fixed. The third relation (4.17*c*) tells us how the flow q_{out} depends on cushion pressure p_c and height y; recall that the height y determines the flow area from cushion to atmosphere. Since the equilibrium cushion pressure $\bar{p}_c$ is determined by (4.17*a*), equations (4.17*b*) and (4.17*c*) serve to define the height $\bar{y}$ which will allow the required flow $q_{\text{out}} = q_{\text{in}}$ to pass from the cushion to the atmosphere. Thus, at equilibrium the vehicle sits a fixed distance $\bar{y}$ off the ground, with $\bar{\dot{y}} = 0$, such that the pressure $\bar{p}_c$ supports the vehicle and the flows q_{out} and q_{in} are in balance. In physical terms the result of our solution makes sense. Generally, it is wise to check the physical meanings implied by a mathematical model, for the equilibrium condition, in order to improve our confidence in the model. This example provides another illustration of a system for which the presence of constant terms in the model, here both mg and q_{in}, result in nonzero equilibrium solutions. The example also illustrates that, even though the system is nonlinear, there is only one equilibrium solution.

Example 4.9 Vertical Descent through the Atmosphere. As a final example, let's consider the vertical descent of a vehicle through the atmosphere (Example 2.7). The general mathematical model obtained was (2.62),

$$\ddot{z} = -g\left(\frac{R_e}{R_e + z}\right)^2 + \left(\frac{\rho_0 A C_D}{2m}\right)e^{-z/h}\dot{z}^2 \tag{4.19}$$

At equilibrium $\dot{z} = \ddot{z} = 0$, so the equilibrium condition is defined by

$$\left[\frac{R_e}{(R_e + z)}\right]^2 = 0$$

This is satisfied at $\bar{z} = \infty$, with $\bar{\dot{z}} = 0$. At equilibrium both aerodynamic drag and gravity forces vanish, and the mass m is at rest an infinite distance from the earth.

We observe that this equilibrium condition is not very interesting from a practical standpoint. The reason for developing this model in the first place is to describe the vehicle motion during descent. This example differs from Examples 4.2 through 4.8 in that it represents a system for which a study of the equilibrium configuration of the system is essentially irrelevant.

Suppose, however, that we consider the version (2.63) of this system model, which applies for motion near the earth's surface:

$$\ddot{z} = -g + \left(\frac{\rho_0 A C_D}{2m}\right)\dot{z}^2 \tag{4.20}$$

We see that the position z does not appear in this mathematical model. We can, therefore, redefine the mathematical model using velocity $v = \dot{z}$ as the dependent variable. This results in a *first-order* nonlinear ODE for the velocity:

$$\dot{v} = -g + \left(\frac{\rho_0 A C_D}{2m}\right)v^2 \tag{4.21}$$

Here the reduction of system order, which should always be made if possible, is analogous to what happened in Example 3.6 when we removed the capacitor.

We can now identify an equilibrium condition for the single state v of the system (4.21) as

$$\bar{v} = \left(\frac{2mg}{\rho_0 A C_D}\right)^{1/2} \tag{4.22}$$

Physically this condition results in a balance between the aerodynamic drag and gravity forces, and the vehicle descends at the constant velocity $\bar{v}$. This condition is referred to as *terminal fall*. In this equilibrium state the vehicle is not at rest, but the system (4.21) is nevertheless in a condition of equilibrium.

Remarks

1. Once we have a system ODE mathematical model, the first step in the analysis is usually to find the equilibrium configurations and to check that they are physically reasonable. We can represent these equilibrium solutions as points in the phase space. From the standpoint of understanding the overall system behavior in the phase space, we have obtained a very limited (but very useful) amount of information. We have found those points $\bar{x}$ in phase space which have the special property that an orbit initiated exactly at $\bar{x}$ will remain at $\bar{x}$ forever. We have, however, said nothing about what will happen to orbits initiated at any point away from equilibrium. We will begin to address this problem in the next subsection, which considers the stability of equilibrium.
2. For some systems, such as Example 4.9 (vertical descent), the existence of equilibrium solutions is largely irrelevant, as our specific concerns are with what happens away from equilibrium. Also, our physical interpretation of equilibrium may be altered if a given model can be converted to lower order, as in Example 4.9.
3. We have to keep in mind that the concept of equilibrium is related to how the system behaves on its own, with no externally imposed effects (inputs) present.

Equilibrium solutions are considered to be basic properties or characteristics of the system, and any basic system property is by definition unrelated to any outside influence (input).

4.1.3 Stability of Equilibrium

The equilibrium solutions studied in the preceding subsection represent a special class of *static* solutions to the state equations (4.5). If a system starts out exactly at an equilibrium condition, then it will remain there forever. In real systems, however, small disturbances often arise which will move the system away from an equilibrium condition. Such disturbances, regardless of their origin, will be considered by us to give rise to initial conditions which do not coincide with an equilibrium condition. Since the system is then not at equilibrium, at least some of the derivatives $\dot{x}_i$ in (4.5) will be nonzero, and the system will exhibit dynamic behavior, which we can monitor by watching orbits evolve in the phase space. We will begin our study of dynamic behavior by considering the following general problem: Suppose we displace the system *slightly* from equilibrium and let it go; what happens? The resulting behavior near equilibrium is often referred to as *local behavior*, to distinguish it from the more general problem of *global behavior*, for which we study systems which start or have moved "far" from equilibrium. In the remainder of this subsection we attack the problem of local behavior qualitatively. Then in Secs. 4.1.4 and 4.2 we'll look at things quantitatively through the process of linearization.

Central to the description of phase space motions near equilibrium is the idea of *stability* of equilibrium, which we now define. Assume we have identified, for a system with n states, an equilibrium solution $\bar{x} = (\bar{x}_1, \ldots, \bar{x}_n)$, the motion near which we wish to study. We next define some "neighborhood" N_1 of $\bar{x}$; this neighborhood N_1 is an n-dimensional volume in phase space which contains $\bar{x}$ in its interior.* Now we picture starting the system out at any point in N_1 and watching the orbit evolve. The stability of equilibrium is then defined as follows (note that the set notation $A \subset B$ means "A is contained in B," or "A is inside B"):

STABILITY OF EQUILIBRIUM. If for *every* n-dimensional neighborhood N, *no matter how small*, of $\bar{x}$, there is *a* neighborhood $N_1 \subset N$, such that *every solution starting out in* N_1 lies in N for all time, then the equilibrium solution $\bar{x}$ is *stable*. Furthermore, if N_1 can be selected so that $x(t) \to \bar{x}$ as $t \to \infty$, then $\bar{x}$ is *asymptotically stable*. See Fig. 4.5, which illustrates the two cases.

If an equilibrium solution is not stable, then it is *unstable*.

As we will see in the examples, the keys to this definition of stability are the words and phrases which are italicized.

---

*It is perhaps easiest to visualize the situation in a space of two or three dimensions.

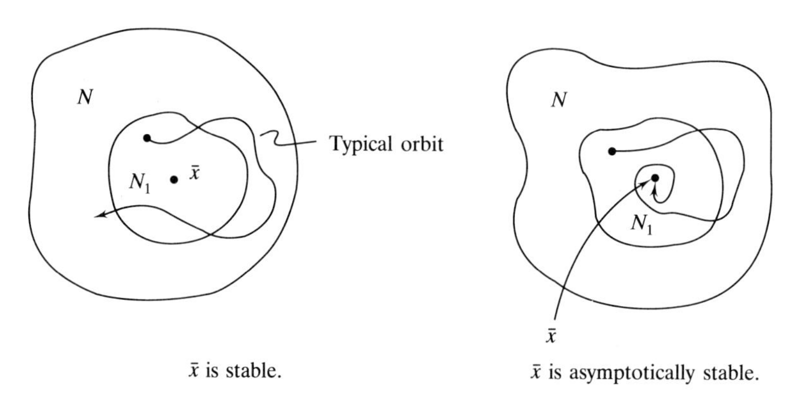

FIGURE 4.5
Illustration of the definitions of *stability* and *asymptotic stability of equilibrium*.

If we were to use the definition of stability literally, we would start by selecting some test neighborhood N containing the equilibrium solution $\bar{x}$. Then we would have to find at least one neighborhood N_1, contained in N, such that *all* orbits started in N_1 stay in N for all time. Then, supposing the stability condition to be met for our first test neighborhood N, we would have to do the test for all such neighborhoods N, no *matter how small*. The phrase "no matter how small" is important because we will see that it is often possible to pass the stability test for large, but not small, N.

Basically, an equilibrium solution is stable if all motions initiated *near* equilibrium remain near equilibrium for all time. For asymptotic stability, in addition, all motions initiated near equilibrium will *approach* equilibrium asymptotically as $t \to \infty$. In the real world any equilibrium solution actually observed is asymptotically stable.

The stability of equilibrium is very important for the following reason. The fact that we can find an equilibrium solution, which by definition satisfies the governing differential equations, is not sufficient to ensure that such a solution can actually be observed in the physical system being modeled. In order to realize such a solution physically, it must be asymptotically stable.

In general we may find any number of different equilibrium solutions for a given system, and the stability of each has to be determined. In nonlinear systems with multiple equilibria, some of the equilibria will be stable, while others will be unstable.

In the remainder of this subsection we will look at some examples for which the stability of equilibrium can be deduced intuitively by sketching the nature of orbits in the phase space, without actually solving the differential equations.

Example 4.10 The Undamped Linear Oscillator. For this linear system, governed by the equation of motion $m\ddot{x} + kx = 0$, the single equilibrium solution is the origin of the phase space, $(\bar{x}, \bar{\dot{x}}) = (0, 0)$. Let us investigate the stability of

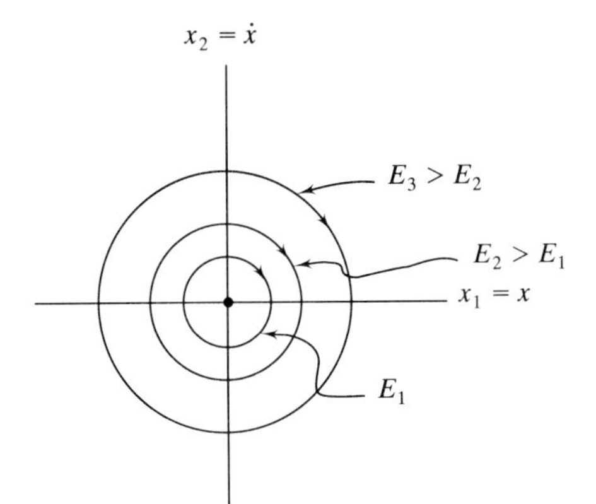

FIGURE 4.6
Phase plane orbits for the undamped oscillator (4.23).

equilibrium by constructing the actual orbits in the phase space. We can do this exactly by noting that for this system the only force is conservative, so the total energy $E = T + V$ is constant. To simplify things, let's set $m = k = 1$, in appropriate units, so that the conservation of energy states

$$\tfrac{1}{2}\dot{x}^2 + \tfrac{1}{2}x^2 = E = \text{constant} \tag{4.23}$$

where the constant E is defined by the initial position x_0 and initial velocity $\dot{x}_0$ as $E = \frac{1}{2}\dot{x}_0^2 + \frac{1}{2}x_0^2$. We observe that (4.23) is the equation of a circle in the phase space. The radius of the circle is $\sqrt{2E}$ and is defined by the energy E. Thus, the phase portraits for this system can be represented schematically as in Fig. 4.6.

Note that the directions indicated for the orbits are clockwise. This is because, when $\dot{x} > 0$, x will be increasing, while when $\dot{x} < 0$, x will be decreasing. We should also note that all orbits form *closed curves* in the phase space. This indicates that the system behavior is periodic; the solutions $x_1(t)$ and $x_2(t)$ for the states repeat exactly with some period T; that is, $x_i(t + T) = x_i(t)$. Since all orbits in the phase plane are closed, we deduce that the equilibrium condition at the origin is merely stable but not asymptotically stable; for any neighborhood N, no matter how large or small, we can always find a second neighborhood $N_1 \subset N$, such that orbits initiated in N_1 will remain in N for all time; but note that the orbits do *not* approach equilibrium asymptotically. It is suggested that the student verify this by drawing various neighborhoods N in the phase space and then showing that, for any N, we can always find some $N_1 \subset N$ for which the stability test is met.

Example 4.11 The Undamped Pendulum. Next we consider a nonlinear system with two equilibrium solutions, the simple pendulum, Example 4.3. Let us visualize what happens if we displace the pendulum from the equilibrium position $\bar{\theta} = \bar{\dot{\theta}} = 0$, for which the pendulum hangs vertically downward. We know intuitively that, if displaced *slightly* from equilibrium, the pendulum will swing back and forth in a periodic manner. The states θ and $\dot{\theta}$ will exhibit behavior similar to that of the states x and $\dot{x}$ of the simple oscillator. The phase portraits, therefore, will be closed curves, as indicated in Fig. 4.7.

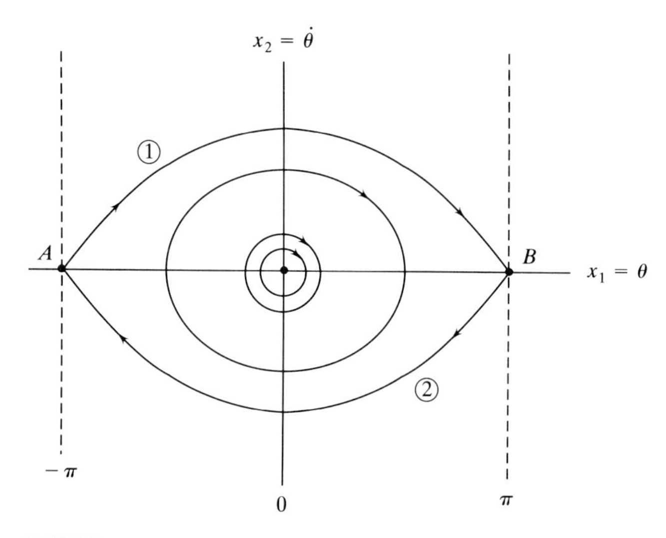

FIGURE 4.7
Phase plane orbits for the simple pendulum.

The same type of reasoning used to assess the stability of the simple oscillator leads to the conclusion that the equilibrium solution $\bar{x} = (0, 0)$ is stable, but not asymptotically stable, for the simple pendulum.

Next, consider the following situation for motions initiated near the equilibrium solution $(\pi, 0)$. Picture the pendulum hanging at rest, vertically upward, and give it an infinitesimal push, barely sufficient to get it moving. The pendulum will then swing all the way around and come back to rest at the equilibrium position, taking a theoretically infinite time to do so. This is indicated by the curves 1 and 2 in Fig. 4.7. Note that the equilibrium points A and B shown correspond physically to the same solution, because $\theta = -\pi$ and $\theta = \pi$ represent identical physical states.*

Now if we apply the stability test to motions initiated near $\bar{x} = (\pi, 0)$ we observe that for any neighborhood N which is small, all motions initiated in N, and hence in any $N_1 \subset N$, will eventually move outside N. This equilibrium solution is, therefore, unstable.

Note that for this system, as for the simple oscillator, the exact phase portraits can be drawn by utilizing the conservation of energy, which here states that

$$\tfrac{1}{2}mL^2\dot{\theta}^2 + mgL(1 - \cos\theta) = E = \text{constant} \tag{4.24}$$

Observe that the phase portraits are not all closed circles, as they are for the linear oscillator, because of the nonlinearity of the system.

*For this system the "natural" phase plane is the surface of the *cylinder* constructed by rotating the diagram shown about the vertical in such a way that the vertical dashed lines passing through A and B coincide.

Example 4.12 The Rolling Ball. As a third example of a system for which the phase portraits can be sketched without much analysis, consider the ball rolling without friction in the double well, as shown in Fig. 4.8.

The three equilibrium solutions were found to be $\bar{x} = (-1, 0)$, $(0, 0)$ and $(1, 0)$. These equilibrium positions are denoted as points 1, 2, and 3, respectively, in Fig. 4.8. We will first outline the construction of the phase portraits and then use these phase portraits to assess the stability of each of the three equilibrium solutions. This is done in a fair amount of detail and it is suggested that obtaining a clear understanding of this example is worth the effort.

To start, consider motions initiated *near* equilibrium $3(1, 0)$ at the bottom of the right-hand well. The ball will roll back and forth in a periodic fashion, and motions will be represented by closed curves which surround the equilibrium point in the phase plane. A similar situation occurs for motions initiated near equilibrium $1(-1, 0)$, at the bottom of the left-hand well. These closed orbits surrounding equilibria 1 and 3 are illustrated in Fig. 4.9.

Next, consider a motion initiated from equilibrium $2(0, 0)$ in the following way: Suppose the mass m to be sitting statically at equilibrium 2 and give it an infinitesimal push in the $+x$ direction to get it going. The mass will travel through the right-hand well and come to rest at point B (Fig. 4.8); then it will travel back through the well and come back to rest at equilibrium 2. During this motion the maximum velocity $\dot{x}$ will occur at the bottom of the well ($x = 1$), at which all of the energy is kinetic. This motion is indicated by curve 1 in the phase portrait (Fig. 4.9). Note that this orbit starts and stops at the same location (the equilibrium solution) in phase space. A similar situation occurs if the mass starts at rest at equilibrium 2 and is given an infinitesimal push in the $-x$ direction; we obtain orbit 2 shown in Fig. 4.9. Make sure the directions of the orbits are clearly understood.

Finally, let's consider an orbit initiated from rest at the point A (Fig. 4.8). In this case, because point A is higher than equilibrium 2, the mass will reach equilibrium 2 with a positive velocity and will continue on through the right-hand well, reaching the point A', at which it will come to rest. Then the mass will travel back to A, completing a cycle, which is continually repeated. This motion is indicated by orbit 3 in Fig. 4.9. The phase plane behavior is pretty interesting. We find a family of closed orbits inside orbit 1, another family inside orbit 2, and a third family outside orbits 1 and 2.

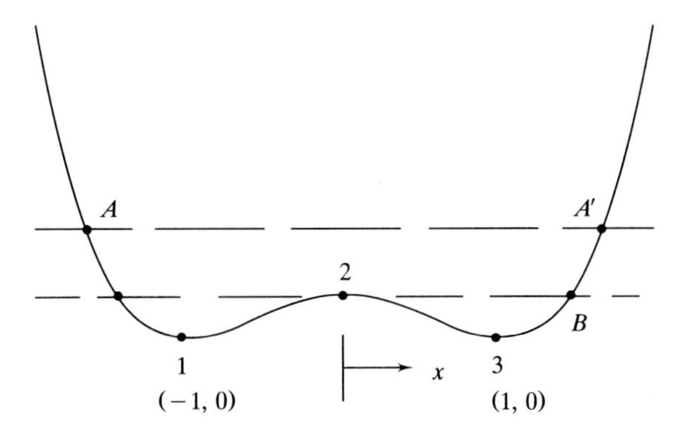

FIGURE 4.8
The rolling ball in a frictionless double well. Equilibrium positions $\bar{x}$ are denoted by the points 1, 2, and 3.

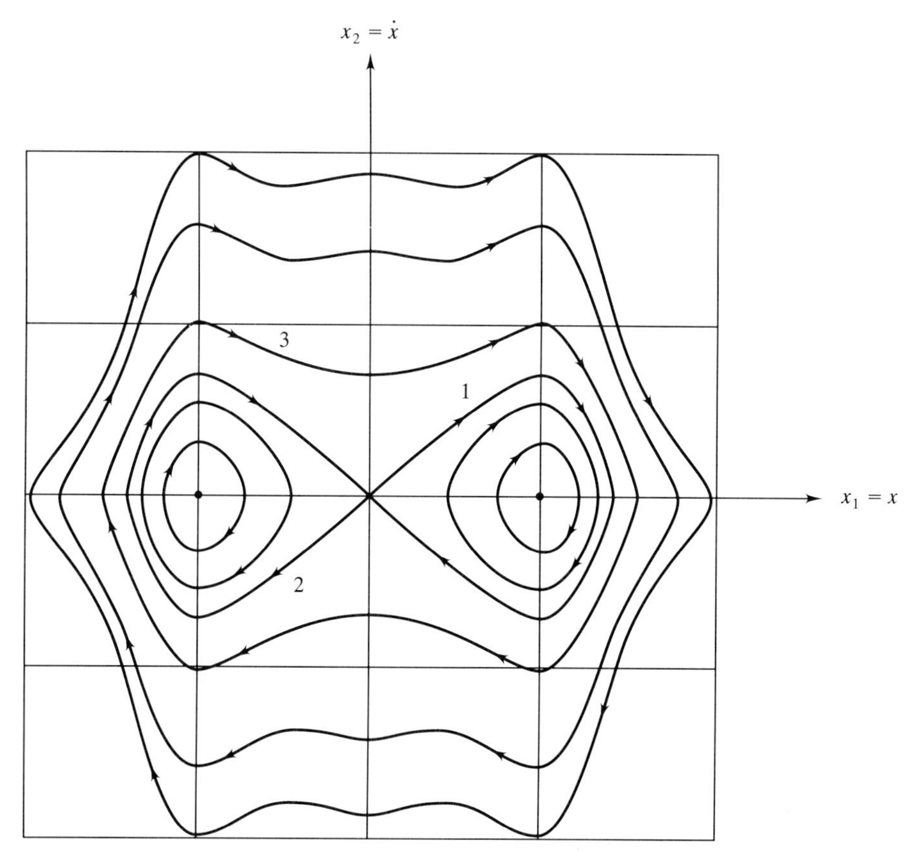

FIGURE 4.9
Phase portraits for the rolling ball (see text for discussion).

We now have to determine the stability of each of the three equilibria. First, the periodic motions near equilibria 1 and 3 are analogous to those of the preceding two examples. We can see that for *any* neighborhood N surrounding either of these equilibria, it is possible to find neighborhoods $N_1 \subset N$ such that motions initiated in N_1 will remain in N forever. Equilibria 1 and 3 are, therefore, stable but not asymptotically stable. The behavior near equilibrium 2 is seen to be similar to that of the pendulum in the vertically upward position. For some neighborhoods N, motions initiated inside N and hence in any N_1 we define, will eventually leave N, so equilibrium 2 is unstable.

It is worthwhile to investigate further the motions near equilibrium 2 from the standpoint of the formal definition of stability. Suppose that we select the (large) neighborhood N of equilibrium as shown in Fig. 4.10. Then it should be clear that we can find some (small) neighborhood N_1, one of which is shown, such that all orbits starting in N_1 will remain in N forever. Thus, the stability test is passed for this particular N. The key to the definition, however, is that the stability test must be passed for *all* N, no matter how small. For example, consider the (small) neighborhood N shown in Fig. 4.11. It is clear now that for any $N_1 \subset N$,

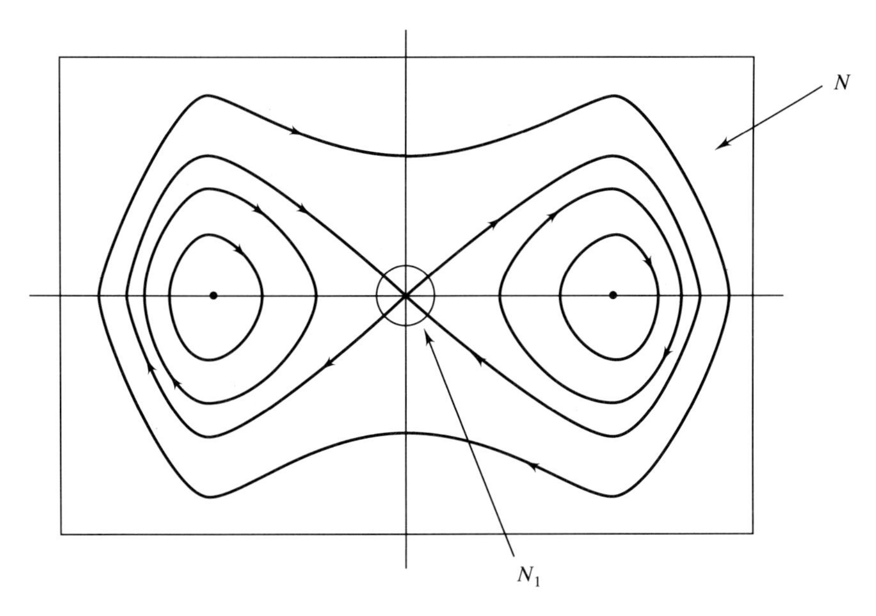

FIGURE 4.10
Stability test for the equilibrium $\bar{x} = (0, 0)$ for the rolling ball for the case in which the neighborhood N is large.

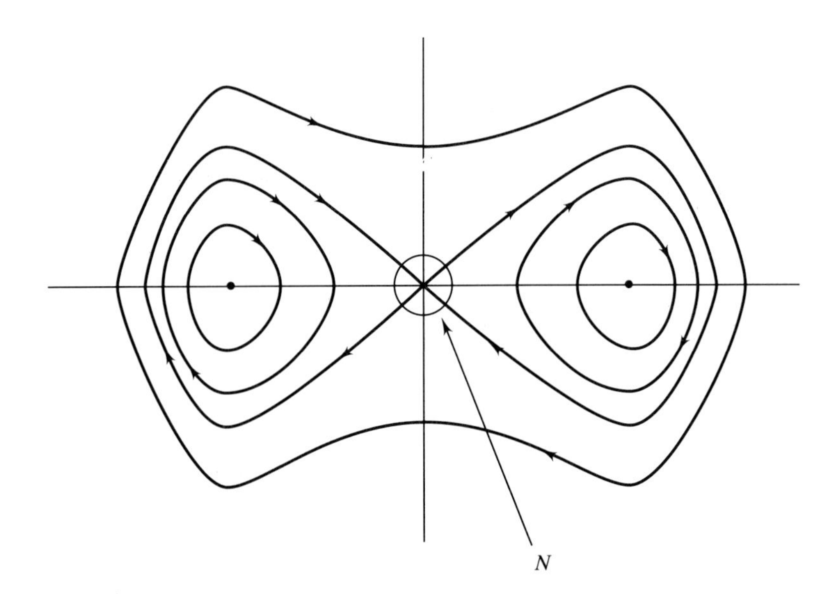

FIGURE 4.11
Stability test for the equilibrium $\bar{x} = (0, 0)$ for the rolling ball for the case in which the neighborhood N is small.

some orbits initiated in N_1 will leave N. In fact, for the situation shown in Fig. 4.11, almost all orbits originating in any N_1 we pick will leave N eventually.

Note that for this system we could, if we wished, have again used conservation of energy to actually construct the phase plane orbits. The kinetic and potential energies of this system are $T = \frac{1}{2}m(\dot{x}^2 + \dot{y}^2)$ and $V = mgy$, so the total energy E, which is constant, is

$$E = \tfrac{1}{2}m\left(\dot{x}^2 + \dot{y}^2\right) + mgy$$

In view of (4.14), which defines y as a function of x, this energy E can be expressed in terms of only x and $\dot{x}$ as

$$E = \tfrac{1}{2}m\left[\dot{x}^2 + \left(-x\dot{x} + x^3\dot{x}\right)^2\right] + mg\left(-\tfrac{1}{2}x^2 + \tfrac{1}{4}x^4\right) \tag{4.25}$$

This result, although a bit messy, will reveal all of the features shown in Fig. 4.9.

In each of the preceding examples we were able to construct the phase plane orbits, using conservation of energy to obtain the exact orbits, or using physical reasoning to sketch qualitatively what the orbits looked like. By investigating the nature of these orbits and invoking the formal definition of stability of equilibrium, we were than able to classify the stability of each equilibrium solution of the system. It is important to note that these examples were selected only to illustrate how the formal definition of stability applies in specific cases. In many situations confronting the engineer in practice, it is difficult, if not impossible, to tell by intuition alone what the phase portraits are going to look like, especially for systems of order 3 or higher.

For example, consider the air cushion landing system Example 3.5. We have seen that at equilibrium the cushion pressure force exactly balances the weight of the vehicle, and the vehicle elevation y is such as to balance the air flows in the system. If we were to picture moving the vehicle to some distance Δy above its equilibrium elevation, while holding the cushion pressure at its equilibrium level $\bar{p}_c$, we could attempt to guess intuitively what to expect in the phase space. Here, however, the phase space is three dimensional, the physical effects are reasonably complex, and our attempts to deduce intuitively the stability of equilibrium will fail.

What we need, then, is a rigorous, quantitative approach which allows us to investigate stability properties in a systematic way. The process by which this is done is *linearization*, followed by analysis of the resulting linearized system model. What we do is to replace the actual, generally nonlinear, ODE model by a system of linear ODEs which describe the dynamic behavior for motions *near* equilibrium. In applying this procedure, we have to be able to do three things: (1) perform the linearization, (2) solve the resulting simultaneous linear ODEs, and (3) interpret the solutions from the standpoint of stability.* These important areas are addressed in Secs. 4.1.4 and 4.2.

*Of course, if the system mathematical model is linear to begin with, as is often the case, we need only be able to solve the linear ODEs and to interpret the results.

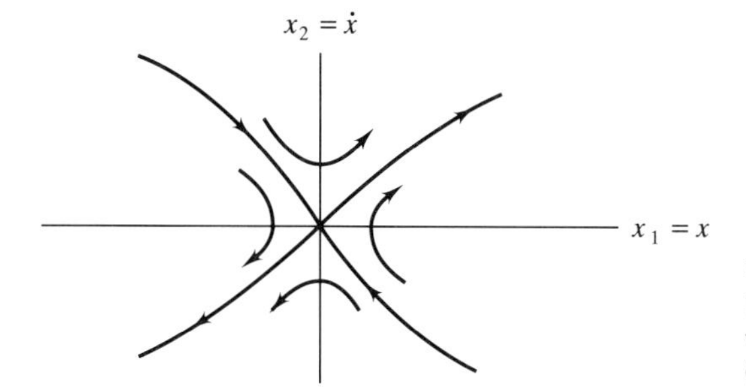

FIGURE 4.12
Phase portrait of the rolling ball in the vicinity of the unstable equilibrium at (0, 0).

Before proceeding to the analysis via linearization, several comments related to the three examples considered here are worthwhile:

1. In each example we found one or more stable equilibria. These equilibria were, however, not *asymptotically* stable. This is due to the absence of any damping in the system. In order that asymptotically stable equilibria exist, there must be some type of dissipation mechanism.
2. In the preceding examples we encountered two types of phase plane motions near equilibrium: (a) closed orbits, which surround equilibrium and signify periodic behavior, and which imply stable, but not asymptotically stable, behavior, and (b) orbits which *intersect* the equilibrium solution as, for example, the motion near equilibrium 2 for the rolling ball, which is shown in more detail in Fig. 4.12. One pair of the intersecting orbits moves *to* equilibrium, while the other pair moves *away* from equilibrium. As we will see, the fact that a pair of equilibrium-intersecting orbits moves away from equilibrium signals the instability of that equilibrium solution.
3. The student is cautioned that the two types of near equilibrium behavior illustrated in the examples are not the only possibilities. In Sec. 4.2 we will explore a number of other types of near equilibrium behavior.
4. The stability of a given equilibrium solution is considered to be a basic property of the system, and, therefore, is unrelated to any external, time-dependent effects (inputs).
5. A note on terminology: practicing engineers often use the term "neutrally stable" to denote a stable, but not asymptotically stable, equilibrium. Also, practicing engineers often use the term "stable" to denote what we have defined here as asymptotically stable. The particular terms used are not important as long as the meaning is clear.

4.1.4 Linearization

We have seen that a key to the definition of *local stability* is what happens in phase space if motions are initiated in arbitrarily small neighborhoods surrounding equilibrium. In order to analyze these near-equilibrium motions, we will use the process of linearization. We suppose that the n-state mathematical model under study is in state variable form

$$\dot{x}_i = f_i(x_1, \ldots, x_n) \qquad i = i, \ldots, n \tag{4.26}$$

that we have found the equilibrium solutions, for which $f_i(\bar{x}_1, \ldots, \bar{x}_n) = 0$, and that we seek to study the motions of the system in the vicinity of each equilibrium solution. The process of linearization involves replacement of any nonlinear functions appearing in the f_i of (4.26) by linear functions that mathematically approach the nonlinear functions as the states approach equilibrium. We will consider first the linearization of single-state models and then extend the analysis to the general, multistate model defined by (4.26).

SINGLE-STATE SYSTEMS. The autonomous mathematical model of a single-state system is

$$\dot{x} = f(x) \tag{4.27}$$

where the unsubscripted variable x is the single state. Suppose that we have found one or more equilibrium solutions $\bar{x}$, defined by $f(\bar{x}) = 0$. We can conveniently represent the single-state problem in graphical form by plotting $f(x)$ versus x, for which a typical graph might be as shown in Fig. 4.13.

The equilibrium solutions $\bar{x}$ are those points at which $f(x)$ crosses the x axis, since $f(x) = 0$ at such points. For the function $f(x)$ shown in Fig. 4.13, there are two equilibrium solutions, labeled A and B. We will focus attention on equilibrium solution A, denoted as $\bar{x}_A$. In order to investigate motions initiated near $\bar{x}_A$ we first introduce a "state deviation" ξ, which measures the

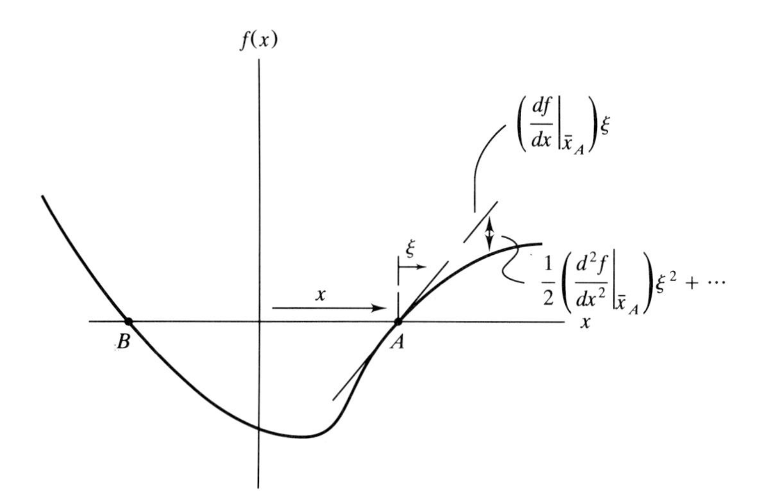

FIGURE 4.13
The hypothetical variation of $f(x)$ in (4.27) with x.

deviation of the state x from its equilibrium value:

$$x \equiv \bar{x} + \xi \tag{4.28}$$

This definition is merely a coordinate translation, as indicated in Fig. 4.13. In terms of the state deviation ξ the mathematical model (4.27) becomes

$$\dot{\bar{x}} + \dot{\xi} = f(\bar{x} + \xi) \tag{4.29}$$

where $\dot{\bar{x}} \equiv 0$ because the equilibrium solution is by definition a constant. Since we want to investigate behavior for small ξ, let us expand (4.29) in a Taylor series about the equilibrium solution $\bar{x}$. This converts (4.29) to

$$\dot{\xi} = f(\bar{x}) + \left.\frac{df}{dx}\right|_{\bar{x}} \xi + \frac{1}{2}\left.\frac{d^2 f}{dx^2}\right|_{\bar{x}} \xi^2 + \cdots \tag{4.30}$$

The first term $f(\bar{x})$ on the right-hand side drops out, as $f(\bar{x}) \equiv 0$ at equilibrium. Note that the derivatives appearing in (4.30) are to be evaluated *at* equilibrium and hence are known constants, since the equilibrium solution $\bar{x}$ is assumed known. The next step in the linearization process is to invoke the smallness condition on the state deviation ξ. Since ξ is arbitrarily small, the nonlinear terms can be considered negligible in comparison to the linear term, so for small motions near equilibrium, (4.30) reduces to

$$\dot{\xi} = a\xi \tag{4.31}$$

where $a \equiv (df/dx)_{\bar{x}}$. Observe that what we have done in going from (4.27) to (4.31) is to replace the actual nonlinear function $f(x)$, in the vicinity of equilibrium, by a straight line approximation, as indicated in Fig. 4.13. By comparing the graphs of the actual function $f(x)$ and the linear approximation near equilibrium, we can obtain some idea of the region of validity of the linear model.

The stability of equilibrium can also be deduced directly from Fig. 4.13. Picture a motion initiated at an initial value $x_0 > \bar{x}_A$, so that $\xi_0 > 0$. Because $f(x_0)$ is positive, x (and ξ) will tend to *increase*, so that x is driven further from equilibrium, in the $+x$ direction. On the other hand, if $x_0 < \bar{x}_A$, so that $\xi_0 < 0$, $f(x_0)$ will be negative, and x will tend to decrease, moving further from equilibrium in the $-x$ direction. Thus, any orbit* initiated near $\bar{x}_A$ will tend to move further away from $\bar{x}_A$, so $\bar{x}_A$ is an unstable equilibrium.

The behavior of the system near $\bar{x}_A$ can be determined quantitatively by solving (4.31), which is a first order, linear homogeneous differential equation with constant coefficients. The student may verify that the solution to (4.31), for an initial deviation from equilibrium ξ_0, is

$$\xi(t) = \xi_0 e^{at} \tag{4.32}$$

*The orbits here are one dimensional; the phase space is the x axis.

This solution shows that, because $a = (df/dx)_{\bar{x}_A} > 0$, motions initiated near $\bar{x}_A$ will tend to move away from $\bar{x}_A$ exponentially, with the exponential rate a.

If we do the same type of analysis for the second equilibrium solution $\bar{x}_B$, we would find $\bar{x}_B$ to be asymptotically stable, as the derivative $(df/dx)_{\bar{x}_B} < 0$, so a in (4.32) would be negative. This indicates that orbits initiated near $\bar{x}_B$ will be driven exponentially *toward* $\bar{x}_B$. We conclude that for a single-state system the stability of equilibrium is determined by the slope $a = (df/dx)_{\bar{x}}$; if $a < 0$, then $\bar{x}$ is asymptotically stable, and if $a > 0$, $\bar{x}$ is unstable. Furthermore, if $a \equiv 0$, then the solution (4.32) says that, if we start a motion at ξ_0, the system, according to the linear model, will stay at ξ_0 forever. In this case the equilibrium $\bar{x}$ is merely stable, according to the linear model.

Let us now consider a couple of examples.

Example 4.13 The Single-Species Population Model. The mathematical model for this single-state system was stated in Example 4.4 as

$$\dot{x} = \alpha x - \beta x^2 \qquad \alpha, \beta > 0 \tag{4.33}$$

The mathematical model is nonlinear, and the two equilibrium solutions are $\bar{x} = 0$ and $\bar{x} = \alpha/\beta$. We seek linearized models to describe behavior near each equilibrium solution. These linear models are obtained by first calculating df/dx, which here is

$$\frac{df}{dx} = \alpha - 2\beta x$$

so that the constant a appearing in (4.31) is

$$a = \left.\frac{df}{dx}\right|_{\bar{x}} = \alpha - 2\beta\bar{x}$$

with $\bar{x} = 0$ or $\bar{x} = \alpha/\beta$. When $\bar{x} = 0, a = \alpha$, whereas when $\bar{x} = \alpha/\beta, a = -\alpha$. We conclude that orbits initiated near $\bar{x} = 0$ will move exponentially away from equilibrium, which is, therefore, unstable. Orbits initiated near $\bar{x} = \alpha/\beta$ will move exponentially toward equilibrium, which is asymptotically stable. We can deduce these types of behaviors by graphing $f(x)$ versus x, which is shown in Fig. 4.14.

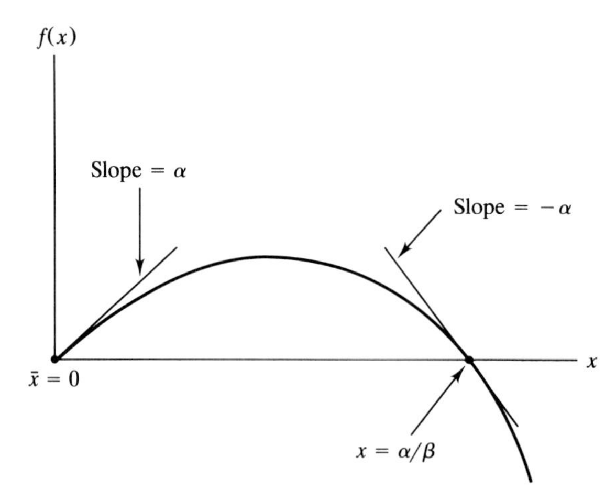

FIGURE 4.14
Graph of $f(x)$ versus x for the single-species population model (4.33).

Example 4.14 The Water Tank. The nonlinear mathematical model governing the height z of water in Example 4.5 was derived in Example 3.1:

$$\dot{z} = \frac{q_0}{A\rho} - \frac{A_0}{A}(2gz)^{1/2} \tag{4.34}$$

and the equilibrium water level $\bar{z}$ was

$$\bar{z} = \frac{1}{2g}\left(\frac{q_0}{\rho A_0}\right)^2 \tag{4.35}$$

In order to obtain the linearized model that governs the water level for small deviations from equilibrium, we compute the derivative df/dz:

$$\frac{df}{dz} = -\frac{A_0 g}{A}(2gz)^{-1/2}$$

Evaluating this derivative at the equilibrium water level $\bar{z}$,

$$a = \left.\frac{df}{dz}\right|_{\bar{z}} = -\frac{A_0^2 g\rho}{Aq_0} \tag{4.36}$$

Because $a < 0$ for this system, the solution (4.32) shows that, for any initial deviation ξ_0 from the equilibrium level $\bar{z}, \xi(t) \to 0$, so that $z(t) \to \bar{z}$, as $t \to \infty$, and we have asymptotic stability. This result makes sense on physical grounds. If the water level starts out slightly above $\bar{z}$, so that $\xi_0 > 0$, then the flow through the orifice A_0 will exceed the constant inflow q_0, and the water level will drop. If, on the other hand, the water level starts out slightly below $\bar{z}(\xi_0 < 0)$, then the inflow q_0 exceeds the flow through A_0, and the water level will rise. Eventually, the equilibrium level will be reached asymptotically for any starting condition near $\bar{z}$. An advantage of having the solution (4.32) to the linear model is that it tells us how fast equilibrium is approached. The graph of $f(z)$ versus z is shown in Fig. 4.15 to further illustrate the behavior.

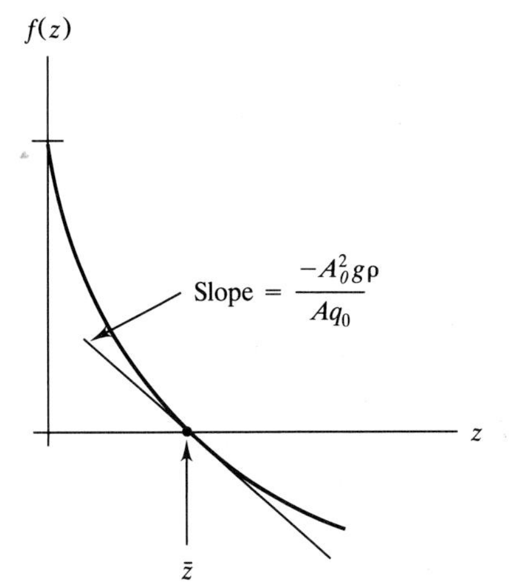

FIGURE 4.15
Graph of $f(z)$ versus z for the water tank model (4.34).

Comments

1. For the single-state system considered in this subsection, the linearization involved calculation of the derivative $(df/dx)|_{\bar{x}}$ to determine the linearized model (4.31). It is also possible to achieve an equivalent formulation using a slightly different sequence of steps. In this alternate procedure we simply plug (4.28) directly into the system equation (4.27), work out the algebra, and delete all nonlinear terms. To illustrate this procedure, consider again the single-species population model (4.33). Let $x = \bar{x} + \xi$ and plug this into (4.33) to obtain

$$\cancel{\dot{\bar{x}}} + \dot{\xi} = \alpha(\bar{x} + \xi) - \beta(\bar{x} + \xi)^2$$

Working out the algebra and moving some terms around enable us to write the above relation as

$$\dot{\xi} = \overset{1}{\left(\alpha\bar{x} - \beta\bar{x}^2\right)} + \overset{2}{\xi(\alpha - 2\beta\bar{x})} - \overset{3}{\beta\xi^2} \tag{4.37}$$

Let us compare the right-hand side of (4.37) to the general form (4.30) that is defined by the Taylor expansion. The term labeled 1 in (4.37) is just $f(\bar{x})$, the first term on the right-hand side of (4.30), which is identically zero. The second term (2), which is linear in the state deviation ξ, is the term $(df/dx)_{\bar{x}}\xi$ appearing in (4.30). Term 3 in (4.37) is analogous to the higher-order terms in the Taylor series expansion (4.30). We see, then, that the linearized system model (4.31) is obtained from (4.37) by deleting terms 1 and 3 in (4.37). We retain only the linear term 2 to obtain the linearized model

$$\dot{\xi} = (\alpha - 2\beta\bar{x})\xi$$

This result is identical to that obtained in Example 4.13. We have performed all of the steps leading to (4.31), but we have done so without formal use of the Taylor series. Both of these procedures are used in practice, and each should be familiar to the student, who should utilize whichever procedure she finds more convenient in a given situation.

2. For a single-state system consider the motion near an unstable equilibrium, $a > 0$ in (4.31). The solution to the linearized model exhibits exponential movement away from equilibrium. After some time, *according to the linear model*, the state deviation ξ will become arbitrarily large. When this happens, the linear model is no longer valid because the neglected terms in the Taylor series (4.30) are no longer small enough to ignore. The linear model has broken down. For motions initiated near an unstable equilibrium, then, the linear model only tells us the initial *tendency* of the system behavior, which is to move exponentially away from equilibrium. The linear model does not allow us to predict what actually happens because it is inherently inaccurate far from equilibrium.

   On the other hand, consider the motion near an asymptotically *stable* equilibrium solution. All motions started near equilibrium will approach equilibrium exponentially, and the linear model remains valid. It describes accurately how the equilibrium condition is approached because the system is always close to equilibrium.

3. The preceding discussion leads naturally to an important practical consideration. This is related to our use of the terms *near equilibrium*, which implies that a linear model is valid, and *far from equilibrium*, which implies that a linear model is not valid. For the engineering analysis of specific systems, we would like to know how far from equilibrium the system can move before we expect a breakdown of the linear model. The way to attack this problem is to compare the actual nonlinear functions appearing in (4.26) with the linear functions used to replace them and to identify a region of the phase space surrounding equilibrium for which the two are close to each other. This is a somewhat subjective process that depends on how we view the phrase *close to each other*. Nevertheless, we can obtain some idea of what to expect regarding the validity of the linear model.
4. For the single-state systems considered in this subsection, it is often possible to solve the autonomous nonlinear, first-order ODE (4.27) exactly. For example, the basic single-state mathematical model (4.27) can be written in differential form as

$$\frac{dx}{f(x)} = dt \tag{4.38}$$

If the initial value of x at $t = 0$ is x_0, then the formal solution to (4.38) is

$$\int_{x_0}^{x} \frac{dx'}{f(x')} = t \tag{4.39}$$

where x' is a dummy variable. If we can perform the integration, perhaps with the aid of integral tables or a symbolic manipulator, then we obtain the exact solution. If we have an exact solution, there is no reason to worry about linearization because the exact solution describes what happens whether close to, or far from, equilibrium. Unfortunately, the existence of exact solutions for autonomous, nonlinear systems of order 2 or greater is rare. For higher-order systems analysis, we usually have to rely on linearization to determine system stability and behavior near equilibrium.*

The purpose of this subsection has been to introduce the basic ideas of linearization by considering a relatively simple class of problems: the single-state nonlinear system. Most engineering dynamic systems of practical importance are of order 2 or greater, and we now turn to the analysis of such systems. The basic ideas regarding linearization are the same.

MULTISTATE SYSTEMS. Assume we have an n-state autonomous system model,

$$\dot{x}_i = f_i(x_1, \ldots, x_n) \qquad i = 1, \ldots, n \tag{4.40}$$

and that we have found one or more equilibrium solutions $\bar{x} = (\bar{x}_1, \ldots, \bar{x}_n)$ from the equilibrium conditions $f_i(\bar{x}_1, \ldots, \bar{x}_n) = 0$. In order to obtain linearized

*An alternative is to find "exact" solutions via numerical integration of the system equations. This topic is considered in Chap. 9.

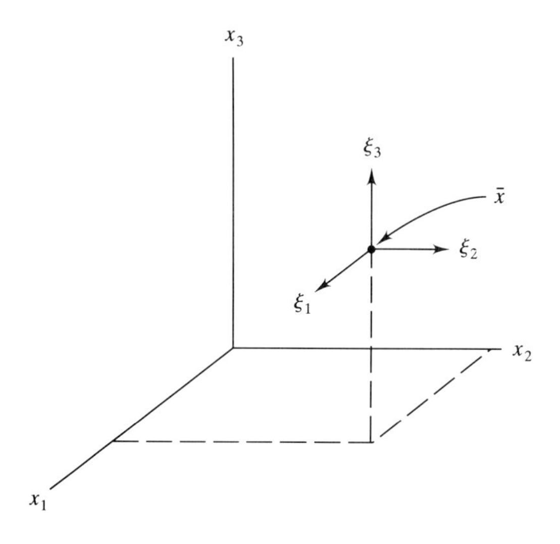

FIGURE 4.16
Definition of the state deviations from equilibrium (ξ_1, ξ_2, ξ_3) for a three-state system.

mathematical models describing system behavior near any of the various equilibria, we introduce the state deviations ξ_i as

$$x_i = \bar{x}_i + \xi_i \qquad i = 1, \ldots, n \tag{4.41}$$

where ξ_i is a small deviation in the ith-state x_i from its equilibrium value $\bar{x}_i$. Equation (4.41) is a translation of coordinates, as shown in Fig. 4.16 for a three-state system. At equilibrium each $\xi_i = 0$. We again employ the Taylor series in order to obtain formally the linearized model. Let us begin by plugging (4.41) into the first of the n-state equations (4.40), for x_1, to obtain

$$\cancel{\dot{\bar{x}}_1} + \dot{\xi}_1 = f_1(\bar{x}_1 + \xi_1, \bar{x}_2 + \xi_2, \ldots, \bar{x}_n + \xi_n) \tag{4.42}$$

Now we expand (4.42) in a Taylor series about the equilibrium solution $\bar{x} = (\bar{x}_1, \ldots, \bar{x}_n)$. Because $f_1(\cdot)$ may be a function of all n states, we need to use the form of the Taylor series appropriate for a function of more than one variable. This leads to the following Taylor expansion of (4.42):

$$\dot{\xi}_1 = f_1(\bar{x}_1, \ldots, \bar{x}_n) + \sum_{i=1}^{n} \left. \frac{\partial f_1}{\partial x_i} \right|_{\bar{x}} \xi_i + \frac{1}{2} \sum_{i=1}^{n} \sum_{j=1}^{n} \left. \frac{\partial^2 f_1}{\partial x_i \, \partial x_j} \right|_{\bar{x}} \xi_i \xi_j + \cdots \tag{4.43}$$

The result, although more complex, is analogous to the single-state version (4.30). All partial derivatives appearing in (4.43) are to be evaluated at equilibrium. The first term $f_1(\bar{x}_1, \ldots, \bar{x}_n)$ in (4.43) drops out from the definition of equilibrium. We also delete the nonlinear terms that appear in the double summation and beyond, as the ξ_i are arbitrarily small. The result is a linear model for the first state deviation ξ_1, as follows:

$$\dot{\xi}_1 = a_{11}\xi_1 + a_{12}\xi_2 + \cdots + a_{1n}\xi_n \tag{4.44}$$

where the constants a_{1j} are the partial derivatives $\partial f_1/\partial x_j$, evaluated at equilibrium:

$$a_{1j} = \left.\frac{\partial f_1}{\partial x_j}\right|_{\bar{x}} \tag{4.45}$$

Now if we apply the Taylor series expansion to each of the original state equations (4.40), we obtain a set of n linear autonomous first-order ODEs for the state deviations ξ_i:

$$\dot{\xi}_1 = a_{11}\xi_1 + a_{12}\xi_2 + \cdots + a_{1n}\xi_n \tag{4.46a}$$

$$\dot{\xi}_2 = a_{21}\xi_1 + a_{22}\xi_2 + \cdots + a_{2n}\xi_n \tag{4.46b}$$

$$\vdots$$

$$\dot{\xi}_n = a_{n1}\xi_1 + a_{n2}\xi_2 + \cdots + a_{nn}\xi_n \tag{4.46c}$$

where now the array of constants a_{ij} are defined by

$$a_{ij} = \left.\frac{\partial f_i}{\partial x_j}\right|_{\bar{x}} \tag{4.47}$$

Our linearized mathematical model (4.46) can be more compactly represented by using vector-matrix notation. First, define the collection of the n-state deviations ξ_i as the "state vector" $\{\xi\}$:

$$\{\xi\} = \begin{Bmatrix} \xi_1 \\ \xi_2 \\ \vdots \\ \xi_n \end{Bmatrix}$$

The braces denote a vector quantity (a vector is considered here to be a matrix with a single column). We next form from the n^2 constants a_{ij} the matrix $[A]$ defined by

$$[A] = \begin{bmatrix} a_{11} & a_{12} & \cdots & a_{1n} \\ a_{21} & a_{22} & \cdots & a_{2n} \\ \vdots & \vdots & & \vdots \\ a_{n1} & a_{n2} & \cdots & a_{nn} \end{bmatrix}$$

This matrix, consisting of the partial derivatives of the $f_i(\cdot)$ in (4.40), evaluated at equilibrium, is usually referred to as the *Jacobian matrix* of the equilibrium solution. It defines all of the basic properties of the linearized model and replaces the single constant $a = (df/dx)_{\bar{x}}$ which appears in the single-state linearized model (4.31). The set of ODEs (4.46) can now be expressed in vector-matrix notation as

$$\{\dot{\xi}\} = [A]\{\xi\} \tag{4.48}$$

where

$$\{\dot{\xi}\} = \begin{Bmatrix} \dot{\xi}_1 \\ \vdots \\ \dot{\xi}_n \end{Bmatrix}$$

In (4.48) the top row reproduces (4.46a), the second row reproduces (4.46b), and so on. The n, first-order linear ODEs defined by (4.48) are to replace the actual, generally nonlinear model (4.40) in the vicinity of equilibrium.

Most of Sec. 4.2 will be devoted to the study of (4.48). Our objective there will be to uncover the nature of near-equilibrium orbits in the phase space and to determine the stability of equilibrium through the analysis of the linearized model (4.48).

We now consider some examples that illustrate the linearization procedure for multistate systems.

Example 4.15 The Airspring. The mathematical model for the airspring with no dissipation was derived in Example 3.4 as

$$m\ddot{x} + Ap_0\left[\left(1 - \frac{x}{d}\right)^{-\gamma} - \left(1 + \frac{x}{d}\right)^{-\gamma}\right] = 0 \tag{4.49}$$

In state variable form, with $x_1 = x$ and $x_2 = \dot{x}$, the mathematical model is

$$\dot{x}_1 = x_2$$

$$\dot{x}_2 = -\frac{Ap_0}{m}\left[\left(1 - \frac{x_1}{d}\right)^{-\gamma} - \left(1 + \frac{x_1}{d}\right)^{-\gamma}\right] \tag{4.50}$$

For this nonlinear two-state system, there is a single equilibrium solution $\bar{x} = (\bar{x}_1, \bar{x}_2) = (0, 0)$. The linearized model is obtained by calculating the four partial derivatives $\partial f_i/\partial x_j$ and evaluating them at equilibrium to obtain the Jacobian matrix $[A]$. The required partial derivatives are listed below:

$$\frac{\partial f_1}{\partial x_1} = 0$$

$$\frac{\partial f_1}{\partial x_2} = 1$$

$$\frac{\partial f_2}{\partial x_1} = \frac{\gamma A p_0}{m}\left[-\frac{1}{d}\left(1 - \frac{x_1}{d}\right)^{-\gamma-1} - \frac{1}{d}\left(1 + \frac{x_1}{d}\right)^{-\gamma-1}\right]$$

$$\frac{\partial f_2}{\partial x_2} = 0$$

Evaluating the above partial derivatives at equilibrium ($\bar{x}_1 = 0, \bar{x}_2 = 0$), we obtain

$$a_{11} = \left.\frac{\partial f_1}{\partial x_1}\right|_{\bar{x}} = 0$$

$$a_{12} = \left.\frac{\partial f_1}{\partial x_2}\right|_{\bar{x}} = 1$$

$$a_{21} = \left.\frac{\partial f_2}{\partial x_1}\right|_{\bar{x}} = -\frac{2\gamma A p_0}{md}$$

$$a_{22} = \left.\frac{\partial f_2}{\partial x_2}\right|_{\bar{x}} = 0$$

The state equations for the deviations ξ_1 and ξ_2 may thus be expressed in the form (4.48) as

$$\begin{Bmatrix} \dot{\xi}_1 \\ \dot{\xi}_2 \end{Bmatrix} = \begin{bmatrix} 0 & 1 \\ -\dfrac{2\gamma A p_0}{md} & 0 \end{bmatrix} \begin{Bmatrix} \xi_1 \\ \xi_2 \end{Bmatrix} \tag{4.51}$$

It is also possible to obtain the linearized model (4.51) by direct substitution of (4.41) into the original state equations (4.50), as we have done for the single-state system (Comment 1, following Example 4.14). In the present example direct substitution of $x_1 = \bar{x}_1 + \xi_1, x_2 = \bar{x}_2 + \xi_2$ into (4.50), with $\bar{x}_1 = \bar{x}_2 = 0$, leads to

$$\cancel{\dot{\bar{x}}_1} + \dot{\xi}_1 = \cancel{\bar{x}_2} + \xi_2 \tag{4.52a}$$

$$\cancel{\dot{\bar{x}}_2} + \dot{\xi}_2 = -\frac{Ap_0}{m}\left[\left(1 - \frac{\xi_1}{d}\right)^{-\gamma} - \left(1 + \frac{\xi_1}{d}\right)^{-\gamma}\right] \tag{4.52b}$$

The next step is to expand in power series the nonlinear terms $(1 - \xi_1/d)^{-\gamma}$ and $(1 + \xi_1/d)^{-\gamma}$ appearing in the second-state equation. To do this, we can make use of the following very useful power series relation, which is commonly used in linearization:

$$(1 \pm z)^n \cong 1 \pm nz + \cdots \tag{4.53}$$

where $|z|$ is small compared to unity, n is any exponent (positive or negative), and the $(+ \cdots)$ indicates a series of terms of degree 2 or higher in z, that is, nonlinear terms. Identifying ξ_1/d with z, (4.53) yields the following linearized forms, valid for $|\xi_1/d|$ small compared to unity:

$$\left(1 - \frac{\xi_1}{d}\right)^{-\gamma} \cong 1 + \frac{\gamma \xi_1}{d}$$

$$\left(1 + \frac{\xi_1}{d}\right)^{-\gamma} \cong 1 - \frac{\gamma \xi_1}{d}$$

Using these linearized relations in (4.52b), we obtain the following linear model for the state deviations ξ_1 and ξ_2:

$$\dot{\xi}_1 = \xi_2$$

$$\dot{\xi}_2 = -\frac{2Ap_0\gamma}{md}\xi_1 \tag{4.54}$$

which is seen to be identical to (4.51). Thus, the linear model (4.54) should describe how the two-state system moves in the vicinity of the equilibrium $\bar{x} = (0, 0)$.

Example 4.16 The System of Example 4.1. The ODE model for this two-state system was stated in (4.3) as

$$\dot{x}_1 = x_1$$

$$\dot{x}_2 = -x_2 + x_1^2$$

so that $f_1(x_1, x_2) = x_1$ and $f_2(x_1, x_2) = -x_2 + x_1^2$. The origin of the phase space is the single equilibrium solution $\bar{x} = (0, 0)$. The partial derivatives defining the system Jacobian matrix $[A]$ are as follows:

$$\frac{\partial f_1}{\partial x_1} = 1 \qquad \frac{\partial f_1}{\partial x_2} = 0$$

$$\frac{\partial f_2}{\partial x_1} = 2x_1 \qquad \frac{\partial f_2}{\partial x_2} = -1$$

Evaluating these partial derivatives at equilibrium ($\bar{x}_1 = \bar{x}_2 = 0$) leads to the linearized model

$$\begin{Bmatrix} \dot{\xi}_1 \\ \dot{\xi}_2 \end{Bmatrix} = \begin{bmatrix} 1 & 0 \\ 0 & -1 \end{bmatrix} \begin{Bmatrix} \xi_1 \\ \xi_2 \end{Bmatrix} \tag{4.55}$$

Example 4.17 Foxes and Rabbits. The mathematical model for the interacting fox and rabbit populations was previously stated in Example 4.7 to be

$$\dot{r} = \alpha r - \gamma rf = f_1(r, f)$$

$$\dot{f} = -\beta f + \delta rf = f_2(r, f) \tag{4.56}$$

Let us determine the linearized model for system motions near the nonzero equilibrium condition obtained in Example 4.7:

$$\bar{r} = \frac{\beta}{\delta}, \qquad \bar{f} = \frac{\alpha}{\gamma}$$

The partial derivatives needed to obtain the linearized model are as follows:

$$\frac{\partial f_1}{\partial r} = \alpha - \gamma f$$

$$\frac{\partial f_1}{\partial f} = -\gamma r$$

$$\frac{\partial f_2}{\partial r} = \delta f$$

$$\frac{\partial f_2}{\partial f} = -\beta + \delta r$$

Next plug in the equilibrium values of $\bar{r}$ and $\bar{f}$ to obtain the elements of the Jacobian matrix:

$$a_{11} = \alpha - \gamma\left(\frac{\alpha}{\gamma}\right) = 0$$

$$a_{12} = -\gamma\left(\frac{\beta}{\delta}\right)$$

$$a_{21} = \delta\left(\frac{\alpha}{\gamma}\right)$$

$$a_{22} = -\beta + \delta\left(\frac{\beta}{\delta}\right) = 0$$

The linearized model is, therefore,

$$\begin{Bmatrix} \dot{\xi}_1 \\ \dot{\xi}_2 \end{Bmatrix} = \begin{bmatrix} 0 & -\gamma\beta/\delta \\ \delta\alpha/\gamma & 0 \end{bmatrix} \begin{Bmatrix} \xi_1 \\ \xi_2 \end{Bmatrix} \tag{4.57}$$

where ξ_1 and ξ_2 represent, respectively, the deviations in rabbit and fox populations from the equilibrium values $\bar{r}, \bar{f}$.

Let us now do the linearization of this system using direct substitution of (4.41) into (4.56). Using $r = \bar{r} + \xi_1, f = \bar{f} + \xi_2$ in (4.56), we obtain

$$\cancel{\dot{\bar{r}}} + \dot{\xi}_1 = \alpha(\bar{r} + \xi_1) - \gamma(\bar{r} + \xi_1)(\bar{f} + \xi_2)$$

$$\cancel{\dot{\bar{f}}} + \dot{\xi}_2 = -\beta(\bar{f} + \xi_2) + \delta(\bar{r} + \xi_1)(\bar{f} + \xi_2)$$

Manipulation of the right-hand sides leads to the following form:

$$\dot{\xi}_1 = (\alpha\bar{r} - \gamma\bar{r}\bar{f}) + \xi_1(\alpha - \gamma\bar{f}) + \xi_2(-\gamma\bar{r}) - \gamma\xi_1\xi_2$$

$$\dot{\xi}_2 = (-\beta\bar{f} + \delta\bar{r}\bar{f}) + \xi_1(\delta\bar{f}) + \xi_2(-\beta + \delta\bar{r}) + \delta\xi_1\xi_2$$

The first term on the right-hand side of each equation is identically zero by the definition of equilibrium. We then delete the nonlinear terms involving the product $\xi_1\xi_2$ and utilize the equilibrium conditions $\bar{r} = \beta/\delta, \bar{f} = \alpha/\gamma$ to obtain

$$\dot{\xi}_1 = -\frac{\gamma\beta}{\delta}\xi_2$$

$$\dot{\xi}_2 = \frac{\alpha\delta}{\gamma}\xi_1$$

which is identical to the result (4.57) obtained via the partial derivative Jacobian calculation.

Having obtained the linearized model (4.57), we would next solve the simultaneous linear ODEs and examine the nature of the solutions in order to determine the stability of equilibrium. This is done in Sec. 4.2.

Comment. The general linearization procedures and examples presented in this section have all been done using the state variable form of the system mathematical model. This does not mean to imply that, in order to do linearization, the model must be in state variable form. The state variable form is merely the most convenient way to develop the general concept. Let us illustrate this point by considering a second-order system for which the linearization is done directly on the governing second-order ODE.

Example 4.18 Simple Pendulum. The equation of motion of the undamped pendulum was found to be Eq. (2.19), Example 2.1,

$$\ddot{\theta} + \frac{g}{L}\sin\theta = 0 \tag{4.58}$$

The two equilibrium solutions are $\bar{x} = (\bar{\theta}, \bar{\dot{\theta}}) = (0, 0), (\pi, 0)$. Let us first examine motions initiated near the equilibrium condition $\bar{\theta} = 0$. Put $\theta = \bar{\theta} + \xi = \xi$, where ξ is a small deviation, and plug into (4.58) to obtain

$$\ddot{\xi} + \frac{g}{L}\sin\xi = 0 \tag{4.59}$$

Now we can utilize the power (Taylor) series expansion of $\sin\xi$ to obtain

$$\ddot{\xi} + \frac{g}{L}\left(\xi - \frac{\xi^3}{6} + \frac{\xi^5}{5!} - \cdots\right) = 0$$

Ignoring the nonlinear terms, we find the linearized model to be

$$\ddot{\xi} + \frac{g}{L}\xi = 0 \tag{4.60}$$

We can if we wish convert this model to state variable form by defining $\xi_1 = \xi$, $\xi_2 = \dot{\xi}$. Notice that the mathematical form of (4.60) is the same as that for the simple linear oscillator (4.7).* This result shows that phase plane motions of the

*Each model is of the form $\ddot{u} + cu = 0$, where u is the dependent variable and c is a positive constant; (4.7) is put into this form through division of the equation by the mass m.

simple pendulum near the equilibrium $\bar{\theta} = 0$ will exhibit the same qualitative behavior as those of the linear oscillator, supporting our conclusion in Example 4.3 that the equilibrium solution $\bar{\theta} = \bar{\dot{\theta}} = 0$ is stable but not asymptotically stable.

Let us continue with the simple pendulum and obtain the linearized model for motions near the unstable equilibrium $\bar{\theta} = \pi$. Here we put $\theta = \pi + \xi$ and plug into (4.58) to obtain

$$\ddot{\xi} + \frac{g}{L}\sin(\pi + \xi) = 0$$

Now expand $\sin(\pi + \xi)$ using the trigonometric identity

$$\sin(\phi + \psi) = \sin\phi\cos\psi + \cos\phi\sin\psi$$

This leads to

$$\ddot{\xi} + \frac{g}{L}(\sin\pi\cos\xi + \cos\pi\sin\xi) = 0$$

Noting that $\sin\pi = 0, \cos\pi = -1$, we have

$$\ddot{\xi} - \frac{g}{L}\sin\xi = 0$$

which is the same as (4.59), except for the sign of the static term. The linearized model is thus

$$\ddot{\xi} - \frac{g}{L}\xi = 0 \tag{4.61}$$

The only difference between (4.60) and (4.61) is the sign of the static term, and we have to conclude that this sign has a profound effect on the stability of equilibrium.

Closing comment. The purpose of Sec. 4.1 has been to introduce the basic concepts of phase space motion, equilibrium, stability of equilibrium, and linearization of the governing ODE model. In the following section we formulate and study the solutions to the autonomous linear ODE model (4.48). Analysis of the solutions to these ODEs will provide a systematic way to study the nature of the near equilibrium phase space orbits and will also enable us to determine the stability of the equilibrium solution for which the linearized model applies.

Note that, frequently, the ODE models that engineers analyze are linear to begin with (as is the case, for example, with many of the ODE models in the examples listed in Tables 2.1 and 3.1). For such models, of course, the process of linearization described in Sec. 4.1.4 is not necessary; one simply deletes any inputs and puts the (already linear) model into state variable form (4.48). It is important to keep in mind that, when we use an ODE model that is linear to begin with, this does not mean that the system cannot exhibit nonlinear behavior. It only means that we choose to ignore nonlinear effects from the outset, generally by employing linear constitutive models to describe system components. But we must be aware that virtually any system will exhibit nonlinear behavior if the system phase space motions are sufficiently far from equilibrium.

4.2 ANALYSIS OF AUTONOMOUS LINEAR SYSTEMS

4.2.1 The Basic Solution: Eigenvalues and Eigenvectors

Let us assume that the system under consideration has n states and is governed by the mathematical model (4.5). Suppose we have found the equilibrium solutions and have obtained linearized models (4.48) for motions in the vicinity of each. Our objective now is to solve the linear ODEs (4.48) to gain an understanding of the near-equilibrium motions and of the stability of equilibrium. The linear problem to be considered, then, is

$$\{\dot{x}\} = [A]\{x\} \tag{4.62}$$

where $[A]$ is the $n \times n$ Jacobian matrix and where we have switched back to x's rather than ξ's; it is henceforth to be understood that in (4.62), $x_1, x_2, \ldots, x_n$ now represent small deviations from the equilibrium values $\bar{x}_1, \ldots, \bar{x}_n$. A given near-equilibrium orbit in phase space is initiated by supplying initial conditions:

$$\{x(0)\} = \begin{Bmatrix} x_1(0) \\ x_2(0) \\ \vdots \\ x_n(0) \end{Bmatrix} \tag{4.63}$$

If $\{x(0)\} = \{0\}$, then the system is at equilibrium.

Our goal here is to develop a systematic way to determine the stability of equilibrium by studying the properties of the solutions to (4.62). Recall that the definition of *stability* entails the study of *all* orbits initiated near equilibrium in the phase space. Thus, we will be interested in the properties of solutions to (4.62) for *all* sets of initial conditions (4.63) in some small n-volume surrounding equilibrium.

Prior to analyzing the multistate version (4.62), let us review the single-state system

$$\dot{x} = ax \qquad x(0) = x_0 \tag{4.64}$$

where a is a scalar. Because the differential equation is linear, homogeneous, and has constant coefficients, the theory of ODEs suggests that we assume a solution of exponential form:

$$x(t) = Ce^{\lambda t} \tag{4.65}$$

where C and λ are constants to be determined. If we plug (4.65) into (4.64), noting that $\dot{x} = \lambda C e^{\lambda t}$, there results

$$\lambda C e^{\lambda t} = aCe^{\lambda t}$$

or

$$Ce^{\lambda t}(\lambda - a) = 0 \tag{4.66}$$

Noting that $e^{\lambda t}$ cannot be zero for all times t, we can cancel it in (4.66) to obtain

$$C(\lambda - a) = 0 \tag{4.67}$$

One solution to (4.67) is $C = 0$, so that $x(t) = 0$. This is the trivial solution, for which the system is at equilibrium. In order that nontrivial solutions exist, we must require that $\lambda = a$. The solution to (4.64) is, therefore,

$$x(t) = Ce^{at} \tag{4.68}$$

This solution is valid for any value of the constant C, as may be verified. Now if we want the solution for a specific initial condition $x(0) = x_0$, we evaluate (4.68) at $t = 0$ to obtain

$$x(0) = x_0 = C$$

so that

$$x(t) = x_0 e^{at} \tag{4.69}$$

We have already encountered the solution (4.69) in Sec. 4.1.4, Eq. (4.32). If $a > 0$, the state x grows exponentially, if $a < 0$, the state x decays exponentially to zero, and, if $a = 0$, the state x remains at its initial value x_0 forever. The stability of equilibrium is thus determined solely by the constant a as follows: (1) If $a > 0$, equilibrium is unstable, as $x(t)$ becomes arbitrarily large for any $x_0 \neq 0$; (2) if $a < 0$, equilibrium is asymptotically stable, as $x(t) \to 0$ as $t \to \infty$ for any x_0; and (3) if $a = 0$, equilibrium is merely stable, as $x(t) = x_0$ for all x_0 and t.

The multistate problem (4.62) consists of a set of n simultaneous, linear, homogeneous ODEs with constant coefficients. This suggests that we try a solution of exponential form for each state. Assume, then, that

$$\begin{aligned} x_1(t) &= c_1 e^{\lambda t} \\ x_2(t) &= c_2 e^{\lambda t} \\ &\vdots \\ x_n(t) &= c_n e^{\lambda t} \end{aligned} \tag{4.70}$$

It is important to note that, in (4.70), the constants, c_i appearing in the solution are assumed generally to differ for the different states, but the exponential term is assumed to be the same for the different states. Thus, all of the states are assumed to be multiples of the same time function $e^{\lambda t}$. The assumed solution (4.70) can be more compactly stated by defining the n-vector $\{C\}$ of constants:

$$\{C\} = \begin{Bmatrix} c_1 \\ c_2 \\ \vdots \\ c_n \end{Bmatrix}$$

so that (4.70) may be written

$$\{x(t)\} = \{C\} e^{\lambda t} \tag{4.71}$$

At this point the n elements of $\{C\}$ and the constant λ are unknown, so we have a total of $n+1$ unknowns in the assumed solution. To proceed with the solution, we plug (4.71) into (4.62), noting that $\{\dot{x}\} = \lambda\{C\}\, e^{\lambda t}$, to obtain

$$\lambda\{C\}\, e^{\lambda t} = [A]\{C\}\, e^{\lambda t}$$

Canceling the $e^{\lambda t}$ as we did in the single-state case, transposing, and inserting the identity matrix $[I]$ between λ and $\{C\}$ result in

$$[[A] - \lambda[I]]\{C\} = \{0\} \tag{4.72}$$

which is the vector-matrix version of (4.67). The identity matrix $[I]$ has ones on the main diagonal and zeroes elsewhere and $\{0\}$ is the "zero vector," having all zero entries. The coefficient matrix in (4.72) consists of the Jacobian matrix $[A]$ with a λ subtracted from each diagonal element of $[A]$, so that in expanded form, (4.72) reads

$$\begin{bmatrix} (a_{11}-\lambda) & a_{12} & \cdots & a_{1n} \\ a_{21} & (a_{22}-\lambda) & \cdots & a_{2n} \\ \vdots & \vdots & & \vdots \\ a_{n1} & a_{n2} & \cdots & (a_{nn}-\lambda) \end{bmatrix} \begin{Bmatrix} c_1 \\ c_2 \\ \vdots \\ c_n \end{Bmatrix} = \begin{Bmatrix} 0 \\ \vdots \\ 0 \end{Bmatrix} \tag{4.73}$$

We call the parameter λ an *eigenvalue*, which in general is any initially unknown constant introduced into a solution, which must later be chosen in a special way. We see that the assumed solution (4.71) has converted the original set of homogeneous linear ordinary *differential equations* (4.62) into a set of homogeneous, linear *algebraic equations*. This is nice, because generally algebraic equations are easier to deal with than are differential equations.

Equation (4.73) is a statement of the ubiquitous "algebraic eigenvalue problem," which describes many phenomena in science and engineering. A few of these familiar to the student include the problem of finding the principal stresses in a loaded solid and the problem of finding the principal moments of inertia and principal axes of a rigid body. Furthermore, problems in differential equations that lead to eigenvalue problems of one form or another include mechanical vibrations (of discrete masses, membranes, plates, beams, built-up structures and so on); wave problems (for example, determining the characteristic frequencies of standing waves in a room or of standing water waves in a harbor); heat transfer in solids; problems in hydrodynamic stability; and buckling phenomena (such as in columns or shells, like an Oly can).

One solution to (4.73) is the trivial solution $\{C\} = \{0\}$. This corresponds to the system being at equilibrium. Since our interest is in motions near, but not at, equilibrium, we want to find nontrivial solutions. Thus, we have to find a nonzero solution $\{C\}$ such that (4.73) is satisfied.

Recall from the theory of linear algebraic equations that a nontrivial solution to a set of homogeneous equations can exist only if the determinant of the coefficient matrix vanishes. We have to require, then, that

$$|[A] - \lambda[I]| = 0 \tag{4.74}$$

where the notation $|\cdot|$ stands for the determinant, which is a scalar. In expanded form, the condition (4.74) is

$$\begin{vmatrix} (a_{11} - \lambda) & a_{12} & \cdots & a_{1n} \\ a_{21} & (a_{22} - \lambda) & \cdots & a_{2n} \\ \vdots & \vdots & & \vdots \\ a_{n1} & a_{n2} & \cdots & (a_{nn} - \lambda) \end{vmatrix} = 0 \tag{4.75}$$

Because the constant λ appears in each diagonal entry of (4.75), the determinant (4.75), when expanded, will be a polynomial of degree n in λ, of the form

$$\lambda^n + a_1\lambda^{n-1} + a_2\lambda^{n-2} + \cdots + a_{n-1}\lambda + a_n = 0 \tag{4.76}$$

Equation (4.76) is called the *characteristic equation* of the system (4.62), and the left-hand side of (4.76) is referred to as the characteristic polynomial.* In (4.76) the constants $a_1, a_2, \ldots, a_n$ will each be functions of the n^2 elements of the Jacobian matrix $[A]$. The result (4.76) says that, in order that (4.71) be a solution to (4.62), we must pick λ so that the characteristic equation is satisfied. Since (4.76) is of nth degree in λ, there will be n values of λ which will solve (4.76). Let us denote these eigenvalues as $\lambda_1, \lambda_2, \ldots, \lambda_n$, and let us assume, at least for now, that the eigenvalues are distinct. Then the characteristic equation (4.76) can also be represented in the factored form:

$$(\lambda - \lambda_1)(\lambda - \lambda_2) \cdots (\lambda - \lambda_n) = 0 \tag{4.77}$$

If we were to multiply out (4.77), we would recover (4.76), with the constants $a_1, \ldots, a_n$ in (4.76) being given in terms of the eigenvalues $\lambda_1, \ldots, \lambda_n$. In practice if $n > 2$ the solutions to the characteristic equation (4.76) are usually found using a computer routine which finds the roots of an nth-degree algebraic equation.

The eigenvalues $\lambda_1, \ldots, \lambda_n$ may be real or complex numbers. If any one of the eigenvalues is complex, then its complex conjugate must also be an eigenvalue, so that complex eigenvalues always occur in conjugate pairs. This happens because the coefficients a_i in the characteristic polynomial (4.76) must be real. Thus, if λ_1 is a complex eigenvalue denoted by $\lambda_1 = \sigma_1 + i\omega_1$, where σ_1 is the real part and ω_1 is the imaginary part, then the complex conjugate $\lambda_2 = \sigma_1 - i\omega_1$ is also an eigenvalue. Only this conjugacy property will ensure that the product $(\lambda - \lambda_1)(\lambda - \lambda_2)$ in (4.77) produce a quadratic polynomial containing real coefficients.

---

*We assume that in (4.76), any constant multiplying λ^n has been divided out.

Let us assume, then, that we have set up and solved the characteristic equation and have found n distinct eigenvalues $\lambda_1, \ldots, \lambda_n$. We conclude that there are n different solutions of the form (4.71), one for each eigenvalue. This is not surprising, as the theory of linear, homogeneous ODEs says we should find n linearly independent solutions for a system of nth order.

Having found the eigenvalues λ_i, we next have to find, for each eigenvalue, the corresponding solution for the vector $\{C\}$, which is referred to as an *eigenvector*. The pair $\lambda_i, \{C\}_i$ defines the ith solution, which we can call $\{x\}_i$. For a given eigenvalue λ_i, the associated eigenvector $\{C\}_i$ is defined by (4.72):

$$[[A] - \lambda_i[I]]\{C\}_i = \{0\} \tag{4.78}$$

Since λ_i is now known, we have n homogeneous algebraic equations for the n unknown entries of $\{C\}_i$. Because the algebraic equations (4.78) are homogeneous, however, we will not be able to find a *unique* solution for the eigenvector $\{C\}_i$. If we find a solution $\{C\}_i$, then any multiple $\alpha\{C\}_i$, where α is any constant, will also be a solution. This can be seen from (4.78); the α simply cancels out.

We can find *a* solution by specifying a value for one of the entries in $\{C\}_i$ and then solving for the remaining entries. Our approach here will be to define the top entry of $\{C\}_i$ to be unity* and then to use $n - 1$ of the equations (4.78) to find the remaining $n - 1$ entries of $\{C\}_i$. Another common procedure is to normalize the eigenvector so that it is of unit length in n space. Conceptually, there is no advantage to doing this.

At this point, then, suppose we have found the complete set of eigenvectors $\{C\}_1, \{C\}_2, \ldots, \{C\}_n$, according to the procedure just described. Each pair $\lambda_i, \{C\}_i$ defines a different, linearly independent solution to (4.62). Denoting this ith solution as $\{x\}_i$, we have

$$\{x\}_i = \{C\}_i e^{\lambda_i t} \tag{4.79}$$

in accordance with (4.71). Keeping in mind that any multiple of (4.79) will also solve (4.62), and that any linear combination of two or more of the solutions (4.79) will solve (4.62), the most general solution to (4.62) will be a linear combination of all of the solutions (4.79), so that the general solution is

$$\{x(t)\} = \alpha_1\{C\}_1 e^{\lambda_1 t} + \alpha_2\{C\}_2 e^{\lambda_2 t} + \cdots + \alpha_n\{C\}_n e^{\lambda_n t} \tag{4.80}$$

or, compactly,

$$\{x(t)\} = \sum_{i=1}^{n} \alpha_i\{C\}_i e^{\lambda_i t} \tag{4.81}$$

Here the n constants $\alpha_1, \ldots, \alpha_n$ are found in specific cases from the initial conditions $x_1(0), \ldots, x_n(0)$. The eigenvalues $\lambda_1, \ldots, \lambda_n$ define the time variation of the response $\{x(t)\}$. For a given system (4.62), some of the eigenvalues may

*Sometimes the top entry of a given eigenvector $\{C\}_i$ turns out to be identically zero. In such cases we have to pick a value for one of the other entries of $\{C\}_i$.

be real and some complex. We now consider these two types of eigenvalues separately, illustrating the basic solution procedure with examples.

4.2.2 Real, Distinct Eigenvalues

Suppose that for an n-state system all of the eigenvalues $\lambda_1, \ldots, \lambda_n$ are real, so that the individual solutions $\{x\}_i$ exhibit purely exponential behavior in time. The eigenvectors $\{C\}_i$ and constants α_i in (4.80) will also be real. The general solution (4.80) is then expressed as a superposition of n exponential functions.

In order to gain insight into the meaning of the eigenvectors, let us consider the relation of the eigenvectors to the vector $\{x(0)\}$ of initial conditions, which we may assume to be given in specific cases. If we evaluate the general solution (4.80) at $t = 0$, we obtain

$$\{x(0)\} = \alpha_1\{C\}_1 + \alpha_2\{C\}_2 + \cdots + \alpha_n\{C\}_n \tag{4.82}$$

This relation says that the initial state vector $\{x(0)\}$ is expressed as a linear combination of the eigenvectors. Because (4.82) must hold for *all* initial state vectors, we deduce that *any* n-dimensional vector can be represented as a linear combination of the eigenvectors. This means that the n eigenvectors are linearly independent and form a basis in n space (or "span" n space), in the same way that the cartesian unit vectors **i**, **j**, and **k** form a basis in ordinary three space. We can "draw" the eigenvectors in the n-dimensional phase space just as we would do with ordinary vectors in three space.

A given initial state vector $\{x(0)\}$ will determine the values of the constants $\alpha_1, \ldots, \alpha_n$ needed to obtain the solution (4.80) for that particular set of initial conditions. To indicate how the α's are found, let us first define the $n \times n$ *matrix of eigenvectors* $[C]$, in which the columns are the eigenvectors $\{C\}_i$, so that

$$[C] \equiv [\{C\}_1 | \{C\}_2 | \cdots | \{C\}_n] \tag{4.83}$$

Let us also form an n vector of the α's, $\{\alpha\} = \lfloor \alpha_1, \alpha_2, \ldots \rfloor^T$.* Then (4.82) may be rewritten as

$$\{x(0)\} = [C]\{\alpha\} \tag{4.84}$$

This constitutes a set of n linear, inhomogeneous algebraic equations which are to be solved for $\alpha_1, \alpha_2, \ldots, \alpha_n$. The solution is guaranteed to be unique because the columns of $[C]$ are linearly independent. Formally, the solution may be written

$$\{\alpha\} = [C]^{-1}\{x(0)\}$$

*When writing out the components of a column vector, it is convenient, in order to save space, to write the column vector as the transpose of a *row vector*, which is a matrix with a single row. This has been done here.

where $[C]^{-1}$ is the matrix inverse of $[C]$, such that $[C]^{-1}[C] = [I]$. The solution tells us the particular linear combination of the n eigenvectors that will produce the initial condition vector in the phase space.

Let us now look at some special cases. Suppose that we pick initial conditions so that only α_1 is nonzero. Then

$$\{x(0)\} = \alpha_1\{C\}_1$$

that is, the initial condition vector is aligned with the first eigenvector. Furthermore, for this special set of initial conditions, the solution (4.80) reduces to

$$\{x(t)\} = \{x(0)\}\, e^{\lambda_1 t}$$

The orbits in phase space will then be straight lines, moving exponentially to the origin of the phase space if $\lambda_1 < 0$, moving exponentially away from the origin if $\lambda_1 > 0$, and staying at $\{x(0)\}$ if $\lambda_1 = 0$. Similar behavior will occur if we pick the initial state vector to be aligned with one of the other eigenvectors $\{C\}_i$. We conclude that there are n different vectors (the eigenvectors) in phase space, each associated with a different solution $\{x\}_i$, such that orbits initiated along these vectors will move directly to or away from the origin of the phase space. This geometric description will be illustrated shortly with some examples.

The stability of the equilibrium solution $\{\bar{x}\} = \{0\}$ can be deduced directly from the eigenvalues. For example, asymptotic stability requires that, for all initial state vectors $\{x(0)\}$, the solution $\{x(t)\} \to \{0\}$ as $t \to \infty$. From (4.80) we see that this condition can be satisfied only if each eigenvalue is negative. Otherwise, we could find some initial state vectors $\{x(0)\}$ for which $\{x(t)\}$ would not approach the origin of the phase space asymptotically.

In order that the equilibrium solution $\{\bar{x}\} = \{0\}$ be unstable, it is only necessary that one of the eigenvalues be positive because we could then find initial conditions $\{x(0)\}$ for which the solution $\{x(t)\}$ would, according to the linear model, move arbitrarily far from equilibrium. Furthermore, if any one of the eigenvalues is zero (but none is positive), then equilibrium is merely stable, as, for some initial conditions the state vector $\{x(t)\}$, while remaining near equilibrium, will not move asymptotically toward it. We can summarize the relation of the eigenvalues to the stability of equilibrium for a system (4.62) with real eigenvalues as in Table 4.1.

TABLE 4.1
Stability of linear systems with real distinct eigenvalues

Eigenvalues	Stability
1. All eigenvalues are negative.	Asymptotically stable
2. One eigenvalue zero; remaining eigenvalues negative.	Stable
3. One or more eigenvalue positive; remaining eigenvalues negative or zero.	Unstable

Let us now consider an example which illustrates the general analysis procedure and the preceding discussion for a linear system with real eigenvalues.

Example 4.19 A Two-State System with Real Eigenvalues. Suppose the linearized model (4.62) for near-equilibrium motions is

$$\dot{x}_1 = 5x_1 + 3x_2$$

$$\dot{x}_2 = -6x_1 - 4x_2$$

or in matrix form,

$$\begin{Bmatrix} \dot{x}_1 \\ \dot{x}_2 \end{Bmatrix} = \begin{bmatrix} 5 & 3 \\ -6 & -4 \end{bmatrix} \begin{Bmatrix} x_1 \\ x_2 \end{Bmatrix}$$

Our objective will be to find the eigenvalues, eigenvectors, and the general solution and to gain geometric insight into the structure of the solutions in the phase space.

The first step is to find the eigenvalues λ_1, λ_2 by forming the characteristic equation (4.75)

$$\begin{vmatrix} (5-\lambda) & 3 \\ -6 & (-4-\lambda) \end{vmatrix} = 0$$

which leads to $(5-\lambda)(-4-\lambda) + 18 = 0$. The characteristic equation in the polynomial form (4.76) is then $\lambda^2 - \lambda - 2 = 0$. The student may verify that the solution to this quadratic equation yields two real eigenvalues $\lambda_1 = -1$ and $\lambda_2 = 2$. Because one of the eigenvalues is positive, the equilibrium solution $\{\bar{x}\} = \{0\}$ is unstable.

The next step is to find the eigenvectors associated with each eigenvalue. We use (4.78) for this purpose, which here is

$$\begin{bmatrix} (5-\lambda_i) & 3 \\ -6 & (-4-\lambda_i) \end{bmatrix} \begin{Bmatrix} c_1 \\ c_2 \end{Bmatrix}_i = \{0\} \tag{4.85}$$

Starting with the first eigenvalue $\lambda_1 = -1$, (4.85) becomes

$$\begin{bmatrix} 6 & 3 \\ -6 & -3 \end{bmatrix} \begin{Bmatrix} c_1 \\ c_2 \end{Bmatrix}_1 = \{0\} \tag{4.86}$$

Observe that the two algebraic equations in (4.86) are not linearly independent, the second merely being the first multiplied by -1. This is consistent with our previous assertion that the solution to (4.78) not be unique. Taking the first of the two equations in (4.86), we have the following relation connecting the two entries c_1 and c_2 of the eigenvector $\{C\}_1$:

$$6c_1 + 3c_2 = 0 \Rightarrow c_2 = -2c_1$$

Let us pick $c_1 = 1$, so that $c_2 = -2$, and the first eigenvector is then

$$\{C\}_1 = \begin{Bmatrix} 1 \\ -2 \end{Bmatrix}$$

Clearly any multiple of this vector will also qualify as an eigenvector, but we follow the convention of defining the top entry to be unity. We would get the same result had we used the second equation of (4.86).

For the second eigenvalue $\lambda_2 = 2$, (4.85) is

$$\begin{bmatrix} 3 & 3 \\ -6 & -6 \end{bmatrix} \begin{Bmatrix} c_1 \\ c_2 \end{Bmatrix}_2 = \{0\}$$

Again we observe that the two algebraic equations defining the entries of the second eigenvector $\{C\}_2$ are not independent. The top equation yields $3c_1 + 3c_2 = 0$, so that if we define $c_1 = 1$, then $c_2 = -1$, and the second eigenvector is

$$\{C\}_2 = \begin{Bmatrix} 1 \\ -1 \end{Bmatrix}$$

Having found the eigenvalues and associated eigenvectors, we now have two solutions which will solve the problem:

$$\text{Solution 1:} \quad \{x\}_1 = \begin{Bmatrix} 1 \\ -2 \end{Bmatrix} e^{-t}$$

$$\text{Solution 2:} \quad \{x\}_2 = \begin{Bmatrix} 1 \\ -1 \end{Bmatrix} e^{2t}$$

The general solution is a linear combination of the two solutions we have found,

$$\{x(t)\} = \alpha_1 \begin{Bmatrix} 1 \\ -2 \end{Bmatrix} e^{-t} + \alpha_2 \begin{Bmatrix} 1 \\ -1 \end{Bmatrix} e^{2t} \tag{4.87}$$

where the constants α_1 and α_2 depend on the initial conditions.

Let us now use the solution (4.87) to construct a picture of the flow of the system in the phase space (Fig. 4.17). First we draw the two lines L_1 and L_2 along

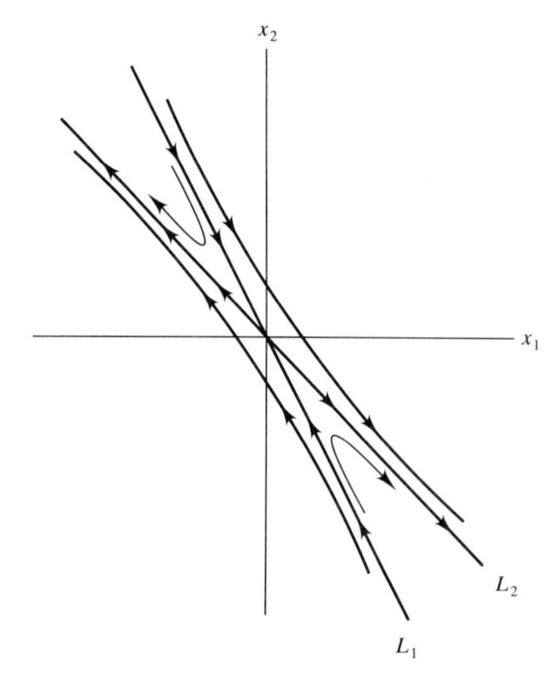

FIGURE 4.17
Phase portraits for the system $\dot{x}_1 = 5x_1 + 3x_2$, $\dot{x}_2 = -6x_1 - 4x_2$

which the eigenvectors $\{C\}_1$ and $\{C\}_2$ are directed. On the line L_1, defined by the eigenvector

$$\{C\}_1 = \begin{Bmatrix} 1 \\ -2 \end{Bmatrix}, \qquad x_2 = -2x_1$$

On the line L_2, defined by the eigenvector

$$\{C\}_2 = \begin{Bmatrix} 1 \\ -1 \end{Bmatrix}, \qquad x_2 = -x_1$$

Suppose we pick initial conditions so that the initial state vector $\{x(0)\}$ lies on $L_1, x_2(0) = -2x_1(0)$. Then $\alpha_2 = 0$ in (4.87), and we observe only solution 1; the orbit approaches the origin of the phase space exponentially, as e^{-t}. The direction of such orbits is indicated by the arrows in Fig. 4.17. If we consider initial conditions on $L_2, x_2(0) = -x_1(0)$, then $\alpha_1 = 0$ in (4.87) and we observe only solution 2; the orbits move exponentially away from the origin, as e^{2t}, indicated by the arrows in the figure.

Now consider initial conditions not on L_1 nor L_2. This situation is what will generally occur, as the probability that a real system, when disturbed from equilibrium, will have initial conditions exactly on one of the eigenvectors, is very remote. In this case both α_1 and α_2 in (4.87) are nonzero, and the response consists of the superposition of the two exponential functions in (4.87). For large times, however, all solutions will asymptote to L_2, because the contribution to the solution of the negative exponential term in (4.87) will approach zero. Thus, orbits initiated very near L_1, so that α_1 is much larger than α_2, will initially move toward the origin, but will eventually be driven along L_2 and, for large times, will move exponentially away from the origin, as sketched in Fig. 4.17. It should be clear that for *almost all* initial conditions, orbits will, according to the linear model, eventually move exponentially away from equilibrium.

Example 4.20 The System of Examples 4.1 and 4.16, Revisited. The linearized mathematical model for this system was shown in Example 4.16 to be (4.55):

$$\begin{Bmatrix} \dot{x}_1 \\ \dot{x}_2 \end{Bmatrix} = \begin{bmatrix} 1 & 0 \\ 0 & -1 \end{bmatrix} \begin{Bmatrix} x_1 \\ x_2 \end{Bmatrix}$$

Notice that this system is in diagonal form; all off-diagonal entries are zero. The algebraic eigenvalue problem (4.73) associated with the system ODE model is

$$\begin{bmatrix} (1-\lambda) & 0 \\ 0 & (-1-\lambda) \end{bmatrix} \begin{Bmatrix} c_1 \\ c_2 \end{Bmatrix} = \bar{0} \tag{4.88}$$

The system characteristic equation is thus $(1-\lambda)(-1-\lambda) = 0$, which yields the two real eigenvalues $\lambda_1 = 1, \lambda_2 = -1$. These are the diagonal entries of $[A]$. If we substitute $\lambda_1 = 1$ into the first equation of (4.88) in order to find the associated eigenvector, we obtain the trivial result $0 = 0$. The second equation in (4.88), however, yields the result $c_2 = 0$, so that the first eigenvector is

$$\{C\}_1 = \begin{Bmatrix} 1 \\ 0 \end{Bmatrix}$$

This eigenvector is directed along the x_1 axis. A similar calculation shows the second eigenvector to be directed along the x_2 axis

$$\{C\}_2 = \begin{Bmatrix} 0 \\ 1 \end{Bmatrix}$$

The general solution is thus

$$\begin{Bmatrix} x_1(t) \\ x_2(t) \end{Bmatrix} = \alpha_1 \begin{Bmatrix} 1 \\ 0 \end{Bmatrix} e^t + \alpha_2 \begin{Bmatrix} 0 \\ 1 \end{Bmatrix} e^{-t} \tag{4.89}$$

where α_1 and α_2 are determined by the initial conditions. Orbits initiated on the x_2 axis [$\alpha_1 = 0$ in (4.89)] will move exponentially to the origin, while orbits initiated on the x_1 axis ($\alpha_2 = 0$) will move exponentially to infinity. All other orbits [α_1, α_2 nonzero in (4.89)] will collapse toward the x_1 axis and will eventually approach infinity along the $\pm x_1$ axes. Equilibrium is unstable because of the positive eigenvalue.

The preceding example illustrates a general result: If the linearized ODE model (4.62) is in diagonal form with real diagonal entries, then the diagonal entries are the eigenvalues, and the eigenvectors are the associated phase space axes.

4.2.3 Complex Eigenvalues and Eigenvectors

The analysis of linear ODE systems having complex eigenvalues and eigenvectors requires familiarity with complex algebra and the complex representation of trigonometric functions. In this subsection we first review briefly some relevant results in these areas. We then consider the special case of a two-state system having eigenvalues which are purely imaginary. Finally, the more general case of complex eigenvalues having nonzero real and imaginary parts will be studied.

COMPLEX ALGEBRA. A complex number α, defined as $\alpha = a + ib$, has a real part a and an imaginary part b and can be represented as a point in the complex plane, as shown in Fig. 4.18, in which the horizontal axis defines the real part and the vertical axis the imaginary part. The complex conjugate of α is denoted as α^* and is defined as $\alpha^* = a - ib$. The product $\alpha\alpha^*$ of a complex number with its conjugate is a real number, $\alpha\alpha^* = a^2 + b^2$. Observe that $(\alpha\alpha^*)^{1/2}$ is the magnitude or length of the complex number α (or of α^*). The sum $\alpha + \alpha^*$ is the real number $2a$, whereas the difference $\alpha - \alpha^*$ is the imaginary number $2ib$.

It is often more convenient to represent complex numbers in polar, rather than cartesian, form. To this end, use is made of DeMoivre's theorem, which expresses the exponential function, with an imaginary argument, in terms of the trigonometric functions:

$$e^{i\theta} = \cos\theta + i \sin\theta \tag{4.90}$$

Thus, we may express the complex number α as $\alpha = Re^{i\theta}$ (see Fig. 4.18), where R is the magnitude of α and θ is its argument, or "angle." DeMoivre's theorem

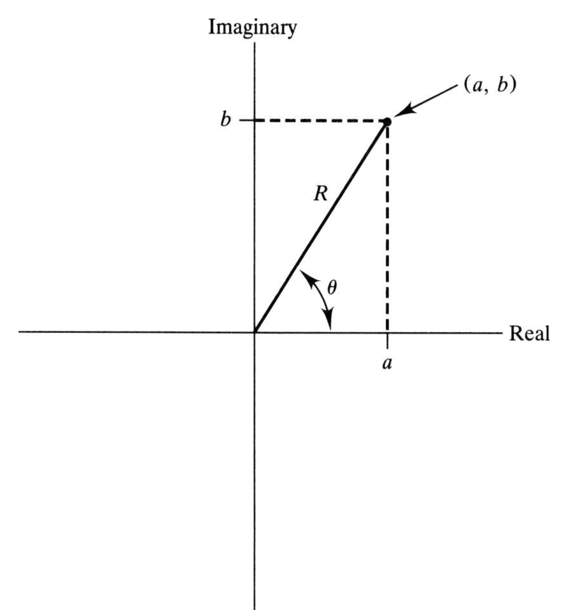

FIGURE 4.18
Representation of a complex number in the complex plane.

then provides the relations $a = R\cos\theta, b = R\sin\theta$, so that $R = (a^2 + b^2)^{1/2}$ and $\tan\theta = b/a$. The polar form is especially useful when multiplying two complex numbers. Defining $\alpha_1 = R_1\,e^{i\theta_1}$ and $\alpha_2 = R_2\,e^{i\theta_2}$, then

$$\alpha_1\alpha_2 = R_1 R_2\,e^{i(\theta_1+\theta_2)} \tag{4.91}$$

so that the arguments are merely added in the complex exponent.

A final useful result is obtained by combining (4.90) with its conjugate, $e^{-i\theta} = \cos\theta - i\sin\theta$, to express the trigonometric functions in terms of the complex exponential,

$$\cos\theta = \frac{1}{2}(e^{i\theta} + e^{-i\theta}) \tag{4.92a}$$

$$\sin\theta = \frac{-i}{2}(e^{i\theta} - e^{-i\theta}) \tag{4.92b}$$

Equation (4.92) is necessary to interpret the solution to ODE models having complex eigenvalues.

ODE MODELS WITH IMAGINARY EIGENVALUES. Here we consider, by example, a two-state system having purely imaginary eigenvalues. The objective will be to see clearly how the complex solution, which arises naturally in this analysis, is equivalent to the solution expressed in terms of the simple harmonic functions.

Example 4.21 A Two-State Linear System with Imaginary Eigenvalues. The autonomous linear system (4.62) to be considered is

$$\dot{x}_1 = 2x_2$$

$$\dot{x}_2 = -4.5x_1$$

or, in matrix-vector form,

$$\begin{Bmatrix} \dot{x}_1 \\ \dot{x}_2 \end{Bmatrix} = \begin{bmatrix} 0 & 2 \\ -4.5 & 0 \end{bmatrix} \begin{Bmatrix} x_1 \\ x_2 \end{Bmatrix} \tag{4.93}$$

The associated eigenvalue problem (4.73) then has the form

$$\begin{bmatrix} -\lambda & 2 \\ -4.5 & -\lambda \end{bmatrix} \begin{Bmatrix} c_1 \\ c_2 \end{Bmatrix} = \bar{0} \tag{4.94}$$

The characteristic equation, obtained by setting the determinant in (4.94) to zero, is

$$\lambda^2 + 9 = 0$$

so that the eigenvalues are $\lambda_{1,2} = \pm 3i$. These eigenvalues are purely imaginary and are the complex conjugates of each other. The associated eigenvectors $\{C\}_1$ and $\{C\}_2$ are obtained from (4.94). For the first eigenvalue $\lambda_1 = 3i$, the first of equations (4.94) yields

$$-3ic_1 + 2c_2 = 0 \Rightarrow c_2 = \tfrac{3}{2}ic_1$$

Following the usual procedure of selecting c_1 to be unity, the first eigenvector is found as

$$\{C\}_1 = \begin{Bmatrix} 1 \\ \dfrac{3}{2}i \end{Bmatrix}$$

The second eigenvector, associated with $\lambda_2 = -3i$, is the complex conjugate of the first, as may be verified directly from (4.94):

$$\{C\}_2 = \begin{Bmatrix} 1 \\ \dfrac{-3}{2}i \end{Bmatrix}$$

The two solutions to (4.93) are thus as follows:

$$\{x\}_1 = \begin{Bmatrix} 1 \\ \dfrac{3}{2}i \end{Bmatrix} e^{3it}$$

$$\{x\}_2 = \begin{Bmatrix} 1 \\ \dfrac{-3}{2}i \end{Bmatrix} e^{-3it} = \{x\}_1^*$$

The second solution is the complex conjugate of the first. The general solution is a linear combination of these two solutions:

$$\{x(t)\} = \alpha_1 \begin{Bmatrix} 1 \\ \dfrac{3}{2}i \end{Bmatrix} e^{3it} + \alpha_2 \begin{Bmatrix} 1 \\ \dfrac{-3}{2}i \end{Bmatrix} e^{-3it}$$

In general, the constants α_1 and α_2 are complex. Furthermore, α_1 and α_2 must be complex conjugates. Otherwise the addition of the two solutions to form the general solution would not result in real solutions for the states $x_1(t)$ and $x_2(t)$, as is required. Letting $\alpha_1 \equiv \alpha$ and $\alpha_2 = \alpha^*$, the general solution is expressed in the form

$$\{x(t)\} = \alpha \begin{Bmatrix} 1 \\ \dfrac{3}{2}i \end{Bmatrix} e^{3it} + \alpha^* \begin{Bmatrix} 1 \\ \dfrac{-3}{2}i \end{Bmatrix} e^{-3it} \tag{4.95}$$

For the two-state system with imaginary eigenvalues the complex form (4.95) arises naturally when one solves the ODE model using the eigenvalue/eigenvector methods of this chapter. In order to relate the result (4.95) to the more familiar trigonometric functions, we first let the constant α be defined in polar form as $\alpha = (A/2)\, e^{i\theta}$, where A and θ are real constants which would be determined by the initial conditions $x_1(0)$ and $x_2(0)$. Thus, (4.95) becomes

$$\{x(t)\} = \frac{A}{2} \begin{Bmatrix} 1 \\ \dfrac{3}{2}i \end{Bmatrix} e^{i(3t+\theta)} + \frac{A}{2} \begin{Bmatrix} 1 \\ \dfrac{-3}{2}i \end{Bmatrix} e^{-i(3t+\theta)} \tag{4.96}$$

Let us now use (4.92) to write out the solutions for the states $x_1(t)$ and $x_2(t)$. For $x_1(t)$, Eq. (4.96) yields

$$x_1(t) = \frac{A}{2}\left[e^{i(3t+\theta)} + e^{-i(3t+\theta)}\right]$$

and in view of (4.92a) this is simply

$$x_1(t) = A \cos(3t + \theta) \tag{4.97}$$

For $x_2(t)$ we obtain

$$x_2(t) = \frac{3A}{2}\left(\frac{i}{2}\right)\left[e^{i(3t+\theta)} - e^{-i(3t+\theta)}\right]$$

In view of (4.92b) this is simply

$$x_2(t) = -\tfrac{3}{2}A \sin(3t + \theta) \tag{4.98}$$

The results (4.97) and (4.98) are equivalent to the complex form (4.96).

The solution for any two-state system for which the eigenvalues are purely imaginary exhibits the following properties:

1. From (4.97) and (4.98) we observe that each state exhibits a harmonic oscillation in time and that the frequency ω of this oscillation is the imaginary part of the eigenvalue; that is, $\lambda_{1,2} = \pm i\omega$.

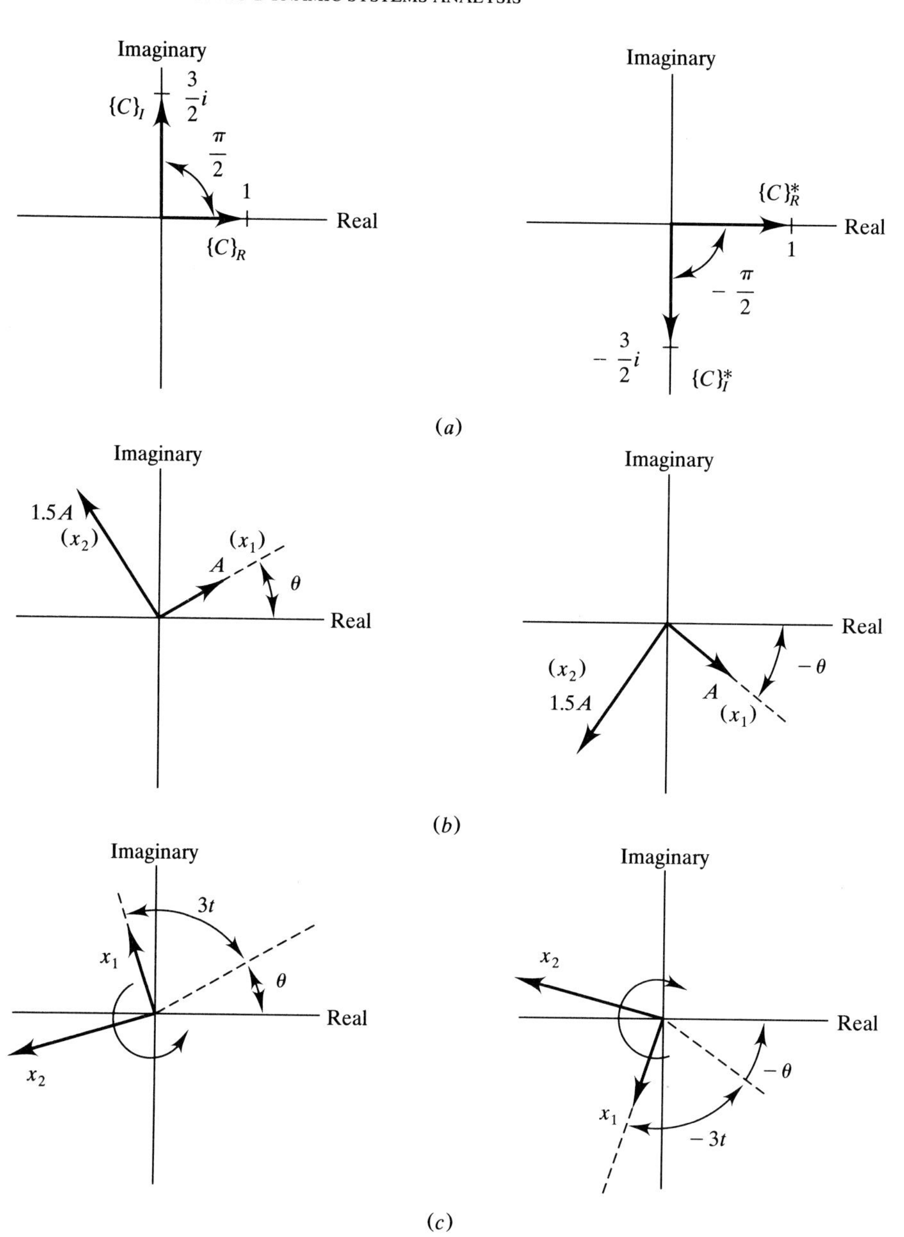

FIGURE 4.19
(a) The conjugate eigenvectors in the complex plane for the system of Example 4.21. When represented as complex numbers in the complex plane, the two entries in the eigenvector, 1 (associated with x_1) and $(\frac{3}{2})i$ (associated with x_2), show that the states x_1 and x_2 will be 90° out of phase and that x_2 has a maximum value $\frac{3}{2}$ that of x_1. (b) Initial conditions. The two parts of the solution (4.96), evaluated at $t = 0$, are shown. Fig. (b) is Fig. (a) multiplied by $(A/2)\,e^{i\theta}$ or $(A/2)\,e^{-i\theta}$. $x_1(0)$ and $x_2(0)$ are obtained by adding the complex vectors in the left and right parts of Fig. (b). (c) Visualization of the motion. The evolution of x_1 and x_2 consists of a rigid body rotation of the vectors of Fig. (b). In the left and right parts of Fig. (c), the vector pairs rotate in opposite directions at a rotation rate of 3 rad/s, so that, at any instant, the vector sums from the left and right figures produce real solutions for x_1 and x_2.

2. The two states x_1 and x_2 have differing amplitudes and phases, and this phase and amplitude information is contained in the eigenvectors. In the preceding example the amplitude of x_1 is A, the amplitude of x_2 is $\frac{3}{2}A$ (where A is determined by the initial conditions), and x_1 and x_2 are 90° out of phase. Figure 4.19 shows how one uses the eigenvectors to deduce this amplitude and phase information.
3. Because the system motions are harmonic in time, orbits in the phase plane will be closed curves, one of which is shown for Example 4.21 in Fig. 4.20. In this figure the initial conditions are $x_1(0) = 1$, $x_2(0) = 0$, so that $A = 1, \theta = 0$. The phase plane orbit is an ellipse with a semimajor axis (along x_2) of $\frac{3}{2}$ and a semiminor axis (along x_1) of unity. The orbit completes one full cycle in a time $T = 2\pi/\omega = 2\pi/3$ s, which is the period of the harmonic motion. Motion along the elliptical orbit is repeated endlessly.
4. In general, a pair of conjugate, purely imaginary eigenvalues $\lambda_{1,2} = \pm i\omega$ signifies a harmonic response of frequency ω for any initial conditions which occur. Thus, the associated equilibrium solution $\bar{x}_1 = \bar{x}_2 = 0$ is merely stable,

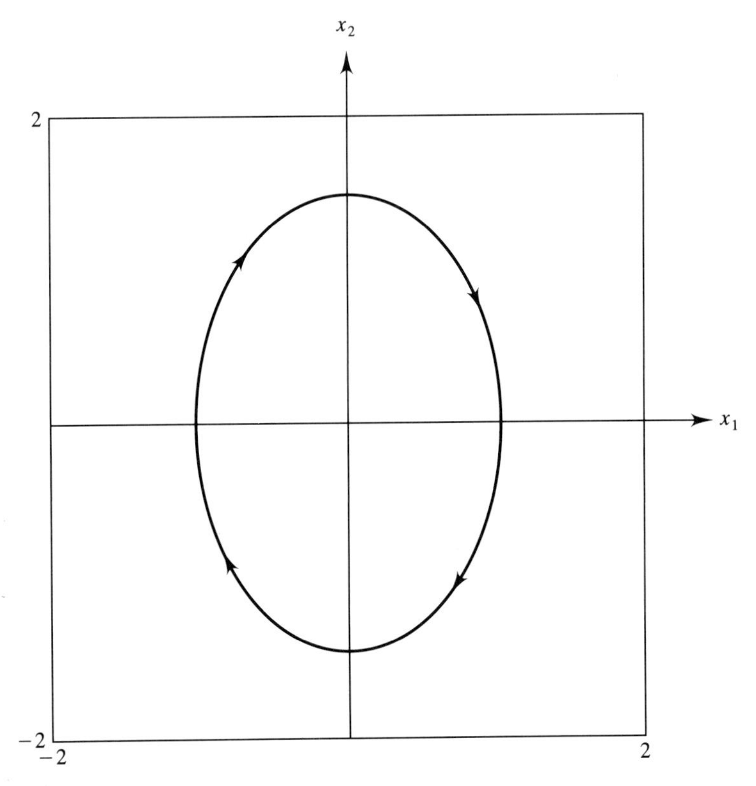

FIGURE 4.20
Phase portrait for the system $\dot{x}_1 = 2x_2$, $\dot{x}_2 = -4.5x_1$, for initial conditions $x_1(0) = 1.0$, $x_2(0) = 0$.

as orbits initiated near equilibrium will remain near equilibrium, but they will not asymptotically approach equilibrium.

A number of the linearized math models developed in Sec. 4.1 and elsewhere model two-state systems with imaginary eigenvalues. Examples include Example 4.17 [foxes and rabbits (4.57)], Example 4.15 [airspring (4.51)], and Example 4.2 [undamped linear oscillator (4.7)]. Each of these systems is characterized by a Jacobian matrix $[A]$ having zeroes on the main diagonal, with the off-diagonal elements nonzero and of opposite sign:

$$[A] = \begin{bmatrix} 0 & k_1 \\ -k_2 & 0 \end{bmatrix} \tag{4.99}$$

where k_1 and k_2 are positive constants. For such system models the eigenvalues are easily shown to be $\lambda_{1,2} = \pm i(k_1 k_2)^{1/2}$, so that these systems will exhibit harmonic motions of frequency $\omega = \sqrt{k_1 k_2}$. For example, the airspring of Example 4.15, Eq. (4.51), will, for small motions about the equilibrium position $\bar{x} = 0$, exhibit harmonic oscillations of frequency $\omega = (2Ap_0\gamma/md)^{1/2}$.

It is useful to note the connection between two-state models with Jacobian matrix (4.99) and the single, second-order ODE model:

$$\ddot{x} + \omega^2 x = 0 \tag{4.100}$$

In state variable form, (4.100) results in the Jacobian matrix:

$$[A] = \begin{bmatrix} 0 & 1 \\ -\omega^2 & 0 \end{bmatrix}$$

where $x_1 \equiv x$ and $x_2 \equiv \dot{x}$. The result is a special case of (4.99), indicating that (4.99) and (4.100) may be viewed as "generic" models of the two-state system with imaginary eigenvalues. In the form (4.100) the oscillation frequency ω appears in squared form as the coefficient of the static term. For example, in Example 4.18 [simple pendulum (4.60)], the oscillation frequency about the static equilibrium $\bar{\theta} = 0$ will be $\omega = \sqrt{g/L}$.

As a final comment note that, in order for a two-state system to have purely imaginary eigenvalues, it is not *necessary* that the main diagonal elements of the Jacobian matrix vanish. It is, however, common, especially in systems modeled by newtonian mechanics.

Example 4.22 A Two-State Linear System with Complex Eigenvalues. The linear system (4.62) to be considered is

$$\dot{x}_1 = \tfrac{3}{2}x_1 + 10x_2$$

$$\dot{x}_2 = -40x_1 + \tfrac{1}{2}x_2$$

or in vector matrix form,

$$\begin{Bmatrix} \dot{x}_1 \\ \dot{x}_2 \end{Bmatrix} = \begin{bmatrix} \frac{3}{2} & 10 \\ -40 & \frac{1}{2} \end{bmatrix} \begin{Bmatrix} x_1 \\ x_2 \end{Bmatrix} \tag{4.101}$$

The eigenvalues for this system are found from the characteristic equation (4.75)

$$\begin{vmatrix} \left(-\lambda + \frac{3}{2}\right) & 10 \\ -40 & \left(-\lambda + \frac{1}{2}\right) \end{vmatrix} = 0$$

which in polynomial form is $\lambda^2 - 2\lambda + 400.75 = 0$. The two solutions to this quadratic equation are the conjugate eigenvalues

$$\lambda_{1,2} = 1 \pm 20i$$

The eigenvectors $\{C\}_1$ and $\{C\}_2$ are found from (4.78):

$$\begin{bmatrix} \left(-\lambda_i + \frac{3}{2}\right) & 10 \\ -40 & \left(-\lambda_i + \frac{1}{2}\right) \end{bmatrix} \begin{Bmatrix} c_1 \\ c_2 \end{Bmatrix}_i = \bar{0}$$

The eigenvector $\{C\}_1$ associated with $\lambda_1 = 1 + 20i$ is then obtained from

$$\begin{bmatrix} \left(\frac{1}{2} - 20i\right) & 10 \\ -40 & \left(-\frac{1}{2} - 20i\right) \end{bmatrix} \begin{Bmatrix} c_1 \\ c_2 \end{Bmatrix}_1 = \bar{0} \tag{4.102}$$

From the first of these equations, we find

$$\left(\tfrac{1}{2} - 20i\right)c_1 + 10c_2 = 0$$

or $c_2 = (-0.05 + 2i)c_1$. Following our usual procedure of selecting the top entry of the eigenvector to be unity, $c_1 = 1$, we then have $c_2 = -0.05 + 2i$, so that the first eigenvector is

$$\{C\}_1 = \begin{Bmatrix} 1 \\ -0.05 + 2i \end{Bmatrix}$$

Note that the second of equations (4.102) would yield the same result; if we multiply the second equation in (4.102) by the factor $-0.0125 + 0.5i$, we obtain the first equation, so that the two equations in (4.102) are really the same.

If we calculate the second eigenvector $\{C\}_2$, associated with $\lambda_2 = 1 - 20i$, we find it to be the complex conjugate of the first:

$$\{C\}_2 = \begin{Bmatrix} 1 \\ -0.05 - 2i \end{Bmatrix}$$

The two solutions to (4.101) are summarized below:

$$\text{Solution 1:} \quad \{x\}_1 = \begin{Bmatrix} 1 \\ -0.05 + 2i \end{Bmatrix} e^{(1+20i)t}$$

$$\text{Solution 2:} \quad \{x\}_2 = \begin{Bmatrix} 1 \\ -0.05 - 2i \end{Bmatrix} e^{(1-20i)t}$$

The two solutions are complex conjugates of each other. The general solution is a linear combination of these two:

$$\{x(t)\} = \alpha_1 \begin{Bmatrix} 1 \\ -0.05 + 2i \end{Bmatrix} e^{(1+20i)t} + \alpha_2 \begin{Bmatrix} 1 \\ -0.05 - 2i \end{Bmatrix} e^{(1-20i)t} \tag{4.103}$$

Now it is clear that the result of summing the two solutions according to (4.103) must result in real solutions for the states $x_1(t)$ and $x_2(t)$. In order that the sum of two complex functions be real, the two complex functions must be complex conjugates. This requires in (4.103) that not only the eigenvalues and eigenvectors be complex conjugates but also the constants α_1 and α_2. Thus, we can write (4.103) as

$$\{x(t)\} = \alpha\begin{Bmatrix} 1 \\ -0.05 + 2i \end{Bmatrix} e^{(1+20i)t} + \alpha^*\begin{Bmatrix} 1 \\ -0.05 - 2i \end{Bmatrix} e^{(1-20i)t} \quad (4.104)$$

The real and imaginary parts of α are determined by the initial conditions.

While (4.104) represents the general solution to the problem of Example 4.22, the complex form may be difficult to interpret physically. In particular, the meaning of the complex eigenvectors may not be obvious, since these eigenvectors cannot be represented as ordinary vectors in the phase space.

Let us investigate the meaning of (4.104) by first evaluating (4.104) at $t = 0$, in order to relate the complex eigenvectors to the initial state vector $\{x(0)\}$. Denoting the complex eigenvector as $\{C\}$, (4.104) at $t = 0$ is

$$\{x(0)\} = \alpha\{C\} + \alpha^*\{C\}^* \quad (4.105)$$

Next, let us express the complex eigenvector $\{C\}$ as the sum of its real and imaginary parts:

$$\{C\} = \{C_R\} + i\{C_I\} \quad (4.106)$$

where, for the example system (4.101), the real part

$$\{C_R\} = \begin{Bmatrix} 1 \\ -0.05 \end{Bmatrix}$$

and the imaginary part

$$\{C_I\} = \begin{Bmatrix} 0 \\ 2 \end{Bmatrix}$$

If we also split the complex constant α into its real and imaginary parts as $\alpha = \frac{1}{2}(a - ib)$, then (4.105) becomes

$$\{x(0)\} = \tfrac{1}{2}(a - ib)\{\{C_R\} + i\{C_I\}\} + \tfrac{1}{2}(a + ib)\{\{C_R\} - i\{C_I\}\}$$

If we multiply out the above expression, we find

$$\{x(0)\} = a\{C_R\} + b\{C_I\} \quad (4.107)$$

This result is seen to be of the same form as (4.82), for the case of two real eigenvalues. We find that, for a two-state system with complex eigenvalues, any initial state vector $\{x(0)\}$, and hence *any* two-dimensional vector, is represented as a linear combination of the real and imaginary parts of the complex eigenvector $\{C\}$. $\{C_R\}$ and $\{C_I\}$ are real, linearly independent vectors that span two space.

Now let us see what a solution of the type (4.104) says about the orbits that occur in the phase space. To keep things general, call the eigenvalues $\lambda_{1,2} = \sigma \pm i\omega$ and the associated eigenvectors $\{C\}$ and $\{C\}^*$, where $\{C\}$ can be expressed as in (4.106). Then the general solution for any two-state system with complex eigenvectors can be expressed in the form

$$\{x(t)\} = \alpha\{C\}\, e^{(\sigma + i\omega t)} + \alpha^*\{C\}^*\, e^{(\sigma - i\omega)t} \tag{4.108}$$

Now let the complex constant α be denoted as $\alpha = (A/2)\, e^{i\theta}$, where A and θ are real constants.* Then the solution (4.108) can be written

$$\{x(t)\} = \frac{A}{2} e^{i\theta}\{\{C_R\} + i\{C_I\}\}\, e^{(\sigma + i\omega)t}$$
$$+ \frac{A}{2} e^{-i\theta}\{\{C_R\} - i\{C_I\}\}\, e^{(\sigma - i\omega)t}$$

Next we can rearrange the exponential terms so as to rewrite this solution as

$$\{x(t)\} = \frac{A}{2} e^{\sigma t}\{\{C_R\}(e^{i(\omega t + \theta)} + e^{-i(\omega t + \theta)}\}$$
$$+ i\{\{C_I\}(e^{i(\omega t + \theta)} - e^{-i(\omega t + \theta)})\}$$

In view of (4.92), we can convert the complex exponential terms to trigonometric form as follows:

$$\{x(t)\} = A\, e^{\sigma t}\{\{C_R\}\cos(\omega t + \theta) - \{C_I\}\sin(\omega t + \theta)\} \tag{4.109}$$

The constants A and θ are determined by the initial conditions.

The solution (4.109) says that each state will consist of a harmonic function having a frequency ω, multiplied by an exponential function $e^{\sigma t}$, which grows or decays depending on whether $\sigma > 0$ or $\sigma < 0$. Thus, each state exhibits *oscillatory* behavior characterized by the imaginary part ω of the eigenvalue, which is the frequency of oscillation. The real part σ of the eigenvalue determines whether the oscillation amplitude of each state increases exponentially ($\sigma > 0$), decreases exponentially ($\sigma < 0$), or remains constant ($\sigma = 0$). The complex eigenvector will determine the relative amplitudes and phases of the two states. For instance, let us write out the solutions for $x_1(t)$ and $x_2(t)$ for the example system (4.101). Noting that

$$\{C_R\} = \begin{Bmatrix} 1 \\ -0.05 \end{Bmatrix}, \{C_I\} = \begin{Bmatrix} 0 \\ 2 \end{Bmatrix} \qquad \sigma = 1, \quad \text{and} \quad \omega = 20$$

Eq. (4.109) yields

$$x_1(t) = A\, e^t \cos(20t + \theta)$$
$$x_2(t) = A\, e^t[-0.05\cos(20t + \theta) - 2\sin(20t + \theta)] \tag{4.110}$$

---

*In the analysis leading to (4.107) we put $\alpha = \frac{1}{2}(a - ib)$, whereas here we are using the polar form $\alpha = (A/2)\, e^{i\theta}$. Use of DeMoivre's theorem yields $\alpha = (A/2)\cos\theta + i(A/2)\sin\theta$, so that $a = A\cos\theta$ and $b = -A\sin\theta$. The two forms are equivalent.

Let us consider the case in which the initial conditions are chosen so that $A = 1, \theta = 0$; that is, $x_1(0) = 1, x_2(0) = -0.05$. The initial state vector $\{x(0)\}$ then is aligned with the real part $\{C_R\}$ of the eigenvector, and the solution (4.110) reduces to

$$x_1(t) = e^t \cos 20t$$

$$x_2(t) = e^t[-0.05 \cos 20t - 2 \sin 20t]$$

The orbit in the phase plane is shown in Fig. 4.21.

The orbit exhibits a spiral character and expands outward because of the exponential growth factor e^t, which is destabilizing.

Based on the preceding discussion and examples, we can make the following observations about the behavior of two-state systems with complex

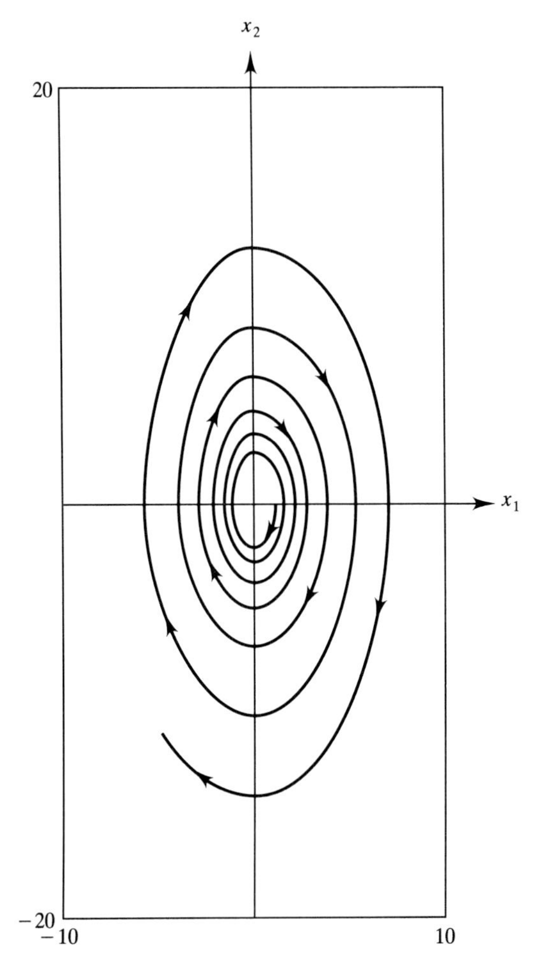

FIGURE 4.21
Phase portrait for the system $\dot{x}_1 = 1.5x_1 + 10x_2$, $\dot{x}_2 = -40x_1 + 0.5x_2$. The initial conditions are $x_1(0) = 1.0$, $x_2(0) = -0.05$.

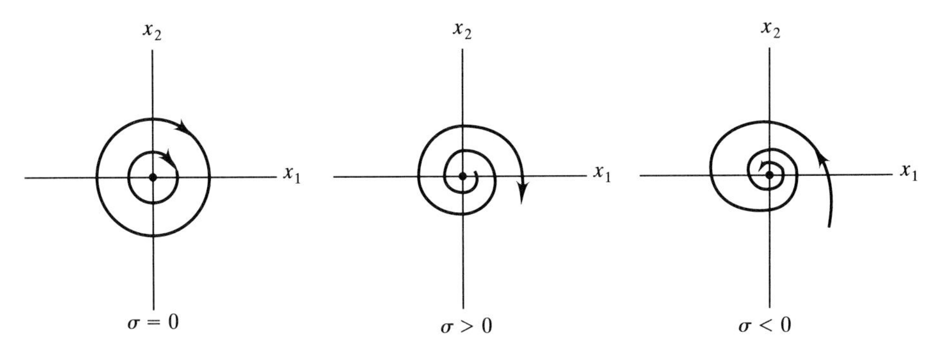

FIGURE 4.22
Schematic of orbits in the phase plane for a two-state system with complex eigenvalues, $\lambda_{1,2} = \sigma \pm i\omega$.

eigenvalues:

1. If $\sigma = 0$, so that the conjugate eigenvalues are imaginary, then (4.109) says that each state will exhibit harmonic behavior at the frequency ω. This periodic behavior will result in *closed curves* enclosing the origin in the phase space.* A complete traversal of such a closed orbit occurs in each full cycle or period T of the response, $T = 2\pi/\omega$. If $\sigma > 0$, then orbits will spiral outward from the origin, with the oscillation amplitude expanding exponentially, as in the Example 4.22. If $\sigma < 0$, then orbits will spiral into the origin of the phase plane, and from (4.109), each state $x_i(t) \to 0$ as $t \to \infty$. These three cases are illustrated qualitatively in Fig. 4.22.
2. Based on the preceding discussion, we can see that the stability of the equilibrium solution $\{\bar{x}\} = \{0\}$ will be governed by the *real part* σ of the eigenvalues. If $\sigma = 0$, then equilibrium is merely stable, as orbits are closed and hence will remain near equilibrium but will not approach equilibrium asymptotically. If $\sigma < 0$, then the equilibrium solution is asymptotically stable, as all orbits spiral into the origin of the phase space. If $\sigma > 0$, then equilibrium is unstable, as any orbit initiated near equilibrium will eventually move arbitrarily far from equilibrium. We can combine these stability characteristics with those for a system with real eigenvalues to deduce the following general condition for asymptotic stability of equilibrium: *An equilibrium solution will be asymptotically stable only if all eigenvalues have a negative real part.*
3. The occurrence of complex eigenvalues in linear models of real systems is very common, as many systems tend to exhibit oscillatory behavior.

---

*The direction of the orbits may be clockwise or counterclockwise depending on the system.

Let us now summarize and classify the general behavior of the two-state system.

4.2.4 Summary: The Two-State System

The general model (4.62) for the near-equilibrium behavior of a two-state system may be written as

$$\begin{Bmatrix} \dot{x}_1 \\ \dot{x}_2 \end{Bmatrix} = \begin{bmatrix} a_{11} & a_{12} \\ a_{21} & a_{22} \end{bmatrix} \begin{Bmatrix} x_1 \\ x_2 \end{Bmatrix} \tag{4.111}$$

Our objective here will be to investigate and classify the motions in phase space and the stability of equilibrium by looking at the eigenvalues of the general two-state model (4.111).

The characteristic equation of the system is

$$\begin{vmatrix} (a_{11} - \lambda) & a_{12} \\ a_{21} & (a_{22} - \lambda) \end{vmatrix} = 0$$

which in polynomial form is

$$\lambda^2 - \lambda(a_{11} + a_{22}) + (a_{11}a_{22} - a_{12}a_{21}) = 0 \tag{4.112}$$

Observe that the coefficient of λ in (4.112) is the negative of the trace T of the matrix $[A]$, the sum of the diagonal elements, while the constant appearing in (4.112) is the determinant of $[A]$, denoted by Δ. Thus, the characteristic equation (4.112) may be expressed as

$$\lambda^2 - T\lambda + \Delta = 0 \tag{4.113}$$

The solution to (4.113) yields the eigenvalues as

$$\lambda_{1,2} = \frac{T}{2} \pm \left[\frac{T^2}{4} - \Delta\right]^{1/2} \tag{4.114}$$

Clearly, then, for a two-state system the eigenvalues will only depend on the trace T and determinant Δ of the system matrix $[A]$. Thus, there will exist an infinite set of different 2×2 matrices which, provided they have the same trace and determinant, will have the same eigenvalues.

Now let us explore the manner in which the eigenvalues depend on T and Δ. A convenient way to do this is to graph Δ versus T and to relate various regions in the $\Delta - T$ plane to the types of eigenvalues which occur. For example, we see from (4.114) that the eigenvalues will be real if $T^2/4 > \Delta$, complex if $T^2/4 < \Delta$, and repeated ($\lambda_{1,2} = T/2$) if $T^2/4 = \Delta$. If $T^2/4 < \Delta$ and $T < 0$, the eigenvalues will be complex conjugates with a negative real part, and the equilibrium solution is asymptotically stable. The eigenvalues for such a

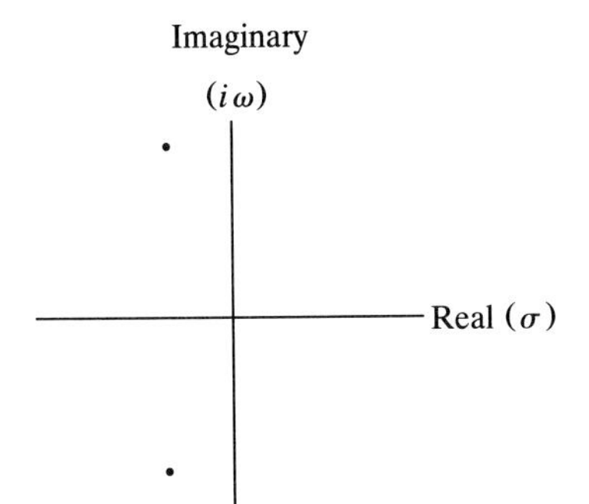

FIGURE 4.23
Representation of eigenvalues as points in the complex plane.

case can be represented pictorially by locating them in the complex plane for which the axes are the real and imaginary parts of $\lambda_{1,2}$, as shown in Fig. 4.23. Note that the conjugate eigenvalues lie equally above and below the real (σ) axis. Based on the preceding discussion, we can identify the types of eigenvalues that characterize various regions of $\Delta - T$ space, and we can represent each eigenpair as in Fig. 4.23. The results are presented in Fig. 4.24. For a given

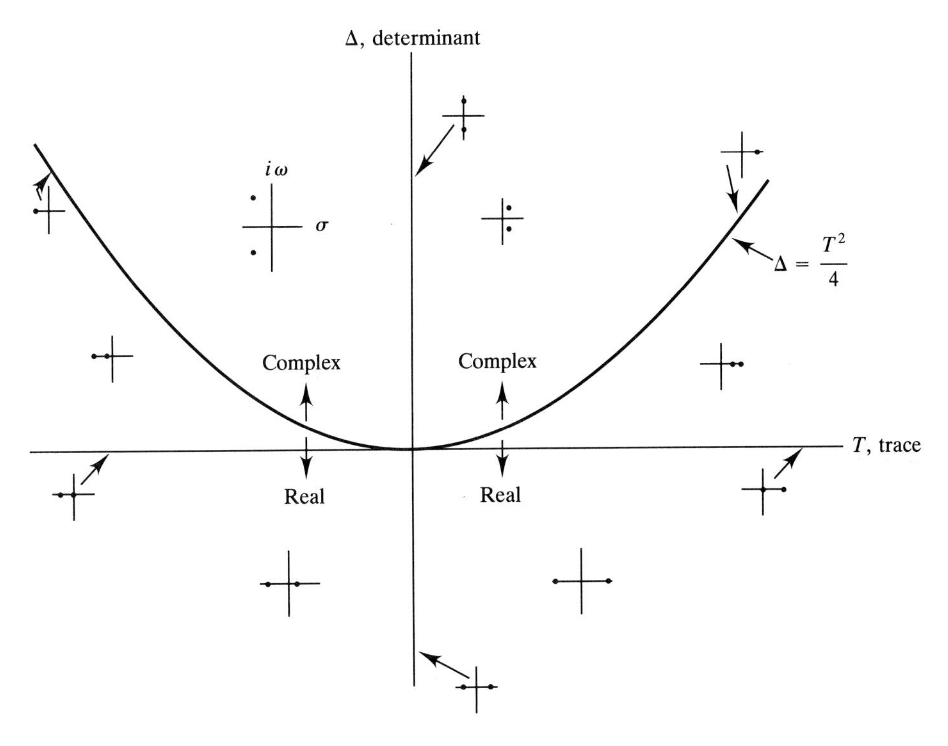

FIGURE 4.24
Types of eigenpairs for two-state systems for various regions of the $\Delta - T$ plane.

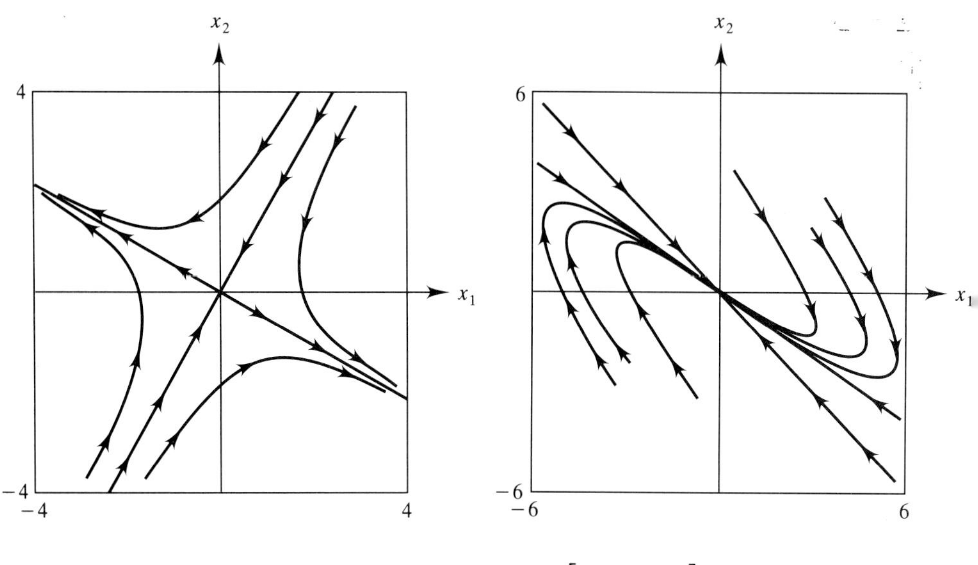

$[A] = \begin{bmatrix} .4 & -1.2 \\ -1.2 & -1.4 \end{bmatrix}$ (Unstable)

$\lambda_1 = -2, \{C\}_1^T = [1 \quad 2]$

$\lambda_2 = +1, \{C\}_2^T = [1 \quad -0.5]$

(a)

$[A] = \begin{bmatrix} 1 & 3 \\ -2 & -4 \end{bmatrix}$ (Asymptotically stable)

$\lambda_1 = -2, \{C\}_1^T = [1 \quad -1]$

$\lambda_2 = -1, \{C\}_2^T = [1 \quad -\frac{2}{3}]$

(b)

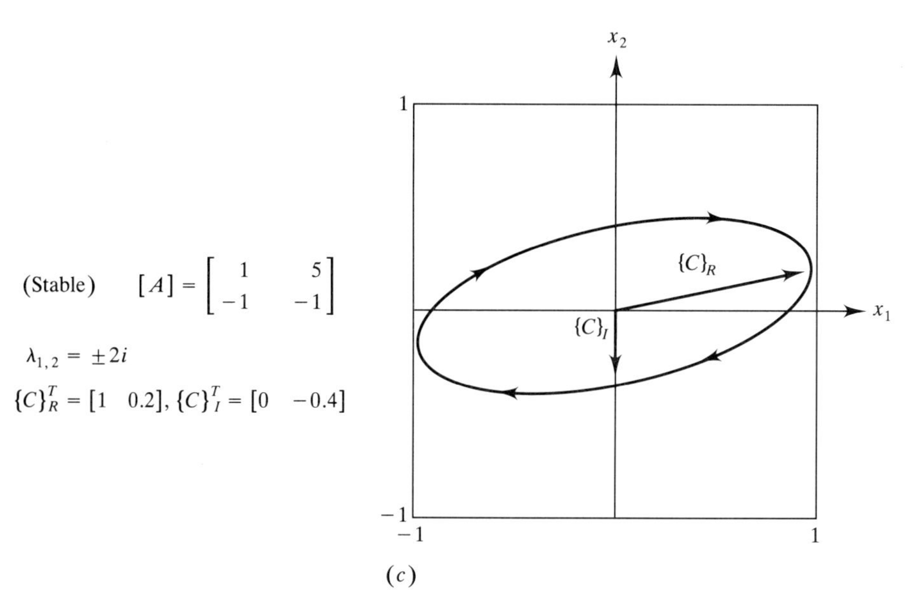

(c)

FIGURE 4.25
Phase portraits for some two-state systems. For each system the Jacobian matrix $[A]$, the eigenvalues, and the eigenvectors are indicated. (a) One positive and one negative eigenvalue (a saddle); (b) two real, distinct negative eigenvalues (a node); (c) a pair of conjugate imaginary eigenvalues (a center); (d) complex conjugate eigenvalues with a negative real part (a spiral sink); (e) one zero and one negative eigenvalue; here orbits asymptote to the eigenvector associated with the zero eigenvalue.

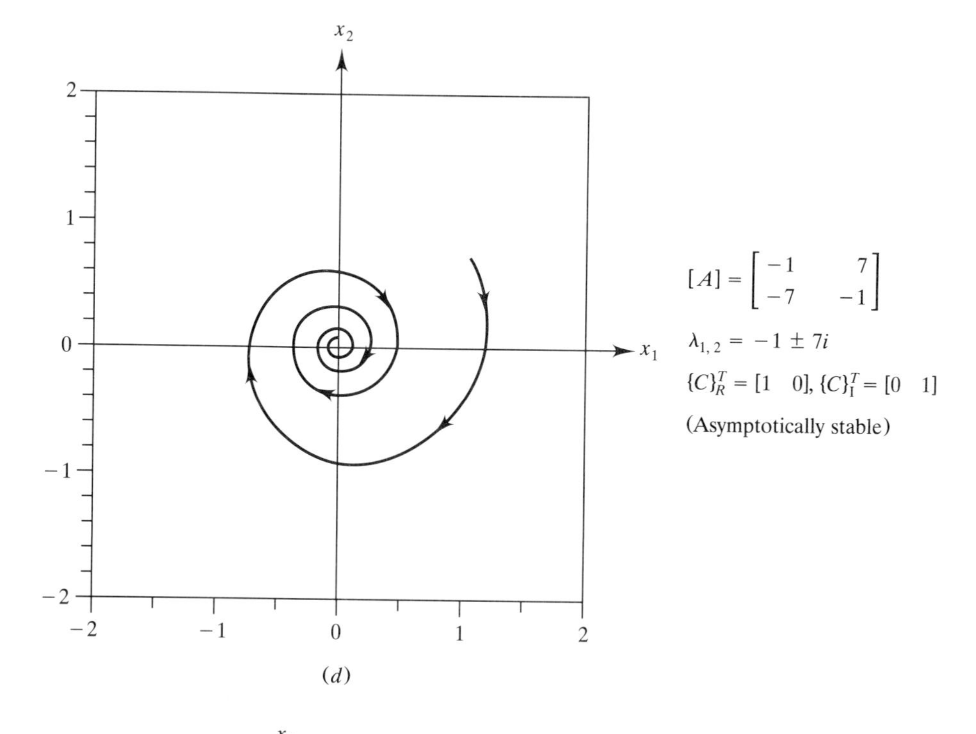

(*d*)

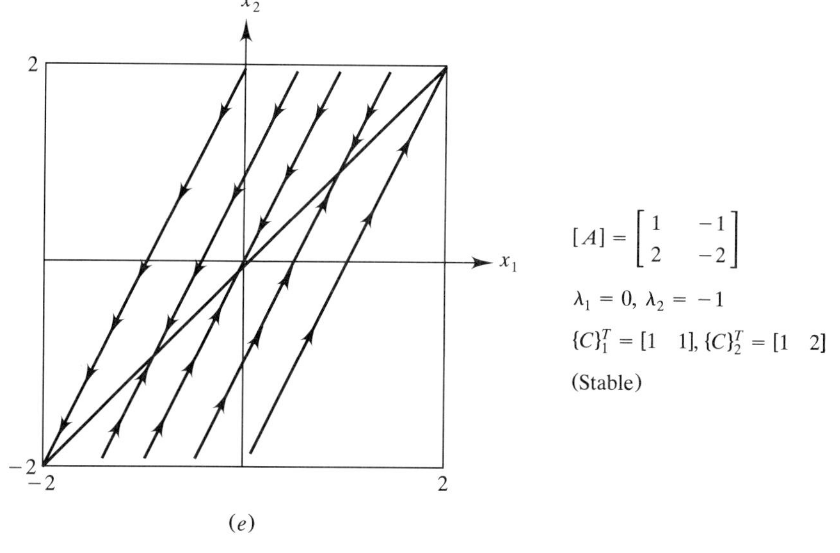

(*e*)

FIGURE 4.25 (*Continued*)

system matrix $[A]$, we calculate the trace T and determinant Δ. A given matrix $[A]$ is, therefore, represented by a point in the $\Delta - T$ plane, and we can immediately deduce the nature of the eigenvalues and the stability of equilibrium. We see that for asymptotic stability it is necessary that $T < 0$ and $\Delta > 0$ (the upper-left quadrant of Fig. 4.24). On the boundaries of this upper-left quadrant ($T = 0$ and $\Delta > 0$, $T < 0$ and $\Delta = 0$) the solutions are merely stable. In the remaining three quadrants of Fig. 4.24 equilibrium solutions are unstable.

In order to provide further insight into the behavior of the phase portraits for two-state systems, several of the cases depicted in Fig. 4.24 are considered in Fig. 4.25. In this figure several system models are given, along with the associated eigenvalues, eigenvectors, stability of equilibrium, classification of equilibrium, and orbits in the phase plane.

4.2.5 Higher-Order Systems

Let us now consider systems of arbitrary order $n > 2$. We suppose, as commonly occurs, that some of the eigenvalues are real and some occur in complex conjugate pairs. Assume we have r real eigenvalues, denoted as $\mu_1, \ldots, \mu_r$, and that we have p pairs of complex conjugate eigenvalues (where $2p + r = n$), denoted by $\sigma_1 \pm i\omega_1, \sigma_2 \pm i\omega_2, \ldots, \sigma_p \pm i\omega_p$. Then by combining the solution forms (4.81) and (4.109), we can write the general solution in the form

$$\{x(t)\} = \sum_{i=1}^{r} \alpha_i \{C\}_i e^{\mu_i t} + \sum_{j=1}^{p} A_j e^{\sigma_j t} \times \left\{ \{C_R\}_j \cos(\omega_j t + \theta_j) - \{C_I\}_j \sin(\omega_j t + \theta_j) \right\} \tag{4.115}$$

where $\{C\}_i$ are the real eigenvectors associated with the real eigenvalues μ_i and $\{C_R\}_j$ and $\{C_I\}_j$ are the real and imaginary parts of the complex eigenvectors associated with the conjugate pair of eigenvalues $\sigma_j \pm i\omega_j$. Then n real constants $\alpha_1, \ldots, \alpha_r, A_1, \ldots, A_p, \theta_1, \ldots, \theta_p$ are determined by the initial conditions. The first summation defines the exponential contributions to the solution due to the real eigenvalues, while the second summation defines the oscillatory contributions to the solution due to the complex eigenvalues. Equation (4.115) shows that, in general, each of the n states will exhibit both exponential and oscillatory behavior. In order that the system be asymptotically stable, all $\mu_i < 0$ and all $\sigma_j < 0$, so that $\{x(t)\} \to \{0\}$ as $t \to \infty$.

In order to gain insight into the structure of the solutions, let us evaluate (4.115) at $t = 0$,

$$\{x(0)\} = \sum_{i=1}^{r} \alpha_i \{C\}_i + \sum_{j=1}^{p} \left\{ A_j \cos\theta_j \{C_R\}_j - A_j \sin\theta_j \{C_I\}_j \right\} \tag{4.116}$$

This relation states that any n-dimensional vector can be expressed as a linear combination of the n vectors $\{C\}_1, \ldots, \{C\}_r, \{C_R\}_1, \ldots, \{C_R\}_p, \{C_I\}_1, \ldots, \{C_I\}_p$. This collection of vectors must, therefore, be linearly independent and must span, or form a basis in, n space. As in the two-state case, each of the eigenvectors $\{C\}_i$ associated with the real eigenvalues μ_i defines a different line, passing through the equilibrium solution at the origin, along which orbits will move exponentially toward or away from the origin, depending on whether $\mu_i < 0$ or $\mu_i > 0$. In a like manner, each *pair* of vectors $\{C_R\}_j, \{C_I\}_j$ defines a *plane* in n space in which orbits will spiral into or away from the origin, depending on whether the real part σ_j of the eigenvalue is negative or positive.

In light of this discussion, let us consider some special cases:

1. Suppose that the initial conditions $\{x(0)\}$ are chosen to lie along one of the real eigenvectors, say, $\{C\}_1$, so that of the n constants in (4.116), only α_1 is nonzero. Then the solution (4.115) reduces to

$$\{x(t)\} = \alpha_1 \{C\}_1 e^{\mu_1 t}$$

and the phase space orbits move along the straight line defined by the eigenvector $\{C\}_1$. Now suppose that the initial conditions are chosen to lie in the *plane* spanned by any two of the eigenvectors, say $\{C\}_1$ and $\{C\}_2$, associated with the real eigenvalues μ_1 and μ_2. In this case only α_1 and α_2 are nonzero, and the solution (4.115) reduces to

$$\{x(t)\} = \alpha_1 \{C\}_1 e^{\mu_1 t} + \alpha_2 \{C\}_2 e^{\mu_2 t}$$

The phase space orbits remain in the $\{C\}_1, \{C\}_2$ plane forever and will exhibit the same behavior as orbits for a two-state system with the real eigenvalues μ_1 and μ_2 and eigenvectors $\{C\}_1$ and $\{C\}_2$. The orbits evolve in a two-dimensional subspace of the n-dimensional phase space. We can extend this reasoning to subspaces of higher dimension.

2. Suppose the initial conditions are chosen to lie in the plane spanned by the vectors $\{C_R\}_1, \{C_I\}_1$ associated with the conjugate pair of eigenvalues $\sigma_1 \pm i\omega_1$, so that only A_1 is nonzero in (4.116). Then the solution (4.115) reduces to

$$\{x(t)\} = A_1 e^{\sigma_1 t}\{\{C_R\}_1 \cos(\omega_1 t + \theta_1) - \{C_I\}_1 \sin(\omega_1 t + \theta_1)\}$$

Orbits will evolve in the $\{C_R\}, \{C_I\}$ plane and will spiral into or away from the origin, depending on whether $\sigma_1 < 0$ or $\sigma_1 > 0$.

As an example, we now consider a three-state linear system having one real and one conjugate pair of eigenvalues.

Example 4.23 A Three-State System. The three-state system to be considered is

$$\begin{Bmatrix} \dot{x}_1 \\ \dot{x}_2 \\ \dot{x}_3 \end{Bmatrix} = \begin{bmatrix} 0 & 1 & 0 \\ -1 & 0 & 0 \\ 1 & 0 & -1 \end{bmatrix} \begin{Bmatrix} x_1 \\ x_2 \\ x_3 \end{Bmatrix} \tag{4.117}$$

The characteristic equation of the system is $\lambda^2(-1-\lambda)+(-1-\lambda)=0$ or $(\lambda^2+1)(\lambda+1)=0$, so that the eigenvalues are $\lambda_1=-1$, $\lambda_2=i$, $\lambda_3=-i$. The associated eigenvectors may be verified to be

$$\{C\}_1 = \begin{Bmatrix} 0 \\ 0 \\ 1 \end{Bmatrix}, \qquad \{C\}_2 = \begin{Bmatrix} 1 \\ i \\ \dfrac{1-i}{2} \end{Bmatrix}, \qquad \{C\}_3 = \begin{Bmatrix} 1 \\ -i \\ \dfrac{1+i}{2} \end{Bmatrix}$$

so that

$$\{C_R\} = \begin{Bmatrix} 1 \\ 0 \\ \dfrac{1}{2} \end{Bmatrix}, \qquad \{C_I\} = \begin{Bmatrix} 0 \\ 1 \\ \dfrac{-1}{2} \end{Bmatrix}$$

The general solution (4.115) is then

$$\{x(t)\} = \alpha_1 \begin{Bmatrix} 0 \\ 0 \\ 1 \end{Bmatrix} e^{-t} + A \left\{ \begin{Bmatrix} 1 \\ 0 \\ \dfrac{1}{2} \end{Bmatrix} \cos(t+\theta) - \begin{Bmatrix} 0 \\ 1 \\ \dfrac{-1}{2} \end{Bmatrix} \sin(t+\theta) \right\} \tag{4.118}$$

We see that the first eigenvector $\{C\}_1$ is directed along the x_3 axis (Fig. 4.26). Motions initiated on this axis will, therefore, move exponentially toward the origin (as e^{-t}). The vectors $\{C_R\}$ and $\{C_I\}$ define a plane, passing through the origin. The "main diagonals" of this plane are defined by the vectors $\pm\{C_R\}$ and $\pm\{C_I\}$, as shown in Fig. 4.26. The $\{C_R\}$ diagonal is a straight line passing through the points $(1, 0, \frac{1}{2})$ and $(-1, 0, \frac{-1}{2})$, while the $\{C_I\}$ diagonal passes through $(0, 1, -\frac{1}{2})$ and $(0, -1, \frac{1}{2})$. Orbits initiated in the $\{C_R\}, \{C_I\}$ plane will be closed curves, exhibiting periodic behavior at unit frequency ($\omega = 1$). The orbits in this plane neither spiral in nor spiral out because the real part of the eigenvalue is zero. A motion not initiated in the $\{C_R\}, \{C_I\}$ plane, nor on the x_3 axis, will exhibit both the exponential and periodic behaviors, as all three terms in (4.118) contribute. Such orbits actually evolve on the surfaces of cylinders, with the cylinder axes along x_3, which intersect the $\{C_R\}, \{C_I\}$ plane. As $t \to \infty$, all such orbits will collapse onto the $\{C_R\}, \{C_I\}$ plane, in which they remain forever, moving about the origin in closed, periodic orbits. For this linear system the equilibrium solution at the origin is stable, but not asymptotically stable, as the complex eigenvalues have a zero real part.

Closing comment. In Sec. 4.2 the objective has been to gain a clear understanding of the physical and geometrical meaning of the eigenvectors and eigenvalues that define the solution to the linearized ODE model which describes motions near equilibrium. The relationship of the eigenvalues to the stability of equilib-

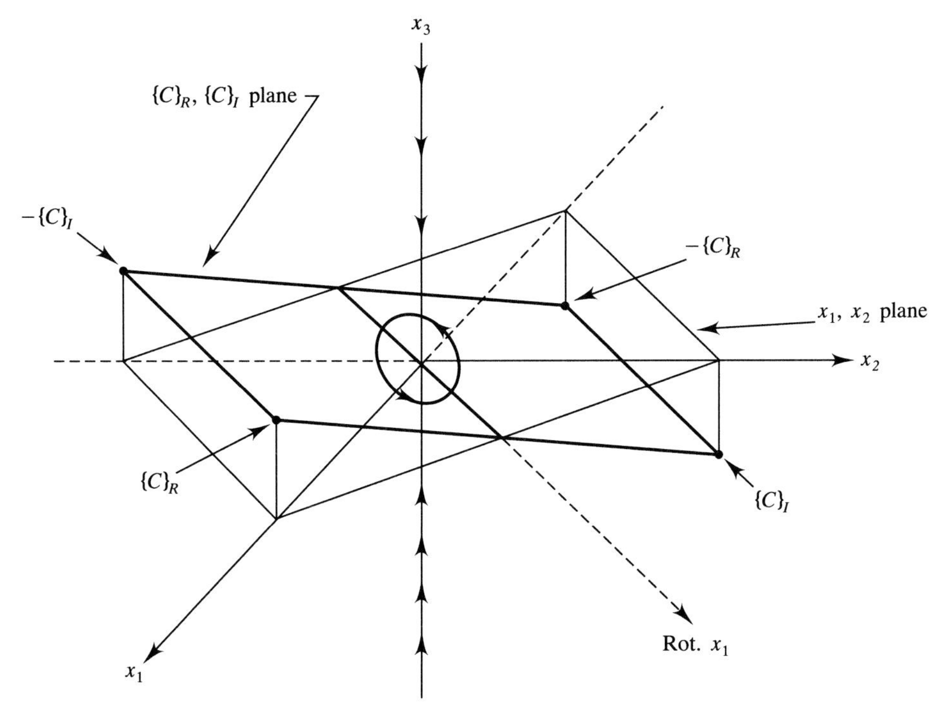

FIGURE 4.26
The phase space for the system of (4.117). Orbits initiated on the x_3 axis will move asymptotically to the original along x_3. The plane defined by the complex eigenvector pair $\{C\}_R, \{C\}_I$ contains closed orbits of frequency $\omega = 1$. This plane is found by plotting the vectors $\pm\{C\}_R$, $\pm\{C\}_I$ and then connecting the four points to form the rectangular plane shown. This plane may be visualized as follows: Rotate the x_1, x_2 plane by 45° about the x_3 axis; then rotate by $-26.6 = \tan^{-1}(0.5)$ about the rotated x_1 axis, which is shown in the figure as "Rot. x_1."

rium is an important aspect of this analysis. The next section contains a series of topics that are intended to fill out and broaden our understanding of the behavior of autonomous linear dynamic systems.

4.3 COMMENTS ON LINEAR SYSTEMS ANALYSIS

4.3.1 *n*th-Order ODE Models

In developing the eigenvalue-eigenvector solution methods of the preceding section, the system ODE model was assumed to be in the state variable form (4.62). The analysis of Sec. 4.2, however, should not be taken to imply that system math models must be in state variable form in order to find solutions. As was discussed in Sec. 1.4, the ODE model of a system of arbitrary order n can often be expressed conveniently as a single nth-order ODE in which the dependent variable $x(t)$ is generally one of the n system states $x_1, x_2, \ldots, x_n$. Normally $x(t)$ would be the state or one of the states in which we are most interested, but any state will do. For an autonomous, time invariant linear system, the nth-order ODE model may be represented in the form

$$x^{(n)} + a_1 x^{(n-1)} + \cdots + a_{n-1}\dot{x} + a_n x = 0 \tag{4.119}$$

where $x^{(n)} = d^n x/dt^n$. The solution to (4.119) may be obtained by assuming the exponential form

$$x(t) = ce^{\lambda t} \tag{4.120}$$

which is the scalar version of (4.71). Substitution of (4.120) into (4.119), followed by cancellation of the constant c, leads directly to the characteristic equation

$$\lambda^n + a_1\lambda^{n-1} + \cdots + a_{n-1}\lambda + a_n = 0$$

which will be the same, whether the system model is in the form (4.62) or (4.119). Denoting the resulting eigenvalues as $\lambda_1, \lambda_2, \ldots, \lambda_n$, the general solution to (4.119) is defined by the linear combination

$$x(t) = \sum_{i=1}^{n} c_i e^{\lambda_i t} \tag{4.121}$$

where the values of the constants c_i are determined by the n initial values $x(0), \dot{x}(0), \ldots, x^{(n-1)}(0)$. The result is equivalent to that of Sec. 4.2, that is, the eigenvalue-eigenvector methods used to solve (4.62).

4.3.2 Modes

In the general solution (4.115), each of the individual solution terms is sometimes referred to as a *mode* of the system response. There are r modes associated with the real eigenvalues, each mode defined by $\alpha_i\{C\}_i e^{\mu_i t}$. Each of these modes exhibits the same type of behavior as a first-order system, and each is associated with a line in phase space defined by the eigenvector $\{C\}_i$. Likewise, each pair of individual solution terms $A_j e^{\sigma_j t}\{\{C_R\}_j \cos(\omega_j t + \theta_j) - \{C_I\}_j \sin(\omega_j t + \theta_j)\}$, associated with the jth pair of conjugate eigenvalues, comprises a mode of the linear system. We get p such modes, each exhibiting growing ($\sigma > 0$) or decaying ($\sigma < 0$) oscillations or periodic behavior ($\sigma = 0$). Each of these modes exhibits the same type of behavior as a second-order system with complex eigenvalues, and each mode is associated with the plane defined by $\{C_R\}_j$ and $\{C_I\}_j$.

The discussion suggests that we consider the overall response (4.115) of the system to be constructed of r, first-order mode responses and p second-order (complex eigenvalue) mode responses. The motion of any linear autonomous system with distinct eigenvalues, no matter how high the system order, consists of combinations of the same types of response we get for first-order systems and for second-order systems with complex eigenvalues. This implies that if we thoroughly understand the first- and second-order systems, the conceptual transition to systems or arbitrary order is straightforward.

4.3.3 Modes and System Time Constants

The real part of a given eigenvalue λ_i governs the growth or decay of the associated mode. If we consider a mode characterized by a real, negative eigenvalue μ_i, the mode will decay exponentially. Because eigenvalues have units of 1/s, the reciprocal of the eigenvalue has units of time. This reciprocal is referred to as a *time constant*, defined by $T_i = 1/\,|\mu_i|$, where it is assumed that $\mu_i < 0$. With this notation the time dependence of a given mode is expressed as $e^{\mu_i t} = e^{-t/T_i}$, where $T_i > 0$ and $\mu_i < 0$. This notation shows that the mode will decay by a factor of e for each T_i seconds of elapsed time. If $|\mu_i|$ is small, then the associated time constant is large and the exponential decay is relatively slow. If $|\mu_i|$ is large, then T_i is small, with relatively fast exponential decay. Time constants are a convenient measure of response characteristics and are often used to characterize system properties. For instance, the phrase, "a first-order linear system with a time constant $T = 100$ ms" conveys the same information as does the single-state mathematical model $\dot{x} = -10x$.

For modes associated with complex eigenvalues $\lambda = \sigma \pm i\omega$, with $\sigma < 0$, the time constant T is defined by the real part σ of the eigenvalue, as this governs the decay of the oscillations, so that $T = 1/\,|\sigma|$.

In describing the decay of the various modes, a common rule of thumb is to say that the given mode has essentially "died out" when it is reduced to 2 percent of its initial magnitude. Although this is a somewhat arbitrary criterion, it is useful in describing modal decay. Because $e^{-4} \approx 0.02$, approximately four time constants are then required for a given mode to decay. This rule of thumb allows the time constant to be used quickly and easily to assess the amount of time that a given mode contributes measurably to the free response. Time constants are also useful in characterizing responses to certain inputs (see Sec. 5.2.2).

4.3.4 Higher-Order Systems and the Least-Damped Mode

We have seen that the free response of an n-state system consists of a superposition of the contributions from all of the modes. Let us suppose that we are analyzing a system (4.62) of fairly high order and that the system is asymptotically stable, so that all eigenvalues have negative real parts. To fix ideas, let us consider the case of a seven-state system for which the eigenvalues are as follows (the values are fairly typical):

$$\lambda_{1,2} = -0.05 \pm 10i \qquad \text{(Mode 1)}$$

$$\lambda_{3,4} = -10 \pm 20i \qquad \text{(Mode 2)}$$

$$\lambda_5 = -20 \qquad \text{(Mode 3)}$$

$$\lambda_6 = -30 \qquad \text{(Mode 4)}$$

$$\lambda_7 = -50 \qquad \text{(Mode 5)}$$

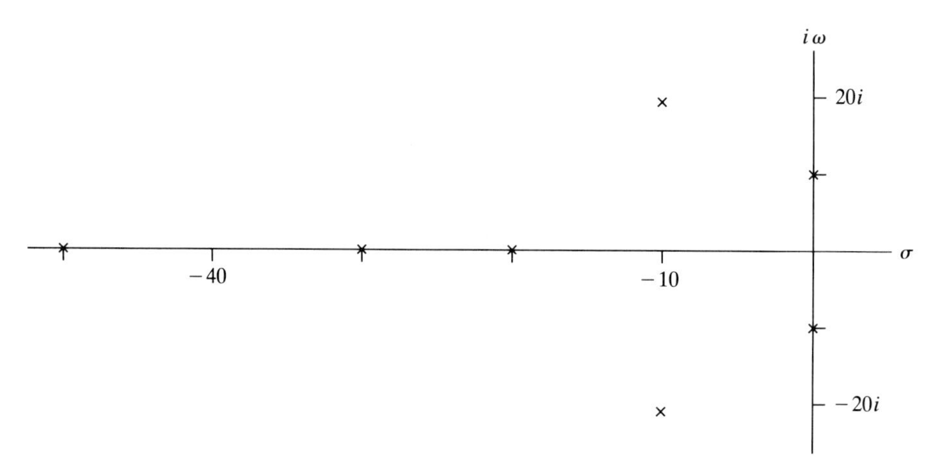

FIGURE 4.27
Complex plane representation of eigenvalues for a hypothetical seven-state system.

These eigenvalues, when plotted in the complex plane, will appear as in Fig. 4.27.

In the seven-dimensional phase space, the eigenvectors associated with modes 1 and 2 will define two planes, in each of which orbits will spiral into the origin (equilibrium). The eigenvectors associated with modes 3 through 5 will each define lines in phase space along which orbits move exponentially toward the origin. The solution is given by (4.115) with $r = 3$ and $p = 2$; there are five different modes involved. Let us examine the nature of the solution (4.115), assuming that at $t = 0$, all five modes are present. We can see that modes 2 through 5 will all quickly die out because the real parts of their eigenvalues are large and negative. On the other hand, mode 1 will decay very slowly. After a short interval of time then, modes 2 through 5 have died out, and all that is left in (4.115) is mode 1. From the phase space perspective, orbits will quickly collapse onto the plane associated with the mode 1 motion.

The point of this is that, except for a brief time interval near $t = 0$, the system response is essentially the same as that of a second-order system having the complex eigenvalues $\lambda_{1,2}$. In situations of this type we call mode 1 the *least-damped mode*; this mode is what we actually observe if we monitor the system behavior for a fairly long time. Situations of the type shown in Fig. 4.27 are fairly common; many, but certainly not all, higher-order systems behave essentially as second-order systems with complex eigenvalues. Graphs such as Fig. 4.27 and visualization of orbits in the phase space are quite helpful in understanding higher-order system behavior. The present discussion is illustrated with an example.

Example 4.24 A Fourth-Order System which Behaves Essentially Like a Second-Order System. Consider a system governed by the single fourth-order ODE:

$$\ddddot{x} + 30.05\dddot{x} + 226.5\ddot{x} + 760\dot{x} + 5000x = 0 \tag{4.122}$$

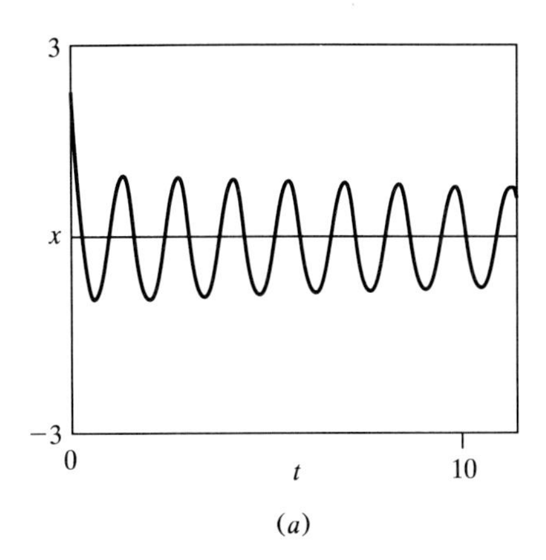

(a)

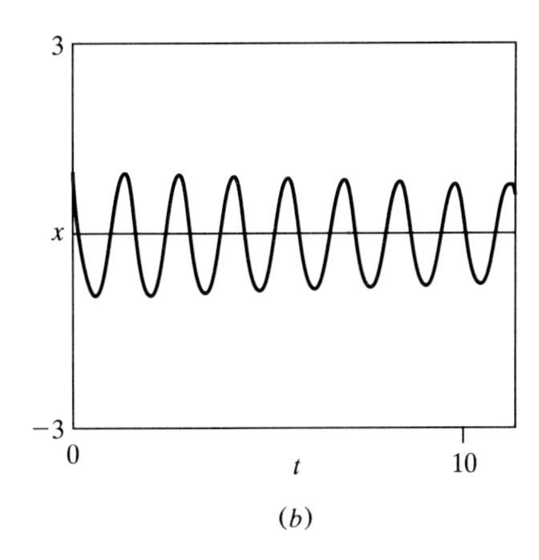

(b)

FIGURE 4.28
(*a*) The solution $x(t)$, (4.124), for the fourth-order system (4.122) for the case $\alpha_1 = \alpha_2 = A_1 = 1$, $\theta = 0$. (*b*) Mode 1 portion of the solution $x(t) = e^{-0.025t}\cos 5t$.

The states of the system are $x_1 = x$, $x_2 = \dot{x}$, $x_3 = \ddot{x}$, $x_4 = \dddot{x}$, and in state variable form the mathematical model is

$$\begin{Bmatrix} \dot{x}_1 \\ \dot{x}_2 \\ \dot{x}_3 \\ \dot{x}_4 \end{Bmatrix} = \begin{bmatrix} 0 & 1 & 0 & 0 \\ 0 & 0 & 1 & 0 \\ 0 & 0 & 0 & 1 \\ -5000 & -760 & -226.5 & -30.05 \end{bmatrix} \begin{Bmatrix} x_1 \\ x_2 \\ x_3 \\ x_4 \end{Bmatrix} \tag{4.123}$$

Note that in (4.123) the last equation, for $\dot{x}_4$, is a rewritten version of (4.122), while the remaining three equations define the other states.

The eigenvalues of this system are found to be

$$\begin{aligned} \lambda_{1,2} &= -0.025 \pm 5i \qquad &\text{(Mode 1)} \\ \lambda_3 &= -10 \qquad &\text{(Mode 2)} \\ \lambda_4 &= -20 \qquad &\text{(Mode 3)} \end{aligned}$$

Let us consider the response of only the first state $x(t) = x_1$. From the general solution (4.115) we have for $x(t)$.

$$x(t) = \alpha_1 e^{-10t} + \alpha_2 e^{-20t} + A_1 e^{-0.025t}\cos(5t + \theta) \tag{4.124}$$

The solution for $x_1 = x(t)$ is shown in Fig. 4.28 for the case $\alpha_1 = \alpha_2 = A_1 = 1$, $\theta = 0$, and we observe that, except for a small time interval near $t = 0$, the behavior is essentially the same as that of a second-order system.

4.3.5 Repeated Eigenvalues and Stability

Table 4.1 shows how the various stability classifications depend on the system eigenvalues: We have asymptotic stability if all eigenvalues have a negative real part, stability if one eigenvalue is zero or if one complex pair has a zero real part, and instability if any eigenvalue has a positive real part. It is to be remembered that these results were developed for the case of *distinct eigenvalues*, for which the characteristic equation of an n-state system has the factored form (4.77)

$$(\lambda - \lambda_1)(\lambda - \lambda_2) \cdots (\lambda - \lambda_n) = 0$$

where $\lambda_1, \lambda_2, \ldots, \lambda_n$ are the system eigenvalues.

We have so far not considered systems that have a repeated eigenvalue, that is, for which the factored form of the characteristic equation is

$$(\lambda - \lambda_1)^2(\lambda - \lambda_3) \cdots (\lambda - \lambda_n) = 0$$

where for convenience the first eigenvalue λ_1 is assumed to be real and repeated. It turns out that the existence of a repeated *nonzero eigenvalue* does not alter the stability of equilibrium. A repeated *zero eigenvalue*, however, will cause equilibrium to be unstable (whereas, if there is a single zero eigenvalue, equilibrium is stable). The interpretation of stability characteristics for systems with repeated eigenvalues is illustrated below by example.

Example 4.25 A Linear System with a Double-Zero Eigenvalue. Consider the linear harmonic oscillator of Example 4.2 in the special case for which both the spring and the damper are removed, so that the mass m is unrestrained in the x direction. The system equation of motion is simply $m\ddot{x} = 0$, or $\ddot{x} = 0$, or in state variable form with $x_1 = x$ and $x_2 = \dot{x}$:

$$\left.\begin{aligned} \dot{x}_1 &= x_2 \\ \dot{x}_2 &= 0 \end{aligned}\right\} \rightarrow \{\dot{x}\} = \begin{bmatrix} 0 & 1 \\ 0 & 0 \end{bmatrix} \{x\}$$

The system characteristic equation is simply $\lambda^2 = 0$, indicating the presence of a repeated zero eigenvalue, $\lambda_{1,2} = 0$. The state equations yield the equilibrium solution $\bar{x}_2 = 0$, $\bar{x}_1$ arbitrary. In physical terms this merely states that the mass m can be placed at rest anywhere along the x axis and the system will be at equilibrium. Suppose we select such an equilibrium position and then displace the mass slightly by imparting a small initial velocity $x_2(0) = \dot{x}(0)$. Then the mass will move away from equilibrium at constant speed, as shown in the phase plane diagram of Fig. 4.29. The solutions for the two states will be $x_1(t) = \dot{x}(0)t$ and $x_2(t) = \dot{x}(0)$, where $x_1(t)$ measures the displacement from the arbitrarily selected equilibrium position. The equilibrium configuration is *unstable* because almost all orbits initiated near equilibrium will move arbitrarily far from equilibrium as $t \rightarrow \infty$.

The result of this example extends to systems of arbitrary order n. If there is a repeated zero eigenvalue, with all other eigenvalues having a negative real part, then the equilibrium solutions will be unstable. Almost all orbits initiated near equilibrium will move away from equilibrium along orbits that asymptote to straight lines with constant speed of motion (in phase space). Furthermore, there will be arbitrariness in the location of equilibrium solutions in the phase space, so that an infinite number of equilibrium solutions, all unstable, is possible.

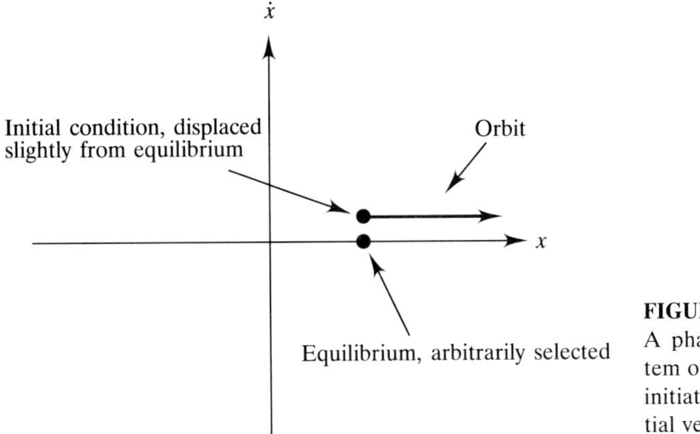

FIGURE 4.29
A phase plane orbit for the system of Example 4.25. The orbit is initiated by imposing a small initial velocity $x_2(0)$.

It is also of interest to explore briefly the case of a real negative eigenvalue which is repeated. This is now done by example.

Example 4.26 A Repeated Negative Eigenvalue. The two-state linear system to be considered is

$$\begin{Bmatrix} \dot{x}_1 \\ \dot{x}_2 \end{Bmatrix} = \begin{bmatrix} 0 & 1 \\ -1 & -2 \end{bmatrix} \begin{Bmatrix} x_1 \\ x_2 \end{Bmatrix} \tag{4.125}$$

The associated algebraic eigenvalue problem (4.73) is

$$\begin{bmatrix} -\lambda & 1 \\ -1 & -2-\lambda \end{bmatrix} \begin{Bmatrix} c_1 \\ c_2 \end{Bmatrix} = \bar{0} \tag{4.126}$$

and the characteristic equation is found to be

$$\lambda^2 + 2\lambda + 1 = 0$$

so that there is a repeated negative eigenvalue $\lambda_{1,2} = -1$. Putting this eigenvalue back into the first of (4.126), we find $c_1 + c_2 = 0$, so that the associated eigenvector is

$$\{C\}_1 = \begin{Bmatrix} 1 \\ -1 \end{Bmatrix}$$

Thus, one solution to the problem is given by

$$\{x\}_1 = \begin{Bmatrix} 1 \\ -1 \end{Bmatrix} e^{-t} \tag{4.127}$$

In the case of distinct eigenvalues, we would next use the second eigenvalue in (4.126) to obtain a second linearly independent solution of the exponential form (4.71). With a repeated eigenvalue, however, we cannot obtain such a second solution. There is only one solution of the assumed form (4.71) that can be used to obtain (4.127) from (4.125). Because the system is of second order, there are two linearly independent solutions. With repeated eigenvalues, however, the second solution has the form of the exponential function contained in the first solution (here e^{-t}), multiplied by time t. For the system of this example the second solution turns out to be as follows:

$$\{x\}_2 = \begin{Bmatrix} 1 \\ -1 \end{Bmatrix} te^{-t} + \begin{Bmatrix} 0 \\ 1 \end{Bmatrix} e^{-t} \tag{4.128}$$

Note that this second linearly independent solution also contains a purely exponential part. The general solution is then expressed as the linear combination

$$\{x(t)\} = \alpha_1 \begin{Bmatrix} 1 \\ -1 \end{Bmatrix} e^{-t} + \alpha_2 \left(\begin{Bmatrix} 1 \\ -1 \end{Bmatrix} te^{-t} + \begin{Bmatrix} 0 \\ 1 \end{Bmatrix} e^{-t} \right) \tag{4.129}$$

where α_1 and α_2 are real constants determined by the initial conditions.

It is not really the purpose here to present in detail methods for finding solutions of the type (4.129). The important thing to notice is the appearance in the response, for systems with repeated, negative eigenvalues $\mu < 0$, of the time function $te^{\mu t}$. We need to assess the effect of this type of function on stability.

The function $te^{\mu t}$ grows linearly with time for times t small enough so that $e^{\mu t} \approx 1$. This indicates that an orbit initiated near equilibrium ($\bar{x}_1 = \bar{x}_2 = 0$) will initially tend to move away from equilibrium. As time elapses, however, the exponential function $e^{\mu t}$ diminishes ($\mu < 0$), and as $t \to \infty$, the product $te^{\mu t} \to 0$ for all $\mu < 0$, no matter how small. Loosely speaking, as $t \to \infty$, $e^{\mu t} \to 0$ more strongly than $t \to \infty$, so that the product $te^{\mu t} \to 0$. Thus, equilibrium is still asymptotically stable, because $\{x(t)\} \to \{\bar{x}\}$ as $t \to \infty$ for all initial conditions. Stability is ensured even though the system initially tends to move away from equilibrium. The result is that the stability classifications for systems with repeated eigenvalues are the same as those of systems with distinct eigenvalues, *except* for the case of a repeated *zero eigenvalue*, which renders an otherwise stable system unstable.

4.3.6 Comparison of Linearized and Actual Phase Portraits for a Nonlinear System

We have seen that if the linear model (4.62) represents the near-equilibrium motion of a nonlinear system, then we expect the linear model to break down if we move sufficiently far from equilibrium. In order to further our understanding of this aspect of the analysis, let us consider a specific example.

Example 4.27 The Rolling Ball. Let us consider the motion of this system in the vicinity of the zero equilibrium solution $\bar{x} = (0, 0)$. We will find the eigenvalues and eigenvectors of the linear model and compare the resulting phase portraits to the actual ones shown in Fig. 4.9. First, for small motions near equilibrium $(0, 0)$, the linearized version of the mathematical model (4.15) is

$$\ddot{x} - gx = 0$$

Letting $g = 1$ to simplify things and denoting the states as $x_1 = x$, $x_2 = \dot{x}$, the

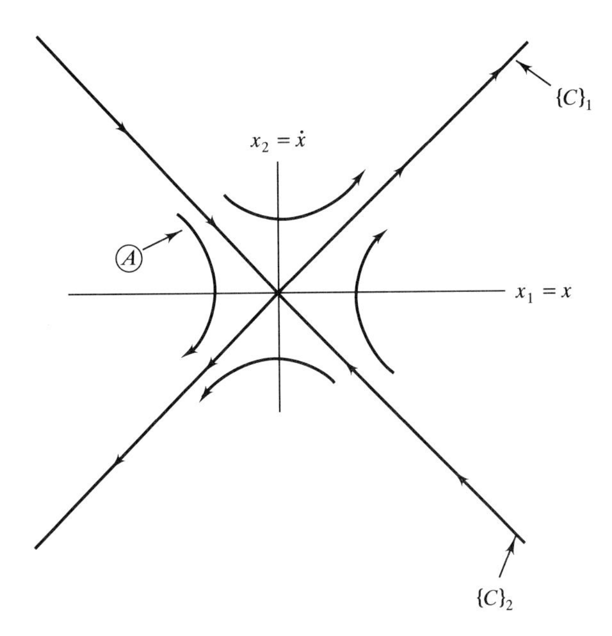

FIGURE 4.30
Linearized motions near the (0, 0) equilibrium.

state variable form of the linearized model is

$$\begin{Bmatrix} \dot{x}_1 \\ \dot{x}_2 \end{Bmatrix} = \begin{bmatrix} 0 & 1 \\ 1 & 0 \end{bmatrix} \begin{Bmatrix} x_1 \\ x_2 \end{Bmatrix}$$

The student may verify that the eigenvalues and eigenvectors are

$$\lambda_1 = 1, \quad \lambda_2 = -1, \quad \{C\}_1 = \begin{Bmatrix} 1 \\ 1 \end{Bmatrix}, \quad \{C\}_2 = \begin{Bmatrix} 1 \\ -1 \end{Bmatrix}$$

Thus, we have an unstable equilibrium (a saddle) for which the eigenvectors determine the phase flow near the origin, as shown in Fig. 4.30. Now let us superimpose the linearized flow shown in Fig. 4.30 on the actual nonlinear flow of Fig. 4.9. The result is shown in Fig. 4.31.

We see that in the vicinity of equilibrium the linearized model accurately depicts the system behavior. It gives, however, a totally erroneous description far from equilibrium, predicting that nearly all orbits initiated near the (0, 0) equilibrium will go to infinity; actually, all of the orbits are bounded. In the actual nonlinear system the orbits leaving the region near (0, 0) will eventually come under the influence of the two outer equilibria at (1, 0) and (−1, 0), and such interactions of multiple equilibria cannot be uncovered by a linear model. For instance, the orbit labeled A in Figs. 4.30 and 4.31 is, according to the linear model, a hyperbola which follows $-\{C\}_1$ to $-\infty$. Actually, however, the orbit A is a closed curve which surrounds the equilibrium at (−1, 0), signifying a periodic motion. Note also that the actual nonlinear orbit B (Fig. 4.31), which starts at (0, 0), encircles the equilibrium at (1, 0), and then comes back to (0, 0), is tangent to the eigenvectors $\{C\}_1$ and $\{C\}_2$ as equilibrium is approached. Thus, in the limit of very small deviations from equilibrium, the linear solutions mathematically

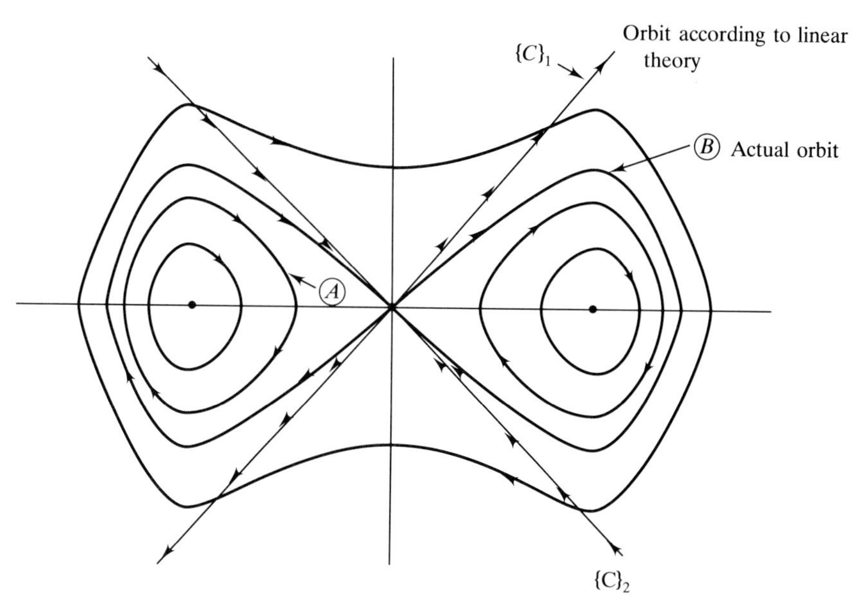

FIGURE 4.31
Comparison of the linearized and actual phase portraits for the rolling ball far from equilibrium.

approach the actual ones. But it is clear that the global (far from equilibrium) motions of the system cannot be analyzed using a linear model. If we were to do a similar analysis for the motion of the simple pendulum in the vicinity of the unstable equilibrium $(\pi, 0)$, we would find very similar behavior, as far as the comparison of the linear and actual phase portraits is concerned.

4.3.7 A Cautionary Note on Stability Determination Using the Linearized Model

Suppose we have a linear model (4.62) which is to represent the small motions of a nonlinear system near equilibrium. We have seen that if the linearized model (4.62) admits one eigenvalue with zero real part, then, if all of the other eigenvalues have negative real parts, we conclude the equilibrium solution to be stable but not asympototically stable. Actually, this is not necessarily true, as we now illustrate with a simple example. Consider the single-state nonlinear system:

$$\dot{x} = -x^3 \qquad x(0) = x_0 \tag{4.130}$$

The only equilibrium solution is $\bar{x} = 0$. The linearized model for motions near this equilibrium is

$$\dot{x} = 0 \qquad x(0) = x_0$$

so the linear model gives a zero eigenvalue and says that the system will stay forever at its initial location x_0, provided that x_0 is close to equilibrium. The actual near equilibrium behavior of the system is different, however. To see this, we can easily find the exact solution by writing (4.130) in the differential form $dx/x^3 = -dt$ and performing the integration, which yields

$$-\frac{1}{2x^2} + \frac{1}{2x_0^2} = -t$$

Solving this equation for $x(t)$, we find

$$x(t) = \pm\left[\frac{1}{(1/x_0^2) + 2t}\right]^{1/2} \tag{4.131}$$

where the $+$ root is used if $x_0 > 0$ and the $-$ root if $x_0 < 0$. Now (4.131) says that as $t \to \infty$, $x(t) \to 0$ for all x_0. We conclude, because (4.131) is the exact solution, that the equilibrium $\bar{x} = 0$ is asymptotically stable, and this is the correct result. The linearized model, then, has given us an incorrect description of the actual behavior near equilibrium. The reason for this in physical terms is as follows: Equation (4.131) shows that $x(t) \to 0$ as $t^{-1/2}$ for large t. This asymptotic approach to zero is "slower than exponential," and any time this occurs we will get an eigenvalue with a zero real part from the linearization.

This example illustrates that, for nonlinear systems for which the linearization implies that the equilibrium solution is merely stable, we have to be aware of the possibility that the linearization *may* give us an incorrect result. On the other hand, a linearization that implies either asymptotic stability or instability is always correct. Finally, the example and discussion do not imply that the linearization *must* fail if we get an eigenvalue with a zero real part, only that it *may* fail, depending on the system. For example, the student can show that the linearization near the equilibrium $(1,0)$ or $(-1,0)$ for the rolling ball (Example 4.6) yields the eigenvalues $\lambda_{1,2} = \pm i$. The periodic orbits surrounding these equilibria (Fig. 4.9), as predicted by the linear model, are what actually occur in the nonlinear system.

4.3.8 Stable, Unstable, and Center Subspaces

We have seen in Sec. 4.2 how the eigenvectors of an n-state system define certain "subspaces" of phase space, such as lines or planes, in which orbits stay for certain initial conditions. The visualization of such subspaces of phase space is useful, and here we introduce another interpretation of subspaces. Consider an n-state system with distinct eigenvalues and separate the eigenvalues (and associated eigenvectors) into three types: (1) those having negative real parts, whether the eigenvalues are real or complex, (2) those having positive real parts and (3) those having zero real parts. Now consider the subspace spanned by the eigenvectors associated with the eigenvalues having negative real parts. This subspace has the property that any orbit initiated in it will asymptotically approach the origin of the phase space. For this reason, we call this subspace the *stable subspace*. In a like manner, the subspace spanned by the eigenvectors associated with the eigenvalues having a positive real part is termed the *unstable subspace*, as orbits initiated in it will move arbitrarily far from the origin, according to the linear model. Finally, the subspace spanned by the eigenvectors associated with the eigenvalues having a zero real part is termed the *center subspace*. Orbits initiated in the center subspace will, according to the linear model, move neither asymptotically toward equilibrium nor arbitrarily far from equilibrium. The union of the stable, unstable, and center subspaces is the phase space.

For the three-state system considered in Example 4.23, the x_3 axis is the stable subspace (it is only one dimensional), and the plane defined by the real and imaginary parts $\{C_R\}$ and $\{C_I\}$ of the complex eigenvector is the center subspace. Notice that for this system all orbits will eventually collapse onto the center subspace.

4.3.9 Attractors

The asymptotically stable equilibrium, for which the stable subspace is the entire phase space, is the simplest example of what is called an *attractor*. Loosely speaking, an attractor is a region of the phase space (in the case of an equilibrium solution, this region is a single point) which all nearby regions of phase space will collapse onto as $t \to \infty$. In order to visualize the idea of an attractor, consider a three-state system with an equilibrium solution at the origin $(0, 0, 0)$ and suppose the eigenvalues are real and negative. Next consider a small sphere centered at the origin and visualize the infinite set of orbits (the flow) generated by taking all of the points on the surface of the sphere as initial conditions. Let the orbits evolve and watch how the surface of the sphere changes as the points starting on the surface move. Because the equilibrium $(0, 0, 0)$ is asymptotically stable, all orbits starting on the surface of the sphere will move asymptotically to the origin. The sphere will deform into an ellipsoid whose volume shrinks continually with time. As $t \to \infty$, all points initially on and inside the sphere will collapse onto the origin, and we have an *attractor*. In a like manner an n-dimensional volume near an asymptotically stable equilibrium of an n-state system will collapse onto the origin; the volume shrinks to zero, and all points in the volume asymptotically approach the attractor.

The region of the phase space that collapses onto the attractor is called the *domain of attraction*, or *basin of attraction*, of the attractor. For nonlinear systems having an asymptotically stable equilibrium, we often find that the domain of attraction is only a portion of the phase space and that orbits initiated outside the domain of attraction will go somewhere else (perhaps to another attractor). We will see some examples of other types of attractors in Chap. 10.

It should be clear that for the autonomous *linear* system (4.62), an asymptotically stable equilibrium solution is the only type of attractor possible and that the entire phase space is the domain of attraction of this attractor, as the general solution (4.115) to the linear problem dictates that $\{x(t)\} \to 0$ as $t \to \infty$ for all initial conditions. Note that the closed periodic orbits of purely imaginary eigenvalues ($\pm i\omega$) are not attractors, because orbits not initiated on a given orbit will not move asymptotically to the orbit.

The concept of an attractor in a dynamical system is very important because attractors represent asymptotically stable motions and hence are what we actually observe over long times in physical processes.

4.3.10 The Modal Matrix and Coordinate Decoupling

At a given time t, a point in the phase space is located by the values of the states $x_1, \ldots, x_n$ at that time. There is, however, nothing magic about the particular set of state variables chosen to describe orbits in phase space. We could, for example, define a new set of state variables $y_1, \ldots, y_n$ according to a *linear coordinate transformation*

$$\{x\} = [T]\{y\} \qquad \{y\} = [T]^{-1}\{x\} \tag{4.132}$$

where the "transformation matrix" $[T]$ must be invertible. This coordinate transformation is unique; if we have all of the x's, then we can find all of the y's, and vice versa, through (4.132).

Any invertible transformation matrix $[T]$ can be used in the study of the dynamical system (4.62). For instance, from (4.62) and (4.132) we could reformulate the mathematical model in terms of the new state variables $y_1, \ldots, y_n$ as follows:

$$\{\dot{x}\} = [T]\{\dot{y}\} = [A]\{x\} = [A][T]\{y\}$$

so that

$$\{\dot{y}\} = [T]^{-1}[A][T]\{y\}$$

or

$$\{\dot{y}\} = [B]\{y\} \tag{4.133}$$

where $[B] = [T]^{-1}[A][T]$. The mathematical model has the same form as (4.62), but the system matrix has changed. If we were to solve (4.133) for the y's, we would get the *same eigenvalues* (but different eigenvectors) as for the original system (4.62). This must be so because the solution to (4.133) for the y's, if put into (4.132) to get the x's, must yield the same result for the x's as we would get by solving (4.62) directly. This can occur only if the individual solutions to (4.62) and (4.133) have the same time dependences, and these are dictated solely by the eigenvalues.

We conclude that there are actually an infinite set of linear systems of the form (4.133) which are "equivalent" to the original one (4.62). The phase space orbits of all of these systems will also be "equivalent" in terms of the stability types of equilibria and the nature of the near-equilibrium orbits. To get phase space orbits for the y's, we can picture taking those for the x's and "stretching" them in a manner consistent with (4.132). Now we normally do our analysis using the original state variables $x_1, \ldots, x_n$ because they have obvious physical meaning. There is no reason to define the transformation (4.132) unless it enables us: (1) to better understand the system behavior and/or (2) to simplify the process of solving the system math model. We will now explore a remarkable transformation that does both of these things.

To see how this special transformation is defined, let us first rewrite the general solution (4.81) to (4.62) by using the previously defined matrix of eigenvectors $[C]$

$$[C] \equiv [\{C\}_1 | \{C\}_2 | \cdots | \{C\}_n] \tag{4.134}$$

which we now call the *modal matrix*. Then the solution (4.81) to (4.62) can be written in the following way:

$$\{x(t)\} = [C]\begin{Bmatrix} \alpha_1 e^{\lambda_1 t} \\ \vdots \\ \alpha_n e^{\lambda_n t} \end{Bmatrix} \tag{4.135}$$

Note that here we are working with the complex eigenvectors associated with any complex conjugate pairs of eigenvalues, rather than with the real and imaginary parts of such eigenvectors, so the matrix $[C]$ is complex, in general. Now each of the terms $\alpha_i e^{\lambda_i t}$ appearing in (4.135) may be considered to be the solution to the first-order equation

$$\dot{y}_i = \lambda_i y_i \Rightarrow y_i(t) = \alpha_i e^{\lambda_i t} \tag{4.136}$$

where the y's are new functions of time defined by (4.133). We view the vector

$$\{y\} = \begin{Bmatrix} \alpha_1 e^{\lambda_i t} \\ \vdots \\ \alpha_n e^{\lambda_n t} \end{Bmatrix} \tag{4.137}$$

as the solution to the problem

$$\{\dot{y}\} = [\lambda]\{y\} \tag{4.138}$$

where $[\lambda]$ stands for the diagonal matrix of eigenvalues:

$$[\lambda] = \begin{bmatrix} \lambda_1 & 0 & \cdots & 0 \\ 0 & \lambda_2 & \cdots & 0 \\ \vdots & \vdots & \ddots & \vdots \\ 0 & 0 & \cdots & \lambda_n \end{bmatrix}$$

We observe that (4.132) and (4.135) combine to form the linear transformation

$$\{x\} = [C]\{y\} \tag{4.139}$$

We view (4.139) as a coordinate transformation that defines the original state variables $x_1, \ldots, x_n$ in terms of a new set of state variables $y_1, \ldots, y_n$, which are governed by the uncoupled equations (4.138).

Now let us express the system mathematical model in terms of the new state variables $y_1, \ldots, y_n$. Following the procedure leading to (4.133), we obtain

$$\{\dot{y}\} = [C]^{-1}[A][C]\{y\}$$

and in view of (4.138) we have the following remarkable result:

$$[C]^{-1}[A][C] = [\lambda] \tag{4.140}$$

The result is this: If we use the modal matrix $[C]$ as the transformation matrix to define a new set of states $y_1, \ldots, y_n$, then the ODEs (4.138) comprising the mathematical model for the new states will be *uncoupled*; we have "diagonalized" the system via the special coordinate transformation (4.139). Each of the new states $y_1, \ldots, y_n$ measures the contribution to the response of a given mode. The equations (4.136) show that each mode behaves independently of all other modes, and the system math model in terms of the modal coordinates $y_1, \ldots, y_n$ consists of a set of n uncoupled first-order equations which can be solved independently of each other.

Thus, rather than viewing a point in phase space as being defined by the states, we could define a "modal phase space," in which the various modal coordinates $y_1, \ldots, y_n$ become the coordinate axes and measure individual modal contributions. We can then say that "this much of mode 1, plus that much of mode 2, . . . , gives us this point in phase space." In this interpretation the basic measures of what the system is doing are not the states but the modal coordinates, each measuring directly the contribution of a given mode.

The clever student now poses the following question: Because we have to have the eigenvectors to construct the coordinate transformation (4.139), this means we should have already solved the problem (4.62) in order to do any of this, so what have we accomplished? The answer is that, for the linear autonomous problem (4.62) that we have been studying, the use of the coordinate transformation (4.139) to uncouple the system equations does not add anything of practical use. We will see, however, when we consider the response of systems to external excitations (inputs), that doing the analysis using the modal coordinates y_i, rather than the state variables x_i, both simplifies the analysis and enables us to better understand why the system responds as it does to a particular input. This is especially true in the case of externally excited multidegree of freedom vibrating systems (Chap. 7), for which the decoupling of the governing ODEs using modal coordinates is *the* basic analytical technique of analysis. The reason for the utility of modal coordinates in general is that, for physical systems subjected to external stimuli, it is the *modes*, rather than the states, which are directly excited. Thus, we should view the response of dynamic systems in two ways: (1) what the states are doing and (2) what the modes are doing.

4.4 CONCLUDING REMARKS

Let us recap what we have done (and have not done) so far in the general analysis of dynamic systems. For the general n-state system (4.5), which may be nonlinear, we have seen how to find the equilibrium solutions, for which the states do not change with time. We introduced orbits in the phase space as a useful way to describe the dynamic behavior of the system. We introduced the general idea of the stability of equilibrium, and we saw how to linearize the general model (4.5) to describe motions of the system near equilibrium, according to (4.62). The eigenvalues and eigenvectors of the linear model (4.62) define the nature of the near-equilibrium motions in phase space, and the eigenvalues define the stability of equilibrium. A thorough understanding of these basic ideas is necessary in the analysis, design, and control of dynamic systems.

There are two important aspects of dynamic systems analysis which we have so far *not* studied. Although we now have a good idea how the linear n-state system (4.62) will behave, we have not said very much about the general behavior of *nonlinear* systems, especially motions far from equilibrium. All we have done is to investigate small neighborhoods of equilibrium using the linear model. We will look briefly at some interesting nonlinear behavior in Chap. 10, where we will see that nonlinear systems can exhibit interesting and practically important phenomena that cannot occur in the linear world. The student should be aware of the fact that, in the past 20 years or so, there has been a tremendous resurgence in the study of dynamic systems. This has been motivated by recent discoveries of new types of nonlinear behavior, specifically, "chaotic motion," discussed briefly in Chap. 10.

The second thing we have to be aware of is that most of the basic concepts introduced in this chapter, including equilibrium, stability of equilibrium, and the linearized model used to study nonequilibrium motions, are *basic properties* of the system (4.5) and hence are associated with the autonomous version (4.5) of the mathematical model (4.4). These basic properties characterize the dynamic behavior of the system for the case in which there are no inputs. Thus, motions of the system are assumed to be initiated by imposing initial conditions, due perhaps to short-lived disturbances, which place the system in a nonequilibrium condition at $t = 0$. For $t > 0$, the system then moves of its own accord in phase space, as dictated by the basic system properties, with no outside influences (inputs) present. In realistic engineering applications, however, we will frequently encounter external influences (inputs) which, although considered to be unrelated to the system itself, will nevertheless influence the system's behavior. In many cases, in fact, the primary concern of the analyst is to determine how the system will respond when acted upon by an input. Chapter 5 is devoted to this important problem. The main point of this comment is that we always have to be aware of the two different types of responses which occur: (1) the "free response" due to nonequilibrium initial conditions, which is governed solely by the basic system properties and (2) the response due to inputs, which is governed by both the basic system properties and the properties of the input.

REFERENCES

4.1. Willems, J. L.: *Stability Theory of Dynamical Systems*, Wiley, New York (1970). Consult for additional material on stability and phase space motions of dynamical systems.

4.2. Hirsch, M. W., and S. Smale: *Differential Equations, Dynamical Systems, and Linear Algebra*, Academic Press, New York (1974). A classic on dynamical systems, this book would generally be considered to be at the graduate level for engineers.

4.3. Meirovitch, L.: *Introduction to Dynamics and Control*, Wiley, New York (1985). See Chap. 8 for the discussion of stability and motion in the phase space.

EXERCISES

1. Consider the three second-order ODEs (2.60) which define the earthquake-building model of Example 2.6. Define state variables and convert the model to state variable form.

2. Consider the nonlinear, fourth-order mathematical model of the two-link planar manipulator, (2.68) and (2.69), of Example 2.10. Identify the four states and convert the model to state variable form. Note that this conversion is a bit involved because of the *inertial coupling* in the equations; that is, both of the accelerations $\ddot{\theta}_1$ and $\ddot{\theta}_2$ appear in each equation.

3. The mathematical model for the airfoil suspended in a wind tunnel (Example 2.9) was found to be (2.65) and (2.66), of the form

$$m\ddot{x} - m\,d\ddot{\theta} + kx = -mg$$

$$I\ddot{\theta} - m\,d\ddot{x} + k_T\theta = mgd \qquad (P4.3)$$

Determine the values of the states x, $\dot{x}$, θ, and $\dot{\theta}$ at equilibrium.

4. a. The mathematical model of the two-link planar manipulator (Example 2.10) is given by equations (2.68) and (2.69). Determine the equilibrium solutions for the states $\theta_1, \dot{\theta}_1, \theta_2, \dot{\theta}_2$, and sketch the configuration of the two links for each of the equilibrium solutions. Recall that the equilibrium solutions are associated with the autonomous version of the mathematical model, so that the inputs $M_1(t)$ and $M_2(t)$ must be set to zero in order to find the equilibrium solutions.

b. Determine, based on physical considerations, the stability of each of the equilibrium solutions found in part *a*.

5. The equations of motion of a certain dynamical system are given by

$$\ddot{r} - (R + r)\dot{\theta}^2 + \omega^2 r + g(R + r)(1 - \cos\theta) = F(t)$$

$$(R + r)\ddot{\theta} + 2\dot{r}\dot{\theta} + g\sin\theta = 0 \qquad (P4.5)$$

Here R, g, and ω^2 are specified constants and $F(t)$ is a specified input function of time.

a. Identify the states in this model.

b. Find at least one equilibrium solution for this model.

c. Obtain a set of linearized equations that govern motions near this equilibrium state.

6. Consider the vertical descent of an object near the surface of the earth (Example 2.7), for which the mathematical model for the velocity v is given by (4.21). Obtain a linearized model that describes small variations of velocity from the equilibrium or "terminal fall" velocity $\bar{v}$ given by (4.22). Is the terminal fall condition stable?

7. Identical spring-damper combinations are connected via a nonlinear spring that produces a restoring force proportional to the cube of the spring deflection (Fig. *P*4.7). Show that the equations of motion are as follows:

$$c\dot{x}_1 + kx_1 + k_1(x_1 - x_2)^3 = 0$$

$$c\dot{x}_2 + kx_2 + k_1(x_2 - x_1)^3 = 0 \qquad (P4.7)$$

For the case $c = k = k_1 = 1.0$ (each in appropriate units), put the model into state variable form. Obtain a linearized model valid for small motions about the equilibrium solution $\bar{x}_1 = \bar{x}_2 = 0$. If the system is displaced slightly from this equilibrium condition, will $\bar{x}_1$ and $\bar{x}_2$ exhibit exponential or oscillatory behavior? Explain.

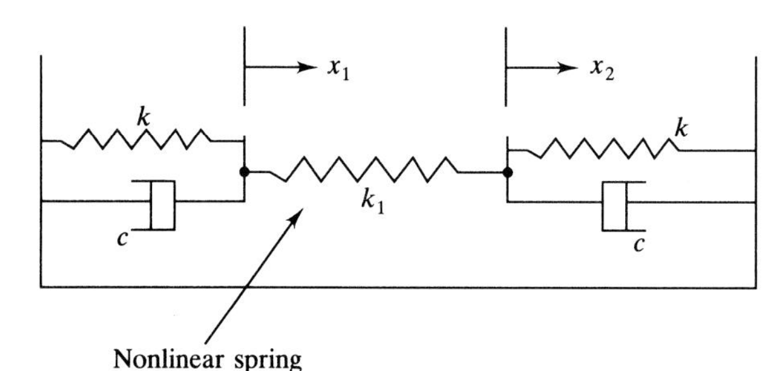

FIGURE *P*4.7

8. The mathematical model for the armature-controlled dc motor was developed in Example 3.8. Assuming the load on the motor to consist only of added inertia, the applicable model is (3.65) with $M_L = 0$. For the case of constant armature voltage $V_A(t) = V_0$, determine the equilibrium value ω of shaft speed $\omega = \dot{\theta}$.

9. A two-state linear system is defined by the mathematical model

$$\{\dot{x}\} = [A]\{x\} \qquad [A] = \begin{bmatrix} 0 & 1 \\ -9 & 0 \end{bmatrix}$$

Determine the eigenvalues and eigenvectors and the stability of equilibrium. Sketch the orbits in the phase plane.

10. Consider the following two-state systems $\{\dot{x}\} = [A]\{x\}$, for which the eigenvalues and eigenvectors are as follows:

a. $\lambda_1 = -\frac{1}{2} \qquad \{C\}_1 = \begin{bmatrix} 1 \\ 2 \end{bmatrix}$

$\lambda_2 = +1 \qquad \{C\}_2 = \begin{bmatrix} 1 \\ \frac{1}{2} \end{bmatrix}$

b. $\lambda_1 = -1 \qquad \{C\}_1 = \begin{bmatrix} 1 \\ -1 \end{bmatrix}$

$\lambda_2 = -2 \qquad \{C\}_2 = \begin{bmatrix} 1 \\ \frac{1}{2} \end{bmatrix}$

Sketch the orbits in the phase plane. Indicate clearly the two directions (lines) along which only one of the two solutions is present.

11. Consider a two-state system for which the eigenvalues and eigenvectors are

$$\lambda_{1,2} = -\tfrac{1}{2} \pm 10i \qquad \{C\}_1 = \begin{Bmatrix} 1 \\ 3 + 4i \end{Bmatrix} \qquad \{C\}_2 = \{C\}_1^*$$

Sketch the orbits of the system in the phase plane, and determine the stability of equilibrium.

12. Consider the motion of a ball rolling without friction on a surface $y(x)$ in the plane, as shown in Fig. P4.12. The points x_1, x_2 shown are given.

a. Identify the equilibrium solutions and their stability and sketch the local (i.e., linearized) phase portraits near each equilibrium.

b. Sketch the global phase portraits (i.e., as in Fig. 4.9), and identify any infinite period orbits.

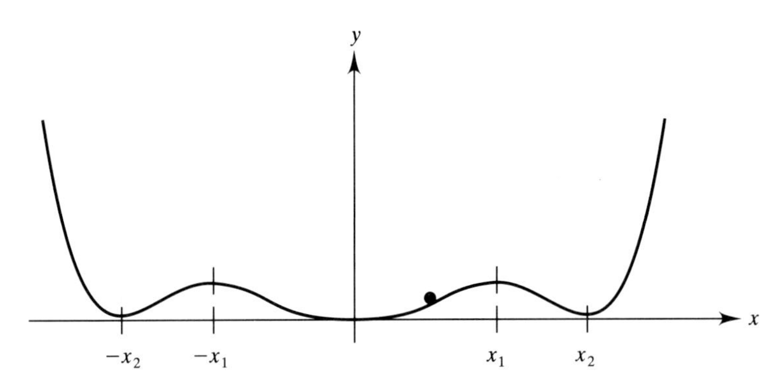

FIGURE P**4.12**

13. Consider the system shown in Fig. P4.13. Derive the system equations in state variable form. Find the eigenvalues and eigenvectors, and determine the stability of the system.

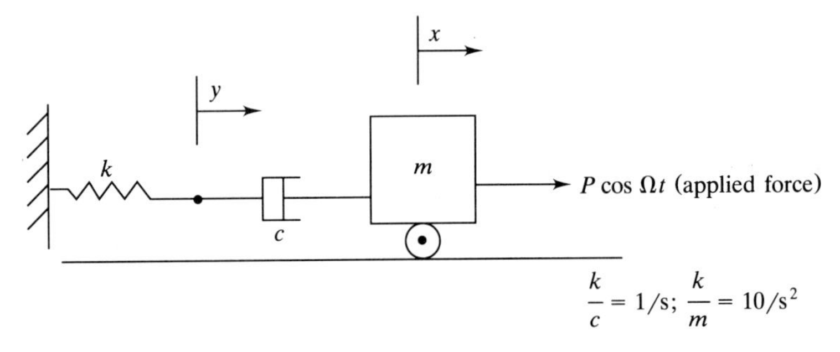

FIGURE P**4.13**

14. The power series expansions for the exponential and trigonometric functions are

$$e^x = 1 + x - \frac{x^2}{2} + \frac{x^3}{6} - \frac{x^4}{24} + \cdots$$

$$\sin x = x - \frac{x^3}{6} + \frac{x^5}{120} + \cdots$$

$$\cos x = 1 - \frac{x^2}{2} + \frac{x^4}{24} + \cdots$$

Use these basic results to verify DeMoirve's theorem directly, $e^{i\theta} = \cos\theta + i\sin\theta$.

For the following systems of equations, write the equations in vector-matrix form, find the eigenvalues, the eigenvectors, and the general solutions, determine the stability of equilibrium, and sketch the orbits in the phase space.

15.
$$\dot{x} = x - 2y$$
$$\dot{y} = -2x + y$$

16.
$$\dot{x}_1 = -7x_1 + 4x_2$$
$$\dot{x}_2 = -8x_1 + x_2$$

17.
$$\dot{x} = -2x - 6y$$
$$\dot{y} = 2x + y$$

18.
$$\dot{x}_1 = -2x_1 + x_2$$
$$\dot{x}_2 = -3x_2$$

19.
$$\dot{x}_1 = x_2$$
$$\dot{x}_2 = -x_1$$
$$\dot{x}_3 = -2x_3$$

20.
$$\dot{x}_1 = x_2$$
$$\dot{x}_2 = -5x_1$$

21.
$$\dot{u} = u + 2v$$
$$\dot{v} = -5u - v$$

22.
$$\dot{x}_1 = -2x_2$$
$$\dot{x}_2 = x_1 + 2x_2$$

23. Consider the following nonlinear state equations:

$$\dot{x}_1 = x_1 - x_1^3 + \alpha x_1 x_2$$
$$\dot{x}_2 = -x_2 + \beta x_1 x_2$$

a. Find all of the equilibrium solutions for this system.
b. Obtain the linearized model which describes the motion near the nonzero equilibrium solution.

24. Consider the system shown in Fig. *P*4.24: A mass m translates vertically and is restrained by a linear spring of stiffness k. A massless bar, pivoted at m, has a static load p_0 applied at one end as shown (as x changes, the load p_0 moves through some displacement, doing work). Ignore gravity, and do not assume that θ has to be small. Find the equilibrium solutions for this system.

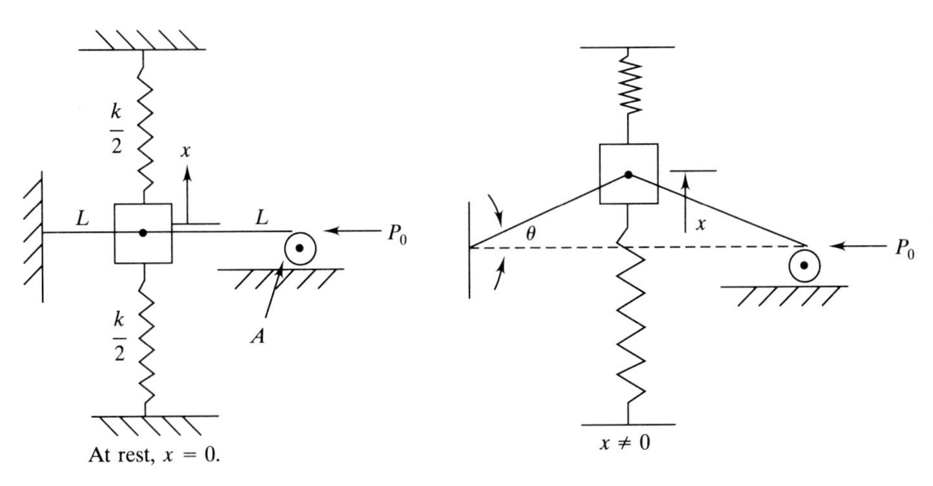

FIGURE ***P*4.24**

25. For the following nonlinear two-state system, find all of the equilibrium solutions. For each, obtain the linearized ODE model valid for motions near equilibrium, find the eigenvalues and eigenvectors, and sketch the phase portraits in the vicinity of the equilibria:

$$\dot{x} = y(1 - x^2)$$
$$\dot{y} = -y + 1$$

26. Consider the two-state nonlinear system defined by the ODE model

$$\dot{x}_1 = x_1$$
$$\dot{x}_2 = -x_2 + x_1^3$$

Plot the orbits in the phase plane, and compare the linearized and actual, nonlinear orbits near and far away from equilibrium. *Hint*: Exact solutions can be found as follows: You can verify that the solutions have the form $x_1(t) = Ae^t$, $x_2(t) = Be^{-t} + Ce^{3t}$, where A and B are found from the initial conditions and C is selected so that the second ODE is satisfied.

27. A three-state linear system (x_1, x_2, x_3) has the following eigenvalues and eigenvectors:

$$\lambda_1 = -1 \qquad \lambda_{2,3} = \pm 4i$$

$$\{C\}_1 = \begin{Bmatrix} 0 \\ 1 \\ 0 \end{Bmatrix} \qquad \{C\}_{2,3} = \begin{Bmatrix} 1 \\ 0 \\ 1 \pm i \end{Bmatrix}$$

Sketch typical orbits in the 3-D phase space, and identify the stable, unstable, and center eigenspaces. Is the equilibrium solution (0, 0, 0) stable? Where (i.e., which subspace) will orbits collapse onto as $t \to \infty$?

28. Discuss the motion of the two-state system modeled as

$$[A] = \begin{bmatrix} 0 & k_1 \\ k_2 & 0 \end{bmatrix}$$

where k_1 and k_2 are nonzero and positive. Specifically, (*a*) will there be oscillatory behavior, (*b*) can such a system be stable?

29. Consider the simple model of the air cushion landing system as described in Example 3.5. The mathematical model for the system was found to be

$$m\ddot{y} = (p_c - p_a)A - mg$$

$$\dot{p}_c = \frac{\gamma R T_c}{V_c}\left(q_{\text{in}} - q_{\text{out}} - \frac{p_c A \dot{y}}{R T_c}\right) \qquad (P4.29)$$

where q_{in} = *Specified constant* mass inflow rate

$$q_{\text{out}} = \frac{CLyp_c}{(RT_c)^{1/2}}\left\{\frac{2\gamma}{\gamma - 1}\left[\left(\frac{p_a}{p_c}\right)^{2/\gamma} - \left(\frac{p_a}{p_c}\right)^{(\gamma+1)/\gamma}\right]\right\}^{1/2} \qquad \gamma = 1.4$$

$$V_c = A(d + y)$$

Assume: (*a*) $p_c > p_a$ always and (*b*) $T_c \approx$ constant.

a. Determine the equilibrium values of the states.

b. Linearize the system model for small motions about equilibrium.

30. An oscillator of mass m is restrained by a linear spring of stiffness k. The mass m slides on a surface which exerts a frictional force F_f. This frictional force is assumed

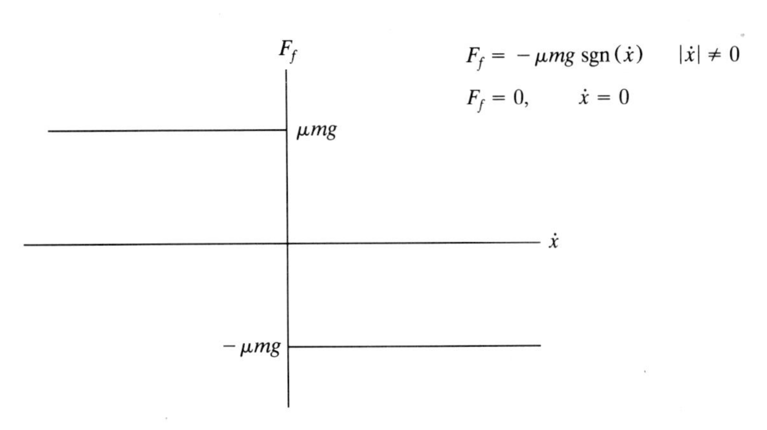

FIGURE ***P*4.30**

to follow the "dry friction" model, for which F_f has constant magnitude and is exerted in the opposite direction in which the mass is moving. The friction force F_f is sketched in Fig. $P4.30$.

Is this a linear damping force? Prove that this force continually dissipates energy if the mass m is in motion. If m is set into motion, will the decay of the resulting oscillations be exponential? Is it possible, according to the linearization procedure discussed in Sec. 4.1.4, to develop a linearized model of this system for small motions near equilibrium? Explain all answers.

CHAPTER 5

RESPONSE OF DYNAMIC SYSTEMS TO INPUTS

5.1 INTRODUCTORY COMMENTS

5.1.1 The Basic Problem

In Chap. 4 we analyzed the autonomous system response governed by the generally nonlinear mathematical model

$$\dot{x}_i = f_i(x_1, \ldots, x_n) \qquad i = 1, \ldots, n \tag{5.1}$$

We found equilibrium solutions $\bar{x}_i$ and developed linearized models, of the form*

$$\{\dot{x}\} = [A]\{x\} \tag{5.2}$$

to study the motions of the system in the vicinity of equilibrium. These motions and the associated stability of equilibrium were governed by the eigenvalues and eigenvectors of (5.2). Since our concern in Chap. 4 was to obtain a thorough understanding of basic system properties, we disallowed any time-dependent outside influences (inputs). Thus, the system was assumed to be set in motion through imposition of initial conditions, often arising in practice from short-lived disturbances, and thereafter to move of its own accord with no outside influences present. We call this type of response "*free*" *response* because the system moves "freely" on its own; external effects (inputs) are not involved.

*In (5.2) it is to be understood that $\{x\}$ represents a small deviation in the state vector from the equilibrium condition.

In this chapter we will consider the response due to externally imposed time-dependent influences (inputs). Our primary objective will be to examine the cause-and-effect relationship between the inputs and the resulting system response, which we term the *output*. Thus, we now return to the more general, nonautonomous mathematical model (4.2),

$$\dot{x}_i = f_i(x_1, \ldots, x_n, t) \qquad i = 1, \ldots, n \tag{5.3}$$

In general there may be one or more inputs involved in the physical problem. Let us suppose that there are m inputs and call them $u_1(t), u_2(t), \ldots, u_m(t)$. For most of the systems we will analyze, the inputs will be assumed to be known functions of time which are imposed on the system *additively*. This means that the state derivatives $f_i(x_1, \ldots, x_n, t)$ in (5.3) can be split into two terms, one involving only the states and one involving only the inputs, in the following way:

$$\dot{x}_i = f_i(x_1, \ldots, x_n) + g_i(t) \tag{5.4}$$

Here the functions $f_i(x_1, \ldots, x_n)$ are the same functions that appear in the autonomous system model (4.5), and the time-dependent functions $g_1(t), \ldots, g_n(t)$ may each be a function of one or more of the inputs $u_j(t)$ and possibly of the derivatives of the inputs. The type of relations typically existing between the inputs $u_j(t)$ and the functions $g_i(t)$ in (5.4) are illustrated in the following example.

Example 5.1 A Mechanical System Excited by a Support Motion. Here we consider the motion of a damped oscillator excited by a prescribed motion $y(t)$ of the support to which the spring and damper are grounded, as shown in Fig. 5.1. The equation of motion of the system is found from Newton's second law to be

$$m\ddot{x} = -c(\dot{x} - \dot{y}) - k(x - y)$$

If we let $x_1 = x$, $x_2 = \dot{x}$, then the state variable form of the system mathematical model is

$$\begin{aligned} \dot{x}_1 &= x_2 \\ \dot{x}_2 &= \frac{1}{m}(-cx_2 - kx_1) + \frac{1}{m}(c\dot{y} + ky) \end{aligned} \tag{5.5}$$

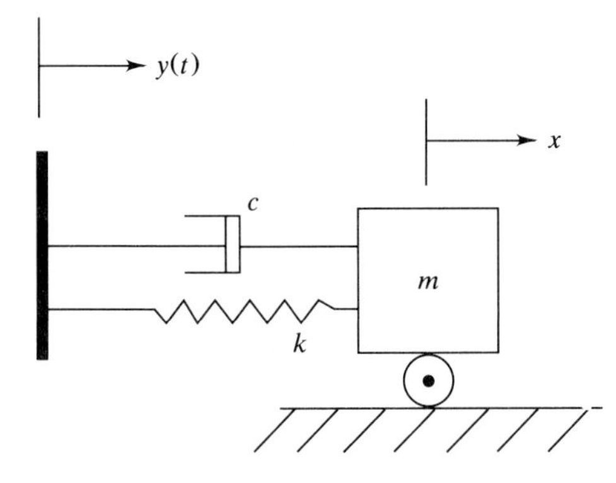

FIGURE 5.1
The linear oscillator excited by a prescribed motion $y(t)$ of the support.

In this case there is a single input $u_1(t) = y(t)$, and the functions $g_i(t)$ appearing in (5.4) are $g_1(t) = 0$, $g_2(t) = (1/m)[c\dot{y}(t) + ky(t)]$. Note that $g_2(t)$ is a function of both the input $y(t)$ and its derivative $\dot{y}(t)$.

The distinction here between the u's and the g's is made only to distinguish the physical inputs $u_j(t)$ from the functions of the inputs $g_i(t)$, which eventually appear in the state equations. Often we just consider the functions $g_i(t)$ to be the inputs.

As another example, consider the automobile-suspension model of Example 2.5, for which the equations of motion for the coordinates x_1, x_2, x_3, and θ are (2.56) through (2.59). The two inputs are $x_f(t)$ and $x_r(t)$, the front and rear road irregularities. The derivatives $\dot{x}_f$ and $\dot{x}_r$ also appear in the equations of motion; we have a 4 degree of freedom, eighth-order system with two support motions.

In our study of the response of systems to inputs, we will often consider, as we did for the autonomous systems of Chap. 4, the effects of inputs on the linear motions of the system near equilibrium. In such cases we will use the linearized version of (5.4):

$$\{\dot{x}\} = [A]\{x\} + \{g(t)\} \tag{5.6}$$

Here $[A]$ is the system Jacobian matrix, the "input vector" $\{g(t)\}^T = \lfloor g_1(t) \cdots g_n(t) \rfloor$, and the state vector $\{x\}$ now represents a deviation from the equilibrium condition.

A final introductory comment relates to the types of phase space orbits that can occur for nonautonomous systems of the type (5.3). We have seen that for the autonomous system (4.5), the vector field defined by the $f_i(x_1, \ldots, x_n)$ is time independent, so that phase space orbits of (4.5) cannot cross each other. The vector field for the nonautonomous system (5.3), however, will be time-varying. Thus, a given phase orbit passing through the point x_0 at time t_0 will be tangent to the vector $f(x_0, t_0)$. At some later time t_1, however, the vector $f(x_0, t_1)$ may differ from $f(x_0, t_0)$, so that an orbit may pass through the same point x_0 at time t_1 in a different direction from that at time t_0. Thus, for the nonautonomous system (5.3), phase space orbits can cross each other, and this can lead to more complex phase space motions than would be observed for an autonomous system of the same order.

5.1.2 Types of Inputs

In the examples presented in Chaps. 2 and 3 we have encountered several types of inputs that commonly occur and can be classified as below:

1. *Support excitation*: Mechanical systems are frequently excited by a motion of the supports to which the system is grounded (the attachment to ground is often modeled as a combination of stiffness and damping elements). Examples of this type of excitation include the following: the 4 degree of freedom

automobile-suspension model (Example 2.5), in which the support motion is due to road irregularities defined by the inputs $x_f(t)$ and $x_r(t)$; the earthquake/building model (Example 2.6), in which the horizontal ground motion $x_g(t)$ is the input; and the pilot ejection model (Example 2.8), in which the input is the motion $y(t)$ of the ejected seat. Support excitations transmit forces to the mechanical objects of the system through the "suspension system" which connects the objects to the support.

2. *Direct load application*: Direct application of forces and moments is also common in mechanical and HPE systems. Examples include the following: the two-link manipulator model (Example 2.10), in which the inputs are the torques $M_1(t)$ and $M_2(t)$ which rotate the manipulator links; engines which provide thrust forces to move or rotate vehicles and spacecraft; and reciprocating or unbalanced rotating parts in machinery and engines, which produce time-varying "inertial forces."
3. *Control position movement*: Many systems incorporate the movement of a positioning device to control gas or liquid flows, electric currents, and so on. An example of this is the hydraulic servovalve (Example 3.3), in which the valve position is adjusted by an operator to control the flow of hydraulic fluid in the system. Note that for the particular system considered in Example 3.3, the input servovalve position $x(t)$ does *not* appear additively as in (5.4). Rather, it multiplies a function of the states p_1 and p_2 [Eq. (3.23)].
4. *Electrical signals*: Voltage and current sources are used to provide inputs to electrical and electromechanical systems. Examples of this are the time-varying input voltage $V(t)$ in the loudspeaker system (Example 3.7) and for the electric motor (Example 3.8).

For our purposes we will always assume, as is implicitly stated in (5.4), that the inputs are not affected by the motions of the system under consideration.

5.1.3 A Comment on the Free and Forced Responses

We have seen that there are two basic types of responses that are of interest in dynamic systems. The first is the free response, in which the system moves of its own accord, unaffected by any inputs. Chapter 4 was devoted entirely to this free response. The second response type is the forced response, which we are now studying. The student should be aware that one or the other (or both) of these types of responses will be important, depending on the nature of the phenomenon being analyzed; for some systems only the free response is of any relevance, while for others, only the response to inputs is important.

Let us discuss some of the examples of Chaps. 2 through 4 in which the free response is most relevant. First, consider the unpowered descent of a vehicle through the atmosphere (as in Example 2.7). There are no inputs involved, and our interest is confined to the free response. This is also true of the fox-rabbit population model (Example 3.10), in which there are no inputs. Our central interest here is in the interaction of the two populations resulting from the nonlinear effects in the system model (3.70). As a final example, we cite the air cushion landing system model (Example 3.5), for which the stability of equilibrium, as it depends on the fan-induced airflow q_{in}, is of primary importance. Thus, we observe that for some systems, only the free response is of interest.

On the other hand, there are also many engineering phenomena for which the primary interest is in the forced response. For example, consider the two-link planar manipulator (Example 2.10) and the loudspeaker (Example 3.7) models. The objective in each case is to provide a specified type of forced response, which is clearly the main interest for these systems. Likewise, for the earthquake/building model (Example 2.6), the pilot ejection model (Example 2.8), and the automobile-suspension model (Example 2.5), the forced response will be of primary concern.

As we will see, the forced response depends on both the basic properties of the system (and hence on the free response) and on the properties of the input. Even if our main interest is in the forced response, we will always have to be aware of the nature of the free response exhibited by the system. Indeed, an understanding of the free response of a given system will often enable us to tell how the system will respond to certain types of inputs. This is especially true in the study of the vibratory systems considered in Chap. 7. Furthermore, in specific cases it is necessary at some point that the system constants, which are basic system properties, be specified. Frequently these constants cannot be determined a priori but must be found through testing; we encounter the problem of system identification. System identification is often performed by monitoring the free response of the system and selecting values of the system constants such that the mathematical model will yield a free response which is very close to that which is measured. More specific examples of use of the free response to perform system identification are contained in Chaps. 7 and 9.

5.1.4 Response of Linear Systems

Let us consider the form of the general solution to the linear version (5.6) of the forced problem. The mathematical model (5.6) consists of a set of n first-order inhomogeneous ODEs with constant coefficients. Thus, the theory of linear ODEs dictates that the general solution $\{x(t)\}$ will consist of the sum of the homogeneous and the particular solutions, which we write as

$$\{x(t)\} = \{x(t)\}_H + \{x(t)\}_P \tag{5.7}$$

The homogeneous solution $\{x(t)\}_H$ is the free response and is by definition the solution to the autonomous problem $\{\dot{x}\}_H = [A]\{x\}_H$ studied in Chap. 4. Let us assume that we have found the eigenvalues λ_i and eigenvectors $\{C\}_i$ which define this solution, so that (5.7) can be written as

$$\{x(t)\} = \sum_{i=1}^{n} \alpha_i \{C\}_i e^{\lambda_i t} + \{x(t)\}_P \tag{5.8}$$

where the form (4.81) for the homogeneous solution has been used. The n constants $\alpha_1, \ldots, \alpha_n$ are once again found from the n initial values $x_1(0), \ldots, x_n(0)$.

We will look at several ways to find the particular solution $\{x(t)\}_P$ for single- and multistate systems. In this chapter we emphasize the method of undetermined coefficients and then in Chap. 8 we introduce Laplace transformation. The method of undetermined coefficients is discussed in Sec. 5.1.7. Here we only note two important characteristics of the particular solution:

1. The particular solution tends to have the same mathematical form as the input; that is, the system tends to "track" the input. Thus, if we subject an n-state linear system to an exponential input $e^{\beta t}$ [i.e., each of the terms $g_i(t)$ in (5.6) is proportional to the function $e^{\beta t}$], then, nearly always, each state $x_i(t)$ will exhibit a response proportional to $e^{\beta t}$. If we subject the same n-state system to a harmonic input of frequency Ω, then the particular solution will also be harmonic, with the same frequency Ω as the input. The "tracking tendency" of linear systems enables us to deduce the mathematical form that the particular solution will assume. It is also an important property in physical terms.
2. The particular solution is *independent* of initial conditions. Thus, consider the response given by (5.8) for an asymptotically stable linear system. For such a system each of the eigenvalues λ_i in (5.8) has a negative real part, so that in the limit $t \to \infty, \{x(t)\}_H \to 0$. The homogeneous solution dies out, and what we actually observe after a sufficiently long time is the particular solution:

$$\lim_{t \to \infty} \{x(t)\} = \{x(t)\}_P \qquad \text{(asymptotically stable linear system)} \tag{5.9}$$

An important consequence of this result is that, for an asymptotically stable linear system, the long-time response to inputs will be independent of initial conditions. This conclusion is, however, not necessarily true in the case of a driven nonlinear system.

If any eigenvalue λ_i in (5.8) has a positive real part, so that the linear system is unstable, then the homogeneous solution $\{x(t)\}_H$ becomes arbitrarily large (according to the linear model). In such cases it doesn't really matter very much what the particular solution is; no matter what the input, the system moves far from equilibrium, and the linear model (5.6) is not valid. Generally we will confine our attention to the forced response of stable or asymptotically stable linear systems.

5.1.5 Transient and Steady-State Responses

The *steady-state response* of the system (5.3) is defined as $\lim_{t\to\infty} x_i(t)$, $i = 1, \ldots, n$, that is, what we observe after a "long time." If the system being analyzed is actually *linear* (5.6), and if the linear system is asymptotically stable, then we see from (5.9) that the steady-state response is governed completely by the particular solution and that the steady-state response is, therefore, independent of the initial conditions. This is one reason why the steady-state response is so important; for an asymptotically stable linear system, we will observe the same steady-state response no matter what the initial conditions.

A common type of steady-state response is simple harmonic motion, which occurs when the input is simple harmonic. For many phenomena that exhibit this type of behavior, a steady-state condition is reached fairly quickly, often in a matter of seconds. In fact, for many engineering systems which are responding to inputs, what we actually observe in the real world is the steady-state response; this is especially true for structural vibratory systems responding to harmonic loads.

The *transient response* is, loosely speaking, what we observe prior to the attainment of steady-state conditions. Often the connotation of this term applies to what happens near the initial time $t = 0$. For linear systems (5.6) the transient response will depend on the initial conditions and on both the homogeneous and particular solutions. When steady-state conditions are reached, we say that the "transients have died out."

In some situations, such as in structural vibrations due to harmonic loads, the steady-state response is usually what is most important. In other situations, such as in design of control systems, both transient and steady-state responses are important.

Unfortunately, for driven *nonlinear* systems, it is not possible to split the solution into homogeneous and particular parts as in (5.7), as the superposition of solutions is not valid for nonlinear systems. For driven nonlinear systems, the idea of transient and steady-state responses is still important, but we often find that the steady-state response depends on the initial conditions (impossible in the linear world). An example of this is provided later in Example 5.2, and further discussion is contained in Chap. 10.

5.1.6 On the Efficacy of Linearized Models of Nonlinear Systems

In studying the near-equilibrium motions of the autonomous nonlinear system (4.5), we determined the Jacobian matrix $[A]$ and used it to define the linearized model (4.62). This linearized model is valid in some neighborhood of equilibrium and can be used to study the near-equilibrium behavior of the nonlinear system. It is thus natural to attempt to use linearized models for the forced problem also.

Suppose we pose the following general problem: We have a forced nonlinear system (5.4). We first find the equilibrium solutions $\bar{x}_i$ for the associated autonomous system (4.5) (no inputs). For each equilibrium solution we obtain the associated autonomous linearized model (4.62), which describes near-equilibrium motions if there were no inputs. Under what circumstances is it valid to study the *forced response* of the system (5.4) using the linearized model (5.6)? The answer is the same as for the no-input case studied in Chap. 4: If the input $\{g(t)\}$ does not drive the system very far from equilibrium, then (5.6) should be a valid representation of (5.4), as the system is then operating in the "linear range."

There is, however, an important distinction between the forced and free problems, insofar as linearization is concerned. In the free problem we can *always** find a small neighborhood of equilibrium such that the linearized model is valid. Hence, we can deduce the stability of equilibrium and the nature of the near-equilibrium orbits. The only question is how small the neighborhood should be, and this depends on the system.

In the forced problem, however, we may have no control over the input and hence no control over whether the input drives the system far from equilibrium. Thus, we may find that for some inputs, the states remain near their equilibrium values, so that the linear model (5.6) is valid. For other inputs, however, the system may be driven far from equilibrium, and the linear model may give a totally erroneous picture of the response. We now illustrate these ideas intuitively with an example.

Example 5.2 The Rolling Ball. Let us return to the problem of the rolling ball (Example 4.6) (Fig. 4.4). We will now change the model in two ways: (1) We add a small amount of linear viscous damping to represent frictional effects, which were ignored in Example 4.6. This addition of damping renders the two equilibrium solutions $\bar{x} = (-1, 0), (1, 0)$ asymptotically stable, rather than merely stable as in Example 4.6. The equilibrium solution $\bar{x} = (0, 0)$ is still unstable (see Fig. 5.2). (2) We will excite the system by moving the entire apparatus in the horizontal direction in some specified manner $x_b(t)$. The input $x_b(t)$, assumed to be specified ahead of time, measures the horizontal location of the top of the "hump," relative to a fixed reference line, as shown. The coordinate x measures the horizontal position of the mass m relative to the top of the hump. Thus, the net horizontal position of m is $x_b(t) + x(t)$.

With the addition of damping and forcing, the equation of motion replacing (4.15) turns out to be as follows (with the mass $m = 1$):

$$\ddot{x}\left[1 + (-x + x^3)^2\right] + c\dot{x} + x\dot{x}^2(-1 + x^2)(-1 + 3x^2) + gx(-1 + x^2) = -\ddot{x}_b(t) \tag{5.10}$$

*Except for the situation described in Sec. 4.3.7, where we saw that, for some systems having an eigenvalue with zero real part in the linearized model, the linearization may not be valid.

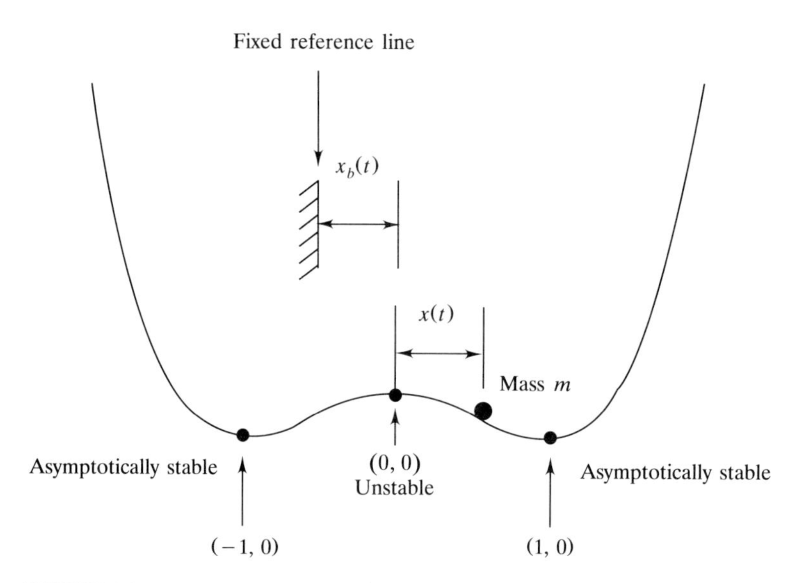

FIGURE 5.2
The rolling ball excited by a "support" motion $x_b(t)$.

The input is due to acceleration $\ddot{x}_b(t)$ of the apparatus as a whole. Now suppose that initially the mass m is at rest at the asymptotically stable equilibrium $(1, 0)$ and that, at $t = 0$, we impose the horizontal motion $x_b(t)$ of the entire apparatus. The linearized model of (5.10), valid for small motions near this equilibrium, is

$$\ddot{x} + c\dot{x} + 2gx = -\ddot{x}_b(t) \tag{5.11}$$

where $x(t)$ in (5.11) now represents the deviation from the equilibrium value $\bar{x} = 1$. Equation (5.11) is seen to be the same as (2.64) for the relative motion in the pilot head acceleration problem. We could, if we wished, convert (5.10) and (5.11) to the state variable forms (5.4) and (5.6), respectively.

Now the question of interest here is whether (5.11) is a valid representation of (5.10). For instance, suppose we shake the apparatus back and forth harmonically, $x_b(t) = x_{b_0} \cos \Omega t$, where Ω is the excitation frequency. Suppose also that the shaking motion is very violent (x_{b_0} is large). Then we would probably expect the mass m to move an appreciable distance from the equilibrium $(1, 0)$; in fact, it is easy to induce a motion in which m executes very large excursions and passes through both wells of the system. If this occurs, then the linearized model (5.11) will fail. On the other hand, for other input functions $x_b(t)$, the mass m may remain near the equilibrium solution $(1, 0)$, and the linear model (5.11) gives a valid representation of the system motion.

So how do we tell, for specific inputs $x_b(t)$, whether the linear model is valid? What we can often do is this: For the input $x_b(t)$ under consideration, we solve (5.11) to determine the particular solution $x_p(t)$ (we ignore the homogeneous solution, as it dies out because of the asymptotic stability of equilibrium). We then examine this solution to find, according to the linear model, the maximum excursions from equilibrium (both x and $\dot{x}$ will be needed). If, according to the linear model, the excursions from equilibrium are always small, so that the magnitudes of the nonlinear terms in (5.10) are small compared to the linear ones, then we can be reasonably sure that the linear model is valid. If, however, the excursions, according to the linear model, are large, so that the nonlinear terms in (5.10) have appreciable magnitudes, then the linear model may give us a totally erroneous view of the behavior.

5.1.7 The Method of Undetermined Coefficients (see also Ref. 5.2)

Constructing the particular solution $\{x(t)\}_P$ to the linear model (5.6) using the method of undetermined coefficients involves the following sequence of steps: (1) selection of the time functions that will comprise the particular solution, (2) formation of the general linear combination of these functions, each with an associated "undetermined coefficient," (3) substitution of the result of (2) into (5.6), (4) equating to zero the coefficients of the various (linearly independent) time functions appearing in the result of (3), and (5) solving for the undetermined coefficients.

The tracking property of linear systems, discussed in Sec. 5.1.4, provides some guidance as to the proper assumed form for the particular solution to (5.6). Here we add a few rules which systematize the process of finding particular solutions. These rules tell us, for a given input function $g_i(t)$ in the vector $\{g(t)\}$ of (5.6), what time dependence we should assume for the particular solution $\{x(t)\}_P$. The rules are based on the following table from Ref. 5.2:

TABLE 5.1

Term in input $g_i(t)$	Terms in assumed particular solution
t^n	$t^n, t^{n-1}, \ldots, t, 1$
$e^{\alpha t}$	$e^{\alpha t}$
$\sin \Omega t$ *or* $\cos \Omega t$	$\sin \Omega t, \cos \Omega t$
Constant	1

The left-hand column denotes the type of term appearing in the input vector $\{g(t)\}$, and the right-hand column lists the associated time functions which should appear in the particular solution. Thus, if $\{g(t)\} = \{g_0\}t^2$, where $\{g_0\}$ is a constant vector, then the assumed form of the particular solution would be $\{x(t)\}_P = \{b_0\} + \{b_1\}t + \{b_2\}t^2$. In this situation, for an n-state system, there will be a total of $3n$ undetermined coefficients: the $3n$ entries of the vectors $\{b_0\}$, $\{b_1\}$, and $\{b_2\}$.

As another example, suppose that for a two-state system, $\{g(t)\} = [t, e^{2t}]^T$. Then the form of the particular solution will be $\{x(t)\}_P = \{b_0\} + \{b_1\}t + \{b_2\}e^{2t}$.

As a final example, suppose that the input vector $\{g(t)\} = \{g_0\}t \cos \Omega t$. Then the particular solution will contain the products of all of the terms appearing separately for the two functions t and $\cos \Omega t$, so that the particular solution will have the form $\{x(t)\}_P = \{b_0\} \cos \Omega t + \{b_1\} \sin \Omega t + \{b_2\}t \cos \Omega t + \{b_3\}t \sin \Omega t$. It is recommended that Refs. 1.1 and/or 5.2 be consulted for a thorough discussion of solution by the method of undetermined coefficients.*

In the remainder of this chapter we will study the response of linear systems of the first, second, and higher orders. An introduction to phenomena occurring when nonlinear effects are important is contained in Chap. 10. There we will see that nonlinear systems can exhibit interesting and practically important behaviors which have no counterpart in the linear world.

5.2 FORCED RESPONSE OF FIRST-ORDER LINEAR SYSTEMS

In this and the following section we will study the response to inputs of *linear* systems of first and second orders. The objectives will be to introduce systematic techniques for the solution of such problems and to develop an understanding of the ways in which such systems respond to various types of inputs. We shall then see (in Sec. 5.4, higher-order systems) that an understanding of the response of higher-order systems is difficult if one does not first obtain a thorough understanding of the responses of first- and second-order systems.

5.2.1 The First-Order Linear System

For single-state systems the linear form (5.6) of the forced problem reduces to the single first-order, inhomogeneous linear ODE with constant coefficients

$$\dot{x} = ax + g(t) \tag{5.12}$$

The input is $g(t)$, and the system is asymptotically stable if $a < 0$, which we will generally assume. We will first develop general relations for the solutions to (5.12), and then we'll apply these results to specific examples.

One way to determine the general solution to (5.12) is as follows: Multiply (5.12) by the function e^{-at} and rewrite the result as

$$e^{-at}[\dot{x} - ax] = e^{-at}g(t)$$

The left-hand side is just $(d/dt)(e^{-at}x)$, so now we have

$$\frac{d}{dt}(xe^{-at}) = e^{-at}g(t)$$

or

$$d(xe^{-at}) = e^{-at}g(t)\,dt \tag{5.13}$$

*If an input function $g_i(t)$ has the *same* time dependence as a homogeneous solution, the procedure discussed here must be amended; see Ref. 5.2.

This relation can be integrated, and we can do the integration in one of two ways: (1) Use the indefinite integral on the right-hand side, with the addition of a constant of integration or (2) integrate (5.13) from the initial time $t = 0$ to arbitrary time t. According to the first of these procedures, the integration yields

$$xe^{-at} = C + \int e^{-a\tau} g(\tau)\, d\tau \tag{5.14}$$

where τ is a dummy variable. Now multiply (5.14) by e^{at} to obtain the solution $x(t)$:

$$x(t) = Ce^{at} + e^{at}\int e^{-a\tau} g(\tau)\, d\tau \tag{5.15}$$

The first term on the right-hand side of (5.15) is the homogeneous solution, while the second term is the particular solution. We automatically get both parts of the solution using this procedure. The constant C is to be evaluated from the initial conditions. Note that in general C will not be the initial value $x(0)$ because the particular solution may be nonzero at $t = 0$. This is illustrated in Sec. 5.2.2.

An alternate way to express the solution is to integrate (5.13) from $t = 0$ to $t = t$, so that we obtain

$$xe^{-a\tau}\Big|_{t=0}^{t} = \int_0^t e^{-a\tau} g(\tau)\, d\tau$$

so that

$$x(t)\, e^{-at} - x(0) = \int_0^t e^{-a\tau} g(\tau)\, d\tau$$

Moving $x(0)$ to the right and multiplying by e^{at} yield

$$x(t) = x(0)\, e^{at} + e^{at}\int_0^t e^{-a\tau} g(\tau)\, d\tau \tag{5.16}$$

The solution (5.16) is the same as (5.15), but it is expressed in a slightly different way. Rather than consisting of the sum of the homogeneous and particular solutions, as in (5.15), the solution (5.16) is expressed as the sum of the "zero input response" and the "zero state response." The zero input response, $x(0)e^{at}$ in (5.16), is the response that would result if there were no input. The zero state response, $e^{at}\int_0^t e^{-a\tau} g(\tau)\, d\tau$ in (5.16), is the response that would occur if the system were initially at rest, $x(0) = 0$. The zero state response actually consists of the particular solution plus a portion of the homogeneous solution, as will be clarified in Sec. 5.2.2.

Clearly (5.15) and (5.16) are merely two different ways to express the same thing, the solution $x(t)$. In some situations, assuming the system to be asymptotically stable ($a < 0$), we are mainly interested in the steady-state response, which is independent of the initial conditions (for an asymptotically stable system). In such cases, (5.15) is convenient because we know that the homogeneous solution will die out for all initial conditions. Thus all we need to look at is the particular solution. In other situations, however, we are interested in both the transient and steady-state responses for the special case of a system initially at rest, that is, all initial conditions zero. In such a case, (5.16) is convenient because the first term $x(0)e^{at}$ will be zero; all we are interested in is the zero state response. These notions are clarified in the following section.

5.2.2 Response of First-Order Systems to a Step Input

The input to be considered is a step function, defined by $g(t) = g_0$ if $t > 0$, $g(t) = 0$ if $t < 0$ (Fig. 5.3).

Thus, (5.12) is

$$\dot{x} = ax + g_0 \qquad (t > 0) \tag{5.17}$$

We'll assume $a < 0$, the system is asymptotically stable, and we'll assume that the system is initially at rest, $x(0) = 0$. In this case, the solution, as given by (5.16), will consist of the zero state response only

$$x(t) = e^{at}\int_0^t e^{-a\tau} g_0 \, d\tau$$

which works out to be

$$x(t) = -\frac{g_0}{a}e^{at}\left(e^{-a\tau}\big|_0^t\right) = -\frac{g_0}{a}(1 - e^{at}) \tag{5.18}$$

The response of the system is shown in Fig. 5.4.

The steady-state response is obtained by evaluating (5.18) as $t \to \infty$ and is $x_{ss}(t) = -g_0/a$; this steady-state response is approached exponentially from below. We observe that the effect of the step input is to shift the equilibrium solution from $\bar{x} = 0$ to $\bar{x} = (-g_0/a)$, and the zero state response shows us how the new equilibrium condition is approached. Step inputs are often used to represent situations in which a constant load is suddenly applied to a system (as when we step on a bathroom scale) or if we wish to alter quickly the equilibrium condition (or "operating point") of the system.

The exponential approach to the steady-state solution $-g_0/a$ is often described in terms of the system time constant $T = 1/|a|$ (Sec. 4.3.3). In terms of the time constant T, the solution (5.18) for the zero state response is

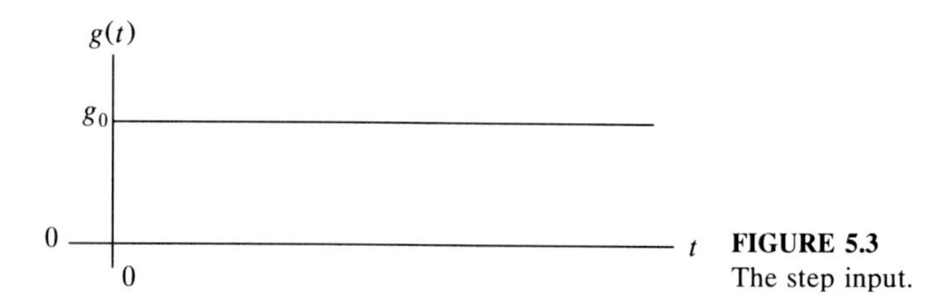

FIGURE 5.3
The step input.

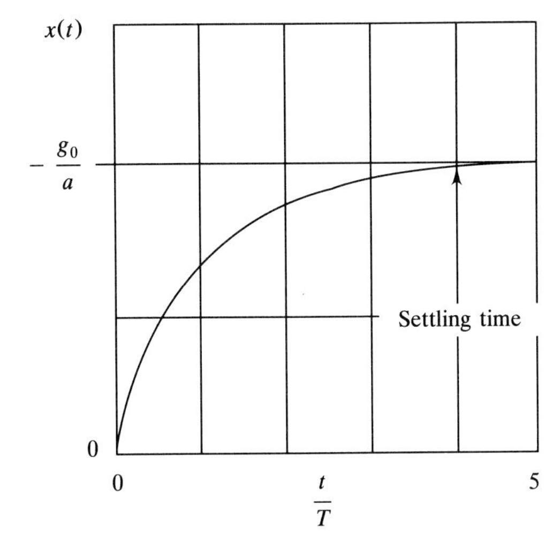

FIGURE 5.4
Zero state response of a first-order system to a step input. Time is measured in units of system time constants.

expressed as

$$x(t) = -\frac{g_0}{a}(1 - e^{-t/T}) \tag{5.19}$$

For a step input it is common (although arbitrary) to consider steady-state conditions to have been reached when the system response $x(t)$ is within 2 percent of the final value $x_{ss}(t)$. This corresponds to approximately four time constants of elapsed time, $t \sim 4T$, at which $(1 - e^{-t/T}) \sim 0.98$. The time constant is thus a useful way to describe how quickly or slowly a first-order system will respond to any input that changes rapidly from one steady value to another (Fig. 5.4).

Example 5.3 Cooling of a Billet (Sec. 3.5). In this example we use θ to represent the temperature, in order to avoid confusion with the time constant T.

A metal block initially at uniform temperature θ_0 is immersed in a hot fluid of higher temperature θ_∞. From experience we know that as $t \to \infty$, the billet temperature $\theta(t) \to \theta_\infty$. If the billet thermal conductivity k is sufficiently high, then a reasonable approximation is that, as the billet heats up, the temperature rises uniformly within the billet. This assumption allows the first-order model (3.67) to be applied

$$\dot{\theta} = \frac{hA}{mc_p}(\theta_\infty - \theta) \tag{5.20}$$

where the various parameters are defined in Sec. 3.5. The time constant T is obtained directly from (5.20) as $T = 1/(hA/mc_p)$, and the temperature response $\theta(t)$ is obtained as follows:

$$\theta(t) = \theta_\infty - (\theta_\infty - \theta_0)\, e^{-t/T} \tag{5.21}$$

Note that this result differs from (5.19) because the initial value of the state is θ_0,

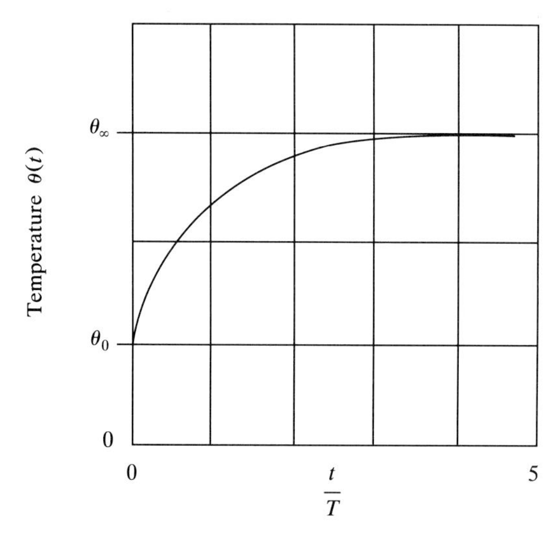

FIGURE 5.5
Temperature response of the billet of Example 5.3 to a sudden immersion in a hot fluid. Time is measured in units of system time constants.

rather than 0, as was assumed in obtaining (5.19). The temperature response is shown in Fig. 5.5, with time measured in units of time constants. The time constants characterizing this type of process are typically in the range of minutes to hours, depending on the material, the geometry, and the flow properties of the heating/cooling fluid medium.

It is worthwhile to comment on the two forms (5.15) and (5.16) of the solution to (5.17). The student may verify that use of (5.15) will result in the solution form

$$x(t) = Ce^{at} - \frac{g_0}{a} \tag{5.22}$$

The constant C is evaluated from the initial condition $x(0)$:

$$x(0) = C - \frac{g_0}{a} \rightarrow C = x(0) + \frac{g_0}{a}$$

Note that C is not simply the initial value $x(0)$ but also contains the initial value of the particular solution. The solution (5.22) is thus

$$x(t) = \left[x(0) + \frac{g_0}{a}\right] e^{at} - \frac{g_0}{a} \tag{5.23}$$

Observe that, if we allow a nonzero initial value $x(0)$ in the other solution form (5.16), we obtain

$$x(t) = x(0)\, e^{at} - \frac{g_0}{a}(1 - e^{at}) \tag{5.24}$$

Here we see clearly the difference between the two solutions forms (5.15) and (5.16). In (5.24), that is, (5.16), a portion of the homogeneous solution appearing in (5.23), that is, (5.15), has been shifted to the second term, so that the first term of (5.24) involves only the initial value $x(0)$, which then drops out in the zero state case.

A second general procedure often used to construct particular solutions to linear, inhomogeneous ODEs with constant coefficients is what we call the *method of undetermined coefficients*. In this procedure we *assume* the mathematical form of the particular solution and then determine the proportionality constants so as to solve the ODE. Normally the particular solution has the same mathematical form as the input, and this guides us in the selection of our assumed solutions. For example, in Example 5.3 the input is a constant for $t > 0$, and this suggests that the particular solution is also a constant. Thus, let $x_p(t) = B$, where the constant B has to be determined. If we plug $x_p(t) = B$ into (5.17), we find

$$0 = aB + g_0$$

so that $B = -(g_0/a)$ is the particular solution. The overall solution is then obtained by adding the particular and homogeneous solutions to obtain

$$x(t) = Ce^{at} - \frac{g_0}{a}$$

which is the same as (5.22).

Sometimes it occurs that the mathematical form of the particular solution differs from that of the input. This occurs when the input has the *same time dependence* as the homogeneous solution (or of one of the homogeneous solutions for an n-state system). For example, consider the single-state problem

$$\dot{x} = ax + be^{at} \tag{5.25}$$

in which the input be^{at} has exactly the same time dependence as the homogeneous solution Ce^{at}. The student may verify that (5.15) gives the particular solution as

$$x_p(t) = bte^{at}$$

which is proportional to te^{at}, not e^{at}.

The advantage of the formulas (5.15) and (5.16) is that they always work, regardless of the input, and one does not have to figure out the form of the particular solution, as in the method of undetermined coefficients. The advantage of the method of undetermined coefficients is that it is tied directly to the physical fact that linear systems track the inputs to which they are subjected, with the exception noted in the preceding paragraph. The student should employ whichever method he or she finds more convenient.

Yet a third method for solving both homogeneous and inhomogeneous linear ODEs with constant coefficients is Laplace transformation, which we will study in Chap. 8 in connection with linear control systems.

Example 5.4 Harmonic Excitation. Let us now consider a second type of input, harmonic excitation, which is very common, especially in systems of second and higher orders and in the vibrations of structural systems. In this case the mathematical model (5.12) is

$$\dot{x} = ax + P \cos \Omega t \qquad (a < 0; P, \Omega \text{ given constants}) \tag{5.26}$$

Here the input $P \cos \Omega t$ is characterized by its amplitude P and its frequency Ω. The solution may be written from (5.15) as

$$x(t) = Ce^{at} + e^{at} \int Pe^{-a\tau} \cos \Omega\tau \, d\tau \tag{5.27}$$

The integral in (5.27) is evaluated from integral tables, enabling the solution to be written as

$$x(t) = Ce^{at} + \frac{P}{a^2 + \Omega^2}(-a \cos \Omega t + \Omega \sin \Omega t) \tag{5.28}$$

The constant C is determined from the initial condition. Evaluating (5.28) at $t = 0$ yields

$$x(0) = C - \frac{aP}{a^2 + \Omega^2}$$

so that

$$C = x(0) + \frac{aP}{a^2 + \Omega^2}$$

The solution is, therefore,

$$x(t) = \left[x(0) + \frac{aP}{a^2 + \Omega^2}\right] e^{at} + \frac{P}{a^2 + \Omega^2}(-a \cos \Omega t + \Omega \sin \Omega t) \tag{5.29}$$

The solution will exhibit both exponential and harmonic character. The exponential behavior, which comes from the homogeneous solution or free response, will die out if the system is asymptotically stable ($a < 0$), and in the steady state the response will be harmonic; only the particular solution is present.

In (5.29) the particular solution is expressed as the sum of sine and cosine contributions. It is usually more convenient to express such forms in terms of a single harmonic function with a phase lag. Thus, the particular solution in (5.29) could also be expressed as*

$$x_p(t) = x_{ss}(t) = \frac{P}{(a^2 + \Omega^2)^{1/2}} \cos(\Omega t - \phi); \qquad \phi = \tan^{-1}\left(\frac{\Omega}{-a}\right) \tag{5.30}$$

The steady-state solution form (5.30) shows that the amplitude of the steady-state motion is $P/(a^2 + \Omega^2)^{1/2}$ and thus depends on both the properties of the input (P, Ω) and the basic system constant a. The phase lag ϕ signifies that, although

*If a function $f(t) = C \cos \Omega t + S \sin \Omega t$, then we can also write $f(t) = A \cos(\Omega t - \phi)$, where $A = (C^2 + S^2)^{1/2}$ and $\tan \phi = S/C$. This can be shown by expanding $\cos(\Omega t - \phi)$ using the formula for the cosine of the difference of two angles, so that $f(t) = A[\cos \Omega t \cos \phi + \sin \Omega t \sin \phi]$. We thus have the identities $A \cos \phi = C$, $A \sin \phi = S$. Squaring and adding these identities yields $A^2 = C^2 + S^2$, while dividing the second identity by the first yields $\tan \phi = S/C$.

the steady-state response is simple harmonic with the *same frequency* as the input, there is a phase difference between the input and the response. Such behavior also occurs in systems of higher order.

Comments on Example 5.4

1. The relation $A = P/(a^2 + \Omega^2)^{1/2}$ for the amplitude A of the steady-state response (5.30) defines what is referred to as the system *frequency response*, which reveals how the steady-state amplitude A will vary with the driving frequency Ω. For the first-order system this frequency response is shown in Fig. 5.6, in which the dimensionless amplitude factor $aA/P = 1/[1 + (\Omega/a)^2]^{1/2}$ is plotted versus frequency ratio (Ω/a). Observe that for small (Ω/a), $(aA/P) \cong 1$. As (Ω/a) is increased through unity, the response amplitude decreases. For large values of (Ω/a), A approaches zero as $1/\Omega$. Further discussion of the first-order system frequency response is contained in Sec. 5.4.
2. Harmonic excitations are often used to model stimuli that persist for long periods of time and have a repeated character. Because of the persistence of the input, the steady-state response $x_{ss}(t)$ is usually of most interest. Physically, the result of Example 5.4 is a very important example of a more general result: **Any asymptotically stable linear system, of any order, when subjected to a harmonic excitation, will exhibit a steady-state response which is also simple harmonic and which has the same frequency as the excitation.**
3. In this example the input was $P \cos \Omega t$, and the particular solution was of the form $C \cos \Omega t + S \sin \Omega t$; that is, both of the trigonometric functions appear in $x_p(t)$, while only one is in the input. Thus, in finding solutions to linear

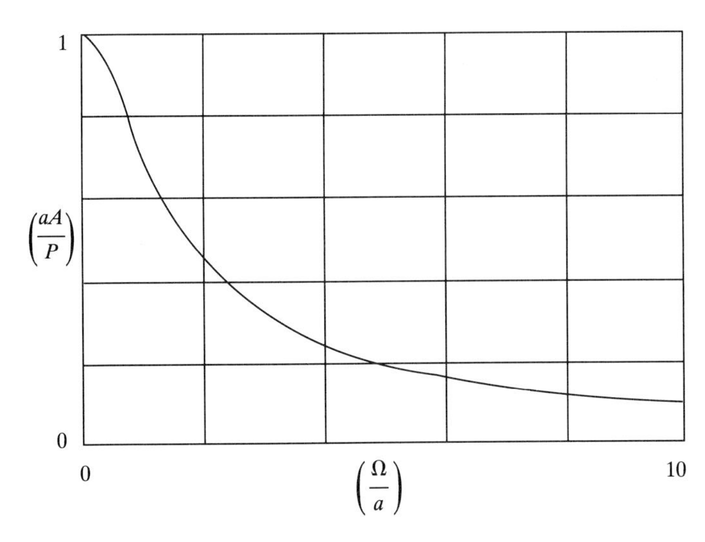

FIGURE 5.6
Frequency response of the first-order system.

problems involving harmonic excitations using the method of undetermined coefficients, we have to remember to include both sines and cosines in the assumed particular solution (see discussion following Table 5.1).

Example 5.5 The Water Tank. Let us return to the water tank system of Examples 3.1 and 4.5. Now we will consider the case in which the water inflow is time dependent and consists of a steady (constant) component q_0 (as in Example 3.1), on which we superpose a harmonically fluctuating component $q_1 \cos \Omega t$. The net water inflow rate is then $q(t) = q_0 + q_1 \cos \Omega t$; we'll assume $q_1 < q_0$ so that water is always flowing into the tank. The situation is shown in Fig. 5.7.

The mathematical model for this example is obtained from (3.6) by merely changing q_0 to $q_0 + q_1 \cos \Omega t$, so we now have

$$\dot{z} = \frac{q_0}{\rho A} - \frac{(2g)^{1/2} A_0}{A} z^{1/2} + \frac{q_1}{\rho A} \cos \Omega t \tag{5.31}$$

This mathematical model is nonlinear, due to the $z^{1/2}$ term, and represents a single-state version of (5.4). Our line of attack will be to use the linearized version of (5.31), analogous to (5.6), to study the forced problem, assuming that the fluctuating input will not drive the system too far from the equilibrium water level $\bar{z}$. We will go through the analysis in a fair amount of detail.

From Example 4.5 the equilibrium water level $\bar{z}$ was found to be

$$\bar{z} = \left(\frac{q_0}{\rho A_0} \right)^2 \frac{1}{2g} \tag{5.32}$$

This equilibrium level is asymptotically stable, so that if $q_1 = 0$ (no input), the water level will approach $\bar{z}$ as $t \to \infty$. The problem now is to determine how the water level will fluctuate as a result of the fluctuation $q_1 \cos \Omega t$ in water inflow rate.

In order to analyze the problem using linear theory, we first have to approximate (5.31) by a linearized model analogous to (5.6). First let $z = \bar{z} + y$,

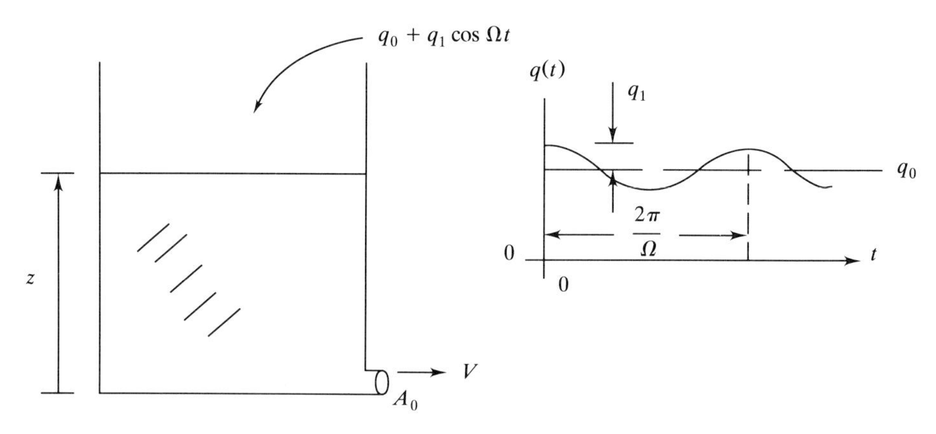

FIGURE 5.7
The water tank problem with variable water inflow rate $q(t)$.

where $y(t)$ is the instantaneous deviation in water level from the equilibrium water level $\bar{z}$. Plugging $z = \bar{z} + y$ into (5.31), we obtain

$$\dot{\bar{z}} + \dot{y} = \frac{q_0}{\rho A} - \frac{(2g)^{1/2}}{A} A_0 (\bar{z} + y)^{1/2} + \frac{q_1}{\rho A} \cos \Omega t \tag{5.33}$$

Next, we rewrite the $(\bar{z} + y)^{1/2}$ term as $\bar{z}^{1/2}[1 + (y/\bar{z})]^{1/2}$, so that (5.33) becomes

$$\dot{y} = \frac{q_0}{\rho A} - \frac{(2g)^{1/2} A_0}{A} \bar{z}^{1/2} \left(1 + \frac{y}{\bar{z}}\right)^{1/2} + \frac{q_1}{\rho A} \cos \Omega t \tag{5.34}$$

To this point we have made no approximations. We now obtain a linearized model by applying the approximation (4.53) to the term $(1 + y/\bar{z})^{1/2}$, assuming that $|y/\bar{z}|$ is small compared to unity; that is, any fluctuations y in water level are small compared to the equilibrium height $\bar{z}$. The approximation (4.53) yields

$$\left(1 + \frac{y}{\bar{z}}\right)^{1/2} \approx 1 + \frac{1}{2} \frac{y}{\bar{z}} \tag{5.35}$$

We will eventually have to determine whether this approximation is valid.

Equation (5.35) is used to rewrite (5.34) as

$$\dot{y} = \frac{q_0}{\rho A} - \frac{(2g)^{1/2} A_0 \bar{z}^{1/2}}{A} - \frac{(2g)^{1/2} A_0}{2A\bar{z}^{1/2}} y + \frac{q_1}{\rho A} \cos \Omega t \tag{5.36}$$

In view of the relation (5.32) for the equilibrium water level $\bar{z}$, the first two terms on the right-hand side of (5.36) drop out, and the linearized model for the forced problem is of the form

$$\dot{y} = ay + P \cos \Omega t \tag{5.37}$$

where $a = -(\rho g A_0^2 / A q_0)$ and $P = q_1/\rho A$. The linear model (5.37) is the same as (5.26) in Example 5.4. The steady-state response of (5.37) is thus given by (5.30):

$$y_{ss}(t) = \frac{P}{(a^2 + \Omega^2)^{1/2}} \cos(\Omega t - \phi)$$

which, using the definitions of a and P for this example, is

$$y_{ss}(t) = \frac{(q_1/\rho A)}{\left[\left(\dfrac{\rho A_0^2 g}{A q_0}\right)^2 + \Omega^2\right]^{1/2}} \cos(\Omega t - \phi) \tag{5.38}$$

Our linearized model predicts that in the steady state the water level will fluctuate harmonically from the equilibrium level $\bar{z}$, and the frequency of this fluctuation is the same as the input frequency Ω. The water fluctuation $y(t)$ will lag the input flow by a phase angle ϕ. From the steady-state solution (5.38), we observe that the maximum steady-state water level fluctuation will be

$$|y_{ss_{\max}}| = \frac{(q_1/\rho A)}{\left[\left(\dfrac{\rho A_0^2 g}{A q_0}\right)^2 + \Omega^2\right]^{1/2}} \tag{5.39}$$

For specific values of q_1, ρ, A, A_0, q_0, and Ω, we can compare this magnitude to the equilibrium level $\bar{z}$ to ascertain whether the linearized model (5.37) is expected to give an accurate representation of the forced response of the actual system (5.31).

5.3 FORCED RESPONSE OF SECOND-ORDER LINEAR SYSTEMS

5.3.1 Introduction and Step Response

We now consider the forced response of linear systems of second order. In analyzing such systems, we will consider two forms of the system mathematical model: (1) The state variable form (5.6) and (2) the single, second-order ODE form, in which we select either of the two states as the "output variable." Historically the latter of these model forms has been used extensively, largely because second-order systems often arise out of the newtonian mechanics problem, so that the mathematical models naturally arise as second-order ODEs, one for each degree of freedom. Either way of expressing the mathematical model is useful for our purposes. Our primary interest, in any event, is to understand how second-order systems respond to inputs. It should be borne in mind that we will compute the same response for the system, regardless of the form of the mathematical model.

Thus, we will suppose that our second-order system model has either of the following forms:

1. *State variable form (5.6):*

$$\begin{Bmatrix} \dot{x}_1 \\ \dot{x}_2 \end{Bmatrix} = \begin{bmatrix} a_{11} & a_{12} \\ a_{21} & a_{22} \end{bmatrix} \begin{Bmatrix} x_1 \\ x_2 \end{Bmatrix} + \begin{Bmatrix} g_1(t) \\ g_2(t) \end{Bmatrix} \tag{5.40}$$

2. *Single second-order ODE form:* The prototype we will use here is the oscillator model

$$m\ddot{x} + c\dot{x} + kx = F(t)$$

This model is typically written in the so-called *standard form* by first dividing by the mass m, so that

$$\ddot{x} + \frac{c}{m}\dot{x} + \frac{k}{m}x = \frac{F(t)}{m}$$

We next define the following system parameters: $\omega^2 = k/m$ and $\xi = c/(2m\omega)$. The parameter $\omega = \sqrt{k/m}$, referred to as the *system natural frequency*, is the oscillation frequency that would occur with no damping and no forcing [see (4.100) and the surrounding discussion]. The parameter ξ is a dimensionless measure of the system damping properties; this parameter governs the rate at which oscillations decay in the unforced system. In terms of this "damping factor" ξ, the term $c/m = 2\xi\omega$. Thus, the standard form of the oscillator model may be written as

$$\ddot{x} + 2\xi\omega\dot{x} + \omega^2 x = g(t) \tag{5.41}$$

where $g(t) = F(t)/m$. The terms in (5.41) have units of acceleration. The importance of the parameters ξ and ω is that two oscillators may have different parameters m, c, and k, but if they are characterized by the same values of the parameters ξ and ω, then they will exhibit free oscillations which have the same frequency and which decay at the same rate.

The two forms (5.40) and (5.41) are equivalent for a given second-order system. For instance, in (5.41) let $x_1 = x$ and $x_2 = \dot{x}$. Then (5.41) can be converted to (5.40) with $a_{11} = 0, a_{12} = 1, g_1(t) = 0, a_{21} = -\omega^2, a_{22} = -2\xi\omega$, $g_2(t) = g(t)$. Likewise, the two first-order equations in (5.40) can be converted to (5.41) following the procedure of Sec. 1.4.1. We select either of the two states as the output variable x appearing in (5.41), and we eliminate the other state from the formulation. It is straightforward to show that, if we select x_1 as our output variable, then (5.40) can be converted to the following second-order ODE for x_1:

$$\ddot{x}_1 - T\dot{x}_1 + \Delta x_1 = \dot{g}_1(t) - a_{22}g_1(t) + a_{12}g_2(t) \equiv g(t) \tag{5.42}$$

where $T = a_{11} + a_{22}$ is the trace of $[A]$ and $\Delta = a_{11}a_{22} - a_{12}a_{21}$ is the determinant of $[A]$. The right-hand side $g(t)$ involves both $g_1(t)$ and $g_2(t)$, as well as $\dot{g}_1(t)$. If we were to select x_2 as our output variable and then eliminate x_1, we would obtain the following second-order ODE for x_2:

$$\ddot{x}_2 - T\dot{x}_2 + \Delta x_2 = \dot{g}_2(t) - a_{11}g_2(t) + a_{21}g_1(t) \equiv g(t) \tag{5.43}$$

It is important to observe that the coefficients appearing on the left-hand side of (5.42) and (5.43) are the same. This must be so, as these coefficients determine the eigenvalues $\lambda_{1,2}$ of the system, and these eigenvalues must be the same, whether we select x_1 or x_2 as the output variable.

We have seen in Sec. 5.2.1 that, for a first-order system, the homogeneous solution is of exponential form. Thus, transient behavior in a first-order linear system will be exponential in character. For a second-order system, however, there will be two possibilities, depending on the eigenvalues $\lambda_{1,2}$ of the autonomous version of the system model: (1) If the eigenvalues λ_1, λ_2 are real and distinct, then the second-order system will also exhibit exponential behavior in the transient response, with two characteristic exponential rates. (2) If the eigenvalues are complex, then the system will exhibit oscillatory transient behavior.

If we consider the form (5.41) of the mathematical model for the case $\omega^2 > 0$, the eigenvalues will be real if the damping factor $\xi > 1$ and will be complex if $\xi < 1$:

$$\lambda_{1,2} = -\xi\omega \pm \omega(\xi^2 - 1)^{1/2} \equiv \mu_1, \mu_2 \qquad (\xi > 1)$$

$$\lambda_{1,2} = -\xi\omega \pm i\omega(1 - \xi^2)^{1/2} \equiv -\xi\omega \pm i\omega_d \qquad (\xi < 1) \tag{5.44}$$

The case $\xi > 1$ is referred to as *overdamped*, as the damping is so strong as to preclude oscillatory motion. The case $\xi < 1$ is referred to as *underdamped*. For reference, the general solutions, including both the homogeneous and particular contributions, for these two cases are summarized below:

Overdamped ($\xi > 1$)

$$x(t) = \alpha_1 e^{\mu_1 t} + \alpha_2 e^{\mu_2 t} + x_p(t) \tag{5.45}$$

where $\mu_1 = -\xi\omega + \omega(\xi^2 - 1)^{1/2}$, and $\mu_2 = -\xi\omega - \omega(\xi^2 - 1)^{1/2}$, and where $x_p(t)$ is the particular solution.

Underdamped ($\xi < 1$)

$$x(t) = Ae^{-\xi\omega t}\cos(\omega_d t - \phi) + x_p(t) \tag{5.46}$$

where $\omega_d = \omega(1 - \xi^2)^{1/2}$ is the frequency of oscillation. In either the overdamped or the underdamped case, for a given input $g(t)$, we first find the particular solution $x_p(t)$. Then the constants α_1 and α_2 in (5.45) or A and ϕ in (5.46) are evaluated from the initial conditions $x(0)$, $\dot{x}(0)$. Let us now consider some examples of the response of second-order systems to inputs.

Example 5.6 Response of the Linear Oscillator to a Step Input. We'll start by considering the response of the linear oscillator (2.52) to the step input shown in Fig. 5.3. In this case the input g_0 is a force that is suddenly applied and persists indefinitely. The mathematical model is (2.52):

$$m\ddot{x} + c\dot{x} + kx = g_0 \qquad (t > 0) \tag{5.47}$$

Division of (5.47) by the mass m allows us to put (5.47) into the standard form (5.41)

$$\ddot{x} + 2\xi\omega\dot{x} + \omega^2 x = \frac{g_0}{m} \tag{5.48}$$

where $c/m = 2\xi\omega$, $k/m = \omega^2$. The particular solution $x_p(t)$ can be obtained easily using the method of undetermined constants. Because the input is a constant, we assume the particular solution also to be a constant, $x_p(t) = B$. The undetermined constant B is then found from (5.48) to be $B = g_0/m\omega^2 = g_0/k$; thus, the particular solution for the displacement x is the input force g_0 divided by the spring constant k. This will be recognized as the displacement that would occur if the load g_0 were applied statically.

Let us examine the response for the underdamped case ($0 < \xi < 1$) and with the system initially at rest, $x(0)$, $\dot{x}(0) = 0$. Then (5.46) yields the general solution as

$$x(t) = Ae^{-\xi\omega t}\cos(\omega_d t - \phi) + \frac{g_0}{k} \tag{5.49}$$

Now we evaluate (5.49) for x and the derivative of (5.49) for $\dot{x}$ at $t = 0$ in order to determine the constants A and ϕ. The results are as follows:

$$x(0) = 0 = A\cos\phi + \frac{g_0}{k} \tag{5.50a}$$

$$\dot{x}(0) = 0 = \omega_d A\sin\phi - \xi\omega A\cos\phi \tag{5.50b}$$

Using the definition $\omega_d = \omega(1 - \xi^2)^{1/2}$, Eq. (5.50$b$) provides the solution for the phase angle ϕ as $\tan\phi = \xi/(1 - \xi^2)^{1/2}$, so that $0 < \phi < \pi/2$ if $0 < \xi < 1$. The initial oscillation amplitude A is then found from (5.50a) to be $A = -g_0/k(1 - \xi^2)^{1/2}$. The solution (5.49) for the zero state response under consideration can then be expressed as

$$x(t) = \frac{g_0}{k}\left[1 - \frac{1}{(1-\xi^2)^{1/2}}e^{-\xi\omega t}\cos(\omega_d t - \phi)\right] \qquad (\xi < 1) \tag{5.51}$$

Now let us investigate what the solution (5.51) reveals in physical terms. We first observe that the response (5.51) consists of a constant or steady term g_0/k, which is the steady-state response, upon which a decaying oscillation (the free or homogeneous response) is superposed. We can gain an understanding of the response by considering first the case of zero damping ($\xi = 0$), for which the mathematics become very simple. For the undamped case, $\omega_d = \omega, \phi = 0$ and the solution (5.51) reduces to the simple form

$$x(t) = \frac{g_0}{k}(1 - \cos\omega t) \qquad (\xi = 0) \tag{5.52}$$

A graph of the solution is presented in Fig. 5.8(a), and the associated orbit in the phase plane is shown in Fig. 5.8(b). Note that the "time scale" in Fig. 5.8(a) is actually the phase angle ωt, so that the solution completes one full period of oscillation every 2π rad. The advantage of using ωt, rather than the actual time t, is that *all* undamped oscillator responses can then be represented by the single solution curve shown in Fig. 5.8(a); if we were to plot x versus t, we would need a separate solution curve for each ω.

The orbit in the phase plane is an ellipse centered at $[(g_0/k), 0]$ and having semimajor axes g_0/k in the x direction and $\omega g_0/k$ in the $\dot{x}$ direction.

For this undamped case, the solution (5.52) is harmonic; the oscillatory (homogeneous) portion of the response does not die out. We observe also in the response an interesting and practically important phenomenon, *overshoot*. This term means that the maximum response reaches a level larger than the steady-state value g_0/k. In fact, the peak displacement is *twice* this steady value. This result is a classic example of the importance of dynamic effects in systems analysis. The load g_0, if applied statically, would result in a static displacement g_0/k. If the same load g_0 is, however, applied suddenly, so that dynamic effects have to be taken into account, the peak displacement is twice the static deflection. Thus, for underdamped mechanical systems of second order, a suddenly applied load will result in larger displacements and stresses than will the same load applied very gradually.

Let us now consider the more realistic situation in which some damping is present ($0 < \xi < 1$). In this case, the response will differ from that shown in Fig. 5.8 in that the oscillatory contribution will decay exponentially, so that the steady-state displacement g_0/k is asymptotically approached. Shown in Fig. 5.9(a)

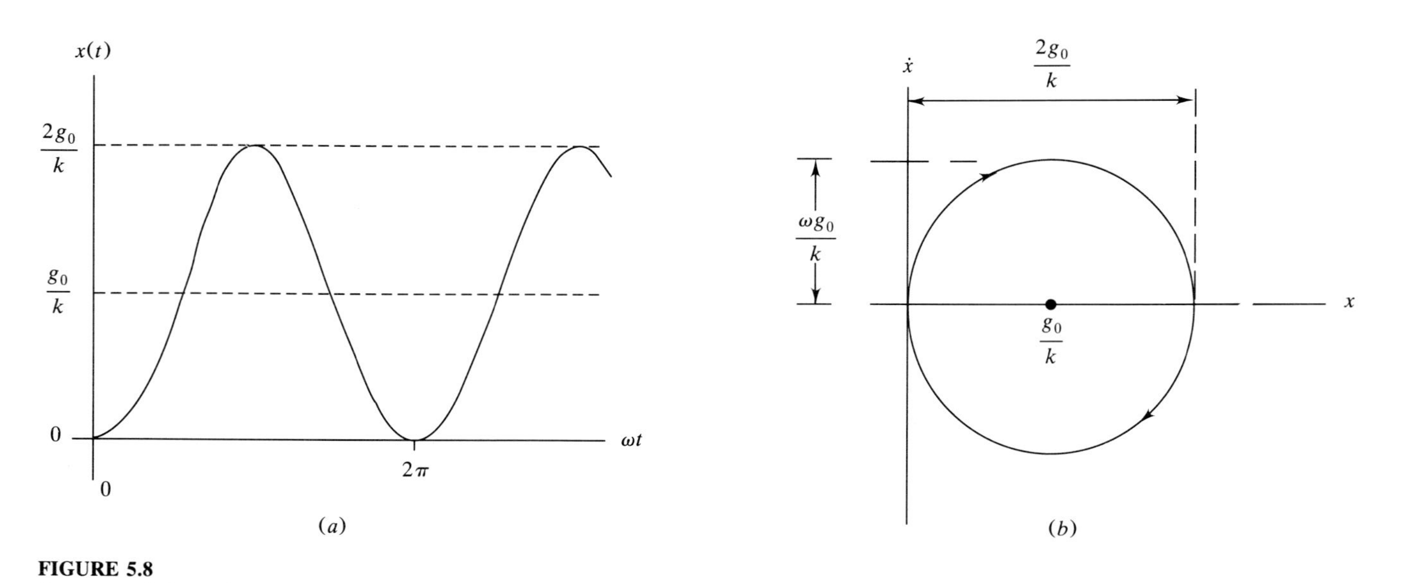

FIGURE 5.8
Response of the undamped second-order system to a step input: (*a*) The solution $x(\omega t)$; (*b*) orbit in the phase plane.

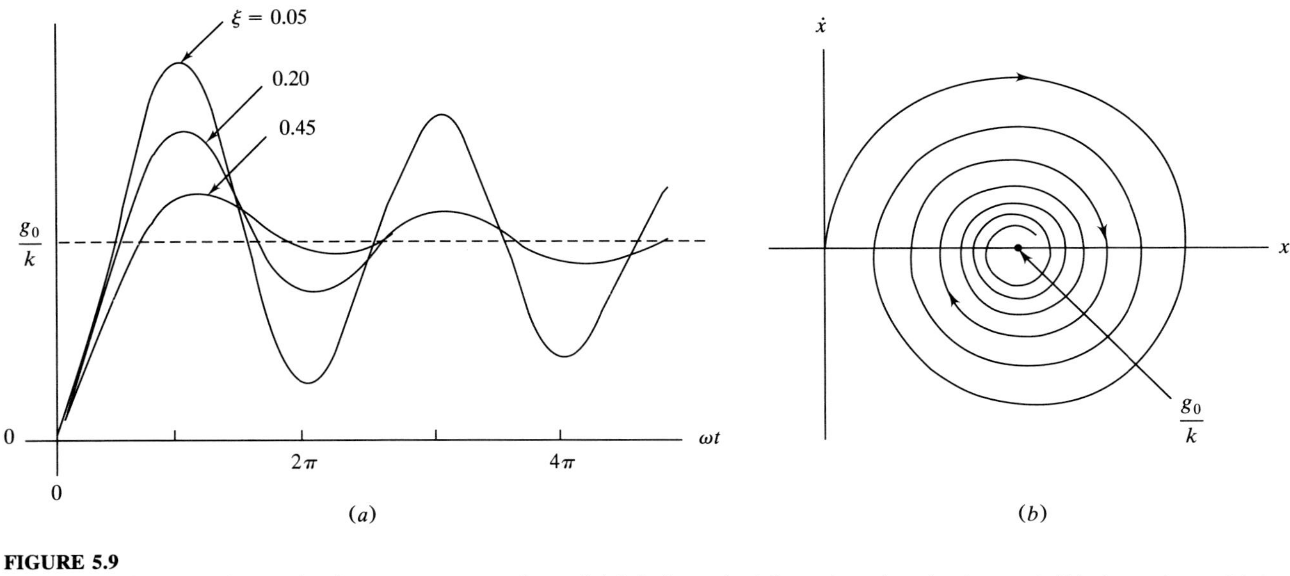

FIGURE 5.9
Response of the damped second-order system to a step input: (*a*) Solutions $x(\omega t)$ for various damping factors ξ; (*b*) phase plane orbit for $\xi = 0.05$.

are graphs of the solution (5.51) for several values of the damping factor ξ. The phase plane orbit is shown in Fig. 5.9(*b*) for the particular value $\xi = 0.05$.

Comments on the step response

1. The maximum overshoot is seen to be a function only of the damping factor ξ. As ξ increases from 0 to 1, the overshoot decreases from 100 to 0 percent. Thus, if a given system exhibits an undesirably large overshoot when subjected to a step input, we can add damping in order to ameliorate the problem. Figure 5.10 shows the variation of peak overshoot as a function of damping factor ξ for the underdamped second-order system ($\xi < 1$). Also shown in Fig. 5.10 is the time T_p to reach the first peak of oscillation.
2. Inspection of the solution (5.51) shows that the frequency ω_d of the oscillatory portion of the response decreases with increasing damping ξ, since $\omega_d = \omega(1 - \xi^2)^{1/2}$. For small damping (say, $\xi < 0.1$) this effect is hardly noticeable, but for larger ξ it is observable in Fig. 5.9. As ξ crosses unity from below, the oscillations cease altogether; the system eigenvalues become real and negative, and the system exhibits only exponential behavior.

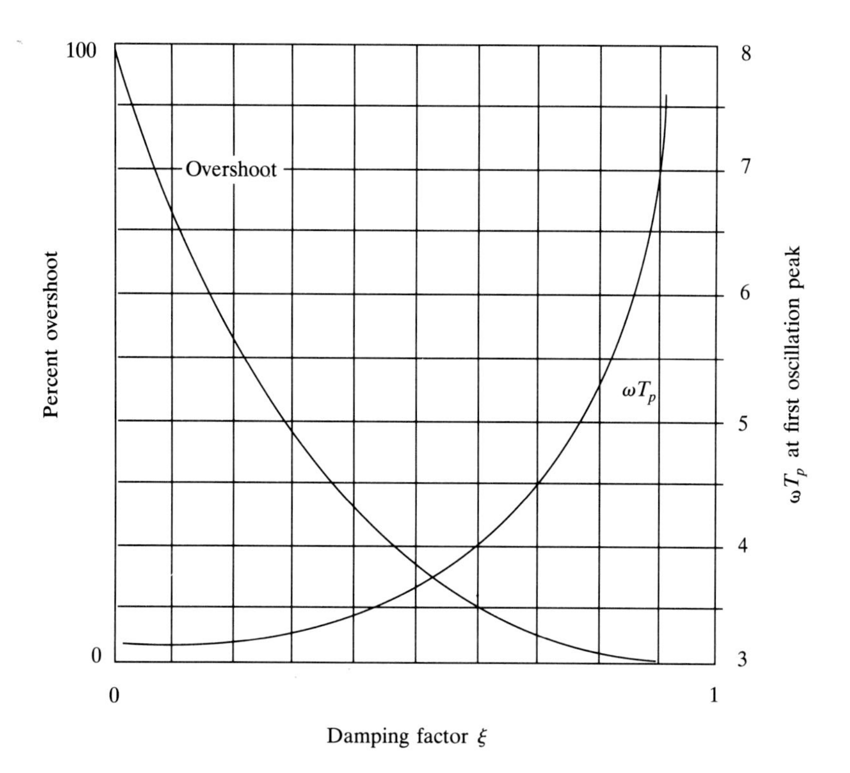

FIGURE 5.10
Step response characteristics of the second-order system: The percent overshoot versus the damping factor ξ and the time T_p to first oscillation peak, expressed as the phase angle ωT_p.

3. In physical terms the imposition of a step input often signifies that the system is being moved to a new equilibrium condition. In the present example the system starts at the equilibrium condition $(\bar{x}, \bar{\dot{x}}) = (0, 0)$. For the damped system, the step input then moves the system to the new equilibrium condition $[(g_0/k), 0]$, as is observed clearly in Fig. 5.9(*b*). Step inputs are often used as simple indicators that allow us to determine how a system responds to a sudden change in the equilibrium or "operating" condition.
4. The dynamic effect of overshoot can be demonstrated easily on an ordinary bathroom scale. If we quickly stand on the scale (a true "step load"), the weight indicator will overshoot our actual weight and then settle back to it. On the other hand, if we apply our weight very gradually, so that the entire process is essentially static, we observe no dynamic effects.

Now let us apply what we have learned to a more realistic engineering problem.

Example 5.7 Pilot Ejection (Example 2.8 Continued). Consider now the pilot ejection example, Example 2.8 (Fig. 2.30), Eq. (2.64). The proposed linear mathematical model (2.64) for the system, rewritten here in the standard form in terms of the damping and frequency parameters ξ and ω, is

$$\ddot{z} + 2\xi\omega\dot{z} + \omega^2 z = -\ddot{y}(t) \tag{5.53}$$

where z is the pilot head displacement relative to the seat being ejected; the actual pilot head displacement $x = z + y$. The input is the seat acceleration $\ddot{y}$ shown in Fig. 2.30. We notice that this input approximates the ideal "truncated step" shown in Fig. 5.11, which consists of a constant input level in the time interval $0 < t < T$ and a zero input level for $t > T$, with $T \approx 0.16$ s for the pilot ejection problem.

Now for $0 < t < T$, the response of the linear second-order system to a truncated step input will be the same as that due to the indefinitely applied step input of Example 5.6. (For $0 < t < T$, the system, which cannot anticipate future values of the input, cannot tell the difference; the responses will differ only for $t > T$, for which the inputs differ.) For $t > T$, the idealized input is zero, and the system then executes a free (autonomous) motion, with "initial conditions" $x(T), \dot{x}(T)$ determined by the response to the truncated step at the instant the input is "removed." The hypothetical experimental results of Fig. 2.30 indicate that the behavior of interest (large pilot head accelerations) occurs during the time interval $(0, T)$, so we can apply the results of Example 5.6 to the present situation.

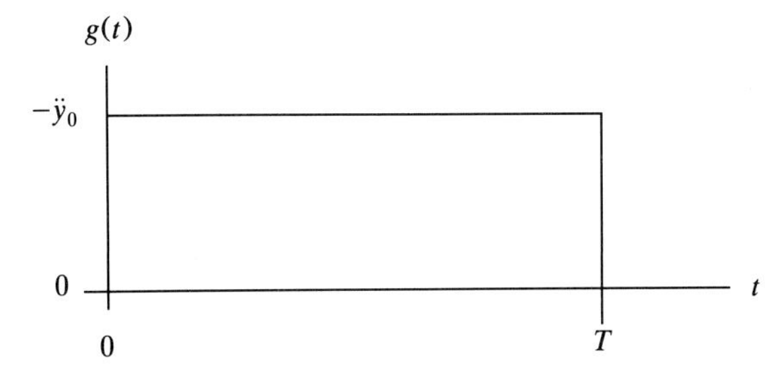

FIGURE 5.11
Idealized truncated step input for Example 5.7.

There are two additional differences between the present problem and that considered in Example 5.6. The first is that the output variable in (5.53) is the *relative* displacement z, rather than the inertially measured displacement x. The second difference is that we are ultimately interested in the *acceleration* $\ddot{x}$, rather than the displacement x. To attack the pilot head acceleration problem using the result of Example 5.6, we note first that the solutions (5.51) and (5.52) for the damped and undamped cases can be applied directly by replacing x by z and that g_0/k becomes $-\ddot{y}_0/\omega^2$. The damped and undamped solutions to (5.53) for the relative displacement $z(t)$ are found to be as follows:

Undamped

$$z(t) = -\frac{\ddot{y}_0}{\omega^2}(1 - \cos \omega t) \tag{5.54}$$

Damped

$$z(t) = -\frac{\ddot{y}_0}{\omega^2}\left[1 - \frac{1}{(1-\xi^2)^{1/2}} e^{-\xi\omega t} \cos(\omega_d t - \phi)\right] \tag{5.55}$$

These solutions apply for $0 < t \le T$.

Now it is clear from Fig. 2.30 that there is an appreciable amount of damping in the pilot/seat system, as the oscillatory behavior induced when the input is applied decays rather rapidly. This should not be too surprising on physical grounds, as the energy dissipation in the human body executing the type of motion being considered is significant. Nevertheless, let us first gain some insight into the problem by using the undamped version (5.54) of the solution for $z(t)$. By differentiating (5.54) twice, we find the relative acceleration $\ddot{z}(t)$ to be

$$\ddot{z}(t) = -\ddot{y}_0 \cos \omega t$$

Then the actual pilot head acceleration $\ddot{x} = \ddot{y} + \ddot{z}$ will be given by

$$\ddot{x}(t) = \ddot{y}_0(1 - \cos \omega t) \qquad (0 < t < T) \tag{5.56}$$

This is the same as the response shown in Fig. 5.8, except that the "response" is now the acceleration, rather than the displacement. The result (5.56) shows that the undamped second-order system, when subjected to a step base *acceleration*, will exhibit a response *acceleration* which is exactly the same as the *displacement* induced by a step *force*, as in Example 5.6.

Now considering the more realistic case in which damping is present, we would expect the output acceleration to behave as in Fig. 5.9. The actual solution for $\ddot{x}(t)$ for the damped case is, however, slightly different from (5.51). If we differentiate (5.55) twice to obtain $\ddot{z}$ and then calculate $\ddot{x}$ from $\ddot{x} = \ddot{y} + \ddot{z}$, we obtain the following result for the pilot head acceleration for the damped case:

$$\ddot{x}(t) = \ddot{y}_0\left\{1 - e^{-\xi\omega t}\left[\frac{(1-2\xi^2)}{(1-\xi^2)^{1/2}}\cos(\omega_d t - \phi) - 2\xi \sin(\omega_d t - \phi)\right]\right\} \tag{5.57}$$

By combining the trigonometric terms into a single harmonic function, this result can be manipulated to the following one:

$$\ddot{x}(t) = \ddot{y}_0\left[1 - \frac{1}{(1-\xi^2)^{1/2}} e^{-\xi\omega t} \cos(\omega_d t - \psi)\right] \qquad (0 < t < T) \tag{5.58}$$

where the phase ψ is now slightly different from the phase ϕ in Example 5.6. Otherwise the results are identical.

Having come this far, let us now assess the validity of the mathematical model, assuming that the data of Fig. 2.30 represent the result of an actual ejection. Inspection of Fig. 2.30 reveals the following discrepancies in comparison to the analytical result (5.58):

1. The mean pilot head acceleration during the ejection maneuver is approximately $13g$, while the seat acceleration is approximately $16g$. According to the result (5.58), these two values should be the same.
2. Comparison of the observed oscillation decay rate (Fig. 2.30) and the result of Fig. 5.9 show that the damping factor ξ is of the order of 0.20 or so.* For this value of damping, the mathematical model predicts a response overshoot of a little less than 60 percent. The observed overshoot, however, is actually close to 100 percent. Thus, the amount of overshoot does not appear to be consistent with the observed oscillation decay rate (Fig. 5.10).

Assuming the "data" to be accurate, there may be several reasons for these discrepancies:

1. The actual input is not the same as the ideal truncated step. Actually, we observe a peak input acceleration of around $17g$ at the beginning of the maneuver. This is larger than the average value of about $16g$ and may account to some extent for the larger than expected overshoot.
2. The system model is highly idealized. The use of a single mass/spring/damper assembly to represent the multiple components of the actual system may represent too much of an idealization to expect exact agreement between the actual and mathematical behaviors. For instance, in addition to the spinal column, the seat/buttocks combination will provide stiffness and damping, and a 2 degree of freedom model may be necessary to account reasonably well for these effects.
3. The effects of nonlinearities may be important. Both the stiffness and damping forces exerted by the spine on the head are undoubtedly nonlinear functions of the relative displacement z and the relative velocity $\dot{z}$, respectively. The inclusion of such nonlinearities in the system model would require some testing, which would be difficult to conduct accurately on human subjects. The step response of a nonlinear second-order system to a step input will certainly differ from the response of a linear system, but the nature of such differences would depend on the nature of the constitutive models used to represent the nonlinear forces. About all we can do here is to note that such effects may be of importance.

In light of the preceding discussion, it would not be too surprising for the measured and model behaviors not to agree exactly in this type of situation.

---

*In Chap. 7 we will consider a more rigorous way to determine the damping from measurement of the rate of decay of oscillations of a second-order system.

Actually, the simple single degree of freedom model does a reasonably good job of modeling the system behavior. The observed and modeled behaviors would typically be sufficiently close that the model could be used if we wished to modify the input seat acceleration so as to reduce the peak pilot head acceleration, which is a primary consideration in the ejection maneuver.

This example also illustrates a situation commonly encountered in the real world of engineering; a simple model describes a fairly complex phenomenon reasonably well, but there are some obvious discrepancies. Based upon the level of accuracy needed and on the purpose of the modeling effort, the engineer must decide whether a more sophisticated model is necessary.

Example 5.8 Foxes and Rabbits. As a final example of a second-order system subjected to a step input, let us consider the following variation of the fox-rabbit problem [Example 3.10, Eqs. (3.70)]. Everything remains the same, except that, at $t = 0$, we impose a fixed "harvesting rate" g_0 on the rabbit population, removing so many rabbits from the population per unit of time. We then have to add a harvesting term to the rabbit evolution equation, so that the math model now reads

$$\dot{r} = \alpha r - \gamma r f - g_0 \qquad (g_0 > 0, t > 0)$$

$$\dot{f} = -\beta f + \delta r f \tag{5.59}$$

Here $g_0 > 0$ is the constant rate of harvesting, and the negative sign indicates that rabbits are being removed. We have a nonlinear mathematical model in the state variable form (5.4) with a step input in one of the state equations.

The system (5.59) is nonlinear, and the nonzero equilibrium condition was found in Example 4.7 to be

$$\bar{r} = \frac{\beta}{\delta}, \qquad \bar{f} = \frac{\alpha}{\gamma} \tag{5.60}$$

In Example 4.17 a linearized model for free motions near this equilibrium condition was developed and was found to be [Eq. (4.57)]

$$\dot{r}_1 = -\left(\frac{\gamma\beta}{\delta}\right) f_1$$

$$\dot{f}_1 = \left(\frac{\alpha\delta}{\gamma}\right) r_1 \tag{5.61}$$

Here r_1 and f_1 replace ξ_1 and ξ_2 in (4.57), and the r_1 and f_1 appearing in (5.61) are now understood to be small deviations from the equilibrium condition (5.60). Since (5.60) applies for the case of no inputs, we now add the step input to obtain the forced linear model

$$\dot{r}_1 = -\left(\frac{\gamma\beta}{\delta}\right) f_1 - g_0 \qquad (t > 0)$$

$$\dot{f}_1 = \left(\frac{\alpha\delta}{\gamma}\right) r_1 \tag{5.62}$$

This linear model (5.62) provides an approximate description of the forced response of the actual nonlinear system (5.59), provided that the input does not cause the system to move too far from equilibrium, as discussed in Sec. 5.1.6.

Now for the linear model (5.62) we have a system initially at rest, $r_1(0) = f_1(0) = 0$; for $t > 0$ the system is then subjected to the step input. The system response is found by superposing the homogeneous and particular solutions as in (5.7).

Let us first examine the nature of the homogeneous solution for this problem. The homogeneous solution is defined by the eigenvalues and eigenvectors of the undriven system (5.61). The student may verify that these are as follows:

$$\lambda_{1,2} = \pm i(\alpha\beta)^{1/2} \equiv \pm i\omega$$

$$\{C\}_1 = \begin{Bmatrix} 1 \\ -i\varepsilon \end{Bmatrix} = \{C\}_2^*; \qquad \varepsilon = \frac{\delta}{\gamma}\left(\frac{\alpha}{\beta}\right)^{1/2}$$

The eigenvalues are purely imaginary, indicating that the homogeneous solution is simple harmonic with frequency $\omega = \sqrt{\alpha\beta}$. According to the solution form (4.109), the homogeneous solution for the two-state deviations r_1 and f_1 may be written as follows:

$$r_{1_H}(t) = A\cos(\omega t + \theta)$$

$$f_{1_H}(t) = \frac{A\varepsilon}{\omega}\sin(\omega t + \theta) \tag{5.63}$$

The constants A and θ are determined by the initial values $r_1(0)$ and $f_1(0)$. The result (5.63) indicates that, if the *undriven* system (5.61) is displaced slightly from equilibrium, then the fox and rabbit populations will each exhibit harmonic behavior, in which the populations are 90° out of phase and of differing amplitudes. This can be seen clearly in the phase plane orbits, as shown in Fig. 5.12. The phase portrait shown in Fig. 5.12 assumes a motion to be initiated with initial conditions

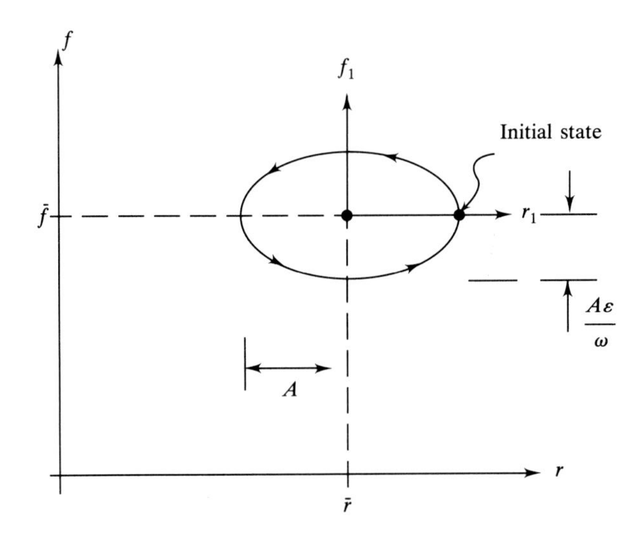

FIGURE 5.12
Phase plane orbit of the fox-rabbit system (5.61) with initial conditions, relative to equilibrium, of $r_1(0) = A$, $f_1(0) = 0$.

$r_1(0) = A$, $f_1(0) = 0$, so that $\theta = 0$ in (5.63). Then $r_1(t) = A\cos\omega t$ and $f_1(t) = (A\varepsilon/\omega)\sin\omega t$. At $t = 0$, the fox population is at its equilibrium level, while the rabbit population is larger than its equilibrium level. The fox population initially tends to grow due to the abundance of rabbits (quadrant 1 in the $r_1 - f_1$ plane). Thereafter, the populations vary cyclically, and the phase plane orbits (for any initial conditions) will be elliptical. Note that the equilibrium condition (5.60) is stable but not asymptotically stable; no "damping" mechanism exists in the model to drive the system asymptotically to equilibrium. According to the linearized model, the cyclic behavior of the two populations will persist forever.

Let us now return to the forced problem and see what happens if we suppose the system to be initially at rest at the equilibrium condition (5.60) and then subject it to the step harvesting rate, according to (5.62). Since for $t > 0$ the input is a constant, we will assume the particular solutions for r_1 and f_1 also to be constants:

$$r_{1_p} = R$$

$$f_{1_p} = F$$

The solutions for R and F are then found to be as follows: $R = 0$, $F = -g_0\delta/\gamma\beta$. Combining these results with the homogeneous solution (5.63), we obtain the general solution as

$$r_1(t) = A\cos(\omega t + \theta)$$

$$f_1(t) = \frac{A\varepsilon}{\omega}\sin(\omega t + \theta) - \frac{g_0\delta}{\gamma\beta} \tag{5.64}$$

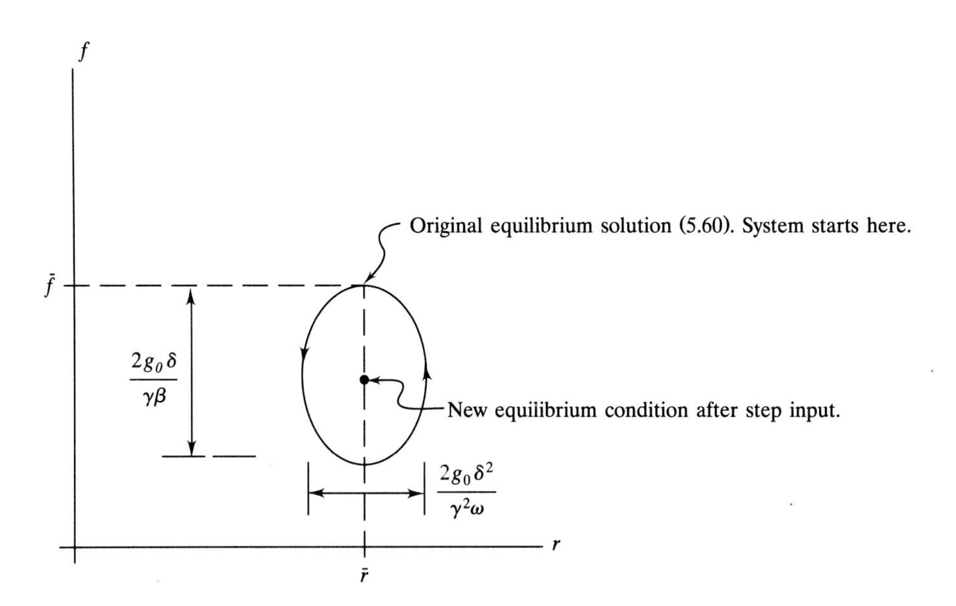

FIGURE 5.13
Response of the linearized fox-rabbit model (5.62) to a step harvesting rate in the rabbit population.

Evaluating these solutions at $t = 0$, for which $r_1(0) = f_1(0) = 0$, we find the constants A and θ to be as follows: $\theta = \pi/2$, $A = g_0\delta^2/\gamma^2\omega$. The zero state response, therefore, is

$$r_1(t) = -\frac{g_0\delta^2}{\gamma^2\omega}\sin\omega t$$

$$f_1(t) = \frac{g_0\delta}{\gamma\beta}\cos\omega t - \frac{g_0\delta}{\gamma\beta} \tag{5.65}$$

The resulting orbit in the phase plane is shown in Fig. 5.13.

We observe that the situation is similar to that studied in Example 5.6 for the undamped case. Essentially, the imposition of the step input changes the equilibrium condition from (5.60) to $\bar{r} = \beta/\delta$ [same as (5.60)] and $\bar{f} = \alpha/\gamma - g_0\delta/(\gamma\beta)$ [different from (5.60)]. Both fox and rabbit populations exhibit 100 percent overshoot because there is no damping in the model, and we see harmonic behavior much the same as in Fig. 5.8(*b*). In physical terms it is also interesting to note that, although the rabbit population is being harvested, it is the fox population that is on average reduced; the average (equilibrium) rabbit population is unaffected.

5.3.2 Response of Second-Order Systems to Harmonic Excitations

We now consider a situation of great practical importance for second-order systems: the harmonic excitation. Engineers frequently encounter excitations that are of a repeated (periodic) character and that persist for long periods of time. We will consider the simplest of such excitations, the simple harmonic input. This input may be used to model excitations due to unbalances in rotating devices such as motors and turbines, motion of reciprocating parts in engines and machinery, and wave effects on ships and submerged structures. Furthermore, system identification is often performed by exciting a system harmonically, measuring the steady-state response, and then determining appropriate values for the system constants.

To fix ideas, we will consider a particular system, the linear single degree of freedom mechanical oscillator, so that the mathematical model is of the form (2.52),

$$m\ddot{x} + c\dot{x} + kx = P\cos\Omega t \tag{5.66}$$

As in Example 5.4 the harmonic excitation is characterized by its amplitude P (in units of force, lb or N) and its frequency Ω (measured in rad/s). We can convert (5.66) to the standard form through division by m and use of the definitions $c/m = 2\xi\omega$ and $k/m \equiv \omega^2$ to obtain

$$\ddot{x} + 2\xi\omega\dot{x} + \omega^2 x = \frac{P}{m}\cos\Omega t \tag{5.67}$$

Let us assume that the system is asymptotically stable and that the free motions (that is, when $P = 0$) consist of exponentially decaying oscillations; the system is underdamped, with $0 < \xi < 1$ and $\omega > 0$. Let us also assume that we are concerned only with the long time or steady-state behavior that is observed after the transient response has died out. Thus, in accordance with (5.9) and the discussion preceding (5.9), we may confine our attention to the particular solution to (5.67), which will give us the desired steady-state response.

The particular solution to (5.67) can be obtained using the method of undetermined coefficients. Because the input is harmonic, we assume a particular solution of the same mathematical form, with both sines and cosines participating (Table 5.1):

$$x_p(t) = a \cos \Omega t + b \sin \Omega t \tag{5.68}$$

An alternative to (5.68) which is often more meaningful physically is to express the particular solution in terms of its amplitude A and phase ϕ by rewriting (5.68) as

$$x_p(t) = A \cos(\Omega t - \phi) \tag{5.69}$$

where $A = (a^2 + b^2)^{1/2}$ and $\tan \phi = b/a$.

Prior to attempting to find the values of A and ϕ which define the particular solution, let us note what the assumed solution says in physical terms. According to (5.69) we expect the steady-state response to be simple harmonic at the *same frequency as the input* and lagging the input in phase by the angle ϕ [supposing that ϕ in (5.69) turns out to be positive], as shown in Fig. 5.14.

In applications to mechanical systems we are primarily concerned with the amplitude A of the steady-state oscillatory motion, as this amplitude will determine maximum displacements, accelerations, and stresses in the mechanical system.

Now in obtaining the particular solution to (5.67), it is convenient, in order to minimize the eventual algebra, to impose a shift in the time t by defining a new time T according to

$$\Omega T \equiv \Omega t - \phi$$

The time T is thus measured from the peak of the steady-state response, as shown in Fig. 5.14. With this device the pair (5.67) and (5.69) are converted to

$$\ddot{x} + 2\xi\omega\dot{x} + \omega^2 x = \frac{P}{m} \cos(\Omega T + \phi) \tag{5.70a}$$

$$x(T) = A \cos \Omega T \tag{5.70b}$$

where dots are now T derivatives. We still have two unknown constants, A and ϕ, to find. The form (5.70) has the advantage that the solution consists of a single harmonic term. Let us now go through the steps of finding the particular solution. Then we will discuss the significance of the results. The particular solution is obtained by first substituting (5.70*b*) into (5.70*a*) to obtain

$$-A\Omega^2 \cos \Omega T - 2\xi\omega A\Omega \sin \Omega T + \omega^2 A \cos \Omega T = \frac{P}{m} \cos(\Omega T + \phi) \tag{5.71}$$

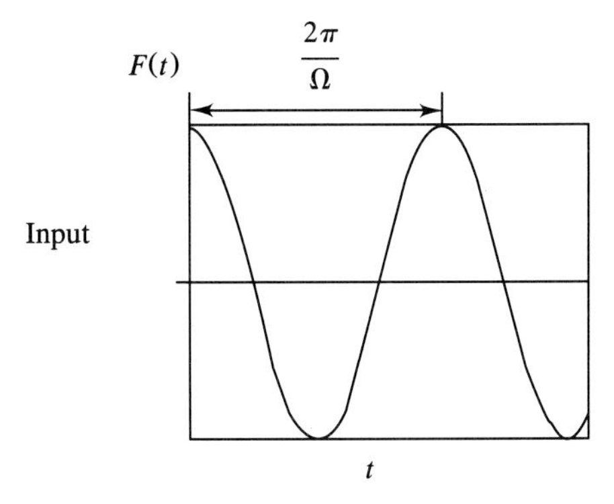

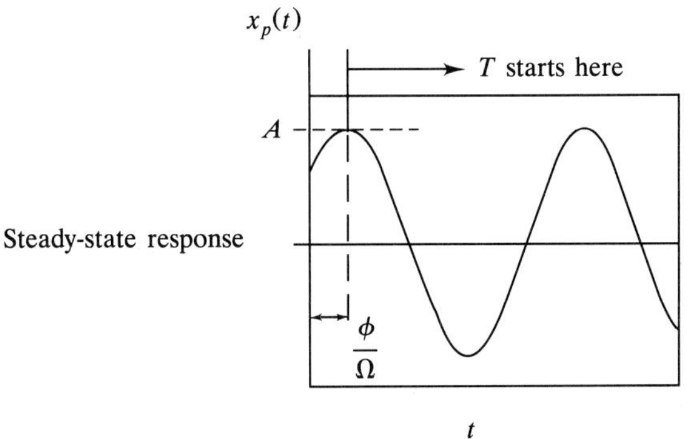

FIGURE 5.14
The input and steady-state response of the harmonically driven second-order system (5.67).

Next we expand the right-hand side using the formula for the cosine of the sum of two angles, whence

$$\cos(\Omega T + \phi) = \cos \Omega T \cos \phi - \sin \Omega T \sin \phi$$

so that (5.71) becomes*

$$(-A\Omega^2 + A\omega^2)\cos \Omega T - 2\xi\omega A\Omega \sin \Omega T = \frac{P}{m}(\cos \Omega T \cos \phi - \sin \Omega T \sin \phi) \tag{5.72}$$

The next step is to separate the terms involving $\cos \Omega t$ from those involving

*Notice that if we had begun with (5.67) and (5.69), we would have had to expand $\cos(\Omega t - \phi)$ twice and $\sin(\Omega t - \phi)$ once on the left-hand side of the resulting Eq. (5.71), complicating the algebra. Use of (5.70) necessitates only one such trigonometric expansion, and this is why (5.70) was introduced.

sin Ωt by writing (5.72) as follows:

$$\cos \Omega T\left(-A\Omega^2 + A\omega^2 - \frac{P}{m}\cos\phi\right) + \sin\Omega T\left(-2\xi\omega\Omega A + \frac{P}{m}\sin\phi\right) = 0 \tag{5.73}$$

Now each bracketed term in (5.73) is a constant, and the functions cos ΩT and sin ΩT are linearly independent. Thus, the only way in which (5.73) can be true for all time T is to require that each bracketed term vanish identically. We thus obtain the following two algebraic equations from which the undetermined constants A and ϕ may be found:

$$-A\Omega^2 + A\omega^2 - \frac{P}{m}\cos\phi = 0$$

$$-2\xi\omega\Omega A + \frac{P}{m}\sin\phi = 0$$

or

$$A(\omega^2 - \Omega^2) = \frac{P}{m}\cos\phi$$

$$A(2\xi\omega\Omega) = \frac{P}{m}\sin\phi \tag{5.74}$$

Observe that there are four different parameters involved in these relations: the two system parameters ω and ξ and the two excitation parameters Ω and P (actually the ratio P/m). This number of parameters can be reduced by dividing both the Eqs. (5.74) by the square ω^2 of the natural frequency ω to obtain

$$A\left(1 - \frac{\Omega^2}{\omega^2}\right) = \frac{P}{m\omega^2}\cos\phi$$

$$A\left(2\xi\frac{\Omega}{\omega}\right) = \frac{P}{m\omega^2}\sin\phi \tag{5.75}$$

We can simplify things a bit further by recognizing that $m\omega^2 = m(k/m) = k$, so that (5.75) can be expressed as

$$A\left(1 - \frac{\Omega^2}{\omega^2}\right) = \frac{P}{k}\cos\phi \tag{5.76a}$$

$$A\left(2\xi\frac{\Omega}{\omega}\right) = \frac{P}{k}\sin\phi \tag{5.76b}$$

We see that in (5.76) there are now only three parameters involved: the system damping factor ξ, the ratio P/k, and the frequency ratio Ω/ω.

The solution for A and ϕ can now be obtained. Divide (5.76b) by (5.76a), canceling A and P/k, to obtain

$$\frac{\sin\phi}{\cos\phi} = \tan\phi = \frac{2\xi\Omega/\omega}{1-\Omega^2/\omega^2} \rightarrow \phi = \tan^{-1}\left(\frac{2\xi\Omega/\omega}{1-\Omega^2/\omega^2}\right) \tag{5.77}$$

The student may verify that if $\Omega/\omega < 1$, then $0 < \phi < 90°$, if $\Omega/\omega = 1$, then $\phi = 90°$, and if $\Omega/\omega > 1, 90° < \phi < 180°$. Therefore, the steady-state response will lag the input by between 0 and 180°, depending on the system damping ξ and the ratio Ω/ω of driving to natural frequency.

The steady-state oscillation amplitude A can be found by squaring and adding (5.76a) and (5.76b) and utilizing the trigonometric identity $\sin^2\phi + \cos^2\phi = 1$ to eliminate the phase angle ϕ. This results in

$$A^2\left(1-\frac{\Omega^2}{\omega^2}\right)^2 + A^2\left(2\xi\frac{\Omega}{\omega}\right)^2 = \left(\frac{P}{k}\right)^2$$

which can be solved for the amplitude A as

$$A = \frac{P/k}{\left[\left(1-\dfrac{\Omega^2}{\omega^2}\right)^2 + \left(2\xi\dfrac{\Omega}{\omega}\right)^2\right]^{1/2}} \tag{5.78}$$

We note that the denominator in (5.78) is dimensionless and that A and P/k each has units of displacement. In fact, P/k is recognized to be the displacement that we would observe if a load of magnitude P were applied *statically*. Thus, we call P/k the *static deflection*. The solution (5.78) for the steady-state amplitude A can be presented more conveniently by expressing A in multiples of the static deflection

$$\frac{A}{P/k} = \frac{1}{\left[\left(1-\dfrac{\Omega^2}{\omega^2}\right)^2 + \left(2\xi\dfrac{\Omega}{\omega}\right)^2\right]^{1/2}} \tag{5.79}$$

Equation (5.79) is the main result of the analysis. It provides a dimensionless relation for the system "frequency response," that is, the steady-state response amplitude A which results when the system is excited by a harmonic excitation of any frequency Ω. The advantage of the dimensionless form is that the result depends on only two parameters, the frequency ratio Ω/ω and the system damping factor ξ. We can, therefore, present the frequency response of all second-order linear time invariant systems in a single family of curves by graphing the amplitude ratio $A/(P/k)$ versus the frequency ratio Ω/ω for various values of the damping factor ξ. Such results are shown in Fig. 5.15, which is one of the most important figures in this book.

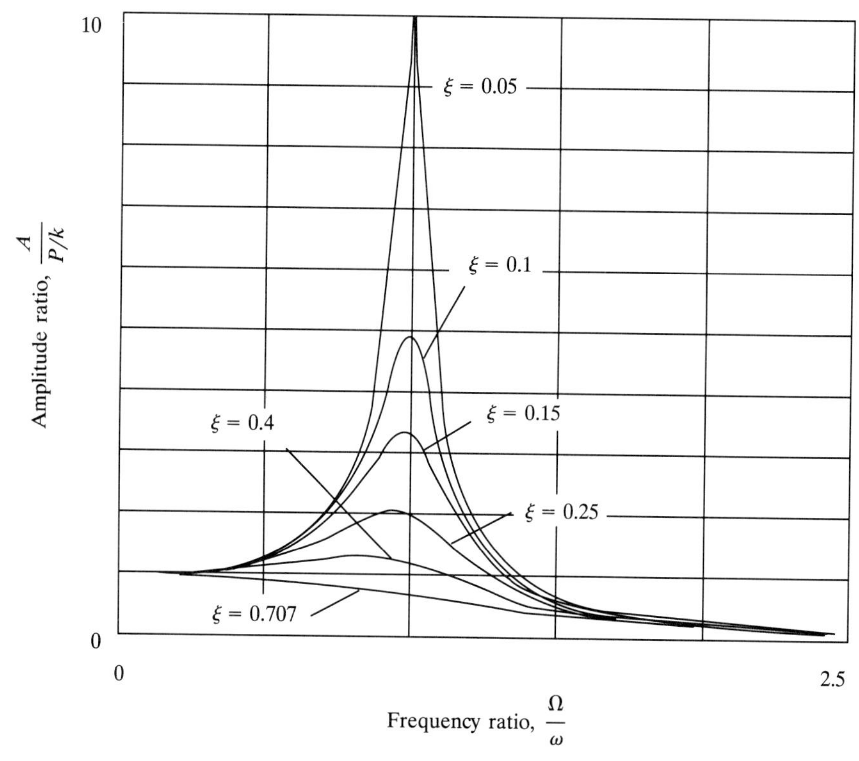

FIGURE 5.15
The steady-state frequency response of the linear second-order system.

Inspection of (5.79) and Fig. 5.15 reveals the following:

1. If $\Omega/\omega \ll 1$, so that the excitation frequency Ω is much lower than the system natural frequency ω, the amplitude ratio $A/(P/k) \cong 1$ for all values of the damping factor $\xi < 1$. Thus, the amplitude A is essentially the same as the static deflection P/k, and the phase angle ϕ from (5.77) is essentially zero. This means that the response is essentially static.
2. If $\Omega/\omega \gg 1$, so that the excitation frequency Ω is much larger than the system natural frequency ω, then the amplitude ratio $A/(P/k) \approx 0$. The system moves very little in response to the input.
3. If $\Omega/\omega = 1$, so that the excitation frequency Ω equals the system natural frequency ω, (5.79) reduces to

$$\frac{A}{P/k} = \frac{1}{2\xi} \qquad \left(\text{when } \frac{\Omega}{\omega} = 1\right) \tag{5.80}$$

so that the response amplitude A is inversely proportional to the system damping factor ξ. This condition, $\Omega/\omega = 1$, is known as *resonance* and results in large response amplitudes if the system damping factor ξ is small, as is often the case in mechanical systems, for which values of $\xi < 0.01$ are not uncommon. Normally, resonance is a condition to be avoided, as large vibration amplitudes can occur even if the exciting force amplitude P is small.

Comment. It is important to keep in mind that there are two frequencies involved in this problem, the system natural frequency ω and the frequency Ω of the external excitation. One of these frequencies (ω) is a basic system property and the other (Ω) characterizes the input, which is assumed to be unrelated to the system itself. The two frequencies ω and Ω, therefore, are physically unrelated to each other; they may be close to each other or they may not, depending on the situation.

Example 5.9 Harmonic Excitation: A Design Problem. A 100-kg machine with reciprocating parts operates at a speed of 1000 rev/min. The machine is mounted on four rubber isolators that provide both stiffness and damping. The rubber isolators are modeled as parallel linear spring damper combinations, each of stiffness $k = 4 \times 10^5$ N/m and damping $c = 100$ Ns/m. At the operating speed of 1000 rev/min, the system exhibits unacceptably large vibration. In order to alleviate the problem, it is proposed to add mass to increase the machine mass to 130 kg. Will the proposed solution work?

First let us determine the reason for the initial problem. We will first assume that the motion of the machine's reciprocating parts causes a (near) harmonic excitation at the operating frequency $\Omega = 1000$ rev/min (105 rad/s). Suspecting that a near-resonance condition is the problem, we next calculate the system natural frequency $\omega = (4k/m)^{1/2}$; the 4 is needed because we have four isolators, each of stiffness k. Calculation yields $\omega^2 = 4(4 \times 10^5\ \text{N/m})/100\ \text{kg} = 1.6 \times 10^4/\text{s}^2$, so that $\omega \cong 126$ rad/s, and the frequency ratio $\Omega/\omega \cong 0.83$. Then the system damping factor $\xi = 4c/2m\omega = c/2\sqrt{mk} = (400\ \text{Ns/m})/2(100 \cdot 1.6 \cdot 10^6)^{1/2}$, $\xi \cong 0.02$. Thus from Fig. 5.15, with $\Omega/\omega = 0.83$ and $\xi \approx 0.02$, we find an amplitude ratio $A/(P/k)$ of approximately 3. Our objective is to reduce this "amplification" by changing the system natural frequency ω, so that the frequency ratio Ω/ω is not as close to unity. Let's see whether the proposed design change will work. After the addition of mass, the system mass is 130 kg, and the system natural frequency ω will be different. The new natural frequency will be $\omega_{\text{new}} = \omega_{\text{old}}/1.3^{1/2}$, because ω varies inversely with the square root of m. Thus we find $\omega_{\text{new}} \cong 110$ rad/s, and the frequency ratio $\Omega/\omega_{\text{new}} \cong 0.95$; we have made things even worse by forcing the system to operate closer to resonance than it did before. Thus, the proposed design solution is a bad one.*

*When going through this type of example in the classroom, I often ask the students, "What do you think happened to the engineer who proposed this design solution?" One wag answered, "He became a university teacher."

5.4 COMPLEX FREQUENCY RESPONSE AND HIGHER-ORDER SYSTEMS

5.4.1 The Complex Frequency Response Function

In Secs. 5.2 and 5.3 we derived the relations (5.30) and (5.79) for the frequency response of the first- and second-order systems (5.26) and (5.67). The resulting graphs of the frequency responses were shown in Figs. 5.6 and 5.15. These frequency response calculations may be made using the method of undetermined coefficients to obtain the steady-state responses according to the assumed solution (5.69).

In this section we consider the frequency response of linear systems of arbitrary order to a harmonic excitation. As in Secs. 5.2 and 5.3, it is assumed that the system under consideration is asymptotically stable and that only the steady-state response to a harmonic excitation is of interest. This general frequency response is of interest for two reasons: (1) In real-world operation, dynamic systems of all types and orders may be acted upon by outside influences (inputs) which have a harmonic or repeated character; thus, it is necessary to understand how a given system will respond to such an influence. (2) It may occur that a given system or system component is difficult to model exactly, so that experimentation is necessary in order to validate and/or obtain an adequate mathematical model. By subjecting the system to harmonic excitations over a sufficiently wide range of exciting frequencies Ω, it is possible to measure the frequency response and from it to identify the system properties. Nowadays it is common to employ sophisticated electronic dynamic analyzers that determine a linear model which will reproduce with reasonable accuracy an experimentally obtained system frequency response.

The objective of this section is to develop an understanding of how linear systems of higher order respond to a harmonic excitation and to understand the close relation between higher-order system frequency response and the frequency responses of first- and second-order systems. To this end we consider linear systems of arbitrary order n, for which the mathematical model is in the form of a single ODE of nth order (as in Sec. 4.3.1):

$$x^{(n)} + a_1 x^{(n-1)} + \cdots + a_{n-1}\dot{x} + a_n x = g(t) \tag{5.81}$$

It is assumed that the system (5.81) is asymptotically stable, so that if the input $g(t)$ is harmonic with frequency Ω, then the steady-state response $x(t)$ will also be harmonic with the same frequency Ω. In order to calculate the frequency response of the system (5.81), we introduce the *complex frequency response* function (CFR), denoted as $H(i\Omega)$ and defined as follows: Let the harmonic input $g(t)$ have unit amplitude and be represented in the complex form $g(t) = e^{i\Omega t}$ (because the physical input is real, not complex, it is to be understood that the physical input is the real part of $e^{i\Omega t}$). Then express the steady-state response $x(t)$ as

$$x(t) = H(i\Omega)e^{i\Omega t} \tag{5.82}$$

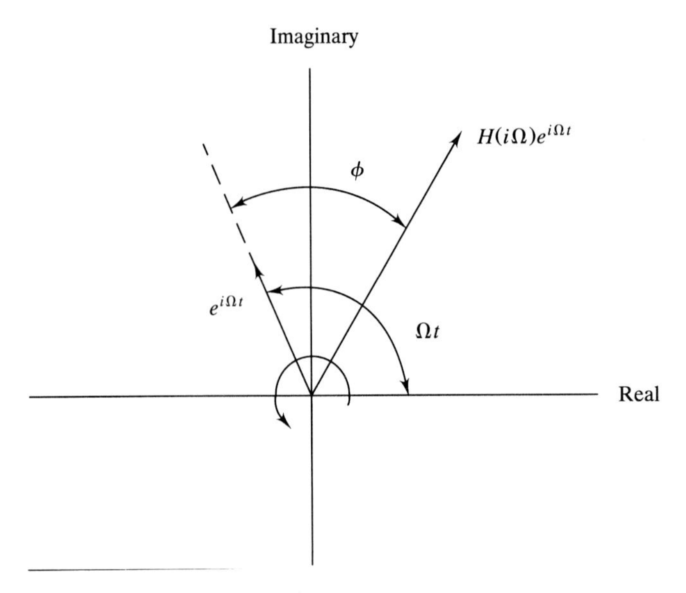

FIGURE 5.16
Representation of the complex frequency response function in the complex plane.

where the complex frequency response function $H(i\Omega)$ is to be determined for the given system. For given excitation frequency Ω, $H(i\Omega)$ will be a complex constant. Equation (5.82) expresses the essential idea of system frequency response: The linear system acts as an amplifier that operates on the harmonic input to produce a harmonic output of the same frequency. The use of complex algebra turns out to be a very direct way to describe this process, as shown in Fig. 5.16 (refer also to the discussion of complex numbers in Sec. 4.2.3).

In Fig. 5.16 the input $e^{i\Omega t}$ is represented as a unit vector in the complex plane. At $t = 0$, this vector is aligned with the real axis. For $t > 0$ the vector $e^{i\Omega t}$ rotates counterclockwise at the rate Ω, completing circuits when $\Omega t = 2\pi, 4\pi, \ldots$. The steady-state response defined by (5.82) is represented by a second rotating vector that is amplified and phase shifted relative to $e^{i\Omega t}$. The input and response vectors rotate as a rigid body such that, at any instant, the physical input and response are the components of the rotating vectors along the real axis. In interpreting the complex frequency response $H(i\Omega)$, it is useful to express $H(i\Omega)$ in the polar form

$$H(i\Omega) = |H(i\Omega)|e^{i\phi} \tag{5.83}$$

where $|H(i\Omega)|$ is the magnitude of $H(i\Omega)$ and ϕ is the phase. Thus, the input and output are expressed as follows:

$$\begin{aligned} \text{Input:} \quad & g(t) = e^{i\Omega t} \\ \text{Output:} \quad & x(t) = |H(i\Omega)|e^{i(\Omega t + \phi)} \end{aligned} \tag{5.84}$$

This form is essentially the same as (5.69) with $-\phi$ changed to ϕ (the angle ϕ depicted in Fig. 5.16 has a negative value; that is, the response lags the input). Thus, $|H(i(\Omega)|$ in (5.83) equals the amplitude A in (5.69) for the case in which the excitation amplitude P is unity. As in Secs. 5.2 and 5.3, our primary concern will be with the magnitude $|H(i\Omega)|$ of the complex frequency response.

Let us now use the CFR to determine the steady-state frequency response of linear systems of arbitrary order. We will begin with systems of first and second order, for which the results are familiar, and we will then consider systems of order greater than 2. We will have two objectives: (1) to become comfortable with the mechanics of finding the steady-state response using the CFR function and (2) to interpret the mathematical results so that an understanding of the frequency response of systems of arbitrary order is obtained. At this point there may not seem to be any advantage in introducing the CFR function. We shall see, however, that the CFR possesses two properties that render it so useful: (1) It can be written down immediately from the system characteristic polynomial, and (2) for higher-order systems the CFR can be easily expressed in a form which allows it to be interpreted in terms of first- and second-order system responses.

FIRST-ORDER SYSTEM. Here we repeat the analysis of Example 5.4 using the CFR. The mathematical model (5.26) is stated as

$$\dot{x} + ax = e^{i\Omega t} \tag{5.85}$$

where the sign of the system constant a has been reversed. To find the CFR function $H(i\Omega)$ for the system (5.85), (5.82) is substituted into (5.85). Noting that if $x(t) = H(i\Omega)\,e^{i\Omega t}$, then $\dot{x} = i\Omega H(i\Omega)\,e^{i\Omega t}$, we obtain from (5.85)

$$(i\Omega + a)He^{i\Omega t} = e^{i\Omega t}$$

Canceling the $e^{i\Omega t}$ and solving for $H(i\Omega)$ yield

$$H(i\Omega) = \frac{1}{a + i\Omega}$$

which is the complex frequency response function for the first-order system (5.85). It is common to express this result in terms of the system time constant $T = 1/a$, which provides the alternative form

$$H(i\Omega) = \frac{1}{(1/T) + i\Omega}$$

or

$$H(i\Omega) = \frac{T}{1 + i\Omega T} \tag{5.86}$$

multiplication of the numerator and denominator of (5.86) by the complex conjugate of the denominator allows (5.86) to be rewritten as

$$H(i\Omega) = \frac{T(1 - i\Omega T)}{1 + (\Omega T)^2}$$

Thus, the magnitude and phase of the first-order system CFR are defined by

$$|H(i\Omega)| = \frac{T}{\left[1 + (\Omega T)^2\right]^{1/2}} \tag{5.87}$$

$$\tan\phi = -\Omega T$$

The result (5.87) for $|H(i\Omega)|$ is the same as (5.30) if P in (5.30) is set to unity.

SECOND-ORDER SYSTEM. Here we determine the CFR for the second-order mechanical oscillator system having the math model similar to (5.67):

$$\ddot{x} + 2\xi\omega\dot{x} + \omega^2 x = \frac{1}{m}e^{i\Omega t} \tag{5.88}$$

Use of (5.82) in (5.88) leads to

$$\left[(i\Omega)^2 + 2\xi\omega(i\Omega) + \omega^2\right]H(i\Omega)\,e^{i\Omega t} = \frac{1}{m}e^{i\Omega t}$$

so that the mechanical oscillator CFR is

$$H(i\Omega) = \frac{1/m}{\left[(\omega^2 - \Omega^2) + 2i\xi\Omega\omega\right]} \tag{5.89}$$

Division of numerator and denominator by the system natural frequency $\omega^2 = k/m$ leads to

$$H(i\Omega) = \frac{1/k}{\left(1 - \dfrac{\Omega^2}{\omega^2}\right) + 2i\xi\dfrac{\Omega}{\omega}} \tag{5.90}$$

The magnitude and phase of $H(i\Omega)$ are then found to be

$$|H(i\Omega)| = \frac{1/k}{\left[\left(1 - \dfrac{\Omega^2}{\omega^2}\right)^2 + \left(2\xi\dfrac{\Omega}{\omega}\right)^2\right]^{1/2}} \tag{5.91a}$$

$$\tan\phi = \frac{-2\xi\Omega/\omega}{1 - \dfrac{\Omega^2}{\omega^2}} \tag{5.91b}$$

The result (5.91a) is the same as the relation (5.78) for the steady-state vibration amplitude A (with $P = 1$) obtained using the method of undetermined coefficients.

The preceding analysis of the CFR for first- and second-order systems was intended to show the mechanics of obtaining the CFR and to show that the results obtained are the same as those found in Secs. 5.2 and 5.3 using the method of undetermined coefficients. Let us now investigate the CFR for a third-order system.

Example 5.10 Complex Frequency Response of a Third-Order System. Suppose the linear system under consideration is of third order with a mathematical model (5.81) of the form

$$\dddot{x} + 10.2\ddot{x} + 3.01\dot{x} + 10.1x = e^{i\Omega t} \tag{5.92}$$

This third-order system has the characteristic equation

$$\lambda^3 + 10.2\lambda^2 + 3.01\lambda + 10.1 = 0 \tag{5.93}$$

and the eigenvalues turn out to be

$$\lambda_{1,2} = -0.1 \pm i$$

$$\lambda_3 = -10$$

so that the system is asymptotically stable. The factored form of the characteristic equation is

$$(\lambda^2 + 0.2\lambda + 1.01)(\lambda + 10) = 0 \tag{5.94}$$

Thus, the autonomous, or free response, of this system will consist of an exponentially decaying oscillation with time constant $T_1 = 10$ s and frequency $\omega = 1$ rad/s (from $\lambda_{1,2} = -0.1 \pm i$), combined with a purely exponential response (from $\lambda_3 = -10$) which decays relatively rapidly with a time constant $T_2 = 0.1$ s.

The complex frequency response $H(i\Omega)$ is found by plugging (5.82) into (5.92) and noting that $x = He^{i\Omega t}$, $\dot{x} = i\Omega He^{i\Omega t}$, $\ddot{x} = (i\Omega)^2He^{i\Omega t}$, $\dddot{x} = (i\Omega)^3He^{i\Omega t}$, which yields

$$\left[(i\Omega)^3 + 10.2(i\Omega)^2 + 3.01i\Omega + 10.1\right]He^{i\Omega t} = e^{i\Omega t} \tag{5.95}$$

Solving for $H(i\Omega)$ results in

$$H(i\Omega) = \frac{1}{(i\Omega)^3 + 10.2(i\Omega)^2 + 3.01(i\Omega) + 10.1} \tag{5.96}$$

We observe from (5.96) the following useful result which allows direct determination of the CFR for systems modeled as (5.81): $H(i\Omega)$ is merely the reciprocal of the system characteristic polynomial, with the eigenvalue λ replaced by $i\Omega$. That is, if we denote the characteristic polynomial as $P(\lambda)$, then $H(i\Omega) = 1/P(i\Omega)$. This occurs because the homogeneous solution to (5.81) is assumed to be proportional to $e^{\lambda t}$, so that the successive differentiations in (5.81) produce successive powers of λ. Similarly, the steady-state response of the system (5.81) is assumed to be (5.82), so that successive differentiations in (5.81) produce successive powers of $(i\Omega)$. Thus, in (5.95) the characteristic polynomial, with $i\Omega$ replacing λ, will always appear as the coefficient of $H(i\Omega)$ on the left-hand side. This result allows the CFR to be determined directly from the system characteristic equation, provided that the mathematical model is of the form (5.81). In view of this result, the calculation of the CFR may also be made by using the factored form (5.94) of the system characteristic polynomial. If this were done in the present example, the result (5.96) could be expressed in the form

$$H(i\Omega) = \frac{1}{(-\Omega^2 + 0.2i\Omega + 1.01)(10 + i\Omega)} \tag{5.97}$$

Because our primary interest here is in the magnitude $|H(i\Omega)|$, use of the factored form (5.97) is convenient, for it allows $|H(i\Omega)|$ to be expressed as the products of the magnitudes of the various factors comprising the system characteristic polynomial. Thus, the magnitude $|H(i\Omega)|$ from (5.97) is

$$|H(i\Omega)| = \frac{1}{|-\Omega^2 + 0.2i\Omega + 1.01|} \cdot \frac{1}{|10 + i\Omega|} \tag{5.98}$$

A little algebra then leads to the result

$$|H(i\Omega)| = \frac{1}{\left[(1.01 - \Omega^2)^2 + (0.2\Omega)^2\right]^{1/2} [10^2 + \Omega^2]^{1/2}} \tag{5.99}$$

In the form (5.99) or (5.98), observe that $|H(i\Omega)|$ comprises the products of two factors, one of which is the same as that of a second-order system [see (5.91)], the other of which is the same as that of a first-order system [see (5.87)].

Generally, for a system (5.81) of arbitrary order, the factored form (5.94) of the system characteristic polynomial allows us to determine directly $|H(i\Omega)|$ as the products of the magnitudes of various first- and second-order factors. A clear understanding of the frequency response of first- and second-order systems, therefore, allows an understanding of higher-order system frequency response to be achieved. As an example, consider the CFR of Example 5.10 shown in Fig. 5.17. Figure 5.17(*a*) depicts the CFR magnitude for the third-order system (5.92), as defined by (5.99). Figures 5.17(*b*) and (*c*), respectively, depict separately the first- and second-order factors appearing in (5.99), so that Fig. 5.17(*a*) is the product of the functions shown in Figs. 5.17(*b*) and (*c*). This figure illustrates the formation of the system CFR from the CFRs associated with the first- and second-order factors appearing in $|H(i\Omega)|$.

In this section (5.4.1) the complex frequency response function $H(i\Omega)$ has been defined and determined for systems of first, second, and third order. The close connection between $H(i\Omega)$ and the system characteristic polynomial was shown. In the following section (5.4.2) we continue study of the CFR with the objective of obtaining a clear understanding of the frequency dependence of higher-order system CFRs. We shall also address the system identification problem: Given an experimentally determined CFR, can we identify properties of the system that produced the experimental data?

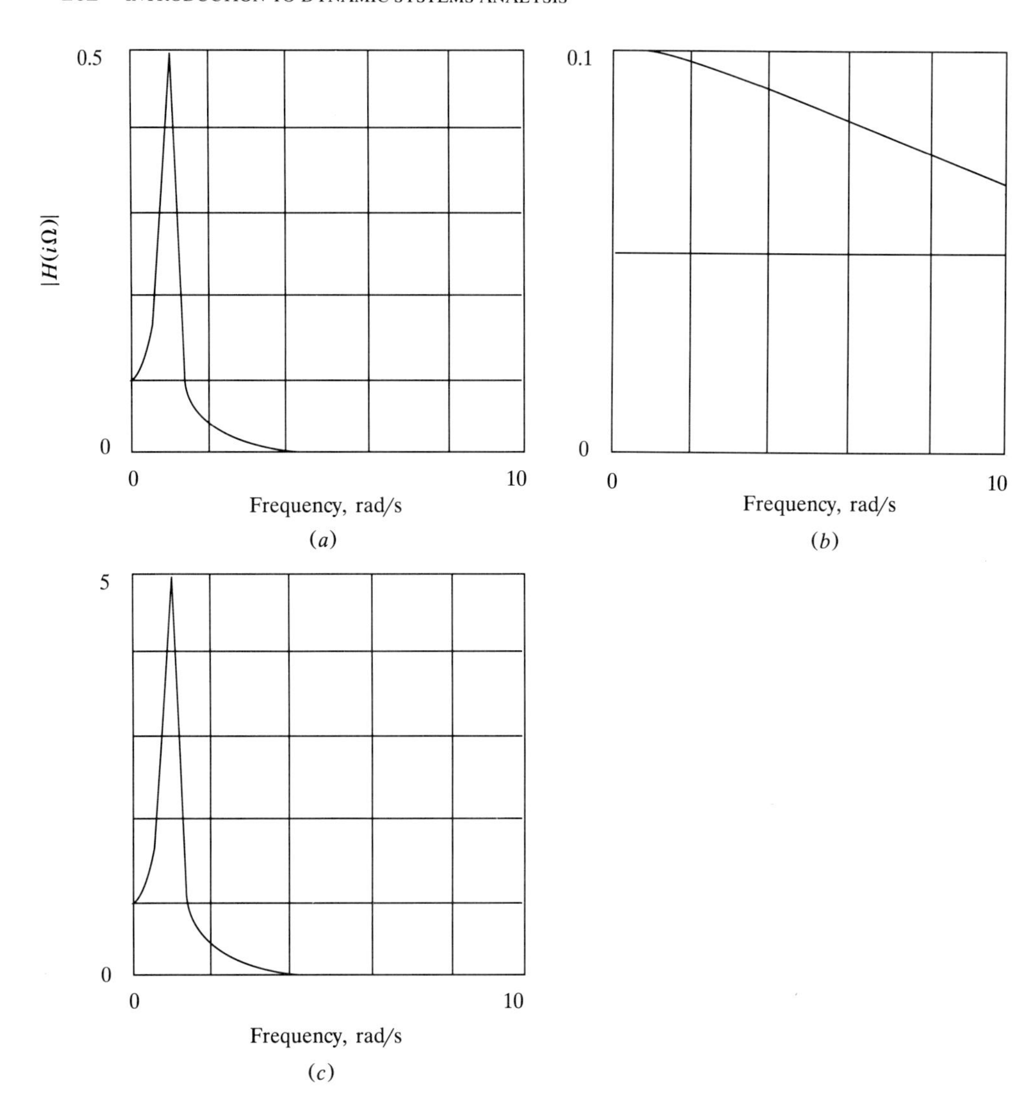

FIGURE 5.17
Frequency response of the third-order system [(5.92) and (5.96)]: (*a*) third-order system frequency response; (*b*) contribution of the first-order factor; (*c*) contribution of the second-order factor. Figure (*a*) is the product of (*b*) and (*c*).

5.4.2 Bode Plot Representation of the CFR

In this section we introduce the Bode plot, which is a log-log plot of $|H(i\Omega)|$ versus driving frequency Ω. This turns out to be a useful way to display and interpret graphically the CFR function $H(i\Omega)$. The practice adopted so far in graphing system frequency response has been to plot $|H(i\Omega)|$ versus Ω on linear scales (for example, see Figs. 5.6, 5.15, and 5.17). It is often more informative, however, to plot this information on a log-log scale, for the following reasons:

1. The use of a log-log scale allows variations in $H(i\Omega)|$ and Ω over several orders of magnitude to be displayed easily. This is of importance in characterizing "high-frequency rolloff" of the system. The term *rolloff* means the tendency of most dynamic system CFRs to approach zero as $\Omega \to \infty$. For instance, the second-order system CFR defined by (5.91) will approach zero as $(1/\Omega^2)$ as $\Omega \to \infty$. The first-order system CFR (5.87) will approach zero as $(1/\Omega)$ as $\Omega \to \infty$. The third-order system CFR (5.99) obtained in Example 5.10 will approach zero as $(1/\Omega^3)$ as $\Omega \to \infty$. The manner in which $|H(i\Omega)|$ decreases through several orders of magnitude as Ω becomes large can be displayed easily on a log-log plot, but it is difficult to display on a linear scale.
2. As noted earlier in this section, the CFR of an nth-order system may be formed as the *products* of various first- and second-order CFRs. Thus, the logarithm of $|H(i\Omega)|$ can be found by *adding* the logarithms of the same first- and second-order CFRs, simplifying the construction and interpretation of the CFR plot.

THE DECIBEL SCALE. The *decibel* is a logarithmic measure of magnitude which is employed in the construction of Bode plots. The decibel, denoted db, is defined as follows: If C is a real number, then C, in decibels, is defined as

$$\text{db} = 20 \log C \tag{5.100}$$

where the logarithm is base 10. The definition (5.100) says that each factor of 10 increase in C corresponds to an increase of 20 db. The table below shows several values of C and the corresponding magnitudes in decibels:

C	Decibels
1000	60
100	40
10	20
1	0
0.1	−20
0.01	−40
0.001	−60

Note that (5.100) can be inverted to obtain the numerical value of C from a given decibel level:

$$C = 10^{(\text{db}/20)} \tag{5.101}$$

Thus a magnitude of 5 db corresponds to $10^{0.25} = 1.778$, and so on.

THE BODE PLOT. The Bode plot is a graph of $|H(i\Omega)|$, measured in decibels, versus Ω, measured logarithmically. In order to illustrate the construction and interpretation of Bode plots, we begin with systems of first and second order.

For a first-order system the CFR magnitude $|H(i\Omega)|$ was defined in terms of the system time constant T in (5.87):

$$|H(i\Omega)| = \frac{T}{(1 + \Omega^2T^2)^{1/2}} \tag{5.102}$$

In decibels $|H(i\Omega)|$ is then

$$\text{db} = 20 \log\left[\frac{T}{(1 + \Omega^2T^2)^{1/2}}\right]$$

or

$$\text{db} = 20 \log T - 10 \log (1 + \Omega^2T^2) \tag{5.103}$$

Notice that the constant T appearing in the numerator of (5.103) merely shifts the entire function (5.103) up or down by an amount $20 \log T$. In the ensuing discussion, we will ignore this term, with the understanding that such a shift in the Bode plot would normally be necessary. Thus, we will actually construct the Bode plot of the function

$$|H(i\Omega)| = [1 + \Omega^2T^2]^{-1/2} \tag{5.104}$$

The resulting first-order system Bode plot is shown in Fig. 5.18. Note that the frequency Ω scale is measured in units of the reciprocal of the system time

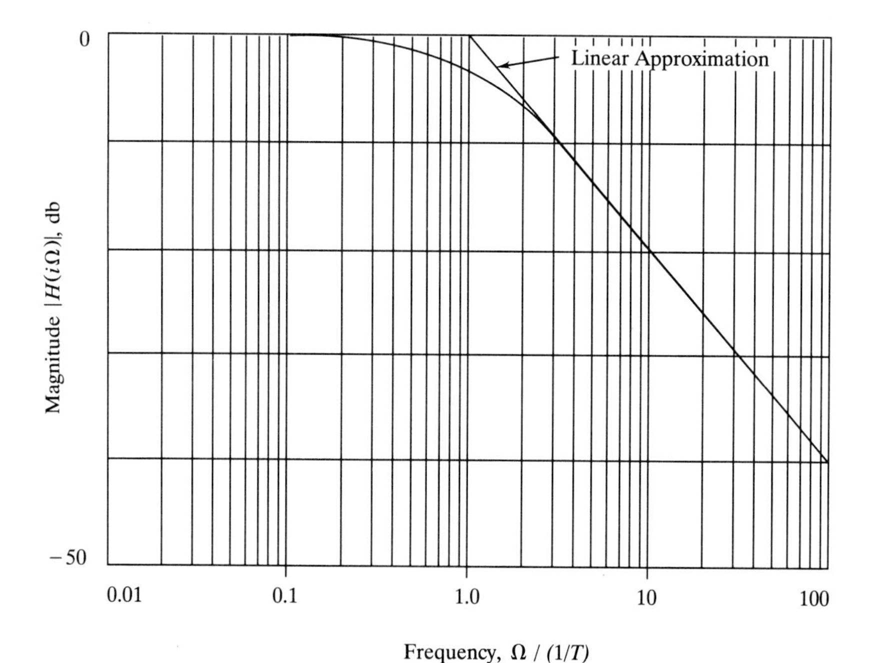

FIGURE 5.18
Bode plot of the magnitude $|H(i\Omega)|$ of the complex frequency response for the first-order system, (5.104). The linear approximation is indicated.

constant T, allowing a single plot to represent all first-order systems (the units of $1/T$ are $1/s$, that is, rad/s, which are the units of the frequency Ω).

Figure 5.18 presents the same information as does Fig. 5.6. The Bode plot, however, allows much wider ranges of frequency and magnitude to be displayed. The following features of the first-order Bode plot of Fig. 5.18 are noted: (1) For driving frequencies $\Omega \ll 1/T$, the magnitude is flat at a level of 0 db, because for this low-frequency range $(1 + \Omega^2 T^2) \approx 1$. (2) As Ω passes through $1/T$ and becomes much larger than $1/T$, the factor $(1 + \Omega^2 T^2) \to \Omega^2 T^2$, so that from (5.104), db $\to -20 \log(\Omega T)$. Thus, for $\Omega \gg 1/T$, the Bode plot asymptotes to a straight line which decreases at the rate of *20 db per decade* of frequency change. (3) If the -20 db per decade asymptote is extended toward lower frequencies, it will intersect the 0-db line at $\Omega = 1/T$, that is, at $\Omega T = 1$ (Fig. 5.18). (4) In characterizing the first-order Bode plot, it is common to approximate the Bode plot by two straight lines with a slope change at $\Omega = 1/T$, as shown in Fig. 5.18. (5) The frequency response is "flat" until $\Omega \to 1/T$, near which frequency rolloff occurs.

The first-order characteristics noted in the preceding paragraph provide a nice way to perform system identification of first-order systems, as demonstrated in the following example.

Example 5.11 System Identification of a Variable Speed Pump. A variable speed pump produces an output flow rate q_0, which, for steady operating conditions, is proportional to the steady input voltage V_0 supplied to the pump. The pump, however, does not respond instantaneously to changes in input voltage. For example, if the pump were initially not operating and were then (at $t = 0$) subjected to a step voltage input V_0, some time would be required for the pump output flow $q(t)$ to reach the steady level q_0. Thus, the pump is a dynamic system for which the state $q(t)$ is governed by some dynamic model (see Exercise 5.14). The mechanics governing the pump flow will be fairly complex and may be difficult to model from first principles, so that experimental characterization of the pump dynamic properties is necessary. By experimentally determining the pump frequency response and then constructing a Bode plot, this system identification can be performed. Figure 5.19 shows the Bode plot obtained experimentally for a small variable speed pump.

The data of Fig. 5.19 were generated by subjecting the pump to an input voltage $V(t) = V_0 + V_1 e^{i\Omega t}$ over a range of exciting frequencies Ω, with V_1 fixed. For this input, the output flow $q(t)$ in the steady state will be given by $q(t) = q_0 + q_1 e^{i(\Omega t + \phi)}$, where q_1 is the steady-state harmonic response amplitude due to the harmonic component $V_1 e^{i\Omega t}$ of the input. The data shown in Fig. 5.19 are the steady-state amplitudes (in decibels) of a voltage signal which is proportional to the shaft speed of the pump, which in turn is proportional to the steady-state output amplitude q_1. Thus, the data shown are proportional to $|H(i\Omega)|$.

For purposes of doing system identification, the data of Fig. 5.19 are interpreted in the following way:

1. In the region of frequency rolloff a straight line of slope -20 db per decade is shown. This line fits the low-frequency rolloff data reasonably well, although there appears to be additional rolloff occurring at about 0.3 Hz ($\log f = -0.5$),

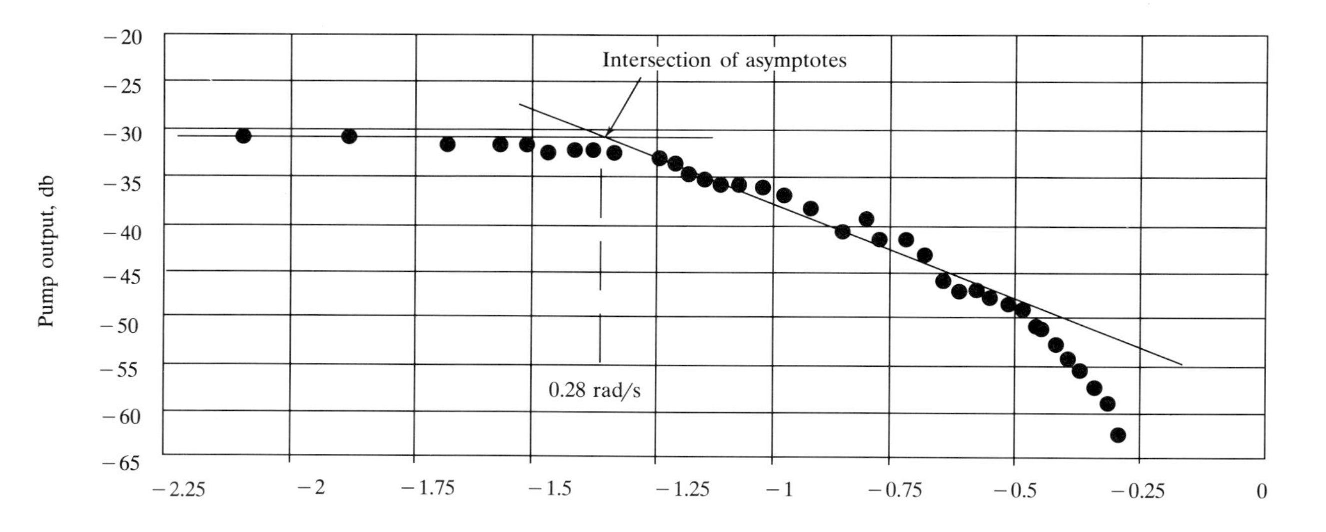

FIGURE 5.19
Bode plot of an actual water pump (Example 5.11). See text for discussion. (Data provided courtesy of Mark Young of Washington State University.)

indicative of higher-order behavior. We will develop here, however, the first-order model which accurately represents the smallest system time constant, since this time constant will essentially govern the transient response.

2. If we extend the -20-db per decade asymptote toward lower frequencies, this asymptote intersects the horizontal line defining the flat, low-frequency portion of the frequency response at a frequency $\Omega = 0.045$ Hz $= 0.28$ rad/s. Thus, the first-order system time constant $T \cong 3.6$ s. We conclude that a reasonable system model for the pump is a first-order system model for which the eigenvalue $\lambda = 1/T = 0.28$/s. The linear system model so determined is then

$$\dot{q} + 0.28q = cV(t) \tag{5.105}$$

where c is a proportionality constant still to be found (see also Exercise 5.14).

Let us now discuss the Bode plot of the second-order system, for which $|H(i\Omega)|$ is defined by (5.91). As in the case of the first-order system, we will set

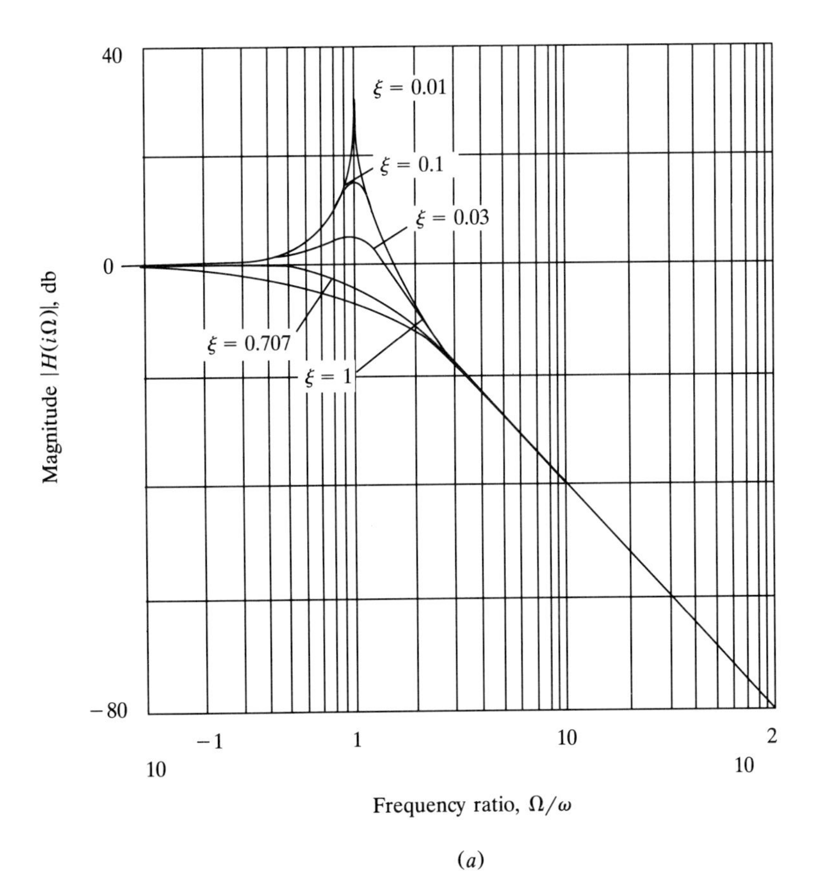

(*a*)

FIGURE 5.20
Bode plot of magnitude $|H(i\Omega)|$ of the complex frequency response function for the second-order system (5.106). Part (*b*) is a blowup of (*a*) near resonance.

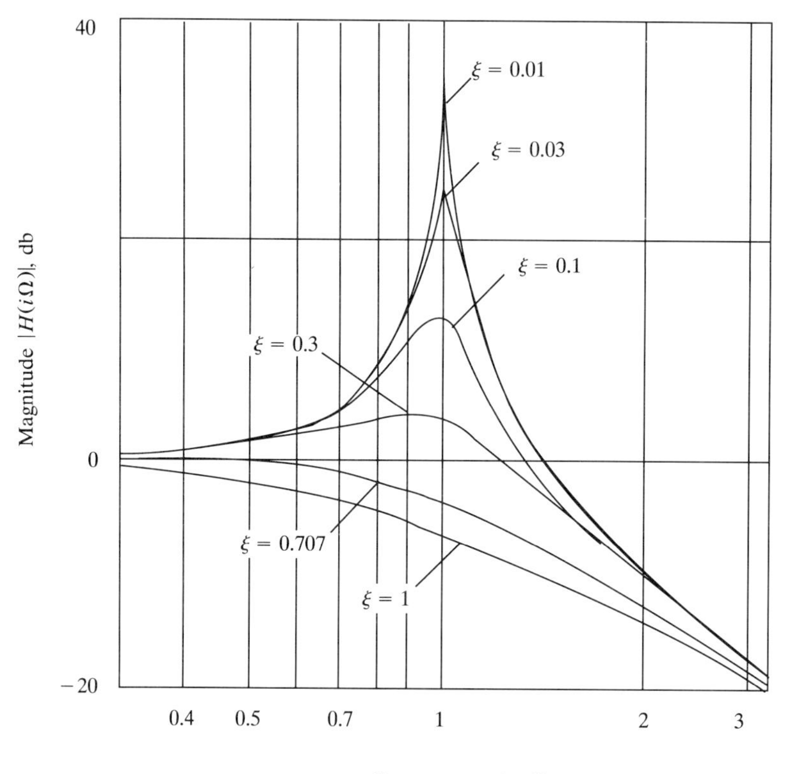

(*b*)

FIGURE 5.20 (*Continued*)

the numerator of $|H(i\Omega)|$ to unity, with the understanding that any non-unit constant multiplier C present for the actual system being considered will merely shift the Bode plot up or down by an amount $20 \log C$. Thus, we investigate the Bode plot of the second-order frequency response function

$$|H(i\Omega)| = \left[\left(1 - \frac{\Omega^2}{\omega^2}\right)^2 + \left(\frac{2\xi\Omega}{\omega}\right)^2\right]^{-1/2} \tag{5.106}$$

On the decibel scale, (5.106) is measured as

$$\text{db} = -10 \log \left[\left(1 - \frac{\Omega^2}{\omega^2}\right)^2 + \left(\frac{2\xi\Omega}{\omega}\right)^2\right] \tag{5.107}$$

The Bode plot based on (5.107) is shown in Fig. 5.20, which contains the same information as Fig. 5.15. The exciting frequency Ω is measured in units of the system natural frequency ω, and results are presented for various values of the damping parameter ξ. The following characteristics of the second-order system Bode plot are observed:

1. As $\Omega/\omega \to 0$, db $\to 0$, and the frequency response is flat for small values of Ω/ω.
2. As $\Omega/\omega \to \infty$, the factor $[(1 - \Omega^2/\omega^2)^2 + (2\xi\Omega/\omega)^2] \to (\Omega/\omega)^4$, so that as $\Omega/\omega \to \infty$, db $\to -10\log(\Omega/\omega)^4 = -40\log(\Omega/\omega)$. This shows that the frequency rolloff of the second-order system will asymptote to a straight line which changes at the rate -40 db per decade of change in the exciting frequency Ω.
3. If the system damping is light ($\xi \ll 1$), we observe the characteristic amplification associated with the region of resonance ($\Omega/\omega \approx 1$). This amplification does not occur in a first-order system.
4. If the -40 db per decade asymptote is extended toward lower frequencies, it intersects the 0-db line at $\Omega = \omega$. This feature is helpful in system identification (which is, nevertheless, just as direct from Fig. 5.15 as it is from the Bode plot for a lightly damped second-order system).

Remark. The "standard form" of the second-order system characteristic polynomial is

$$\lambda^2 + 2\xi\omega\lambda + \omega^2 = 0$$

and the eigenvalues are $\lambda_{1,2} = -\xi\omega \pm \omega(\xi^2 - 1)^{1/2}$. We often tacitly assume that the eigenvalues are complex, and if this is so, we obtain the Bode plot shown in Fig. 5.20. If, however, the eigenvalues are real ($\xi > 1$), which may occur if the damping mechanism is sufficiently strong, then we may view the second-order system characteristic polynomial as the product of two first-order polynomials, as illustrated in the following example.

Example 5.12 A Second-Order System with Real Eigenvalues. Suppose we investigate a hypothetical, heavily damped second-order system for which $\omega = 1$ and $\xi = 3$. Then the eigenvalues are found to be $\lambda_1 = -0.17, \lambda_2 = -5.83$, and the system characteristic polynomial may be written as $(\lambda + 0.17)(\lambda + 5.83)$. The system time constants are $T_1 = 1/|\lambda_1| = 5.83$ s and $T_2 = 1/|\lambda_2| = 0.17$ s, and the CFR magnitude is proportional to

$$|H(i\Omega)| \sim \frac{1}{\left[\left(1 + \Omega^2 T_1^2\right)\left(1 + \Omega^2 T_2^2\right)\right]^{1/2}} \tag{5.108}$$

In decibels $|H(i\Omega)|$ is

$$\text{db} = -10\log\left[\left(1 + \Omega^2 T_1^2\right)\left(1 + \Omega^2 T_2^2\right)\right]$$

or, separating the factors,

$$\text{db} = -10\log\left(1 + \Omega^2 T_1^2\right) - 10\log\left(1 + \Omega^2 T_2^2\right) \tag{5.109}$$

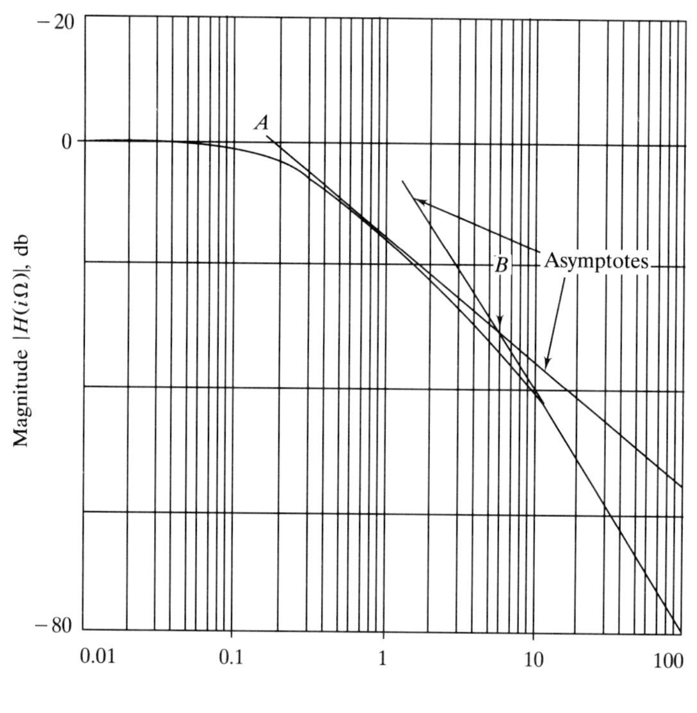

FIGURE 5.21
Bode plot for the system of Example 5.12, a second-order system with real eigenvalues [Eq. (5.108)]. The intersection A of the -20 db per decade asymptote with the zero db line identifies the first (largest) system time constant T_1. The intersection B of the -20 db per decade and -40 db per decade asymptotes identifies the second time constant T_2.

The associated Bode plot is shown in Fig. 5.21 and has the following features:

1. As $\Omega \to 0$, db $\to 0$ and the frequency response is flat for sufficiently small Ω.
2. As Ω passes through $1/T_1 = |\lambda_1|$, frequency rolloff occurs. This rolloff, however, tends toward a -20 db per decade asymptote because the factor $(1 + \Omega^2 T_2^2) \approx 1$ for these lower frequencies.
3. As Ω increases past $1/T_2 = |\lambda_2|$, additional frequency rolloff occurs, as each of the first-order factors is diminishing at -20 db per decade when $\Omega \gg 1/T_2$. At high frequencies Ω the rolloff asymptotes to -40 db per decade, as it will do for any second-order system (5.88), whether the eigenvalues are real or complex.
4. If the two real eigenvalues differ significantly in magnitude, as they do in this example, then it is possible to identify two "breaks" in the frequency rolloff and to estimate the largest system time constant (and associated eigenvalue) by extending the asymptote to the 0-db level, as shown in Fig. 5.21. The second time constant is then estimated by locating the intersection of the high-frequency, -40 db per decade asymptote with the lower frequency, -20 db per decade asymptote, as shown in Fig. 5.21. If, however, a second-order system has

real, negative eigenvalues that are close to each other in magnitude, it is difficult to isolate the frequency rolloffs due to the individual eigenvalues. Available are computer routines that will determine appropriate time constants which will produce a close match with a measured Bode plot. Based on the preceding results for systems of the first and second order, we now study the third-order system considered in Example 5.10.

Example 5.13 The System of Example 5.10 Revisited. The third-order system considered is given by (5.92) and (5.99), and the CFR magnitude is shown on linear scales in Fig. 5.17. From (5.99) the magnitude $|H(i\Omega)|$, measured in decibels, is

$$\text{db} = -10\log\left[(1.01-\Omega^2)^2 + (0.2\Omega)^2\right] - 10\log(100+\Omega^2) \quad (5.110)$$

Equation (5.110) consists of the sum of a second-order response factor and a first-order response factor. The complex eigenvalues associated with the second-order factor were found in Example 5.10 to be $\lambda_{1,2} = -0.1 \pm i$, and the associated second-order parameters ξ, ω are $\omega = \sqrt{1.01}$, $\xi = 0.1/\sqrt{1.01} = 0.0995$. The real eigenvalue λ_3 associated with the first-order factor is $\lambda_3 = -10$, with an associated time constant $T_3 = 0.1$ s. The system Bode plot, from (5.110), is shown in Fig. 5.22.

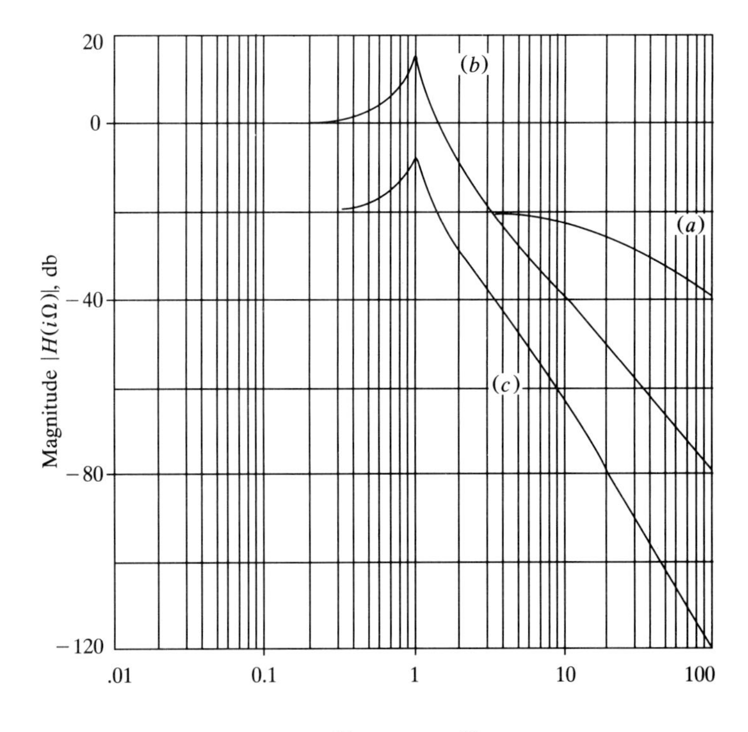

FIGURE 5.22
Bode plot for the third-order system of Examples 5.10 and 5.13, Eq. (5.99). The first- and second-order factors are shown, along with the resultant third-order system plot. (*a*) First-order factor; (*b*) second-order factor; (*c*) third-order system.

Also shown in Fig. 5.22 are the Bode plots of the second- and first-order factors appearing in (5.110). These two factors add to form the system Bode plot. From Fig. 5.22 the following behavior is observed:

1. As $\Omega \to 0$, db $\to -10\log(1.02) - 10\log(100) = -20.04$ db. Here we have not adjusted $|H(i\Omega)|$ to pass through 0 db as $\Omega \to 0$.
2. As Ω increases from very small values, we observe a flat frequency response, followed by the second-order resonant amplification occurring near $\Omega = \omega = \sqrt{1.01}$, followed by the second-order frequency rolloff which tends toward a -40 db per decade decrease.
3. As Ω passes through $\omega = \sqrt{1.01}$, the first-order factor $-10\log(100 + \Omega^2)$ remains flat. The frequency rolloff due to this first-order factor does not become evident until Ω approaches and passes through $1/T_3 = 10$. For $\Omega \gg 10$, the system frequency rolloff asymptotes to -60 db per decade, which is characteristic of the third-order system.

We close this section with a final example.

Example 5.14. The measured frequency response of a dynamic system produces the Bode plot shown in Fig. 5.23. From this information it is desired to determine how quickly the system will respond if it is subjected to a step input.

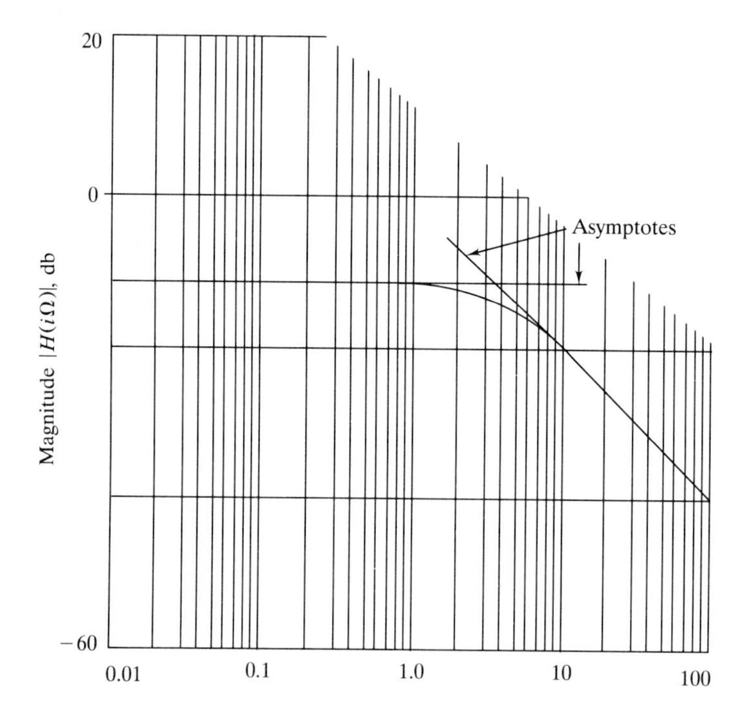

FIGURE 5.23
Bode plot for the system of Example 5.14. See text for discussion.

The Bode plot exhibits the character expected of a first-order system: The frequency response is flat at low frequencies and then rolls off at -20 db per decade. Determination of the intersection of the -20 db per decade asymptote with the extension of the flat portion of the response curve, as shown in Fig. 5.23, yields a first-order system time constant $T = 0.278$ s. From the analysis of first-order system response to a step input, Sec. 5.2.2, we conclude that after four time constants, that is, when $t \cong 1.1$ s, the system response will have reached to within 2 percent of its steady-state level.

5.4.3 Comments on the Frequency Response

1. *Bode plots and eigenvalues:* The way in which a linear system responds to a given input is governed by the system eigenvalues. The Bode plot is thus an effective way to represent the system frequency response (important in its own right) and to display indirectly the system eigenvalues.
2. *Bode plots:* In the examples of Bode plots presented so far, the CFR magnitude has been of the form $|H(i\Omega)| = C/\,|P(i\Omega)|$, where C is a constant (sometimes set to unity) and $P(\lambda)$ is the system characteristic polynomial. For such systems the Bode plot is flat at low frequencies Ω and exhibits frequency rolloff dictated by the eigenvalues obtained from $P(\lambda) = 0$. It is fairly common, however, to encounter systems for which the magnitude $|H(i\Omega)|$ has the form $|H(i\Omega)| = |f(i\Omega)|/\,|P(i\Omega)|$, that is, for which the numerator is also a function of the exciting frequency Ω. In such cases the nature of the Bode plot is altered, as illustrated in the following example.

Example 5.15 Base Excitation of a Mechanical Oscillator (Fig. 5.1). We suppose that the linear oscillator of Fig. 5.1 is subjected to a unit harmonic motion of the base to which the stiffness and damping elements are attached, so that $y(t) = e^{i\Omega t}$. The system equation of motion is

$$m\ddot{x} + c\dot{x} + kx = ky + c\dot{y} = (l + ic\Omega)\,e^{i\Omega t} \tag{5.111}$$

Note that, while the harmonic input is $y(t)$, both the input and its derivative appear in the equation of motion. Division of (5.111) by m and introduction of the parameters $\xi = c/2m\omega$ and $\omega^2 = k/m$ lead to the standard form

$$\ddot{x} + 2\xi\omega\dot{x} + \omega^2 x = \left(\omega^2 + 2i\xi\omega\Omega\right) e^{i\Omega t} \tag{5.112}$$

The magnitude of the complex frequency response function then works out to be

$$|H(i\Omega)| = \frac{\left[1 + \left(2\xi\dfrac{\Omega}{\omega}\right)^2\right]^{1/2}}{\left[\left(1 - \dfrac{\Omega^2}{\omega^2}\right)^2 + \left(2\xi\dfrac{\Omega}{\omega}\right)^2\right]^{1/2}} \tag{5.113}$$

The denominator is the same as that appearing in (5.91) for the second-order system considered in the preceding section (5.4.2.). The numerator, however, is now a function of the exciting frequency Ω because of the presence of the derivative of the input in the system mathematical model. The Bode plot is obtained by converting (5.113) to the decibel scale:

$$\text{db} = 10\log\left[1 + \left(\frac{2\xi\Omega}{\omega}\right)^2\right] - 10\log\left[\left(1 - \frac{\Omega^2}{\omega^2}\right)^2 + \left(\frac{2\xi\Omega}{\omega}\right)^2\right] \quad (5.114)$$

The second term in (5.114) is simply (5.91) (Fig. 5.20). To this we add the first term arising from the numerator of (5.114). The Bode plot of this first term of (5.114) is shown in Fig. 5.24(a) and is simply the negative of the first-order Bode plot of Fig. 5.18, with a system time constant $T = \frac{\omega}{2\xi}$. The noteworthy feature of the numerator shown in Fig. 5.24(a) is that, as Ω passes beyond $2\xi/\omega$, the function shown *increases* with frequency at a rate which asymptotes to +20 db per decade.

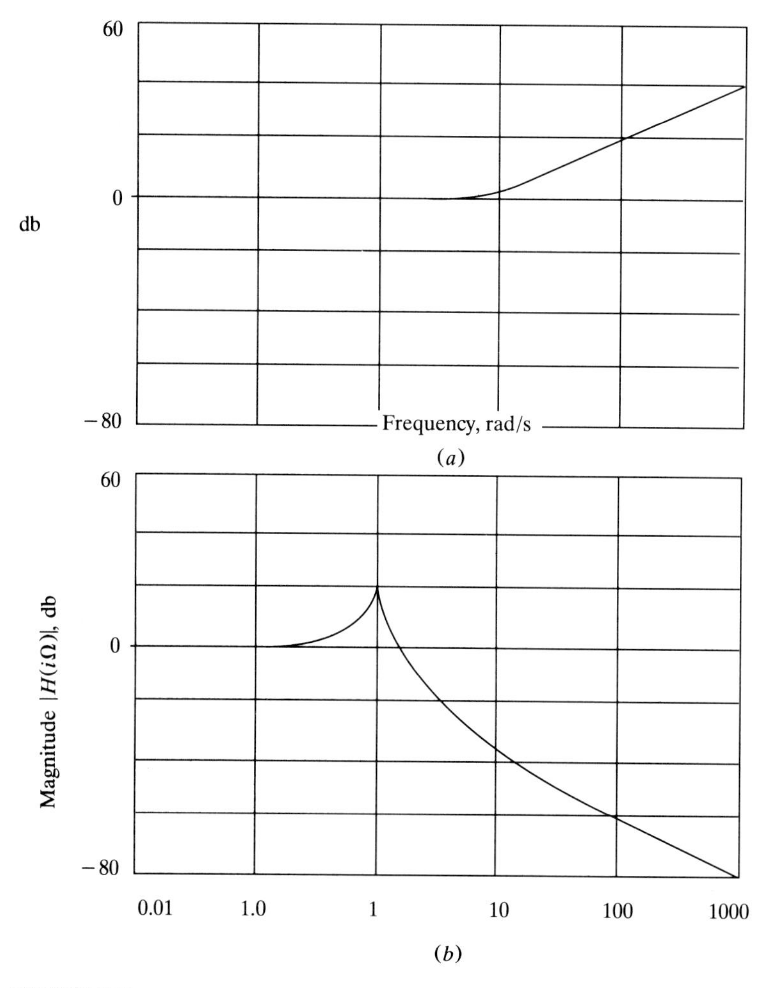

FIGURE 5.24
(a) Bode plot of the numerator of Eq. (5.113) in Example 5.15. (b) Bode plot for the system of Example 5.15, Eq. (5.113).

The overall system Bode plot from (5.114) is shown in Fig. 5.24(b) for a value $\xi = 0.05$. One observes flat frequency response for very low frequencies, followed by the "spike" associated with resonance when $\Omega/\omega = 1$, followed by an increase in $|H(i\Omega)|$ as Ω/ω passes through $\frac{1}{2\xi}$, followed by frequency rolloff at the asymptotic rate of -20 db per decade [-40 db per decade from the denominator of (5.114), combined with $+20$ db per decade from the numerator].

The preceding example illustrates the effect of a frequency-dependent numerator in $H(i\Omega)$. From this example it should be clear that, if, as $\Omega \to \infty$, the numerator $|f(i\Omega)| \to \Omega^m$, while the denominator $|P(i\Omega)| \to \Omega^n$, with $n > m$, then the eventual frequency rolloff of the Bode plot as $\Omega \to \infty$ will asymptote to $-20(n - m)$ db per decade.

3. In this section (5.4) the Bode plot has been defined as a log-log plot of $|H(i\Omega)|$ versus exciting frequency Ω. Actually, the term *Bode plot* implies, in addition, a plot of the phase ϕ (linear scale) versus Ω (log scale), since the two quantities $|H(i\Omega)|$ and ϕ are necessary in order to define completely the CFR $H(i\Omega) = |H(i\Omega)|e^{i\phi}$. Here we have not studied the behavior of the phase ϕ, the magnitude $|H(i\Omega)|$ being considered of primary importance. In Sec. 8.6, which presents the use of Bode plots to design control systems, the Bode plot consisting of both $|H(i\Omega)|$ and ϕ will be necessary. The use of Bode plots in control systems design is a standard, classical technique that historically has been used widely by control engineers.

5.5 CONCLUDING REMARKS

5.5.1 Solution Methods and Types of Inputs

We have so far in this chapter considered the response of first-, second-, and higher-order systems to some simple inputs such as the step and harmonic excitation. For first- and second-order systems the mathematical simplicity of the inputs considered allows one to obtain closed-form analytical solutions for the response. It frequently occurs, however, that the inputs that drive the system are themselves the result of complex processes. Often such inputs cannot be represented in simple functional form. In such cases, one of which is illustrated below in Example 5.16, a common line of attack is to obtain numerical solutions to the governing mathematical model by doing numerical integration using a digital computer. An example of this type is now presented.

Example 5.16 Excitations That Cannot Be Modeled Exactly. Suppose we consider a single degree of freedom oscillator system which is subjected to a time-dependent input force $F(t)$, so that the system motion is governed by the mathematical model

$$\ddot{x} + 2\xi\omega\dot{x} + \omega^2 x = \frac{F(t)}{m} \tag{5.115}$$

As we have seen, this simple model may be used to represent approximately the motion of numerous systems. Let us now suppose, however, that the input force $F(t)$ arises itself from complex processes, which make it difficult to predict ahead of time what the force will be. Examples of this type of situation might include the

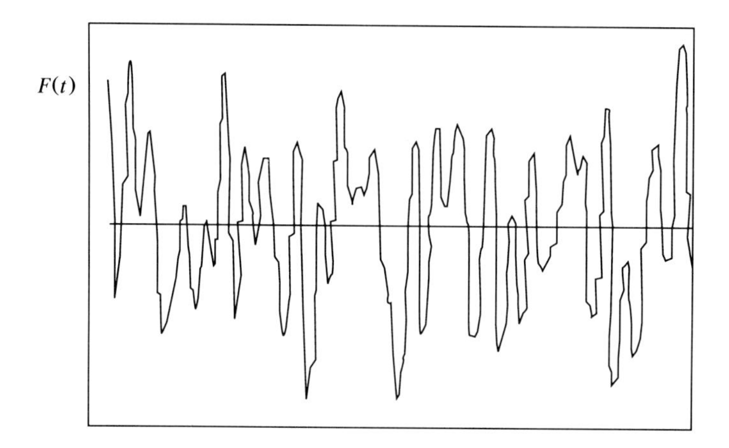

FIGURE 5.25
Hypothetical forcing function $F(t)$ of Example 5.16.

following: (1) the ground motion during an earthquake, which excites vibratory motions of buildings and other structures, (2) the force exerted by waves on ships or offshore structures, and (3) the force exerted by the wind on buildings, transmission lines, towers, and so on. In the preceding situations the excitations typically cannot be easily predicted ahead of time; furthermore, these types of excitations often cannot be represented in simple functional form and may vary irregularly with time. Shown in Fig. 5.25 is a hypothetical input force of the type discussed. With the time units properly scaled, this function would be typical of the input forces in any of the above situations. Suppose the function of Fig. 5.25 represented the recorded ground acceleration during an earthquake or the measured wave force on an offshore structure. In using these data to determine how the oscillator system of (5.115) responds, the usual procedure would be to store the input data on tape or disk and have the computer read it as needed, solving (5.115) using a numerical integration routine. In realistic engineering applications of this type, numerical integration of ODEs, discussed in Chap. 9, is very common.

The purpose of this example has been to illustrate that our ability to find simple analytical solutions to driven ODE models will depend not only on the nature of the system but also on the nature of the input.

5.5.2 Closing Comments on the Forced Response

In this chapter we have considered the transient and steady-state responses of stable linear systems subjected to effects arising outside the system (inputs). A linear analysis is expected to be valid provided that the inputs do not drive the system far enough from equilibrium that nonlinearities come into play. An important physical characteristic of linear systems is the "tracking property": the system steady-state response tends to follow the same functional form as the input, and this is the basis of the method of undetermined coefficients which we used to determine the particular solutions to the ODE models considered.

We reiterate here the fact that the transient and steady-state responses of linear systems are governed not only by the properties of the input but also by the system eigenvalues, which arise in the autonomous (unforced) problem. Thus, even if the forced response is of primary interest for a given linear system, a competent engineer will always analyze the autonomous problem first in order to determine the system eigenvalues, a knowledge of which enables the forced response to be better understood.

On the other hand, if the math model of a given system or system component cannot be well or completely defined from first principles, then the system or component response to simple inputs such as steps and harmonic excitations can be determined experimentally. These data can then be used to determine system properties and a system mathematical model which provides an adequate representation of the measured system response.

REFERENCES

5.1. Meirovitch, L., *Introduction to Dynamics and Control*, Wiley, New York (1985). See in particular Chap. 4.

See References 3.1–3.3 (Chap. 3) for additional and complementary material on system response to inputs.

For the response of the second-order oscillator system to step, harmonic, and other excitations, see References 7.1–7.5, listed at the end of Chap. 7.

5.2. Hildebrand, F. B., *Advanced Calculus for Applications*, Prentice-Hall, Englewood Cliffs, N.J. (1962). Chapter 1 contains some useful information on finding particular solutions to linear ODEs.

EXERCISES

1. Consider the system shown in Fig. *P*5.1. There are two inputs: (a) a force $F(t)$ applied directly to mass 2 and (b) a known motion $y(t)$ of the support to which the damper is attached. Determine the system mathematical model in the state variable form

$$\{\dot{x}\} = [A]\{x\} + \{g(t)\}$$

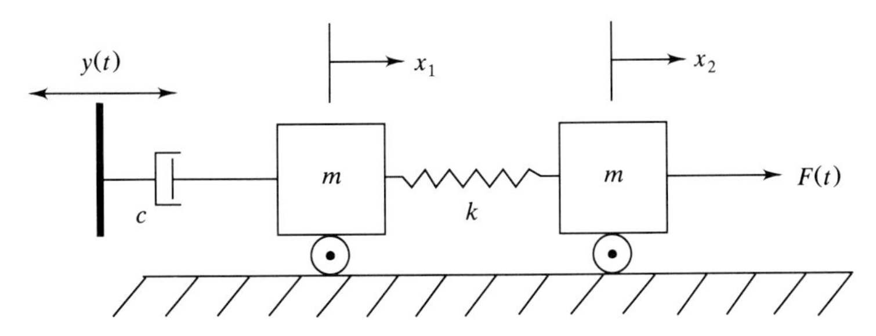

FIGURE *P*5.1

2. An object moves vertically downward through a viscous liquid under the action of gravity and a drag force F_D, which is proportional to the speed v of the object ($F_D = -\alpha v$, where the negative sign indicates that the drag force is opposite the direction of travel). Determine the steady-state velocity v_{ss} of the object, assuming the mass m of the object and the positive constant α are known.

3. A mass m is attached to a shock absorber as shown in Fig. $P5.3$. At $t = 0$, the mass is released and moves downward under the action of gravity. Assuming the shock absorber to be a linear damping device with damping constant c, determine the histories of velocity $v(t) = \dot{x}(t)$ and position $x(t)$.

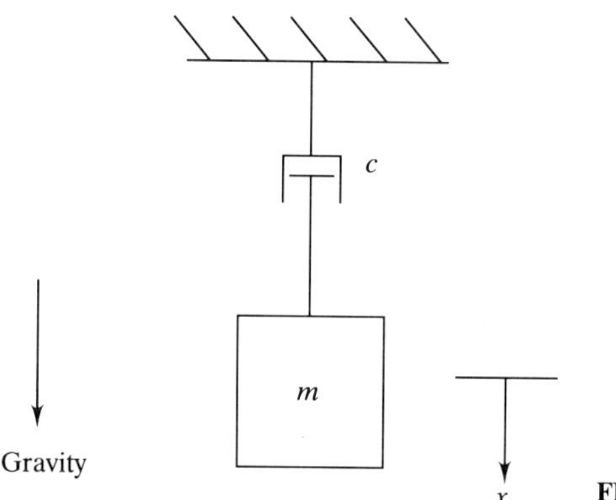

FIGURE $P5.3$

4. A commonly encountered problem in system design is the response of a system to a "ramp" input. For a first-order system the mathematical model would be

$$\dot{x} = -ax + bt \qquad (a, b \text{ each} > 0)$$

For the initial condition $x(0) = 0$, determine the system response $x(t)$.

5. Consider the linear system having n states and subjected to a harmonic excitation according to the mathematical model

$$\{\dot{x}\} = [A]\{x\} + \{P\} \cos \Omega t$$

Assume the system to be asymptotically stable and assume the particular solution to be of the form

$$\{x_p\} = \{S\} \sin \Omega t + \{C\} \cos \Omega t$$

where the n-dimensional vectors $\{S\}$ and $\{C\}$ are to be found. Show that the solutions for $\{S\}$ and $\{C\}$ can be expressed as

$$\begin{Bmatrix} S \\ \cdots \\ C \end{Bmatrix} = [B]^{-1} \begin{Bmatrix} P \\ \cdots \\ 0 \end{Bmatrix}$$

i.e., find the $2n \times 2n$ matrix $[B]$. An important result of this problem is that *any* asymptotically stable linear system, when excited harmonically, will respond (in the steady state) harmonically with the same frequency as the excitation.

6. A single degree of freedom oscillator with $m = 10$ kg, $k = 10^3$ N/m, and $c = 20$ N-s/m is subjected to a step force input which moves the mass m from its rest position to a new nonzero position. The system is observed to exhibit too much overshoot and to respond too slowly. Redesign the system (i.e., change k and c) so that the maximum overshoot is no more than 10% and the time T_p to reach the first peak of oscillation is $T_p \leq 0.1$ s (see Fig. 5.10).

7. A single degree of freedom oscillator, when subjected to a step input, exhibits an overshoot of 50%. Determine the system damping factor ξ. Determine the percent overshoot which would result if the system stiffness were doubled, with the mass m and damping constant c unchanged.

8. A machine and the suspension which supports it are modeled as a single degree of freedom oscillator. The system damping factor $\xi = 0.02$ and the natural frequency $\omega = 10$ Hz. During operation of the machine the moving parts cause a harmonic excitation at a frequency $\Omega = 9$ Hz. The resulting steady-state vibration amplitude is unacceptably large. The following design solutions are proposed:

a. Add a damping device which doubles the system damping factor ξ.
b. Increase the system mass by 40%.
c. Increase the suspension stiffness by 40%.

Which of these strategies will produce the greatest reduction in vibration amplitude? *Explain.*

9. A third-order linear system subjected to a harmonic excitation is governed by the mathematical model

$$\dddot{x} + a_2\ddot{x} + a_1\dot{x} + a_0 x = e^{i\Omega t}$$

Determine the magnitude of the complex frequency response $H(i\Omega)$, where $x(t) = H(i\Omega)\, e^{i\Omega t}$. Is it possible to have a resonance condition for this system (that is, can $|H(i\Omega)|$ become very large for certain values of Ω, a_0, a_1, and a_2)? Explain.

10. A two-state system is subjected to a step input according to the mathematical model

$$\begin{Bmatrix} \dot{x}_1 \\ \dot{x}_2 \end{Bmatrix} = \begin{bmatrix} 0 & 1 \\ -25 & -1 \end{bmatrix} \begin{Bmatrix} x_1 \\ x_2 \end{Bmatrix} + \begin{Bmatrix} 0 \\ 1 \end{Bmatrix} \qquad (t > 0)$$

If the system is initially at rest, $x_1(0) = x_2(0) = 0$, determine the histories $x_1(t)$ and $x_2(t)$ and sketch the resulting phase portrait.

11. A pendulum is induced to rotate by moving the pivot A in the horizontal direction in a known manner, i.e., the pivot position $x(t)$, velocity $\dot{x}(t)$, and acceleration $\ddot{x}(t)$ are

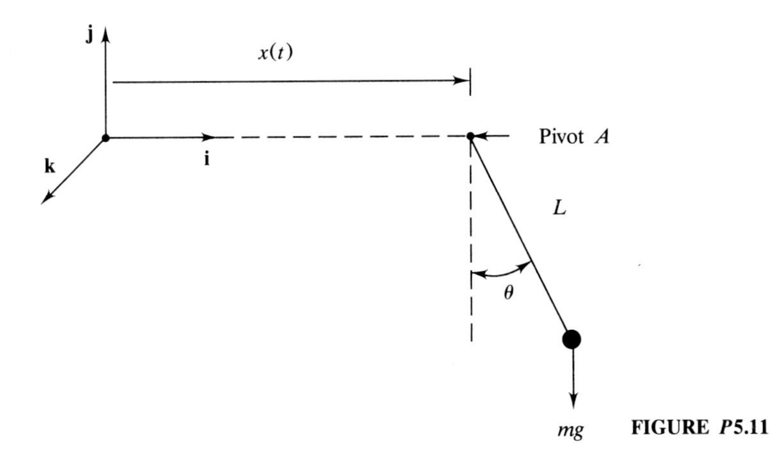

FIGURE P5.11

known functions of time (Fig. $P5.11$). Assume that for $t \leq 0$, the pendulum is at rest such that $\theta = 0$, $\dot{\theta} = 0$ and $x(t) = 0$. For $t > 0$, the pivot A moves to the right with constant acceleration, $\ddot{x}(t) = a_0$, where a_0 is a known constant. Assuming θ remains small, find the maximum pendulum rotation that occurs. The system damping is small enough to be ignored.
Hint: Start with Eq. (2.23) and simplify it for the case of small θ to obtain the system equation of motion.

12. A two-state linear system, initially at rest, is subjected to an input. The resulting phase portrait is as shown in Fig. $P5.12$. Is the system stable? Determine the steady-state responses $x_{1_{ss}}$ and $x_{2_{ss}}$. Determine the type of input (e.g., ramp, step, harmonic?).

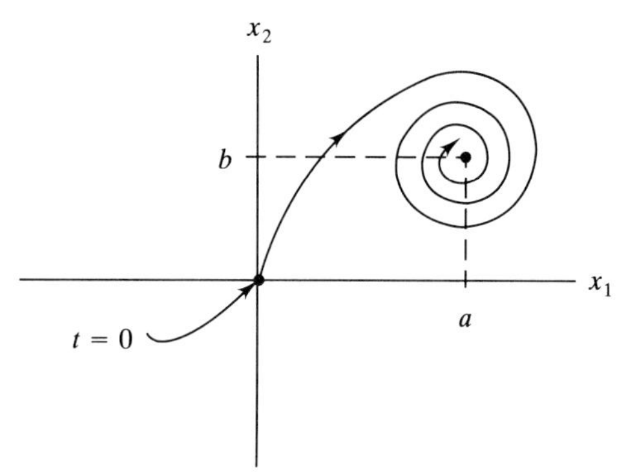

FIGURE $P5.12$

13. The damped linear oscillator shown in Fig. $P5.13$ is subjected to an input motion $y(t)$ of the base to which the spring k is grounded. The motion $y(t)$ is a ramp input, $y(t) = at$, where a is a specified constant. Determine the steady-state response $x_{ss}(t)$ and find the steady-state "error" $e(t) \equiv y(t) - x_{ss}(t)$, which is the difference between the input and the response. Discuss ways to reduce this error.

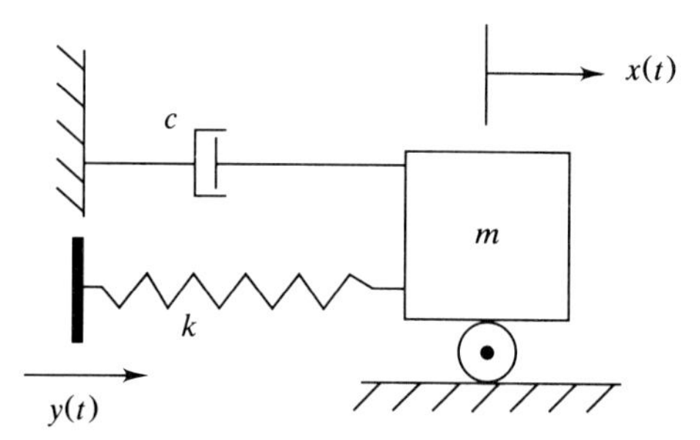

FIGURE $P5.13$

14. Consider the pump system of Example 5.11. Show from first principles that the first-order model (5.105) is reasonable by (a) deriving the equation of rotational motion for the pump shaft rotation $\theta(t)$ and (b) assuming that pump output q is proportional to shaft speed $\dot{\theta}$. Also assume that the torque which rotates the pump shaft is proportional to the input voltage $V(t)$.

For the following complex frequency response magnitude functions $|H(i\Omega)|$, identify the system eigenvalues and construct the approximate Bode plots.

15.

$$|H(i\Omega)| = \frac{1}{(1 + 0.01\Omega^2)^{1/2}}$$

16.

$$|H(i\Omega)| = \frac{1}{(1 + 0.01\Omega^2)^{1/2}(1 + \Omega^2)^{1/2}}$$

17.

$$|H(i\Omega)| = \frac{1}{(1 + \Omega^2)^{1/2}(1 + 100\Omega^2)^{1/2}(1 + 10\Omega^2)^{1/2}}$$

18.

$$|H(i\Omega)| = \frac{1}{(1 + \Omega^2)^{1/2}\left[(1 - 0.01\Omega^2)^2 + (0.01\Omega)^2\right]^{1/2}}$$

For each of the following Bode plots, determine the order or the system, and estimate the system time constants and eigenvalues.

19. See Fig. $P5.19$.

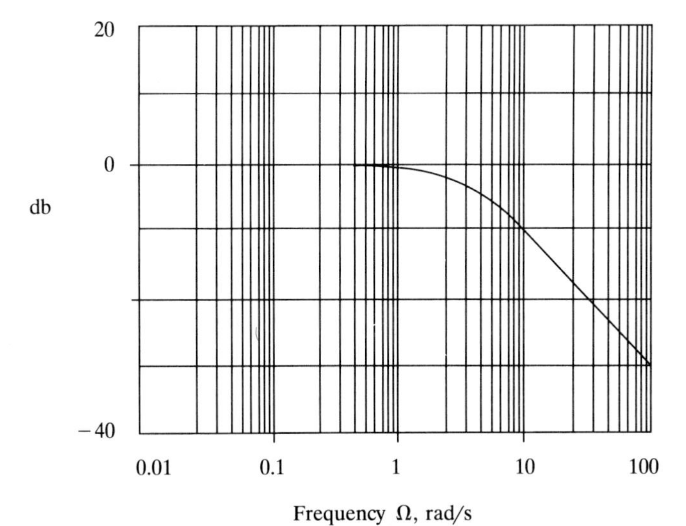

FIGURE ***P*5.19**

20. See Fig. $P5.20$.

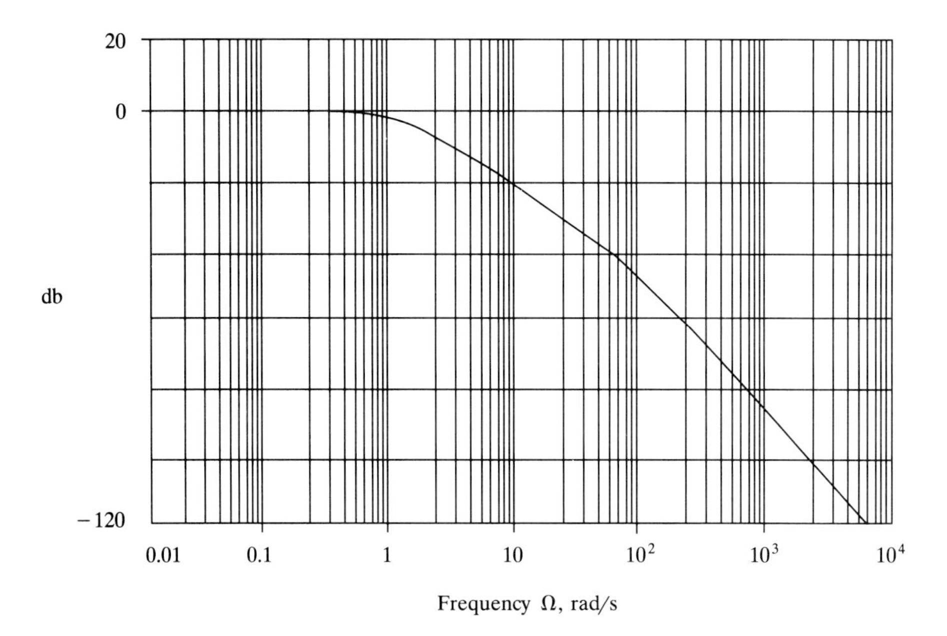

FIGURE ***P*5.20**

CHAPTER 6

FURTHER TOPICS IN DYNAMICS

6.1 INTRODUCTION

In this chapter we'll study two topics that will extend considerably our ability to model mechanical systems: (1) the use of rotating coordinate systems and (2) lagrangian mechanics. An introductory discussion of these topics is given in this section. Rotating coordinate systems are then considered in detail in Sec. 6.2 and lagrangian mechanics in Sec. 6.3.

6.1.1 Rotating Coordinate Systems

In Chap. 2 we applied the basic laws of newtonian mechanics to develop mathematical models for two generic situations: (1) the translation of particles (or of the center of mass of a rigid body) and (2) the plane rotation of a rigid body. In order to formulate the equations of motion, we have to define the appropriate vector quantities (position, velocity, acceleration, force, moment, angular momentum, and so on), which are then used in the laws of motion to obtain the ODE model of the system. The required vectors may be defined in terms of any coordinate system which we find convenient to use. Such coordinate systems may be nonrotating, such as the cartesian (x, y, z) system, or rotating, such as the polar coordinate (r, θ) system discussed for planar motions in Chap. 2.

The example systems considered in Sec. 2.5 were confined to situations in which all motions were planar. In engineering practice, we can get a lot of mileage out of planar models of mechanical systems because many systems exhibit, at least approximately, planar motions. There are, however, numerous practical situations that involve independent rotations about two or more axes, so that the associated motions of the system are three dimensional. In addition, such three-dimensional motions are often all but impossible to analyze unless one employs a rotating coordinate system. It is the construction, understanding, and use of rotating coordinate systems that will occupy our attention in Sec. 6.2.

Examples of engineering systems requiring the use of rotating coordinate systems include spacecraft and missile systems, rail vehicle wheelsets and trucks, gyroscopes and instrumentation based on the gyroscopic effect, robotic manipulators, and rotating machinery. As a specific example, consider the use of three orthogonal accelerometers, mounted in a rotating, translating vehicle and used to measure the accelerations experienced by (and hence the loads exerted on) the vehicle. The measured accelerations are naturally resolved along the three axes of a rotating coordinate system. An understanding of rotating coordinate systems is clearly required in order to interpret the data.

In practice, rotating coordinate systems are used in the analysis of both particle motions and nonplanar rigid body dynamics. In either case the rotating coordinate systems are set up in the same way. In the examples considered in Sec. 6.2 we will confine our attention to the analysis of particle motions. The student should be aware, however, that an understanding of rotating coordinate systems is absolutely essential if one is to successfully analyze three-dimensional rotational dynamics of rigid bodies.

6.1.2 Lagrangian Mechanics

The newtonian laws used to derive the equations of motion of mechanical systems are stated as *vector* relations. The fundamental quantity describing the state of motion is the linear (or angular) momentum. The essential problem is to determine how the application of a force (or moment) will cause the momentum to change with time. For a system having several degrees of freedom, such as in Examples 2.4 through 2.6, and 2.10, we apply the appropriate vector laws for each particle and/or rigid body of the system.

In the lagrangian formulation of mechanics the fundamental quantity describing the state of motion is the *kinetic energy* T. The essential problem is to determine how the *work* W done by the applied forces causes the kinetic energy to change with time. Thus, force and momentum are replaced by work and kinetic energy as the basic quantities of interest. The lagrangian formulation thus utilizes *scalar* quantities (T, W) to describe the motion of the system. Furthermore, the kinetic energy and work done for the *system as a whole* are used. This is in contrast to the newtonian formulation, in which the appropriate vector law must be applied separately to each of the bodies comprising the system.

As we will see, the lagrangian formulation offers a much easier way to obtain system equations of motion for certain types of systems. Moreover, the essential underlying concept of "virtual change" finds application in the analysis of nearly every kind of engineering system. Thus, the underlying ideas of lagrangian mechanics are not confined to the mechanics of particles and rigid bodies, as are those of newtonian mechanics.

6.2 ROTATING COORDINATE SYSTEMS

In this section we consider the class of problems which involve general (nonplanar) rotations and for which the use of rotating coordinate systems facilitates the analysis. Keep in mind that the use of rotating coordinate systems involves nothing new conceptually. We still employ the same basic laws of motion. Rotating coordinate systems are just the best way to describe the *geometry* of motion in certain situations. In using rotating coordinate systems, we have to be careful to distinguish between *absolute*, or actual quantities, which are measured relative to an inertial frame of reference, and *relative* quantities, which are measured relative to an observer fixed in and rotating with the rotating coordinate system.

6.2.1 Angular Velocity

The kinematics of translation and of rotation differ in an important way: In formulating the kinematics of translation, we monitor the motion of a *point* as it moves in space. This point generally represents the location of a particle or of the center of mass of a rigid body. In formulating the kinematics of rotation, however, we monitor the *orientation of a straight line*. This line may be conceptually "painted on" a rigid body, it may be one of the rotating coordinate axes, or it may be some other rotating line. The way in which the orientation of this reference line changes with time defines the angular velocity of the object or coordinate system in which the line is fixed. The ability to formulate mathematically the angular velocity of a coordinate system or of a rigid body is an important requirement in any analysis involving nonplanar rotation.

The angular velocity $\boldsymbol{\omega}$ is a vector defined with reference to Fig. 6.1. We suppose that at some time t a reference line L is oriented as indicated. An infinitesimal time Δt later, the orientation of the line has changed slightly. In the limit $\Delta t \to 0$, the surface swept out by the line L in the time interval Δt will be a plane P, as indicated in Fig. 6.1. In this plane P the angle $\Delta\theta$ defines the change in orientation that has occurred in the time interval Δt. The angular velocity $\boldsymbol{\omega}$ of the line L is a vector defined as follows: (1) the direction of $\boldsymbol{\omega}$ is perpendicular to the plane P in accordance with the right-hand rule (Fig. 6.1); (2) the magnitude of $\boldsymbol{\omega}$, denoted as $|\boldsymbol{\omega}|$, is defined as

$$|\boldsymbol{\omega}| = \lim_{\Delta t \to 0} \left| \frac{\Delta\theta}{\Delta t} \right| = \left| \frac{d\theta}{dt} \right| = |\dot{\theta}|$$

For nonplanar rotation the visualization of Fig. 6.1 may be difficult, because the magnitude and direction of $\boldsymbol{\omega}$ will be continually changing, so that,

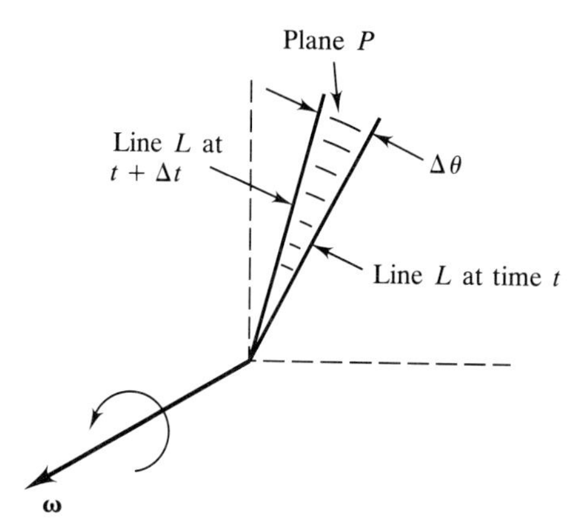

FIGURE 6.1
The angular velocity vector.

over finite time intervals, the surface swept out by the line L will be curved, rather than planar. This is what makes plane rotation so simple. We know that $\boldsymbol{\omega}$ is always oriented normal to a fixed plane in which the rotation occurs.

In setting up rotating coordinate systems, we will generally use the notation of Fig. 6.2. Here the (x, y, z) coordinate system, with associated unit vectors $(\mathbf{i}, \mathbf{j}, \mathbf{k})$, is fixed relative to an inertial frame of reference. A second coordinate system, characterized by the unit vectors $(\mathbf{e}_1, \mathbf{e}_2, \mathbf{e}_3)$, originates at the same point $\mathbf{O}$ as does the fixed system. The $(\mathbf{e}_1, \mathbf{e}_2, \mathbf{e}_3)$ triad, however, rotates with some angular velocity $\boldsymbol{\omega}$ relative to the fixed system. Generally, we will use $(\mathbf{i}, \mathbf{j}, \mathbf{k})$ to represent a fixed, nonrotating coordinate system and $(\mathbf{e}_1, \mathbf{e}_2, \mathbf{e}_3)$ to represent the rotating unit vectors defining the rotating coordinate system. The manner in which the rotating coordinate system is defined will depend on the physical nature of the rotations in given situations. In this construction a useful property of the angular velocity $\boldsymbol{\omega}$ is its vectorial nature, which allows physically

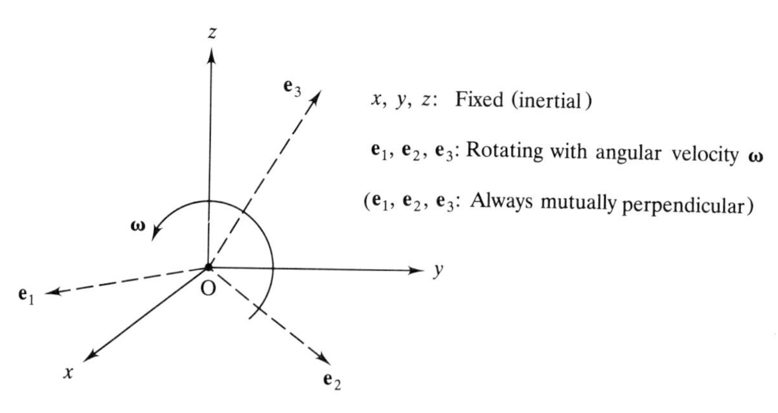

FIGURE 6.2
Fixed and rotating coordinate systems, each originating at **O**.

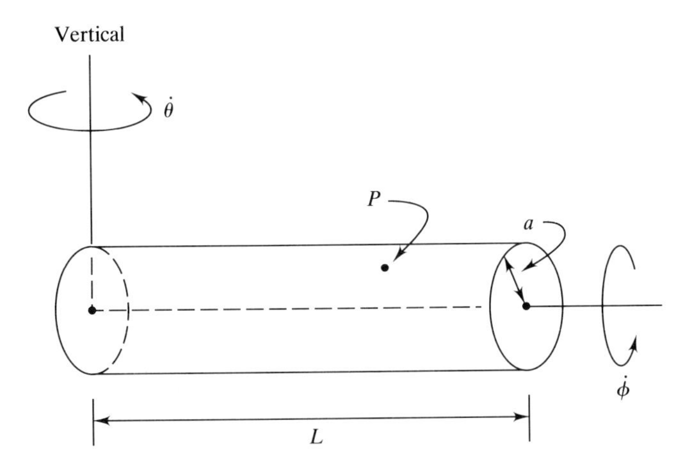

FIGURE 6.3
A cylinder rotating simultaneously about two axes.

separate rotations to be characterized as vectors and to then be summed to form the overall angular velocity vector. We now illustrate the construction and characterization of rotating coordinate systems with an example.

Example 6.1 A Cylinder Rotating Simultaneously about Two Axes. The system is shown in Fig. 6.3: A cylinder of radius a and length L spins about its longitudinal axis at the rate $\dot{\phi}$; meanwhile, the cylinder is rotated about the vertical axis at the rate $\dot{\theta}$. Here $\dot{\theta}$ and $\dot{\phi}$ need not be constants. Suppose we want to find the velocity and acceleration of some point P on the surface of the cylinder (try to visualize this motion). We might even complicate the problem by letting P be a bug which is wandering around on the cylinder while it is rotating. Now let's set up rotating coordinate systems for this problem and determine the coordinate system angular velocity. We'll do the problem in two ways.

Case 1. In the first case we'll set up an $\mathbf{e}_1, \mathbf{e}_2, \mathbf{e}_3$ system which is oriented so that the $\mathbf{e}_2$ axis is always coincident with the spin (longitudinal) axis of the cylinder (Fig. 6.4). Thus, the $\mathbf{e}_1, \mathbf{e}_2$ duet rotates in the horizontal plane at the angular rate $\dot{\theta}$ about the vertical, which becomes the third axis $\mathbf{e}_3$ of our rotating triad. In this construction the rotating coordinate system does *not* spin with the cylinder about its longitudinal axis at the rate $\dot{\phi}$, so that the $\mathbf{e}_3$ axis of our rotating system is fixed.

Thus, the angular velocity of the rotating coordinate system is simply

$$\boldsymbol{\omega}_b = \dot{\theta}\mathbf{e}_3 \tag{6.1}$$

The $(\ )_b$ subscript will be used hereafter to represent any quantity that is specifically related to, or measured relative to, the rotating coordinate system. Here read "$\boldsymbol{\omega}_b$ = the angular velocity of the rotating coordinate system." This distinction is necessary in order to distinguish the angular velocity of the *coordinate system* from that of the *cylinder*. If we were going to analyze the 3-D rotational

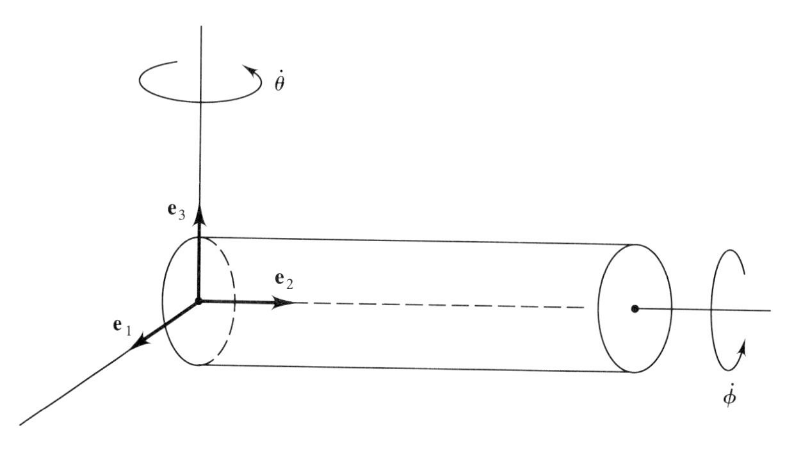

FIGURE 6.4
The rotating coordinate system rotates with $\dot{\theta}$ but not with $\dot{\phi}$.

motion of the cylinder as a problem in 3-D rigid body dynamics, we also would have to calculate the angular velocity of the cylinder itself. Let's do this to see where the utility of the vectorial nature of angular velocity comes into play.

The angular velocity of the cylinder consists of two independent rotations, each of which can easily be written down in vector form:

Rotation 1: $\dot{\phi}\mathbf{e}_2$ Spin about longitudinal axis

Rotation 2: $\dot{\theta}\mathbf{e}_3$ Rotation of cylinder about vertical

Now, since $\boldsymbol{\omega}$ is a vector, the *combined* case where both rotations occur simultaneously is just the vector sum of the two rotations, so we get

$$\boldsymbol{\omega} = \dot{\phi}\mathbf{e}_2 + \dot{\theta}\mathbf{e}_3 \tag{6.2}$$

This is the angular velocity of the *cylinder*. We haven't used a subscript on $\boldsymbol{\omega}$ here because we want to make sure to distinguish between the angular velocity of the rotating coordinate system and the angular velocity of the cylinder. Now let's look at a second way to set up the rotating coordinate system.

Case 2. In this case we'll define a rotating coordinate system to be fixed rigidly in the cylinder; that is, the rotating coordinate system is now to have the same angular velocity as the cylinder. The rotating unit vectors will now be denoted by $\mathbf{g}_1$, $\mathbf{g}_2$, and $\mathbf{g}_3$ in order to distinguish them from the vectors $\mathbf{e}_1$, $\mathbf{e}_2$, and $\mathbf{e}_3$ defined in case 1. In defining this rotating coordinate system, we have to be careful because now the $\mathbf{g}_1, \mathbf{g}_3$ duet in the rotating system will be rotating in the vertical plane at the rate $\dot{\phi}$, while meanwhile this vertical plane is itself rotating at the rate $\dot{\theta}$ (see Fig. 6.5). We want to resolve $\boldsymbol{\omega}_b$ into the rotating system, and we can do this in two steps: First, we know that the cylinder (hence also the rotating coordinate system) angular velocity is still given by (6.2) with $\mathbf{e}_2$ replaced by $\mathbf{g}_2$ and $\mathbf{e}_3$ replaced by the fixed vertical unit vector $\mathbf{u}$. Now, however, the fixed vertical vector $\mathbf{u}$ must be resolved along the rotating $\mathbf{g}_1$ and $\mathbf{g}_3$ axes. In order to do this, we have to define

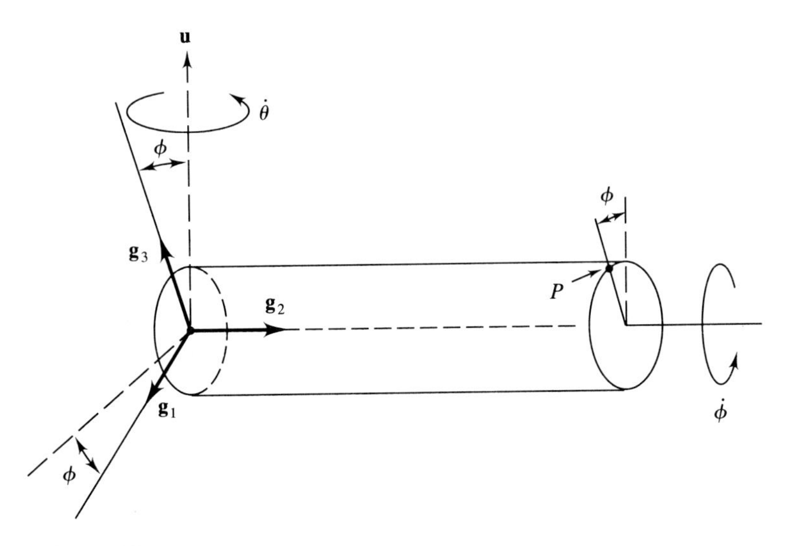

FIGURE 6.5
The rotating coordinate system is now fixed in the cylinder and rotates with both $\dot{\theta}$ and $\dot{\phi}$. The rotating unit vectors are denoted by $\mathbf{g}_1$, $\mathbf{g}_2$, and $\mathbf{g}_3$. The vector $\mathbf{u}$ is fixed and points in the vertical direction.

the angle $\phi = \int \dot{\phi}\, dt$, as shown in Fig. 6.5. We thus resolve $\mathbf{u}$ as follows:

$$\mathbf{u} = -\sin\phi\mathbf{g}_1 + \cos\phi\mathbf{g}_3$$

The angular velocity of the rotating coordinate system and of the cylinder in which the coordinate system is fixed is thus

$$\boldsymbol{\omega}_b = \boldsymbol{\omega} = -\dot{\theta}\sin\phi\mathbf{g}_1 + \dot{\phi}\mathbf{g}_2 + \dot{\theta}\cos\phi\mathbf{g}_3 \qquad (6.3)$$

Comment. Equations (6.2) and (6.3) define the *same* vector in inertial space. The two expressions merely resolve this vector into two different coordinate systems. We can picture how this vector looks in inertial space by the following construction, valid in the special case for which $\dot{\theta}$ and $\dot{\phi}$ are constant (Fig. 6.6). We add vectorially the two orthogonal components to get $\boldsymbol{\omega}_b$, and we see that $\boldsymbol{\omega}_b$ will rotate in space so that it lies along the surface of a cone, as shown. Even though $\dot{\theta}$ and $\dot{\phi}$ are assumed constant here, $\boldsymbol{\omega}_b$ is *not* constant, since its direction is continually changing. The angular acceleration vector $\dot{\boldsymbol{\omega}}_b$ lies in the horizontal plane A and is tangent to the curve on which the tip of $\boldsymbol{\omega}_b$ lies (Fig. 6.7). Looking down from above, here is what we see (Fig. 6.7): The projection of $\boldsymbol{\omega}_b$ onto the horizontal plane rotates at the rate $\dot{\theta}$; the tip of this projection traces out a circular path. Thus, the angular acceleration $\dot{\boldsymbol{\omega}}_b$ is directed tangentially as shown.

Determination of the rotating coordinate system angular velocity $\boldsymbol{\omega}_b$, resolved along the unit vectors that define the rotating system, is an important step in the analysis of dynamic phenomena using rotating coordinate systems.

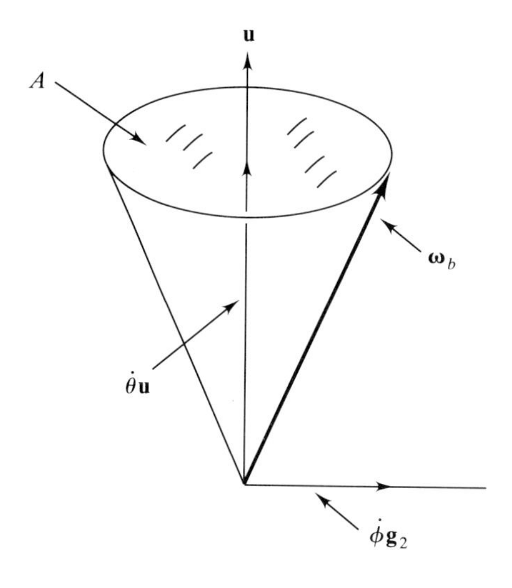

FIGURE 6.6
Visualization of the angular velocity vector $\boldsymbol{\omega}_b$ for the spinning, rotating cylinder.

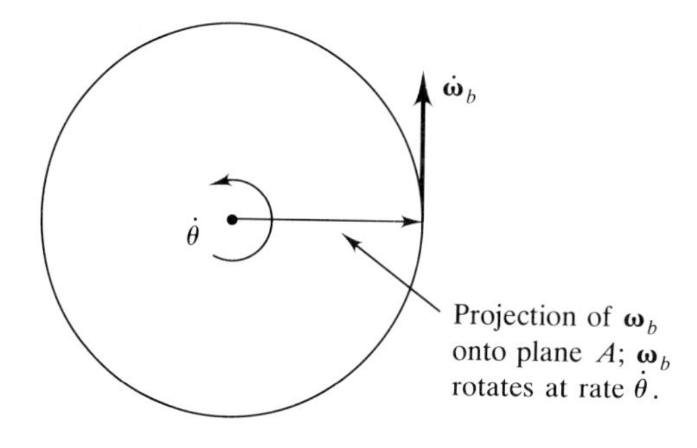

FIGURE 6.7
The projection of the angular velocity vector $\boldsymbol{\omega}_b$ onto the horizontal plane (for the case in which the rotation rates $\dot{\theta}$ and $\dot{\phi}$ are constants).

6.2.2 Inertial and Relative Velocity and Acceleration

The acceleration **a** appearing in Newton's second law must be the acceleration measured by an observer in an inertial frame of reference. We will denote this acceleration here as $\mathbf{a}_B$, where the B subscript will generally denote a quantity defined relative to an inertial frame of reference. Thus, Newton's second law states $\mathbf{F} = m\mathbf{a}_B$. The notation $(\ )_B$ is necessary in order to distinguish the inertially measured acceleration $\mathbf{a}_B$ from the acceleration relative to an observer fixed at the origin and rotating with the rotating coordinate system. We will denote this relative acceleration by $\mathbf{a}_b$. In a similar manner the inertial and relative velocities will be denoted by $\mathbf{v}_B$ and $\mathbf{v}_b$, respectively. The difference between the inertial and relative quantities is illustrated by the following example of a plane motion described by polar coordinates.

Example 6.2 Particle Moving on a Rotating Turntable. The motion to be considered is shown in Fig. 6.8. A turntable has a radial line L painted on it as shown. A particle of mass m starts ($t = 0$) at a distance r_0 from the turntable

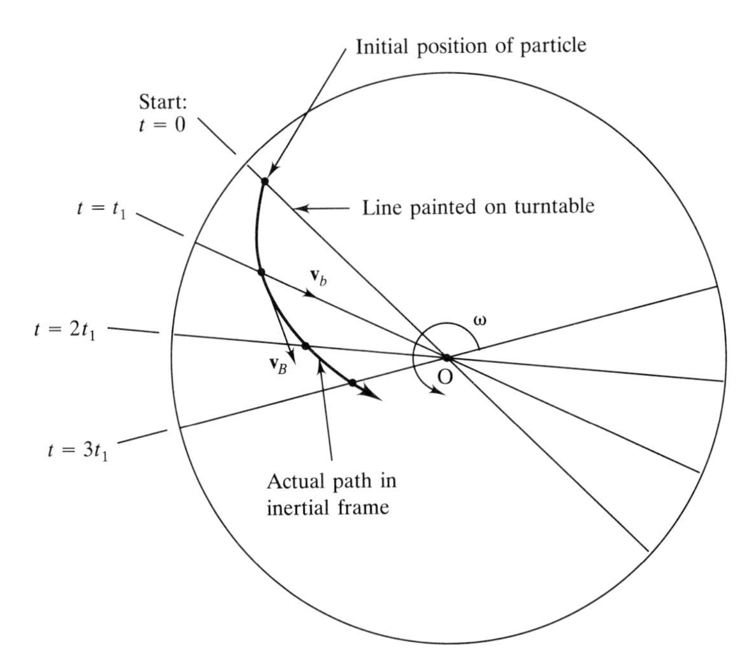

FIGURE 6.8
Particle on a rotating turntable.

center **O**, about which the turntable rotates at the fixed rate ω. For $t > 0$, the mass m moves radially inward, remaining on the line L, such that its distance $r(t)$ from the origin decreases at the constant rate $-v_0(v_0 > 0)$.

An observer at **O** rotating with the turntable sees a purely radial motion as the particle moves inward along the line L. The velocity $\mathbf{v}_b$ relative to the rotating observer is thus v_0 in the radially inward direction. The actual path of motion, observed by someone fixed relative to an inertial frame, however, is curved as shown. The inertial velocity $\mathbf{v}_B$ is tangent to this curved path. The vectors $\mathbf{v}_b$ and $\mathbf{v}_B$ are indicated at the time $t = t_1$. The difference between $\mathbf{v}_B$ and $\mathbf{v}_b$ is the tangential component of velocity $r\omega$, which the rotating observer does not see.

The preceding qualitative description is now quantified by setting up a rotating coordinate system and analyzing the inertial and relative velocity vectors $\mathbf{v}_B$ and $\mathbf{v}_b$. Shown in Fig. 6.9 is the rotating coordinate system employed. This rotating coordinate system is fixed in the turntable and rotates with the angular velocity $\boldsymbol{\omega}_b = \dot{\theta}\mathbf{k}$, where $\mathbf{k}$ is an axis out of the paper. The radial unit vector $\mathbf{e}_r$ is always directed toward the particle; i.e., it is along the line L. The tangential unit vector $\mathbf{e}_\theta$ is normal to $\mathbf{e}_r$ and points in the direction of increasing θ. An inertially based (i.e., nonrotating) cartesian xy system is also shown for reference. The two coordinates used to define the position of the particle are the distance r of the particle from the origin **O** and the angle θ, which defines the instantaneous angular displacement of $\mathbf{e}_r$ relative to the x axis.

In terms of the polar coordinates the position vector $\mathbf{r}$ is given by

$$\mathbf{r} = r\mathbf{e}_r \tag{6.4}$$

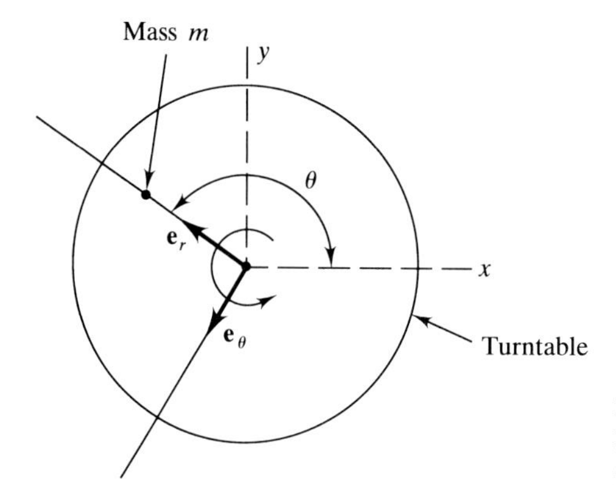

FIGURE 6.9
Polar coordinates for the system of Example 6.2.

The position vector is not subscripted because $\mathbf{r}_B = \mathbf{r}_b$; the inertial and relative position vectors are the *same*. Based on the preceding discussion, the relative velocity $\mathbf{v}_b$ is simply

$$\mathbf{v}_b = -v_0\,\mathbf{e}_r \tag{6.5}$$

The inertial velocity, however, includes the component $r\omega$ in the tangential ($\mathbf{e}_\theta$) direction, so that

$$\mathbf{v}_B = -v_0\mathbf{e}_r + r\omega\mathbf{e}_\theta \tag{6.6}$$

where $r(t) = r_0 - v_0 t$. Furthermore, the relative acceleration $\mathbf{a}_b = 0$ because the mass m appears, to an observer rotating with the turntable, to be moving toward the observer with constant speed and direction.

In the preceding example the polar coordinates r and θ were used to specify the location of a particle in the plane. Let us now investigate in detail the formulation of the general, planar motion of a particle using polar coordinates. This formulation represents one of the simplest situations in which a rotating coordinate system is used. Particular attention will be devoted to the difference between the inertial and relative velocities and accelerations. The general setup is shown in Fig. 6.10.

The rotating unit vectors are $\mathbf{e}_r$ and $\mathbf{e}_\theta$ and are defined so that the radial unit vector $\mathbf{e}_r$ always points directly at the particle. The particle moves around and $\mathbf{e}_r$ follows it, necessitating changes with time of the orientation angle θ. The distance from the origin $\mathbf{O}$ to the particle is denoted by the scalar r, so that the position vector $\mathbf{r}$ is given, by definition, as

$$\mathbf{r} = r\mathbf{e}_r \tag{6.7}$$

The orientation of the position vector $\mathbf{r}$ is defined by the angle θ. The planar rotating unit vectors $(\mathbf{e}_r, \mathbf{e}_\theta)$ and fixed unit vectors $(\mathbf{i}, \mathbf{j})$ are related by the transformation

$$\begin{aligned}\mathbf{e}_r &= \cos\theta\mathbf{i} + \sin\theta\mathbf{j}\\ \mathbf{e}_\theta &= -\sin\theta\mathbf{i} + \cos\theta\mathbf{j}\end{aligned} \tag{6.8}$$

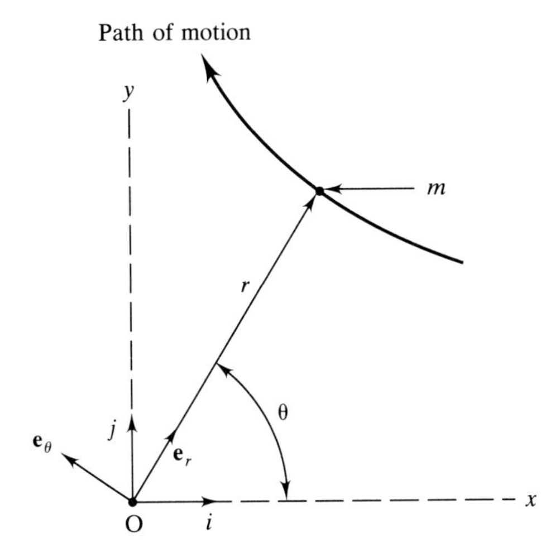

FIGURE 6.10
Polar coordinates; the particle m is confined to the xy plane.

Here the rotating coordinate system triad consists of $(\mathbf{e}_r, \mathbf{e}_\theta, \mathbf{k})$, with rotation always to occur about the $\mathbf{k}$ axis, so that the angular velocity $\boldsymbol{\omega}_b$ of the rotating coordinate system is

$$\boldsymbol{\omega}_b = \dot{\theta}\mathbf{k} \tag{6.9}$$

The inertial velocity $\mathbf{v}_B$ is obtained by differentiating (6.7):

$$\mathbf{v}_B = \frac{d\mathbf{r}}{dt} = \dot{r}\mathbf{e}_r + r\dot{\mathbf{e}}_r \tag{6.10}$$

Because the coordinate system is rotating, the unit vector $\mathbf{e}_r$ will have a nonzero rate of change: Although the magnitude of $\mathbf{e}_r$ is by definition unity, the direction of $\mathbf{e}_r$ is changing if $\dot{\theta} \neq 0$, and this change in direction must be included in differentiating $\mathbf{r}$. In order to determine $\dot{\mathbf{e}}_r$ (as well as $\dot{\mathbf{e}}_\theta$), let's look at the following picture (Fig. 6.11) of the changes in $\mathbf{e}_r$ and $\mathbf{e}_\theta$ which take place during an incremental time interval dt, during which an incremental rotation $d\theta$ of the coordinate system is assumed to have occurred.

First, we observe that the incremental vector $d\mathbf{e}_r$ is directed along $\mathbf{e}_\theta$, while the incremental vector $d\mathbf{e}_\theta$ is directed along $-\mathbf{e}_r$. Because the unit vectors $\mathbf{e}_r$ and $\mathbf{e}_\theta$ have unit length, the magnitudes of $d\mathbf{e}_r$ and $d\mathbf{e}_\theta$ are equal to $d\theta$, so that

$$\begin{aligned} d\mathbf{e}_r &= d\theta\,\mathbf{e}_\theta \\ d\mathbf{e}_\theta &= -d\theta\mathbf{e}_r \end{aligned} \tag{6.11}$$

Dividing (6.11) by the time increment dt, we find

$$\begin{aligned} \dot{\mathbf{e}}_r &= \dot{\theta}\mathbf{e}_\theta \\ \dot{\mathbf{e}}_\theta &= -\dot{\theta}\mathbf{e}_r \end{aligned} \tag{6.12}$$

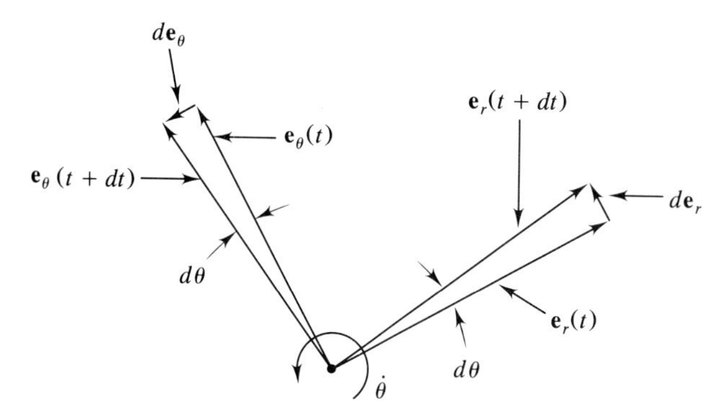

FIGURE 6.11
Determination of $\dot{\mathbf{e}}_r$ and $\dot{\mathbf{e}}_\theta$.

These are the results we need in (6.10) and later on for computing the inertial acceleration $\mathbf{a}_B$. We can anticipate the general results for three-dimensional motion by computing the cross product $\boldsymbol{\omega}_b \times \mathbf{e}_r$, as follows:

$$\boldsymbol{\omega}_b \times \mathbf{e}_r = \begin{vmatrix} \mathbf{e}_r & \mathbf{e}_\theta & \mathbf{k} \\ 0 & 0 & \dot{\theta} \\ 1 & 0 & 0 \end{vmatrix} = \dot{\theta}\mathbf{e}_\theta = \dot{\mathbf{e}}_r$$

Thus, we conclude that

$$\dot{\mathbf{e}}_r = \boldsymbol{\omega}_b \times \mathbf{e}_r \tag{6.13a}$$

and a similar calculation shows that

$$\dot{\mathbf{e}}_\theta = \boldsymbol{\omega}_b \times \mathbf{e}_\theta \tag{6.13b}$$

Returning to (6.10), we can now write out the inertial velocity as

$$\mathbf{v}_B = \dot{r}\mathbf{e}_r + r\dot{\theta}\mathbf{e}_\theta \tag{6.14}$$

which of course comes as no surprise, as we know from previous experience that the tangential velocity (along $\mathbf{e}_\theta$) is just $r\dot{\theta}$. In (6.14) the first term on the right-hand side, $\dot{r}\mathbf{e}_r$, is the relative velocity $\mathbf{v}_b$ seen by an observer rotating with the rotating coordinate system. The rotating observer does not see the portion of the inertial velocity $\mathbf{v}_B$ due to rate of change of the unit vector $\mathbf{e}_r$.

From (6.14) we obtain the inertial acceleration $\mathbf{a}_B$ as $\mathbf{a}_B = d\mathbf{v}_B/dt$, or

$$\mathbf{a}_B = \ddot{r}\mathbf{e}_r + \dot{r}\dot{\mathbf{e}}_r + \dot{r}\dot{\theta}\mathbf{e}_\theta + r\ddot{\theta}\mathbf{e}_\theta + r\dot{\theta}\dot{\mathbf{e}}_\theta$$

Use of (6.12) for $\dot{\mathbf{e}}_r$ and $\dot{\mathbf{e}}_\theta$ leads to the following result for the inertial acceleration $\mathbf{a}_B$:

$$\mathbf{a}_B = (\ddot{r} - r\dot{\theta}^2)\mathbf{e}_r + \left(r\ddot{\theta} + 2\dot{r}\dot{\theta}\right)\mathbf{e}_\theta \tag{6.15}$$

The relative acceleration $\mathbf{a}_b$, however, is simply

$$\mathbf{a}_b = \ddot{r}\mathbf{e}_r$$

because, relative to the rotating observer, only a linear, translational motion in the radial direction occurs, defined by the rate at which $\dot{r}$ is changing.

If we assume a resultant force $\mathbf{F}$ to be acting on the particle m, resolution of $\mathbf{F}$ into its radial and tangential components, $\mathbf{F} = F_r\mathbf{e}_r + F_\theta\mathbf{e}_\theta$, allows us to determine the two equations of motion in polar coordinates. Use of (6.15) in Newton's law $\mathbf{F} = m\mathbf{a}_B$ leads to

$$F_r = m(\ddot{r} - r\dot{\theta}^2)$$

$$F_\theta = m(r\ddot{\theta} + 2\dot{r}\dot{\theta}) \tag{6.16}$$

These relations were stated previously as (2.12).

Study of Example 6.2 and the general development of polar coordinates reveals that the relative velocity $\mathbf{v}_b$ may be obtained mathematically by time differentiating the position vector $\mathbf{r}$, but *not including* the rates of change of the rotating unit vectors. In a similar fashion the relative acceleration $\mathbf{a}_b$ is obtained by time differentiating the relative velocity $\mathbf{v}_b$ but not including the rate of change of the unit vectors. These results show that the "relative time derivatives" measured relative to the rotating observer differ from inertial or actual time derivatives. We now explore this difference in the operation of time differentiation.

6.2.3 The Rate of Change of a Vector in a Rotating Coordinate System

We normally use a dot to denote the time derivative of a scalar or a vector. This dot notation is understood here to represent the time derivative measured by an observer fixed in an inertial frame of reference. In order to indicate this explicitly, we use the following notation to represent the inertial time derivative of a vector $\mathbf{s}$:

$$\dot{\mathbf{s}} \equiv \left.\frac{d\mathbf{s}}{dt}\right|_B$$

This notation is necessary in order to distinguish the inertial (actual) time derivative from the relative time derivative, which is the rate of change of the vector $\mathbf{s}$ relative to a rotating observer. This relative rate of change will be denoted by

$$\left.\frac{d\mathbf{s}}{dt}\right|_b \equiv \text{rate of change of } \mathbf{s} \text{ relative to a rotating observer}$$

Note that the inertial and relative rates of change of a scalar c are equal:

$$\left.\frac{dc}{dt}\right|_B = \left.\frac{dc}{dt}\right|_b$$

because the value of a scalar is independent of the coordinate system. Thus, we will use the notation $\dot{c}$ to denote the time derivative of a scalar.

In terms of the derivative notation introduced, the relative velocity $\mathbf{v}_b$ and the relative acceleration $\mathbf{a}_b$ are defined by

$$\mathbf{v}_b = \left.\frac{d\mathbf{r}}{dt}\right|_b \qquad \mathbf{a}_b = \left.\frac{d\mathbf{v}_b}{dt}\right|_b \tag{6.17}$$

This notation can be used to express the inertial velocity $\mathbf{v}_B$ in polar coordinates, Eq. (6.10), as

$$\mathbf{v}_B = \left.\frac{d\mathbf{r}}{dt}\right|_B = \left.\frac{d\mathbf{r}}{dt}\right|_b + r\dot{\mathbf{e}}_r$$

where we have used the previous result

$$\mathbf{v}_b = \left.\frac{d\mathbf{r}}{dt}\right|_b = \dot{r}\mathbf{e}_r$$

If we also utilize (6.13a), the above becomes

$$\mathbf{v}_B = \left.\frac{d\mathbf{r}}{dt}\right|_B = \left.\frac{d\mathbf{r}}{dt}\right|_b + r(\boldsymbol{\omega}_b \times \mathbf{e}_r)$$

Because r is a scalar, the last term may be expressed as $r(\boldsymbol{\omega}_b \times \mathbf{e}_r) = \boldsymbol{\omega}_b \times (r\mathbf{e}_r) = \boldsymbol{\omega}_b \times \mathbf{r}$. The inertial velocity $\mathbf{v}_B$ may then be written as

$$\mathbf{v}_B = \left.\frac{d\mathbf{r}}{dt}\right|_B = \left.\frac{d\mathbf{r}}{dt}\right|_b + \boldsymbol{\omega}_b \times \mathbf{r} \tag{6.18a}$$

or

$$\mathbf{v}_B = \mathbf{v}_b + \boldsymbol{\omega}_b \times \mathbf{r} \tag{6.18b}$$

The physical interpretation of (6.18a) should be clear: The term $\mathbf{v}_b = (d\mathbf{r}/dt)|_b$ represents the inertial velocity which *would* exist if the coordinate system (and the particle which it is following) were not rotating, so that $\boldsymbol{\omega}_b = 0$. The motion in this case is purely radial.

We conclude this section by developing a general relation, applicable to *any* vector $\mathbf{s}$ resolved into a 3-D rotating coordinate system $(\mathbf{e}_1, \mathbf{e}_2, \mathbf{e}_3)$, which connects the inertial and relative time derivatives. Any vector $\mathbf{s}$ may be expressed in the general component form

$$\mathbf{s} = s_1\mathbf{e}_1 + s_2\mathbf{e}_2 + s_3\mathbf{e}_3$$

where the scalar components s_1, s_2, and s_3 are functions of time. Computing the inertial time derivative of $\mathbf{s}$, we have

$$\dot{\mathbf{s}} = \left.\frac{d\mathbf{s}}{dt}\right|_B = \underbrace{\left(\dot{s}_1\mathbf{e}_1 + \dot{s}_2\mathbf{e}_2 + \dot{s}_3\mathbf{e}_3\right)}_{\left.\frac{d\mathbf{s}}{dt}\right|_b} + \left(s_1\dot{\mathbf{e}}_1 + s_2\dot{\mathbf{e}}_2 + s_3\dot{\mathbf{e}}_3\right) \tag{6.19}$$

As indicated, the first set of terms, which involves only the rates of change of the scalar components of $\mathbf{s}$, gives the rate of change of $\mathbf{s}$ relative to the rotating observer. The second set of terms can be rewritten by noting, without proof, that the relations (6.13*a* and *b*) turn out to be true also for the general case of arbitrary three-dimensional rotation of the coordinate system; that is, the rate of change of any of the three rotating unit vectors $\mathbf{e}_i (i = 1, 2, \text{ or } 3)$ is given by

$$\dot{\mathbf{e}}_i = \boldsymbol{\omega}_b \times \mathbf{e}_i \qquad (i = 1, 2, 3) \tag{6.20}$$

where $\boldsymbol{\omega}_b$ is the angular velocity of the rotating coordinate system. Thus, the second set of terms in (6.19) may be expressed as

$$\sum_{i=1}^{3} s_i \dot{\mathbf{e}}_i = \sum_{i=1}^{3} s_i (\boldsymbol{\omega}_b \times \mathbf{e}_i) = \sum_{i=1}^{3} \boldsymbol{\omega}_b \times (s_i \mathbf{e}_i)$$

$$= \boldsymbol{\omega}_b \times \sum_{i=1} s_i \mathbf{e}_i = \boldsymbol{\omega}_b \times \mathbf{s}$$

The result is that (6.19) may be written in the compact form

$$\left.\frac{d\mathbf{s}}{dt}\right|_B = \left.\frac{d\mathbf{s}}{dt}\right|_b + \boldsymbol{\omega}_b \times \mathbf{s} \tag{6.21}$$

This relation, which applies to *any* vector $\mathbf{s}$, is one of the most useful relations in dynamics. The relation splits the inertial time derivative of a vector into two contributions: (1) rates of change of the scalar components s_1, s_2, and s_3 of $\mathbf{s}$, which yield the rate of change $(d\mathbf{s}/dt)|_b$ relative to the rotating observer, and (2) rates of change of the rotating unit vectors, $\boldsymbol{\omega}_b \times \mathbf{s}$.

A final comment is necessary before moving on to the next section. Let's use (6.21) to compute the relation between the coordinate system angular acceleration that would be measured in an inertial frame and the angular acceleration measured relative to the rotating frame. Thus

$$\left.\frac{d\boldsymbol{\omega}_b}{dt}\right|_B = \left.\frac{d\boldsymbol{\omega}_b}{dt}\right|_b + \boldsymbol{\omega}_b \times \boldsymbol{\omega}_b$$

Because the cross product of any vector with itself is identically zero, the result shows that fixed and rotating observers *agree* on the angular acceleration. This special property is enjoyed only by the angular velocity vector.

We'll next use the relation (6.21) to develop Newton's second law for 3-D particle dynamics, formulating the law in terms of a rotating coordinate system.

6.2.4 Particle Dynamics in Rotating Coordinate Systems

We now want to study particle dynamics using Newton's law $\mathbf{F} = m\mathbf{a}_B$. In this vector law we have to use the *inertial acceleration* $\mathbf{a}_B$, but we'll want to resolve this acceleration along a set of coordinate axes that are rotating with angular velocity $\boldsymbol{\omega}_b$. The situation is shown in Fig. 6.12, where the unit vectors are

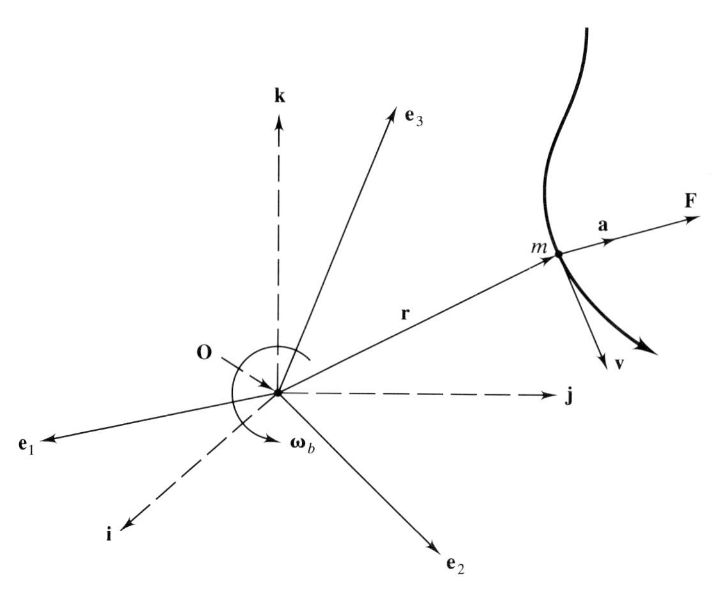

FIGURE 6.12
Inertial and rotating coordinate systems for the newtonian mechanics problem.

defined as follows: $\mathbf{i}, \mathbf{j}, \mathbf{k}$; fixed in inertial frame; $\mathbf{e}_1, \mathbf{e}_2, \mathbf{e}_3$: rotating with angular velocity $\boldsymbol{\omega}_b$ and originating at same point $\mathbf{O}$ as the $\mathbf{i}, \mathbf{j}, \mathbf{k}$ system.

First, note that in Fig. 6.12 $\mathbf{r}_b$ and $\mathbf{r}_B$ represent the *same* vector; when we resolve this vector into the rotating system, we call it $\mathbf{r}_b$, and when we resolve it into the $\mathbf{i}, \mathbf{j}, \mathbf{k}$ system, we call it $\mathbf{r}_B$. To find the inertial acceleration $\mathbf{a}_B$, we differentiate the position vector $\mathbf{r}$ twice with respect to time:

$$\mathbf{a}_B = \left.\frac{d^2\mathbf{r}_b}{dt^2}\right|_B$$

The first differentiation gives the inertial velocity $\mathbf{v}_B = (d\mathbf{r}_b/dt)|_B$, which from (6.21) may be written as follows:

$$\mathbf{v}_B = \left.\frac{d\mathbf{r}_b}{dt}\right|_b + \boldsymbol{\omega}_b \times \mathbf{r}_b$$

Noting that $(d\mathbf{r}_b/dt)|_b = \mathbf{v}_b$ is the velocity relative to an observer fixed in the rotating system, we can rewrite this as

$$\mathbf{v}_B = \mathbf{v}_b + \boldsymbol{\omega}_b \times \mathbf{r}_b \tag{6.22}$$

Next, we differentiate (6.22) to find $\mathbf{a}_B$:

$$\mathbf{a}_B = \left.\frac{d\mathbf{v}_B}{dt}\right|_B = \left.\frac{d\mathbf{v}_b}{dt}\right|_B + \frac{d}{dt}(\boldsymbol{\omega}_b \times \mathbf{r}_b)|_B$$

The derivative of the cross product can be expanded following the standard rule for differentiation of the product of two vectors, so we obtain

$$\mathbf{a}_B = \overbrace{\left.\frac{d\mathbf{v}_b}{dt}\right|_B}^{(1)} + \overbrace{\left.\frac{d\boldsymbol{\omega}_b}{dt}\right|_B \times \mathbf{r}_b}^{(2)} + \overbrace{\boldsymbol{\omega}_b \times \left.\frac{d\mathbf{r}_b}{dt}\right|_B}^{(3)}$$

We now use (6.21) again to convert this result to one involving the relative quantities $\mathbf{r}_b$, $\mathbf{v}_b$, and we obtain the following, with the terms labeled 1, 2, and 3 above also labeled below:

$$\mathbf{a}_B = \overbrace{\left.\frac{d\mathbf{v}_b}{dt}\right|_b + \boldsymbol{\omega}_b \times \mathbf{v}_b}^{(1)} + \overbrace{\dot{\boldsymbol{\omega}}_b \times r_b}^{(2)} + \overbrace{\boldsymbol{\omega}_b \times \left(\left.\frac{d\mathbf{r}_b}{dt}\right|_b + \boldsymbol{\omega}_b \times \mathbf{r}_b\right)}^{(3)}$$

Noting that $(d\mathbf{v}_b/dt)|_b$ is the relative acceleration $\mathbf{a}_b$, the above leads to the following useful result:

$$\mathbf{a}_B = \mathbf{a}_b + 2\boldsymbol{\omega}_b \times \mathbf{v}_b + \boldsymbol{\omega}_b \times (\boldsymbol{\omega}_b \times \mathbf{r}_b) + \dot{\boldsymbol{\omega}}_b \times \mathbf{r}_b \tag{6.23}$$

The term $2\boldsymbol{\omega}_b \times \mathbf{v}_b$ is the so-called Coriolis acceleration, while the $\boldsymbol{\omega}_b \times (\boldsymbol{\omega}_b \times \mathbf{r}_b)$ is the centripetal acceleration. Since all the terms on the right-hand side of (6.23) are defined by vectors that are expressed in the rotating coordinate system, we have accomplished our objective: We've calculated the inertial acceleration $\mathbf{a}_B$ but in such a way that this vector $\mathbf{a}_B$ is automatically resolved along the rotating coordinate axes.

In general, we would execute the following steps in analyzing a particle dynamics problem using a rotating coordinate system:

1. Define the rotating coordinate system according to convenience and personal preference (often there are several reasonable choices).
2. Determine the angular velocity $\boldsymbol{\omega}_b$ of the rotating coordinate system in terms of one or more individual rotations occurring in the given problem; since the rotations involved do not necessarily occur naturally about one of the three rotating axes, we may have to do a little work to resolve $\boldsymbol{\omega}_b$ along the rotating axes (as in the case 2 formulation for the spinning, rotating cylinder).
3. Set up the position vector $\mathbf{r}_b$ in the rotating system.
4. Compute the relative velocity $\mathbf{v}_b$ and the relative acceleration $\mathbf{a}_b$; to do this, one merely differentiates $\mathbf{r}_b$ (and then $\mathbf{v}_b$) as if the rotating unit vectors were not changing with time.
5. Compute the angular acceleration $\dot{\boldsymbol{\omega}}_b$. Recall that

$$\left.\frac{d\boldsymbol{\omega}_b}{dt}\right|_b = \left.\frac{d\boldsymbol{\omega}_b}{dt}\right|_B$$

6. Plug the results of 2 through 5 into (6.23) and do the vector algebra.
7. Analyze the equations of motion to see what the particle does.

To illustrate these procedures, we'll now consider several examples.

Example 6.3 A Gyro/Gimbal Assembly. The problem considered is to find the inertial acceleration of the point P located in the position shown in the gyro/gimbal assembly of Fig. 6.13. The system operation is detailed as follows: The gyro disk D spins about an axis ($\mathbf{e}_3$) through its center at the fixed angular rate $\dot{\psi}$. This disk is mounted in a gimbal (frame), which is itself free to rotate about the horizontal at the rate $\dot{\theta}$. This assembly is bolted rigidly to a structure, and we will assume that the structure is itself rotating at the rate $\dot{\phi}$ about an axis out of the paper (to make the problem fit into our current framework, we'll also require that the point **O** always remain fixed).

We'll set up our rotating coordinate system ($\mathbf{e}_1, \mathbf{e}_2, \mathbf{e}_3$) so that the $\mathbf{e}_1, \mathbf{e}_2$ axes lie in the plane of the disk D, with the $\mathbf{e}_3$ axis normal to the disk as shown. These axes, however, do *not* spin with the disk. Thus, the angular velocity of the rotating

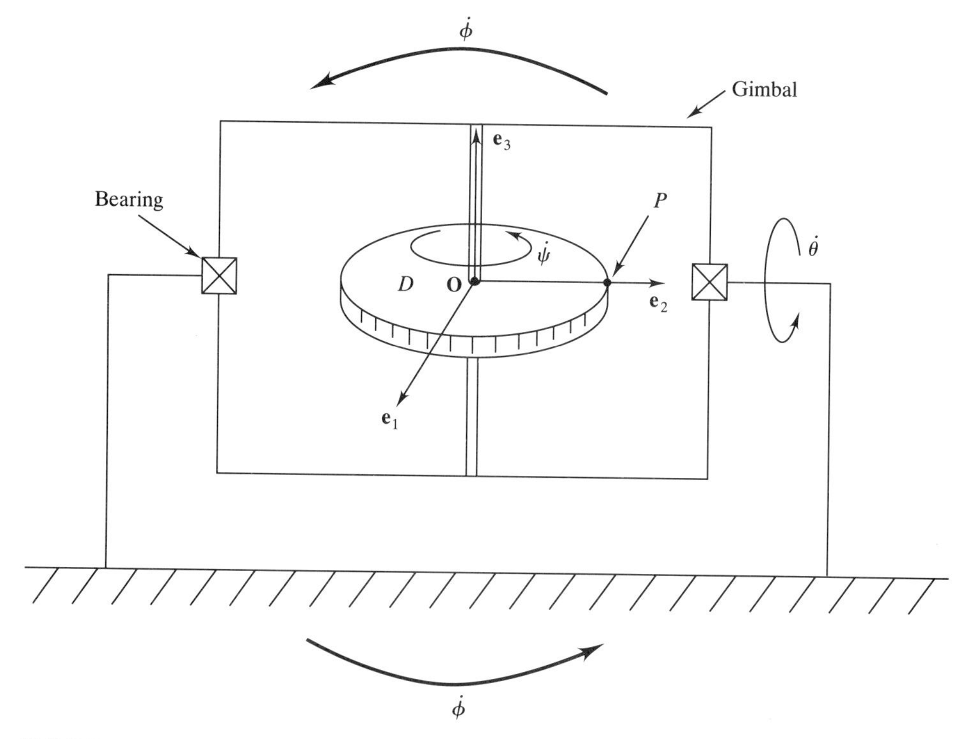

FIGURE 6.13
The gyro/gimbal assembly. At $t = 0$ the gimbal is in the plane of the paper; for $t \geq 0$ let θ be the angle between the instantaneous gimbal plane and the plane of the paper. Assumes $\dot{\psi}$ = constant but $\dot{\theta}$ and $\dot{\phi}$ may be varying with time.

coordinate system is due only to the rotation rates $\dot{\theta}$ and $\dot{\phi}$ (the reason for doing it this way will emerge). As the gimbal rotates ($\dot{\theta}$), the $\mathbf{e}_1$ and $\mathbf{e}_3$ axes rotate around with it. The $\mathbf{e}_2$ axis is always the axis about which this gimbal rotation takes place. In view of this description, we can write out the angular velocity $\boldsymbol{\omega}_b$ of the rotating coordinate system as

$$\boldsymbol{\omega}_b = \dot{\theta}\mathbf{e}_2 + \dot{\phi}\mathbf{n}$$

where $\mathbf{n}$ is a unit vector that always points out the paper (note that if we had been calculating the angular velocity $\boldsymbol{\omega}$ of the *disk D*, we would have to add a term $\dot{\psi}\mathbf{e}_3$ to the above formula). The next step is to resolve $\boldsymbol{\omega}_b$ along the three rotating axes (this works out automatically for $\dot{\theta}$ but not for $\dot{\phi}$). To do this, we have to express $\mathbf{n}$ in terms of $\mathbf{e}_1$ and $\mathbf{e}_3$. The needed relation is

$$\mathbf{n} = \mathbf{e}_1 \cos\theta + \mathbf{e}_3 \sin\theta$$

where the angle θ is the angle between the instantaneous gimbal plane and the plane of the paper. Substitution of this relation into the equation for $\boldsymbol{\omega}_b$ yields

$$\boldsymbol{\omega}_b = \dot{\phi}\cos\theta\,\mathbf{e}_1 + \dot{\theta}\mathbf{e}_2 + \dot{\phi}\sin\theta\,\mathbf{e}_3 \qquad (6.24)$$

Now we can move on to the position vector $\mathbf{r}_b$. We'll define this in terms of the angle ψ, which measures the orientation of the point P relative to the $\mathbf{e}_2$ axis. The result is

$$\mathbf{r}_b = -a\sin\psi\,\mathbf{e}_1 + a\cos\psi\,\mathbf{e}_2 \qquad (6.25)$$

Now let's pause a minute and see what would have happened if we had defined the rotating coordinate system to be fixed rigidly in the disk (i.e., rotating with it). Then the position vector would have been simply $\mathbf{r}_b = a\mathbf{e}_2$. We see, however, that the relation for the angular velocity of the coordinate system would be quite complicated; we would have to start out with

$$\boldsymbol{\omega}_b = \dot{\psi}\mathbf{e}_3 + \dot{\theta}\mathbf{n}_1 + \dot{\phi}\mathbf{n}_2$$

[here $\mathbf{n}_2 = \mathbf{n}$ is normal to the paper, and $\mathbf{n}_1$ is the horizontal gimbal axis of rotation (for $\dot{\theta}$)].

We would then have to resolve both $\mathbf{n}_1$ and $\mathbf{n}_2$ into the rotating system, and each of these transformations would involve the two angles θ and ϕ. It's quite messy, although certainly doable. The rotating coordinate system that we are actually using is an intermediate one in terms of which it is reasonably straightforward to write out both $\mathbf{r}_b$ and $\boldsymbol{\omega}_b$.

Now, using (6.24) and (6.25), we can calculate the remaining vectors we need. The results are given below:

$$\mathbf{v}_b = -a\dot{\psi}\cos\psi\,\mathbf{e}_1 - a\dot{\psi}\sin\psi\,\mathbf{e}_2$$

$$\mathbf{a}_b = a\dot{\psi}^2\sin\psi\,\mathbf{e}_1 - a\dot{\psi}^2\cos\psi\,\mathbf{e}_2$$

$$\dot{\boldsymbol{\omega}}_b = \left(\ddot{\phi}\cos\theta - \dot{\phi}\dot{\theta}\sin\theta\right)\mathbf{e}_1 + \ddot{\theta}\mathbf{e}_2 + \left(\ddot{\phi}\sin\theta + \dot{\phi}\dot{\theta}\cos\theta\right)\mathbf{e}_3 \qquad (6.26)$$

The acceleration of the point P is determined by performing the indicated operations given by (6.23), using (6.24) through (6.26). This is a strongly suggested exercise for the student. The algebra is a bit messy, but the practice is worthwhile.

Example 6.4 A Rotating, Telescoping Boom (Fig. 6.14). A very light (assumed massless) boom with a mass m attached to its end "telescopes," so that the distance s increases (or decreases) in some specified manner [that is, $s(t)$ is assumed given]. Meanwhile the entire apparatus is rotated about the vertical at the fixed rate Ω. The angle α is constant, so that the mass m will move on the surface of a cone of vertex angle α. The objective is to find the force $\mathbf{F}$ which must be exerted on the mass m in order to produce this motion. This force, which is exerted by the boom on the mass, needs to be known in order to ensure that the stresses that result in the boom are acceptably small.

In the analysis of this problem, our first task is to define a convenient rotating coordinate system. We'll look at two ways to do this and will then compare the merits of each setup. First, we'll define the $\mathbf{e}_1, \mathbf{e}_2, \mathbf{e}_3$ system as shown in Fig. 6.14; the rotating system rotates with the telescoping boom. The axes will be oriented so that the boom always lies in the $\mathbf{e}_2 - \mathbf{e}_3$ plane, and $\mathbf{e}_3$ is always vertical (along the direction of the angular velocity vector $\boldsymbol{\omega}_b$ of the rotating coordinate system). We next tabulate the vectors needed to determine the inertial acceleration $\mathbf{a}_B$ of the mass m:

$$\mathbf{r}_b = -s \sin \alpha \mathbf{e}_2 + s \cos \alpha \mathbf{e}_3$$

$$\mathbf{v}_b = -\dot{s} \sin \alpha \mathbf{e}_2 + \dot{s} \cos \alpha \mathbf{e}_3$$

$$\mathbf{a}_b = -\ddot{s} \sin \alpha \mathbf{e}_2 + \ddot{s} \cos \alpha \mathbf{e}_3$$

$$\boldsymbol{\omega}_b = \Omega \mathbf{e}_3 \qquad \dot{\boldsymbol{\omega}}_b = 0$$

$$\boldsymbol{\omega}_b \times \mathbf{v}_b = \Omega \dot{s} \sin \alpha \mathbf{e}_1$$

$$\boldsymbol{\omega}_b \times (\boldsymbol{\omega}_b \times \mathbf{r}_b) = s\Omega^2 \sin \alpha \mathbf{e}_2$$

Now we put these results into (6.23) to determine the inertial acceleration $\mathbf{a}_B$,

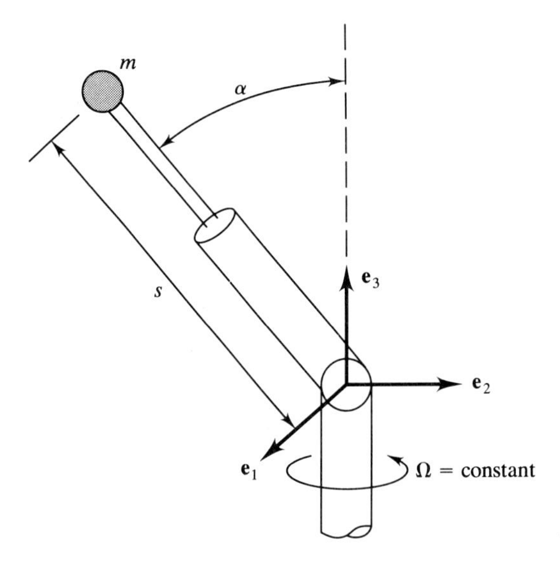

FIGURE 6.14
A rotating, telescoping boom.

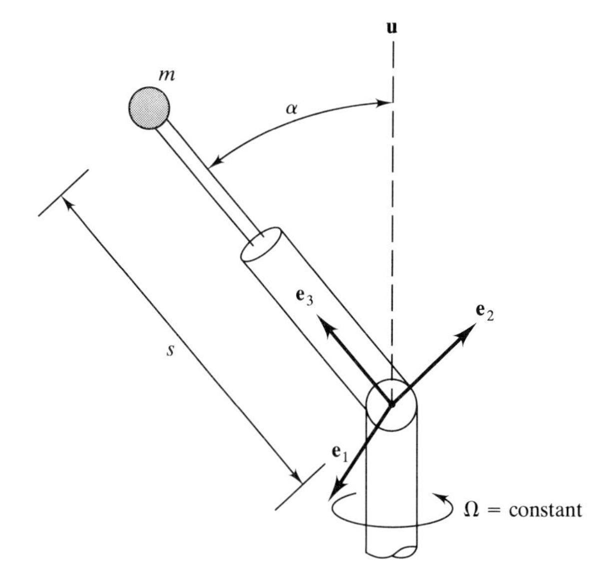

FIGURE 6.15
The rotating telescoping boom, using a rotating coordinate system with one unit vector ($\mathbf{e}_3$) aligned with the boom.

resolved along the rotating system unit vectors, so that Newton's law comes out as follows (verify calculations!):

$$\mathbf{F} = m\left[2\Omega\dot{s}\sin\alpha\mathbf{e}_1 + (s\Omega^2 - \ddot{s})\sin\alpha\mathbf{e}_2 + \ddot{s}\cos\alpha\mathbf{e}_3\right] \tag{6.27}$$

Since we know s, $\dot{s}$, $\ddot{s}$, and Ω, the force $\mathbf{F}$ can be calculated.

Next, let's look at a second way to set up the rotating coordinate axes (Fig. 6.15). What we'll now do is to align the $\mathbf{e}_3$ axis along the boom (the $\mathbf{e}_2$ axis also changes, but $\mathbf{e}_1$ doesn't), so we now have the setup shown in Fig. 6.15. Now as far as setting up the requisite vectors is concerned, note the following differences between this setup and the first one: In the first setup it was easy to write out the angular velocity $\boldsymbol{\omega}_b$ of the rotating coordinate system, since $\boldsymbol{\omega}_b$ was along one of the three axes ($\mathbf{e}_3$). On the other hand, the relative position vector $\mathbf{r}_b$ had to be resolved using the $\mathbf{e}_3$ and $\mathbf{e}_2$ axes. In the second setup this situation has been reversed: Now the relative position vector $\mathbf{r}_b$ is always aligned along the $\mathbf{e}_3$ axis, but the angular velocity $\boldsymbol{\omega}_b$ of the rotating coordinate system (along the $\mathbf{u}$, or vertical direction) must be resolved along the new $\mathbf{e}_3$ and $\mathbf{e}_2$ axes. The required vectors for the second setup are summarized below:

$$\mathbf{r}_b = s\mathbf{e}_3$$

$$\mathbf{v}_b = \dot{s}\mathbf{e}_3$$

$$\mathbf{a}_b = \ddot{s}\mathbf{e}_3$$

$$\boldsymbol{\omega}_b = \Omega\mathbf{u} = \Omega\sin\alpha\mathbf{e}_2 + \Omega\cos\alpha\mathbf{e}_3$$

$$\dot{\boldsymbol{\omega}}_b = 0$$

$$\boldsymbol{\omega}_b \times \mathbf{v}_b = \dot{s}\Omega\sin\alpha\mathbf{e}_1 \qquad \text{(same as before; } \mathbf{e}_1 \text{ components don't change)}$$

$$\boldsymbol{\omega}_b \times (\boldsymbol{\omega}_b \times \mathbf{r}_b) = s\Omega^2\cos\alpha\sin\alpha\mathbf{e}_2 - s\Omega^2\sin^2\alpha\mathbf{e}_3$$

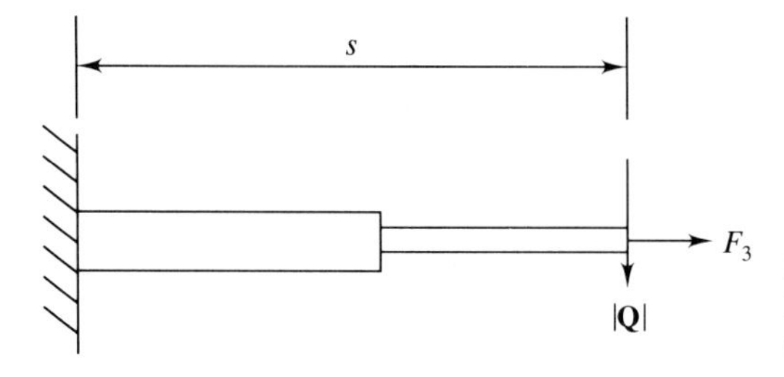

FIGURE 6.16
Boom configuration for structural analysis.

The statement of Newton's law is thus

$$\mathbf{F} = m\left[2\Omega\dot{s}\sin\alpha\mathbf{e}_1 + s\Omega^2\cos\alpha\sin\alpha\mathbf{e}_2 + (\ddot{s} - s\Omega^2\sin^2\alpha)\mathbf{e}_3\right] \tag{6.28}$$

It is important to keep in mind that the force $\mathbf{F}$ appearing in (6.27) and (6.28) is the *same*; it has simply been resolved into two different coordinate systems.

From a practical standpoint the second coordinate setup may be the more useful. For instance, if we take the $\mathbf{e}_3$ component of $\mathbf{F}$ from (6.28), $F_3 = m(\ddot{s} - \Omega^2 s\sin^2\alpha)$, we have the component of the force directed *along* the boom; thus, this force will result directly in a tensile stress in the boom. Likewise, the resultant force in the $\mathbf{e}_1$-$\mathbf{e}_2$ plane results in a shear $\mathbf{Q}$ applied at the end of the boom; the magnitude of the shear is given by

$$|\mathbf{Q}| = m\left[(2\dot{s}\Omega\sin\alpha)^2 + (s\Omega^2\cos\alpha\sin\alpha)^2\right]^{1/2}$$

In order to analyze the stresses in the boom, then, we may as a first approximation consider the boom to be a beam of variable cross section having the loadings shown in Fig. 6.16. So we can compute the tensile and bending stresses in the boom.* We see that for purposes of this type of calculation, the coordinate system of Fig. 6.15 is a "natural" one.

Now let's see what happens if we make the system of this example a little more complicated. We'll assume that the angle α may also be changing with time in a specified manner $\alpha(t)$. Thus, we have two rotations, and $\boldsymbol{\omega}_b$ will generally be changing with time. To analyze this version of the problem, we'll use the set of rotating coordinates defined in Fig. 6.15 ($\mathbf{e}_3$ aligned along the boom and the rotating coordinates rotating with both the Ω and $\dot{\alpha}$ rotations). The relevant vectors turn out to be as follows:

$$\mathbf{r}_b = s\mathbf{e}_3$$

$$\mathbf{v}_b = \dot{s}\mathbf{e}_3$$

$$\mathbf{a}_b = \ddot{s}\mathbf{e}_3$$

$$\boldsymbol{\omega}_b = \Omega\mathbf{u} + \dot{\alpha}\mathbf{e}_1 = \dot{\alpha}\mathbf{e}_1 + \Omega\sin\alpha\mathbf{e}_2 + \Omega\cos\alpha\mathbf{e}_3$$

$$\dot{\boldsymbol{\omega}}_b = \ddot{\alpha}\mathbf{e}_1 + \Omega\dot{\alpha}\cos\alpha\mathbf{e}_2 - \Omega\dot{\alpha}\sin\alpha\mathbf{e}_3$$

*Actually, if the boom is light, then it may undergo fairly large deflections and may also vibrate. Such effects are ignored here, as the boom is assumed to move as a massless rigid body.

Here the algebra involved in computing $\mathbf{a}_B$ is messier, but straightforward. Once we obtain $\mathbf{F} = m\mathbf{a}_B$, we can again resolve the forces acting on m into tensile and shear components. As we can see, the key to analyzing mechanical systems in terms of rotating coordinates is to be able to (1) define a convenient and useful coordinate system and (2) calculate the relative position vector $\mathbf{r}_b$ and the angular velocity $\boldsymbol{\omega}_b$ of the coordinate system. Once these two vectors have been determined, calculation of the other vectors needed to find the inertial acceleration $\mathbf{a}_B$ is straightforward. In practice, we usually pick a coordinate system in which either $\mathbf{r}_b$ or $\boldsymbol{\omega}_b$ is (relatively) easy to determine.

Comment. The inertial acceleration $\mathbf{a}_B$ that we have been calculating via (6.23) is of course the *same vector* we would get if we had done the problem using a set of fixed (nonrotating) cartesians, $\mathbf{a}_B = \ddot{x}\mathbf{i} + \ddot{y}\mathbf{j} + \ddot{z}\mathbf{k}$. In the rotating coordinate formulation, we are simply resolving this vector along axes which are rotating, because it turns out to be both more convenient and more natural (consider the difficulty in trying to set up either of the preceding example problems directly using fixed cartesians).

Example 6.5 The Rotating Cylinder (Example 6.1) Revisited. In this example we will determine the inertial acceleration of the point P on the surface of the spinning, rotating cylinder of Fig. 6.5. The rotating coordinate system is fixed rigidly in the cylinder, and the angular velocity $\boldsymbol{\omega}_b$ of the rotating coordinate system is given by (6.3), with **e**'s replacing **g**'s:

$$\boldsymbol{\omega}_b = -\dot{\theta}\sin\phi\,\mathbf{e}_1 + \dot{\phi}\mathbf{e}_2 + \dot{\theta}\cos\phi\,\mathbf{e}_3 \tag{6.29}$$

The position vector $\mathbf{r}_b$ is seen from Fig. 6.5 to be

$$\mathbf{r}_b = L\mathbf{e}_2 + a\mathbf{e}_3$$

Because the point P is fixed on the cylinder, it undergoes no motion relative to an observer rotating with the rotating coordinate system. Thus $\mathbf{v}_b = \mathbf{a}_b = 0$ and the inertial acceleration $\mathbf{a}_B$ defined by (6.23) reduces to

$$\mathbf{a}_B = \boldsymbol{\omega}_b \times (\boldsymbol{\omega}_b \times \mathbf{r}_b) + \dot{\boldsymbol{\omega}}_b \times \mathbf{r}_b \tag{6.30}$$

where the coordinate system angular acceleration $\dot{\boldsymbol{\omega}}_b$ is given by

$$\dot{\boldsymbol{\omega}}_b = \left(-\ddot{\theta}\sin\phi - \dot{\theta}\dot{\phi}\cos\phi\right)\mathbf{e}_1 + \ddot{\phi}\mathbf{e}_2 + \left(\ddot{\theta}\cos\phi - \dot{\theta}\dot{\phi}\sin\phi\right)\mathbf{e}_3$$

This result was obtained by time differentiating (6.29), ignoring the contribution due to the rate of change of the rotating unit vectors. This is valid because of the previous result $(d\boldsymbol{\omega}_b/dt)|_B = (d\boldsymbol{\omega}_b/dt)|_b$, developed at the end of Sec. 6.2.3.

Combining the relevant vectors according to (6.30), we obtain the following inertial acceleration:

$$\begin{aligned}\mathbf{a}_B = &\left(a\ddot{\phi} - L\ddot{\theta}\cos\phi - a\dot{\theta}^2\sin\phi\cos\phi\right)\mathbf{e}_1 \\ &+\left(-L\dot{\theta}^2 + 2a\dot{\theta}\dot{\phi}\cos\phi + a\ddot{\theta}\sin\phi\right)\mathbf{e}_2 \\ &+\left(-L\ddot{\theta}\sin\phi - a\dot{\phi}^2 - a\dot{\theta}^2\sin^2\phi\right)\mathbf{e}_3\end{aligned} \tag{6.31}$$

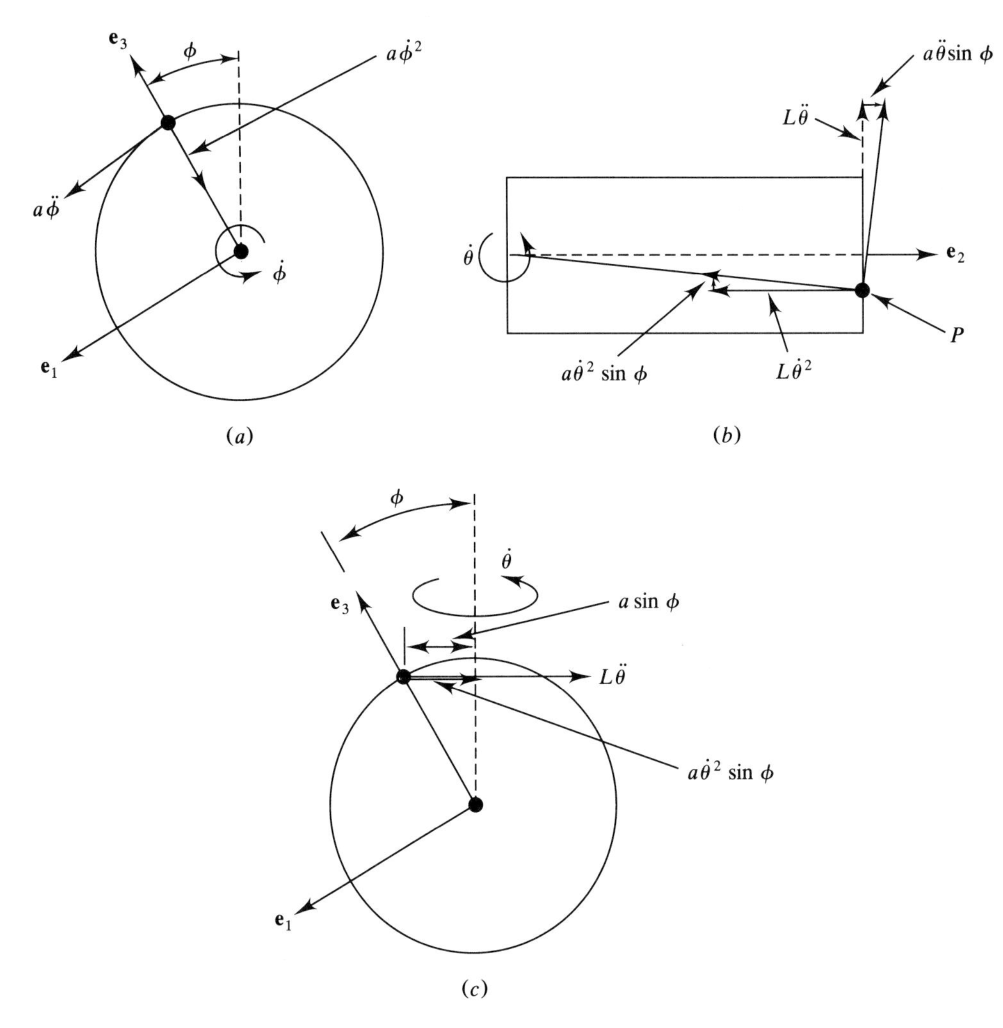

FIGURE 6.17
(*a*) View looking down the $\mathbf{e}_2$ axis to describe the portion of the acceleration $\mathbf{a}_B$ due to spin only ($\dot{\phi}$, $\ddot{\phi}$); (*b*) top view, looking down the inertially fixed $\mathbf{u}$ axis; (*c*) same view as (*a*). (*b*) and (*c*) show the portion of the acceleration $\mathbf{a}_B$ due to rotation ($\dot{\theta}, \theta$) only.

While (6.31) appears fairly complicated, it is not too difficult to visualize most of the terms that appear. This visualization is shown in Fig. 6.17; Fig. 6.17(*a*) depicts the accelerations which are due to the spin motion ($\dot{\phi}, \ddot{\phi}$) only. The terms $a\ddot{\phi}\mathbf{e}_1 - a\dot{\phi}^2\mathbf{e}_3$ in (6.31) describe the acceleration which would exist if $\dot{\theta}$ were zero. We observe in Fig. 6.17(*a*) the tangential ($a\ddot{\phi}\mathbf{e}_1$) and radial ($-a\dot{\phi}^2\mathbf{e}_3$) accelerations which characterize the circular motion of a particle. Figures 6.17(*b*) and (*c*) show the acceleration components that are due to rotation ($\dot{\theta}, \ddot{\theta}$) only. This rotational

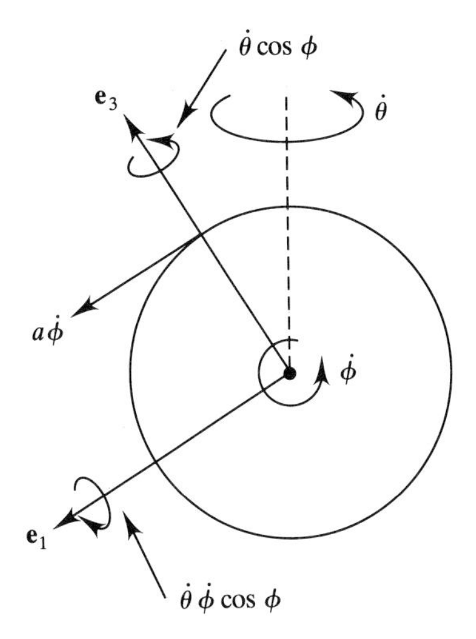

FIGURE 6.18
Acceleration component due to interaction of the two rotations.

motion produces the following terms in (6.31):

$$\overbrace{L\ddot{\theta}(-\cos\phi\mathbf{e}_1 - \sin\phi\mathbf{e}_3)}^{(1)} \underset{}{\overset{(2)}{-\,L\dot{\theta}^2\mathbf{e}_2}} - \overbrace{a\dot{\theta}^2\sin\phi(\cos\phi\mathbf{e}_1 + \sin\phi\mathbf{e}_3)}^{(3)} + \overbrace{a\ddot{\theta}\sin\phi\mathbf{e}_2}^{(4)}$$

These terms describe the acceleration which would exist if $\dot{\phi}$ were zero and if the $\mathbf{e}_3$ axis were rotated a fixed amount ϕ relative to the vertical $\mathbf{u}$. The first term is the tangential acceleration $L\ddot{\theta}$, which is in the horizontal plane and normal to $\mathbf{e}_2$ and $\mathbf{u}$. Term 1 above is the resolution of this acceleration component along the $\mathbf{e}_1$ and $\mathbf{e}_3$ axes. Term 2 is the centripetal acceleration directed radially inward (along $-\mathbf{e}_2$). Term 3 represents the acceleration component $a\dot{\theta}^2 \sin\phi$, resolved along the $\mathbf{e}_1$ and $\mathbf{e}_3$ directions. Finally, term 4 is due to rotation of the position vector component $a \sin\phi$ (in the horizontal plane) about the vertical.

Careful examination of Fig. 6.17 enables a physical interpretation of all of the terms in (6.31) except the term that involves the *interaction* of the two rotations: $2a\dot{\phi}\dot{\theta}\cos\phi\mathbf{e}_2$. The origin of this term is more difficult to see and is depicted in Fig. 6.18. A portion $a\dot{\phi}\dot{\theta}\cos\phi\mathbf{e}_2$ of this acceleration is due to the angular velocity component $\dot{\theta}\cos\phi\mathbf{e}_3$ rotating the velocity component $a\dot{\phi}\mathbf{e}_1$, which produces an $\mathbf{e}_2$ acceleration. An equal contribution arises from the angular acceleration component $-\dot{\theta}\dot{\phi}\cos\phi\mathbf{e}_1$ rotating the position vector component $a\mathbf{e}_3$ in the $\mathbf{e}_2$ direction.

It should be evident from this example that in the general case of rotation about two or even three axes, the vector formulation culminating in (6.23) is the only reliable way to proceed. Attempts to determine the acceleration components directly using visualizations as in Figs. 6.17 and 6.18 require extreme care and often lead to errors.

Example 6.6 Bead on a Rotating Hoop. As a final example, consider a bead of mass m which is constrained to slide along a circular wire of radius a; this wire is rotated at the fixed rate Ω about a vertical axis, as shown in Fig. 6.19. At any instant the location of the mass m around the hoop periphery is defined by the coordinate θ. A rotating coordinate system is defined such that $\mathbf{e}_1$ and $\mathbf{e}_2$ always lie in the plane of the hoop, with $\mathbf{e}_1$ always pointing directly at the particle; $\mathbf{e}_3$ is normal to the hoop plane. The rotating coordinate system, therefore, rotates with both the Ω and $\dot{\theta}$ motions. Our objective will be to develop the equations of motion for the mass m. Notice that the system possesses a single degree of freedom, defined by the coordinate θ; if $\theta(t)$ is known for all time t, then the exact location in space of the bead m is also known.

In order to determine the inertial acceleration $\mathbf{a}_B$ of the mass m, we first determine the angular velocity $\boldsymbol{\omega}_b$ of the rotating coordinate system, which is

$$\boldsymbol{\omega}_b = \Omega\mathbf{u} + \dot{\theta}\mathbf{e}_3$$

The vertical unit vector $\mathbf{u}$ is defined by

$$\mathbf{u} = -\mathbf{e}_1 \cos\theta + \mathbf{e}_2 \sin\theta$$

so that

$$\boldsymbol{\omega}_b = -\Omega \cos\theta\,\mathbf{e}_1 + \Omega \sin\theta\,\mathbf{e}_2 + \dot{\theta}\mathbf{e}_3 \tag{6.32}$$

The position vector $\mathbf{r}_b = a\mathbf{e}_1$; furthermore, $\mathbf{v}_b = \mathbf{a}_b = 0$ because the mass m does not move relative to a rotating observer. Thus, (6.30) once again applies. The angular acceleration $\dot{\boldsymbol{\omega}}_b$ is obtained from (6.32), noting that $\dot{\Omega} = 0$, as

$$\dot{\boldsymbol{\omega}}_b = \Omega\dot{\theta} \sin\theta\,\mathbf{e}_1 + \Omega\dot{\theta} \cos\theta\,\mathbf{e}_2 + \ddot{\theta}\mathbf{e}_3$$

Combining the relevant vectors according to (6.30), the inertial acceleration $\mathbf{a}_B$ is

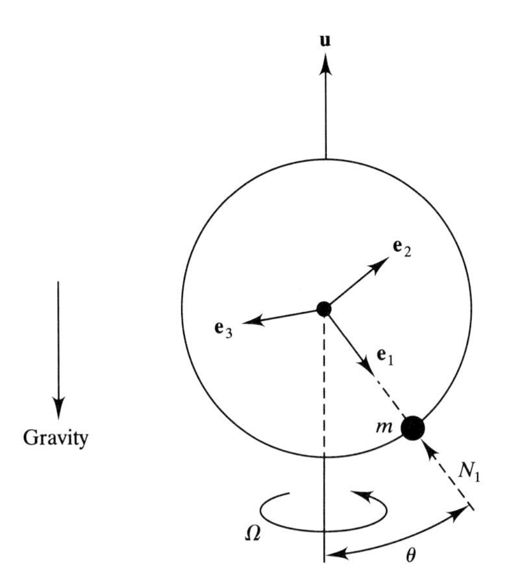

FIGURE 6.19
Bead on a rotating hoop.

determined as

$$\mathbf{a}_B = (-a\dot{\theta}^2 - a\Omega^2 \sin^2\theta)\mathbf{e}_1 + (a\ddot{\theta} - a\Omega^2 \sin\theta\cos\theta)\mathbf{e}_2 + (-2a\Omega\dot{\theta}\cos\theta)\mathbf{e}_3 \tag{6.33}$$

Next let us consider the forces acting on the particle m. The force due to gravity is shown in Fig. 6.19. In addition to this force, there will exist two normal forces N_1 and N_3 (defined to be positive along $-\mathbf{e}_1$ and $-\mathbf{e}_3$; N_1 is shown in Fig. 6.19). These constraint forces, exerted by the hoop on the mass m, arise so as to maintain the postulated motion (i.e., to keep the bead on the hoop). Because there is no friction, the hoop exerts no force on the bead in the tangential ($\mathbf{e}_2$) direction. Resolving the force due to gravity along the $\mathbf{e}_1$ and $\mathbf{e}_2$ axes, the resultant force vector $\mathbf{F}$ is given by

$$\mathbf{F} = (-N_1 + mg\cos\theta)\mathbf{e}_1 - mg\sin\theta\,\mathbf{e}_2 - N_3\mathbf{e}_3$$

Equating this force to the product $m\mathbf{a}_B$, with $\mathbf{a}_B$ given by (6.33), leads to the following three scalar equations:

$$-N_1 + mg\cos\theta = m(-a\dot{\theta}^2 - a\Omega^2\sin^2\theta) \tag{6.34a}$$

$$-mg\sin\theta = m(a\ddot{\theta} - a\Omega^2\sin\theta\cos\theta) \tag{6.34b}$$

$$-N_3 = -2ma\Omega\dot{\theta}\cos\theta \tag{6.34c}$$

Equations (6.34*a*, *b*, and *c*) are a set of three differential equations in the three unknowns θ, N_1, and N_3. Equation (6.34*b*), which involves the unknown coordinate θ, but not the constraint forces N_1 and N_2, can theoretically be solved to determine $\theta(t)$. The solution will show how the combined effects of gravity and hoop rotation affect the bead motion for any set of initial conditions $\theta(0), \dot{\theta}(0)$. Once the solution $\theta(t)$ has been found, it can be substituted into (6.34*a* and *c*) in order to determine the time dependence of the constraint forces N_1 and N_3. But notice that these constraint forces do not need to be found in order to determine the motion $\theta(t)$ of the mass m.

Now a few comments are in order:

1. Example 6.3 outlined above may seem artificial, since we were only attempting to calculate the acceleration of a point on a rigid body, the disk D. The gyro assembly described is really a phenomenon in the province of rigid body gyroscopic mechanics. A rapidly spinning (large $\dot{\psi}$) rotor mounted on a gimbal capable of rotation forms the basic element of many useful devices: the gyrocompass, angular velocity measuring instrumentation, and gyroscopic devices for guidance and stabilization, to mention a few. It turns out that in order to work out the rigid body dynamics for such phenomena, rotating coordinate systems are indispensable. Thus, analysis using rotating coordinate systems is certainly not confined to 3-D particle phenomena.
2. An editorial comment on the free body diagram. When you first learned how to analyze phenomena in particle mechanics, you undoubtedly learned to work with the free body diagram, in which the directions of the various forces

and accelerations are displayed pictorially, then calculated and finally put into equation form. This is a useful tool of analysis for problems in two dimensions and for simple problems in three dimensions. For more difficult problems in three dimensions, however, especially those involving rotations, it is often exceedingly difficult to utilize the free body diagram: The only realistic way to do such problems is to set up an appropriate coordinate system, along with the requisite vectors, and to then proceed using the *vector* formulation of mechanics described here. The point to be made is the following: The general vector formulation *always* provides a systematic way to attack the problem. If you understand its use thoroughly in the 3-D cases we've been working on, then you will have little trouble in analyzing simpler problems. The greatest utility of the free body diagram as applied to relatively difficult 3-D problems is probably this: It is useful as a visualization *after* you analyze the problem using vectorial mechanics in that it allows you to picture what the equations are telling you.

6.3 LAGRANGE'S EQUATIONS

6.3.1 Introduction

Our objective in this section will be to develop an alternate formulation of the basic laws of mechanics. The results, Lagrange's equations, will provide a different (from Newton's laws) line of attack for development of the equations of motion of mechanical systems. What we will see is that the equations of motion are obtained directly from the following *scalar* quantities for the *system as a whole*, no matter how complex the system happens to be: (1) the kinetic energy T, (2) the potential energy V, which we will use to characterize those forces which are conservative, and (3) the work W_{nc} done by nonconservative forces. (For a summary of the basic ideas of work, kinetic energy, and conservative forces, see Sec. 2.3.) In order to develop Lagrange's equations, we will eventually need to introduce a new concept, the idea of virtual displacement and accompanying virtual work, or, more generally, the idea of virtual change in a quantity.

We will develop Lagrange's equations for the special case of a single particle of mass m moving in three dimensions. The results we obtain, however, will turn out to be general and will be applicable to any mechanical system. Thus, let us begin by stating the equations of motion for the single particle in terms of a fixed (nonrotating) cartesian (x, y, z) system. The three scalar equations of motion are given by (2.7):

$$m\ddot{x} = F_x \tag{6.35a}$$

$$m\ddot{y} = F_y \tag{6.35b}$$

$$m\ddot{z} = F_z \tag{6.35c}$$

The resultant force components F_x, F_y, and F_z may depend on position, velocity, and time, as in (2.10). The particle kinetic energy T is defined by

$$T = \tfrac{1}{2}m\left(\dot{x}^2 + \dot{y}^2 + \dot{z}^2\right) \tag{6.36}$$

and is seen to be a quadratic function of the velocity components. Recall that the change in kinetic energy during some interval of time is equal to the work done on the particle in that interval.

Now let us assume that the force components F_x, F_y, and F_z can be split into conservative and nonconservative portions, $F_x = F_{x_c} + F_{x_{nc}}$, and so on. The conservative forces F_{x_c}, F_{y_c}, and F_{z_c} can be characterized by a potential energy function V, which we will assume to be a function of position, $V = V(x, y, z)$. The conservative forces are then defined in terms of the partial derivatives of V as follows (Sec. 2.3):

$$F_{x_c} = -\frac{\partial V}{\partial x}, \qquad F_{y_c} = -\frac{\partial V}{\partial y}, \qquad F_{z_c} = -\frac{\partial V}{\partial z} \tag{6.37}$$

Recall that the change in potential energy during some interval of time is equal to the negative of the work done on the particle by the conservative forces during that interval. The nonconservative force components $F_{x_{nc}}$, $F_{y_{nc}}$, and $F_{z_{nc}}$, arising perhaps from damping or external, time-dependent effects, are assumed to be known functions of position, velocity, and time.

6.3.2 Degrees of Freedom and Constraints

In Sec. 6.3.1 the particle is assumed to move freely in three dimensions under the action of the forces that act upon it. The system has 3 *degrees of freedom*; a minimum of three position coordinates is needed to characterize unambiguously the configuration of the system. Frequently, however, the particle motion is *constrained* to lie on certain surfaces (such as the surface of a sphere) or to lie along certain curves in space. In such cases the allowed motions, and hence the number of degrees of freedom, are reduced, and the original coordinates used to describe the motion are not independent of each other. The idea of constrained motions is illustrated by some examples.

Example 6.7 The Simple Pendulum (Example 2.1). The planar motion of the simple pendulum (Fig 2.4) can be described by the cartesian coordinates x and y. These coordinates, however, are not independent but are related by the geometric constraint relation

$$x^2 + y^2 = L^2 \tag{6.38}$$

Clearly x and y cannot change independently of each other. Any change in x (or y) must be accompanied by an appropriate change in y (or x), according to (6.38). We also note that the assumption of planar motion provides another constraint, $z = 0$. Normally, we do not state this type of constraint mathematically; it is just understood. The pendulum, therefore, has only 1 degree of freedom.

Example 6.8 The Rolling Ball (Examples 4.6 and 4.12). The rolling ball, Fig. 4.4, is constrained to move in the x-y plane and to lie on the curve defined by

$$y(x) = -\tfrac{1}{2}x^2 + \tfrac{1}{4}x^4 \tag{6.39}$$

which is the geometric constraint relation for this system (along with $z = 0$, understood). Observe once again that the coordinates x and y are not independent, and the system has a single degree of freedom.

In general, for systems in which all of the constraints are of the form of algebraic equations, such as (6.38) and (6.39), the number of degrees of freedom will equal the theoretical maximum (e.g., three for a single particle) less the number of constraint relations.*

In physical terms constraints are always accompanied by forces of constraint, which must arise so as to maintain the motion of the system along the constrained path or on the constrained surface. For example, for the rolling ball of Example 6.8, a constraint force is exerted by the surface on the ball. This constraint force is always *normal* to the instantaneous direction of travel. Constraint forces are real forces that must be included in a newtonian analysis. But, as we will see, they do not appear in the lagrangian formulation.

6.3.3 Virtual Displacement and Virtual Work

We have so far introduced the ideas of constraints and degrees of freedom for the single particle whose motion is defined by the cartesian coordinates x, y, and z. We now come to the central concept underlying lagrangian mechanics, the idea of "virtual change" in a quantity. We will introduce this idea by considering first the *virtual displacements* for the cartesian particle of Sec. 6.3.1. Suppose that at some time t the particle coordinates are $x(t)$, $y(t)$, and $z(t)$; there may or may not be constraints. We define a set of virtual displacements, denoted by δx, δy, and δz, as a set of infinitesimal displacements, consistent with any constraints but otherwise completely arbitrary, which we impose on the system in a kind of mathematical experiment. Furthermore, the virtual displacements are imposed with the time t held fixed ($\delta t \equiv 0$). What we are doing is to "pretend" that at time t the system is in a configuration slightly different from the actual one. Because the set of virtual displacements is arbitrary, as long as they do not violate any constraints, we see that the number of allowable sets of virtual displacements is infinite. What we will do is to address the question, "By considering all possible sets of virtual displacements, can we deduce anything about the actual motion of the system?"

It is important to note that the operation δ of virtual change is distinctly different from the differential operator d of calculus. Specifically, when we write

*Algebraic equality constraints are not the only possible type. For instance, a particle constrained to move inside a sphere of radius R is constrained according to the *inequality* $x^2 + y^2 + z^2 < R^2$, but the particle still has 3 degrees of freedom. We will consider only constraints which are of the form of algebraic equations.

out the differentials dx, dy, and dz, we are referring to *actual* changes in the coordinates x, y, and z occurring in a time interval dt. The virtual displacements, on the other hand, are *imagined*, *arbitrary* displacements imposed with the time held fixed.

Comment. Even though the δ and d operators signify much different things in physical terms, the virtual change operator δ follows the same mathematical rules as the differential operator. Thus, just as the differential of x^2 is $d(x^2) = 2x\,dx$, the virtual change of x^2 is $\delta(x^2) = 2x\,\delta x$. Furthermore, the operations of differentiation and virtual change commute, so that

$$\delta\left(\frac{dx}{dt}\right) = \frac{d}{dt}(\delta x)$$

As stated above, the virtual displacements must not violate any constraints. Thus, if we consider the three-dimensional motion of an unconstrained particle, then δx, δy, and δz are independent of each other. But if a constraint exists, then δx, δy, and δz are not independent. For example, consider the rolling ball, constrained according to (6.39). Applying the virtual change operator δ to this equation, we see that the permissible variations for δx and δy must satisfy

$$\delta y = -x\delta x + x^3\delta x = (-x + x^3)\,\delta x \tag{6.40}$$

Satisfaction of (6.40) will guarantee that the imposed virtual displacement will move the particle (in thought only, of course!) to a location on the curve of constraint; a pair $(\delta x, \delta y)$ not satisfying (6.40) will move the mass to a location *not on* the surface (6.39), and this is not allowed.

Now let us consider the *virtual work* done by allowing all of the applied forces existing at time t to move through the imposed virtual displacements. Because the time is held fixed as we impose our virtual displacement, the applied forces are also to be held fixed; these forces merely assume whatever their values happen to be at the time in question. We then define the virtual work δW as the work done by all of the applied forces in a virtual displacement $(\delta x, \delta y, \delta z)$. For the single particle we have then

$$\delta W = F_x\delta x + F_y\delta y + F_z\delta z \tag{6.41}$$

This quantity is a scalar and involves the effects of all of the applied forces. Note, however, that constraint forces will do no work in a virtual displacement, because such forces are always *normal* to the net virtual displacement. (In fact, constraint forces may be *defined* as forces which do no work in a virtual displacement.)

The virtual work δW can be related to the particle motion by using (6.35) as follows: Multiply (6.35*a*) by δx, (6.35*b*) by δy, and (6.35*c*) by δz and add the results to obtain

$$m(\ddot{x}\delta x + \ddot{y}\delta y + \ddot{z}\delta z) = F_x\delta x + F_y\delta y + F_z\delta z \tag{6.42}$$

Equation (6.42) is "Lagrange's form of D'Alembert's principle." In the *static* case, for which no motion occurs and the accelerations $\ddot{x}$, $\ddot{y}$, $\ddot{z}$ are zero, (6.42) becomes the principle of virtual work of statics, which may be stated as follows: For a system in static equilibrium, the virtual work done by the applied forces in a virtual displacement from the equilibrium configuration will be zero.

Equation (6.42) will provide the starting point for our development of Lagrange's equations. Before going through this development, however, we need to broaden our view of the coordinates used to describe the motion of a mechanical system.

6.3.4 Generalized Coordinates

So far in this section we have restricted attention to the motion of a single particle described by cartesian coordinates x, y, and z. It is clear, however, that we may use other types of coordinates to describe the particle motion. For example, we developed the newtonian laws for planar motion in terms of the polar coordinates r and θ in Secs. 2.1.1 and 6.2.2. Many other possibilities exist, such as cylindrical coordinates and spherical coordinates (for instance, the most convenient way to represent motion constrained to lie on the surface of a sphere is through use of the angles defining latitude and longitude). In fact, any set of quantities that allow us to specify unambiguously the system configuration will qualify as usable coordinates. Let us define such a set of quantities as *generalized coordinates* and denote them, for the particle in 3-D, as q_1, q_2, and q_3 (q's are historically used to denote the generalized coordinates, and this practice is followed here).

Example 6.9 A Particle Moving in the Plane. An alien spaceship moves in the plane over terrain which consists of a slope of fixed angle α (Fig. 6.20). An observer at **O** might monitor the position of the spaceship by measuring the distance q_1 along the slope to the point S directly underneath the spaceship and then measuring the vertical distance q_2 of the spaceship above the point S (Fig. 6.20).

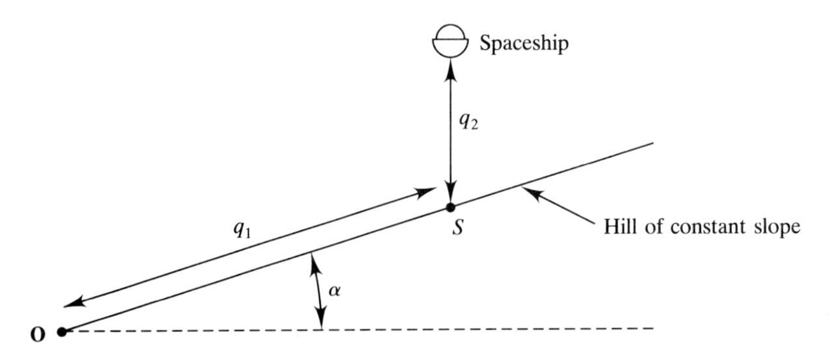

FIGURE 6.20
Generalized coordinates for the alien spaceship.

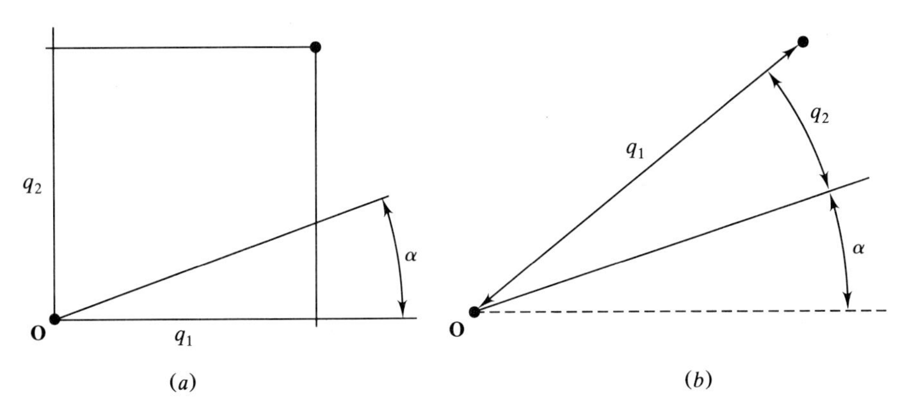

FIGURE 6.21
(a) Cartesian and (b) polar coordinates for the alien spaceship.

Other obvious choices of generalized coordinates include the cartesian set [Fig. 6.21(a)] and polar coordinates [Fig. 6.21(b)].

Clearly, for a given system, there may be numerous sets of generalized coordinates which can be used to describe the motion. We want to select the one which is most convenient for our purposes.

In the ensuing development of Lagrange's equations for a single particle, we will take (6.42) in cartesians as the starting point, but we will want to develop a result which is applicable for any set of generalized coordinates that we decide to use. We will, therefore, need to know the manner in which the cartesians and the generalized coordinates are related. Specifically, we'll assume that we can determine a coordinate transformation that gives us the cartesian coordinates x, y, and z if we specify the generalized coordinates q_1, q_2, and q_3:

$$
\begin{aligned}
x &= x(q_1, q_2, q_3) \\
y &= y(q_1, q_2, q_3) \\
z &= z(q_1, q_2, q_3)
\end{aligned}
\tag{6.43}
$$

For instance, in plane polar coordinates, with $r = q_1$ and $\theta = q_2$, the coordinate transformation would be

$$
\begin{aligned}
x &= q_1 \cos q_2 \\
y &= q_1 \sin q_2 \\
z &= 0
\end{aligned}
\tag{6.44}
$$

whereas for the "spaceship coordinates" of Fig. 6.20, we would have

$$
\begin{aligned}
x &= q_1 \cos \alpha \\
y &= q_1 \sin \alpha + q_2 \\
z &= 0
\end{aligned}
\tag{6.45}
$$

Using coordinate transformations of this type, we can relate the virtual displacements δx, δy, and δz of the cartesians to the virtual displacements δq_1, δq_2, and δq_3 of the generalized coordinates. For example, applying the virtual change operator to (6.44), we have

$$\delta x = \delta q_1 \cos q_2 - \delta q_2 q_1 \sin q_2$$

$$\delta y = \delta q_1 \sin q_2 + \delta q_2 q_1 \cos q_2$$

$$\delta z = 0$$

For the general case (6.43), application of the chain rule leads to the result

$$\delta x = \frac{\partial x}{\partial q_1}\delta q_1 + \frac{\partial x}{\partial q_2}\delta q_2 + \frac{\partial x}{\partial q_3}\delta q_3$$

$$\delta y = \frac{\partial y}{\partial q_1}\delta q_1 + \frac{\partial y}{\partial q_2}\delta q_2 + \frac{\partial y}{\partial q_3}\delta q_3$$

$$\delta z = \frac{\partial z}{\partial q_1}\delta q_1 + \frac{\partial z}{\partial q_2}\delta q_2 + \frac{\partial z}{\partial q_3}\delta q_3 \tag{6.46}$$

6.3.5 Lagrange's Equations

Next we will use the ideas introduced here to develop Lagrange's equations for the single particle from (6.42). The strategy will be to reexpress (6.42) in terms of the generalized coordinates and their virtual displacements and to manipulate the terms so that they involve the particle kinetic energy, potential energy, and work done by nonconservative forces.

Let us begin by substituting (6.46) into (6.42), so that all virtual displacements involve the generalized coordinates. We then obtain

$$m\left[\ddot{x}\left(\frac{\partial x}{\partial q_1}\delta q_1 + \frac{\partial x}{\partial q_2}\delta q_2 + \frac{\partial x}{\partial q_3}\delta q_3\right) + \ddot{y}\left(\frac{\partial y}{\partial q_1}\delta q_1 + \frac{\partial y}{\partial q_2}\delta q_2 + \frac{\partial y}{\partial q_3}\delta q_3\right)\right.$$
$$\left. + \ddot{z}\left(\frac{\partial z}{\partial q_1}\delta q_1 + \frac{\partial z}{\partial q_2}\delta q_2 + \frac{\partial z}{\partial q_3}\delta q_3\right)\right]$$
$$= F_x\left(\frac{\partial x}{\partial q_1}\delta q_1 + \frac{\partial x}{\partial q_2}\delta q_2 + \frac{\partial x}{\partial q_3}\delta q_3\right) + F_y\left(\frac{\partial y}{\partial q_1}\delta q_1 + \frac{\partial y}{\partial q_2}\delta q_2 + \frac{\partial y}{\partial q_3}\delta q_3\right)$$
$$+ F_z\left(\frac{\partial z}{\partial q_1}\delta q_1 + \frac{\partial z}{\partial q_2}\delta q_2 + \frac{\partial z}{\partial q_3}\delta q_3\right)$$

Now group together the terms involving δq_1, the terms involving δq_2, and the terms involving δq_3 to obtain the following:

$$\delta q_1\left\{m\left(\ddot{x}\frac{\partial x}{\partial q_1}+\ddot{y}\frac{\partial y}{\partial q_1}+\ddot{z}\frac{\partial z}{\partial q_1}\right)-F_x\frac{\partial x}{\partial q_1}-F_y\frac{\partial y}{\partial q_1}-F_z\frac{\partial z}{\partial q_1}\right\}$$

$$+\delta q_2\left\{m\left(\ddot{x}\frac{\partial x}{\partial q_2}+\ddot{y}\frac{\partial y}{\partial q_2}+\ddot{z}\frac{\partial z}{\partial q_2}\right)-F_x\frac{\partial x}{\partial q_2}-F_y\frac{\partial y}{\partial q_2}-F_z\frac{\partial z}{\partial q_2}\right\}$$

$$+\delta q_3\left\{m\left(\ddot{x}\frac{\partial x}{\partial q_3}+\ddot{y}\frac{\partial y}{\partial q_3}+\ddot{z}\frac{\partial z}{\partial q_3}\right)-F_x\frac{\partial x}{\partial q_3}-F_y\frac{\partial y}{\partial q_3}-F_z\frac{\partial z}{\partial q_3}\right\}=0 \tag{6.47}$$

We now consider the coefficient of δq_1 only. We will manipulate this coefficient into a useful result, and then the analogous results for the coefficients of δq_2 and δq_3 will be evident. First we split the applied forces F_x, F_y, and F_z into conservative and nonconservative forces, $F_x = F_{x_c} + F_{x_{nc}}$ and so on. The conservative forces F_{x_c}, F_{y_c}, and F_{z_c} are defined by the partial derivatives of the potential energy (6.37), so that the coefficient of δq_1, denoted as c_1, may be written

$$c_1 = m\left(\ddot{x}\frac{\partial x}{\partial q_1}+\ddot{y}\frac{\partial y}{\partial q_1}+\ddot{z}\frac{\partial z}{\partial q_1}\right)+\left(\frac{\partial V}{\partial x}\frac{\partial x}{\partial q_1}+\frac{\partial V}{\partial y}\frac{\partial y}{\partial q_1}+\frac{\partial V}{\partial z}\frac{\partial z}{\partial q_1}\right)$$
$$-F_{x_{nc}}\frac{\partial x}{\partial q_1}-F_{y_{nc}}\frac{\partial y}{\partial q_1}-F_{z_{nc}}\frac{\partial z}{\partial q_1} \tag{6.48}$$

The potential energy V is assumed to have been originally expressed as a function of the cartesian coordinates x, y, and z. In view of the coordinate transformation (6.43), however, V may also be expressed as a function of the generalized coordinates q_1, q_2, and q_3. If we write this dependence as $V[x(q_1, q_2, q_3), y(q_1, q_2, q_3), z(q_1, q_2, q_3)]$, then we observe that the second term of c_1 is simply the partial derivative of V with respect to q_1:

$$\frac{\partial V}{\partial q_1}=\frac{\partial V}{\partial x}\frac{\partial x}{\partial q_1}+\frac{\partial V}{\partial y}\frac{\partial y}{\partial q_1}+\frac{\partial V}{\partial z}\frac{\partial z}{\partial q_1} \tag{6.49}$$

according to the chain rule. Thus, (6.48) simplifies to

$$c_1 = m\left(\ddot{x}\frac{\partial x}{\partial q_1}+\ddot{y}\frac{\partial y}{\partial q_1}+\ddot{z}\frac{\partial z}{\partial q_1}\right)+\frac{\partial V}{\partial q_1}$$
$$-F_{x_{nc}}\frac{\partial x}{\partial q_1}-F_{y_{nc}}\frac{\partial y}{\partial q_1}-F_{z_{nc}}\frac{\partial z}{\partial q_1} \tag{6.50}$$

Let us next focus attention on the nonconservative force terms in (6.51). The product

$$\left(F_{x_{nc}} \frac{\partial x}{\partial q_1} + F_{y_{nc}} \frac{\partial y}{\partial q_1} + F_{z_{nc}} \frac{\partial z}{\partial q_1} \right) \delta q_1$$

appearing in (6.47) is the virtual work δW_{nc} done by all nonconservative forces in a virtual displacement δq_1, with $\delta q_2 = \delta q_3 = 0$. Let us denote this virtual work as

$$\delta W_{nc_1} \equiv Q_1(t)\, \delta q_1 \tag{6.51}$$

The quantity $Q_1(t)$ is a "generalized force," associated with the generalized coordinate q_1:

$$Q_1(t) = F_{x_{nc}} \frac{\partial x}{\partial q_1} + F_{y_{nc}} \frac{\partial y}{\partial q_1} + F_{z_{nc}} \frac{\partial z}{\partial q_1} \tag{6.52}$$

Thus $Q_1(t)$ is simply the function of proportionality defining the amount of virtual work done by all nonconservative forces in a virtual displacement δq_1. With this definition, (6.50) simplifies to

$$c_1 = m\left(\ddot{x} \frac{\partial x}{\partial q_1} + \ddot{y} \frac{\partial y}{\partial q_1} + \ddot{z} \frac{\partial z}{\partial q_1} \right) + \frac{\partial V}{\partial q_1} - Q_1(t) \tag{6.53}$$

Finally, let us consider the acceleration terms in (6.53). The next step is to manipulate these terms so that they involve the system kinetic energy. First, consider the term $\ddot{x}\partial x/\partial q_1$. This term may be written as (verify!)

$$\ddot{x} \frac{\partial x}{\partial q_1} = \frac{d}{dt}\left(\dot{x} \frac{\partial x}{\partial q_1} \right) - \dot{x} \frac{d}{dt}\left(\frac{\partial x}{\partial q_1} \right) \tag{6.54}$$

Switching the order of differentiation in the last term in (6.54) yields

$$\ddot{x} \frac{\partial x}{\partial q_1} = \frac{d}{dt}\left(\dot{x} \frac{\partial x}{\partial q_1} \right) - \dot{x} \frac{\partial}{\partial q_1}\left(\frac{dx}{dt} \right) \tag{6.55}$$

We can also express the first term on the right-hand side of (6.55) differently by first noting that the cartesian velocity component $\dot{x}$ may be expressed, according to the chain rule, as

$$\dot{x} = \frac{\partial x}{\partial q_1}\dot{q}_1 + \frac{\partial x}{\partial q_2}\dot{q}_2 + \frac{\partial x}{\partial q_3}\dot{q}_3$$

Now if we compute the partial derivative of this equation with respect to the generalized velocity $\dot{q}_1$, we obtain the useful result

$$\frac{\partial \dot{x}}{\partial \dot{q}_1} = \frac{\partial x}{\partial q_1} \tag{6.56}$$

This result can be used to rewrite the first term on the right-hand side of (6.55) so that

$$\ddot{x}\frac{\partial x}{\partial q_1} = \frac{d}{dt}\left(\dot{x}\frac{\partial \dot{x}}{\partial \dot{q}_1}\right) - \dot{x}\frac{\partial \dot{x}}{\partial q_1} \tag{6.57}$$

Next we note that the term $\dot{x}\partial\dot{x}/\partial\dot{q}_1$ may be expressed as

$$\dot{x}\frac{\partial \dot{x}}{\partial \dot{q}_1} = \frac{\partial}{\partial \dot{q}_1}\left(\frac{1}{2}\dot{x}^2\right) \tag{6.58}$$

and, similarly, that the term $\dot{x}\partial\dot{x}/\partial q_1$ may be expressed as

$$\dot{x}\frac{\partial \dot{x}}{\partial q_1} = \frac{\partial}{\partial q_1}\left(\frac{1}{2}\dot{x}^2\right) \tag{6.59}$$

Use of (6.58) and (6.59) in (6.57) results in

$$\ddot{x}\frac{\partial x}{\partial q_1} = \frac{d}{dt}\left[\frac{\partial}{\partial \dot{q}_1}\left(\frac{1}{2}\dot{x}^2\right)\right] - \frac{\partial}{\partial q_1}\left(\frac{1}{2}\dot{x}^2\right) \tag{6.60}$$

Thus, we have now converted one of the three acceleration terms appearing in (6.53) into the form (6.60). If we execute the same steps [(6.54) through (6.60)] for the remaining two acceleration terms in (6.53), then (6.53) may be reexpressed as

$$c_1 = \frac{d}{dt}\left\{\frac{\partial}{\partial \dot{q}_1}\left[\frac{1}{2}m\left(\dot{x}^2 + \dot{y}^2 + \dot{z}^2\right)\right]\right\} - \frac{\partial}{\partial q_1}\left[\frac{1}{2}m\left(\dot{x}^2 + \dot{y}^2 + \dot{z}^2\right)\right] + \frac{\partial V}{\partial q_1} - Q_1(t) \tag{6.61}$$

where the mass m has been moved inside the differential operators. We have now achieved our objective of reformulating the inertia terms in (6.53) in terms of the kinetic energy $T = \frac{1}{2}m(\dot{x}^2 + \dot{y}^2 + \dot{z}^2)$, and (6.61) may be expressed in the concise form

$$c_1 = \frac{d}{dt}\left(\frac{\partial T}{\partial \dot{q}_1}\right) - \frac{\partial T}{\partial q_1} + \frac{\partial V}{\partial q_1} - Q_1(t) \tag{6.62}$$

Recalling that c_1 is the coefficient of δq_1 in (6.47) and recognizing that the coefficients of δq_2 and δq_3 in (6.47) will be given by relations analogous to

(6.62), we may rewrite (6.47) as follows:

$$\begin{aligned}
&\delta q_1\left[\frac{d}{dt}\left(\frac{\partial T}{\partial \dot{q}_1}\right) - \frac{\partial T}{\partial q_1} + \frac{\partial V}{\partial q_1} - Q_1(t)\right] \\
&+\delta q_2\left[\frac{d}{dt}\left(\frac{\partial T}{\partial \dot{q}_2}\right) - \frac{\partial T}{\partial q_2} + \frac{\partial V}{\partial q_2} - Q_2(t)\right] \\
&+\delta q_3\left[\frac{d}{dt}\left(\frac{\partial T}{\partial \dot{q}_3}\right) - \frac{\partial T}{\partial q_3} + \frac{\partial V}{\partial q_3} - Q_3(t)\right] = 0
\end{aligned} \tag{6.63}$$

Equation (6.63), which involves only the generalized coordinates, now replaces the cartesian version (6.42) of Lagrange's form of D'Alembert's principle. Equation (6.63) must be valid for any set of virtual displacements $\delta q_1, \delta q_2, \delta q_3$ which we impose. Because the virtual displacements δq_1, δq_2, and δq_3 are *independent* and *arbitrary* (so long as infinitesimal), the only way in which (6.63) can be satisfied generally is to require that the coefficients of δq_1, δq_2, and δq_3 vanish identically. If we did not invoke this requirement, then we could always find a set of virtual displacements δq_i for which (6.63) would be violated. We conclude that (6.63) is equivalent to the following set of three differential equations, one for each of the generalized coordinates:

$$\frac{d}{dt}\left(\frac{\partial T}{\partial \dot{q}_1}\right) - \frac{\partial T}{\partial q_1} + \frac{\partial V}{\partial q_1} = Q_1(t) \tag{6.64a}$$

$$\frac{d}{dt}\left(\frac{\partial T}{\partial \dot{q}_2}\right) - \frac{\partial T}{\partial q_2} + \frac{\partial V}{\partial q_2} = Q_2(t) \tag{6.64b}$$

$$\frac{d}{dt}\left(\frac{\partial T}{\partial \dot{q}_3}\right) - \frac{\partial T}{\partial q_3} + \frac{\partial V}{\partial q_3} = Q_3(t) \tag{6.64c}$$

These three equations, termed *Lagrange's equations*, are the equations of motion of the particle, in terms of the generalized coordinates q_1, q_2, and q_3, and they replace the cartesian newtonian equations (6.35) as the starting point for the analysis of particle motion. Note that the equations are identical except for the subscripts, so that the results (6.64) may be compactly expressed as the single equation

$$\frac{d}{dt}\left(\frac{\partial T}{\partial \dot{q}_i}\right) - \frac{\partial T}{\partial q_i} + \frac{\partial V}{\partial q_i} = Q_i(t) \qquad i = 1, 2, 3 \tag{6.65}$$

The equations of motion (6.65) for the particle will consist of a set of three second-order ODEs; the second-order terms arise in $(d/dt)(\partial T/\partial \dot{q}_i)$.

Comment. In deriving (6.65) from (6.42), we assumed the existence of the coordinate transformation (6.43) connecting the cartesian and generalized coordinates. In the derivation we also made use of the nine partial derivatives $(\partial x/\partial q_1) \cdots (\partial z/\partial q_3)$. In actually using Lagrange's equations to develop a mathematical model of particle motion, however, we will never actually have to define the transformation (6.43), and we will never have to calculate the partial derivatives $(\partial x/\partial q_1) \cdots (\partial z/\partial q_3)$. These were introduced only to facilitate the derivation of (6.65) from (6.42) in general form.

In applying Lagrange's equations for a single unconstrained particle, we would implement the following steps:

1. Define a set of (independent) generalized coordinates q_1, q_2, and q_3.
2. Determine the particle kinetic energy T as a function of the generalized velocities $\dot{q}_1$, $\dot{q}_2$, and $\dot{q}_3$ and the generalized coordinates q_1, q_2, and $q_3, T = T(q_i, \dot{q}_i)$.
3. Determine the potential energy V as a function of the generalized coordinates q_1, q_2, and q_3. This function $V = V(q_i)$ will characterize the influence of all conservative forces.
4. Determine the generalized forces $Q_1(t)$, $Q_2(t)$, and $Q_3(t)$, which characterize the influence of all nonconservative forces. This can be most easily done as follows: Impose a virtual displacement δq_1, holding δq_2 and δq_3 equal to zero. Then determine the virtual work δW_{nc}, done by all nonconservative forces, as these forces are moved through the virtual displacement δq_1. The result will be of the form (6.51), and $Q_1(t)$ is then easily identified. Then do similar operations for δq_2 and δq_3 to obtain $Q_2(t)$ and $Q_3(t)$.
5. Perform the mathematical operations indicated in (6.65) to obtain the equations of motion.

The basic result, Eq. (6.65), has been derived for the single particle. It turns out, however, that this result applies for *any* mechanical system, no matter how complex. In the general case in which we define the system motion in terms of a set of n independent generalized coordinates, T and V are the kinetic and potential energies of the *system as a whole* and we apply Lagrange's equation (6.65) for each of the n generalized coordinates.

Let us now consider some examples. The objectives in presenting these examples will be to: (1) illustrate the basic steps outlined above in applying Lagrange's equations to develop mathematical models of mechanical systems and (2) show that for certain types of systems, notably those with geometric constraints, Lagrange's equations provide a much more direct route to the system equations of motion than do the newtonian mechanics.

Example 6.10 The Harmonic Oscillator (Example 2.3). We begin with the undamped, unforced linear oscillator shown in Fig. 6.22. The single generalized coordinate is the displacement x (we can denote x by q if we want). There are no nonconservative forces, so $Q_1(t) = 0$ and the single Lagrange's equation for the

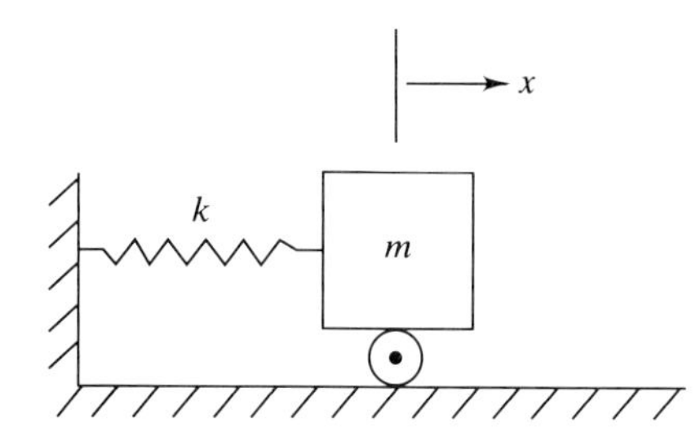

FIGURE 6.22
The undamped, unforced linear oscillator.

system is

$$\frac{d}{dt}\left(\frac{\partial T}{\partial \dot{x}}\right) - \frac{\partial T}{\partial x} + \frac{\partial V}{\partial x} = 0 \tag{6.66}$$

The kinetic and potential energies are

$$T = \tfrac{1}{2}m\dot{x}^2; \qquad V = \tfrac{1}{2}kx^2$$

We may then list the required derivatives appearing in Lagrange's equation (6.66) as follows:

$$\frac{\partial T}{\partial \dot{x}} = m\dot{x}$$

$$\frac{d}{dt}\left(\frac{\partial T}{\partial \dot{x}}\right) = m\ddot{x}$$

$$\frac{\partial V}{\partial x} = kx$$

$$\frac{\partial T}{\partial x} = 0$$

Combining these terms according to (6.66), we obtain the familiar equation of motion

$$m\ddot{x} + kx = 0$$

as would also be obtained via Newton's second law. Note that the term $\partial T/\partial x$ does not contribute anything to the equation of motion because the kinetic energy T is a function of velocity only (this will always occur when we use cartesian coordinates which are independent).

Now let us add some linear damping and a time-dependent force $F(t)$, assumed specified, so that we have the system shown in Fig. 6.23.

The single generalized coordinate is x, and we now have two nonconservative forces. The kinetic and potential energies will be the same as for the undamped, unforced oscillator, so we need only consider the virtual work done by the nonconservative forces. These are determined by imposing a virtual displacement δx and computing the resulting virtual work due to the nonconservative forces. The imposed virtual displacement, as well as the nonconservative forces, are illustrated in Fig. 6.24.

The virtual work δW_{nc} done by the nonconservative forces is given by

$$\delta W_{nc} = F(t)\delta x - c\dot{x}\delta x = (F(t) - c\dot{x})\delta x$$

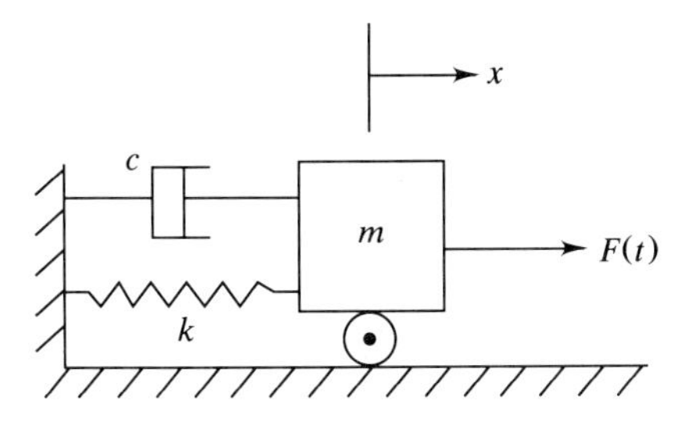

FIGURE 6.23
The damped, forced linear oscillator.

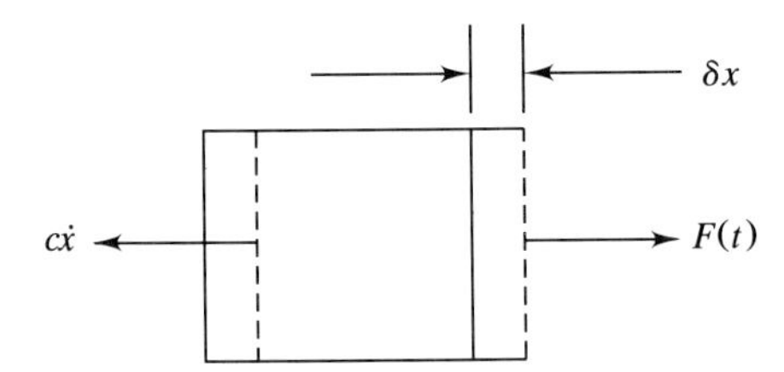

FIGURE 6.24
The virtual displacement δx. The solid lines represent the system position x at any time t. The dashed lines represent the hypothetical displacement $x + \delta x$, at the same time, after the imposition of the virtual displacement δx.

so the generalized force Q_x is the coefficient of δx, in accordance with (6.51)

$$Q_x(t) = F(t) - c\dot{x}$$

and this is seen to be merely the sum of the two nonconservative forces. It is important to keep in mind that all forces are held *fixed* as the virtual displacement is imposed. The equation of motion (6.65) for the system is

$$m\ddot{x} + kx = F(t) - c\dot{x}$$

or, in the usual form,

$$m\ddot{x} + c\dot{x} + kx = F(t)$$

Example 6.10 is intended only to introduce the basic steps in the use of Lagrange's equation. There is no advantage, as compared to the newtonian formulation, in using Lagrange's equation for this system.

Example 6.11 Particle Motion in Polar Coordinates (Sec. 6.2.2). Let us now use Lagrange's equations to derive the general equations of motion (6.16) for the plane motion of a particle using polar coordinates, as shown in Fig. 6.25.

The two generalized coordinates are r and θ. We have included the force due to gravity, and we assume that the nonconservative forces have been resolved into the radial and tangential components F_r and F_θ.

In order to obtain the system kinetic energy, recall that the inertial velocity vector $\mathbf{v}$ is given by (6.14):

$$\mathbf{v} = \dot{r}\mathbf{e}_r + r\dot{\theta}\mathbf{e}_\theta$$

expressed in terms of the rotating unit vectors $\mathbf{e}_r$ and $\mathbf{e}_\theta$. The kinetic energy T of the particle is then given by (2.34):

$$T = \tfrac{1}{2}m\mathbf{v}\cdot\mathbf{v} = \tfrac{1}{2}m(\dot{r}^2 + r^2\dot{\theta}^2) \qquad (6.67)$$

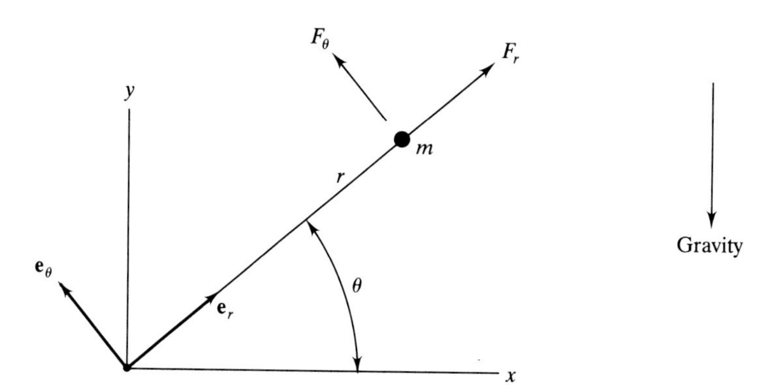

FIGURE 6.25
Planar motion of a particle in terms of polar coordinates r and θ.

Note that T depends not only on the generalized velocities $\dot{r}$ and $\dot{\theta}$, but also on the generalized coordinate r. The potential energy V due to gravity is

$$V = mgr \sin \theta \tag{6.68}$$

where we have arbitrarily assigned V to be zero when $\theta = 0$, that is, when the particle is on the x axis.

The generalized force Q_r is obtained by imposing a virtual displacement δr [Fig. 6.26(*a*)], with $\delta\theta = 0$.

The virtual work δW_{nc_r} done by the nonconservative forces is then

$$\delta W_{nc_r} = F_r \delta_r$$

Note that the tangential force F_θ does no virtual work in a virtual displacement δr, as F_θ is normal to the direction of δr. The generalized force Q_r is then $Q_r = F_r$.

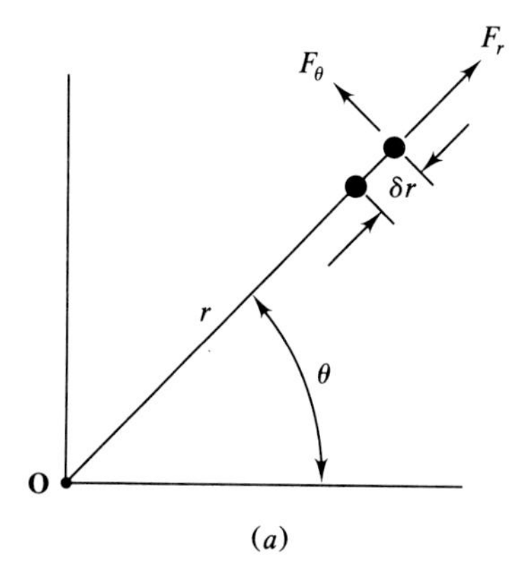

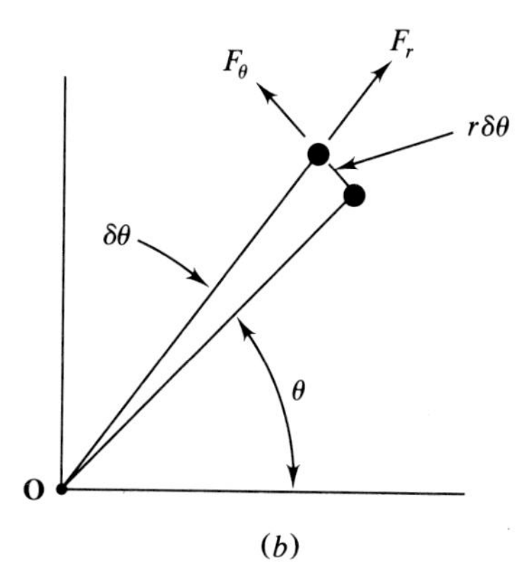

FIGURE 6.26
Virtual displacements: (*a*) δr; (*b*) $\delta\theta$.

Next we impose a virtual displacement $\delta\theta$, with $\delta r = 0$, as shown in Fig. 6.26(*b*). In this case the tangential force F_θ is moved through a distance $r\delta\theta$, so that the associated virtual work δW_{nc_θ} is

$$\delta W_{nc_\theta} = F_\theta r \delta\theta$$

so the generalized force $Q_\theta = rF_\theta$. This is actually a moment, exerted about the origin, because the associated generalized coordinate is a rotation. Note that F_r does no virtual work in a virtual displacement $\delta\theta$.

We now perform the required differential operations on T and V, according to (6.65), to obtain the equations of motion. For the generalized coordinate $q_1 = r$, the required terms are listed below:

$$\frac{\partial T}{\partial \dot{r}} = m\dot{r}$$

$$\frac{d}{dt}\left(\frac{\partial T}{\partial \dot{r}}\right) = m\ddot{r}$$

$$\frac{\partial T}{\partial r} = mr\dot{\theta}^2$$

$$\frac{\partial V}{\partial r} = mg \sin\theta$$

The first equation of motion is, therefore,

$$m\ddot{r} - mr\dot{\theta}^2 + mg\sin\theta = F_r(t) \tag{6.69}$$

Note that the centrifugal term $-mr\dot{\theta}^2$ arises in the $\partial T/\partial r$ term in Lagrange's equation for r and thus has its source in the system kinetic energy.

For the θ coordinate the required terms in Lagrange's equation are listed below:

$$\frac{\partial T}{\partial \dot{\theta}} = mr^2\dot{\theta}$$

$$\frac{d}{dt}\left(\frac{\partial T}{\partial \dot{\theta}}\right) = mr^2\ddot{\theta} + 2mr\dot{r}\dot{\theta}$$

$$\frac{\partial T}{\partial \theta} = 0$$

$$\frac{\partial V}{\partial \theta} = mgr\cos\theta$$

The second equation of motion is thus

$$mr^2\ddot{\theta} + 2mr\dot{r}\dot{\theta} + mgr\cos\theta = rF_\theta(t) \tag{6.70}$$

If one divides (6.70) by r and eliminates gravity in (6.69) and (6.70), one obtains the same equations of motion (2.12) and (6.16) derived previously using the newtonian formulation.

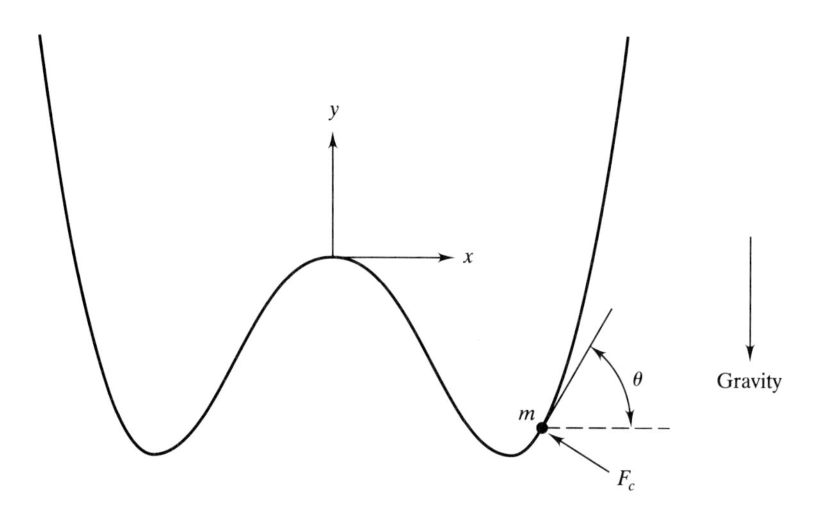

FIGURE 6.27
The rolling ball in a double wall.

Example 6.12 The Rolling Ball (Example 4.6). We now derive the equation of motion (4.15) of the rolling ball in a double well (Fig. 6.27).

This example was introduced in Chap. 4, but we did not derive the equation of motion. The ball is assumed always to lie on the surface defined by

$$y = -\tfrac{1}{2}x^2 + \tfrac{1}{4}x^4 \tag{6.71}$$

We thus have a geometric constraint, and, although the particle moves in the plane, there is only 1 degree of freedom, as the motion is constrained to lie on the particular curve in the plane defined by (6.71). Because there is a single degree of freedom, we need only one generalized coordinate, which we take to be the horizontal position x. The strategy will be to express the kinetic and potential energies T and V as functions of only $\dot{x}$ and x and then to apply Lagrange's equation for the single generalized coordinate x.

First, the potential energy V due to gravity is defined by

$$V = mgy$$

where V is assigned the value zero when $y = 0$. The constraint relation (6.71) allows us to express V as a function of x as

$$V = mg\left(-\tfrac{1}{2}x^2 + \tfrac{1}{4}x^4\right) \tag{6.72}$$

The kinetic energy T of a particle moving in the plane is defined by

$$T = \tfrac{1}{2}m\left(\dot{x}^2 + \dot{y}^2\right) \tag{6.73}$$

This relation is true generally, whether or not the particle is constrained. In the present case, however, x and y are not independent, so we will use the constraint relation to express T as a function of $\dot{x}$ and x only. This is done by differentiating the constraint relation (6.71) with respect to time, which yields

$$\dot{y} = -x\dot{x} + x^3\dot{x} = \dot{x}(x^3 - x)$$

so that

$$\dot{y}^2 = \dot{x}^2(x^3 - x)^2$$

Use of this result to eliminate $\dot{y}^2$ in (6.73) then gives the kinetic energy T as

$$T = \tfrac{1}{2}m\left[\dot{x}^2 + \dot{x}^2(x^3 - x)^2\right] \tag{6.74}$$

Now there are assumed to be no nonconservative forces present, so that (6.66) applies. The terms in Lagrange's equation are listed below:

$$\frac{\partial T}{\partial \dot{x}} = m\dot{x}\left[1 + (x^3 - x)^2\right]$$

$$\frac{d}{dt}\left(\frac{\partial T}{\partial \dot{x}}\right) = m\ddot{x}\left[1 + (x^3 - x)^2\right] + m\dot{x}\left[2(x^3 - x)(3x^2 - 1)\dot{x}\right]$$

$$\frac{\partial T}{\partial x} = m\dot{x}^2(x^3 - x)(3x^2 - 1)$$

$$\frac{\partial V}{\partial x} = mgx(-1 + x^2)$$

Combining these terms according to (6.66) and then dividing the result by m, which appears in all terms, lead to

$$\ddot{x}\left[1 + (x^3 - x)^2\right] + \dot{x}^2x\left[(x^2 - 1)(3x^2 - 1)\right] + gx(-1 + x^2) = 0 \tag{6.75}$$

which has been given previously as (4.15).

Comments on Example 6.12

1. The system considered in this example is one for which the lagrangian formulation provides a straightforward route to the equation of motion. This is because the force of constraint F_c does not explicitly have to be taken into account, as F_c does no virtual work in any admissible virtual displacement. In the newtonian formulation, on the other hand, the constraint force F_c has to be included. If we were to derive the equation of motion using the newtonian formulation, we would start with (6.35a and b), which for this system are (see Fig. 6.27)

$$m\ddot{x} = -F_c \sin\theta \tag{6.76a}$$

$$m\ddot{y} = -mg + F_c \cos\theta \tag{6.76b}$$

where the angle θ defines the line of action of the constraint force F_c and is given by

$$\tan\theta = \frac{dy}{dx} = -x + x^3$$

We now have to eliminate both the constraint force F_c and the coordinate y, to obtain a single equation of motion for x. In order to do this, we first rewrite (6.76a) as

$$m\ddot{x} = -F_c \cos\theta \tan\theta \tag{6.77}$$

Next, solving (6.76b) for $F_c \cos\theta$, we obtain

$$F_c \cos\theta = m(\ddot{y} + g) \tag{6.78}$$

Equation (6.78) is then used to replace $F_c \cos\theta$ in (6.77); we also substitute $\tan\theta = -x + x^3$ to obtain

$$m\ddot{x} = -m(\ddot{y} + g)(-x + x^3) \tag{6.79}$$

Finally, we can calculate $\ddot{y}$ by twice differentiating the constraint relation (6.71) with respect to time. The result, after some rearrangement, will be the equation of motion (6.75).

2. It is important to note that we cannot derive the equation of motion by applying Lagrange's equation for both the x and y coordinates. Specifically, suppose we were to try this: Let $T = \frac{1}{2}m(\dot{x}^2 + \dot{y}^2)$, $V = mgy$, and apply Lagrange's equation for x and then for y. We would then obtain the following results:

$$\begin{aligned} m\ddot{x} &= 0 \\ m\ddot{y} &= -mg \end{aligned} \tag{6.80}$$

These equations are incorrect [cf. (6.76)]. The reason for this is clear when we consider the statement (6.63), applied to the present problem, which is (without the generalized forces, which are zero here)

$$\left[\frac{d}{dt}\left(\frac{\partial T}{\partial \dot{x}}\right) - \frac{\partial T}{\partial x} + \frac{\partial V}{\partial x}\right]\delta x + \left[\frac{d}{dt}\left(\frac{\partial T}{\partial \dot{y}}\right) - \frac{\partial T}{\partial y} + \frac{\partial v}{\partial y}\right]\delta y = 0 \tag{6.81}$$

Equation (6.81) is true as stated; however, because δx and δy are *not independent* [$\delta y = (-x + x^3)\delta x$ from (6.71)], we cannot set the coefficients of δx and δy separately to zero, and this is exactly what we have done to obtain (6.80).

3. Based on the previous comment, it should be clear that we must proceed as follows in applying (6.65): Suppose we have a mechanical system for which the motions are described by a total of N coordinates, and suppose that there exist m algebraic constraint relations. The system then has $n = N - m$ degrees of freedom, and we may select any set of n independent coordinates to use as our generalized coordinates $q_1, \ldots, q_n$. The kinetic and potential energies must then be expressed in terms of only these coordinates q_i and their derivatives $\dot{q}_i$. This is done by using the algebraic constraints to eliminate the m "excess coordinates" wherever they appear in T and V, as we did in obtaining (6.74) and (6.72). Once this has been done, we apply Lagrange's equation (6.65) for each of $q_1, \ldots, q_n$, obtaining the n equations of motion.*

---

*It is possible to derive Lagrange's equations in such a way that we do not have to eliminate the m "excess coordinates." This necessitates the use of Lagrange multipliers, resulting in a different form of Lagrange's equations than (6.65). This formulation is beyond the scope of the present exposition.

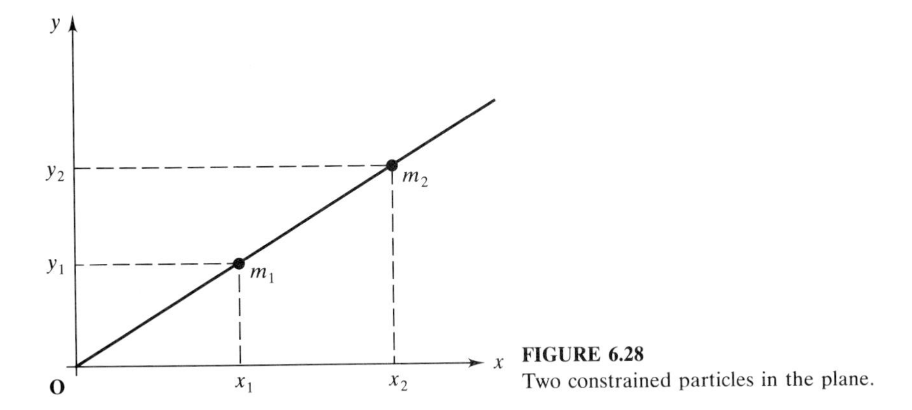

FIGURE 6.28
Two constrained particles in the plane.

Example 6.13 Two Constrained Particles in the Plane. In order to illustrate further the preceding comment 3, consider the motion of two particles m_1 and m_2 in the plane. The configuration of the system is defined by the four coordinates x_1, y_1, x_2, and y_2 (Fig. 6.28).

Suppose there are two algebraic constraints, as follows:

$$x_2 = 2x_1$$
$$y_2 = 2y_1$$

The two particles are restricted to lie on a straight line emanating from the origin **O**, such that m_2 is always twice as far from **O** as is m_1. There are 2 degrees of freedom, and we may select the two generalized coordinates in a number of ways. The following pairs will qualify as generalized coordinates: (x_1, y_1), (x_2, y_2), (x_1, y_2), and (x_2, y_1); the coordinates in each of these pairs are independent of each other. Note, however, that we *cannot* select the following pairs as generalized coordinates because they are not independent: $(x_1, x_2), (y_1, y_2)$.

Example 6.14 Pendulum with Moving Support. Here we consider a variation of the system analyzed in Example 2.2. In the present example a motion of the pendulum support is specified, but this support motion is now vertical rather than horizontal. In addition, we suppose a time-dependent force $F(t)$ to be applied directly to the pendulum mass m, with $F(t)$ always directed in the horizontal direction (Fig. 6.29).

The system has a single degree of freedom; the rotation angle θ will be used as the generalized coordinate. Let us first develop relations for the kinetic and potential energies. First, in terms of the cartesian unit vectors **i** and **j**, the velocity **v** of m may be expressed as

$$\mathbf{v} = (L\dot{\theta}\cos\theta)\mathbf{i} + \left(L\dot{\theta}\sin\theta + \dot{y}_p\right)\mathbf{j}$$

as indicated in Fig. 6.29. The kinetic energy is then $T = \frac{1}{2}m\mathbf{v}\cdot\mathbf{v}$:

$$T = \tfrac{1}{2}m\left[(L\dot{\theta}\cos\theta)^2 + \left(L\dot{\theta}\sin\theta + \dot{y}_p\right)^2\right]$$

or

$$T = \tfrac{1}{2}m\left(L^2\dot{\theta}^2 + 2L\dot{\theta}\dot{y}_p\sin\theta + \dot{y}_p^2\right) \tag{6.82}$$

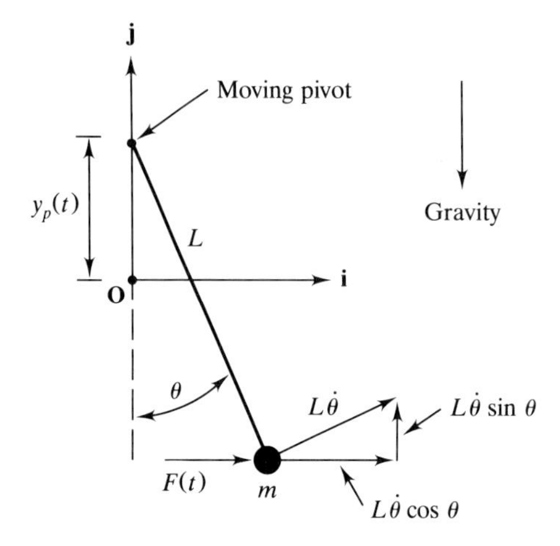

FIGURE 6.29
Pendulum with a specified vertical motion $y_p(t)$ of the support.

The potential energy V is due only to gravity. If we define V to be zero when $\theta = y_p = 0$, then

$$V = mg\left[y_p + L(1 - \cos\theta)\right] \tag{6.83}$$

Observe that both T and V now depend explicitly on time, due to the appearance in T and V of the specified support motion $y_p(t)$.

The generalized force $Q_\theta(t)$ is due to the applied nonconservative force $F(t)$ and is calculated by imposing a virtual displacement $\delta\theta$, as shown in Fig. 6.30.

The virtual work δW_{nc} done by $F(t)$ due to the imposed virtual displacement $\delta\theta$ is seen to be

$$\delta W_{nc} = F(t)L\cos\theta\,\delta\theta$$

Note that, while the mass m moves a distance $L\delta\theta$, only a portion $L\cos\theta\,\delta\theta$ of this distance is along the line of action of the nonconservative force $F(t)$. The

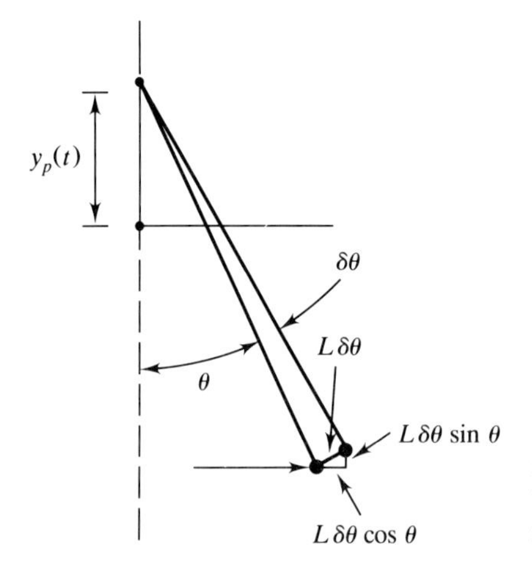

FIGURE 6.30
Virtual displacement $\delta\theta$ for the pendulum with moving support.

generalized force $Q_\theta(t)$ is thus

$$Q_\theta(t) = L\cos\theta F(t)$$

Physically, this is the moment exerted by $F(t)$ about the moving support.

By now performing the differential operations of Lagrange's equation (6.65), we find

$$\frac{\partial T}{\partial \dot{\theta}} = m\left(L^2\dot{\theta} + L\dot{y}_p \sin\theta\right)$$

$$\frac{d}{dt}\left(\frac{\partial T}{\partial \dot{\theta}}\right) = m\left(L^2\ddot{\theta} + L\ddot{y}_p \sin\theta + L\dot{y}_p\dot{\theta}\cos\theta\right)$$

$$\frac{\partial T}{\partial \theta} = mL\dot{\theta}\dot{y}_p \cos\theta$$

$$\frac{\partial V}{\partial \theta} = mgL\sin\theta$$

The equation of motion is, therefore,

$$mL^2\ddot{\theta} + mL\ddot{y}_p \sin\theta + mgL\sin\theta = L\cos\theta F(t) \tag{6.84}$$

This equation of motion is observed to be nonlinear. If we consider the case for which θ remains small, then (6.84) reduces, after linearization and division by mL^2, to

$$\ddot{\theta} + \theta\left(\frac{g}{L} + \frac{\ddot{y}_p}{L}\right) = \frac{F(t)}{mL} \tag{6.85}$$

This linear system (6.85) represents one for which one of the inputs, the support motion $y_p(t)$ does not appear *additively*, as we assumed according to (5.6). Rather, the input multiplies the coordinate θ, producing a time-dependent coefficient.

Example 6.15 Bead on a Rotating Hoop (Example 6.6 Revisited). Here we consider the system of Example 6.6, Fig. 6.19. A bead is free to move without friction on the surface of a circular hoop of radius a. The hoop, meanwhile, rotates at the fixed rate Ω about a vertical axis through the hoop center. Because the hoop rotation rate Ω is constant, the orientation of the plane of the hoop with respect to the plane of the paper is known. Thus, the system has a single degree of freedom, defined by the coordinate θ. If at any time θ is known, then the exact position of the bead m in space is determined. The objective here is to use Lagrange's equation to derive the equation of motion for the θ coordinate.

The kinetic energy T is obtained from the definition $T = \frac{1}{2}m\mathbf{v}_B \cdot \mathbf{v}_B$, where $\mathbf{v}_B$ is the inertial velocity of the bead m. We adopt the rotating coordinate system of Fig. 6.19 and recall the following results of Example 6.6:

$$\mathbf{r}_b = a\mathbf{e}_1$$

$$\boldsymbol{\omega}_b = -\Omega\cos\theta\mathbf{e}_1 + \Omega\sin\theta\mathbf{e}_2 + \dot{\theta}\mathbf{e}_3$$

Noting that $\mathbf{v}_b = 0$, since the bead does not move relative to a rotating observer, the inertial velocity $\mathbf{v}_B = \mathbf{v}_b + \boldsymbol{\omega}_b \times \mathbf{r}_b$ is obtained as

$$\mathbf{v}_B = a\dot{\theta}\mathbf{e}_2 - a\Omega \sin\theta\,\mathbf{e}_3 \tag{6.86}$$

The kinetic energy is then found to be

$$T = \tfrac{1}{2}m[a^2\dot{\theta}^2 + a^2\Omega^2 \sin^2\theta]$$

The potential energy V, due to gravity, is defined as

$$V = mga(1 - \cos\theta)$$

where $V = 0$ when $\theta = 0$. There are no generalized forces here, so Lagrange's equation (6.65) produces the following terms:

$$\left(\frac{\partial T}{\partial \dot{\theta}}\right) = ma^2\dot{\theta}$$

$$\frac{d}{dt}\left(\frac{\partial T}{\partial \dot{\theta}}\right) = ma^2\ddot{\theta}$$

$$\frac{\partial T}{\partial \theta} = ma^2\Omega^2 \sin\theta \cos\theta$$

$$\frac{\partial V}{\partial \theta} = mga \sin\theta$$

Combining these terms according to (6.65) yields the equation of motion

$$ma^2\ddot{\theta} - ma^2\Omega^2 \sin\theta\cos\theta + mga\sin\theta = 0 \tag{6.87}$$

Observe that if (6.87) were divided by a, it is the same as (6.34*b*) derived via newtonian mechanics. Newtonian mechanics, however, also produces two additional equations which define the constraint forces exerted normal to the hoop. These additional equations are not forthcoming from the lagrangian formulation. We merely obtain a single equation for the lone generalized coordinate θ.

6.3.6 Comments on Lagrangian Mechanics

In this section, only some of the more basic aspects of lagrangian mechanics have been introduced for the analysis of particle motions. We close this section with some remarks on the results developed, as well as on some aspects of the analysis which have not been considered here.

1. In lagrangian mechanics the quantities that describe the system motion are scalars: work, potential energy, and kinetic energy. Scalars are independent of coordinate system. Thus, the work, the kinetic energy, and the potential energy are unique quantities and must be the same, regardless of the coordinate system used to calculate them. For example, if the rotating hoop problem were redone using a coordinate system rotating only with Ω but not with θ (see Exercise 6.21), the dependence of T on θ and $\dot{\theta}$, and of V on θ, must be identical to those found in Example 6.15. Thus, once the generalized

coordinates q_i are defined, the functional dependencies $T(q_i, \dot{q}_i)$ and $V(q_i)$ are uniquely determined, independently of the coordinate system used.*

2. In spite of the fact that the basic quantities of the analysis are scalars, we must still be conversant with vectorial kinematics and rotating coordinate systems in order to formulate the system kinetic energy. For a particle, $T = \frac{1}{2}m\mathbf{v}_B \cdot \mathbf{v}_B$, so that the velocity relative to an inertial frame of reference must be determined, and this is often done most conveniently by employing a rotating coordinate system.

3. We have considered simple systems for which the generalized coordinates are either (*a*) unconstrained and so truly independent of each other or (*b*) constrained by algebraic relations which dictate, for example, that a particle is only allowed to move along a certain curve or on a certain surface (e.g., as in Example 6.8). If m algebraic constraints exist, the n coordinates are no longer independent, and our procedure has been to eliminate, via the constraint equations, m of the coordinates from T and V. Thus, we end up with a reduced number $(n - m)$ of generalized coordinates which are independent. The forces of constraint are automatically eliminated from the analysis. This is a primary advantage of lagrangian mechanics. Forces of constraint are real forces which always appear in the newtonian formulation. But forces of constraint do no virtual work in any virtual displacement of the system coordinates and thus cannot appear in the lagrangian formulation described here. Thus, if the system motions, but not the forces of constraint, are of main interest, then lagrangian mechanics is often superior to newtonian mechanics as a direct route to the system mathematical model. On the other hand, if it is necessary to determine the forces of constraint as part of the analysis, then newtonian mechanics may be more direct. A good example of the difference in the two formulations for constrained motions is afforded by the rolling ball problem of Example 6.12.

4. In the development of Lagrange's equation in this section, a basic result of the analysis is (6.63), which actually applies for any system having 3 degrees of freedom, with generalized coordinates q_1, q_2, q_3. In the more general case of a system with n degrees of freedom, (6.63) has the form

$$\sum_{i=1}^{n}\left[\frac{d}{dt}\left(\frac{\partial T}{\partial \dot{q}_i}\right) - \frac{\partial T}{\partial q_i} + \frac{\partial V}{\partial q_i} - Q_i(t)\right]\delta q_i = 0 \tag{6.88}$$

If we consider the *static* case, for which $T \equiv 0$ and for which the Q_i are static rather than time-dependent forces, we obtain the principle of virtual work

*In this comment it is assumed that the various coordinate systems have the same origin **O**. If one coordinate system is translating *uniformly* with respect to the other, then the kinetic energies as defined in the two systems will differ. The equations of motion obtained from Lagrange's equation will, however, be identical, as a uniform translation has no effect on the dynamics (see Exercise 6.22 for a simple demonstration of this fact).

from statics, which may be stated as follows:

$$\delta V = \sum \left(\frac{\partial V}{\partial q_i} \right) \delta q_i = \sum Q_i \, \delta q_i = \delta W_{nc}$$

In words: If a system is in static equilibrium, then a set of virtual displacements δq_i imposed on the equilibrium configuration will result in a virtual change δV in potential energy that must equal the virtual work δW_{nc} done by the (nonconservative) applied forces. This principle is widely used to formulate mathematical models to solve static problems in solid mechanics.

5. The analysis of this section was done for a single particle in 3-D. The result (6.65), however, turns out to be applicable to any mechanical system, regardless of the number of degrees of freedom, with the following proviso: The system must be unconstrained, or, if constrained, the m constraint relations must be of algebraic type, so that they can be used to eliminate m of the coordinates from the analysis. Other (nonalgebraic) types of constraints can be handled in lagrangian mechanics using the Lagrange multiplier, but we have not considered this possibility.

6. In the development of Sec. 6.3.5 leading to Lagrange's equations (6.64), the potential energy V was assumed to be a function of the particle coordinates only, but not of the velocities nor of the time t. Then in Example 6.14 the potential energy V was formulated as a function of both time and position [see (6.83) and note that $y_p(t)$ is a given function of time]. This turns out to be OK, provided that the potential energy, when expressed as a function of the cartesians (x, y, z) measured relative to a fixed origin, is not a function of time. In Example 6.14 what has happened is that the coordinate transformation (6.43) is time dependent. From Fig. 6.29 the transformation (6.43) is seen to be (with $\theta = q$)

$$x = L \cos q$$

$$y = L(-1 + \sin q) + y_p(t)$$

where x and y are the fixed cartesian coordinates of the particle, in terms of which $V = V(y)$, that is, is a function of y only. Although (6.43) was assumed to be time independent, the development in Sec. 6.3.5 leading to Lagrange's equations (6.64) is unaffected by a time dependence in the transformation (6.43), because the time is not involved in any virtual displacement. The functional form $V = V(q_1, q_2, q_3, t)$ is common in situations involving a support motion, as in Example 6.14.

7. We emphasize that in this section only some of the basic ideas of lagrangian mechanics have been introduced. A number of important aspects were not addressed. A thorough treatment can be found in the references listed at the end of this chapter.

REFERENCES

The following references contain additional material on rotating coordinate systems, Lagrange's equations, and other topics in dynamics:

6.1. Meirovitch, L.: *Introduction to Dynamics and Control*, Wiley, New York (1985).
6.2. Greenwood, D. T.: *Principles of Dynamics*, 2d, ed., Prentice-Hall, Englewood Cliffs, N.J. (1988).
6.3. Greenwood, D. T.: *Classical Dynamics*, Prentice-Hall, Englewood Cliffs, N.J. (1977).
6.4. Meirovitch, L.: *Methods of Analytical Dynamics*, McGraw-Hill, New York (1970).
6.5. Groesberg, S. W.: *Advanced Mechanics*, Wiley, New York, (1968).
6.6. Ginsberg, J. H.: *Advanced Engineering Dynamics*, Harper & Row, New York (1988).

EXERCISES

1. Use Lagrange's equations to derive the equations of motion for the system shown in Fig. *P*6.1 assuming the angle θ to be small. The spring k is unstretched when $x = \theta = 0$.

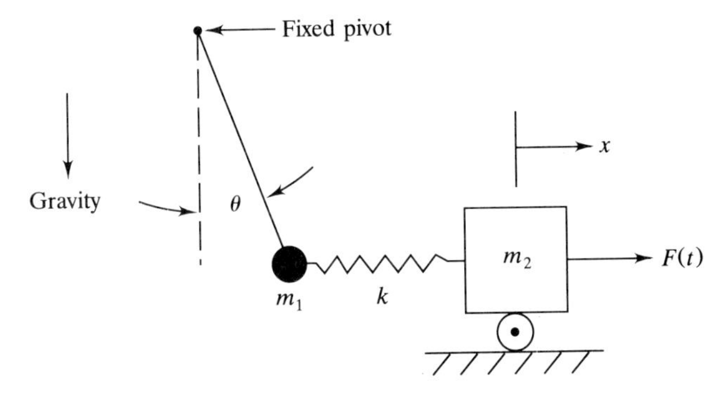

FIGURE *P*6.1

2. Use Lagrange's equations to derive the equations of motion for the system of Exercise 2.17 (see Fig. *P*2.17).
3. Use Lagrange's equations to derive the equations of motion (2.68) and (2.69) for the two-link manipulator of Example 2.10. Care must be taken in the determination of the system kinetic energy T.
4. Use Lagrange's equation to derive the equation of motion of the system of Exercise 2.2 (see Fig. *P*2.2).
5. Use Lagrange's equation to derive the equation of motion for the system of Exercise 2.9 (see Fig. *P*2.9).
6. A small bead of mass m slides without friction on the surface $y = x^2$, as shown in Fig. *P*6.6. Determine the equation of motion, using x as the single generalized coordinate.

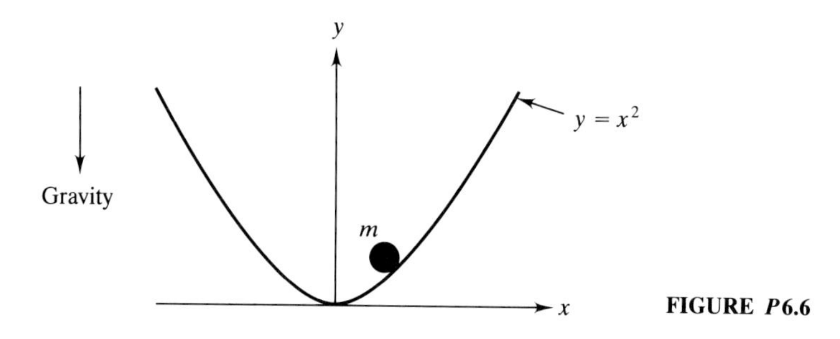

FIGURE P6.6

7. Consider the system of Exercise 4.24 (see Fig. *P*4.24). Assume the static force P_0 to be replaced by a known, time-varying force $p(t)$. Use Lagrange's equation to derive the equation of motion for the generalized coordinate x. Care must be taken in determining the virtual work δW done by the force $p(t)$ in a virtual displacement δx. As in *P*4.24, the angle θ need not be small.

8. Consider the rotating, telescoping boom shown in Fig. *P*6.8. A force $F(t)$ acts along the boom axis on the mass m, which moves in and out as measured by the coordinate s, which is an unknown function of time. A torque $M(t)$ rotates the boom; this rotation is measured by the (unknown) angle α. Meanwhile, the entire assembly rotates about the vertical at the fixed rate Ω. Use Lagrange's equations to derive the equations of motion for the s and α coordinates. Assume the boom to be massless. *Hint:* Resolve the velocity of m along the $\mathbf{e}_1, \mathbf{e}_2, \mathbf{e}_3$ directions shown. These axes are fixed in the boom and rotate with it.

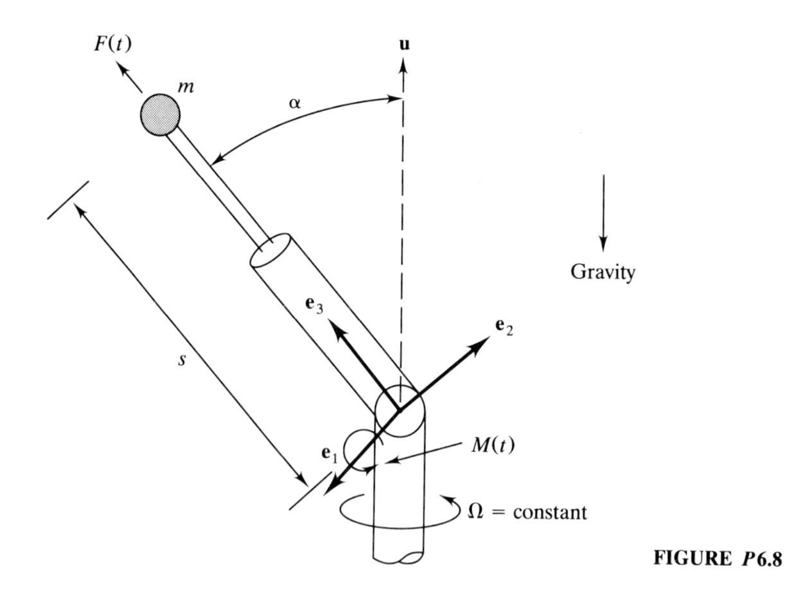

FIGURE P6.8

9. A bar B with symmetrically located masses is mounted on the shaft S as shown in Fig. *P*6.9. The shaft S rotates about the vertical at the fixed rate Ω. Meanwhile, the

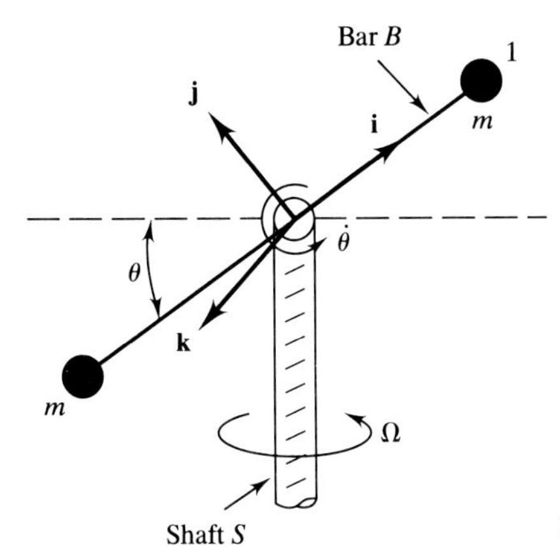

FIGURE *P*6.9

bar B rotates about an axis in the horizontal plane at the fixed rate $\dot{\theta}$. The orientation of the bar B in the vertical plane is measured by the angle θ. A rotating coordinate system $\mathbf{i}, \mathbf{j}, \mathbf{k}$ is fixed in the bar B as shown. Determine the following kinematic vectors, resolved along the $\mathbf{i}, \mathbf{j}, \mathbf{k}$ axes:

a. The rotating coordinate system angular velocity, $\boldsymbol{\omega}_b$
b. The absolute (inertial) velocity $\mathbf{v}_B$ of the mass labeled 1
c. The acceleration $\mathbf{a}_B$ of mass 1

10. A shaft rotates at the fixed rate Ω about an axis $\mathbf{n}$ out of the paper (Fig. *P*6.10). Attached to the shaft is a hoop of radius a which rotates at the fixed rate $\dot{\phi}$ about the axis shown. A bead of mass m slides without friction around the hoop at the fixed rate $\dot{\theta}$; the location of the bead is defined by the angle θ. An angle ϕ measures the orientation of the hoop relative to the plane of the paper (when $\phi = 0$ the hoop lies in the plane of the paper). Determine the angular velocity of the rotating $\mathbf{e}_1, \mathbf{e}_2, \mathbf{e}_3$ system and the acceleration of the mass m. *Note:* The coordinates $\mathbf{e}_1, \mathbf{e}_2, \mathbf{e}_3$ rotate with the Ω and $\dot{\phi}$ rotations, but not with the $\dot{\theta}$ rotation.

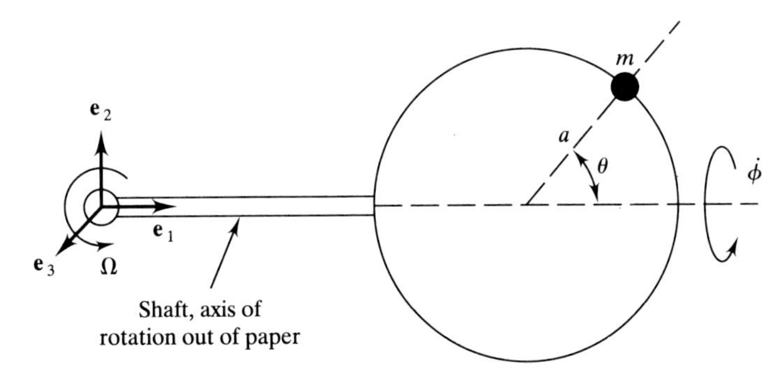

FIGURE *P*6.10

11. A 2-DOF system consists of a uniform bar, of moment of inertia I_C about its center of mass, which rotates about a fixed, frictionless pivot A at the center of mass, and a

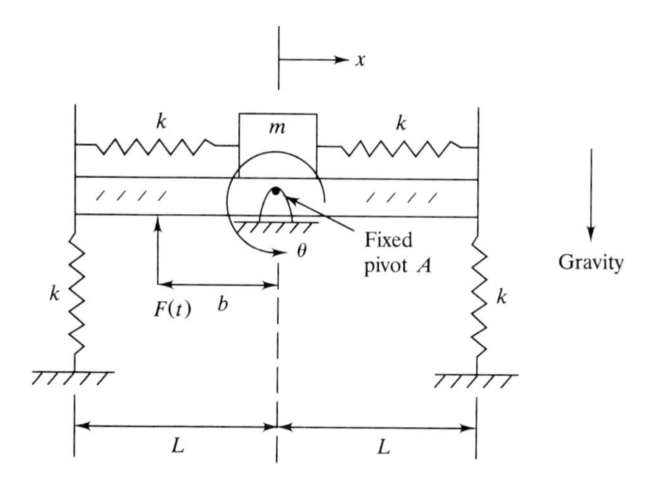

FIGURE ***P*6.11**

mass m which may slide without friction along the top of the bar. Four equal springs restrain the system motions as shown in Fig. $P6.11$. The coordinate θ measures the bar rotation, and the coordinate x measures the position of the sliding mass m ($x = 0$ when m is over the bar center of mass at A). A known vertical force $F(t)$ is applied to the bar at the location shown.

a. Determine the kinetic and potential energies of the system.
b. Determine the equations of motion, assuming x and θ to be small.

12. The Bar OA rotates about a vertical axis at the *constant* rate $\dot{\phi}$. The boom AB rotates about a horizontal axis through A. The boom rotation rate $\dot{\theta}(t)$ is given, as is $\ddot{\theta}(t)$. A rotating coordinate system is defined as shown in Fig. $P6.12$. This coordinate system rotates with OA but not with the boom AB.

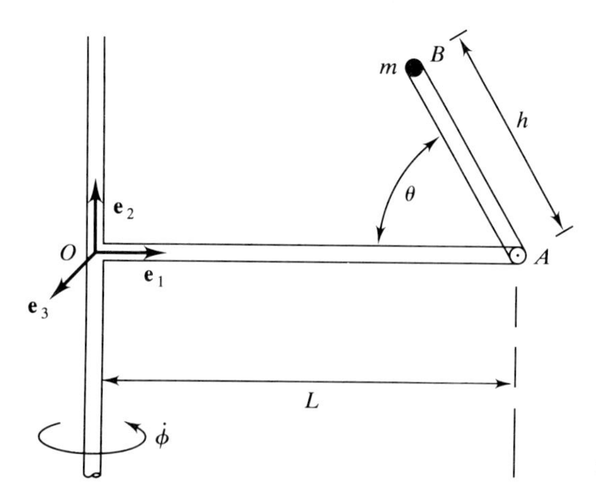

FIGURE ***P*6.12**

Determine the following quantities:

a. The angular velocity $\boldsymbol{\omega}_b$ of the rotating coordinate system
b. The angular velocity $\boldsymbol{\omega}$ of the boom AB
c. The relative velocity $\mathbf{v}_b$ of a mass located at B
d. The relative acceleration $\mathbf{a}_b$ of the mass m at B
e. The angular accelerations $\dot{\boldsymbol{\omega}}$ and $\dot{\boldsymbol{\omega}}_b$

13. A pendulum consists of a uniform bar of mass m and length L, as shown in Fig. $P6.13$. The pivot A of the pendulum is made to move along a 45° incline with known velocity $v(t)$ and acceleration $a(t)$. Determine the kinetic and potential energies of the system and the equation of motion for the single generalized coordinate θ.

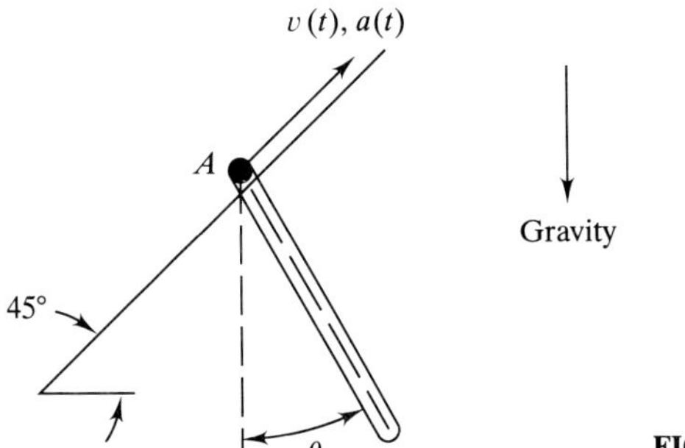

FIGURE $P6.13$

14. A bug crawls along a straight line painted on a cylinder as shown in Fig. $P6.14$. The bug moves in such a way that the coordinate r increases at the constant rate v_0. The cylinder meanwhile rotates about its axis ($\bar{e}_2$) at the rate $\dot{\phi}$, which is constant.

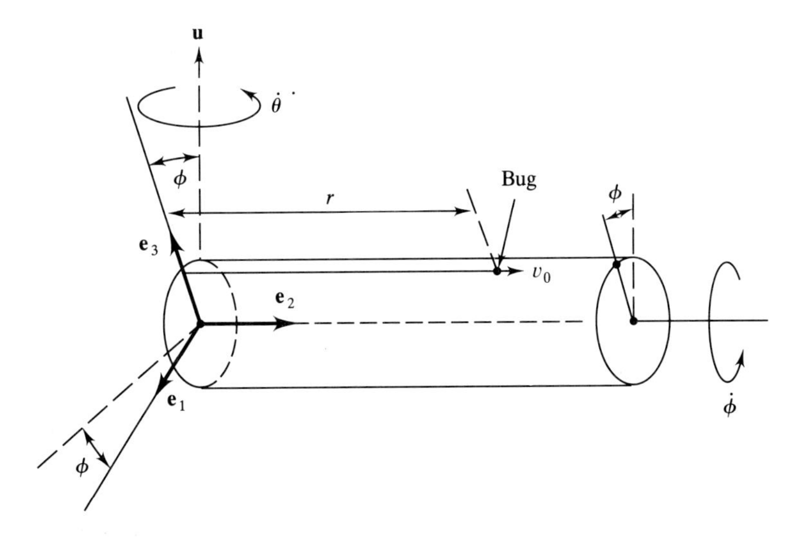

FIGURE $P6.14$

The cylinder is rotated about the vertical **u** at the rate $\dot{\theta}$, also constant. The rotating coordinates $\mathbf{e}_1, \mathbf{e}_2, \mathbf{e}_3$ are fixed in and rotate with the cylinder. An angle ϕ defines the instantaneous angle between **u** and $\mathbf{e}_3$. Determine the acceleration of the bug.

15. Two particles, each of mass m, are attached via a rigid massless rod of length $2a$, as shown in Fig. $P6.15$. This mass/rod system spins at the *constant* rate $\dot{\theta}$ about the axis $\mathbf{e}_2$. Meanwhile the gimbal G is rotated $(\phi, \dot{\phi})$ about the horizontal axis $\mathbf{e}_1$ by the application of a moment M_0. The coordinate system rotates with the gimbal (ϕ) rotation but not with the θ rotation. Thus $\mathbf{e}_1$ is fixed in space and $\mathbf{e}_2, \mathbf{e}_3$ rotate in a vertical plane.

Determine the system kinetic energy and the equation of motion for the ϕ coordinate.

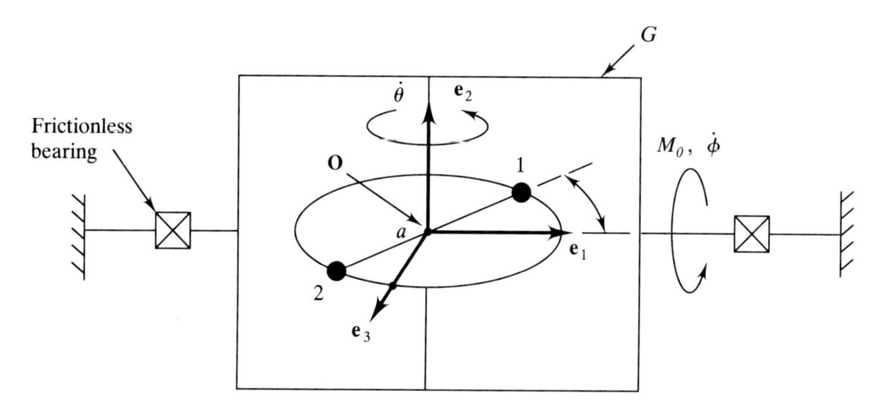

FIGURE $P6.15$

16. Determine the inertial acceleration $\mathbf{a}_B$ for the system of Example 6.3 by using (6.24) through (6.26) in (6.23).

17. Determine the inertial acceleration $\mathbf{a}_B$ of the telescoping boom mass of Example 6.4 for the case in which both Ω and $\dot{\alpha}$ rotations occur simultaneously (the analysis has been partially completed at the end of Example 6.4). Use the rotating coordinate system defined in Fig. 6.15, for which the rotating $\mathbf{e}_1, \mathbf{e}_2, \mathbf{e}_3$ system now rotates with both the Ω and $\dot{\alpha}$ rotations.

18. Determine the inertial acceleration $\mathbf{a}_B$ of the point P on the spinning, rotating cylinder of Example 6.1, using a coordinate system which rotates with $\dot{\theta}$, but not with $\dot{\phi}$, as in Fig. 6.4 (in the analysis done in Example 6.5, $\mathbf{a}_B$ was determined in terms of a rotating coordinate system which rotates with both the $\dot{\theta}$ and $\dot{\phi}$ rotations, as in Fig. 6.5).

19. Determine the transformation analogous to (6.8) which expresses the cartesian unit vectors **i** and **j** in terms of $\mathbf{e}_r$ and $\mathbf{e}_\theta$.

20. A uniform bar of mass M and length L is free to pivot in the plane about its center of mass at **O**, as shown in Fig. $P6.20$. A bead of mass m slides without friction on the surface of the bar. The coordinates defining bar orientation and mass m position are

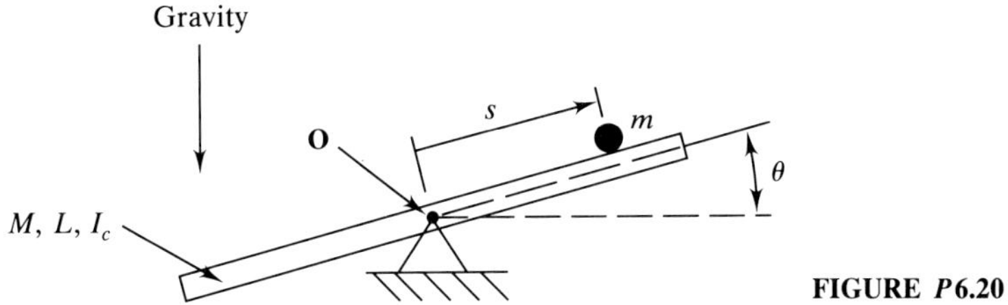

FIGURE $P6.20$

θ and s, respectively. Use Lagrange's equations to develop the equations of motion for s and θ. Include the effects of gravity.

21. A version of the bead on the rotating hoop problem (Examples 6.6 and 6.15) is shown in Fig. $P6.21$. The bead slides without friction on the surface of a circular wire of radius a. The hoop also rotates about a vertical axis through the center **O**, but the rotation rate is *not constant*, as it was in Example 6.6 and in Example 6.15. Rather, a specified, time-dependent moment $M(t)$ is exerted in order to rotate the hoop. Thus, the hoop angular velocity $\dot{\phi}$ varies and ϕ becomes a second degree of freedom (along with θ). Derive the two equations of motion for the θ and ϕ coordinates. For the case $M(t) = 0$, is it true that $\dot{\phi} = \text{constant}$? Explain.

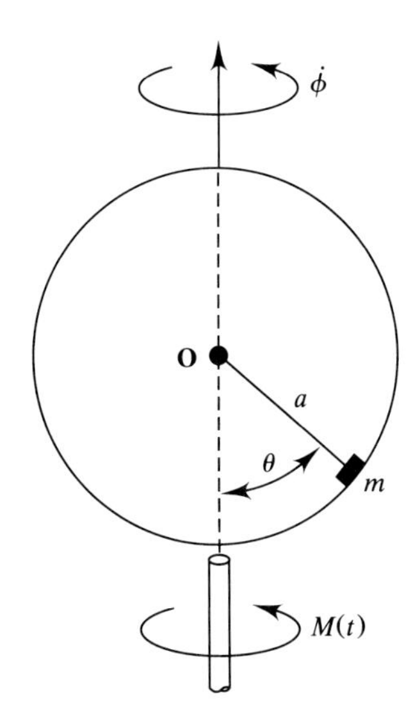

FIGURE $P6.21$

22. A particle m moves in one dimension. Consider the following two cases: (1) A coordinate x measures the position of the particle relative to a fixed origin **O**; (2) a coordinate x measures the position of the particle relative to the origin of a coordinate system which translates at the uniform speed V in the x direction. The force F exerted on the particle is a function of x, $\dot{x}$, and t but not of V. Use Lagrange's equation to prove that the equations of motion for the two cases are identical; that is, a uniform translation of coordinates has no effect on the equation of motion, even though the kinetic energy in case 2 is a function of V.

CHAPTER 7

LINEAR MECHANICAL VIBRATIONS

7.1 INTRODUCTION

In this chapter we consider those mechanical systems that tend naturally to exhibit oscillatory motions. We refer to this oscillatory behavior as *mechanical vibration*. In physical terms a mechanical system must possess the two properties of mass and stiffness in order to exhibit vibratory motion. As discussed in Sec. 2.4.1, *stiffness* refers to conservative, restoring forces (and/or moments) which are generated as a result of the system motions. Because stiffness forces are conservative, they can be characterized by a potential energy function. In this chapter we will restrict attention to small motions in the vicinity of a static equilibrium configuration. We will, therefore, assume such motions to be described by linear mathematical models. Effects of nonlinearities are considered in Chap. 10.

Some vibrational phenomena of practical importance to engineers include the following (this list is by no means exhaustive):

1. Devices that have rotating or reciprocating parts, such as engines, turbines, fans, motors, and machinery. For these systems the motions of the moving parts are accompanied by time-dependent loads that excite vibratory motion of the system as a whole.
2. Vibrations of large structures such as buildings, bridges, cooling towers, offshore oil structures, transmission lines, and so on. Here the vibrations may be excited by wind (or water) loadings or by ground motion during an earthquake.

3. Vibratory motions exhibited by vehicles of all types. Examples include the rolling, pitching, and yawing oscillations of ships, space vehicles, and airplanes, vibration of automobiles traversing uneven roads, and the lateral motions of railway vehicles at high speed or on low-quality rail.

Usually, vibratory motions are unwanted, as is the case in all of the situations just described. Vibrations may lead to undesirably large motions, loads, and/or stresses in the system, leading to a degradation of performance or to physical discomfort. In extreme cases the system may fail catastrophically. Furthermore, even relatively small amplitude vibrations can lead to fatigue failure of structures if the vibrations persist over long periods of time. The job of the design engineer is to ensure, through proper system design, that vibrations are limited to acceptably small levels.

Sometimes, however, mechanical vibrations are produced intentionally to perform some function. An example is the loudspeaker, in which mechanical vibrations of the speaker cone produce sound. The music produced by stringed instruments such as violins and pianos and by reed instruments is another example. In addition, various types of equipment utilize vibrational motion to induce mixing or material separation, as in paint mixers and wood chip separators.

We now discuss the various ways in which vibrational phenomena will be classified in our study of vibrations. Four types of classifications will be relevant: (1) linear versus nonlinear, (2) damped versus undamped, (3) free versus forced, and (4) single degree of freedom versus multi-degree-of-freedom. Each of these is discussed briefly below:

1. *Linear versus nonlinear*: We will study only linear vibratory systems in this chapter. For such systems all forces and inertia terms appearing in the mathematical model will be linear functions of position, velocity, and acceleration. In physical terms we are assuming that the system movement from the static equilibrium configuration is sufficiently small that a linear model is valid. If the system movement from equilibrium is sufficiently large, then nonlinear effects will become important. Such effects are discussed in Chap. 10. It must be noted that *any* vibratory system, if driven by a sufficiently strong input, is capable of exhibiting nonlinear behavior. Nevertheless, linear models work well in a surprisingly large number of situations, partly because one objective in designing systems which exhibit vibratory motion is to ensure that the vibratory motions are small. Thus, if such a system is properly designed, it will tend naturally to exhibit linear behavior. (This comment does not necessarily imply that in all situations we can produce a system exhibiting linear behavior; sometimes nonlinear effects are unavoidable.)
2. *Damped versus undamped*: All of the mechanical systems that we will analyze possess some inherent damping. Nevertheless, it is often useful to analyze models of undamped systems in order to understand the basic

oscillatory behavior. This is especially true in the case of multi-degree-of-freedom systems, for which the analysis of the undamped version of the system math model is considerably simpler than the analysis with damping included. This will become clear in Sec. 7.3.

3. *Free versus forced vibration*: In free vibration the system is displaced from equilibrium at $t = 0$ and thereafter moves of its own accord with no outside influences. This is the problem considered at length in Chap. 4, and, in state variable form, an autonomous mathematical model of the type (4.62) would describe the linear behavior of the system states. In forced vibration, the system vibrates in response to some externally imposed input. This is the problem considered in Chap. 5, and, in state variable form, a nonautonomous math model of the type (5.6) would describe the linear response.

   In practical terms it is the forced vibration problem that is usually of most interest. The forced response, however, will depend on both the nature of the input and on the basic properties of the system (that is, the free response). Thus, we must always have an understanding of the free response, even if the forced response is of primary concern.

4. *Single-DOF versus multi-DOF*: As discussed in Sec. 6.3.2, the number of degrees of freedom is the minimum number of independent position coordinates needed to model the system motion. Thus, the mathematical model of a single-DOF mechanical system, obtained through application of either Newton's law or Lagrange's equation, will arise naturally as a second-order, linear ordinary differential equation. This ODE will be autonomous for free vibration and nonautonomous for forced vibration. A multi-degree-of-freedom system having some number n of degrees of freedom will be characterized by a mathematical model consisting of n, second-order ODEs.

In the first six chapters we have already encountered numerous systems which may exhibit vibratory motion. Some of these are summarized in Table 7.1, in which information related to the previous four classification types is displayed. In this chapter we will extend our understanding of vibrations by considering three primary topics: (1) a continuation of work already done for the free and forced vibration of the single degree of freedom oscillator, with special attention to the more practical aspects, (2) an introduction to multi-degree-of-freedom vibrations and modal analysis, and (3) an introduction to spectrum analysis, which is an indispensable tool in the experimental analysis of the response of mechanical and many other oscillatory systems.

As in Chaps. 4 and 5, our primary concern here will be to understand the fundamentals of the system motions through analysis of the governing mathematical models. In Chaps. 4 and 5 we analyzed system models using the state variable approach; the solutions we obtained gave us the time dependence of each of the states of the system. As noted previously, all of the vibratory models we will study in this chapter will arise as sets of second-order ODEs. Thus, a

TABLE 7.1
Classification of previously studied vibratory systems

System	No. DOF	Linear/ nonlinear	Source of damping in model	Source of stiffness	Source of input
1. Example 2.1, pendulum (2.19)	1	Nonlinear (Linear: $\sin\theta \to \theta$)	Air resistance, pivot friction (not included in model)	Gravity	None
2. Example 2.2, pendulum w/moving support	1	Nonlinear (Linear: $\sin\theta \to \theta$ $\cos\theta \to 1$)	Air resistance, pivot friction (not included in model)	Gravity	Moving pivot
3. Example 2.3, harmonic oscillator	1	Linear	Dashpot	Spring	Externally applied force
4. Example 2.4, auto suspension (2.55)	2	Linear	Shock absorber, tire damping	Coil spring tires	None
5. Example 2.5, auto suspension (2.56) to (2.59)	4	Linear	Shock absorbers, tire damping	Coil springs tires	Road irregularities
6. Example 2.6, earthquake building model	3	Linear	Air resistance, dissipation in structure (not included in model)	Building walls	Ground motion
7. Example 5.2, rolling ball (5.10)	1	Nonlinear	Friction between ball/surface	Gravity	Support motion

vibratory system having n degrees of freedom will be of order $2n$ and will possess $2n$ states: the n system coordinates $q_1, q_2, \ldots, q_n$ and their derivatives: $\dot{q}_1, \dot{q}_2, \ldots, \dot{q}_n$. Rather than converting the system math model to the state variable form ($2n$ first-order ODEs), we will just find the n coordinate functions $q_1(t), \ldots, q_n(t)$ which solve the n second-order ODEs comprising the model. This solution procedure is the approach most often used by vibrations engineers. The procedure is simpler than the state variable approach (we only solve for n quantities, rather than $2n$), but it yields the same amount of information, because n of the states (the n velocities $\dot{q}_1$) are merely derivatives of the other n (the q_i); so, if we solve for the $q_i(t)$, the $\dot{q}_1(t)$ are easily found by differentiation.

7.2 SINGLE DEGREE OF FREEDOM VIBRATION

7.2.1 Damped Free Vibration

We'll start by considering free vibration of the prototype oscillator model of Fig. 2.23 (Example 2.3), repeated here as Fig. 7.1 with the input force $F(t)$ removed.

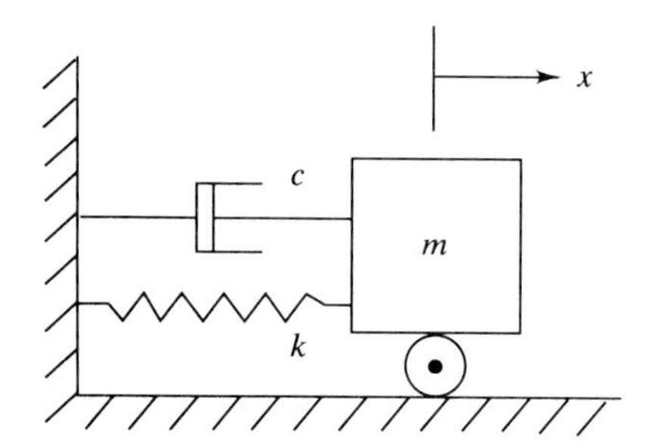

FIGURE 7.1
The damped linear oscillator with no forcing.

The inertia, damping, and stiffness effects combine to yield the equation of motion (2.52), with $F(t) = 0$,

$$m\ddot{x} + c\dot{x} + kx = 0 \tag{7.1}$$

Division by the mass m and utilization of the definitions $\omega^2 \equiv k/m$ and $\xi \equiv c/2m\omega$ then yield the "standard form"

$$\ddot{x} + 2\xi\omega\dot{x} + \omega^2 x = 0 \tag{7.2}$$

At equilibrium the states x and $\dot{x}$ are zero. Motions of the system are initiated by imposing nonzero initial conditions $x(0)$ and $\dot{x}(0)$. We assume the linear model (7.2) to provide a valid description of motions that remain close to equilibrium.

Notice that the standard form (7.2) contains only two system parameters, the dimensionless damping factor ξ and the undamped natural frequency ω, whereas the model (7.1) contains the three parameters m, c, and k. Thus, two oscillators having different m, c, and k values will exhibit the same response to given initial conditions, provided the two oscillators have the same values of ξ and ω. The natural parameters that define free vibrational motion are thus ξ and ω, not m, c, and k. Furthermore, a reduction in the number of system parameters (from three to two in this case) appearing in a system mathematical model should always be sought.

We'll now investigate the properties of the solution to (7.2) with the objectives of (1) understanding the system response and (2) utilizing the solutions as a means to interpret experimental data of freely oscillating single degree of freedom systems.

To obtain the solution $x(t)$ to (7.2), we may use basic ideas from the theory of ordinary differential equations. For a linear, homogeneous ODE with constant coefficients, one may assume a solution of the form

$$x(t) = Ce^{\lambda t} \tag{7.3}$$

where C will turn out to be arbitrary and λ is an eigenvalue which is to be determined.

Substitution of (7.3) into (7.2), followed by cancellation of the $e^{\lambda t}$, leads to the algebraic equation

$$C(\lambda^2 + 2\xi\omega\lambda + \omega^2) = 0 \tag{7.4}$$

The trivial solution $x(t) = 0$ results if $C = 0$. For nontrivial solutions we have to require that λ satisfy

$$\lambda^2 + 2\xi\omega\lambda + \omega^2 = 0 \tag{7.5}$$

with the constant C arbitrary. This is the system characteristic equation, exactly the same as we would obtain by converting (7.2) to state variable form (as was done in Sec. 4.2). The two solutions to the quadratic equation (7.5) are

$$\lambda_{1,2} = -\xi\omega \pm (\xi^2\omega^2 - \omega^2)^{1/2} \tag{7.6}$$

We observe that if the dimensionless damping factor $\xi > 1$, then both eigenvalues are real and negative and oscillations do not occur. This case is referred to as *overdamped*. In the remainder of this section we'll assume that $\xi < 1$, so that (7.6) may be expressed as

$$\lambda_{1,2} = -\xi\omega \pm i\omega(1 - \xi^2)^{1/2} \qquad (\xi < 1) \tag{7.7}$$

The eigenvalues are complex conjugates with a negative real part, signifying an exponentially decaying oscillatory motion. The imaginary part of the eigenvalues is the oscillation frequency; let us define this frequency as ω_d:

$$\omega_d = \omega(1 - \xi^2)^{1/2} \tag{7.8}$$

We call ω_d the *damped natural frequency* to distinguish it from the undamped natural frequency ω. Observe that, for fixed ω, increases in the damping factor ξ will reduce the oscillation frequency ω_d. If, however, the damping is "light," say $\xi \leq 0.05$, then, from (7.8), ω_d and ω are essentially the same, and the effect of damping on the observed oscillation frequency would be observable only over a large number of cycles of oscillation.

The two linearly independent solutions $x_1(t)$ and $x_2(t)$ implied by (7.7) are

$$x_1(t) = e^{(-\xi\omega + i\omega_d)t}$$

$$x_2(t) = e^{(-\xi\omega - i\omega_d)t}$$

The general solution $x(t)$ is then a linear combination of these two, which may be expressed as

$$x(t) = e^{-\xi\omega t}(C_1 e^{i\omega_d t} + C_1^* e^{-i\omega_d t}) \tag{7.9}$$

where the common exponential term $e^{-\xi\omega t}$ has been factored, and it has been noted that the constants associated with the two solutions will be complex conjugates (Sec. 4.2.3) in order to ensure that $x(t)$ be real. From previous work [for example, the analysis leading to (4.109)], (7.9) may be expressed in terms of the trigonometric functions as follows:

$$x(t) = A_0 e^{-\xi \omega t} \cos(\omega_d t - \phi) \tag{7.10}$$

Equation (7.10) is the basic result of the analysis. The constants A_0 and ϕ are determined by the initial values $x(0)$ and $\dot{x}(0)$ and may be shown to be (verify!)

$$A_0 = \left[x_0^2 + \left(\frac{\dot{x}_0 + \xi \omega x_0}{\omega_d} \right)^2 \right]^{1/2}$$
$$\tan \phi = \frac{\dot{x}_0 + \xi \omega x_0}{\omega_d x_0} \tag{7.11}$$

The oscillator executes an exponentially decaying oscillation with frequency ω_d. This is illustrated in Fig. 7.2(a), which shows the history of $x(t)$ for the case $\xi = 0.02$, $\omega = 1$. The initial conditions used to generate Fig. 7.2 were $x(0) = \dot{x}(0) = 1$, so that $A_0 \cong 1.43$ and $\phi \cong 45.57°$. Figure 7.2(b) shows a portion of the associated orbit in the phase plane (this orbit will eventually spiral asymptotically into the origin).

Shown in Fig. 7.2(a), along with the solution $x(t)$, is the *amplitude envelope*, that is, the exponential function $A_0 e^{-\xi \omega t}$ which bounds the oscillation. This function intersects the x axis ($t = 0$) at A_0. The "period of oscillation"* is $\tau = 2\pi/\omega_d$. Note that in Fig. 7.2(a) this period is essentially the same as the period $2\pi/\omega$ the oscillator would exhibit if there were no damping, since from (7.8), $\omega_d \approx \omega$ if $\xi = 0.02$. The phase portrait, Figure 7.2(b), exhibits the inward spiraling character typical of damped oscillatory systems.

We'll now look more closely at the manner in which the oscillations decay with time. This decay is governed by the amplitude envelope, which we call $A(t)$ and define according to

$$A(t) = A_0 e^{-\xi \omega t} \tag{7.12}$$

where A_0 is the initial value of the amplitude. Let us begin the investigation by examining the dependence of the amplitude of successive oscillation peaks on the damping factor ξ. Suppose we pick any oscillation peak, say the nth, which occurs at some time t_n. Calling the amplitude of this peak A_n, we have from (7.12)

$$A(t_n) = A_n = A_0 e^{-\xi \omega t_n} \tag{7.13}$$

The peak which follows the nth will be defined as A_{n+1}, and it occurs at time t_{n+1}, so that

$$A(t_{n+1}) = A_{n+1} = A_0 e^{-\xi \omega t_{n+1}} \tag{7.14}$$

---

*Because the response $x(t)$ is not periodic (it does not repeat itself every so many seconds), we cannot properly speak of a "period of oscillation." Nevertheless, this term is typically used and is taken to mean the time between successive oscillation peaks.

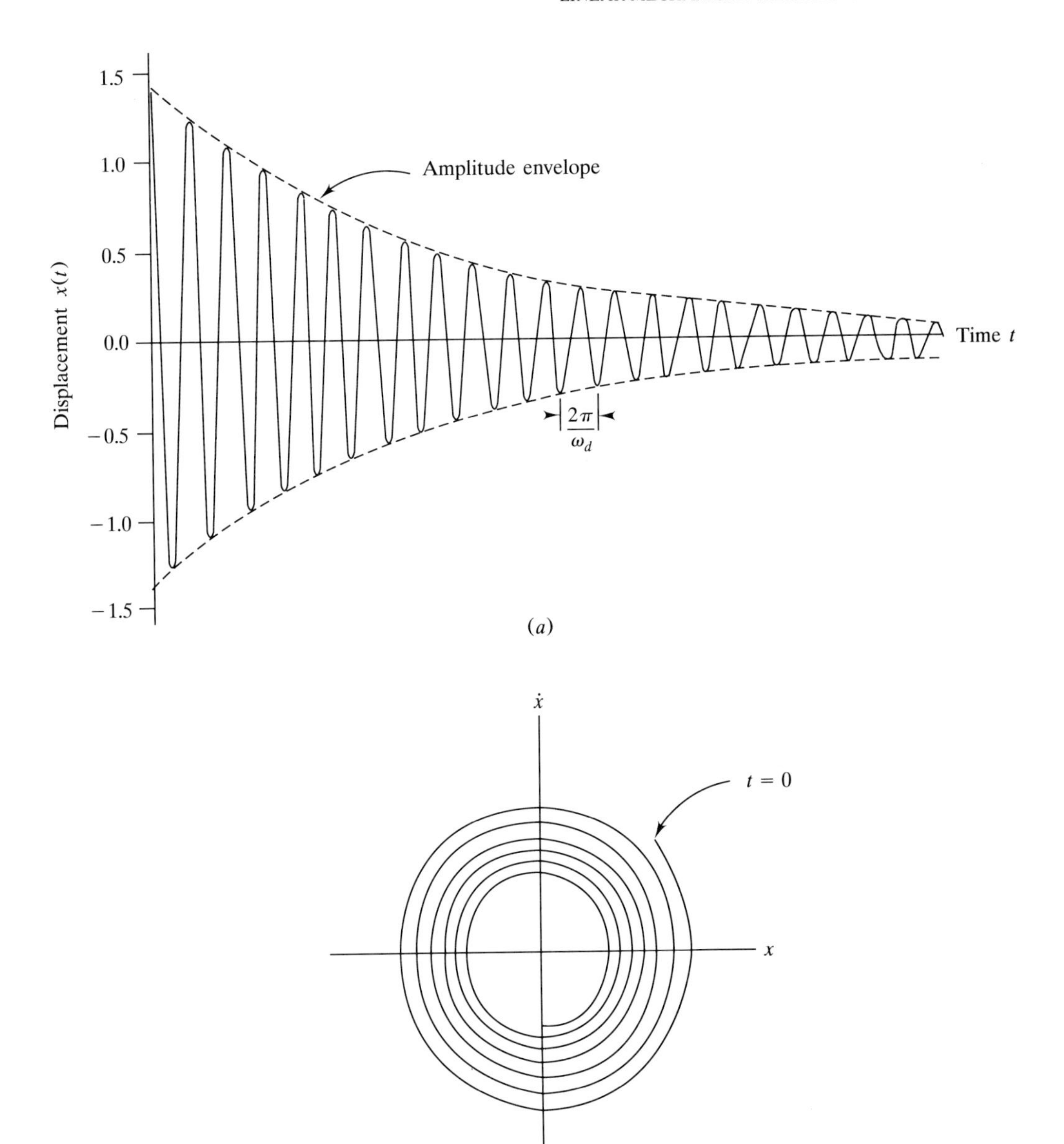

FIGURE 7.2
(a) The solution $x(t)$, Eq. (7.10); (b) associated phase portrait for the damped oscillator: $\xi = 0.02$, $\omega = 1.0$, $x(0) = \dot{x}(0) = 1$.

Let us now determine the ratio of successive peaks, which is

$$\frac{A_n}{A_{n+1}} = \frac{A_0 e^{-\xi\omega t_n}}{A_0 e^{-\xi\omega t_{n+1}}} = e^{\xi\omega(t_{n+1}-t_n)} \tag{7.15}$$

Notice that $A_n/A_{n+1} > 1$, so that the exponent in (7.15) is positive. Next we note that the time between successive peaks, $t_{n+1} - t_n$, is just the oscillation period $2\pi/\omega_d$, so that

$$\frac{A_n}{A_{n+1}} = e^{[2\pi\xi/(1-\xi^2)^{1/2}]} \tag{7.16}$$

This important relation shows that the ratio of successive amplitudes is a function *only* of the damping factor ξ and does not depend on the system natural frequency ω. Furthermore, (7.16) is valid for any n, so that the ratio of *any* two successive peaks will be the same. In systems for which the damping is light, say, $\xi \leq 0.05$, the factor $(1 - \xi^2)^{1/2}$ appearing in (7.16) is approximately unity, and (7.16) reduces to the simple form

$$\frac{A_n}{A_{n+1}} \cong e^{2\pi\xi} \qquad \text{(light damping approximation)} \tag{7.17}$$

Equation (7.16) or (7.17) forms the basis of one experimental technique for determination of the system damping factor ξ. As noted in Sec. 2.4, the damping factor ξ in real vibratory systems is usually very difficult to determine a priori because most often the damping is due to a combination of several effects, each of which is generally difficult to predict. For instance, if we take a little car, attach it to a real spring, and let it vibrate freely as in Fig. 7.1, there will be energy dissipation due to a number of sources: (1) air resistance, (2) friction in the wheel bearings and between the wheels and the surface on which the car moves, and (3) energy dissipation (heating) in the spring as it deforms. What we do is to put in a linear viscous damper to approximate these combined effects. If we then induce a free oscillation of the system and measure the response, we can determine a representative system damping factor ξ by computing the ratio of successive peaks from (7.16) or (7.17).

Most often this "decay testing" technique is employed by measuring a large number of successive amplitudes and analyzing the data in the following way (cf. Fig. 7.3): First we select a reference oscillation peak, usually as near to $t = 0$ as possible (but any peak will do), and we define this reference amplitude as A_0. Then, we note that the ratio A_0/A_n, where A_n is the amplitude n cycles later than A_0, will be given by

$$\frac{A_0}{A_n} = e^{2\pi n\xi/(1-\xi^2)^{1/2}} \tag{7.18}$$

This relation was obtained by replacing the oscillation period $\tau = t_{n+1} - t_n$ in (7.15) by $n\tau$, since n, rather than 1, cycles have elapsed. We next take the

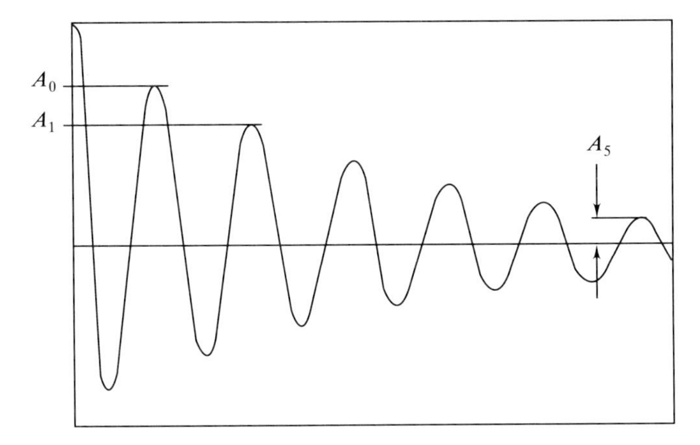

FIGURE 7.3
Definition of successive amplitudes in free vibration decay testing.

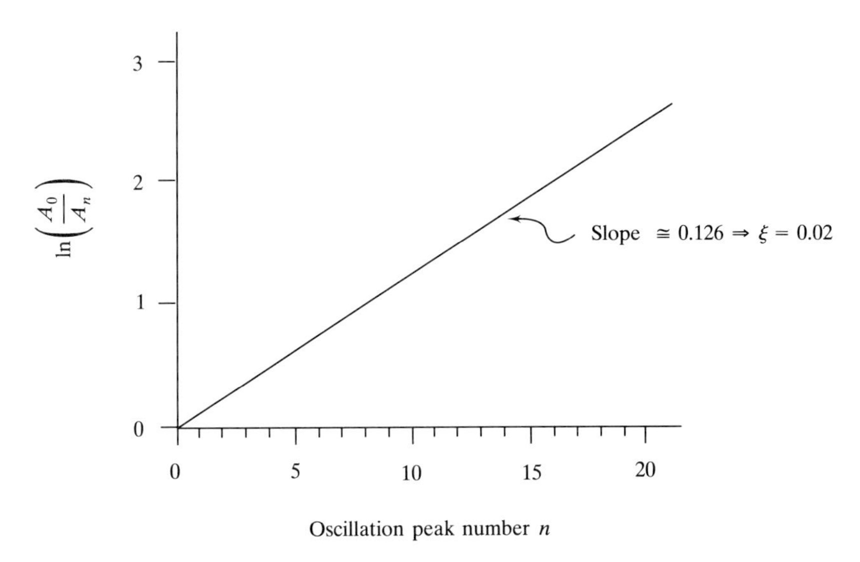

FIGURE 7.4
Logarithmic plot of amplitude ratio versus cycle number n, for the response shown in Fig. 7.2(a).

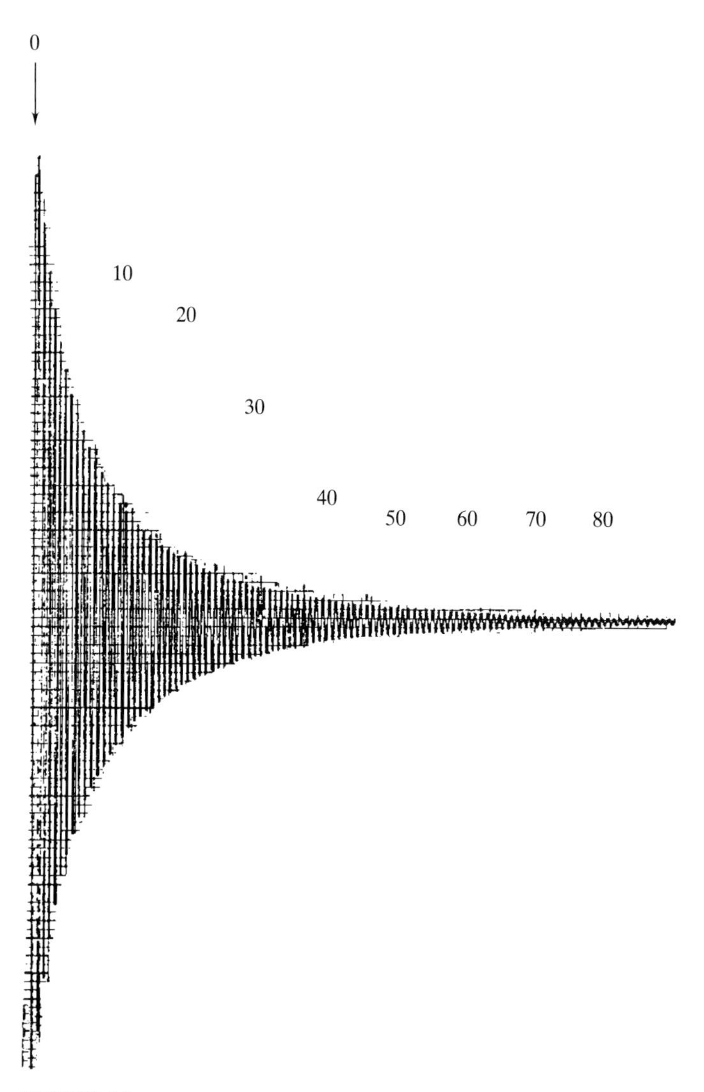

FIGURE 7.5
Free vibration data for a cantilever beam. The data shown are measured bending strain versus time. The bending strain is proportional to vibrational displacement. The total elapsed time for the 80 cycles indicated is 76.5 s, yielding a damped natural frequency of 1.046 Hz. The undamped natural frequency is essentially the same because of the small damping factor ξ. (Data provided courtesy of Carl Baker, Battelle Pacific Northwest Laboratory.)

natural logarithm of (7.18):

$$\ln\left(\frac{A_0}{A_n}\right) = \frac{2\pi n\xi}{(1-\xi^2)^{1/2}} \tag{7.19}$$

This relation shows that, if we construct a graph of $\ln(A_0/A_n)$ versus n, the slope of this graph will be $2\pi\xi/(1-\xi^2)^{1/2}$, or just $2\pi\xi$ if the damping is light. This graph is illustrated in Fig. 7.4, for which the data of Fig. 7.2(a) were used. Notice that this graph will always start at the origin, because $\ln(A_0/A_0) = \ln(1) = 0$. We note that, while any two peaks, if accurately measured, can be used to determine ξ according to (7.19), the technique based on Fig. 7.4 is much preferable, as it allows us to find an average slope over many cycles of data, thus minimizing errors in the measurement of individual peaks.

In order to illustrate the decay testing method further, actual data from a free vibration test of a vibrating cantilever beam are shown in Fig. 7.5. The vibrational motion was monitored with a strain gage mounted near the root of the beam. The strain gage measures the history of bending strain, which in this experiment is proportional to the displacement of the beam. The damping in this system is due to air resistance and to the dissipation (heating) in the metal as it deforms.

TABLE 7.2
Amplitude data from Fig. 7.5

n	Double amplitude (cm)	$\ln(A_0/A_n)$
0	24.1	0
1	21.1	0.133
2	19.7	0.202
3	17.4	0.326
4	15.6	0.435
5	14.0	0.543
7	11.9	0.740
10	8.65	1.025
15	5.9	1.407
20	4.35	1.712
25	3.35	1.973
30	2.55	2.246
35	1.97	2.504
40	1.6	2.712
45	1.29	2.928
50	1.04	3.143
55	0.83	3.369
60	0.70	3.539
70	0.50	3.875
80	0.35	4.232

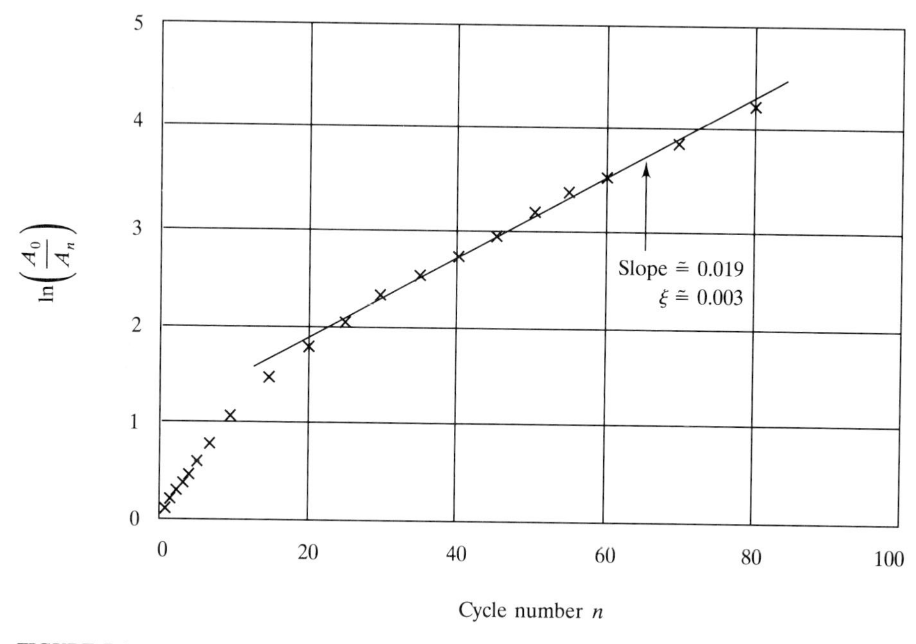

FIGURE 7.6
Plot of $\ln(A_0/A_n)$ versus cycle number n for the data of Fig. 7.5, as tabulated in Table 7.2. See text for discussion.

Selecting the indicated oscillation peak as the reference amplitude A_0, the successive peaks were then measured. (The scale of the measured quantity is arbitrary because the data are eventually used in ratio form.) The amplitudes are tabulated in Table 7.2, along with additional information needed to construct the type of plot shown in Fig. 7.4.

Figure 7.6 shows the resulting plot of $\ln(A_0/A_n)$ versus n. As noted in the figure, the damping factor ξ was determined as $\xi \cong 0.003$. This small value is typical of metals deforming in the elastic range. We also observe in Fig. 7.6 that the graph of $\ln(A_0/A_n)$ versus n does not follow a straight line. For small n, corresponding to the relatively larger amplitudes at the start of the test, there is noticeable nonlinearity in the graph. For large n, corresponding to relatively small amplitude motion, the data exhibit a linear variation, and this linear region was used to measure the damping factor ξ.

Comments

1. The nonlinear variation of $\ln(A_0/A_n)$ with n indicates that nonlinear damping effects are appreciable at large amplitude, and the oscillation decay at large amplitude is, therefore, not exponential. If the damping were truly linear, that is, characterized by a damping force given by $F_D = -c\dot{x}$, then $\ln(A_0/A_n)$ versus n would have to vary linearly with n. For the experiment considered

here, the nonlinear damping appears to be due to the effect of air drag on the vibrating beam; this effect produces forces that are approximately proportional to the square of the velocity, that is, $F_{D_{\text{air}}} \cong -c|\dot{x}|\dot{x}$.

2. If we are confronted with data similar to that of Fig. 7.6, then we should measure the system linear damping factor in the small amplitude range (large n) of the data. The small amplitude range is associated with motions near equilibrium, for which the linear model is to be defined.
3. The linear range of the data will tell us the amplitudes for which the linear viscous damping model works. In Fig. 7.6 the data are linear beyond about $n = 25$, with a corresponding amplitude denoted as A. Thus, if the system oscillates with amplitudes below A, the linear model should be valid.
4. By measuring the elapsed time T_n for the system to execute a number n of full cycles of motion, we can also determine the oscillator natural frequency ω_d as $\omega_d = 2\pi n/T_n$; the undamped natural frequency ω can then be found from (7.8). From Fig. 7.5 we observe a system natural frequency of $\omega \cong 1.046$ Hz. The decay test, therefore, allows us to identify both of the basic linear system constants ξ and ω. Free vibration decay testing is thus seen to be a useful technique to perform system identification and is commonly used for this purpose.

In order to gain further insight into the damping factors that typically occur in mechanical systems, representative ranges of the damping factor ξ for various systems are summarized in Table 7.3.

Some references that provide more detailed information on typical system damping factors are listed at the end of this chapter. In particular, see Ref. 7.1, which contains a lot of useful information on damping properties. One observes that for many mechanical vibratory systems the damping is light unless some type of damping component (such as an automobile shock absorber or the large dampers used in the recoil mechanisms of artillery) is added with the intention of producing large damping.

TABLE 7.3
Representative damping factors (ξ)

System	Typical range of damping factor
Metals (in elastic range)	< 0.01
Metal structures with joints	~ 0.03
Aluminum/steel transmission lines	$\sim 4 \times 10^{-4}$
Auto shock absorbers	~ 0.3
Rubber	~ 0.05
Large buildings vibrating during earthquake	$\sim 0.01\text{–}0.05$
Field artillery recoil mechanism	~ 1.0

Now let's continue the study of single degree of freedom vibrations by considering the effect of external excitations (inputs). As we shall see, the response to inputs will be strongly dependent on the system parameters ξ and ω, and this is a main reason why careful system identification is necessary.

7.2.2 Forced Vibration of Single Degree of Freedom Linear Oscillators

We now consider the forced vibration of the single degree of freedom oscillator model of Fig. 7.1. We assume that a known, time-dependent external force $F(t)$ acts on the system, so that the equation of motion is (2.52)

$$m\ddot{x} + c\dot{x} + kx = F(t)$$

Division by the mass m allows us to put the equation of motion in the standard form, in terms of the system parameters ξ and ω:

$$\ddot{x} + 2\xi\omega\dot{x} + \omega^2 x = \frac{F(t)}{m} \tag{7.20}$$

In Sec. 5.1.2 we discussed various types of excitations $F(t)$ in physical terms. Those most relevant to vibratory systems are support excitations, electromechanical excitations, and direct load application. Direct load application, wherein a known time-dependent force $F(t)$ of mechanical origin is assumed to be exerted directly on the mass m, commonly arises in two ways: (1) due to the effects of rotating or reciprocating parts in engines or machinery and (2) the time-dependent loads exerted by fluid pressure in a variety of circumstances (an example is the rolling moments produced by transverse waves on ships, inducing the ship to execute an oscillatory rolling motion). In practical situations involving fluid pressure loadings, the most challenging part of the analysis may be the development of an adequate model to characterize the loading itself.

In mathematical terms we can categorize the input force as one of four basic types, as described below:

1. *Harmonic excitation*: In this case the loading is of the form $F(t) = P\cos\Omega t$ and is characterized by its amplitude P and frequency Ω. We have considered the response to this type of loading in Sec. 5.3.2, in which the basic results of the analysis are the steady-state response amplitude relation (5.79) and the accompanying Fig. 5.15.
2. *Periodic excitation*: Some excitations, although not simple harmonic, nevertheless repeat themselves indefinitely and so are termed *periodic*. A periodic function $x(t)$ has the property that $x(t + \tau) = x(t)$ for all t, where τ is the period of the function. Shown in Fig. 7.7 is a periodic function that is distinctly nonsinusoidal. The period τ of this function is noted in the figure.

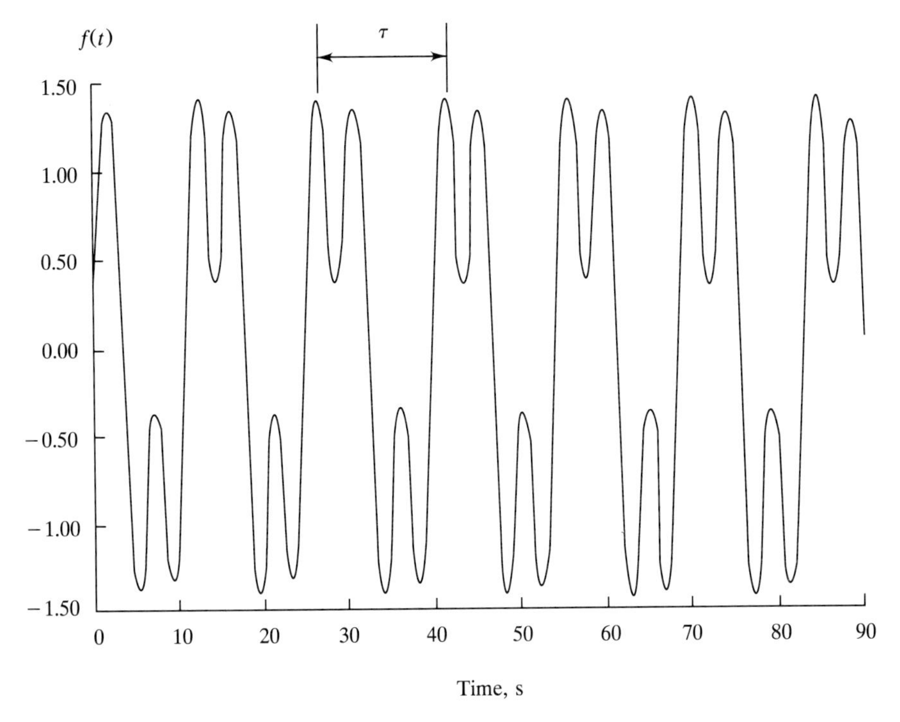

FIGURE 7.7
A periodic function $f(t + \tau) = f(t)$. The period τ is noted. (Provided courtesy of Dr. Zahidul Rahman, Ref. 7.8.)

We also define the "fundamental frequency" ω of the periodic function* as $\omega = 2\pi/\tau$. Examples of periodic excitations include the force exerted by a person pushing a child on a swing, the dynamic loading on an engine due to motion of reciprocating parts, and the electrochemical stimulus that causes contractions of muscle in the heart.

3. *Aperiodic excitations*: Excitations that do not have a repeated or periodic character include the step and truncated step inputs considered in Sec. 5.2.2 and in Examples 5.3 and 5.7, and the "ramp input," which increases linearly with time. These types of idealized excitations are often used to represent actual loadings that are applied suddenly and persist for a certain length of time (steps), or loadings which increase monotonically from zero (ramps).

*This ω is in general unrelated to the harmonic oscillator $\omega = \sqrt{k/m}$. Here ω is simply the frequency one obtains from $\omega = 2\pi/\tau$, that is, it is merely another way to express the period of the periodic function.

4. *"Random" excitation*: This is actually another example of an excitation which is aperiodic, but the so-called random excitation represents those situations in which the loading is irregular and cannot be predicted ahead of time. Examples include the ground motion during an earthquake, the wind loadings exerted on buildings, and the contour of rough road over which a vehicle moves. Excitations that are characterized as being random generally exhibit some type of oscillatory behavior, as in Fig. 5.25, but do not repeat themselves. Random excitations and the responses which they produce (which are also "random") are generally analyzed statistically.

Regardless of the type of excitation, we will be concerned with both the transient and steady-state responses. In cases for which the excitation is harmonic or periodic or persists for a long period of time, the steady-state response is usually of most interest, as it is what we actually observe if we monitor the behavior of the system. In many cases involving aperiodic forcing, which may be exerted only over limited time intervals (for example, the truncated step, Example 5.7, or for a sudden wind gust or blast load on a structure), we are more interested in the transient response; indeed for excitations of short duration, steady-state conditions may not even be reached.

In Examples 5.3 and 5.7 we considered the response to simple aperiodic excitations (the step and truncated step). Further examples of such idealized aperiodic inputs will be considered in Chap. 8, control systems. In Secs. 5.2 and 5.3 we studied the steady-state response to harmonic excitation. The remainder of this section will be devoted to extending these results for the harmonically excited case. The references 7.1 through 7.7 should be consulted for material on response of oscillators to periodic and random excitations, which are not considered here.

7.2.3 Response to Harmonic Excitation

If a harmonic input load is applied to the mass m of Fig. 7.1, the equation of motion (7.20) of the linear oscillator is

$$\ddot{x} + 2\xi\omega\dot{x} + \omega^2 x = \frac{P}{m}\cos\Omega t \tag{7.21}$$

In Sec. 5.3.2 the steady-state solution, given by Eqs. (5.69), (5.77), and (5.78), was determined, and Fig. 5.15 presented the system frequency response. Let us repeat these results below with the inclusion of the homogeneous or free response solution (7.10), which defines the transient part of the response:

$$x(t) = A_0 e^{-\xi\omega t}\cos(\omega_d t - \psi) + \frac{\left(\dfrac{P}{k}\right)\cos(\Omega t - \phi)}{\left[\left(1 - \dfrac{\Omega^2}{\omega^2}\right)^2 + \left(2\xi\dfrac{\Omega}{\omega}\right)^2\right]^{1/2}} \tag{7.22}$$

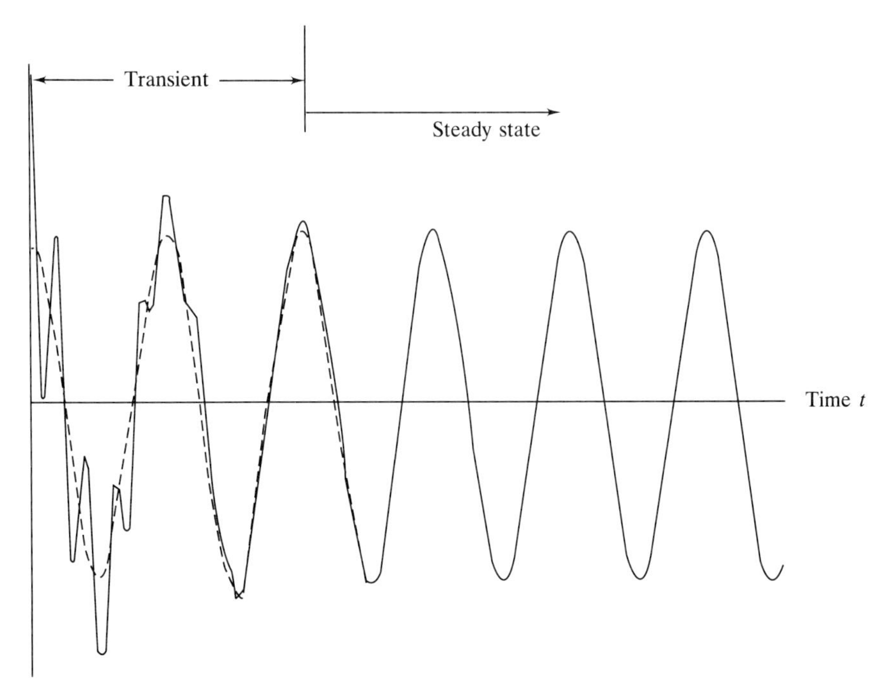

FIGURE 7.8
The solution $x(t)$, Eq. (7.22), for $\Omega = 1.0$, $\omega = 5.0$, $\xi = 0.05$, $P/k = 1.0$, $x(0) = 2$, $\dot{x}(0) = 0$. The dashed line near $t = 0$ is the steady component, on which the actual solution is superposed.

The constants A_0 and ψ appearing in the homogeneous solution are to be determined from the initial conditions. The homogeneous solution will die out exponentially, and the long-time behavior (steady-state response) is defined by the particular solution. The complete solution will thus consist of the steady oscillation at the driving frequency Ω, on which is superposed a decaying oscillation of a generally different frequency ω_d. Shown in Fig. 7.8 is a complete solution for the case $\Omega = 1.0$, $\omega = 5$ and with other parameters as noted in the figure caption.

The transient and steady-state regimes are noted in the figure. In the transient regime the homogeneous and particular solutions can each be seen easily, because their frequencies ($\Omega = 1.0$, $\omega_d \approx 5$) are widely separated. Five cycles of free oscillation occur over each cycle of the steady-state component of the response. The duration of the transient regime will clearly depend on the system damping factor ξ.

The associated orbit in the phase plane is shown in Fig. 7.9, and the large transient component is evident; the steady-state orbit, which is realized for all

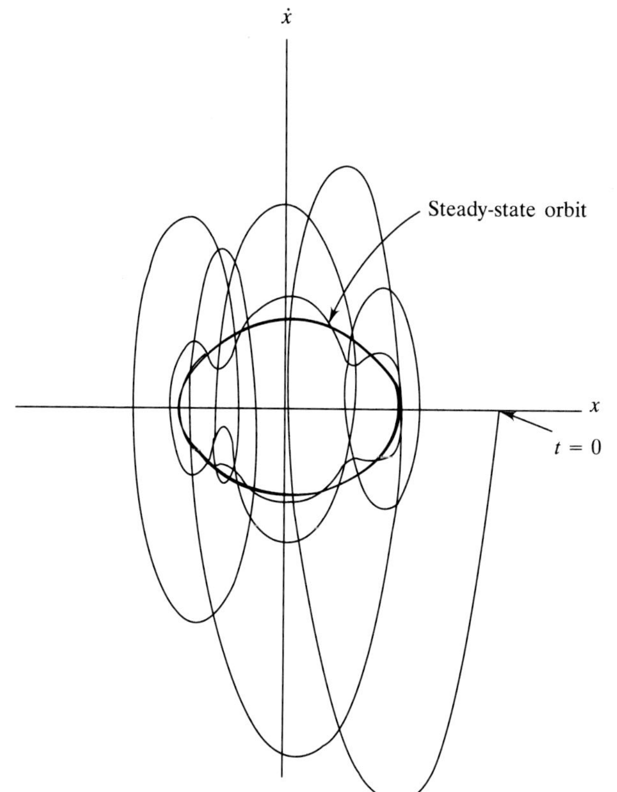

FIGURE 7.9
Phase plane orbit for the solution of Fig. 7.8. The steady-state solution is indicated.

initial conditions, is also noted in the figure. Note here that, unlike in the autonomous case, for which orbits in the phase space cannot cross each other (Sec. 4.1.1), the orbits of Fig. 7.9 cross many times because the vector field that defines the tangent to the orbit at any time is time dependent.

The steady-state orbit shown in Fig. 7.9 is another example of an *attractor* (first defined in Sec. 4.3.9), a set of points in the phase space (in this case a closed curve) onto which all nearby regions of the phase space will collapse asymptotically. For this driven linear system, the orbits generated for *any* initial conditions will end up on the attractor. We thus observe that the domain or basin of attraction, the set of points in the phase space that will asymptote to the attractor, is the entire phase space. Because the attractor consists of a closed curve, signifying a periodic steady-state response, we refer to the attractor as a *periodic attractor*. Periodic attractors cannot occur in an autonomous linear system, for which the only type of attractor is an equilibrium solution. Periodic attractors will always occur, however, in an asymptotically stable linear system driven by a periodic stimulus.

Shown in Fig. 7.10 are the transient and steady-state responses for an oscillator excited at resonance, $\Omega = \omega = 5$. Notice that in the transient regime

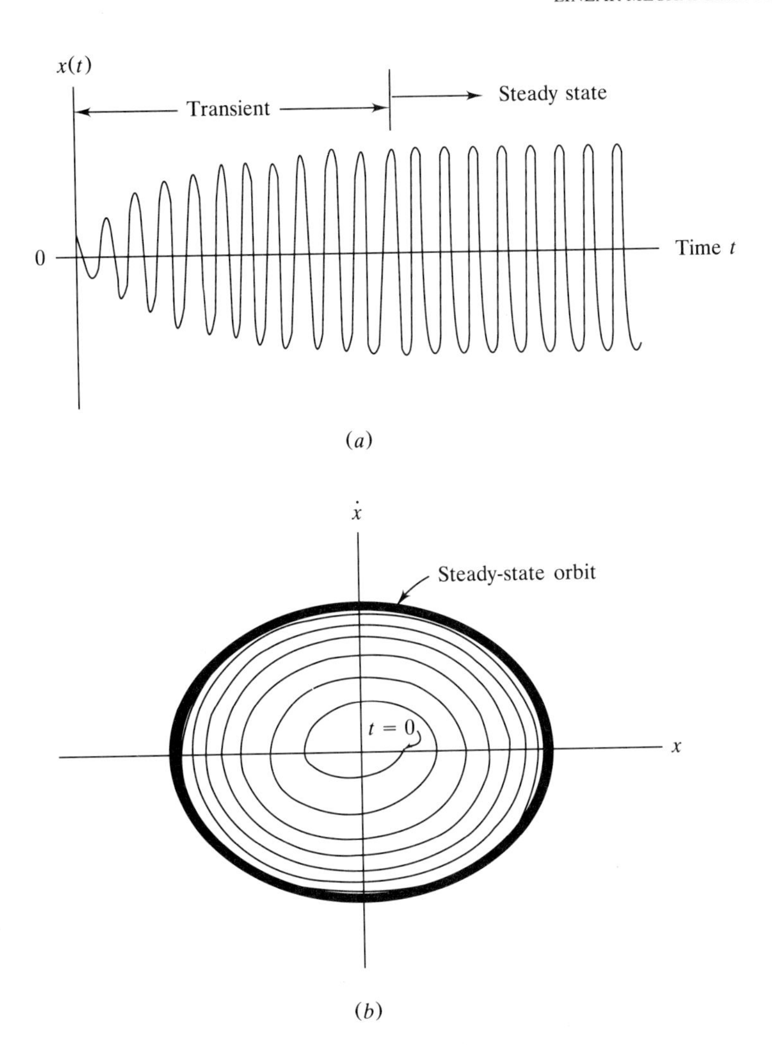

FIGURE 7.10
The response of the driven oscillator for the case $\Omega = \omega = 5$, for the same parameters and initial conditions as in Figs. 7.8 and 7.9: (*a*) The solution $x(t)$; (*b*) the phase portrait. The attractor is the outermost orbit, to which any orbit asymptotes.

it is not possible visually to separate the homogeneous and particular contributions to the solution, as they have the same frequency.

Let us now consider two common situations of practical interest in which harmonic loadings arise. The first is the so-called rotating unbalance, in which unbalance of a rotating component such as a shaft, flywheel, turbine fan, wheel, or rotor can produce harmonic excitation. The second case is the base or support excitation, in which the support to which the stiffness and damping elements are attached is moved harmonically. We'll consider these cases in turn.

7.2.4 Rotating Unbalance

An idealized model of the single degree of freedom rotating unbalance phenomenon is shown in Fig. 7.11.

The rotating device is modeled as a disk that rotates at the constant angular rate Ω. The mass unbalance is represented by a small mass m_0, which is offset by a distance e from the axis of rotation. The total mass of the system, including the unbalanced mass m_0, is M. The system is assumed to be very stiff in the horizontal direction, so that only vertical vibration can occur. In physical terms the "centrifugal force" $m_0 e\Omega^2$ due to the unbalanced mass m_0 will possess a vertical component which varies harmonically, and this is the source of the harmonic excitation.

The equation of motion of the system is conveniently derived using Lagrange's equation. The kinetic energy T of the system is, from Fig. 7.11,

$$T = \tfrac{1}{2}(M - m_0)\dot{x}^2 + \tfrac{1}{2}m_0(\dot{x} + e\Omega \cos \Omega t)^2 + \tfrac{1}{2}m_0(-e\Omega \sin \Omega t)^2$$

where $\dot{x} + e\Omega \cos \Omega t$ and $-e\Omega \sin \Omega t$ are the vertical and horizontal components of velocity, respectively, of the unbalanced mass m_0. The potential energy V due to the stiffness element is $V = \tfrac{1}{2}kx^2$ and the generalized force of damping is $-c\dot{x}$. Thus, with

$$\frac{\partial T}{\partial \dot{x}} = (M - m_0)\dot{x} + m_0(\dot{x} + e\Omega \cos \Omega t) = M\dot{x} + m_0 e\Omega \cos \Omega t$$

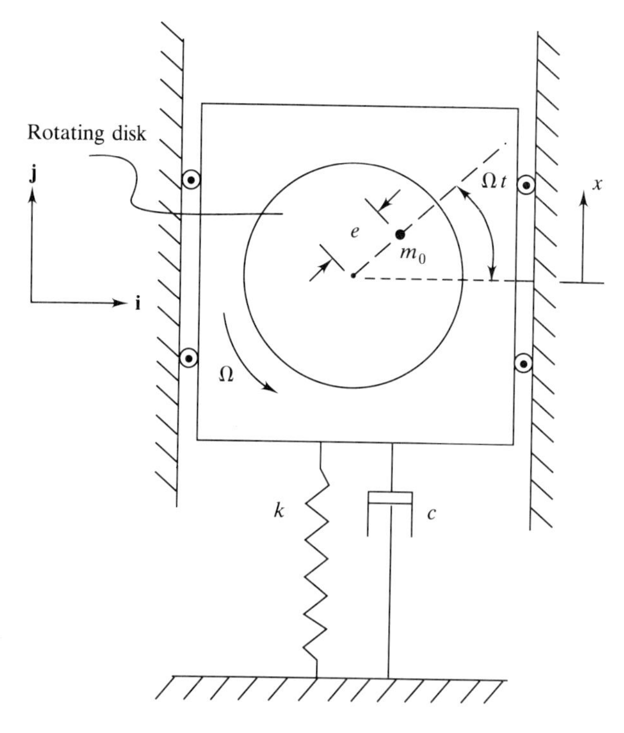

FIGURE 7.11
Idealized model for 1-DOF rotating unbalance excitation.

and

$$\frac{d}{dt}\left(\frac{\partial T}{\partial \dot{x}}\right) = M\ddot{x} - m_0 e \Omega^2 \sin \Omega t$$

Lagrange's equation yields the equation of motion as

$$M\ddot{x} + c\dot{x} + kx = m_0 e \Omega^2 \sin \Omega t \tag{7.23}$$

or, in standard form,

$$\ddot{x} + 2\xi\omega\dot{x} + \omega^2 x = \frac{m_0 e}{M}\Omega^2 \sin \Omega t \tag{7.24}$$

Observe that the harmonic load appearing in (7.23) is merely the vertical component of the centrifugal "force" arising from the unbalanced mass m_0. We also note that the amplitude $m_0 e \Omega^2$ of the excitation depends on the product $m_0 e$; thus, in an analysis we do not need to know m_0 and e individually, but only their product.

There is a basic difference between the excitation amplitude here and that assumed in Sec. 5.3.2. In Sec. 5.3.2 the excitation was taken as $P \cos \Omega t$. It was assumed that if we were to excite the system a number of times at different frequencies Ω (for example, in order to generate the system frequency response as in Fig. 5.15), the driving amplitude P would remain constant in those different cases. Here, however, we see that the driving amplitude arises naturally as a function of the driving frequency Ω, increasing as Ω^2. Thus, if we operate the device of Fig. 7.11 at several different driving frequencies, we have to account for the dependence of driving amplitude on frequency, and the results will differ from those obtained in Sec. 5.3.2. Nevertheless, the steady-state solution (5.78) previously found can be used directly, as follows: For rotating unbalance the amplitude P of the excitation is $P = m_0 e \Omega^2$. Substitution of this relation for P into the relation (5.78) for steady-state response amplitude A yields

$$A = \frac{\dfrac{m_0 e \Omega^2}{k}}{\left[\left(1 - \dfrac{\Omega^2}{\omega^2}\right)^2 + \left(2\xi\dfrac{\Omega}{\omega}\right)^2\right]^{1/2}}$$

This relation can be nondimensionalized by multiplying the numerator and denominator of the right-hand side by the system mass M, so that

$$A = \frac{\left(m_0 e \Omega^2 \dfrac{M}{k}\right)\dfrac{1}{M}}{\left[\left(1 - \dfrac{\Omega^2}{\omega^2}\right)^2 + \left(2\xi\dfrac{\Omega}{\omega}\right)^2\right]^{1/2}}$$

Next, note that $M/k = 1/\omega^2$ and rewrite the numerator of the preceding equation as

$$\left(m_0 e \Omega^2 \frac{M}{k}\right)\frac{1}{M} = \frac{m_0 e}{M}\frac{\Omega^2}{\omega^2}$$

Finally, move the $m_0 e/M$ to the left-hand side to obtain the dimensionless result

$$\frac{MA}{m_0 e} = \frac{\dfrac{\Omega^2}{\omega^2}}{\left[\left(1 - \dfrac{\Omega^2}{\omega^2}\right)^2 + \left(2\xi\dfrac{\Omega}{\omega}\right)^2\right]^{1/2}} \tag{7.25}$$

This is the basic result of the analysis, analogous to (5.79). The dimensionless amplitude ratio is now $MA/m_0 e$ rather than $A/(P/k)$. The right-hand side of

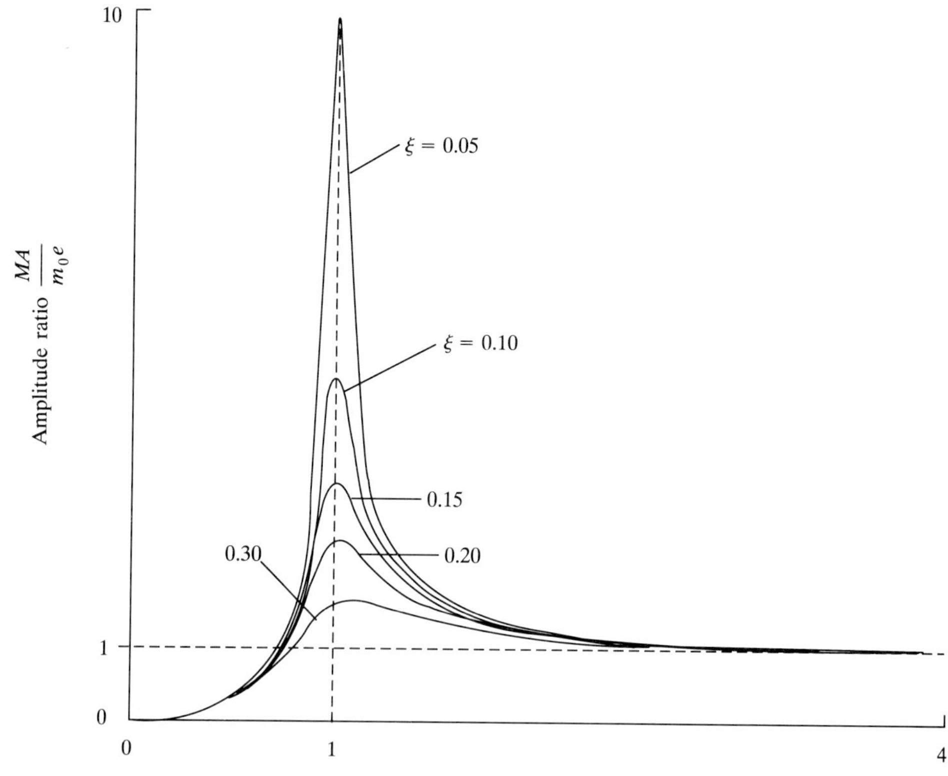

FIGURE 7.12
Frequency response curves for rotating unbalance excitation.

(7.25) is seen to be (Ω^2/ω^2) times the right-hand side of (5.79). Thus, to obtain the frequency response curves for rotating unbalance, we multiply the response curves of Fig. 5.15 by (Ω^2/ω^2). The resulting frequency response is shown in Fig. 7.12 for several values of the system damping factor ξ. Observe the following important features of Fig. 7.12 in comparison to Fig. 5.15:

1. As $\Omega/\omega \to 0$, the vibration amplitude A for rotating unbalance also approaches zero, whereas $A \to P/k$ if P is not a function of the driving frequency. This reflects the fact that, as $\Omega \to 0$, the excitation force amplitude $P = m_0 e \Omega^2$ also approaches zero.
2. As $\Omega/\omega \to \infty$, $MA/m_0 e \to 1$, whereas $A \to 0$ in Fig. 5.15. This reflects the fact that the increase in driving amplitude with Ω is balanced by the high-frequency "filtering" property of the system, which tends to reduce the amplitude as $1/\Omega^2$.
3. At resonance $\Omega/\omega = 1$, $MA/m_0 e = 1/2\xi$, exactly as before. Thus, the vibration behavior of the system in a near-resonant condition will be essentially the same as in the case for which P does not depend on Ω.

From a practical standpoint the behavior exhibited in Fig. 7.12 should be expected any time that vibration of a device with rotating or moving parts is excited.

7.2.5 Harmonic Base Excitation

We now consider a second common way of practical importance in which harmonic excitations can occur: the harmonic support motion depicted in Fig. 7.13. Here the base to which the stiffness and damping elements are grounded is assumed to move in the known harmonic manner indicated. The equation of motion is easily determined via newtonian mechanics to be

$$m\ddot{x} + c[\dot{x} - \dot{x}_b(t)] + k[x - x_b(t)] = 0$$

or

$$m\ddot{x} + c\dot{x} + kx = c\dot{x}_b(t) + kx_b(t) \tag{7.26}$$

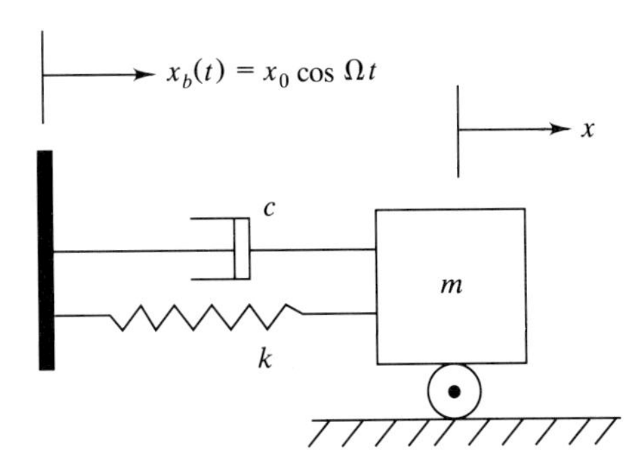

FIGURE 7.13
The harmonic base excitation model.

From this result we see that different input forces are transmitted through the spring and damper. Division of (7.26) by m yields the standard form

$$\ddot{x} + 2\xi\omega\dot{x} + \omega^2 x = \omega^2 x_b(t) + 2\xi\omega\dot{x}_b(t) \tag{7.27}$$

Use of the harmonic base excitation $x_b(t) = x_0 \cos \Omega t$ allows (7.27) to be expressed as

$$\ddot{x} + 2\xi\omega\dot{x} + \omega^2 x = \omega^2 x_0 \cos \Omega t - 2\xi\omega\Omega x_0 \sin \Omega t$$

The right-hand side may be written as a single harmonic term, so that

$$\ddot{x} + 2\xi\omega\dot{x} + \omega^2 x = x_0\omega^2\left[1 + \left(2\xi\frac{\Omega}{\omega}\right)^2\right]^{1/2} \cos(\Omega t - \psi) \tag{7.28}$$

where the phase ψ is defined by $\tan \psi = -2\xi\Omega/\omega$. The result (7.28) shows the actual force transmitted to the mass m to be out of phase with the base motion $x_b(t)$.

Comparison of (7.28) and (7.21) shows the exciting force amplitude P to be defined according to

$$\frac{P}{m} = x_0\omega^2\left[1 + \left(2\xi\frac{\Omega}{\omega}\right)^2\right]^{1/2}$$

so that the ratio $P/k = (P/m)(m/k) = (P/m)(1/\omega^2)$ will be

$$\frac{P}{k} = x_0\left[1 + \left(2\xi\frac{\Omega}{\omega}\right)^2\right]^{1/2} \tag{7.29}$$

We may, therefore, determine the steady-state vibration amplitude A by using (7.29) for P/k in (5.78). This yields the result

$$A = \frac{x_0\left[1 + \left(2\xi\frac{\Omega}{\omega}\right)^2\right]^{1/2}}{\left[\left(1 - \frac{\Omega^2}{\omega^2}\right)^2 + \left(2\xi\frac{\Omega}{\omega}\right)^2\right]^{1/2}}$$

Division by the base motion amplitude x_0 yields the following dimensionless amplitude:

$$\frac{A}{x_0} = \frac{\left[1 + \left(2\xi\frac{\Omega}{\omega}\right)^2\right]^{1/2}}{\left[\left(1 - \frac{\Omega^2}{\omega^2}\right)^2 + \left(2\xi\frac{\Omega}{\omega}\right)^2\right]^{1/2}} \tag{7.30}$$

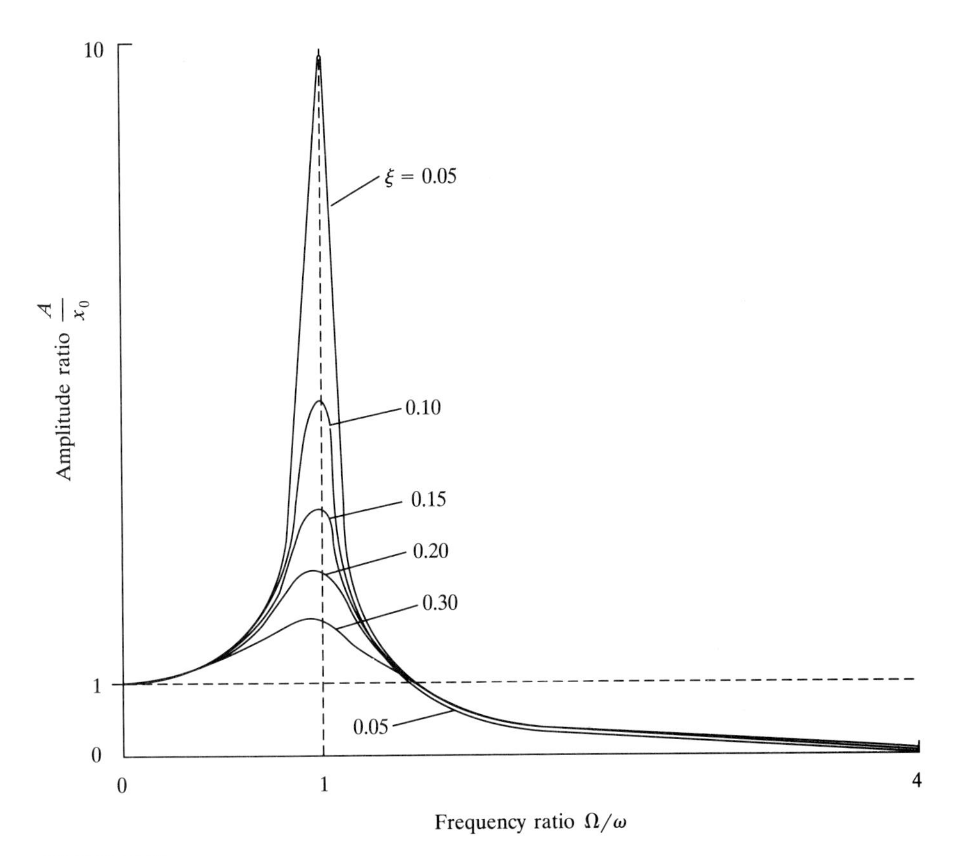

FIGURE 7.14
Frequency response for a harmonic base excitation.

which is the basic result of the analysis. Once again there is a difference between (7.30) and (5.79), because the exciting force defined by (7.29) is a function of the excitation frequency Ω, as well as of the system damping factor ξ. The frequency response for a base excitation of the type shown in Fig. 7.13, will, therefore, differ from that depicted in both Figs. 5.15 and 7.12. The frequency response given by (7.30) is shown in Fig. 7.14. Note that in (7.30) the vibration amplitude A is measured in multiples of the amplitude x_0 of the base motion. For small values of (Ω/ω), the term $(2\xi\Omega/\omega)^2$ in the numerator of (7.30) will be small compared to unity, so that the amplitude ratio A/x_0 will be essentially the same as that of (5.79). Thus, at low frequencies of excitation the response will be very close to that shown in Fig. 5.15. In physical terms the force transmitted through the damper is negligibly small.

In the high-frequency limit, $\Omega/\omega \to \infty$, (7.30) asymptotes to

$$\lim_{\Omega/\omega \to \infty} \frac{A}{x_0} = \frac{2\xi\Omega/\omega}{(\Omega/\omega)^2} = \frac{2\xi}{(\Omega/\omega)} \tag{7.31}$$

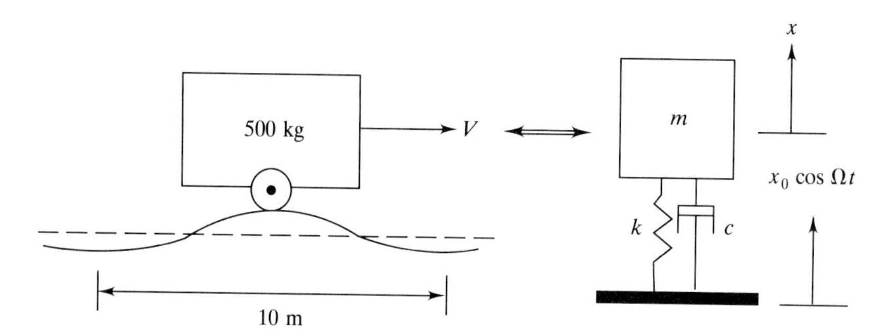

FIGURE 7.15
Trailer on a sinusoidal roadway modeled as a mass/spring/damper system subjected to a base excitation.

This function approaches zero as $1/(\Omega/\omega)$, whereas (5.79) shows $A/(P/k)$ to approach zero as $1/(\Omega/\omega)^2$ at high frequencies. At high excitation frequencies, therefore, the amplitude diminishes more slowly for a base excitation as the exciting frequency is increased.

An interesting feature of the response shown in Fig. 7.14 is that, for $\Omega/\omega > \sqrt{2}$, the larger the system damping factor ξ, the larger the vibration amplitude A. This trend may seem unusual, as we would normally expect damping to suppress vibration. In this case, however, the force transmitted through the damper increases linearly with ξ.

Example 7.1 A Trailer Traveling over a Sinusoidal Road. A loaded trailer of mass 500 kg travels at 25 m/s over an undulating road which is modeled by a sinusoidal profile of amplitude $x_0 = 1$ cm and wavelength $\lambda = 10$ m (Fig. 7.15). The combined stiffness of the coil springs is $k = 10^5$ N/m, and the system damping factor is $\xi = 0.1$. We assume that the tire-axle assembly moves rigidly,* so that the road undulations have the effect of moving the axles up and down harmonically to produce a base excitation. The vibratory motion of the trailer body is to be studied.

We first determine the parameters needed to define the steady-state response. The system natural frequency $\omega = \sqrt{k/m} \cong 14.1$ rad/s. The frequency Ω of the exciting base motion is determined by the trailer speed V and road wavelength λ as $\Omega = 2\pi V/\lambda \cong 15.7$ rad/s. Thus, the frequency ratio $\Omega/\omega \cong 1.11$; the system is being driven slightly above the resonant condition. From Fig. 7.14 or from (7.30), we determine the amplitude ratio A/x_0 to be $A/x_0 \cong 3$, so that the trailer will execute a steady oscillation of amplitude $A \cong 3$ cm, which is a significant motion. If the vehicle were to slow down a bit, so that $\Omega/\omega \approx 1.0$, then

*A more realistic model of this system would include the effects of tire stiffness and damping and would consider the motions of both the axle and the trailer body, as in Example 2.4. An example of this type is considered in Example 7.2, Sec. 7.3.1.

the oscillation amplitude would increase to the resonance value given by (7.30) with $\Omega/\omega = 1$,

$$\left.\frac{A}{x_0}\right|_{\Omega/\omega=1} = \frac{\left(1+4\xi^2\right)^{1/2}}{2\xi}$$

which yields $A \approx 5.2$ cm. Of course, this analysis assumes that the trailer maintains contact with the road and that the stiffness and damping are linear. The latter assumption is suspect if the vibration amplitude is actually as large as we have calculated.

Comments on the harmonically forced oscillator

1. Inspection of Figs. 5.15 and 7.12 shows that, if the system damping factor ξ is less than about 0.20, then the steady-state response far from resonance (Ω/ω not near unity) will be essentially independent of the damping. Suppose, for example, that a system driven by a rotating unbalance is operated at an excitation condition $\Omega/\omega = 3$, with a system damping factor $\xi = 0.05$. If we were to triple the damping, so that $\xi = 0.15$, we see from Fig. 7.12 that the steady-state response would be virtually the same as before. On the other hand, for systems operating at or near resonance ($\Omega/\omega \approx 1$), the steady-state vibration amplitude is inversely proportional to ξ, so that the amount of damping is crucial in determining the severity of the vibration. We see then that the *main effect of damping is to limit the response at or near resonance*. This is also true in the case of a base excitation (Fig. 7.14); however, Fig. 7.14 shows that, well above resonance, the amplitude increases linearly with ξ, as previously noted.
2. The simple rotating unbalance model of Fig. 7.11, and the associated analysis, provide an adequate model of many vibrational phenomena involving systems with rotating parts. Furthermore, the essential features of vibration of devices with reciprocating parts, such as an automobile engine, can be understood in the context of our simple rotating unbalance model, although reciprocating parts generally lead to more complicated "inertial loadings." Thus, nearly all linear, single degree of freedom vibrations excited by rotating or reciprocating parts can be described by the analysis presented here.

   On the other hand, base excitations often arise naturally in such a way that the excitation differs significantly from the simple harmonic model used here. Examples include the ground motion during an earthquake and the travel of vehicles over rough roads. Thus, many base excitation phenomena of practical importance cannot be described by the results obtained here.
3. In Sec. 7.2.1 free vibration testing was presented as a technique for system identification. Harmonically forced vibration can also be employed for this purpose. It is common in engineering practice to determine the steady-state frequency response of systems by exciting them harmonically over a range of frequencies and to use the results to determine the system damping factor ξ and undamped natural frequency ω. For example, suppose that we employ a "counterrotating eccentric weight exciter," as shown in Fig. 7.16, to excite a vibratory motion. The exciter produces a "centrifugal force" along the vertical direction only (the horizontal components of the excitation cancel). Suppose we attach this device rigidly to the single degree of freedom system under

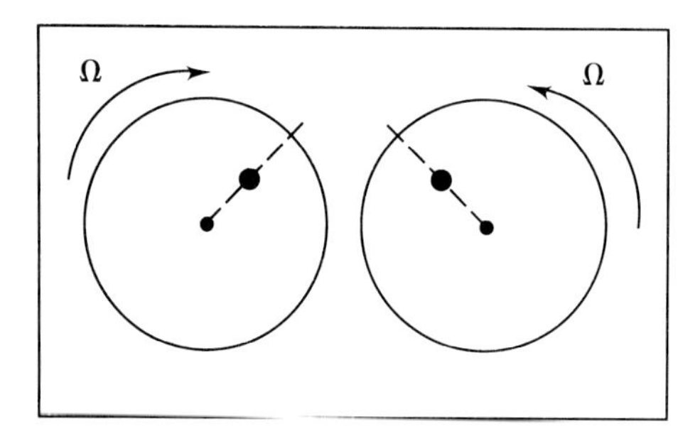

FIGURE 7.16
A counterrotating eccentric weight exciter.

consideration and that we excite the system over a range of driving frequencies Ω. The resulting steady-state frequency response, characteristic of a rotating unbalance excitation, might then look like Fig. 7.17. Note that the data will consist of a plot of actual vibration amplitude A versus driving frequency Ω, rather than in the dimensionless form of Fig. 7.12. If the system damping is light, the peak vibration amplitude A_R occurs essentially at resonance ($\Omega \approx \omega$), so the system natural frequency ω can be determined by measuring the exciting frequency at which the amplitude is a maximum. The damping factor ξ is then found as follows: We know from (7.25) that, at resonance,

$$\frac{MA_R}{m_0 e} = \frac{1}{2\xi}$$

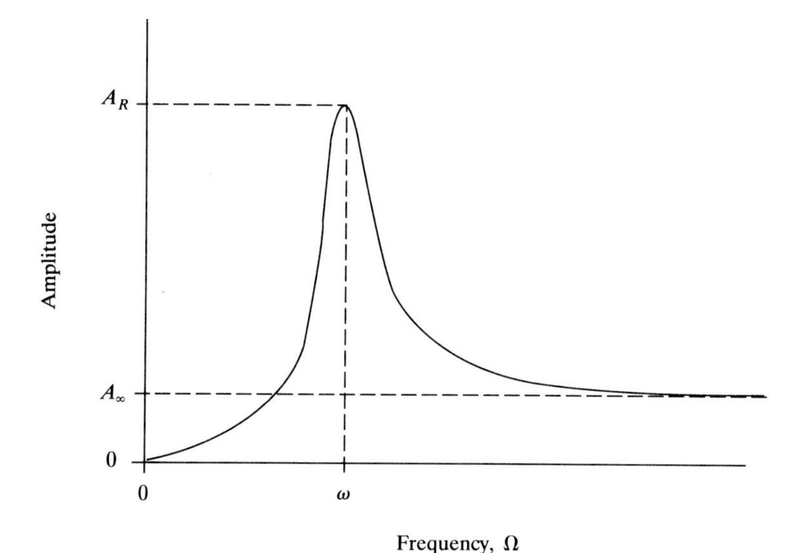

FIGURE 7.17
Hypothetical data obtained via a rotating unbalance excitation.

and that as $\Omega/\omega \to \infty$,

$$\frac{MA_\infty}{m_0 e} = 1$$

By measuring A_R and A_∞, as shown in Fig. 7.17, the above relations can be divided by each other to eliminate the constant $M/m_0 e$, and one obtains

$$\frac{A_R}{A_\infty} = \frac{1}{2\xi} \to \xi = \frac{A_\infty}{2A_R}$$

Similar procedures can be used if some other type of harmonic excitation, such as an electromagnetic shaker, is employed. Additional methods of forced vibration testing are discussed in several of the references cited at the end of this chapter.

7.3 MULTI-DEGREE-OF-FREEDOM LINEAR VIBRATION

In many situations the use of a single position coordinate does not suffice to describe the vibratory motion of a mechanical system. Some previously considered examples in which multiple degrees of freedom are necessary include the auto-suspension system models of Examples 2.4 and 2.5, the earthquake building model of Example 2.6, and the airfoil in wind tunnel model of Example 2.9. Each of these multi-degree-of-freedom systems possesses mass, stiffness, and damping properties, although the damping was not always incorporated into the model. A common feature of these example systems is that the mass, stiffness, and damping properties are modeled by *discrete* elements: The masses are considered to be particles or rigid bodies, and the damping and stiffness effects are described by linear springs and dashpots. The mathematical models naturally arise as sets of coupled second-order ODEs. From a mathematical viewpoint the analysis of such systems of second-order ODEs is the main topic of this section.

There are many vibrational phenomena of practical importance for which the mathematical models do *not* arise naturally as sets of second-order ODEs. Specifically, consider the class of problems involving the vibration of *nonrigid bodies*. Examples include the vibration of large space structures, cooling towers, buildings, bridges, airplane structures such as wings and body panels, and automobile body panels. The vibrational motion in these systems consists of time-dependent deformations of the bodies themselves. The mass, damping, and stiffness properties of these systems are distributed *continuously* in space, and the vibrational displacements are functions of both space and time. The mathematical models of such systems arise naturally as *partial differential equations*, which model any phenomena for which the dependent variables depend on more than one independent variable. A simple example of a vibrational motion governed by a partial differential equation is provided in Sec. 9.6.1, the vibrating cantilever beam.

The vibrations of nonrigid bodies are often analyzed by first replacing the actual structure with an approximate version in which the displacements and the mass, stiffness, and damping properties are associated with a finite number of isolated *points* in the structure, thus forming a discrete model of the type considered in this section. This approximation is usually made using the finite element method, which is introduced in Chap. 9. Often such approximate discrete models of complex structures require hundreds of degrees of freedom in order to account adequately for the often complex geometries involved. The intent of this and the preceding paragraph is to point out that the techniques discussed here in Sec. 7.3 have a range of applicability well beyond particle and rigid body vibrational dynamics.

We'll begin our study of multi-degree-of-freedom linear vibrations by considering the free vibration of a simple, undamped system with 2 degrees of freedom. As we shall see, the transition from 1 to 2 degrees of freedom leads to notable differences in the way the analysis is conducted. On the other hand, the transition from 2 to an arbitrarily large number of degrees of freedom is conceptually straightforward. The basic concepts needed to describe and understand the system motions are the same, whether the system has 2 or a very large number of degrees of freedom.

In analyzing multi-degree-of-freedom vibratory systems, we will use vector-matrix methods which are very similar to those employed for the multi-state systems considered in Chaps. 4 and 5. The solutions will be defined by eigenvalues and eigenvectors, as in those chapters. The multi-degree-of-freedom vibration problem is another example of the utility of vector-matrix and eigenvalue-eigenvector methods for the solution of linear ODE models.

7.3.1 Undamped Free Vibration of 2-DOF Systems

We now introduce the basic ideas by considering the vibration of a specific system with 2 degrees of freedom, as shown in Fig. 7.18. Here the springs are equal, as are the masses, and the coordinates are x_1 and x_2. This model is actually the same as Example 2.6 (earthquake/building) except there are 2 instead of 3 degrees of freedom. This model, therefore, might represent the free vibration of a two-story structure. The model is also similar to the auto-suspension model of Example 2.4, with the dampers removed.

The system equations of motion may be derived using Lagrange's equations or Newton's second law. Using Lagrange's equations here, we first determine the kinetic and potential energies to be

$$T = \tfrac{1}{2}m\dot{x}_1^2 + \tfrac{1}{2}m\dot{x}_2^2$$

$$V = \tfrac{1}{2}kx_1^2 + \tfrac{1}{2}k(x_1 - x_2)^2$$

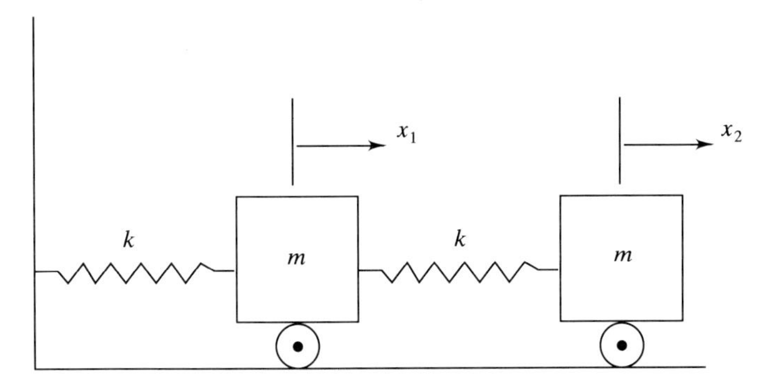

FIGURE 7.18
Undamped 2-DOF system with equal masses and equal springs.

Application of (6.65) provides the equations of motion as

$$
\begin{aligned}
m\ddot{x}_1 + 2kx_1 - kx_2 &= 0 \\
m\ddot{x}_2 + kx_2 - kx_1 &= 0
\end{aligned}
\tag{7.32}
$$

The equations of motion are next expressed in vector-matrix form by defining displacement and acceleration vectors as follows:

Position or displacement vector $\{x\} = \begin{Bmatrix} x_1 \\ x_2 \end{Bmatrix}$ Acceleration vector $\{\ddot{x}\} = \begin{Bmatrix} \ddot{x}_1 \\ \ddot{x}_2 \end{Bmatrix}$

We may then express the equations of motion in the following vector-matrix form:

$$
\begin{bmatrix} m & 0 \\ 0 & m \end{bmatrix} \begin{Bmatrix} \ddot{x}_1 \\ \ddot{x}_2 \end{Bmatrix} + \begin{bmatrix} 2k & -k \\ -k & k \end{bmatrix} \begin{Bmatrix} x_1 \\ x_2 \end{Bmatrix} = \mathbf{0} \tag{7.33}
$$

where $\mathbf{0}$ is the zero vector

$$
\mathbf{0} = \begin{Bmatrix} 0 \\ 0 \end{Bmatrix}
$$

We refer to the matrix coefficient of the acceleration vector as the *mass matrix* and the matrix coefficient of the displacement vector as the *stiffness matrix*. These matrices define the basic system properties.

By analogy with the undamped single degree of freedom system, for which the free oscillation solution is simple harmonic, we will seek solutions to (7.33) by assuming that simple harmonic motion is possible. Let us assume a solution of the form

$$
\begin{Bmatrix} x_1 \\ x_2 \end{Bmatrix} = \begin{Bmatrix} c_1 \\ c_2 \end{Bmatrix} \cos(\omega t - \phi) \tag{7.34}
$$

where c_1, c_2, ϕ, and ω are initially unknown. This assumed solution is seen to be similar to the solution (4.71) used in Chap. 4 to solve the autonomous state variable problem. In the present case, however, we are assuming at the outset that the time dependence will be harmonic, whereas the more general form (4.71) may produce harmonic (imaginary eigenvalues), damped oscillatory (complex eigenvalues), or exponential (real eigenvalues) behavior.

Noting that the acceleration vector $\{\ddot{x}\}$ will be given from (7.34) by

$$\{\ddot{x}\} = -\omega^2 \begin{Bmatrix} c_1 \\ c_2 \end{Bmatrix} \cos(\omega t - \phi) \tag{7.35}$$

the substitution of (7.34) and (7.35) into (7.33) results in

$$-\omega^2 \begin{bmatrix} m & 0 \\ 0 & m \end{bmatrix} \begin{Bmatrix} c_1 \\ c_2 \end{Bmatrix} \cos(\omega t - \phi) + \begin{bmatrix} 2k & -k \\ -k & k \end{bmatrix} \begin{Bmatrix} c_1 \\ c_2 \end{Bmatrix} \cos(\omega t - \phi) = \mathbf{0}$$

The two terms can be combined into one as follows:

$$\left[\begin{bmatrix} 2k & -k \\ -k & k \end{bmatrix} - \omega^2 \begin{bmatrix} m & 0 \\ 0 & m \end{bmatrix} \right] \begin{Bmatrix} c_1 \\ c_2 \end{Bmatrix} \cos(\omega t - \phi) = \mathbf{0}$$

Observe that the matrix which multiplies $\{C\}$ is of the form $[[k] - \omega^2[m]]$. Because the solution has to be valid for all time, $\cos(\omega t - \phi)$ is in general nonzero, and we therefore require that

$$\left[\begin{bmatrix} 2k & -k \\ -k & k \end{bmatrix} - \omega^2 \begin{bmatrix} m & 0 \\ 0 & m \end{bmatrix} \right] \begin{Bmatrix} c_1 \\ c_2 \end{Bmatrix} = \mathbf{0} \tag{7.36}$$

or combining terms from the mass and stiffness matrices:

$$\begin{bmatrix} (2k - m\omega^2) & -k \\ -k & (k - m\omega^2) \end{bmatrix} \begin{Bmatrix} c_1 \\ c_2 \end{Bmatrix} = \mathbf{0} \tag{7.37}$$

This result is similar to (4.73) in the autonomous state variable problem. We have obtained a pair of linear, homogeneous algebraic equations. These will possess a nontrivial solution (c_1 and/or c_2 nonzero) only if the determinant of the coefficient matrix vanishes. Thus, we require that

$$\begin{vmatrix} (2k - m\omega^2) & -k \\ -k & (k - m\omega^2) \end{vmatrix} = 0 \tag{7.38}$$

We see that in order that the assumed solution (7.34) be valid, we must pick the oscillation frequency ω so that (7.38) is satisfied. The frequency ω (actually ω^2) appears in the formulation as an eigenvalue, just as the eigenvalue λ appears in (4.72) and (4.73). Equation (7.38) is the characteristic equation for this free vibration problem. If we expand the determinant (7.38), we obtain

$$(2k - m\omega^2)(k - m\omega^2) - k^2 = 0$$

or, multiplying out the terms,

$$m^2\omega^4 - 3km\omega^2 + k^2 = 0$$

Division by m^2 then yields

$$\omega^4 - 3\frac{k}{m}\omega^2 + \left(\frac{k}{m}\right)^2 = 0 \tag{7.39}$$

We observe that the characteristic equation (7.39) is an algebraic equation of the fourth degree and defines the oscillation frequencies that are possible. Note, however, that odd powers of ω do not appear in (7.39), so that the equation is quadratic in ω^2. We may, therefore, solve (7.39) for ω^2 using the quadratic formula. The solution is given by

$$\omega^2 = \frac{3k}{2m} \pm \left[\left(\frac{3k}{2m}\right)^2 - \left(\frac{k}{m}\right)^2\right]^{1/2}$$

or

$$\omega^2 = \frac{k}{m}\left(\frac{3 \pm \sqrt{5}}{2}\right) \tag{7.40}$$

Because we have obtained two solutions for ω^2, we conclude that two different motions of the type assumed in (7.34) are possible, each having its own oscillation frequency. Defining these frequencies in ascending order, we then find

$$\omega_1 = \sqrt{\frac{k}{m}}\left(\frac{3 - \sqrt{5}}{2}\right)^{1/2} = 0.618\sqrt{\frac{k}{m}}$$

$$\omega_2 = \sqrt{\frac{k}{m}}\left(\frac{3 + \sqrt{5}}{2}\right)^{1/2} = 1.618\sqrt{\frac{k}{m}}$$

The two natural frequencies are expressed as multiples of $\sqrt{k/m}$.

We have found that there are two harmonic solutions to the problem, each characterized by its natural frequency ω_1 or ω_2. Each of these frequencies will have associated with it an eigenvector $\begin{Bmatrix} c_1 \\ c_2 \end{Bmatrix}$. In order to find the eigenvectors, we proceed exactly as in Sec. 4.2.1. We substitute the first frequency squared ω_1^2 into (7.37) to obtain the eigenvector $\begin{Bmatrix} c_1 \\ c_2 \end{Bmatrix}_1$ associated with the first solution:

$$\begin{bmatrix} (2k - m\omega_1^2) & -k \\ -k & (k - m\omega_1^2) \end{bmatrix} \begin{Bmatrix} c_1 \\ c_2 \end{Bmatrix}_1 = \mathbf{0} \tag{7.41}$$

Note the similarity of (7.41) to (4.78), which defines the eigenvectors of the autonomous state variable problem. Substitution of the value $\omega_1^2 = 0.382\ k/m$

into (7.41) yields

$$\begin{bmatrix} (2k - 0.382k) & -k \\ -k & (k - 0.382k) \end{bmatrix} \begin{Bmatrix} c_1 \\ c_2 \end{Bmatrix}_1 = \mathbf{0} \tag{7.42}$$

The first row of (7.42) is the equation

$$(1.618k)c_1 - kc_2 = 0$$

which says that $c_2 = 1.618c_1$. The second of Eqs. (7.42) also yields this result. As for the state variable analysis of Sec. 4.2, the solution for the eigenvector $\begin{Bmatrix} c_1 \\ c_2 \end{Bmatrix}$ is arbitrary to within a multiplicative constant, and, in order to find a solution, we will define the top entry c_1 of the eigenvector to be unity. Thus, the first eigenvector, associated with the frequency ω_1, is

$$\begin{Bmatrix} c_1 \\ c_2 \end{Bmatrix}_1 = \begin{Bmatrix} 1 \\ 1.618 \end{Bmatrix}$$

The first of the two harmonic solutions to the problem then has the mathematical form

$$\begin{Bmatrix} x_1 \\ x_2 \end{Bmatrix}_1 = \begin{Bmatrix} 1 \\ 1.618 \end{Bmatrix} \cos(\omega_1 t - \phi) \tag{7.43}$$

where $\omega_1 = 0.682\sqrt{k/m}$.

To obtain the second eigenvector, substitute $\omega_2^2 = 2.618\ k/m$ into (7.37) to obtain

$$\begin{bmatrix} (2k - 2.618k) & -k \\ -k & (k - 2.618k) \end{bmatrix} \begin{Bmatrix} c_1 \\ c_2 \end{Bmatrix}_2 = \mathbf{0}$$

The first of these two equations yields

$$-0.618kc_1 - kc_2 = 0$$

or

$$c_2 = -0.618c_1$$

Thus the second eigenvector is

$$\begin{Bmatrix} c_1 \\ c_2 \end{Bmatrix}_2 = \begin{Bmatrix} 1 \\ -0.618 \end{Bmatrix}$$

so that the second of the two harmonic solutions has the form

$$\begin{Bmatrix} x_1 \\ x_2 \end{Bmatrix}_2 = \begin{Bmatrix} 1 \\ -0.618 \end{Bmatrix} \cos(\omega_2 t - \phi_2) \tag{7.44}$$

where $\omega_2 = 1.618\sqrt{k/m}$.

The general solution to the problem is a linear combination of the two solutions (7.43) and (7.44):

$$\begin{Bmatrix} x_1(t) \\ x_2(t) \end{Bmatrix} = A_1 \begin{Bmatrix} 1 \\ 1.618 \end{Bmatrix} \cos(\omega_1 t - \phi_1) + A_2 \begin{Bmatrix} 1 \\ -0.618 \end{Bmatrix} \cos(\omega_2 t - \phi_2) \tag{7.45}$$

This general solution contains four undetermined constants, A_1, ϕ_1, A_2, and ϕ_2. In specific cases these constants would be evaluated from the initial conditions: $x_1(0)$, $\dot{x}_1(0)$, $x_2(0)$, and $\dot{x}_2(0)$.

Now let us explore the physical meaning of the result (7.45). We first note that, depending on the initial conditions, the motion of the system will be one of the following three types:

1. If we pick initial conditions such that $A_2 = 0$, $A_1 \neq 0$, then we will observe in the response only solution 1, Eq. (7.43), in which each mass moves harmonically at the lower frequency ω_1.
2. If we pick initial conditions so that $A_1 = 0$, $A_2 \neq 0$, then we will observe in the response only solution 2, Eq. (7.44), in which each mass moves harmonically at the higher frequency ω_2.
3. If the initial conditions are such that neither A_1 nor A_2 is zero (this is what will normally occur), then both solutions are present, and (7.45) shows that we will observe a response in which the position of each mass consists of a superposition of two sinusoids of differing frequencies ω_1 and ω_2.

A Comment on Terminology. Vibration analysts refer to the individual solutions such as (7.43) and (7.44) as the *modes* of the system. Thus, (7.43) defines the mode 1 or first-mode solution, and case 1 above ($A_1 \neq 0$, $A_2 = 0$) would result in a "mode 1 motion" (only mode 1 is present). The eigenvectors associated with a given mode are referred to as *mode shapes* because they define how the masses of the system move in relation to each other if the system is oscillating in a particular mode.

We now want to look closely at the physical meaning of the eigenvectors or mode shapes. First consider case 1 above ($A_1 \neq 0$, $A_2 = 0$), which is a single-mode motion in which only mode 1 is present. From (7.45) the solutions for $x_1(t)$ and $x_2(t)$ will then be as follows:

$$x_1(t) = A_1 \cos(\omega_1 t - \phi_1)$$

$$x_2(t) = 1.618 A_1 \cos(\omega_1 t - \phi_1)$$

It is evident that in a mode 1 motion $x_2(t) = 1.618 x_1(t)$, for all times t. Thus, the two masses will move harmonically in such a way that their displacements are always in the ratio 1.618. This motion is illustrated in Fig. 7.19 (here the masses are represented as points). In Fig. 7.19(a) the system is at rest ($x_1, x_2 = 0$). In Fig. 7.19(b) (at some time t_1) both displacements are positive and the

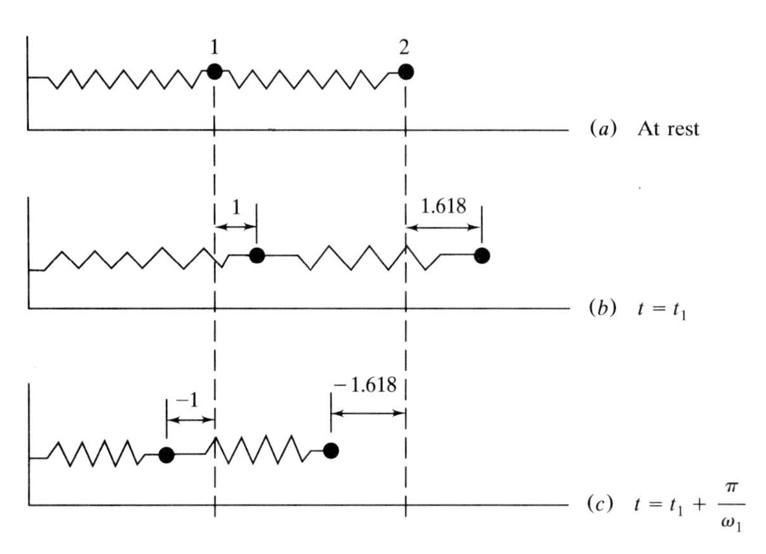

FIGURE 7.19
The displacement patterns in a mode 1 motion of the system of Fig. 7.18.

ratio of displacements is $x_2/x_1 = 1.618$. In Fig. 7.19(*c*), a half-cycle after t_1, both displacements are negative. The masses move together (that is, in phase) but mass 2 moves 1.618 times as far as mass 1. This information is contained in the mode shape.

Now consider mode 2, for which the mode shape

$$\begin{Bmatrix} c_1 \\ c_2 \end{Bmatrix}_2 = \begin{Bmatrix} 1 \\ -0.618 \end{Bmatrix}$$

tells us that the masses move such that $x_2(t) = -0.618x_1(t)$, for all t. Thus, in a mode 2 motion the masses move in opposition; one mass is moving to the right, while the other mass is moving to the left, and vice versa. This is illustrated in Fig. 7.20. The physical meaning of the modes and mode shapes must be clearly understood if one is to understand multi-degree-of-freedom vibratory phenomena. This is also true in the case of forced vibration of multi-DOF systems, considered later in this section. There we will see that, in physical terms, it is the individual *modes* of the system that tend to be excited, rather than the individual masses or coordinates.

Example 7.2 Auto / Suspension System (After Steidel, Ref. 7.4). In this example we consider the idealized 2-degree-of-freedom auto-suspension model of Example 2.4 and make the simplification that the damping is removed. The resulting idealized model is shown in Fig. 7.21. We consider the free vibration of this system, and our objective will be to find the natural frequencies and mode shapes. Although this model may appear simplistic, it actually provides the basis for the understanding of the essential behavior of automobile-axle vibratory motions.

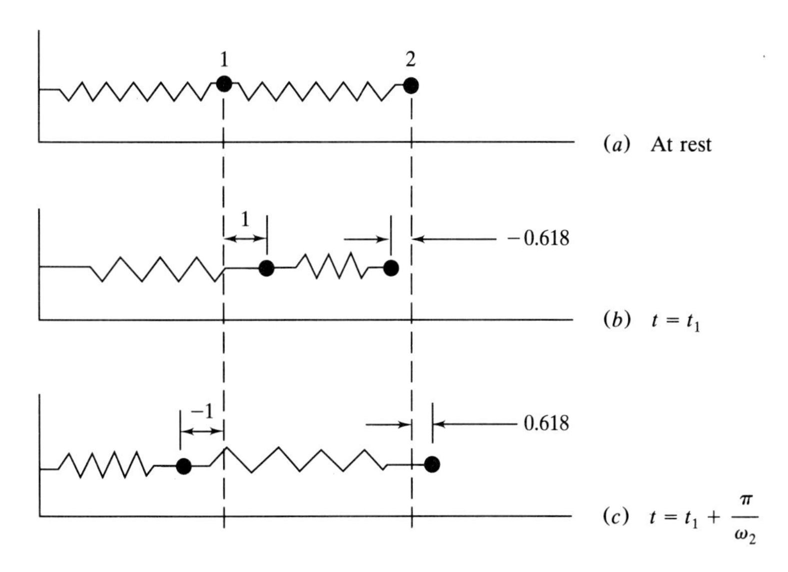

FIGURE 7.20
The displacement pattern in a mode 2 motion of the system of Fig. 7.18.

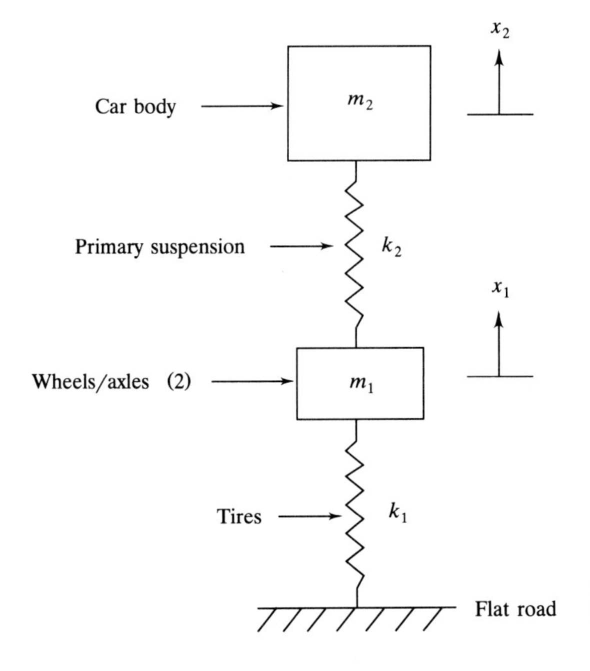

FIGURE 7.21
Idealized 2-DOF car-axle model.

Furthermore, when we consider the case of forced vibration later in this section, we'll see why the suspension system is able to accomplish its main function of isolating the car body from the axles (this means that, even if the axles are undergoing relatively large motions, the car body vibrates very little in most circumstances).

The following values of the system parameters, representative of an actual automobile, are to be used: $m_1 = 200$ kg, $m_2 = 800$ kg, $k_1 = 6(10^5)$ N/m, and $k_2 = 5(10^4)$ N/m. Notice that the tire stiffness k_1 is an order of magnitude larger than the primary suspension stiffness k_2; this is typical. The system is similar in configuration to that of Fig. 7.18, except that the masses and stiffnesses differ. Applying Lagrange's equations to determine the equations of motion, we first determine the kinetic and potential energies as

$$T = \tfrac{1}{2}m_1\dot{x}_1^2 + \tfrac{1}{2}m_2\dot{x}_2^2$$

$$V = \tfrac{1}{2}k_1x_1^2 + \tfrac{1}{2}k_2(x_2 - x_1)^2$$

Application of (6.65) leads to the equations of motion,

$$\begin{aligned} m_1\ddot{x}_1 + (k_1 + k_2)x_1 - k_2x_2 &= 0 \\ m_2\ddot{x}_2 + k_2x_2 - k_2x_1 &= 0 \end{aligned} \tag{7.46}$$

In vector-matrix form (7.46) is

$$\begin{bmatrix} m_1 & 0 \\ 0 & m_2 \end{bmatrix}\begin{Bmatrix} \ddot{x}_1 \\ \ddot{x}_2 \end{Bmatrix} + \begin{bmatrix} (k_1 + k_2) & -k_2 \\ -k_2 & k_2 \end{bmatrix}\begin{Bmatrix} x_1 \\ x_2 \end{Bmatrix} = \mathbf{0} \tag{7.47}$$

Following the steps defined by (7.34) through (7.37), (7.47) is converted into the algebraic eigenvalue problem

$$\begin{bmatrix} (k_1 + k_2 - m_1\omega^2) & -k_2 \\ -k_2 & (k_2 - m_2\omega^2) \end{bmatrix}\begin{Bmatrix} c_1 \\ c_2 \end{Bmatrix} = 0 \tag{7.48}$$

Recall that the coefficient matrix in (7.48) consists of the stiffness matrix minus ω^2 times the mass matrix. The system natural frequencies are obtained by setting the determinant of the coefficient matrix to zero. This leads to the following characteristic equation:

$$\omega^4 - \left(\frac{k_1 + k_2}{m_1} + \frac{k_2}{m_2}\right)\omega^2 + \frac{k_1k_2}{m_1m_2} = 0 \tag{7.49}$$

The solutions for the two values of ω^2 which satisfy (7.49) are obtained from the quadratic formula as

$$\omega_{1,2}^2 = \frac{1}{2}\left(\frac{k_1 + k_2}{m_1} + \frac{k_2}{m_2}\right) \pm \frac{1}{2}\left[\left(\frac{k_1 + k_2}{m_1} + \frac{k_2}{m_2}\right)^2 - \frac{4k_1k_2}{m_1m_2}\right]^{1/2}$$

Substituting the given system parameters into (7.49), we find the system natural frequencies to be

$$\omega_1^2 = 57.6 \Rightarrow \omega_1 = 7.6 \text{ rad/s}$$

$$\omega_2^2 = 3255 \Rightarrow \omega_2 = 57 \text{ rad/s}$$

These frequencies are widely separated: ω_2 is approximately eight times ω_1. The next step in the analysis is to find the mode shapes or eigenvectors. For the lower natural frequency ω_1, the first of the two equations in (7.48) yields

$$\left(k_1 + k_2 - m_1\omega_1^2\right)c_1 - k_2c_2 = 0 \tag{7.50}$$

Substituting the known values of k_1, k_2, m_1, and ω_1^2 into (7.50), we find $c_2 = 12.8c_1$, so that the first mode shape is

$$\{C\}_1 = \begin{Bmatrix} 1 \\ 12.8 \end{Bmatrix}$$

A similar procedure leads to the second mode shape:

$$\{C\}_2 = \begin{Bmatrix} 1 \\ -0.02 \end{Bmatrix}$$

Now let's look at the physical meaning of the results. Recalling that the top entry c_1 of a given mode shape defines the motion of the axles m_1, while the bottom entry c_2 of a given mode shape defines the motion of the car body, we see that the two modes of motion exhibit the following behavior: (1) In the lower-frequency (ω_1) mode the car body and axles move together (that is, in phase), with the car body moving 12.8 units of displacement for each unit moved by the axles. Notice that the oscillation frequency of mode 1 is quite low, slightly greater than 1 cycle/s. This mode 1 motion is shown in Fig. 7.22(*a*). (2) In the higher-frequency

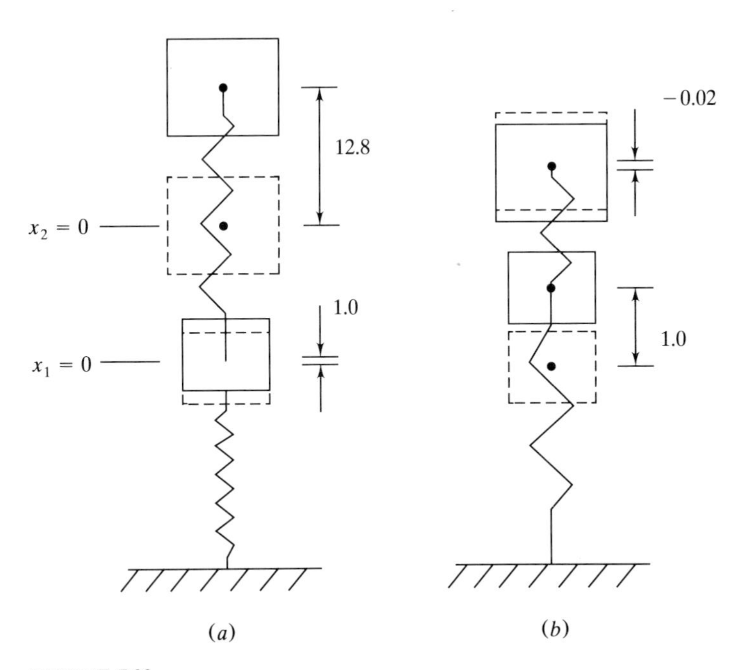

FIGURE 7.22
(*a*) Mode 1, low-frequency motion consisting mainly of motion of the car body; (*b*) higher-frequency mode 2 motion consisting mainly of motion of the axles (not to scale).

mode the two masses move in opposition, with the car body moving up 0.02 units of displacement for each unit moved down by the axles. [Fig. 7.22(*b*)].

The overall motion consists of a superposition of the two modes, as in (7.45), so there is a possibility of both the car body and the axles undergoing appreciable motion. We can see, however, that from the standpoint of passenger comfort, the mode 2 motion is a much more desirable circumstance because the car body moves only a small amount, even though the axles may be vibrating fairly violently. One objective in the design is to ensure that, for road conditions which the vehicle will likely encounter, any motion that is induced consists mainly of mode 2.

7.3.2 Undamped Free Vibration of Systems with an Arbitrary Number of Degrees of Freedom

We now generalize the results of the preceding section to linear vibratory systems having an arbitrary number n of degrees of freedom. Conceptually, the methods of analysis are the same whether there are 2 or any number $n > 2$ of degrees of freedom. For an n-DOF system the mass and stiffness matrices may be reprcsented in the general forms

$$[m] = \begin{bmatrix} m_{11} & m_{12} & \cdots & m_{1n} \\ m_{21} & m_{22} & \cdots & m_{2n} \\ \vdots & \vdots & & \vdots \\ m_{n1} & m_{n2} & \cdots & m_{nn} \end{bmatrix} \qquad n \times n \text{ mass matrix}$$

$$[k] = \begin{bmatrix} k_{11} & k_{12} & \cdots & k_{1n} \\ k_{21} & k_{22} & \cdots & k_{2n} \\ \vdots & \vdots & & \vdots \\ k_{n1} & k_{n2} & \cdots & k_{nn} \end{bmatrix} \qquad n \times n \text{ stiffness matrix}$$

The $(n \times 1)$ displacement and acceleration vectors are defined as

$$\{x\} = \begin{Bmatrix} x_1 \\ x_2 \\ \vdots \\ x_n \end{Bmatrix}; \qquad \{\ddot{x}\} = \begin{Bmatrix} \ddot{x}_1 \\ \ddot{x}_2 \\ \vdots \\ \ddot{x}_n \end{Bmatrix}$$

This notation does not mean to imply that all of the displacements will involve translation of a point. Some of the displacement variables x_i in $\{x\}$ may be rotations, and some of the m_{ij} in the mass matrix $[m]$ would then be moments of inertia. Furthermore, for the systems of Fig. 7.18 and Example 7.2 the mass matrices were diagonal. This occurs if the mass elements are considered as particles moving in one direction. But generally, both $[m]$ and $[k]$ may be fully populated matrices.

Use of the preceding notation enables the matrix-vector equations of motion for linear, undamped free vibration to be written in the form

$$\begin{bmatrix} m_{11} & m_{12} & \cdots & m_{1n} \\ m_{21} & m_{22} & \cdots & m_{2n} \\ \vdots & \vdots & & \vdots \\ m_{n1} & m_{n2} & \cdots & m_{nn} \end{bmatrix} \begin{Bmatrix} \ddot{x}_1 \\ \ddot{x}_2 \\ \vdots \\ \ddot{x}_n \end{Bmatrix} + \begin{bmatrix} k_{11} & k_{12} & \cdots & k_{1n} \\ k_{21} & k_{22} & \cdots & k_{2n} \\ \vdots & \vdots & & \vdots \\ k_{n1} & k_{n2} & \cdots & k_{nn} \end{bmatrix} \begin{Bmatrix} x_1 \\ x_2 \\ \vdots \\ x_n \end{Bmatrix} = \mathbf{0} \tag{7.51}$$

This form represents the n second-order equations of motion. For example, the first equation of motion would read $m_{11}\ddot{x}_1 + m_{12}\ddot{x}_2 + \cdots + m_{1n}\ddot{x}_n + k_{11}x_1 + k_{12}x_2 + \cdots + k_{1n}x_n = 0$, and so on. We will develop the general analysis using the following compact form of (7.51):

$$[m]\{\ddot{x}\} + [k]\{x\} = \mathbf{0} \tag{7.52}$$

with the understanding that $[m]$ and $[k]$ are $n \times n$ matrices and $\{x\}$ and $\{\ddot{x}\}$ are $n \times 1$ matrices, often referred to as *column vectors*. The solution to (7.52) is obtained as in Sec. 7.3.1. We assume a harmonic solution of the form (7.34):

$$\{x(t)\} = \{C\} \cos(\omega t - \phi) \tag{7.53}$$

where $\{C\}$ is the $n \times 1$ mode shape or eigenvector. Substitution of (7.53) into (7.52) and cancellation of the $\cos(\omega t - \phi)$ then lead to

$$-\omega^2[m]\{C\} + [k]\{C\} = \mathbf{0}$$

or

$$[[k] - \omega^2[m]]\{C\} = \mathbf{0} \tag{7.54}$$

which is analogous to (7.37). Equation (7.54) defines the algebraic eigenvalue problem which is equivalent to the matrix-vector differential equation (7.52). To obtain nontrivial solutions to (7.54), we must require that the determinant of the coefficient matrix vanish, so that we obtain the characteristic equation

$$|[k] - \omega^2[m]| = 0 \tag{7.55}$$

If we were to expand this determinant, the result would be a polynomial in ω of degree $2n$, containing only even powers of ω:

$$\omega^{2n} + a_{n-1}\omega^{2n-2} + a_{n-2}\omega^{2n-4} + \cdots + a_1\omega^2 + a_0 = 0 \tag{7.56}$$

Equation (7.56) admits n solutions for the frequency squared; we will assume these frequencies to be distinct and order them in ascending order $\omega_1^2 < \omega_2^2 < \omega_3^2 < \cdots < \omega_n^2$. If $n \geq 3$, the eigenvalue problem (7.54) would normally be solved on a digital computer.

The results indicate that there will be n system natural frequencies and hence n modes of motion. For each mode the mode shape must be calculated from (7.54). The ith mode shape $\{C\}_i$, where i is any of $1, 2, \ldots, n$, is obtained by placing the (assumed known) frequency squared ω_i^2 into (7.54):

$$\left[[k] - \omega_i^2[m]\right]\{C\}_i = \mathbf{0} \tag{7.57}$$

Recall that the n homogeneous algebraic equations in (7.57) are not independent, so that the solution for the mode shape $\{C\}_i$ is not unique. Any multiple of any solution we find will also solve (7.57). To find a solution, we will follow the convention used in Chap. 4 and in Sec. 7.3.1: We set the top entry of $\{C\}_i$ to unity and then solve for the remaining $n - 1$ entries using $n - 1$ of the n algebraic equations in (7.57).*

Once we have found the n mode shapes, we form the general solution as a superposition of the individual modal solutions

$$\{x(t)\} = \sum_{i=1}^{n} A_i\{C\}_i \cos(\omega_i t - \phi_i) \tag{7.58}$$

Here the $2n$ constants $A_1, \ldots, A_n, \phi_1, \ldots, \phi_n$ would be determined by the initial conditions $x_1(0), \ldots, x_n(0), \dot{x}_1(0), \ldots, \dot{x}_n(0)$. Observe that if the initial conditions are such that each of the constants A_i in (7.58) is appreciable, so that all of the modes make a significant contribution to the overall motion, then each displacement $x_i(t)$ will consist of a superposition of n harmonic functions of differing frequencies. Such motions may be quite complex and difficult to interpret visually.

---

*A word of caution is needed here. It sometimes occurs that an entry in one of the mode shapes $\{C\}_i$ is identically zero, indicating that, in that particular mode, the mass associated with the zero entry does not move at all. If we were to define that entry to be unity, then the solution for the remaining entries would be undefined, indicating that something is incorrect. An example is afforded by the following system of three equal masses and four equal springs:

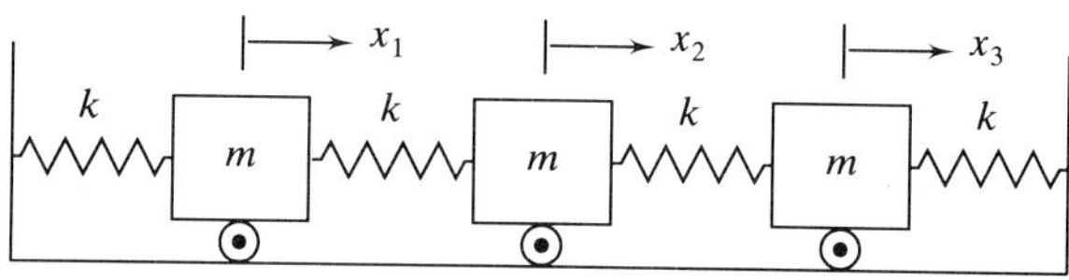

On physical grounds it is clear that one of the modes will consist of equal and opposite motions of masses 1 and 3, with mass 2 remaining stationary (the resultant force on m_2 is zero). Thus, the associated mode shape would be

$$\{C\} = \begin{Bmatrix} 1 \\ 0 \\ -1 \end{Bmatrix}$$

In this case we would run into computational difficulties if we attempted to set the second entry of $\{C\}$ to unity and then to solve for the first and third entries.

Example 7.3 A 3-DOF Structure (After Clough and Penzien, Ref. 7.2). In this example we consider the 3-DOF structure shown in Fig. 7.23. The values of the masses and stiffnesses are as indicated. The objective of this example is to clarify further the physical meaning of the mode shapes. This is the same system as considered in Example 2.6. The free vibrational motion is governed by (2.60) with the input ground motion $x_g(t)$ deleted. In vector-matrix form the equations of motion are

$$\begin{bmatrix} m & 0 & 0 \\ 0 & m & 0 \\ 0 & 0 & m \end{bmatrix} \begin{Bmatrix} \ddot{x}_1 \\ \ddot{x}_2 \\ \ddot{x}_3 \end{Bmatrix} + \begin{bmatrix} (k_1 + k_2) & -k_2 & 0 \\ -k_2 & (k_2 + k_3) & -k_3 \\ 0 & -k_3 & k_3 \end{bmatrix} \begin{Bmatrix} x_1 \\ x_2 \\ x_3 \end{Bmatrix} = \mathbf{0}$$

If we convert this version of (7.51) to the algebraic eigenvalue problem (7.54) and then find the natural frequencies and mode shapes, the following results would be obtained for the three modes of motion:

$$\text{Mode 1:} \quad \omega_1 = 12.17 \text{ rad/s}, \qquad \{C\}_1 = \begin{Bmatrix} 1 \\ 2.66 \\ 4.78 \end{Bmatrix}$$

$$\text{Mode 2:} \quad \omega_2 = 24.35 \text{ rad/s}, \qquad \{C\}_2 = \begin{Bmatrix} 1 \\ 1.62 \\ -2.08 \end{Bmatrix}$$

$$\text{Mode 3:} \quad \omega_3 = 50.95 \text{ rad/s}, \qquad \{C\}_3 = \begin{Bmatrix} 1 \\ -1.40 \\ 0.20 \end{Bmatrix}$$

In order to visualize the individual modal motions, Fig. 7.24 is presented to show how the individual masses move in each mode. Here the masses are represented as points, and the actual displacements are greatly exaggerated and not to exact scale. In mode 1 all three masses move in phase, with the displacements increasing as we

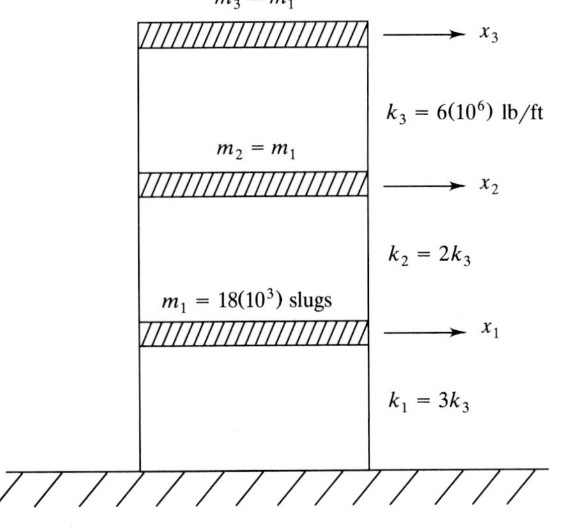

FIGURE 7.23
A three-story structure with mass and stiffness properties as noted (damping is neglected).

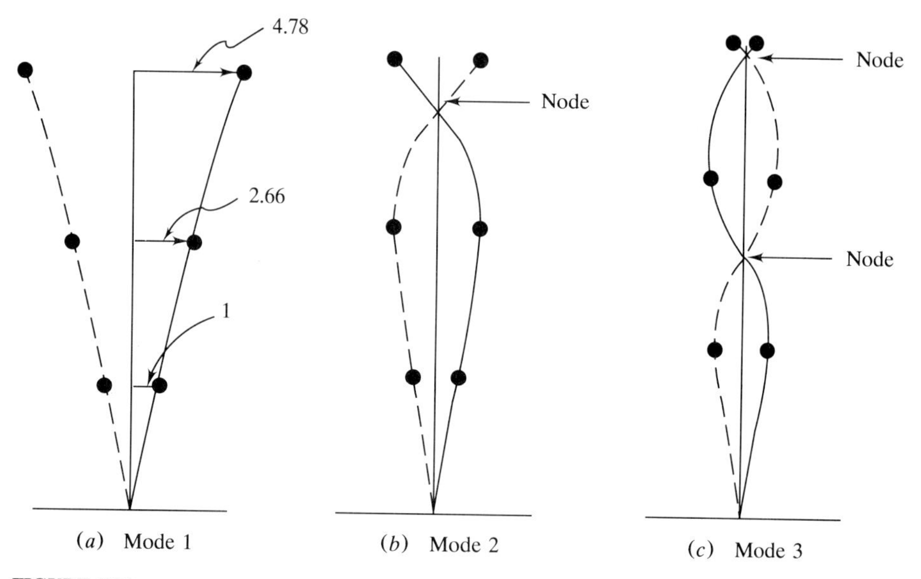

FIGURE 7.24
The three modes of motion for the system of Fig. 7.23. The solid-lined shape represents the displacement pattern at some time t, while the dashed line represents the displacement pattern one-half cycle later (i.e., at time $t + \pi/\omega_i$).

move up the structure. In mode 2 m_1 and m_2 move in one direction, while m_3 moves in the other. At some point between m_2 and m_3, identified as a "node" in Fig. 7.24(*b*), there is zero displacement. In mode 3 masses m_1 and m_3 move in one direction, while m_2 moves in the other. There are now two locations of zero displacement (nodes), as noted in Fig. 7.24(*c*). If the system is vibrating such that only a single mode is present, the displacement pattern is identical at all times; the pattern at any time is just some multiple of the mode shape.

From this example we can infer what would happen if the structure had any number n of stories. The mode shape for the ith mode would exhibit $(i - 1)$ nodal points. Thus, for relatively large i the mode shapes will be quite sinuous. Furthermore, the natural frequencies ω_i for large i tend to be much higher than the lowest few.

Comments

1. Let us consider, for an n-DOF system, the initial conditions necessary to produce a motion in which only a single mode is present. For simplicity we'll assume throughout this discussion that all initial velocities $\dot{x}_1(0), \ldots, \dot{x}_n(0)$ are zero. Thus, a motion is produced by imposing a displacement pattern defined by the initial displacements $x_1(0), \ldots, x_n(0)$. The system is then released at $t = 0$ from this initial displacement pattern and allowed to vibrate freely. Evaluation of the general solution (7.58) at $t = 0$ yields

$$\{x(0)\} = A_1\{C\}_1 \cos \phi_1 + A_2\{C\}_2 \cos \phi_2 + \cdots + A_n\{C\}_n \cos \phi_n$$

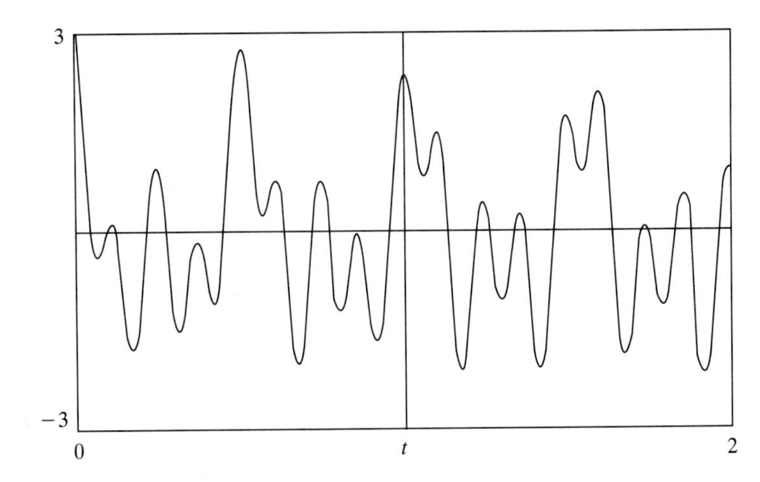

FIGURE 7.25
x_1 versus time as given by (7.60) for the system of Fig. 7.23.

Now each term $A_i \cos \phi_i$ may be defined as another constant a_i, so that we may write

$$\{x(0)\} = a_1\{C\}_1 + a_2\{C\}_2 + \cdots + a_n\{C\}_n \tag{7.59}$$

Suppose we want to produce a free vibration in which only mode 3 is present. Then from (7.59), we have to impose an initial displacement pattern $\{x(0)\}$ such that $a_1, a_2, a_4, \ldots, a_n$ are all zero, with a_3 nonzero. It is then evident from (7.59) that the initial displacement vector $\{x(0)\}$ must be proportional to the third mode shape $\{C\}_3$. This certainly makes sense on physical grounds. In general the degree of participation of a given mode will depend on how "close" the initial condition vector $\{x(0)\}$ is to that particular mode shape.

2. Because (7.59) must be valid for any of the infinite possible initial condition vectors $\{x(0)\}$, the result (7.59) implies that *any* n vector must be representable as a linear combination of the n mode shapes. Thus, the mode shapes must be linearly independent and must form a basis in n space, just as do the eigenvectors in the state variable problem [see Eq. (4.82)].
3. In practice initial conditions generally arise in such a way that no single mode is preferred; typically, numerous modes will participate in the motion.* Thus the individual displacement coordinate histories $x_i(t)$, as given by the solution (7.58), will each consist of a superposition of numerous sinusoids of differing frequencies. These individual displacements, if plotted as functions of time, may appear very irregular, and the time histories difficult to interpret. An example is shown in Fig. 7.25, which is a graph of $x_1(t)$ for the system of Fig. 7.23, with

*The chance that an arbitrarily generated initial condition vector $\{x(0)\}$ will lie exactly along the line in n space defined by a mode shape is extremely remote.

initial conditions such that $A_1 = A_2 = A_3 = 1$ and $\phi_1 = \phi_2 = \phi_3 = 0$, that is

$$x_1(t) = \cos(12.17t) + \cos(24.35t) + \cos(50.95t) \tag{7.60}$$

Although the function (7.60) is simple, the response shown exhibits a complicated structure and is, by itself, difficult to interpret. Time histories of response such as the one shown above are best interpreted using spectrum analysis, which is discussed in Sec. 7.4.

THE SYMMETRY OF THE MASS AND STIFFNESS MATRICES. In idealized, freely vibrating mechanical systems governed by (7.51), an important system property is that the mass and stiffness matrices $[m]$ and $[k]$ turn out to be *symmetric*, that is $m_{ji} = m_{ij}$ and $k_{ji} = k_{ij}$. We now explore the reason for this important property, which turns out to be crucial in the analysis of forced vibrations using modal analysis (Sec. 7.3.3). For simplicity let us consider a system of 2 degrees of freedom. The free oscillation equations of motion may be obtained using Lagrange's equations. Denoting the two displacement coordinates as q_1 and q_2, we have from (6.65)

$$\frac{d}{dt}\left(\frac{\partial T}{\partial \dot{q}_1}\right) - \frac{\partial T}{\partial q_1} + \frac{\partial V}{\partial q_1} = 0 \tag{7.61a}$$

$$\frac{d}{dt}\left(\frac{\partial T}{\partial \dot{q}_2}\right) - \frac{\partial T}{\partial q_2} + \frac{\partial V}{\partial q_2} = 0 \tag{7.61b}$$

Next, suppose the kinetic energy T and the potential energy V have the following quadratic forms:

$$\begin{aligned} T &= \tfrac{1}{2}\alpha_1\dot{q}_1^2 + \tfrac{1}{2}\alpha_2\dot{q}_2^2 + \beta\dot{q}_1\dot{q}_2 \\ V &= \tfrac{1}{2}\gamma_1 q_1^2 + \tfrac{1}{2}\gamma_2 q_2^2 + \delta q_1 q_2 \end{aligned} \tag{7.62}$$

where α_1, α_2, β, γ_1, γ_2, and δ are constants. Lagrange's equation (7.61a) then produces the following terms for q_1:

$$\frac{\partial T}{\partial \dot{q}_1} = \alpha_1\dot{q}_1 + \beta\dot{q}_2$$

$$\frac{d}{dt}\left(\frac{\partial T}{\partial \dot{q}_1}\right) = \alpha_1\ddot{q}_1 + \beta\ddot{q}_2$$

$$\frac{\partial T}{\partial q_1} = 0$$

$$\frac{\partial V}{\partial q_1} = \gamma_1 q_1 + \delta q_2$$

Combining these terms according to (7.61a), we obtain the first equation of motion as

$$\alpha_1\ddot{q}_1 + \beta\ddot{q}_2 + \gamma_1 q_1 + \delta q_2 = 0$$

In a similar fashion the second equation of motion (7.61*b*) yields

$$\beta \ddot{q}_1 + \alpha_2 \ddot{q}_2 + \delta q_1 + \gamma_2 q_2 = 0$$

In vector-matrix form the two equations of motion may be written as follows:

$$\begin{bmatrix} \alpha_1 & \beta \\ \beta & \alpha_2 \end{bmatrix} \begin{Bmatrix} \ddot{q}_1 \\ \ddot{q}_2 \end{Bmatrix} + \begin{bmatrix} \gamma_1 & \delta \\ \delta & \gamma_2 \end{bmatrix} \begin{Bmatrix} q_1 \\ q_2 \end{Bmatrix} = \mathbf{0} \tag{7.63}$$

We see that $[m]$ and $[k]$ are automatically symmetric if the kinetic and potential energies have the form (7.62). The coordinate coupling (i.e., off-diagonal) terms in (7.63) arise from the terms $\dot{q}_1 \dot{q}_2$ and $q_1 q_2$ in T and V, respectively. The above analysis is easily extended to a system with an arbitrary number of degrees of freedom.

Lagrange's equations (7.61) show clearly the intimate connection between the system energies T and V and the resulting equations of motion. The equations of motion (7.51) are homogeneous and *linear* with *constant coefficients*. Each term in (7.61) involves a differentiation with respect to either $\dot{q}_i$ or q_i; therefore, only *quadratic* terms with constant coefficients can appear in T and V. This shows that the system energies *must* have the form (7.62) if the equations of motion are to have the form (7.51). Symmetry of $[m]$ and $[k]$ follows naturally.

Further comments on mode shapes. The natural frequencies and mode shapes are fundamental properties of a freely vibrating, undamped multi-DOF system. For such systems it is always possible to define the displacement coordinates in several different ways. If we choose to analyze a given system using two different sets of coordinates, we will of course obtain the same set of natural frequencies, as these frequencies are fundamental properties which are independent of the choice of coordinates. The mode shapes, however, will be different, *mathematically*, in the two cases, because the mode shapes or eigenvectors do depend on the choice of coordinates. The physical system motions that are defined by the mode shapes, however, must be identical in the two cases. We illustrate this point with the following example.

Example 7.4 The System of Fig. 7.18 Revisited. Here we consider the two-mass, two-spring system of Fig. 7.18, which was used to introduce the basic analysis methods for multi-degree-of-freedom systems. We now redo the analysis with the following variation: x_1 will still measure the position of mass 1 relative to a fixed reference, but now x_2 will measure the position of mass 2 *relative* to mass 1, with $x_2 = 0$ when the spring connecting the two masses is unstretched (x_2 is in fact the amount of stretch in this connecting spring). Thus the (inertial) displacement of mass 2 from the fixed reference is $x_1 + x_2$, as shown in Fig. 7.26. The use of one or more coordinates that measure relative displacement is common (see Example 2.10; the coordinate θ_2 measures a relative angular displacement). With this choice of coordinates the kinetic and potential energies are

$$\begin{aligned} T &= \tfrac{1}{2} m \dot{x}_1^2 + \tfrac{1}{2} m (\dot{x}_1 + \dot{x}_2)^2 \\ V &= \tfrac{1}{2} k x_1^2 + \tfrac{1}{2} k x_2^2 \end{aligned} \tag{7.64}$$

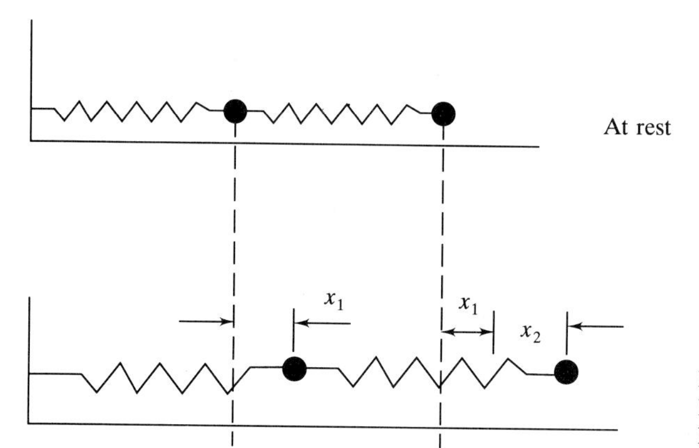

FIGURE 7.26
Definition of coordinates for Example 7.4.

This choice of coordinates simplifies the potential energy but complicates the kinetic energy; note that the kinetic energy of mass 2 is defined by the inertial velocity of mass 2, which is $\dot{x}_1 + \dot{x}_2$. Further, (7.64) is seen to be of the same mathematical form as (7.62).

Use of Lagrange's equations yields the following equations of motion:

$$\begin{bmatrix} 2m & m \\ m & m \end{bmatrix} \begin{Bmatrix} \ddot{x}_1 \\ \ddot{x}_2 \end{Bmatrix} + \begin{bmatrix} k & 0 \\ 0 & k \end{bmatrix} \begin{Bmatrix} x_1 \\ x_2 \end{Bmatrix} = \mathbf{0} \tag{7.65}$$

which differs from the original equations (7.33). The stiffness matrix is now diagonal, whereas the mass matrix contains off-diagonal or coupling terms. We

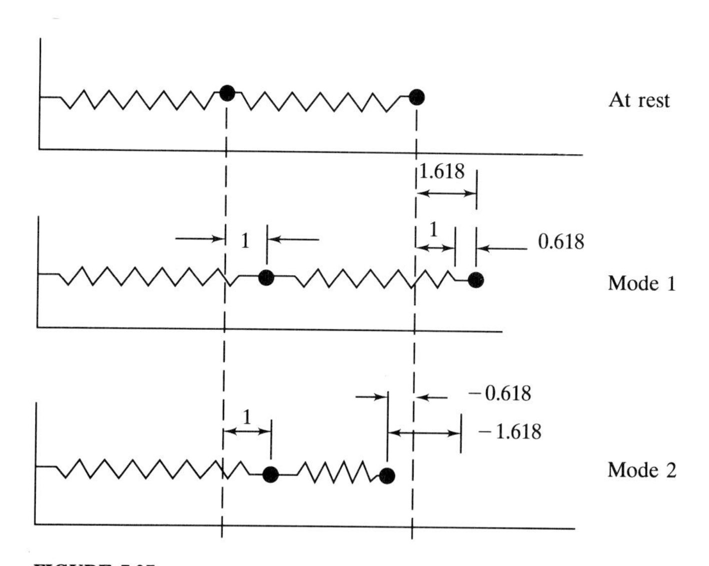

FIGURE 7.27
Physical interpretation of mode shapes based on (7.66) and the definition of the coordinates in Example 7.4.

now have "dynamic coupling" or "inertial coupling," whereas in the original version (7.33), we have "static coupling."

The characteristic equation (7.54) for (7.65) may be shown to be

$$\omega^4 - 3\frac{k}{m}\omega^2 + \frac{k^2}{m^2} = 0$$

which is identical to (7.39). Thus, the two natural frequencies are the same as those obtained previously. Calculation of the mode shapes, however, shows them to be mathematically different: The results are

$$\{C\}_1 = \begin{Bmatrix} 1 \\ 0.618 \end{Bmatrix}, \qquad \{C\}_2 = \begin{Bmatrix} 1 \\ -1.618 \end{Bmatrix} \tag{7.66}$$

which differ from the mode shapes previously obtained [see (7.43) and (7.44)]. By comparison with Fig. 7.19, however, the student may verify that the physical motions in each mode are the same as before. The modal motions defined by (7.66) are shown in Fig. 7.27. The point to be noted here is that, although the mode shapes will *mathematically* depend on the particular choice of coordinates, the *physical* motions implied will be the same, regardless of how the coordinates are defined. The physical modal motions are fundamental properties of the system.

7.3.3 Forced Multi-DOF Vibration via Modal Analysis

We now consider the effect of externally imposed excitations (inputs) for the undamped, multi-degree-of-freedom system having an arbitrary number n of degrees of freedom. The influence of damping will be addressed later in this section. Thus, we suppose that a set of time-dependent forces and/or moments is applied, so that the system equations of motion take the form

$$[m]\{\ddot{x}\} + [k]\{x\} = \{F(t)\} \tag{7.67}$$

where $\{F(t)\}$ is the input vector, arising possibly from a number of different sources, as discussed in Sec. 7.1.

The solution to (7.67) will consist of the superposition of the homogeneous solution (7.58) and the particular solution $\{x(t)\}_p$. Because there is no damping in the model (7.67), the homogeneous solution will not die out with time.

In real systems, of course, some damping will be present, so that we would expect the homogeneous solution or free response actually to decay, leaving only the steady-state response defined by the particular solution. We shall, therefore, focus attention only on the particular solution to (7.67). In any event, our objective in studying the undamped model (7.67) is to achieve an understanding of the types of forced response which can occur, and this task is best begun by considering the undamped system model.

There are a number of methods available for finding the particular solution $\{x\}_p$ to (7.67) for a given input vector $\{F(t)\}$. These include the methods of undetermined coefficients, Laplace transformation (discussed in Chap. 8), numerical integration (Chap. 9), and modal analysis. The method of undetermined coefficients was used extensively in Chap. 5 and is illustrated briefly here. As we have done throughout this chapter, we will seek a solution to the "standard vibrational form" (7.67) of the system math model, rather than first converting the model to state variable form. To fix ideas, let us consider the case for which the excitation $\{F(t)\}$ is a harmonic excitation of the form $\{F_0\}\cos \Omega t$, where $\{F_0\}$ is a known constant vector. Then (7.67) becomes

$$[m]\{\ddot{x}\} + [k]\{x\} = \{F_0\}\cos \Omega t \tag{7.68}$$

In keeping with the ideas of Chap. 5, we assume the particular solution to have the same mathematical form as the input, so we try an assumed particular solution of the form*

$$\{x(t)\}_p = \{B\}\cos \Omega t \tag{7.69}$$

where the n vector $\{B\}$ is to be determined. Substitution of (7.69) into (7.68) and cancellation of the $\cos \Omega t$ in the result lead to the following set of linear, inhomogeneous algebraic equations:

$$\left[[k] - \Omega^2[m]\right]\{B\} = \{F_0\} \tag{7.70}$$

The solution formally is then given by

$$\{B\} = \left[[k] - \Omega^2[m]\right]^{-1}\{F_0\}$$

where the inverse of the matrix $[A] = [[k] - \Omega^2[m]]$ is the transpose of the matrix of cofactors of $[A]$, divided by the determinant of $[A]$. Notice that the coefficient matrix $[A]$ in (7.70) has the same mathematical form as the coefficient matrix in (7.54) for the free vibration problem, but with the square of the driving frequency Ω^2 replacing the square of the system natural frequency ω^2. We know that the determinant of $[[k] - \omega_i^2[m]]$ vanishes. If, therefore, the driving frequency Ω is equal to any of the system natural frequencies ω_i, then the inverse of $[A]$ will not exist and the solution $\{B\}$ is unbounded. The response of an *undamped* multi-DOF system to a harmonic excitation Ω which is equal to any of the system natural frequencies ω_i will thus produce a mathematically unbounded response. We should expect, then, that for the case of harmonic excitation of the n-DOF system, there will be n resonance conditions possible, that is, whenever $\Omega = \omega_i$, and this is in fact the case.

While the method of undetermined coefficients is conceptually straightforward, it may not reveal in easily perceived physical terms why the system

---

*If damping were present in the system math model, we would have to use an assumed solution of the form

$$\{x(t)\}_p = \{B\}\cos \Omega t + \{D\}\sin \Omega t$$

Equation (7.69) reflects the fact that, without damping, the excitation and the response will be in (or 180° out of) phase with each other.

responds as it does to certain excitations. We now turn, therefore, to the technique of *modal analysis*, which is the most common analytical method for study of the vibration of driven multi-DOF linear mechanical systems. The utility of modal analysis is based on the fact that, in physical terms, it is the individual *modes* of the system that tend naturally to be excited, rather than the individual displacement coordinates.

The strategy of modal analysis is as follows: In the basic linear mathematical model of the problem (7.67), the variables that describe the system response are the $\hbar$ displacement coordinates $x_i(t)$. What we do via modal analysis is to seek a new set of n coordinates, say, $y_i(t)$, which are connected to the original displacement coordinates $x_i(t)$ by a linear coordinate transformation, such that the new coordinates $y_i(t)$ have the following properties: (1) Each $y_i(t)$ measures the contribution to the response of a particular *mode* (the ith), rather than the contribution of a particular displacement coordinate. (2) The mathematical model governing the behavior of a given modal coordinate $y_i(t)$ is *independent* of all the other modal coordinates. Thus the various modal responses can be analyzed independently of each other. These ideas will become clear as we work our way through the analysis.

Modal analysis is based on certain orthogonality properties enjoyed by the undamped free vibration mode shapes, and it is necessary to understand these properties in order to develop the basic results of modal analysis. These orthogonality conditions are now discussed.

ORTHOGONALITY CONDITIONS FOR UNDAMPED FREE VIBRATION MODE SHAPES. The orthogonality conditions involve the undamped free vibration mode shapes and are hence a property of the *unforced* system (7.52). Let us assume, for the n-DOF system (7.67) of interest, that we have solved (7.52) to obtain the n natural frequencies $\omega_1, \omega_2, \ldots, \omega_n$ and the associated mode shapes $\{C\}_1, \{C\}_2, \ldots, \{C\}_n$. Further, we assume that the natural frequencies are distinct, so that $\omega_s \neq \omega_p$ if $s \neq p$.

To establish the orthogonality properties, consider the algebraic form (7.54) of the free vibration problem for any two modes s and p not the same:

$$\text{Mode } p\text{:} \quad [k]\{C\}_p - \omega_p^2[m]\{C\}_p = \mathbf{0} \tag{7.71a}$$

$$\text{Mode } s\text{:} \quad [k]\{C\}_s - \omega_s^2[m]\{C\}_s = \mathbf{0} \tag{7.71b}$$

Now premultiply (7.71a) by the transpose $\{C\}_s^T$ and premultiply (7.71b) by the transpose $\{C\}_p^T$ to obtain

$$\{C\}_s^T[k]\{C\}_p - \omega_p^2\{C\}_s^T[m]\{C\}_p = 0 \tag{7.72a}$$

$$\{C\}_p^T[k]\{C\}_s - \omega_s^2\{C\}_p^T[m]\{C\}_s = 0 \tag{7.72b}$$

Note that (7.72) are now scalar equations, whereas (7.71) are vector equations. The next step is to take the transpose of (7.72b), noting that the transpose of the product of two or more matrices is the product of transposes in the reverse order [for example, $(ABC)^T = C^T B^T A^T$, where A, B, and C are matrices] and that the column vectors are just a special type of matrix ($n \times 1$). The transpose of (7.72b) is

$$\{C\}_s^T[k]^T\{C\}_p - \omega_s^2\{C\}_s^T[m]^T\{C\}_p = 0 \tag{7.73}$$

We next invoke the symmetry property of the system mass and stiffness matrices, $[k]^T = [k], [m]^T = [m]$, as discussed in Sec. 7.3.2, to rewrite (7.73) as

$$\{C\}_s^T[k]\{C\}_p - \omega_s^2\{C\}_s^T[m]\{C\}_p = 0 \tag{7.74}$$

We now subtract (7.74) from (7.72a), noting that the first term involving the stiffness matrix cancels out, with the result

$$\left(\omega_s^2 - \omega_p^2\right)\{C\}_s^T[m]\{C\}_p = 0 \tag{7.75}$$

Now since $\omega_s^2 \neq \omega_p^2$ by supposition, we have to conclude that the two mode shapes $\{C\}_s$ and $\{C\}_p$ satisfy the following relation:

$$\{C\}_s^T[m]\{C\}_p = 0 \qquad \text{any } s, p \text{ not the same} \tag{7.76}$$

This is the first orthogonality condition. We say that the "mode shapes are orthogonal with respect to the mass matrix." In the special case for which the mass matrix $[m]$ is diagonal and for which all diagonal entries are equal, $[m]$ is proportional to the identity matrix $[I]$, so that (7.76) reduces to $\{C\}_s^T\{C\}_p = 0$; that is, the mode shapes themselves are orthogonal in the ordinary vector sense (their dot product vanishes).

A second orthogonality condition is obtained by substituting (7.76) into (7.74), which shows that we must also have

$$\{C\}_s^T[k]\{C\}_p = 0 \qquad (s \neq p) \tag{7.77}$$

so that the mode shapes are also orthogonal with respect to the stiffness matrix. Any two mode shapes not the same satisfy both (7.76) and (7.77). We should note that a given mode is not orthogonal to itself; that is, (7.76) and (7.77) do not hold if $s = p$. Armed with the orthogonality conditions (7.76) and (7.77), we are now ready to attack the forced problem via modal analysis.

MODAL ANALYSIS. The strategy of modal analysis is to define a linear coordinate transformation connecting the original time-dependent displacement coordinates $x_1, x_2, \ldots, x_n$ and a new set of time-dependent coordinates $y_1, y_2, \ldots, y_n$, such that the new coordinate $y_i(t)$ represents the contribution to the overall response of the ith mode.

A general linear coordinate transformation may be defined by

$$\{x\} = [T]\{y\} \tag{7.78}$$

where $[T]$ is an invertible $n \times n$ transformation matrix. The inverse transformation is then

$$\{y\} = [T]^{-1}\{x\}$$

It should be clear that there are an infinite number of possible linear coordinate transformations; in (7.78) any invertible $n \times n$ matrix will do. We want to define the particular coordinate transformation (7.78) for which the new coordinates $y_1, y_2, \ldots, y_n$ are each associated with a given mode. The transformation matrix which accomplishes this feat is the *modal matrix* $[C]$, which is constructed by lining up the n mode shapes side by each to form an $n \times n$ matrix:

$$[C] \equiv [\{C\}_1 \ \{C\}_2 \ \cdots \ \{C\}_n] \tag{7.79}$$

The ith column of the modal matrix $[C]$ is the ith mode shape. Using $[C]$ as the transformation matrix in (7.78), we have

$$\{x\} = [C]\{y\} \tag{7.80}$$

In order to understand clearly the physical motivation for (7.80), let us rewrite (7.80) in the following way:

$$\{x(t)\} = \{C\}_1 y_1(t) + \{C\}_2 y_2(t) + \cdots + \{C\}_n y_n(t) \tag{7.81}$$

Equation (7.81) says that at any time t, the displacement pattern $\{x(t)\}$ is represented as a superposition of the n mode shapes. The contribution of a given mode $\{C\}_i$ to the overall displacement pattern is defined by the instantaneous value of the associated modal coordinate $y_i(t)$.

Let us now plug (7.80) into the system equation of motion (7.67) and determine the differential equations that govern the modal coordinates $y_i(t)$. We shall see that a remarkable simplification occurs. Plugging (7.80) into (7.67), we have

$$[m][C]\{\ddot{y}\} + [k][C]\{y\} = \{F(t)\} \tag{7.82}$$

Next, let us multiply (7.82) by the transpose $\{C\}_1^T$ of the first mode shape, so that

$$\{C\}_1^T[m][C]\{\ddot{y}\} + \{C\}_1^T[k][C]\{y\} = \{C\}_1^T\{F(t)\} \tag{7.83}$$

In order to simplify (7.83), let's write out the first term in expanded form, with the aid of (7.81):

$$\begin{aligned}\{C\}_1^T[m][C]\{\ddot{y}\} &= \{C\}_1^T[m]\{\{C\}_1\ddot{y}_1 + \{C\}_2\ddot{y}_2 + \cdots + \{C\}_n\ddot{y}_n\} \\ &= \{C\}_1^T[m]\{C\}_1\ddot{y}_1 + \{C\}_1^T[m]\{C\}_2\ddot{y}_2 + \cdots \\ &\quad + \{C\}_1^T[m]\{C\}_n\ddot{y}_n\end{aligned} \tag{7.84}$$

At this point we utilize the orthogonality condition (7.76), by which all terms in (7.84) except the first will vanish, and we are left with only

$$\{C\}_1^T[m][C]\{\ddot{y}\} = \underbrace{\{C\}_1^T[m]\{C\}_1\ddot{y}_1}_{\text{Scalar}}$$

which involves the single modal coordinate y_1. A similar result obtains for the second term of (7.83), so we now have

$$\underbrace{\{C\}_1^T[m]\{C\}_1}_{\text{Scalar}}\ddot{y}_1 + \underbrace{\{C\}_1^T[k]\{C\}_1}_{\text{Scalar}}y_1 = \{C\}_1^T\{F(t)\} \tag{7.85}$$

The coefficients of $\ddot{y}_1$ and y_1 in (7.85) are scalars which can be calculated, since $\{C\}_1$, $[m]$, and $[k]$ are known. Let us define these coefficients as

$$\begin{aligned} M_1 &\equiv \{C\}_1^T[m]\{C\}_1 \\ K_1 &\equiv \{C\}_1^T[k]\{C\}_1 \end{aligned} \tag{7.86}$$

The scalar M_1 is a mass like quantity that we refer to as the *modal mass*, while we call K_1 the *modal stiffness*. Furthermore, consider the ratio K_1/M_1:

$$\frac{K_1}{M_1} = \frac{\{C\}_1^T[k]\{C\}_1}{\{C\}_1^T[m]\{C\}_1}$$

Inspection of (7.72*a* or 7.72*b*) with $s = p = 1$ shows this ratio to be the square of the first natural frequency ω_1^2. Thus, if we write (7.85) in the compact form

$$M_1\ddot{y}_1 + K_1 y_1 = \{C\}_1^T\{F(t)\}$$

and divide by the mode 1 mass M_1, we obtain the standard form for a single degree of freedom undamped oscillator:

$$\ddot{y}_1 + \omega_1^2 y_1 = \frac{f_1(t)}{M_1} \tag{7.87}$$

where $f_1(t) = \{C\}_1^T\{F(t)\}$ is the portion of the applied load vector $\{F(t)\}$ which excites mode 1.

Equation (7.87) is a remarkable result, for it shows that the behavior of mode 1 is independent of what any of the other modes are doing.

We would obtain similar results for each of the other modal coordinates: To obtain the modal equation of motion for an arbitrary modal coordinate y_i, we premultiply (7.82) by $\{C\}_i^T$ and use the orthogonality conditions (7.76) and (7.77) to obtain the equation of motion

$$\ddot{y}_i + \omega_i^2 y_i = \frac{f_i(t)}{M_i} \qquad (i = 1, \ldots, n) \tag{7.88a}$$

with

$$M_i = \{C\}_i^T[m]\{C\}_i \tag{7.88b}$$

$$K_i = M_i\omega_i^2 \tag{7.88c}$$

$$f_i(t) = \{C\}_i^T\{F(t)\} \tag{7.88d}$$

The importance of the result (7.88) is that each mode behaves exactly like a single degree of freedom undamped oscillator; each mode has been decoupled from every other mode.

Let us now summarize the procedures to be followed in analyzing the forced vibration problem (7.67) using modal analysis:

1. Solve the free vibration problem to find the natural frequencies ω_i and mode shapes $\{C\}_i$.
2. For each mode compute M_i and $f_i(t)$ according to (7.88*b* and *d*).
3. Form (7.88*a*) for each mode, and solve each modal equation for the particular solutions $y_{1p}, y_{2p}, \ldots, y_{np}$.
4. Obtain the particular solutions for the original displacement coordinates $x_{1p}, x_{2p}, \ldots, x_{np}$ using the coordinate transformation (7.80),

$$\{x(t)\}_p = [C]\{y(t)\}_p$$

We now illustrate the modal analysis procedure with an example.

Example 7.5 Forced Vibration of a 2-DOF System. The system to be considered is shown in Fig. 7.28. A harmonic force is applied directly to mass 1. The equations of motion of this system are

$$\begin{bmatrix} m & 0 \\ 0 & m \end{bmatrix} \begin{Bmatrix} \ddot{x}_1 \\ \ddot{x}_2 \end{Bmatrix} + \begin{bmatrix} 2k & -k \\ -k & 2k \end{bmatrix} \begin{Bmatrix} x_1 \\ x_2 \end{Bmatrix} = \begin{Bmatrix} F_0 \\ 0 \end{Bmatrix} \cos \Omega t$$

The first step in the analysis is to solve the free vibration problem ($\{F(t)\} = \mathbf{0}$) to obtain the natural frequencies and mode shapes. The results of this calculation are stated below:

$$\text{Mode 1:} \quad \omega_1^2 = \frac{k}{m}, \qquad \{C\}_1 = \begin{Bmatrix} 1 \\ 1 \end{Bmatrix}$$

$$\text{Mode 2:} \quad \omega_2^2 = \frac{3k}{m}, \qquad \{C\}_2 = \begin{Bmatrix} 1 \\ -1 \end{Bmatrix}$$

In a mode 1 motion the masses move together, in phase and with equal amplitudes. In a mode 2 motion, the masses move in opposition, with equal amplitudes but 180° out of phase.

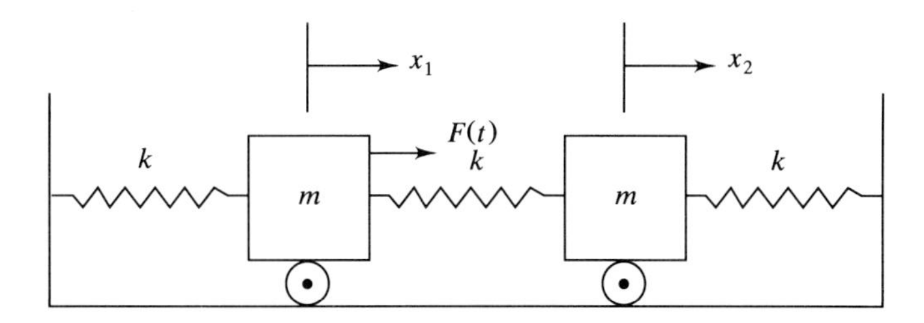

FIGURE 7.28
A driven two-mass, three-spring system.

We next form the two modal equations (7.88). For the first mode we have

$$\ddot{y}_1 + \omega_1^2 y_1 = \frac{f_1(t)}{M_1} \tag{7.89}$$

where
$$M_1 = \{C\}_1^T[m]\{C\}_1 = \lfloor 1 \quad 1 \rfloor \begin{bmatrix} m & 0 \\ 0 & m \end{bmatrix} \begin{Bmatrix} 1 \\ 1 \end{Bmatrix} = 2m$$

$$f_1(t) = \{C\}_1^T\{F(t)\} = \lfloor 1 \quad 1 \rfloor \begin{Bmatrix} F_0 \\ 0 \end{Bmatrix} \cos \Omega t = F_0 \cos \Omega t$$

The modal equation is, therefore,

$$\ddot{y}_1 + \omega_1^2 y_1 = \frac{F_0}{2m} \cos \Omega t \tag{7.90}$$

The particular solution to (7.90) may be assumed in the form $y_{1p}(t) = B \cos \Omega t$. The solution turns out to be

$$y_{1p}(t) = \frac{F_0}{2m\omega_1^2} \frac{\cos \Omega t}{(1 - \Omega^2/\omega_1^2)} = \frac{F_0}{2k} \frac{\cos \Omega t}{(1 - \Omega^2/\omega_1^2)} \tag{7.91}$$

The modal equation for mode 2 is

$$\ddot{y}_2 + \omega_2^2 y_2 = \frac{f_2(t)}{M_2}$$

where
$$M_2 = \lfloor 1 \quad -1 \rfloor \begin{bmatrix} m & 0 \\ 0 & m \end{bmatrix} \begin{Bmatrix} 1 \\ -1 \end{Bmatrix} = 2m$$

$$f_2(t) = \lfloor 1 \quad -1 \rfloor \begin{Bmatrix} F_0 \\ 0 \end{Bmatrix} \cos \Omega t = F_0 \cos \Omega t$$

so that
$$\ddot{y}_2 + \omega_2^2 y_2 = \frac{F_0}{2m} \cos \Omega t \tag{7.92}$$

The particular solution y_{2p} is the same as (7.91) with 1 subscripts replaced by 2 subscripts, so that

$$y_{2p}(t) = \frac{F_0}{2m\omega_2^2} \frac{\cos \Omega t}{(1 - \Omega^2/\omega_2^2)} = \frac{F_0}{6k} \frac{\cos \Omega t}{(1 - \Omega^2/\omega_2^2)} \tag{7.93}$$

The particular solutions (7.91) and (7.93) for y_{1p} and y_{2p} can now be substituted into the coordinate transformation (7.80), which here is

$$\begin{Bmatrix} x_{1p} \\ x_{2p} \end{Bmatrix} = \begin{bmatrix} 1 & 1 \\ 1 & -1 \end{bmatrix} \begin{Bmatrix} y_{1p} \\ y_{2p} \end{Bmatrix}$$

to obtain the particular solutions for the original displacement coordinates x_1 and x_2 as

$$x_{1p}(t) = \frac{F_0}{2k} \cos \Omega t \left(\frac{1}{1 - \Omega^2/\omega_1^2} + \frac{1/3}{1 - \Omega^2/\omega_2^2} \right)$$

$$x_{2p}(t) = \frac{F_0}{2k} \cos \Omega t \left(\frac{1}{1 - \Omega^2/\omega_1^2} - \frac{1/3}{1 - \Omega^2/\omega_2^2} \right) \tag{7.94}$$

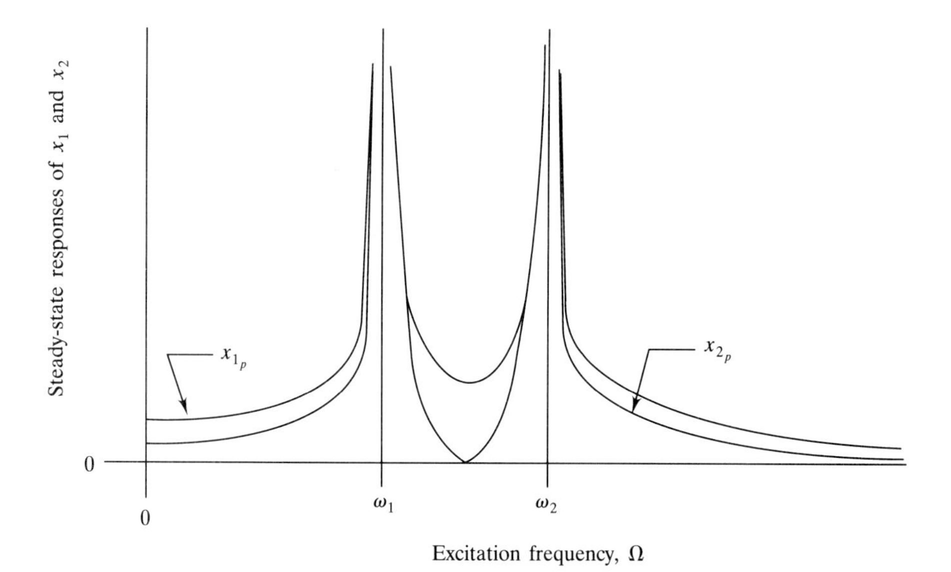

FIGURE 7.29
The dimensionless amplitudes a_1 and a_2 from (7.94). The amplitudes of x_1 and x_2 have been nondimensionalized by the factor $F_0/2k$.

There are two important features of the steady-state response given by (7.94) which should be noted. The first is that both masses move harmonically at the same frequency Ω as the excitation. The second is that there are now two conditions of resonance possible; these resonances occur when $\Omega = \omega_1$, or when $\Omega = \omega_2$. This is illustrated in Fig. 7.29, which shows the variations of the dimensionless amplitudes of x_1 and x_2 as a function of the driving frequency Ω, with the natural frequencies ω_1 and ω_2 noted. We expect large amplitude vibrational response if Ω is very near either of these two natural frequencies.

MODAL ANALYSIS: SYSTEMS WITH $n > 2$ DEGREES OF FREEDOM. Let us now extend our basic results to the case of a system with an arbitrary number n of degrees of freedom, excited by a harmonic excitation. The system math model (7.68) then leads to the following set of uncoupled second-order ODEs (7.88) for the individual modes:

$$\ddot{y}_i + \omega_i^2 y_i = \frac{f_i}{M_i} \cos \Omega t \tag{7.95}$$

where $f_i = \{C\}_i^T\{F_0\}$ and $M_i = \{C\}_i^T[m]\{C\}_i$. One may verify that the particular solution for the ith mode response is

$$y_{ip}(t) = \frac{f_i}{M_i \omega_i^2}\left(\frac{1}{1 - \Omega^2/\omega_i^2}\right) \cos \Omega t \tag{7.96}$$

We now observe that there are n conditions of resonance possible, and they occur when the exciting frequency Ω equals any of the n natural frequencies $\omega_1, \omega_2, \ldots, \omega_n$. Thus, a frequency response plot like Fig. 7.29 for any of the displacement coordinates would exhibit n peaks. In real engineering systems with many degrees of freedom, however, it is often difficult to excite the higher-frequency modes, and we now explore the reason for this.

Examination of (7.96) shows that the tendency of a given mode to be excited by a harmonic excitation depends on two things: (1) the magnitude of the factor $f_i/(M_i\omega_i^2)$ and (2) the nearness of the exciting frequency Ω to the modal frequency ω_i. We now explore these in turn. In this discussion we will suppose that the mode shapes $\{C\}_i$ have been normalized so that the length of each is unity. This will eliminate any effects of the arbitrary lengths of the mode shapes that occur with our usual procedure of defining the top entry of the mode shape to be unity [such effects will cancel when we convert (7.96) into solutions for the displacement coordinates via (7.80)].

Consider first the factor $f_i/(M_i\omega_i^2)$. The ith mode forcing amplitude f_i, defined as $f_i = \{C\}_i^T\{F_0\}$, is the projection of the original forcing vector $\{F_0\}$ along the (now) unit vector $\{C\}_i$. The magnitude of f_i will, therefore, depend on the extent to which $\{F_0\}$ is aligned in n space with the ith mode shape $\{C\}_i$. For example, for the system of Example 7.5 the normalized mode shapes and the load vector $\{F_0\}$ are shown in Fig. 7.30. This figure shows that the modal forcing amplitudes f_1 and f_2 will be equal. Suppose, however, that we consider

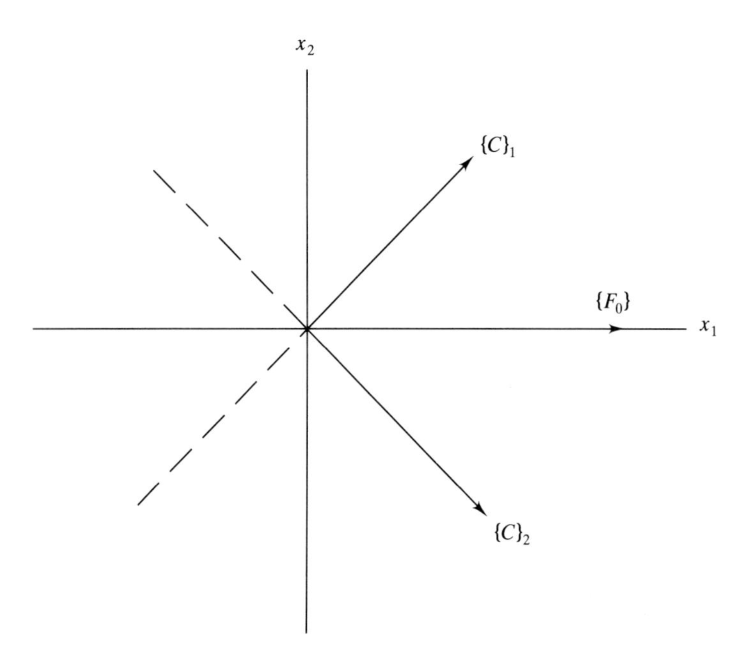

FIGURE 7.30
The mode shapes and load vector for the system of Example 7.5.

excitations that differ from that applied in Fig. 7.28, as follows: (1) Suppose equal driving forces $F_0 \cos \Omega t$ are applied to each mass in Fig. 7.28, so that the input vector $\{F(t)\} = F_0 \begin{Bmatrix} 1 \\ 1 \end{Bmatrix} \cos \Omega t$. Because $\{F_0\}$ is directed along the first mode shape, this excitation will excite only mode 1:

$$f_1 = \frac{1}{\sqrt{2}} \lfloor 1 \quad 1 \rfloor \begin{Bmatrix} 1 \\ 1 \end{Bmatrix} F_0 = \frac{2F_0}{\sqrt{2}} \qquad \text{and} \qquad f_2 = \frac{1}{\sqrt{2}} \lfloor 1 \quad -1 \rfloor \begin{Bmatrix} 1 \\ 1 \end{Bmatrix} F_0 = 0.$$

Likewise, if we consider equal and opposite driving forces, so that $\{F_0\} = F_0 \begin{Bmatrix} 1 \\ -1 \end{Bmatrix}$, then the vector $\{F_0\}$ will be aligned with the second mode shape $\{C\}_2$, and only this mode will be excited. For harmonic excitations occurring in practice, it seldom occurs that $\{F_0\}$ is aligned closely with a given mode shape; nevertheless, it does often occur that $\{F_0\}$ is more closely aligned with the lower than higher mode shapes, so the tendency is usually for f_i to diminish for the higher modes.

We next observe that in the factor $f_i/(M_i\omega_i^2)$ the square of the ith mode natural frequency appears in the denominator. For most structural systems ω_i^2 increases rapidly with mode number i, and this tends to cause a significant reduction in the associated modal amplitude. Thus, in general, the factor $f_i/(M_i\omega_i^2)$ will be smaller for the higher modes.

The second major factor in the excitation of a given mode due to a harmonic input is the nearness of the exciting frequency Ω to the modal frequency ω_i. Engineering systems may be subjected to driving frequencies Ω which differ for different operating conditions of the system. It is usually possible, however, to determine the expected range of driving frequencies Ω that may be encountered. The vibration modes having natural frequencies ω_i well above this range are usually not observed during system operation.

The difficulty in exciting the higher modes of vibration is very important from the standpoint of analysis. It allows us effectively to *ignore* these higher modes altogether. Thus, a system model having a large number of degrees of freedom may be reduced substantially in size by considering only those modes that are important. This reduction may be visualized in the phase space as follows: The phase space for an n-DOF system will be of dimension $N = 2n$. Each pair of modal states y_i and $\dot{y}_i$ will have associated with it a plane p_i in N space, such that a motion initiated in p_i will consist only of that mode. Suppose that only a relatively small number m of the modes is actually observable in practice. Then in the phase space the relevant dynamics actually occur in an $M = 2m$ dimensional *subspace* of the phase space. From both conceptual and practical standpoints, this "collapsing" of N-dimensional phase space orbits onto (often much) lower dimensional subspaces is important and occurs to some extent in most systems having some type of dissipation mechanism. In any event, to understand multi-degree-of-freedom vibrational phenomena, the behavior of the modes, rather than of the individual displacement coordinates, is what is important.

We close this subsection with an example in which the excitation of a 2-degree-of-freedom system is interpreted from the standpoint of modal analysis.

Example 7.6 Auto / Suspension System (Example 7.2) Revisited. In the analysis of the automobile-suspension model of Example 7.2, we obtained the two natural frequencies and mode shapes as

$$\omega_1 \cong 7.6 \text{ rad/s } (\approx 1.2 \text{ Hz}) \qquad \lfloor C \rfloor_1^T = \lfloor .0779 \quad 0.997 \rfloor$$

$$\omega_2 \cong 57 \text{ rad/s } (\approx 9.1 \text{ Hz}) \qquad \lfloor C \rfloor_2^T = \lfloor 0.9998 \quad -0.02 \rfloor,$$

where the mode shapes have been normalized so that each is of length unity. These numbers are typical for actual automobiles. The low-frequency mode consists mainly of a translation of the car body, while the higher-frequency mode is essentially a translation of the axles. Here the objective is to understand, via modal analysis, why the auto suspension is effective in isolating the car body from road irregularities of the type normally encountered. We consider an idealized version of the forced vibration problem for which damping is ignored and for which a base excitation is caused by travel at a fixed speed V_0 over a roadway having a sinusoidal shape (see the figure accompanying Exercise 7.7). The amplitude and wavelength of the road irregularity are denoted by λ and x_0. The equations of motion are then (7.47), amended to include the effect of the base excitation:

$$\begin{bmatrix} m_1 & 0 \\ 0 & m_2 \end{bmatrix} \begin{Bmatrix} \ddot{x}_1 \\ \ddot{x}_2 \end{Bmatrix} + \begin{bmatrix} (k_1 + k_2) & -k_2 \\ -k_2 & k_2 \end{bmatrix} \begin{Bmatrix} x_1 \\ x_2 \end{Bmatrix} = \begin{Bmatrix} k_1 x_0 \\ 0 \end{Bmatrix} \cos \Omega t$$

where the driving frequency $\Omega = V_0/\lambda$ (in hertz).

To determine the equations for the individual modal coordinates y_1 and y_2 according to (7.88), we first use the physical parameters given in Example 7.2 to obtain the following: $M_1 = 796.4 \text{ kg} \cdot \text{m}^2$, $M_2 = 200.2 \text{ kg} \cdot \text{m}^2$, $f_1(t) = 0.467(10^5)x_0 \times \cos \Omega t$ N-m, $f_2(t) = 6(10^5)x_0 \cos \Omega t$ N-m (with x_0 in meters). Note the units of these parameters: The entries in the mode shapes $\{C\}_i$ have dimensions of length, so that the product (7.88*b*) defining M_i will have dimensions of mass-length squared, and so on. The equations (7.88*a*) for the modal coordinates are then found to be

$$\ddot{y}_1 + \omega_1^2 y_1 = 58.7 x_0 \cos \Omega t$$

$$\ddot{y}_2 + \omega_2^2 y_2 = 2996 x_0 \cos \Omega t$$

where y_1 (mainly car body motion) and y_2 (mainly axle motion) define the (dimensionless) contributions of the two modes. From the preceding modal differential equations, we observe that mode 1 will tend to be excited when the driving frequency $\Omega \approx \omega_1$, while mode 2 will tend to be excited when $\Omega \approx \omega_2$. The former case requires either a very slow forward speed or a road irregularity of very long wavelength. The latter case is much more common in practice. For instance, at a moderate forward speed of $V_0 = 20$ m/s over a road characterized by an irregularity wavelength $\lambda = 2m$, the driving frequency $\Omega = 10$ Hz, which is close enough to the mode 2 natural frequency to excite this mode noticeably. The analysis also indicates why drivers often increase speed while passing over the "corduroy" irregularities common on back roads of the western United States.

These irregularities are typically characterized by very short wavelengths, of the order of a meter or less, so that increasing speed can cause the exciting frequency Ω to be well above the *higher* system natural frequency ω_2, resulting in a lessening of the tendency to excite the second mode. In summary, although it is not too difficult to find road conditions for which the axles vibrate significantly due to a second mode excitation, in this condition the car body motion is usually of much smaller amplitude. Occasionally, as on certain roads in southeastern Pennsylvania, one encounters very long wavelength irregularities. If the speed of the car is chosen with care, a mode 1 motion can be excited, and this can be a very exhilarating experience.

We close this example with some cautionary notes regarding this analysis. We have ignored damping in order to be able to draw some conclusions from a simple analysis. Any realistic analysis of an actual automobile-suspension system would include this important effect. We have also based the analysis on the highly idealized, 2-degree-of-freedom model for which the axles are lumped together in a single mass. Models actually used to design suspension systems would include numerous degrees of freedom, for example, roll, pitch, and yaw of the car body and rotation, as well as translation of the axles, and so on. Nevertheless, the simple undamped, 2-degree-of-freedom model considered here goes a long way toward describing the basic mechanism by which the car body is isolated from irregularity in the roadway.

INCLUSION OF DAMPING EFFECTS. The preceding material on modal analysis was done for multi-degree-of-freedom systems with no damping, in order to illustrate the basic ideas of modal analysis. All real-world structural systems possess damping, and here we summarize a conceptually simple way to include the effects of damping in the modal analysis. What we do is the following: we first *ignore* damping and do the undamped analysis exactly as described in the preceding section, leading to the basic modal analysis equations (7.88). Then we assume that, for each mode, there exists a modal damping factor ξ_i, such that (7.88*a*) is amended to include the effects of damping as follows:

$$\ddot{y}_i + 2\xi_i\omega_i\dot{y}_i + \omega_i^2 y_i = f_i(t)/M_i \tag{7.97}$$

The form (7.97) is the standard form of the damped oscillator model. The factor ξ_i will describe the manner in which a motion consisting of only the ith mode would decay in a free oscillation.

Values of the modal damping factors ξ_i might be determined through testing. For example, measurement of the modal amplitude in a forced vibration test conducted near resonance ($\Omega = \omega_i$) allows the ith mode damping factor ξ_i to be estimated (see Chap. 8 of Ref. 7.7 for more detail on experimental methods used in "modal testing").

Assuming that reasonable estimates for the modal damping factors ξ_i can be found, (7.97) allows each mode to be analyzed as a single-DOF damped, driven oscillator, uncoupled from all of the other modes as in the undamped case.

Remarks on damping

1. The inclusion of linear damping in the matrix-vector equation of motion (7.67) would lead to the following model, which would replace (7.67):

$$[m]\{\ddot{x}\} + [c]\{\dot{x}\} + [k]\{x\} = \{F(t)\} \tag{7.98}$$

where $[c]$ is the damping matrix, assumed to be symmetric. Suppose that we calculate the *undamped* free vibration frequencies and mode shapes and then attempt to apply the modal decoupling procedure, done in equations (7.82) through (7.88), on the damped system (7.98). This procedure will result in (7.97) *only if* the undamped mode shapes turn out to be orthogonal with respect to the damping matrix $[c]$. This may occur if $[c]$ is *proportional* to the mass matrix $[m]$ or to the stiffness matrix $[k]$, in which case a modal damping constant C_i may be defined as $C_i = \{C\}_i^T[c]\{C\}_i$ (other possibilities also exist, as discussed in Ref. 7.2). The modal damping factor is then calculated as $\xi_i = C_i/(2M_i\omega_i)$.
2. In general, the damping matrix $[c]$ may be such that the mode shapes $\{C\}_i$ are not orthogonal with respect to the damping matrix. In such cases, the attempt to use the undamped mode shapes to apply the procedure of equations (7.82) through (7.88) will not decouple the equations of motion (7.98). In this situation normal modes as we have defined them here do not exist, and the use of (7.97) to analyze the system motions is not rigorously correct.
3. For many (if not most) real-world structural dynamic systems, the damping mechanisms in the system are sufficiently diverse and difficult to model that the damping matrix $[c]$ is essentially impossible to determine. In such cases, comment 2 above becomes somewhat moot, and one really has little recourse but to assume that it is possible to define modal damping ratios and to apply (7.97).

7.4 INTRODUCTION TO SPECTRUM ANALYSIS

7.4.1 Introduction

In the analysis of system response, whether free or forced, we have used two ways to display the response graphically: by plotting one or more of the states as a function of time and by constructing phase portraits of the response. These methods of response presentation are used in both analysis and experimentation and are often referred to as *time domain methods*, because the way in which the response evolves in time is what is analyzed. If the system responses exhibit a reasonably simple character, such as sinusoidal or exponential decay, then time domain results are easily and directly interpreted. In many situations, however, the responses and often the inputs that cause these responses exhibit time variations which are difficult to interpret. This problem of time domain response interpretation may be especially difficult if the response is measured in an experiment on a complex, possibly nonlinear, system for which the mathematical model is only partially complete; the objective of the experiment might be to aid in model development and/or parameter identification.

What is usually done in order to analyze "complicated" responses is to consider them to be a superposition of a possibly large number of different sinusoids, each sinusoid defined by its amplitude, phase, and frequency. The objective of the response interpretation then becomes the determination of the *frequency domain characteristics* (amplitude, phase, frequency) of the contributing sinusoids. This interpretation from the frequency domain perspective requires the use of Fourier analysis, to which this section is devoted.

In order to illustrate the typical complexity of time responses in vibratory systems, we consider two examples. The first is Example 7.3, the freely vibrating three-story structure, for the particular initial conditions used to generate the history $x_1(t)$ shown in Fig. 7.25. We observe that, even though $x_1(t)$ is defined by the simple relation (7.60), the graph of x_1 versus time exhibits a fairly complex structure. It is evident that several frequencies are contributing to $x_1(t)$. It is also clear, however, that if we were given only the "data" of Fig. 7.25, it would be difficult to infer from the graph shown the type of mathematical relation that defines $x_1(t)$. What is needed is a technique that allows us to "decompose" the given time series into the individual harmonic constituents which appear in (7.60). This technique, Fourier analysis, is the topic of Secs. 7.4.2 through 7.4.4.

Example 7.7 Forced Vibration of a Cantilever Beam (a "Complex" Motion). In this example we consider the bending vibration of a uniform cantilever beam which is mounted on an electromagnetic shaker and excited by a harmonic motion of the base, as shown in Fig. 7.31. For certain excitation conditions the beam can be made to vibrate at large amplitude, so that nonlinear effects cannot be ignored.

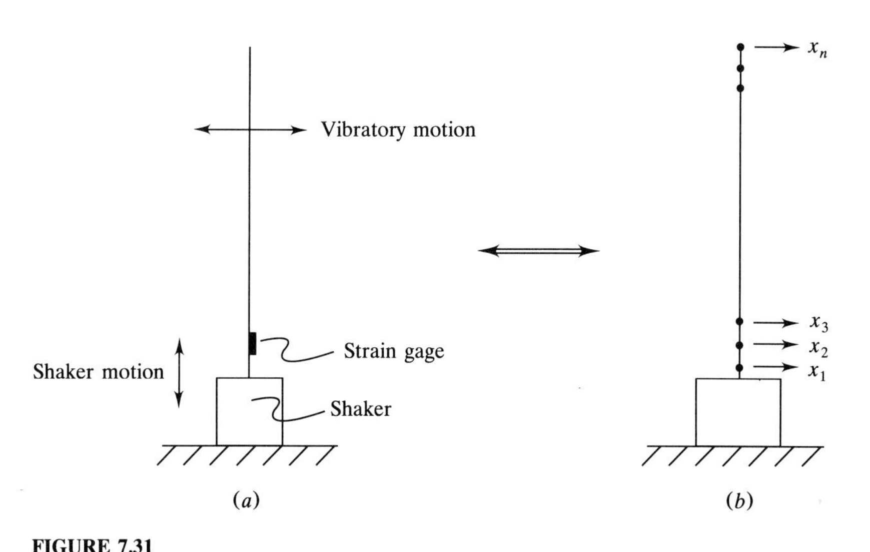

FIGURE 7.31
Schematic of the experimental setup for the harmonically driven cantilever beam. (a) The experimental setup; (b) conceptual model of the system as a large number of masses connected by stiffness elements (see text for discussion).

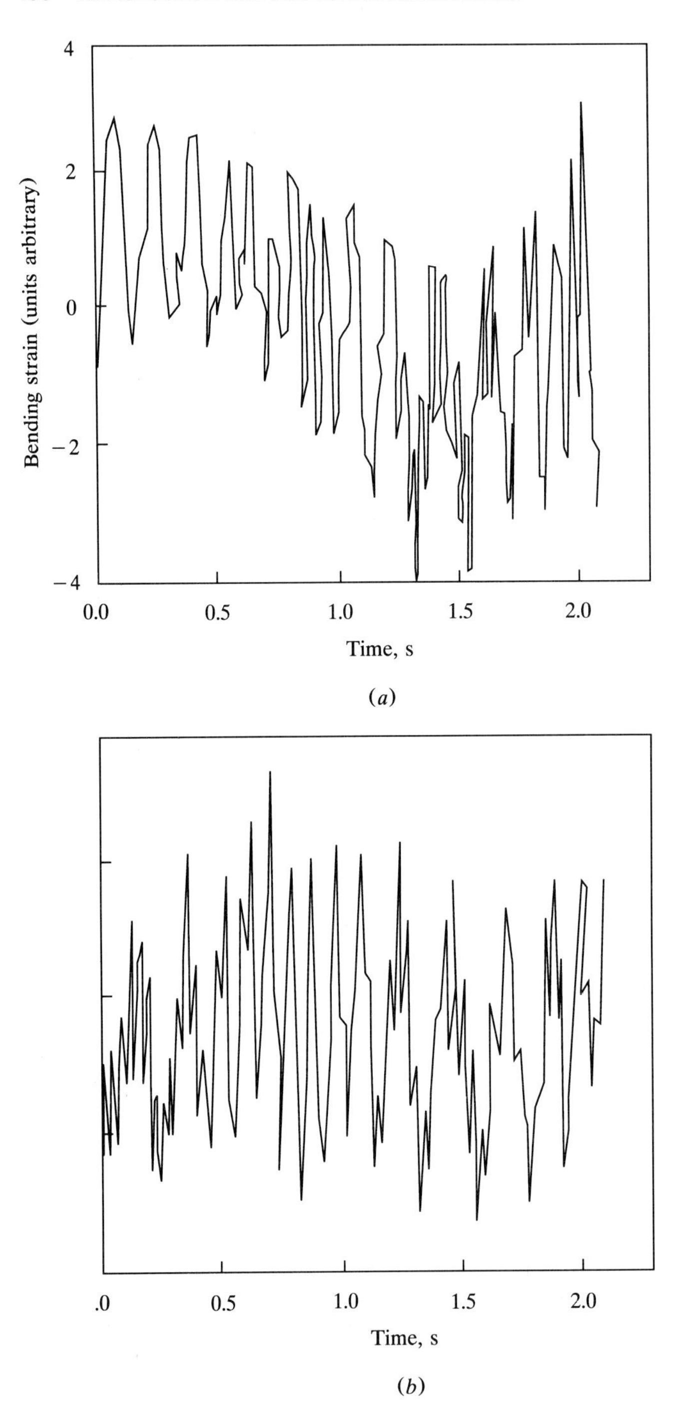

FIGURE 7.32
Steady-state bending strain versus time for the harmonically excited cantilever of Fig. 7.31. (a) The response during a 2-s interval; (b) the response during a different 2-s interval. (Data courtesy of Maia Genaux, Ref. 7.9.)

Because the mass and stiffness properties of the beam are distributed continuously, this system does not conform to our discrete, multi-degree-of-freedom system model. We can, however, consider the beam motion approximately in terms of the lumped model shown in Fig. 7.31(*b*), which consists of a large number n of discrete masses with stiffness elements between them; this is just the building model of Fig. 7.23, extended to n stories. The actual motion is, however, nonlinear, so we have a nonlinear system with a large number of degrees of freedom, excited by a harmonic input.* In this experiment the system response was monitored by a strain gage placed near the root of the cantilever. Thus the response graph will consist of the bending strain ε, at a specific location along the beam, versus time. Typical steady-state responses during two time intervals in one experimental run are shown in Fig. 7.32. The measured bending strain $\varepsilon(t)$, although oscillatory, is very irregular. The time history shown is difficult to interpret in a meaningful way.

We will begin the study of spectrum analysis by considering first (Sec. 7.4.2) the relatively simple analysis of functions which are periodic. Then in Sec. 7.4.3 we will define the discrete frequency spectrum, whereby the characteristics of a function are displayed in the frequency domain. The discrete frequency spectrum is used to display response characteristics of periodic functions and of aperiodic functions that are composed of a finite number of sinusoids. In Sec. 7.4.4 we define the continuous frequency spectrum, which is used to characterize aperiodic functions that cannot be represented as a finite superposition of sinusoids.

Although spectrum analysis is presented here as a way to analyze the response of vibratory systems, the student should be aware that spectrum analysis is one of the most useful techniques of characterizing a wide variety of phenomena in engineering and science. In particular, in the interpretation of experimental data for nearly all time-varying processes, spectrum analysis is indispensable.

7.4.2 Periodic Functions and the Fourier Series

A periodic function $f(t)$ is one that repeats itself exactly every so many seconds, so that for all time t,

$$f(t+T) = f(t) \tag{7.99}$$

where T is the period of the function $f(t)$. The trigonometric functions $\sin \omega t$ and $\cos \omega t$, for which $T = 2\pi/\omega$, are the simplest examples of periodic functions.

*The actual mathematical model one would use to analyze this system is beyond our capabilities (for a simple version of this type of problem, the linear motion of a *freely* vibrating cantilever, see Sec. 9.6). In any event the objective of this example is only to illustrate the type of time series one may be confronted with when measuring the response of a "complex" system (here a nonlinear system with many degrees of freedom).

Suppose we consider a function $f(t)$ which is a superposition of two sinusoids according to

$$f(t) = A_1 \cos \omega_1 t + A_2 \cos \omega_2 t \tag{7.100}$$

Here for simplicity both sinusoids are assumed to have the same phase. Now consider the ratio ω_2/ω_1 of the two frequencies. If ω_2/ω_1 is a rational number, then this ratio can be written as the ratio of two integers. For example, if $\omega_1 = 1/3$ and $\omega_2 = 5/9$, then $\omega_2/\omega_1 = 5/3$. Functions of this type will always be periodic; if $\omega_1 = 1/3$, then the period T_1 of $\cos \omega_1 t$ is $2\pi/(1/3) = 6\pi$, while if $\omega_2 = 5/9$, the period T_2 of $\cos \omega_2 t$ is $2\pi/(5/9) = 18\pi/5$. The period of the function (7.100), or of any periodic function which is a superposition of sinusoids, is equal to the minimum time needed for each contributing sinusoid to complete an integer number of cycles. For $\omega_1 = 1/3$ and $\omega_2 = 5/9$, $\cos \omega_1 t$ completes 3 cycles in $T = 18\pi$, $\cos \omega_2 t$ completes 5 cycles in $T = 18\pi$, and the period of the function $f(t)$ is thus 18π. Note that in general the period of the function $f(t)$ need not coincide with the period of any of the individual sinusoids comprising $f(t)$.

Next let us consider the function $f(t)$ defined by

$$f(t) = A_1 \cos t + A_2 \cos\left(\sqrt{3}\, t\right) \tag{7.101}$$

In this case the two frequencies are $\omega_1 = 1$ and $\omega_2 = \sqrt{3}$. The ratio ω_2/ω_1 is now irrational, and the function $f(t)$ is not periodic. Although it nearly repeats itself, it does not actually do so; no time interval T can be found such that both sinusoids in (7.101) complete exactly an integer number of cycles every T seconds. A function such as (7.101) is referred to as *quasi-periodic or doubly periodic*.

In the remainder of this subsection we will consider only periodic functions. We suppose that the periodic function $f(t)$ under consideration has a period T which we can determine or which is known a priori. For example, a hypothetical periodic function is shown in Fig. 7.33, and the period T is noted. We associate with this period the *fundamental frequency* $\omega = 2\pi/T$ of the periodic function. What we would now like to do is to develop a systematic way to obtain a mathematical representation of any periodic function $f(t)$. Let us try the following: Assume that $f(t)$ can be represented formally as a superposition of sinusoids, each of which has a frequency that is an integral multiple of the fundamental frequency ω. The integral multiple condition is necessary in order to ensure that the function so produced is periodic and has the correct period. In order to get our superposition of sinusoids to converge to the actual function $f(t)$, we take an infinite series of terms and write

$$f(t) = a_0 + \sum_{n=1}^{\infty} \left(a_n \cos n\omega t + b_n \sin n\omega t\right) \tag{7.102}$$

This representation (7.102) is the *Fourier series*. The frequencies of the contributing sinusoids are seen to be $\omega, 2\omega, 3\omega, \ldots,$ that is, integral multiples of

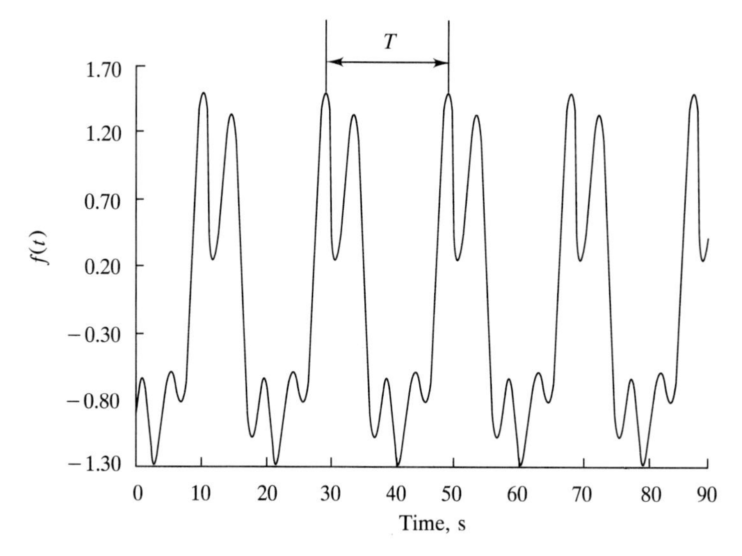

FIGURE 7.33
A periodic function, with the period T noted. (Courtesy Dr. Zahidul Rahman, Ref. 7.8.)

the fundamental frequency ω. The constant a_0, which allows us to shift the function along the vertical axis, does not affect periodicity.

For a given periodic function $f(t)$, we will assume the fundamental frequency ω to be known through precise measurement or from advance knowledge of the period. Our job then is to find the constants or "Fourier coefficients" a_0, a_n, b_n such that when we form the summation (7.102), we reproduce the actual function $f(t)$.

In order to determine the Fourier coefficients, we will need the following relations involving certain integrals of the trigonometric functions; each of the following integrals is defined over one period $T = 2\pi/\omega$ of the periodic function:

$$\int_0^{T=2\pi/\omega} \cos n\omega t\, dt = \int_0^{T=2\pi/\omega} \sin n\omega t\, dt = 0 \qquad (\text{all } n \geq 1) \quad (7.103)$$

$$\int_0^T \cos^2 n\omega t\, dt = \int_0^T \sin^2 n\omega t\, dt = \frac{T}{2} \qquad (\text{all } n \geq 1) \quad (7.104)$$

$$\int_0^T \sin n\omega t \sin m\omega t\, dt = \int_0^T \cos n\omega t \cos m\omega t\, dt = 0 \qquad (m \neq n) \quad (7.105)$$

$$\int_0^T \sin n\omega t \cos m\omega t\, dt = 0 \qquad (\text{all } m, n) \quad (7.106)$$

The relations (7.103) are evident if we graph $\cos n\omega t$ or $\sin n\omega t$ for any multiple of the period: The positive and negative areas merely cancel each other. Relations (7.105) and (7.106) are orthogonality conditions enjoyed by the trigonometric functions. In the same way that two vectors are orthogonal if their dot product vanishes, two *functions* are said to be orthogonal over an interval $[0, T]$ if the integral of their product vanishes.

We first determine the constant a_0 in (7.102) by integrating (7.102) over one period, $t = 0$ to $t = T$, and switching the order of summation and integration in the result:

$$\int_0^T f(t)\,dt = \int_0^T a_0\,dt + \sum_{n=1}^{\infty}\left[a_n\int_0^T \cos n\omega t\,dt + b_n\int_0^T \sin n\omega t\,dt\right]$$

In view of (7.103), the integrals in the summation vanish and we have

$$a_0 = \frac{1}{T}\int_0^T f(t)\,dt$$

Thus a_0 is seen to be the *average value* of the function $f(t)$ over one period. We next find the constant a_1 by multiplying (7.102) by $\cos \omega t$ and then integrating from $t = 0$ to $t = T$:

$$\int_0^T f(t)\cos \omega t\,dt = a_0\int_0^T \cos \omega t\,dt + \sum_{n=1}^{\infty}\left[a_n\int_0^T \cos n\omega t \cos \omega t\,dt + b_n\int_0^T \cos \omega t \sin n\omega t\,dt\right]$$

The a_0 term vanishes because of (7.103) and, in view of the orthogonality conditions (7.105) and (7.106), the summation reduces to the single term $a_1 \int_0^T \cos^2 \omega t\,dt$, so that

$$\int_0^T f(t)\cos \omega t\,dt = a_1\int_0^T \cos^2 \omega t\,dt = a_1\frac{T}{2} \qquad \text{(from 7.104)}$$

Thus, the Fourier coefficient a_1 is obtained as

$$a_1 = \frac{2}{T}\int_0^T f(t)\cos \omega t\,dt$$

The remaining Fourier coefficients a_n, b_n are found by a similar procedure. For example, to find b_3, multiply (7.102) by $\sin 3\omega t$ and integrate the result from 0 to T. The resulting relations which define the Fourier coefficients are summarized below:

$$a_0 = \frac{1}{T}\int_0^T f(t)\,dt \tag{7.107a}$$

$$a_n = \frac{2}{T}\int_0^T f(t)\cos(n\omega t)\,dt \tag{7.107b}$$

$$b_n = \frac{2}{T}\int_0^T f(t)\sin(n\omega t)\,dt \tag{7.107c}$$

Thus, any periodic function $f(t)$ can be "decomposed" into its individual harmonic contributions. For periodic functions $f(t)$ which we encounter in practice, the integrals (7.107) cannot usually be done in closed form; rather, they would be done on a digital computer. Let us now consider a simple example that illustrates the representation of a periodic function as a superposition of harmonic functions according to (7.102).

Example 7.8 The Square Wave. The square wave function $f(t)$ is shown in Fig. 7.34. This function has the property that $f(t) = A$ in $0 < t < T/2$ and $f(t) = -A$ in $T/2 < t < T$. The function is discontinuous (hence undefined) at $t = 0, T/2, T, \ldots$. For this "simple" function the Fourier coefficients can be determined analytically, as follows: First, the average value of $f(t)$ over one period is clearly zero, $a_0 = 0$. The cosine and sine coefficients are then defined by

$$a_n = \frac{2}{T}\left(\int_0^{T/2} A \cos n\omega t \, dt - \int_{T/2}^{T} A \cos n\omega t \, dt\right)$$

$$b_n = \frac{2}{T}\left(\int_0^{T/2} A \sin n\omega t \, dt - \int_{T/2}^{T} A \sin n\omega t \, dt\right)$$

so that

$$a_n = \frac{2A}{n\omega T}\left(\sin n\omega t \,\Big|_0^{\pi/\omega} - \sin n\omega t \,\Big|_{\pi/\omega}^{2\pi/\omega}\right)$$

$$b_n = \frac{2A}{n\omega T}\left(-\cos n\omega t \,\Big|_0^{\pi/\omega} + \cos n\omega t \,\Big|_{\pi/\omega}^{2\pi/\omega}\right)$$

Evaluating a_n, we find that $a_n = 0$ for all n; there are no cosine terms in the Fourier series representing the square wave shown in Fig. 7.34. Evaluating the sine

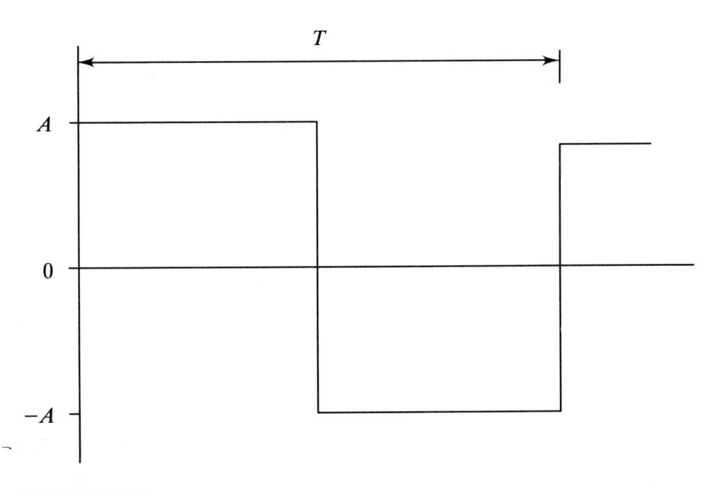

FIGURE 7.34
The square wave function.

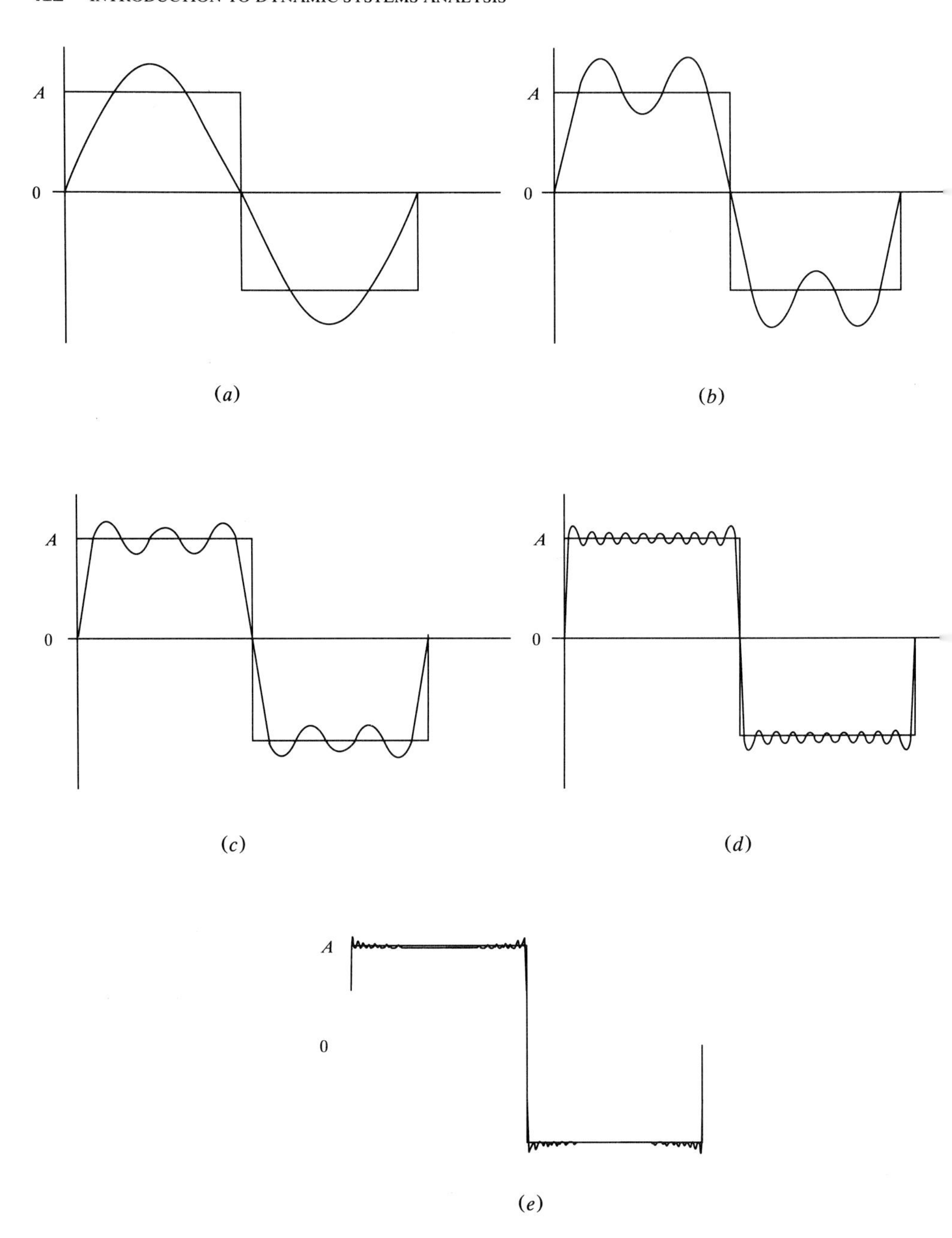

FIGURE 7.35
Truncated Fourier series representations of the square wave: (*a*) 1-term approximation, $f(t) \equiv (4A/\pi)\sin \omega t$; (*b*) 2-term approximation, $f(t) = (4A/\pi)\sin \omega t + (4A/3\pi)\sin 3\omega t$; (*c*) 3-term approximation; (*d*) 10-term approximation; (*e*) 50-term approximation.

coefficients b_n and noting that $\omega T = 2\pi$, we find

$$b_n = \frac{4A}{n\pi} \qquad \text{if } n \text{ is odd}$$

$$b_n = 0 \qquad \text{if } n \text{ is even}$$

Thus the Fourier series for the square wave of Fig. 7.34 is an "odd sine series" given by

$$f(t) = \frac{4A}{\pi} \sin \omega t + \frac{4A}{3\pi} \sin 3\omega t + \frac{4A}{5\pi} \sin 5\omega t + \cdots$$

or

$$f(t) = \sum_{n=1}^{\infty} \frac{4A}{n\pi} \sin n\omega t \qquad (n \text{ odd}) \tag{7.108}$$

The infinite series (7.108) will converge to $f(t)$ at every time t for which the function $f(t)$ is continuous. If we take only a finite number of terms in the series (a "truncated" Fourier series), then this truncated series will define an approximate representation of $f(t)$. An example is shown in Fig. 7.35, which displays various truncated Fourier series representations of the square wave $f(t)$.

Notice how the addition of terms improves the approximation. Furthermore, the approximation deteriorates somewhat near the points of discontinuity of the function $f(t)$. This is to be expected because we are attempting to reconstruct a discontinuous function using a set ($\sin n\omega t$) of continuous functions.

The first harmonic term in the Fourier series, $a_1 \cos \omega t + b_1 \sin \omega_1 t$ [$(4A/\pi) \sin \omega t$ in Example 7.8], is called the *fundamental harmonic*, as it is defined by the lowest frequency $\omega = 2\pi/T$ used to construct the Fourier series. The remaining terms in the series, having frequencies of $2\omega, 3\omega, 4\omega, \ldots,$ are called *higher harmonics*, or *overtones*, as their frequencies are integral multiples of the fundamental frequency ω. Generally, in analyzing arbitrary periodic functions $f(t)$ from a practical standpoint, it is possible to establish a "cutoff" frequency beyond which the harmonics are small enough to ignore.

7.4.3 The Discrete Frequency Spectrum

Let us suppose that we have determined the Fourier series representation of a given periodic function $f(t)$. We have found a_0, a_n, b_n for $1 \leq n \leq N$. We now present a useful way to visualize the harmonic content of the periodic function using the *discrete frequency spectrum*. In order to define the discrete frequency spectrum, we first note that the general Fourier series (7.102) may also be expressed by writing each harmonic in terms of its amplitude A_n and phase ϕ_n, rather than in terms of its separate sine and cosine contributions. Noting that

$$a_n \cos n\omega t + b_n \sin n\omega t = A_n \cos(n\omega t - \phi_n)$$

where

$$A_n = \left(a_n^2 + b_n^2\right)^{1/2}$$
$$\tan \phi_n = \frac{b_n}{a_n} \tag{7.109}$$

The Fourier series (7.102) may be written as

$$f(t) = a_0 + \sum_{n=1}^{\infty} A_n \cos(n\omega t - \phi_n) \tag{7.110}$$

The advantage of the form (7.110) is that we are usually interested in the amplitudes A_n of the individual harmonics, and only secondarily in their phases. When we speak of the "harmonic content" of a periodic function, we refer mainly to the amplitudes of the individual harmonics.

The discrete frequency spectrum is defined by plotting the absolute value of the amplitudes A_n versus frequency Ω. Because the amplitudes A_n are defined only for certain *discrete* frequencies $(\omega, 2\omega, \ldots,)$, the graph will consist of a series of "spikes," each one measuring the amplitude of a given harmonic. The discrete frequency spectrum for the square wave function of Example 7.8 is shown in Fig. 7.36. At a glance we are able to discern the frequency content of the function. The discrete frequency spectrum shows in an easily perceived way which harmonics are important contributors to the periodic response. Let us now look at the discrete frequency spectra of some other periodic functions.

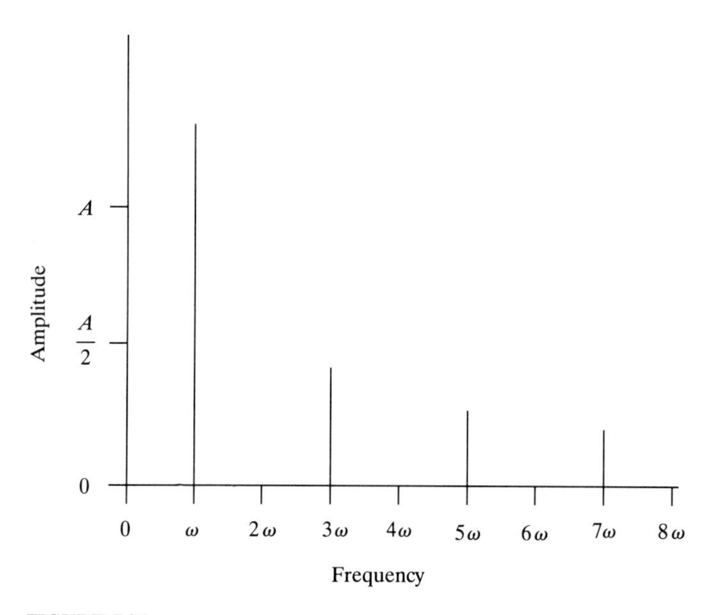

FIGURE 7.36
The discrete frequency spectrum for the square wave function of Fig. 7.34.

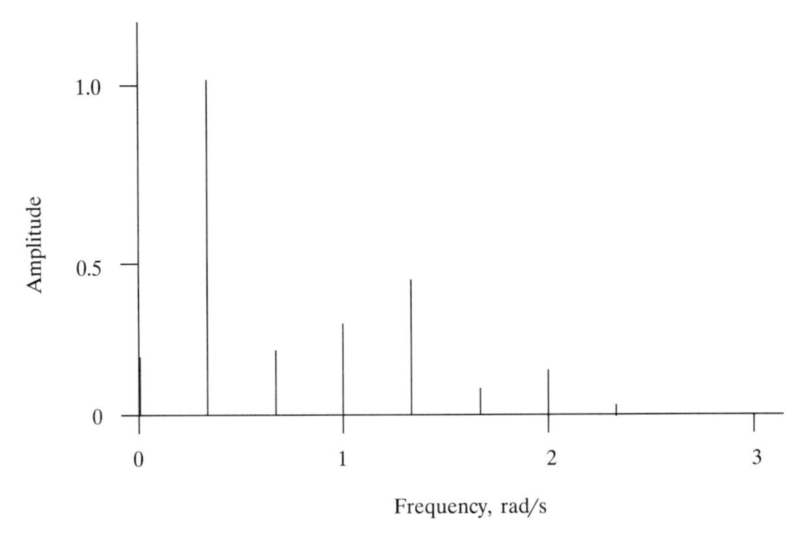

FIGURE 7.37
The discrete frequency spectrum of the periodic function of Fig. 7.33.

Example 7.9 The Periodic Function of Fig. 7.33. A Fourier analysis of this periodic function shows the amplitudes of the individual harmonics to be $a_0 = -0.16$, $A_1 = 1.03$, $A_2 = 0.19$, $A_3 = 0.27$, $A_4 = 0.42$, $A_5 = 0.07$, $A_6 = 0.12$, $A_7 = 0.02, \ldots$. The discrete frequency spectrum is shown in Fig. 7.37. For this function both odd and even harmonics appear. While this function is dominated by the fundamental harmonic A_1, which is clearly visible in Fig. 7.33, the second, third, fourth, and sixth harmonics are also significant.

Example 7.10 The Periodic Function of Fig. 7.7. The amplitudes A_n for this periodic function are $a_0 = 0$, $A_1 = 1.08$, $A_2 = 0$, $A_3 = 0.67$, $A_4 = 0$, $A_5 = 0.10$, $A_6 = 0$, $A_7 = 0.05, \ldots$. The associated discrete frequency spectrum is shown in Fig. 7.38. The periodic function considered here is dominated by the fundamental and third harmonics, each of which is visible in the time series of Fig. 7.7. There are no even harmonics, and this results in the function f being symmetric with respect to $f = 0$.

Example 7.11 Pressure in the Beating Heart. Shown in Fig. 7.39(*a*) are actual data of one period of the pressure $p(t)$ in the root aorta of a beating dog heart. The time history shown repeats itself every 0.7 s. This pressure drives blood through the circulatory system, and to understand the operation of the heart, we must understand why this pressure exhibits the particular variation with time which is shown. The function $p(t)$ shown in Fig. 7.39(*a*) has a rich harmonic content [Fig. 7.39(*b*)], with many harmonics contributing to the pressure function.

We close this section by considering the discrete frequency spectrum of functions $f(t)$ which are composed of sinusoids whose frequencies are in the ratio of irrational numbers, so that the functions are aperiodic. The frequency

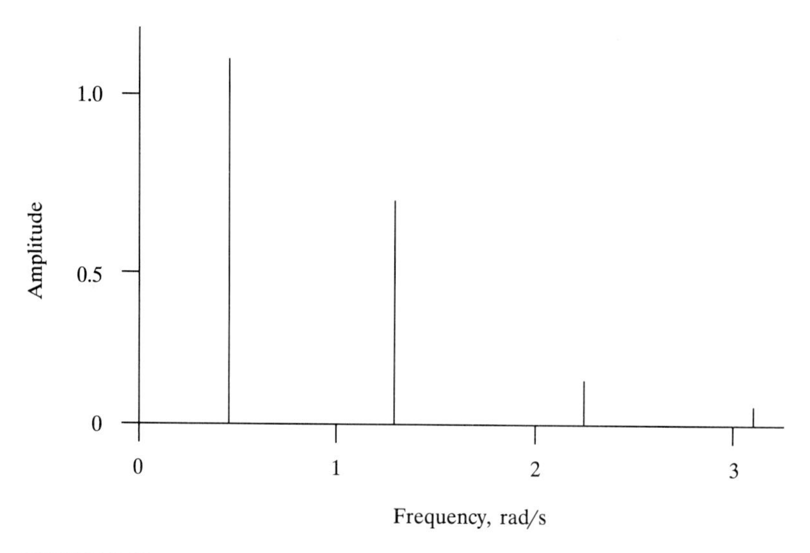

FIGURE 7.38
The discrete frequency spectrum of the periodic function of Fig. 7.7.

spectra for such functions will appear as spikes at the frequencies of the individual sinusoids, but these frequencies will not be equally spaced as in the case of periodic functions.*

Example 7.12 The Three-Story Building of Example 7.3. Consider one possible free vibration response of this model as defined previously in (7.60) and in Fig. 7.25. The contributing frequencies are $\omega_1 \cong 12.17/\text{s}$, $\omega_2 \cong 24.35/\text{s}$, and $\omega_3 \cong 50.95/\text{s}$; actually, all of these frequencies are irrational, and the preceding numerical values have been rounded off. The discrete frequency spectrum of the time series of Fig. 7.25 is shown in Fig. 7.40. The spectrum shown in Fig. 7.40 is very simple and provides directly the frequency content of the function $x_1(t)$, whereas the direct interpretation of the time series $x_1(t)$ of Fig. 7.25 is quite difficult. Furthermore, suppose that $x_1(t)$ represents the response measured during a free vibration experiment in which the system is set into motion in some fashion, perhaps by hitting it with a hammer to impart an initial velocity. Then we know that the observed response will consist of a superposition of one, several, or all of the free vibration modes; in Fig. 7.25 the three modes of the system appear in equal proportions. Thus, provided the induced motion is linear, the frequencies which appear in the discrete frequency spectrum will be (at least some of) the system natural frequencies $(\omega_1, \omega_2, \ldots, \omega_n)$. The result is that for a free vibration test of a multi-degree-of-freedom system, the system natural frequencies, which we may not be able to calculate accurately in a real system, will be the frequencies

*The discrete frequency spectrum of multifrequency aperiodic functions cannot be determined using the Fourier analysis presented in this subsection, that is, (7.102) and (7.107). The spectrum is determined using the Fourier *transform*, which is defined in Sec. 7.4.4.

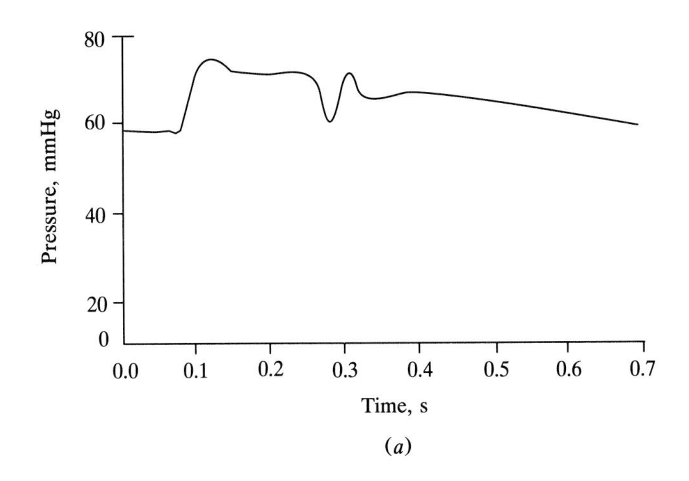

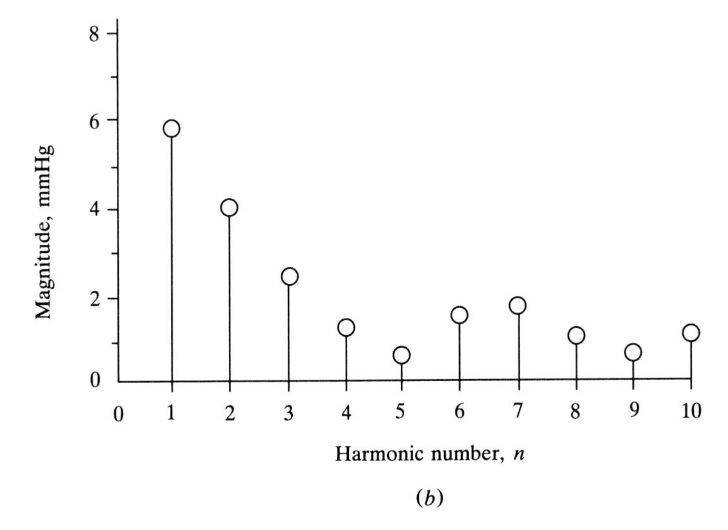

FIGURE 7.39
(*a*) One cycle of the pressure in the root of the aorta of a 25-kg dog. The periodic waveform shown repeats itself at 0.7-s intervals. (*b*) The discrete frequency spectrum of the periodic function of (*a*). Numerous harmonics are important in this distinctly nonsinusoidal periodic function. In addition, the average value a_0, not shown in the figure, is quite large, $a_0 \approx 65$ mm. [After Kenneth B. Campbell *et al.*, "Pulse Reflection Sites and Effective Length of the Arterial System," Fig. 2, *American Journal of Physiology*, **256** (*Heart Circulation Physiology*, **25**), pp. H1684–H1689 (1989). Used with permission of Kenneth B. Campbell, Washington State University.]

that appear in the discrete frequency spectrum. Free vibration testing and subsequent determination of the discrete frequency spectrum of the data using a spectrum analyzer or by direct calculation of the Fourier transform using a digital computer then allow the system natural frequencies to be determined experimentally. If we have a mathematical model of the system of the type (7.51), then the calculated and experimentally determined natural frequencies can be

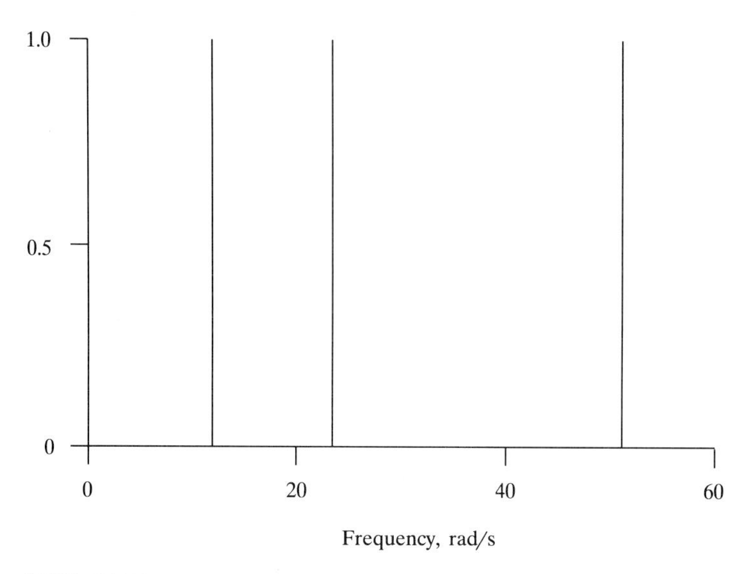

FIGURE 7.40
Discrete frequency spectrum of the function $x_1(t)$ shown in Fig. 7.25.

compared in order to validate the model. This process, which is used extensively in engineering practice, would not be possible without viewing the response from the frequency domain perspective.

Example 7.13 Natural Frequency Determination for the Cantilever Beam of Example 7.7. Shown in Fig. 7.41 is the discrete frequency spectrum, as measured by a spectrum analyzer, for a free vibration test of the cantilever beam of Fig. 7.31. The test conditions are as follows: The shaker is not operating, and the beam is initially in the static, vertical position. At $t = 0$, the beam is hit impulsively at a certain position along its length. The resulting free vibrational motion contains many individual modes, several of which are prominent enough to be measured. The individual modes whose frequencies appear in the discrete frequency spectrum are noted in Fig. 7.41. The measured values of the natural frequencies of the cantilever were found to be approximately as follows: $\omega_1 \cong 0.5$ Hz, $\omega_2 \cong 4.8$ Hz, $\omega_3 \cong 14.0$ Hz, $\omega_4 \cong 27.8$ Hz, and $\omega_5 \cong 46.0$ Hz. Measurement of natural frequencies of structures as described in this example is very common and is a familiar process to the practicing engineer.

Comment. In this subsection we have considered those functions that can be represented as a superposition of sinusoids of certain discrete frequencies. This includes *all* periodic functions and *some* aperiodic functions, for which the functional representations may be written

$$\text{Periodic:} \quad f(t) = a_0 + \sum_{n=1}^{\infty} A_n \cos(n\omega t - \phi_n) \tag{7.111}$$

$$\text{Aperiodic:} \quad f(t) = a_0 + \sum_{n=1}^{N} A_n \cos(\omega_n t - \phi_n) \tag{7.112}$$

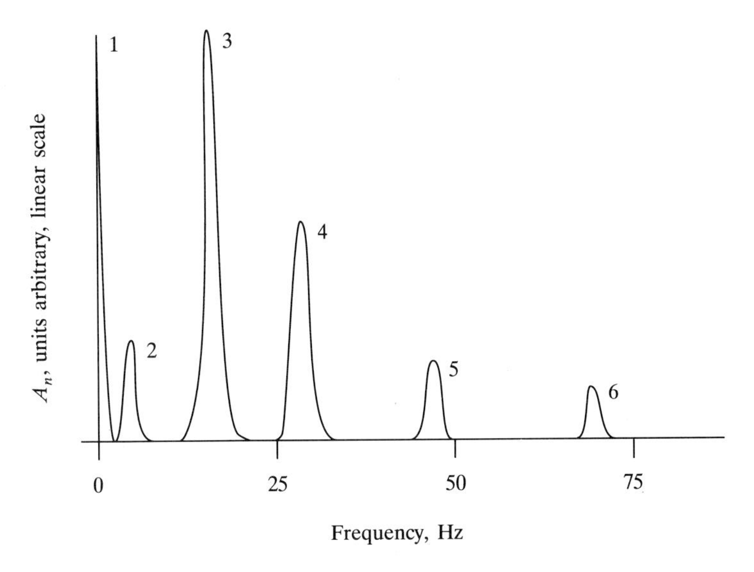

FIGURE 7.41
The measured discrete frequency spectrum for a freely vibrating cantilever beam. The first six modes of vibration are visible in the spectrum. The mode numbers are noted next to the associated peaks. See text for discussion. (Data courtesy of Maia Genaux, Ref. 7.9.)

In (7.111) the frequencies are integral multiples of the fundamental frequency ω, ensuring periodicity, whereas in (7.112) the frequencies ω_n are unrelated. It turns out that many, if not most, aperiodic functions cannot be represented as a summation of discrete frequency sinusoids as in (7.112). Rather, the frequency spectra for most aperiodic functions are *continuous*, meaning that *all* frequencies contribute something to the aperiodic function. The representation of such functions in the frequency domain requires, therefore, that the summation of (7.112) over discrete frequencies be replaced by an integral over all frequencies. This is discussed in the next subsection.

7.4.4 The Continuous Frequency Spectrum: The Fourier Transform

In this section we develop the continuous frequency spectrum of aperiodic functions in terms of the Fourier transform. We will consider a general aperiodic function $f(t)$ to be a periodic function for which the period approaches infinity. The summation (7.102), which involves only the discrete frequencies $\omega, 2\omega, 3\omega, \ldots$, will then turn into an *integral* over *all* frequencies. The following development follows Clough and Penzien, Ref. 7.2.

In order to develop the Fourier transform, it is convenient to express the Fourier series relations (7.102) and (7.107) in complex exponential form. Noting that the trigonometric terms in (7.102) may be written

$$\cos n\omega t = \tfrac{1}{2}(e^{in\omega t} + e^{-in\omega t})$$

$$\sin n\omega t = -\frac{i}{2}(e^{in\omega t} - e^{-in\omega t})$$

(7.102) may be written in complex exponential form as

$$f(t) = a_0 + \sum_{n-1}^{\infty}\left[\frac{a_n}{2}(e^{in\omega t} + e^{-in\omega t}) - \frac{ib_n}{2}(e^{in\omega t} - e^{-in\omega t})\right]$$

Collecting coefficients of $e^{in\omega t}$ and of $e^{-in\omega t}$ in the above, we have

$$f(t) = a_0 + \sum_{n=1}^{\infty}\left[\tfrac{1}{2}(a_n - ib_n)e^{in\omega t} + \tfrac{1}{2}(a_n + ib_n)e^{-in\omega t}\right] \qquad (7.113)$$

For convenience let us define the complex constant c_n as

$$c_n = \tfrac{1}{2}(a_n - ib_n)$$

Then, noting that $\frac{1}{2}(a_n + ib_n) = c_n^*$, the complex conjugate of c_n, (7.113) becomes

$$f(t) = a_0 + \sum_{n=1}^{\infty}(c_n e^{in\omega t} + c_n^* e^{-in\omega t}) \qquad (7.114)$$

The above expression can be combined into the summation of a single term by noting that

$$\sum_{n=1}^{\infty} c_n^* e^{-in\omega t} = \sum_{n=-\infty}^{-1} c_n e^{in\omega t}$$

where $c_{-n} = c_n^*$. Furthermore, the constant a_0 can be incorporated into the summation as an $n = 0$ term, so that (7.114) may be written as

$$f(t) = \sum_{n=-\infty}^{\infty} c_n e^{in\omega t} \qquad (c_{-n} = c_n^*) \qquad (7.115)$$

For any positive integer m, the pair of terms $c_m e^{im\omega t} + c_{-m}e^{-im\omega t}$ in the summation (7.115) will reproduce the $\sin m\omega t$ and $\cos m\omega t$ terms in the original series (7.102). For example, the student may verify that

$$c_3 e^{3i\omega t} + c_{-3}e^{-3i\omega t} = a_3 \cos 3\omega t + b_3 \sin 3\omega t$$

The equation (7.115) is thus an alternate way to express the Fourier series (7.102). We can also derive the relations, analogous to (7.107*a*, *b*, and *c*) for the complex Fourier coefficient c_n as follows: First, recalling that $c_n = \frac{1}{2}(a_n - ib_n)$, and using (7.107*b* and *c*), we have

$$c_n = \frac{1}{2}\left(\frac{2}{T}\int_0^T f(t)\cos n\omega t\,dt - \frac{2i}{T}\int_0^T f(t)\sin n\omega t\,dt\right)$$

or

$$c_n = \frac{1}{T}\int_0^T f(t)\underbrace{(\cos n\omega t - i\sin n\omega t)}_{e^{-in\omega t}}\,dt$$

so that in complex exponential form

$$c_n = \frac{1}{T}\int_0^T f(t)e^{-in\omega t}\,dt \tag{7.116}$$

Notice that when $n = 0$, (7.116) also provides the correct relation (7.107*a*) for the constant a_0:

$$a_0 = c_0 = \frac{1}{T}\int_0^T f(t)\,dt$$

The pair (7.115), (7.116) now replace (7.102), (7.107) as the basic relations defining the construction of the periodic function $f(t)$.

We now note the following in (7.115) and (7.116):

1. The complex constant c_n is associated with the frequency $n\omega$. Let us call this frequency $n\omega \equiv \omega_n$. Then c_n may be viewed as a function of ω_n, and, to note this, we write

$$Tc_n \equiv c(\omega_n) \qquad \text{or} \qquad c_n \equiv \frac{1}{T}c(\omega_n)$$

   where the constant $1/T$ has been absorbed into c_n.
2. The fundamental frequency of the periodic function is ω, which is also equal to the *separation* of the succeeding frequencies $2\omega, 3\omega, \ldots,$ which contribute to $f(t)$. Thus let us define $\omega \equiv \Delta\omega$. We then think of $\Delta\omega$ as the separation in successive frequencies.
3. The period T can be written as $T = 2\pi/\omega = 2\pi/\Delta\omega$.

In view of these comments and definitions the relations (7.115) and (7.116) may be written as

$$f(t) = \sum_{n=-\infty}^{\infty} c(\omega_n)e^{i\omega_n t}\frac{\Delta\omega}{2\pi} \qquad [c(-\omega_n) = c^*(\omega_n)] \tag{7.117a}$$

$$c(\omega_n) = \int_0^{2\pi/\Delta\omega} f(t)e^{-in\omega_n t}\,dt \tag{7.117b}$$

These relations still define the Fourier series for the periodic function $f(t)$. The Fourier representation for an arbitrary *aperiodic* function $f(t)$ is obtained in the limit as the period $T \to \infty, \Delta\omega \to 0$. In physical terms, as T becomes very large, the separation $\Delta\omega$ of the frequencies contributing to $f(t)$ becomes very small, and in the limit $T \to \infty$, then $\Delta\omega \to 0$, and the frequency spectrum becomes continuous. Thus, in the above relations the frequency ω_n becomes a continuous variable (call it ω), $\Delta\omega$ becomes the differential $d\omega$, and $c(\omega_n)$ becomes a continuous function of the frequency, $c(\omega_n) \to c(\omega)$. Finally, the summation over the discrete frequencies ω_n becomes an integral over the continuous frequency ω. Thus, the relations (7.117) become

$$f(t) = \frac{1}{2\pi}\int_{-\infty}^{\infty} c(\omega)e^{i\omega t}\,d\omega \tag{7.118a}$$

$$c(\omega) = \int_{-\infty}^{\infty} f(t)e^{-i\omega t}\,dt \tag{7.118b}$$

The lower limit of integration in (7.117*b*) has been extended to $-\infty$ in (7.118*b*) to allow for functions $f(t)$ defined for $t < 0$. (7.118*a*) is analogous to (7.102), while (7.118*b*) is analogous to (7.107). The complex function $c(\omega)$, which is a continuous function of the frequency ω, is the *Fourier transform* of the function $f(t)$.

The quantity $(1/2\pi)[c(\omega)\,d\omega + c(-\omega)\,d\omega]$ defines the contribution to the aperiodic function $f(t)$ of frequencies in a small band of width $d\omega$, centered about the frequencies $\pm\omega$. Note that, because $c(-\omega) = c^*(\omega)$, the magnitudes $|c(\omega)|$ and $|c(-\omega)|$ must be equal.

THE CONTINUOUS FREQUENCY SPECTRUM. The Fourier transform $c(\omega)$ is a complex function and can, therefore, be expressed in terms of its magnitude and phase ϕ as

$$c(\omega) = |c(\omega)|e^{i\phi}$$

We now define the continuous frequency spectrum to be the magnitude of the Fourier transform $c(\omega)$, just as the magnitudes of the Fourier coefficient A_n in (7.111) defined the discrete frequency spectrum. Thus, a graph of the continuous frequency spectrum will now be a continuous curve, $|c(\omega)|$ versus ω, rather than a set of discrete lines at certain frequencies.

Let us now see, in conceptual terms, what we would need to do to calculate the continuous frequency spectrum. For illustration, suppose that we have a set of measured data for which the Fourier transform $c(\omega)$ is to be determined. We see, according to (7.118*b*), that, theoretically, we would require data over an infinite time interval. In practice, we take data over a finite time interval which is "sufficiently long"; usually, data from many such finite time intervals are analyzed individually and the results averaged. Next, pick a frequency, say $\omega = 2\pi$ (1 Hz). To obtain the value of $|c(\omega)|$ at this frequency, we have to do the integral (7.118*b*) over the entire time interval for which we have data, with ω fixed at the value selected. This integration would be done numerically on a digital computer. The result of this integration is the value of $c(\omega)$ at the frequency we have selected, so we have obtained a single point on the continuous frequency spectrum curve. Next we pick another frequency and repeat the calculation to obtain a second point on the continuous frequency spectrum curve. This process is continued until we have enough points to define to our satisfaction the variation of $|c(\omega)|$ with ω. In general, adequate determination of $|c(\omega)|$ requires the calculation to be done for a large number of frequencies.

One observes that the calculation, as just described, is going to eat up large amounts of CPU time. Fortunately, in the mid-1960s a method, now referred to as the *Fast Fourier Transform*, or FFT, was devised for the rapid calculation of Fourier transforms. Such a technique was a necessity at that time because digital computers were many times slower than those of today. The FFT revolutionized numerical Fourier transform calculation and is today the backbone of spectrum analysis. We will not go into the details of this method; see several of the references at the end of this chapter for the details. We merely note here that the FFT can be done rapidly on modern digital computers and by spectrum analyzers which operate directly on measured data.

We will now consider some examples of aperiodic functions $f(t)$ and their Fourier transforms. For our purposes two categories of aperiodic functions will be considered: (1) those functions $f(t)$ for which $f(t) \to 0$ as $t \to \infty$, that is, functions which die out with time (any such function will be aperiodic); and (2) those functions $f(t)$ which do not die out in the steady state and which, furthermore, exhibit an oscillatory character that cannot be represented by (7.111) or (7.112). From a practical standpoint the latter class of functions is more important. This is because many engineering systems exhibit steady-state, aperiodic oscillatory behavior that persists over (effectively) infinite time. Examples include the motions of structures in a wind or in the ocean, vehicles traveling over rough terrain, the velocity fluctuations in a turbulent fluid flow, the temperatures in or near a flame, or the response of mechanical systems having nonlinearities. In all of these situations the responses have the properties that (1) they are defined over long periods of time, (2) they are aperiodic and cannot be represented by (7.112), and (3) they exhibit some type of oscillatory behavior.

Example 7.14 The Exponential Function. We consider the function $f(t) = e^{-at}$, $t \geq 0$, $f(t) = 0$ for $t < 0$, where $a > 0$. In this case, the Fourier transform can be calculated analytically:

$$c(\omega) = \int_0^\infty e^{-at} e^{-i\omega t}\, dt = \int_0^\infty e^{-(a+i\omega)t}\, dt$$

$$c(\omega) = \frac{1}{-(a+i\omega)} e^{-(a+i\omega)t} \Big|_{t=0}^{\infty} = \frac{1}{a+i\omega}$$

The function $f(t)$ and its frequency spectrum are shown in Fig. 7.42. Notice that $|c(\omega)|$ has the property that $|c(\omega)| = |c(-\omega)|$. For this reason we often present $|c(\omega)|$ for positive frequencies only, with the understanding that the function $|c(-\omega)|$ is the same. Because the exponential function looks nothing like a sinusoid and does not oscillate, the continuous frequency content is spread over a wide range of frequencies, the main contributions occurring in the range $-a < \omega < a$.

Example 7.15 An Oscillatory Aperiodic Function of Long Duration. Shown in Fig. 7.43(a) is a portion of the history of a function $f(t)$ which is aperiodic.* This

*The "data" shown were generated via numerical integration of a harmonically excited nonlinear oscillator model, considered further in Chap. 10.

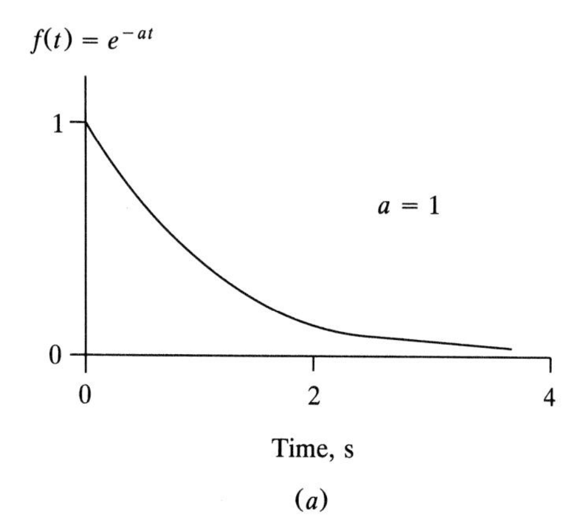

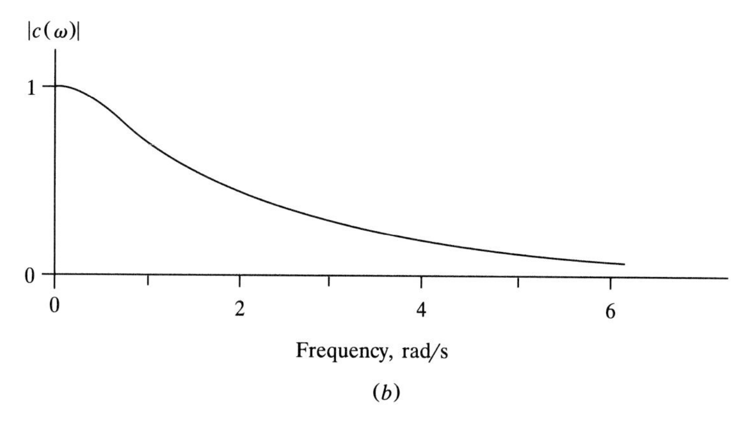

FIGURE 7.42
(a) The exponential function $f(t)$; (b) the magnitude $|c(\omega)|$ of the Fourier transform.

time history may look substantially different in other time intervals. The continuous frequency spectrum is shown in Fig. 7.43(b). This spectrum was obtained by taking data samples of 32-s duration, applying the FFT algorithm on a digital computer, and averaging the result of 50 different samples. One observes both discrete and continuous characteristics in the frequency spectrum. Notice the large spike at $\omega_1 \cong 0.07$ Hz, the large spike at $\omega_3 \approx 0.021$ Hz ($= 3\omega_1$), and the smaller spike at $\omega_5 \approx 0.35$ Hz ($= 5\omega_1$). These discrete frequency components are integer multiples of ω_1, indicating that a portion of the function $f(t)$ is composed of sinusoids of the form $A_n \cos(n\omega_1 t - \phi_n)$, $n = 1, 3, 5$. We also observe regions in which the frequency spectrum is continuous. Thus, mathematically the response shown in Fig. 7.43 may be decomposed into periodic and aperiodic components as

$$f(t) = \sum_{\substack{n=1 \\ n \text{ odd}}}^{5} A_n \cos(n\omega_1 t - \phi_n) + f_1(t)$$

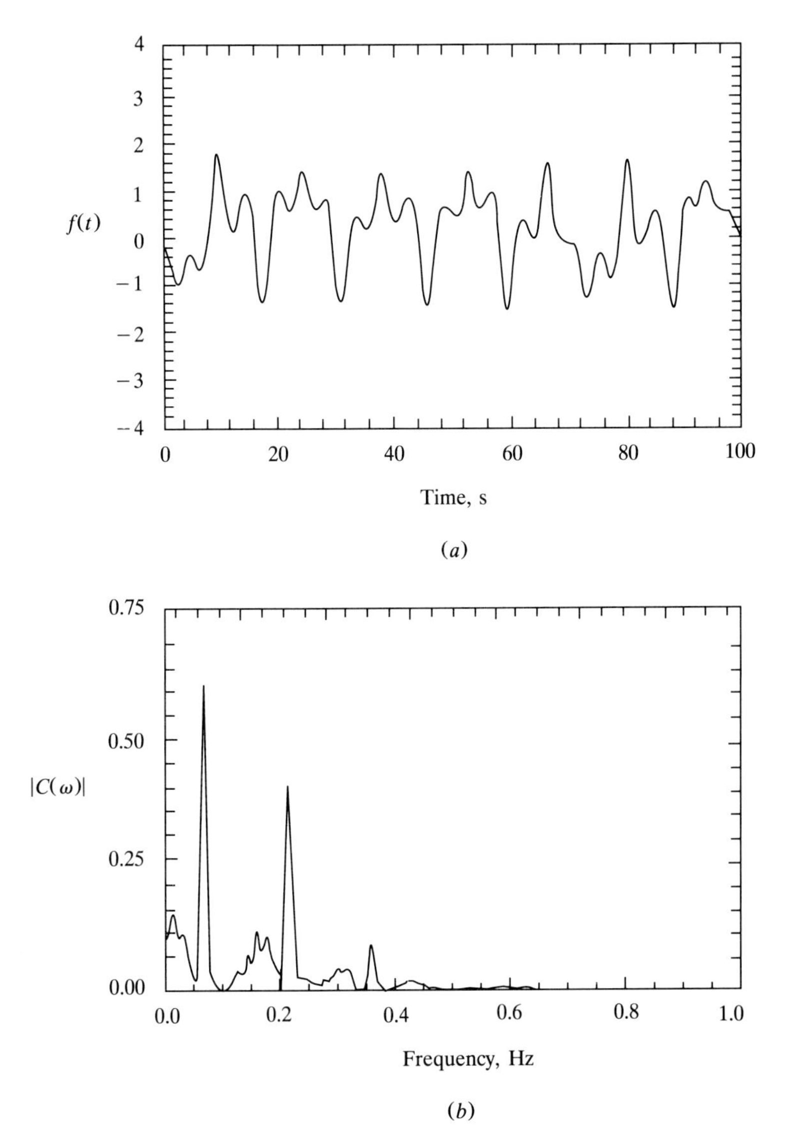

FIGURE 7.43
(*a*) An aperiodic function $f(t)$; (*b*) the continuous frequency spectrum. (Reproduced from Ref. 7.10 with permission of the *Journal of Sound and Vibration*.)

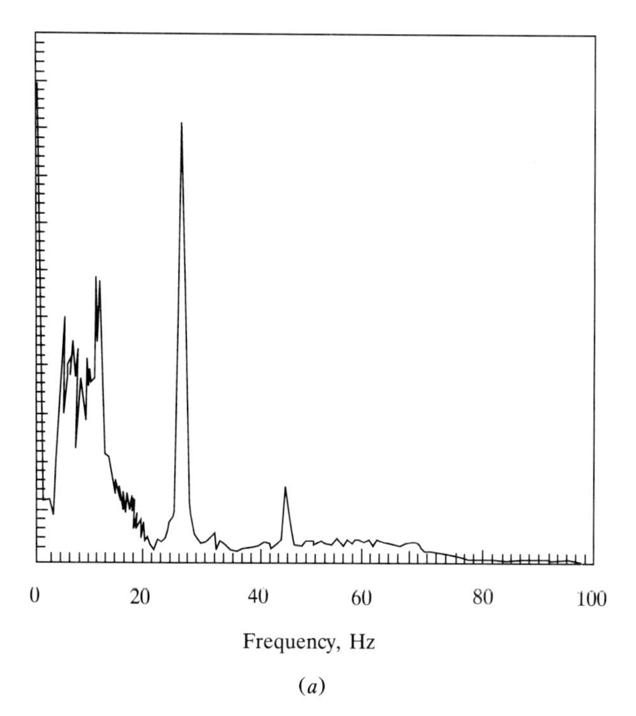

(a)

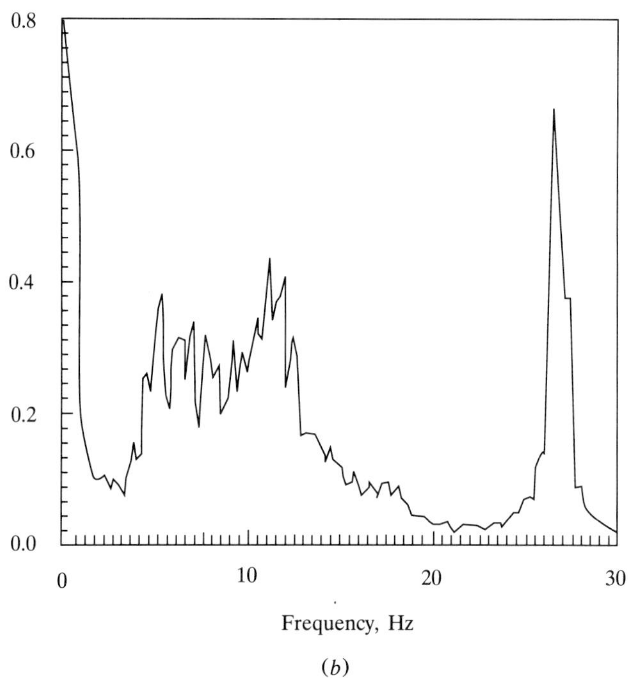

(b)

FIGURE 7.44
(a) $|c(\omega)|$ versus ω for the data of Fig. 7.32; (b) detail of the frequency range $0 \leq \omega \leq 30$ Hz. (Data courtesy of Maia Genaux, Ref. 7.9.)

where $f_1(t)$ may be viewed as a small contribution which possesses elements of unpredictability or "randomness."

Example 7.16 The Harmonically Forced Cantilever. We now consider the response of the harmonically excited cantilever beam, for which representative steady-state response data have been shown in Fig. 7.32, as described in Example 7.7. The frequency spectrum obtained from these data are shown in Fig. 7.44. Here we once again observe both discrete and continuous frequency content. The spikes that occur at $\omega_1 \approx 27$ Hz and at $\omega_2 \approx 45$ Hz represent the fourth and fifth natural frequencies of the beam (see Fig. 7.41). The presence of these spikes indicates that these two modes are excited by the harmonic input. In addition, there appear significant continuous frequency components in the range $4 < \omega < 20$ Hz. A spectrum of the type shown is typical of processes for which the time series appears to be erratic or "unpredictable."

Comment / Disclaimer. The purpose of Sec. 7.4 has only been to introduce some, but not all, of the ways in which the responses of engineering systems may be characterized from the frequency domain perspective. For a given function $f(t)$, the same information is contained in the time series $f(t)$ and in the Fourier transform $c(\omega)$. It turns out, though, that the frequency spectrum often reveals more directly the underlying mechanisms or phenomena involved in the process being analyzed. Moreover, the techniques of spectrum analysis are heavily used in engineering and scientific practice in a wide variety of different fields. The arsenal of the modern engineer should include these techniques.

In our brief introduction to spectrum analysis we have discussed only a few of the basic ideas. The student should be aware that there are numerous aspects of spectrum analysis which have been omitted here. Several of the references provided at the end of this chapter provide valuable discussions which are much more comprehensive.

REFERENCES

7.1. Blevins, R. D.: *Flow-Induced Vibration*, van Nostrand Reinhold, New York (1977). See Ch. 8, which contains a nice description of damping properties of various vibratory systems.
7.2. Clough, R. W., and J. Penzien: *Dynamics of Structures*, McGraw-Hill, New York (1975). This is an excellent book and is recommended especially for the material on multi-degree-of-freedom vibratory systems.
7.3. Thomson, W. T.: *Theory of Vibration with Application*, 4th ed., Prentice-Hall, Englewood Cliffs, N.J. (1993). This is another excellent book with numerous practical examples.
7.4. Steidel, R. F.: *An Introduction to Mechanical Vibrations*,. 2d ed., Wiley, New York (1979). Contains numerous real-world examples and homework problems.
7.5. Rao, S. S.: *Mechanical Vibrations*, 2d ed., Addison-Wesley, Reading, Mass. (1993). See especially the chapter on instrumentation and measurement.
7.6. Meirovitch, L.: *Elements of Vibration Analysis*, McGraw-Hill, New York (1975). See especially Ch. 4 on modal analysis.
7.7. Inman, D. J.: *Vibration, with Control, Measurement and Stability*, Prentice-Hall, Englewood Cliffs, N.J. (1989). Contains useful information on modal analysis, experimental methods, and other topics from a modern perspective.

The following references are cited in the figures:

7.8. Rahman, Z. H. M.: *Higher Order Perturbation Methods in Nonlinear Oscillations*, Ph.D. thesis, Department of Mechanical Engineering, Washington State University (1986).
7.9. Genaux, M. E.: *Chaos: An Investigation of Beam Vibration*, MSME Project Report, Department of Mechanical and Materials Engineering, Washington State University (1989).
7.10. Burton, T. D., and M. Anderson: "On Asymptotic Behavior in Cascaded Chaotically Excited Non-Linear Oscillators," *Journal of Sound and Vibration*, **133**(2), 353–358 (1989).

EXERCISES

1. A uniform bar of mass m and length L rotates freely about the fixed, frictionless pivot **O**. An equal mass m, assumed to behave as a particle, is attached to the bar at its end, as shown in Fig. $P7.1$. Determine the natural frequency ω of small oscillations about the equilibrium configuration $\theta = \dot{\theta} = 0$.

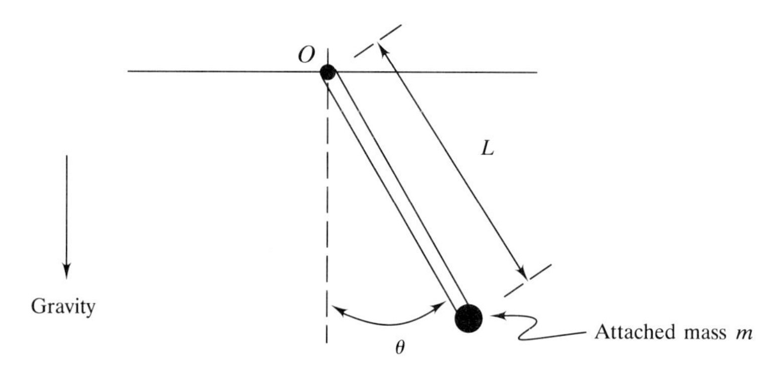

FIGURE $P7.1$

2. Two pendula are coupled via the linear spring as shown in Fig. $P7.2$. Assume that each pendulum undergoes small angle oscillations. Derive the equations of motion and find the natural frequencies and mode shapes for this system. The spring is unstretched when $\theta_1 = \theta_2 = 0$.

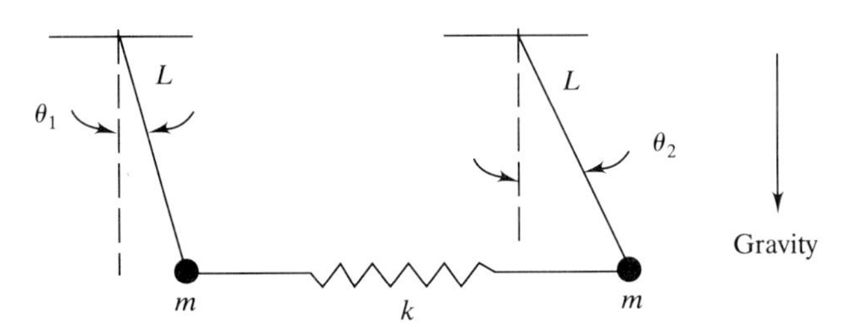

FIGURE $P7.2$

3. For the system shown in Fig. $P7.3$, determine the natural frequencies for the special case $I_c = m = L = k = 1.0$ (each in appropriate units). I_c is the moment of inertia of the bar about its center of mass, which is located at the fixed pivot A. Ignore gravity.

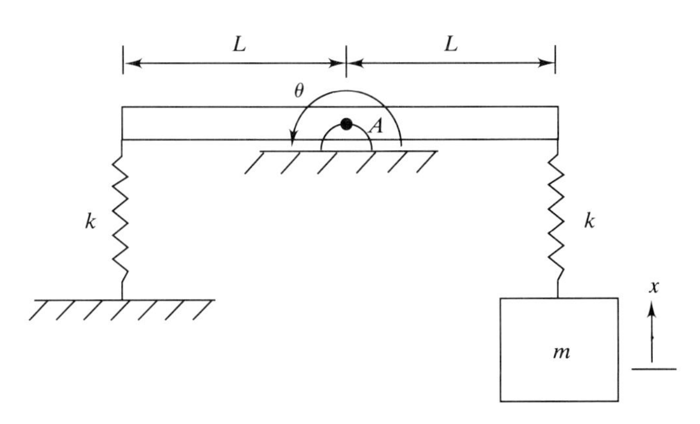

FIGURE $P7.3$

4. The natural frequencies and mode shapes for a certain 2-degree-of-freedom system are given below:

$$\text{Mode 1:}\quad \omega_1 = 10\ \text{Hz},\qquad \{C\}_1 = \begin{Bmatrix} 1 \\ 2 \end{Bmatrix}$$

$$\text{Mode 2:}\quad \omega_2 = 20\ \text{Hz};\qquad \{C\}_2 = \begin{Bmatrix} 1 \\ -\frac{1}{2} \end{Bmatrix}$$

a. For a free vibrational motion with initial conditions $\{x(0)\} = \begin{Bmatrix} -0.5 \\ -0.9 \end{Bmatrix}$, $\{\dot{x}(0)\} = \begin{Bmatrix} 0 \\ 0 \end{Bmatrix}$, which mode will dominate the response? Explain.

b. For a harmonically forced vibration, which mode would you expect to dominate the response in each of the following cases:

$$(1)\quad \{F(t)\} = \begin{Bmatrix} F_0 \\ 1.8F_0 \end{Bmatrix} \cos \Omega t; \qquad \Omega = 15\ \text{Hz}$$

$$(2)\quad \{F(t)\} = \begin{Bmatrix} -F_0 \\ 0.4F_0 \end{Bmatrix} \cos \Omega t; \qquad \Omega = 25\ \text{Hz}$$

$$(3)\quad \{F(t)\} = \begin{Bmatrix} F_0 \\ 0.4F_0 \end{Bmatrix} \cos \Omega t; \qquad \Omega = 25\ \text{Hz}$$

5. A curved manipulator arm is subjected to a free oscillation test in order to determine its moment of inertia I_c with respect to the center of mass, the location of which is known. The test is conducted by allowing the arm to rotate freely as a pendulum. If

47 oscillations are observed to occur during 1 min of free oscillation, find the moment of inertia I_c. The mass of the manipulator arm is 3 kg, and the center of mass is located 10 cm from the pivot about which the arm rotates.

6. A car of mass m supports a cylinder, also of mass m; the cylinder center is connected to the car by a linear spring of stiffness k, as shown in Fig. P7.6. The cylinder may roll without slipping on the car surface. Determine the natural frequencies of the system. Care is required to determine the kinetic energy of the cylinder.

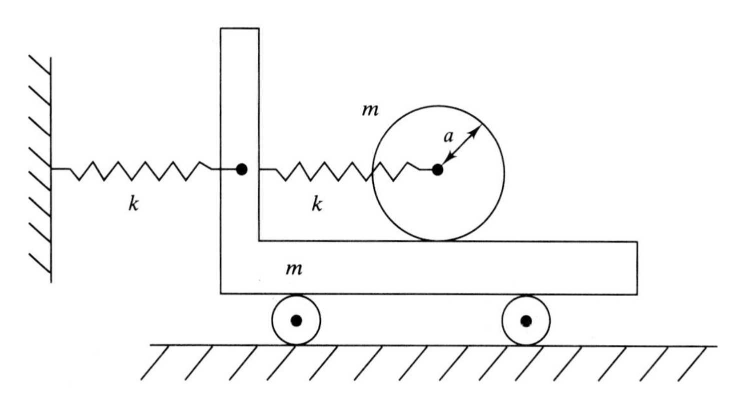

FIGURE *P*7.6

7. An automobile, modeled as a 2-DOF system as shown in Fig. P7.7, travels over a rough road, which is modeled as a sinusoid with wavelength λ. The car travels at speed V. m_1 = 600 kg and m_2 = 150 kg.

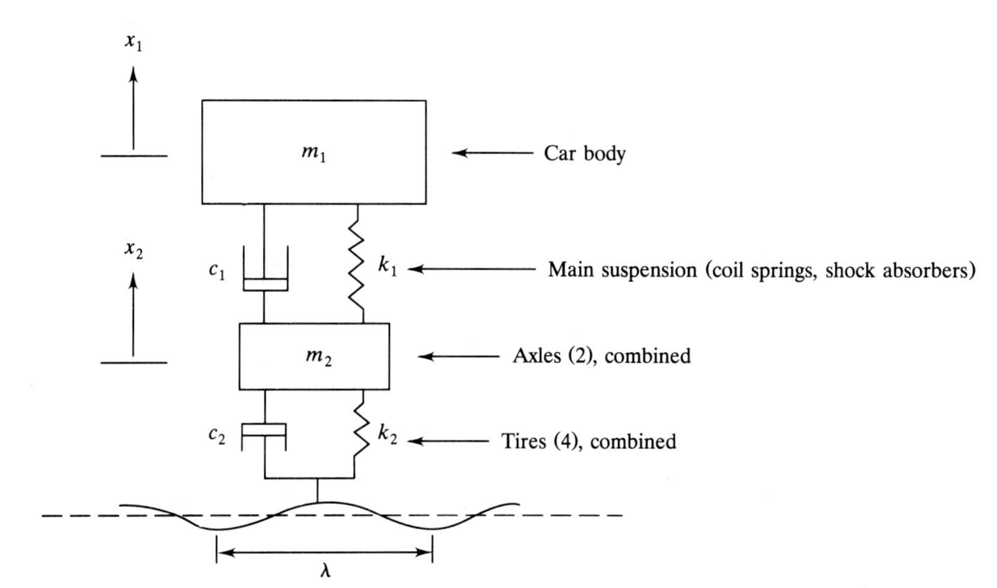

FIGURE *P*7.7

a. Determine the equations of motion in the form $[m]\{\ddot{x}\} + [c]\{\dot{x}\} + [k]\{x\} = \{F(t)\}$. Assume that the *undamped*, free vibration natural frequencies and mode shapes are defined by

$$\text{Mode 1: } \omega_1 = 1 \text{ Hz}, \qquad \{C\}_1 = \begin{Bmatrix} 1 \\ 0.1 \end{Bmatrix}$$

$$\text{Mode 2: } \omega_2 = 10 \text{ Hz}, \qquad \{C\}_2 = \begin{Bmatrix} 1 \\ -10 \end{Bmatrix}$$

To include the effect of damping on the forced response, assume that the equations for the modal coordinates y_1 and y_2 (where $\{x\} = [C]\{y\}$) are

$$\ddot{y}_i + 2\xi_i \omega_i \dot{y}_i + \omega_i^2 y_i = \frac{F_i(t)}{M_i} \qquad (i = 1, 2)$$

where $\xi_1 = 0.25$, $\xi_2 = 0.10$ are damping factors associated with the two modes.

b. Determine the steady-state amplitudes of the modal coordinates y_1 and y_2 and the resulting steady-state amplitudes for x_1 and x_2 for the following conditions:

$$\lambda = 1 \text{ m}, \qquad V = 10 \text{ m/s}$$

and

$$\lambda = 1 \text{ m}, \qquad V = 20 \text{ m/s}$$

8. A three-story building is modeled as the 3-DOF system shown in Fig. *P*7.8. The mass and stiffness properties, along with the free vibration mode shapes and frequencies, are indicated:

$$m_1 = m_2 = m_3 = 18{,}000 \text{ lb s}^2/\text{ft}$$

$$k_1 = 6(10^6) \text{ lb/ft}; \qquad k_2 = 2k_1, \qquad k_3 = 3k_1$$

$$\omega_1 = 12.17/\text{s}; \qquad \omega_2 = 24.35/\text{s}; \qquad \omega_3 = 50.95/\text{s}$$

$$[C] = \begin{bmatrix} 1 & 1 & 1 \\ 0.556 & -0.78 & -6.80 \\ 0.209 & -0.48 & 4.87 \end{bmatrix}$$

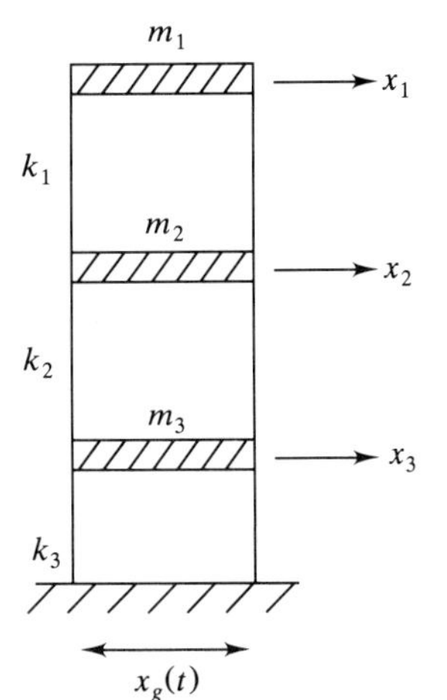

FIGURE *P*7.8

Assume that during an earthquake* a harmonic motion of the ground occurs, $x_g = x_{g_0} \cos \Omega t$, with $x_{g_0} = 0.15$ in, $\Omega = 4$ Hz. This ground motion applies a load on mass 3 through the "spring" k_3 of magnitude $F_3(t) = k_3 x_{g_0} \cos \Omega t$, so the load vector is $\{F(t)\}^T = \lfloor 0 \quad 0 \quad k_3 x_{g_0} \cos \Omega t \rfloor$.

a. Determine the equations for the three normal or modal coordinates y_i, where $\{x\} = [C]\{y\}$.
b. Find the particular responses for y_1, y_2, y_3.
c. Determine the displacement history of mass 3, and find the maximum "spring force" k_3 developed during the motion.

9. A single degree of freedom oscillator is excited by imparting a harmonic motion $y(t) = y_0 \cos \Omega t$ to the base to which the damper c is grounded as shown in Fig. $P7.9$. Determine the steady-state vibration amplitude ratio A/y_0 (A is the steady-state vibration amplitude) as a function of the frequency ratio Ω/ω and the system damping factor ξ. Sketch the frequency response.

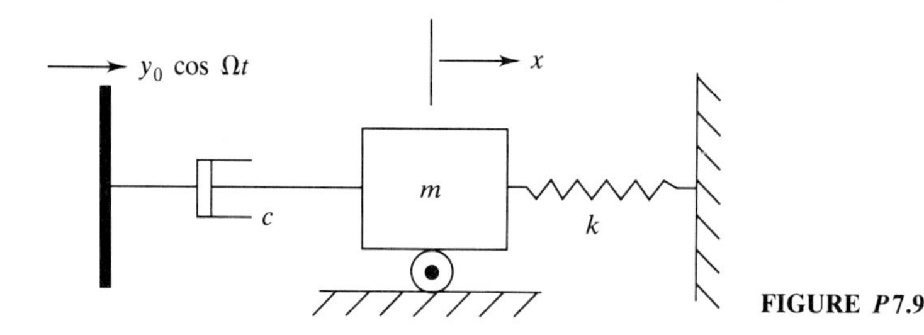

FIGURE $P7.9$

10. A function $f(t)$ has the mathematical form

$$f(t) = 2 \cos t + \sin t + 3 \cos \sqrt{5}\, t$$

Is this function periodic? Determine and plot the discrete frequency spectrum of $f(t)$.

11. Calculate the Fourier series and plot the discrete frequency spectrum of the function $f(t)$, one period of which is shown in Fig. $P7.11$.

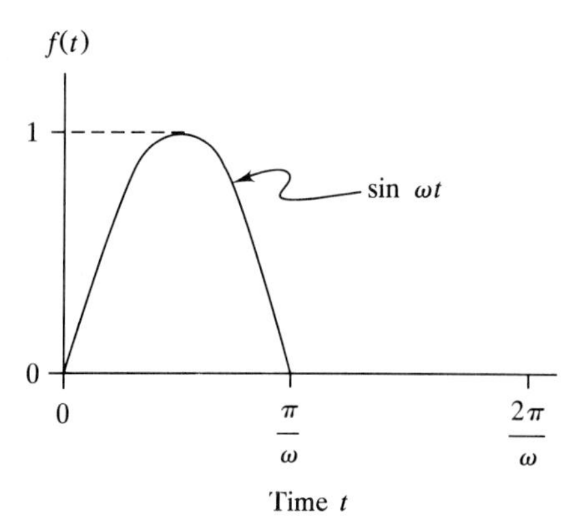

FIGURE $P7.11$

*A crude estimate based on the El Centro earthquake of 1940. Actually, the ground motion during an earthquake is not simple harmonic.

12. The data shown in Fig. $P7.12$ were obtained during a free vibration test of a single degree of freedom system. Determine the system damping factor ξ.

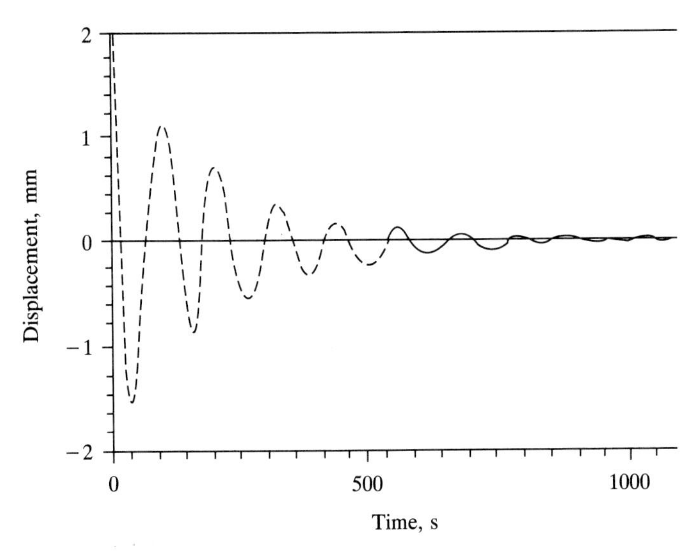

FIGURE P**7.12**

13. A fan mounted on a light frame exhibits steady vibration due to a slight imbalance in the rotating parts. The measured steady-state vibration amplitude A as a function of

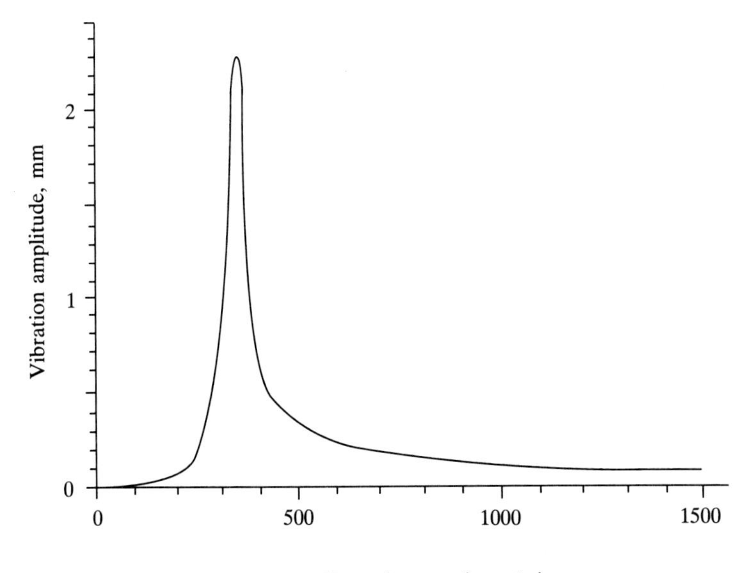

FIGURE P**7.13**

fan operating speed Ω is shown in Fig. $P7.13$. Estimate the system damping factor ξ. The desired operating speed of the fan is 400 rev/min, at which the vibration amplitude is unacceptably large. It is proposed to ameliorate the problem by stiffening the frame so as to increase the system stiffness by 20%. Discuss the merits of this proposed solution.

14. Consider the system of Example 7.5, Fig. 7.28. Verify that the natural frequencies and mode shapes are as stated in Example 7.5.

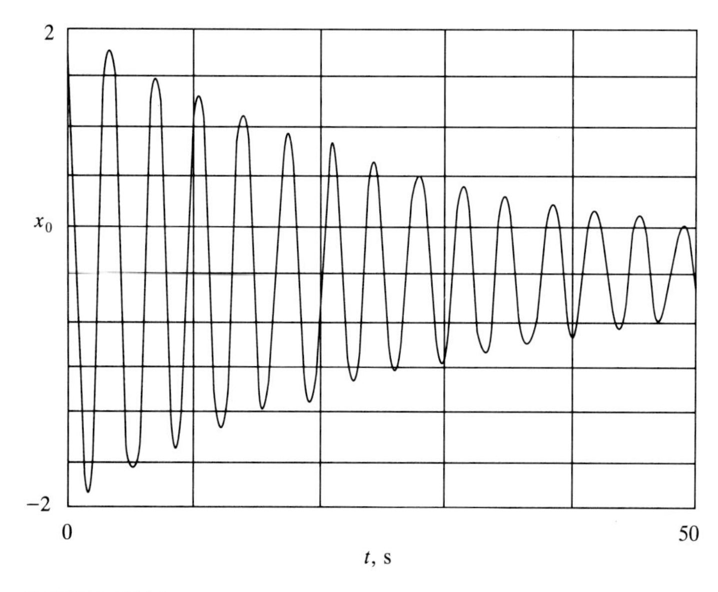

FIGURE $P7.15$

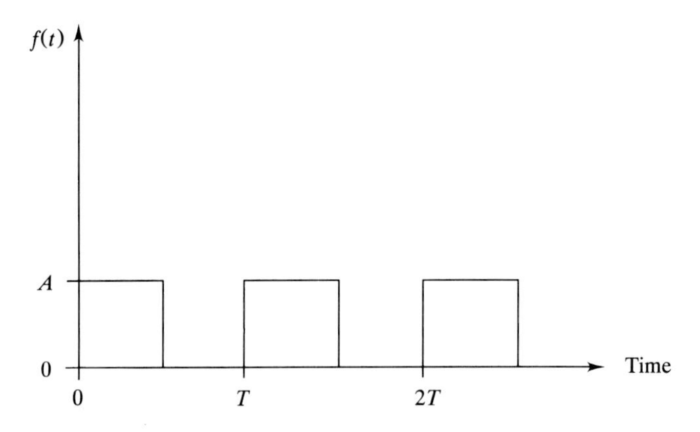

FIGURE $P7.16$

15. Shown in Fig. P7.15 is the result of a free vibration test of a single degree of freedom oscillator. Determine the system damping factor and the undamped natural frequency ω.

16. Use the results of Example 7.8 to determine the Fourier series representation of the function $f(t)$ shown in Fig. $P7.16$.

CHAPTER 8

LINEAR CONTROL SYSTEMS

8.1 INTRODUCTION

In the preceding chapters we've studied *uncontrolled* systems: A given system may be set in motion as a result of initial conditions or as a result of some external stimulus (input). The system then responds according to its basic properties and to those of the input, if present. We will now consider those systems which are controlled. By *controlled* we mean that additional causal effects are made to act on the system in such a way as to make the dynamic behavior exhibit certain preordained characteristics. The additional causal effects are produced by the control system or controller. From the dynamic systems point of view, the addition of the controller to the uncontrolled system creates a new and more complex dynamic system. Nevertheless, all of the basic systems analysis fundamentals we have seen previously will apply to both the uncontrolled and the controlled systems.

Control systems may be classified in two ways: (1) open-loop and (2) closed-loop. An *open-loop control system* executes a set of preprogrammed actions or maneuvers, but there is no means of applying corrective action if the maneuvers do not turn out to be close to the intended ones. For instance, if you attempted to drive home from your favorite watering hole with your eyes closed, based on your memory of the route, this would be an example of open-loop control. In *closed-loop control*, the actual state of the system is continually compared to the desired state of the system, and corrections are made to try to move the system to the desired state. Normal automobile operation involves this type of control. Because nearly all control systems are of the closed-loop type, we will study only closed-loop control here.

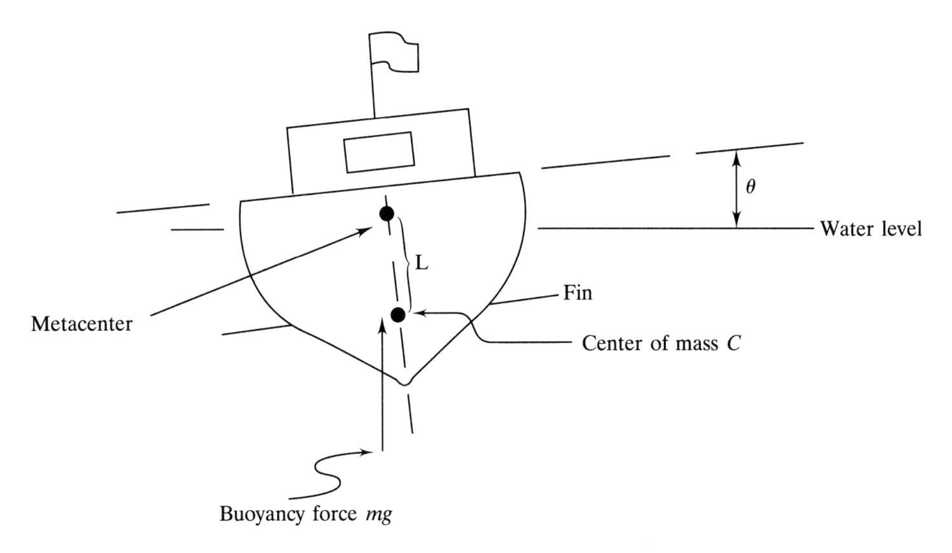

FIGURE 8.1
Rolling motion of a ship.

We now present some of the essential ideas involved in control systems analysis by considering several examples in some detail. A main purpose of these examples is to introduce those aspects of control systems which are new to us, that is, they are not involved in the analysis of uncontrolled systems.

Example 8.1 Ship Roll Control. A common problem in ship dynamics is that of rolling motions induced by the action of waves. In this example we'll illustrate conceptually how such rolling motions might be controlled. We'll start by developing a simple mathematical model to describe the rolling motion of a ship, as shown in Fig. 8.1.

The coordinate θ measures the roll orientation of the ship. The mass of the ship is m and its roll moment of inertia with respect to the center of mass C is I_C. Thus, considering rolling as the only rotational motion which occurs, the equation of motion for roll will be (2.30):

$$I_C\ddot{\theta} = M_C \tag{8.1}$$

where M_C is the sum of all roll moments about the center of mass C. Without as yet worrying about the control problem, we can identify three sources of roll torque about the center of mass: (1) A torque due to buoyancy. A buoyancy force mg acts through the metacenter, which is the point, on the vertical axis through the center of mass C, located directly above the center of mass of the water displaced by the ship (the buoyancy force is applied at the center of mass of the displaced water). (2) Torques due to the action of wave motion of the surrounding water. These torques are usually somewhat unpredictable and will be assumed to cause the rolling motion we wish to control. (3) A fluid damping torque, which we'll assume to be proportional to roll rate $\dot{\theta}$. With these three torques written out, (8.1) becomes, for small θ,

$$I_C\ddot{\theta} = \underset{\substack{\uparrow \\ \text{Buoyancy}}}{-mgL\theta} + \underset{\substack{\uparrow \\ \text{wave} \\ \text{disturbance}}}{M_d(t)} - \underset{\substack{\uparrow \\ \text{damping}}}{c\dot{\theta}}$$

or

$$I_C\ddot{\theta} + c\dot{\theta} + mgL\theta = M_d(t) \tag{8.2}$$

Observe that (8.2) is of the form (2.52) of the single DOF damped oscillator, with external forcing which arises from the disturbing waves. Thus, the uncontrolled system will act like a mechanical oscillator subjected to an (often) irregular, time-dependent input. Let's assume that, for typical disturbance torques $M_d(t)$, the resulting rolling motion is unacceptably large and that we want to add a control system to keep the roll excursions within acceptable limits. One way to do this is as follows:

1. Mount control fins on either side of the ship, as shown in Fig. 8.1. Assume that these fins can be rotated, equally and oppositely, to produce positive or negative roll torques of desired magnitude as the ship moves through the water.
2. Define the *desired* roll orientation, which we'll call $\theta_d(t)$. In this example, $\theta_d(t)$ will be zero, but for generality we'll leave it as $\theta_d(t)$.
3. Continually *measure* the actual roll orientation $\theta(t)$ (this measurement might be made with some type of gyroscopic sensor). Compare the actual to the desired roll orientation to form the *error* $e(t) = \theta_d(t) - \theta(t)$. We want to keep this error close to zero.
4. We want to arrange the action of the roll fins so as to continually drive the error to zero. A possible control strategy is to have the roll fins produce a control torque, denoted as $M_C(t)$, which is proportional to the instantaneous error $e(t)$; thus, let

$$M_C(t) = k_p(\theta_d - \theta) \tag{8.3}$$

The "control constant" k_p will be a function of the size of the fins and the amount of fin deflection per unit error. We assume that k_p can be selected by us in order to achieve acceptable performance. A "control law" of the type (8.3) is referred to as *proportional control* because the control action is proportional to the error. In simple, intuitive terms the control law (8.3) makes sense: If the instantaneous error is large, a large corrective action is applied; if the instantaneous error is small, a small corrective action is applied.

How is the control action actually accomplished? The error signal is fed to a servomotor or hydraulic actuator which keeps the fins deflected so as to produce a roll torque proportional to the error. (If the relation between fin orientation and roll torque produced is linear, then the fin deflection will also be proportional to the error.) Our mathematical model now looks like this:

$$I_C\ddot{\theta} + c\dot{\theta} + mgL\theta = M_d(t) + \underset{\substack{\uparrow \\ \text{Control torque}}}{k_p(\theta_d - \theta)} \tag{8.4}$$

Now in physical terms we can see what the controller is doing by rewriting (8.4) as follows:

$$I_C\ddot{\theta} + c\dot{\theta} + (mgL + k_p)\theta = M_d(t) + k_p\theta_d(t) \qquad (8.5)$$

We see that a portion of the control torque $(k_p\theta_d)$ is an input which defines the desired operating condition, while a portion $k_p\theta$ increases the stiffness of the system. Whereas the uncontrolled system has an undamped natural frequency equal to $\sqrt{mgL/I_C}$, the controlled system now has the higher undamped natural frequency $\sqrt{(mgL + k_p)/I_C}$, and the controlled system will exhibit quicker response. Thus, the addition of the proportional controller has altered the basic properties of the system, creating a new system with different and presumably better dynamic characteristics.

Let us compare the controlled and uncontrolled systems in Example 8.1. We observe several new things which are involved in the closed-loop control problem:

1. *Continual measurement of the actual state of the system:* Here we measure $\theta(t)$, the variable we wish to control.
2. *Continual determination of the error $e(t)$, the difference between the desired state and the actual state:* The process of error determination involves *feedback*, that is, the measured value of the actual state of the system is "fed back" to a device which compares the actual state to the desired state to determine the error. Because feedback is essential in the controls process, closed-loop control systems are also known as feedback control systems.
3. *Actuation:* A part of the control system consists of devices which exert some corrective action designed to drive the error to zero. In this example the actuators might be hydraulic or electromagnetic; the actuators exert moments which rotate the control fins, in turn causing a roll moment to be exerted on the ship.

The operations of measurement, feedback, and actuation are new to us and are central to the design and analysis of closed-loop control systems. In practical terms the addition of the controller requires the use of instrumentation for measurement, electronic devices such as microprocessors for error determination, and load-generating actuators which apply corrective action.

In this example a typical design sequence might go something like this:

1. Determine the relevant properties of the uncontrolled system: I_C, mg, L, and the center of mass location C.
2. Obtain a representative sampling of typical disturbance torques $M_d(t)$. These may be based on existing data for ships of similar configuration, model tests, and so on. The actual disturbance torques to which the ship will be subjected cannot be predicted exactly ahead of time, so we would have to use a representative sample previously determined.

3. Establish requirements which the controlled system should meet. For instance, we might require the magnitude of roll angle $|\theta(t)|$ always to be less than some acceptable value for the class of disturbances in 2.
4. Using the system mathematical model, find values of the control constant k_p which will produce acceptable performance for typical disturbances.
5. Design a measurement/fin/actuator system which will produce the required k_p. This means we will need to know the dependence of control torque $M_C(t)$ on fin size and deflection and on the forward speed of the ship. Then we need to design actuators to act on the measured error to keep the fins deflected in the proper manner.
6. Test everything and modify as needed. If the mathematical modeling has been done carefully, the effort involved in testing/model verification can be reduced.

In practice some of the above steps will be fairly involved. In addition, the student should realize that the mathematical model described here is highly idealized. For instance, the roll fin/actuator system will possess inertia, damping, and possibly stiffness, so it would be necessary to construct a mathematical model to describe the dynamic behavior of the actuation system. Keep in mind that the intent of the example is only to present some of the basic ideas involved in control system design and analysis, rather than to present a realistic analysis of ship roll control.

Example 8.2 Jet Airplane Roll Control. We next consider the roll control of a high-performance jet airplane, as shown in Fig. 8.2. The basic problem is similar to ship roll control: We desire to maintain a specified roll orientation, determined by the pilot, in the presence of aerodynamic disturbance torques. For simplicity we'll assume the pilot desires to maintain level flight at zero roll angle. If we consider uncontrolled rolling, the system equation of motion is (8.1). Roll moments arise due to disturbances $M_d(t)$ and due to roll damping, so that the linear mathematical

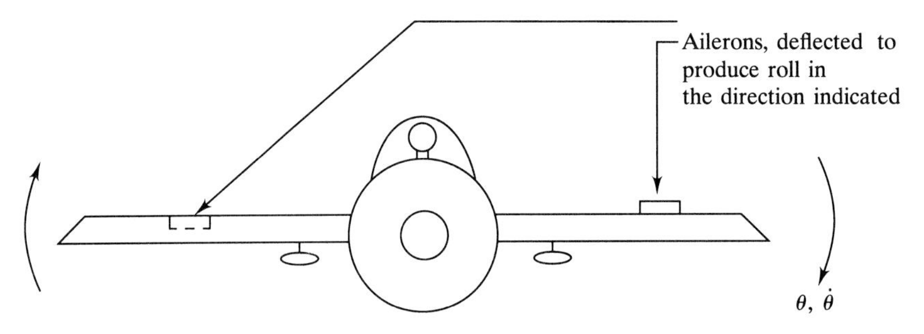

FIGURE 8.2
Jet airplane roll control.

model analogous to (8.2) is

$$I_C\ddot{\theta} + c\dot{\theta} = M_d(t) \tag{8.6}$$

Observe that this system model possesses no "stiffness," that is, there is no "self-righting" tendency as in the ship problem, where buoyancy creates such an effect. The autonomous version of the math model (8.6) yields the eigenvalues $\lambda_1 = 0$, $\lambda_2 = -c/I_C$, so that the uncontrolled system is stable but not asymptotically stable. Furthermore, the equilibrium solution ($M_d = 0$) for the two states θ and $\dot{\theta}$ is $\dot{\theta} = 0$, θ arbitrary. Thus, suppose that a disturbance torque $M_d(t)$ is applied over a short time interval, imparting some initial roll velocity $\dot{\theta}_0$ to the airplane. This roll velocity will decay exponentially to zero, but the eventual roll orientation θ may take on any value (depending on $\dot{\theta}_0$).

In this example a control strategy similar to that used for the ship roll control problem may be implemented by designing a measurement/actuator system which deflects the ailerons so that errors in desired roll orientation are driven to zero. Here, however, the roll control system must accomplish several tasks: It must alter the system properties so that there is a single equilibrium solution $\theta = \dot{\theta} = 0$, it must render this equilibrium solution asymptotically stable, and it must then keep any disturbance induced rolling motions sufficiently close to this equilibrium. For the ship problem, only the latter function needs to be performed.

Example 8.3 Automobile Cruise Control. Here we consider the automobile cruise control, which is supposed to maintain a desired forward speed V_0 of the car, regardless of winds, grades, and changes in road conditions. In order to simplify things, we'll consider only the effect of changes in grade, as shown in Fig. 8.3; zero wind and unvarying road conditions will be assumed. At any instant the grade is defined by the angle α. Thus, if the car is traveling along a level road at the desired speed V_0 and suddenly encounters a steep grade, the actual speed V will tend to drop below V_0. The cruise control is supposed to increase fuel flow to increase the speed back to V_0.

A mathematical model of the car motion along the road is obtained by application of Newton's second law along the instantaneous direction of travel,

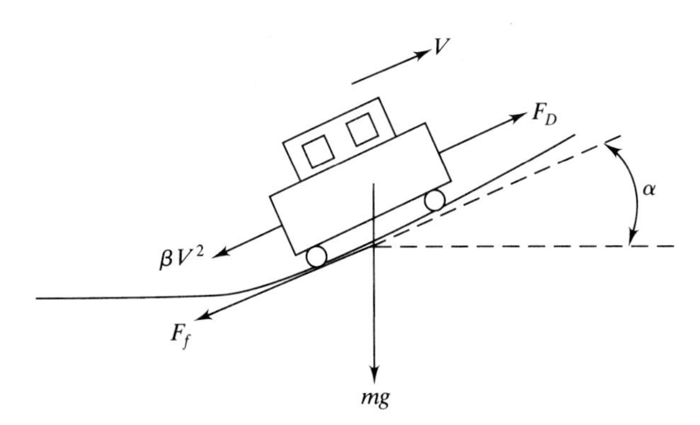

FIGURE 8.3
Auto moving in the plane. The various forces acting are shown.

resulting in

$$m\dot{V} = -\beta V^2 + F_D - F_f - mg \sin \alpha \tag{8.7}$$

where V = Actual vehicle speed (desired to be V_0, which is a specified constant)
m = Mass of car
β = Coefficient defining aerodynamic drag force βV^2
F_D = Tractive or drive force provided by engine
F_f = Retarding force due to friction between wheels and road
$\alpha(t)$ = Instantaneous grade angle

We have a first-order nonlinear system in which the single state is the speed V. When cruising on level road ($\alpha = 0$) at the desired speed ($V = V_0$, $\dot{V} = 0$), (8.7) reduces to

$$F_{D_0} = F_f + \beta V_0^2 \tag{8.8}$$

which defines the tractive force F_{D_0} (and the throttle setting) needed to maintain the desired speed on level highway.

Now in order to view this problem from the controls standpoint, let's modify the model (8.7) in the following way: Let the forward speed $V = V_0 + v$, where v is the deviation from the desired speed V_0. Likewise, let $F_D = F_{D_0} + f_D$, where F_{D_0} is the steady value defined by (8.8) and f_D is the deviation in tractive force. Here f_D will be produced by the cruise controller as it alters the throttle setting. We will assume F_f to be constant. With the new notation, the math model (8.7) may be written as (with $\dot{V}_0 = 0$)

$$m\dot{v} = -\beta(V_0 + v)^2 + F_{D_0} + f_D - F_f - mg\alpha(t)$$

or

$$m\dot{v} = -\beta(2V_0 v + v^2) + f_D - mg\alpha(t) + F_{D_0} - F_f - \beta V_0^2 \tag{8.9}$$

where α is assumed small enough that $\sin \alpha \approx \alpha$. This model is next linearized by assuming $v^2 \ll 2V_0 v$, that is, $v \ll 2V_0$. Furthermore, the equilibrium condition (8.8) allows us to eliminate the final three terms in (8.9). We then arrive at the following nonautonomous linear first-order model for the deviation v from the desired speed V_0:

$$m\dot{v} = -2\beta V_0 v + f_D - mg\alpha(t) \tag{8.10}$$

Note that the steps followed in obtaining the linearized model (8.10) from (8.7) are the same as those followed in the linearization procedure described in Sec. 4.1.4. Here V_0 is the equilibrium value of the state V, and the definition $V = V_0 + v$ is just (4.28). The job of the cruise control is to maintain the actual state V sufficiently close to the equilibrium value V_0. In the context of controls, equilibrium conditions at which the system is desired to be are sometimes referred to as *operating points*.

The objective now is to design the cruise controller to provide a corrective traction force f_D (by altering the throttle setting) to maintain the speed deviation v sufficiently close to zero in the presence of changes in grade $\alpha(t)$. Many actual cruise controllers employ a proportional control strategy, so that

$$f_D = k_p(v_d - v)$$

where v_d, the desired velocity deviation, is zero here. The ODE model of the

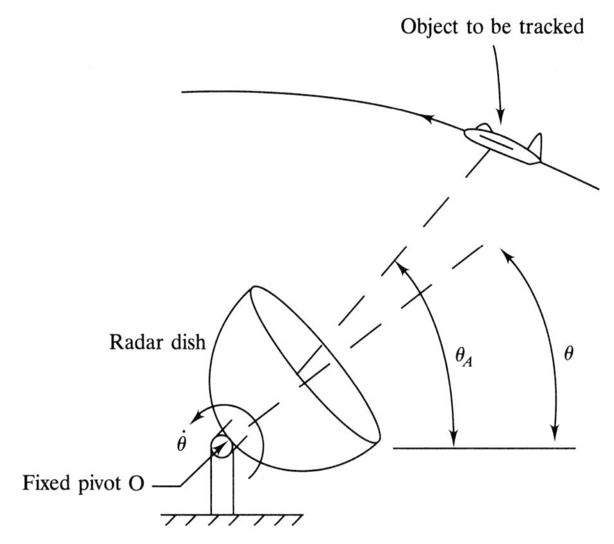

FIGURE 8.4
Radar tracking system.

controlled system with proportional control is then

$$m\dot{v} = -2\beta V_0 v + k_p(v_d - v) - mg\alpha(t) \tag{8.11}$$

The control constant k_p must be such that the speed deviation v is acceptably small (say, a few miles per hour) for the range of grade disturbances $\alpha(t)$ likely to be encountered. We note, however, that actual cruise controllers do not necessarily maintain the desired speed V_0 even in the presence of steady grade disturbances. For instance, if one encounters a long, uphill 6 percent grade ($\tan \alpha = 6/100$) and if the desired speed $V_0 = 55$ mi/h, the actual steady-state speed reached on the grade is likely to be nearer 53 mi/h: There is some "steady-state error." This error depends on the control constant k_p and on the system constants m, β, V_0, g in the model (8.11).

Example 8.4 Radar Tracker. In the preceding examples the main purpose of the control system is to ensure that the system dynamics remain sufficiently close to an asymptotically stable equilibrium (or "operating") condition in the presence of disturbances which tend to move the system away from this equilibrium. In the present example the purpose of the controller will be to cause the system motion to follow or "track" an input function to within a specified accuracy. The system considered is shown in Fig. 8.4.

We assume that an airplane, flying in the plane of the paper, is to be tracked by radar as it flies over. The instantaneous orientation angles of the tracker and airplane are $\theta(t)$ and $\theta_A(t)$, respectively. The radar tracker is rotated by a motor which must provide a torque which causes $\theta(t)$ always to be sufficiently close to $\theta_A(t)$. Thus, the purpose of the controller is to cause the system state $\theta(t)$ to follow closely the input functions $\theta_A(t)$ likely to be encountered. An idealized math model, including some linear damping to represent friction, would be

$$I\ddot{\theta} + c\dot{\theta} = M_C(t) \tag{8.12}$$

where $M_C(t)$ is the control torque, c the damping constant, and I the moment of inertia of the radar dish with respect to the fixed pivot **O**. If we adopt a proportional control strategy, so that $M_C(t) = k_p[\theta_A(t) - \theta(t)]$, then the model (8.12) becomes

$$I\ddot{\theta} + c\dot{\theta} + k_p\theta = k_p\theta_A(t) \tag{8.13}$$

Mathematically, this is the same problem as considered in Chap. 5: the response of a second-order system to an input. Here, however, we have the added proviso that the state $\theta(t)$ should track the input $\theta_A(t)$ closely at all times.

Examples 8.1 through 8.4 have one important characteristic in common: They belong to the class of *single input/single output*, or SISO, systems. For our purposes *single output* means that there is a single variable which is to be controlled,* and associated with this controlled variable is a single input which specifies the desired operating value or history of the controlled variable. In general, even if the system is complex and of high order mathematically, our viewpoint is that, if we can make a single variable, say, one of the states, behave in the manner we wish, then the control is successful.

A large number of control problems fit into the SISO category, and we will spend most of the rest of this chapter studying it. Example 8.5 is provided below, however, in order to illustrate a fairly complex control problem involving multiple inputs and multiple outputs (MIMO).

Example 8.5 Missile Guidance System. Here we consider an idealized targeting scenario for a long-range strategic missile such as the Minuteman III or SS-18. The operation is described briefly with reference to Fig. 8.5.

These systems are equipped with multiple warheads which are targeted independently. We consider the targeting of a single reentry vehicle (RV). The multiple RVs are mounted on a platform (bus) which separates from the booster (not shown in Fig. 8.5) following launch. The RVs are then released one at a time from the bus and spend most of their flight outside the atmosphere, eventually reentering the atmosphere as shown. In order for a given RV (for example, the one shaded in Fig. 8.5) to hit as close as possible to the target, that RV must reenter the atmosphere at the correct location and with the correct velocity (the ensuing flight through the atmosphere to the target has previously been calculated as accurately as possible). In order to hit the correct reentry point with the correct velocity, the RV must be released from the bus at an appropriate location and with the appropriate velocity. The correct RV release conditions are achieved by the bus, which is equipped with a controller and small thrusters, which allow the bus to maneuver. The actual position and velocity are continually measured and are compared to the (preprogrammed) desired position and velocity, and the bus maneuvers to try to drive any errors to zero prior to RV release.

---

*In Examples 8.1 through 8.4 the controlled variable was one of the system states. In general, however, the controlled variable, or output, might be some linear combination of several or even all of the states. For simplicity we will assume in this chapter that the variable to be controlled is one of the states.

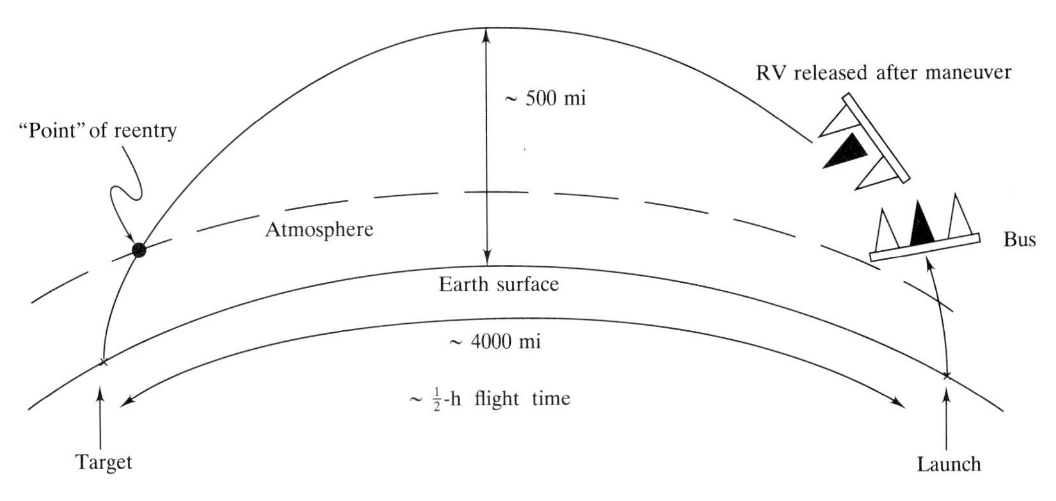

FIGURE 8.5
Strategic missile targeting.

This fairly complex control problem is an example of a *multiple input/multiple output*, or MIMO, system. Because position and velocity are vectors, a total of *six* variables must be controlled: the three components of position and the three components of velocity. Thus, the controller must attempt to zero-out six errors prior to RV release from the bus.

In the older ballistic systems such as the Minuteman, the bulk of the targeting error is due to errors in RV release conditions, so that the overall targeting problem is dominated by this guidance/control problem. The actual design, analysis, and implementation of such a guidance system involved a tremendous expenditure of effort and is well beyond our scope. This example has been presented only to illustrate a representative MIMO situation.

Comment. In the preceding examples, 8.1, 8.3, and 8.4, only one type of control strategy was discussed: proportional control, for which the corrective action is proportional to the instantaneous error. This is not meant to imply that proportional control is the only or the best type of control strategy. Actually, there are numerous other types of control strategies which are used instead of, or in addition to, proportional control. In general, each of these strategies will utilize an error to determine a control action, but the control action need not be proportional to the error. (It might be proportional to the derivative or integral of the error, for instance.) A few of these other strategies are discussed in Sec. 8.4, et sequence.

In the preceding SISO examples, 8.1 through 8.4, the addition of the controller will, from a dynamic systems point of view, change the system dynamic model in two ways:

1. It will alter the basic properties of the system itself. For example, the proportional controller, when added to the systems in Examples 8.1 and 8.4, will alter the overall system stiffness.

2. The controller will provide as input the desired history of the variable to be controlled (for example, θ in Example 8.1, v_d in Example 8.3, and θ_A in Example 8.4).

With these two things and Examples 8.1 through 8.5 in mind, we state the three primary functions which the controller is designed to perform:

1. *Maintain or produce adequate stability:* The controlled system must be asymptotically stable (if it is not, all other considerations are moot). Thus, if the uncontrolled system is unstable or neutrally stable (as in Example 8.2), the controller must stabilize the system. Furthermore, for most controlled systems it is required (often, for reasons 2 and 3 which follow) that reasonably large damping be present. This means that the eigenvalues of the autonomous version of the controlled system math model should lie sufficiently to the left of the imaginary axis when plotted in the complex plane as in Fig. 4.23. Thus, the controller must alter the overall system properties in such a way that the eigenvalues of the controlled system result in adequate overall damping.
2. *Close tracking of the input:* For SISO systems the input specifies the desired history of the controlled variable. The controller has to be designed to keep the output sufficiently close to the input and to respond quickly enough to relatively sudden changes in the input. Both transient and steady-state operations are important. For instance, if a mechanical system such as a robot is commanded to maneuver its end effector quickly from one position to another, the system should move quickly and smoothly toward the new position (transient regime) and should eventually reach a steady-state position close to that desired.
3. *Disturbance rejection:* The controller must maintain the system near the desired operating point (as in Examples 8.1 through 8.3) or must track the input closely (as in Example 8.4), in the presence of disturbances.

While related to performance goals of the controller, the preceding descriptions are not very specific (notice that the phrase "sufficiently close" appears frequently). In specific applications the performance goals have to be quantified; we might have to specify how much damping is desirable, what steady-state error is tolerable, and so on. These considerations are addressed in Sec. 8.4.

In the remainder of this chapter we will study the single input/single output (SISO) linear control problem. We have seen in the modeling examples of Chaps. 2 and 3 that many engineering systems can exhibit nonlinear behavior. The use of linear models to study control systems, however, is often justified. For example, consider problems of the type Examples 8.1 through 8.3, for which the objective is to keep the system close to the desired equilibrium or operating condition. If the control system is properly designed, it will cause the system to remain close to equilibrium. If the system stays close to equilibrium, nonlinear effects should not be significant, and a linear model is valid. For information on nonlinear control systems, the student is referred to the references at the end of the chapter.

In addition to the restriction of linearity, we will consider only SISO systems: a single variable is to be controlled, and the desired value or history of this variable is supplied as input. For information on the MIMO problem, consult one of the references listed at the end of this chapter.

The primary mathematical tool we will utilize to characterize and study the SISO linear control problem is Laplace transformation. Laplace transforms are widely used to solve linear differential equations. Furthermore, in the controls context, the basic properties of systems and their components, as well as of controllers, are often specified in terms of Laplace transforms. Section 8.2 which follows provides a summary of the essentials of Laplace transformation, which we will use heavily in the remainder of this chapter.

8.2 BASICS OF LAPLACE TRANSFORMATION

8.2.1 Introduction

We have so far looked at several ways to solve the ordinary differential equations which comprise our system math models: the eigenvalue-eigenvector techniques associated with the state variable formulation (Chap. 4) and the method of undetermined coefficients to solve nonautonomous problems (Chap. 5). Now we'll look at yet another way: Laplace transformation. The reason for doing this is that the Laplace transform is the language of classical control theory. Systems, control components, and inputs are characterized by their Laplace transforms. In this subsection we'll define the Laplace transform (LT), calculate a few simple transforms, and then see how the LT is applied to the study of linear differential equations (the Laplace transform is applicable only to linear systems). Use of LTs to characterize and analyze linear control systems is made in Secs. 8.3 et sequence.

DEFINITION. Consider any function of time $f(t)$ [$f(t)$ could be a state or an input in a dynamic system, for example]. The Laplace transform of this function is denoted as $\mathscr{L}[f(t)]$ and is defined by the following integral:

$$\mathscr{L}[f(t)] = \int_{t=0}^{\infty} e^{-st} f(t)\, dt \tag{8.14}$$

Here the "Laplace variable" s is generally a complex number. We do not need to say anything more about s, except that we assume that the real part of s is positive, so that the above integral actually exists, at least for most functions of interest to us. [One can see that (8.14) will exist provided that $f(t)$ is a function of "exponential order," that is, $f(t)$ is bounded by an exponential of the form $e^{\alpha t}$, $\alpha > 0$, such that the integral in (8.14) exists.]

We observe that the integral defined in (8.14) is a function of the Laplace variable s, the time dependence having been "integrated out." To indicate this functional dependence on s, we also denote the Laplace transform of $f(t)$ as $\mathscr{L}[f(t)] = \bar{f}(s)$ or $\mathscr{L}[f(t)] = F(s)$. Thus, either $\bar{f}(s)$ or $F(s)$ is understood to be the Laplace transform of $f(t)$. The student should verify that the Laplace transform is a linear operator:

$$\mathscr{L}[a_1 f_1(t) + a_2 f_2(t)] = a_1 \mathscr{L}[f_1(t)] + a_2 \mathscr{L}[f_2(t)] \tag{8.15}$$

where a_1 and a_2 are constants and $f_1(t)$ and $f_2(t)$ arbitrary functions of time. To illustrate the calculation (8.14), we'll now determine the transforms of a few simple functions.

Example 8.6 The Unit Step Function. The unit step function $u(t)$ is defined as $u(t) = 0$ if $t < 0$ and $u(t) = 1$ if $t > 0$. Equation (8.14) then yields

$$\begin{aligned} \mathscr{L}[u(t)] &= \int_0^\infty e^{-st}(1)\,dt = -\frac{1}{s}e^{-st}\Big|_0^\infty = \frac{1}{s} \\ \mathscr{L}[u(t)] &= \frac{1}{s} \end{aligned} \tag{8.16}$$

where use has been made of the fact that e^{-st}, evaluated at $t = \infty$, is zero because the real part of s is positive. It should be clear that the transform of an arbitrary step function $f(t) = cu(t)$, where c is a constant, will be just $\mathscr{L}[cu(t)] = c/s$, in view of (8.15).

Example 8.7 Exponential Function. We consider the exponential function $f(t) = e^{at}$, where a is any constant, real or complex. In this case (8.14) yields

$$\mathscr{L}[e^{at}] = \int_0^\infty e^{-st}e^{at}\,dt = \int_0^\infty e^{-(s-a)t}\,dt = -\frac{1}{s-a}e^{-(s-a)t}\Big|_0^\infty$$

The exponential term vanishes at the upper limit and is unity at the lower limit, so that

$$\mathscr{L}[e^{at}] = \frac{1}{s-a} \tag{8.17}$$

The result (8.17) turns out to be very useful.

Example 8.8 The Trigonometric Functions ($\sin \omega t, \cos \omega t$). The easiest way to compute the Laplace transforms of these functions is first to express the trigonometric functions in complex exponential form and then to use (8.17). Thus, put $\cos \omega t = \frac{1}{2}(e^{i\omega t} + e^{-i\omega t})$, hence

$$\begin{aligned} \mathscr{L}[\cos \omega t] &= \frac{1}{2}\mathscr{L}(e^{i\omega t}) + \frac{1}{2}\mathscr{L}(e^{-i\omega t}) \\ &= \frac{1}{2}\left[\frac{1}{(s-i\omega)} + \frac{1}{(s+i\omega)}\right] \\ &= \frac{s}{s^2+\omega^2} = \mathscr{L}[\cos \omega t] \end{aligned} \tag{8.18a}$$

A similar calculation shows that

$$\mathscr{L}[\sin \omega t] = \frac{\omega}{s^2 + \omega^2} \tag{8.18b}$$

Table 8.1 provides the Laplace transforms of a number of functions.

TABLE 8.1
Some Laplace transforms

	Time function, $f(t)$, $t > 0$	Laplace transform, $F(s)$
1	Unit impulse	1
2	Unit step function	$\frac{1}{s}$
3	t	$\frac{1}{s^2}$
4	t^2	$\frac{2}{s^3}$
5	t^n	$\frac{n!}{s^{n+1}}$
6	e^{at}	$\frac{1}{s-a}$
7	$\frac{e^{-at} - e^{-bt}}{b-a}$	$\frac{1}{(s+a)(s+b)}$
8	$\frac{1}{(n-1)!} t^{n-1} e^{-at}$	$\frac{1}{(s+a)^n}$
9	$\frac{e^{-at}}{(b-a)(c-a)} + \frac{e^{-bt}}{(c-b)(a-b)} + \frac{e^{-ct}}{(a-c)(b-c)}$	$\frac{1}{(s+a)(s+b)(s+c)}$
10	$\sin \omega t$	$\frac{\omega}{s^2 + \omega^2}$
11	$\cos \omega t$	$\frac{s}{s^2 + \omega^2}$
12	$1 - e^{-t/T}$	$\frac{1}{s(1 + Ts)}$
13	$1 - \frac{t+T}{T} e^{-t/T}$	$\frac{1}{s(1 + Ts)^2}$
14	$\frac{t^{n-1} e^{-t/T}}{T^n (n-1)!}$	$\frac{1}{(1 + Ts)^n}$
15	$e^{-at} \sin \omega t$	$\frac{\omega}{(s+a)^2 + \omega^2}$
16	$e^{-at} \cos \omega t$	$\frac{(s+a)}{(s+a)^2 + \omega^2}$
17	$1 + \frac{e^{-\xi \omega t}}{(1 - \xi^2)^{1/2}} \sin(\omega(1 - \xi^2)^{1/2} t - \psi)$; $\cos \psi = -\xi$	$\frac{\omega^2}{s(s^2 + 2\xi\omega s + \omega^2)}$

8.2.2 Transformation of Derivatives and Integrals

Since we will apply Laplace transformation to solve ordinary differential equations, we will need to determine the transform of the derivatives of any function $f(t)$. Thus, consider the transform of the first derivative of a function $f(t)$:

$$\mathscr{L}\left[\frac{df}{dt}\right] = \int_0^\infty e^{-st}\left(\frac{df}{dt}\right) dt$$

We can integrate this equation by parts ($\int u\,dv = uv - \int v\,du$); let $u = e^{-st}$, $dv = (df/dt)\,dt$, leading to

$$\mathscr{L}\left[\frac{df}{dt}\right] = f(t)e^{-st}\Big|_0^\infty - \int_0^\infty (-se^{-st})\, f(t)\, dt$$

$$= -f(0) + s\int_0^\infty e^{-st} f(t)\, dt$$

so that

$$\mathscr{L}\left[\frac{df}{dt}\right] = s\mathscr{L}[f(t)] - f(0) \tag{8.19}$$

In words, the transform of the derivative of $f(t)$ equals s times the transform of $f(t)$, minus the initial value of $f(t)$. Essentially, the operation of time differentiation in the time domain is equivalent to multiplication by s in the Laplace domain.

The result (8.19) can be applied successively to determine the transform of the derivative of any order. For example, to determine the transform of the second derivative, $\mathscr{L}[d^2f/dt^2]$, let $g = df/dt$. Then $\mathscr{L}[d^2f/dt^2] = \mathscr{L}[dg/dt]$, so that

$$\mathscr{L}\left[\frac{d^2f}{dt^2}\right] = s\bar{g}(s) - g(0)$$

$$\mathscr{L}\left[\frac{df}{dt}\right] = s\bar{f}(s) - f(0) = \mathscr{L}[g(t)]$$

$$\mathscr{L}\left[\frac{d^2f}{dt^2}\right] = s^2\bar{f}(s) - sf(0) - \dot{f}(0) \tag{8.20}$$

In applying the LT to differential equations, we'll see that the LT method automatically gets the required initial conditions into the formulation. The student should be able to show that the transform of the nth derivative of $f(t)$ will be

$$\mathscr{L}\left[\frac{d^nf}{dt^n}\right] = s^n\bar{f}(s) - s^{n-1}f(0) - s^{n-2}\dot{f}(0) - \cdots - \frac{d^{n-1}f}{dt^{n-1}}(0) \tag{8.21}$$

Next, we'll determine the transform of the integral of a function $f(t)$:

$$\mathscr{L}\left[\int_0^t f(\tau)\, d\tau\right]$$

To do this define $g(t) = \int_0^t f(\tau)\, d\tau$, so that $dg/dt = f(t)$. Then, use

$$\mathscr{L}\left[\frac{dg}{dt}\right] = s\bar{g}(s) - g(0)$$

Next, replace dg/dt by $f(t)$, note that $g(0) = 0$, and solve for $\bar{g}(s)$, which is the desired transform; thus,

$$\mathscr{L}\left[\int_0^t f(\tau)\, d\tau\right] = \frac{\bar{f}(s)}{s} \tag{8.22}$$

The process of integration in the time domain is equivalent to division by s in the Laplace domain.

8.2.3 Inversion of the Laplace Transform

In the above examples we have addressed the question "given some function $f(t)$, what is its Laplace transform $\bar{f}(s)$?" In applying LTs to ordinary differential equations, we will need also to address the question "given a transform $\bar{f}(s)$, what is the associated function of time $f(t)$?" We speak of "inverting" the transform. Although there is a formal procedure for computing $f(t)$ given $\bar{f}(s)$, it is a bit beyond us, and we will simply use the transform tables to find the function $f(t)$, which has the given transform $\bar{f}(s)$. For example, if we know that $\bar{f}(s) = 1/(s^2 + \omega^2)$, then we can say from (8.18b) that $f(t) = (1/\omega)\sin \omega t$.

8.2.4 Solving ODEs Using Laplace Transforms

Suppose we have a single linear differential equation in a dependent variable $x(t)$. The approach used to solve such ODEs using Laplace transformation is to take the Laplace transform of the entire differential equation, determine the transform $\bar{x}(s)$, and then use the tables to find $x(t)$ from $\bar{x}(s)$. We'll illustrate this procedure with a few examples.

Example 8.9 A First-Order ODE. Consider the homogeneous, linear first-order differential equation

$$\dot{x} = ax; \qquad x(0) = x_0$$

First we take the Laplace transform of the differential equation

$$\mathscr{L}[\dot{x}] = s\bar{x}(s) - x(0) = a\bar{x}(s)$$

Next, solve the above relation to obtain the transform $\bar{x}(s)$:

$$\bar{x}(s) = \frac{x_0}{s - a} = x_0\left(\frac{1}{s - a}\right)$$

Now answer the question: What function $x(t)$ has the above as its Laplace transform? Referring to (8.17), we conclude that

$$x(t) = x_0 e^{at}$$

Observe that the initial value x_0 appears directly using this technique; we do not end up with a solution which contains an undetermined constant which then has to be evaluated from the initial condition, as occurs with the eigenvalue-eigenvector or assumed solution methods.

Example 8.10 A Second-Order ODE. As a second example consider the homogeneous, linear second-order equation

$$\ddot{x} + \omega^2 x = 0; \qquad x(0) = 0; \qquad \dot{x}(0) = \dot{x}_0$$

which would represent an undamped oscillator with a nonzero initial velocity. Taking the Laplace transform of the equation, we find, with $\mathscr{L}[\ddot{x}] + \mathscr{L}[\omega^2 x] = 0$,

$$s^2\bar{x}(s) - sx(0) - \dot{x}(0) + \omega^2\bar{x}(s) = 0$$

Rearranging and substituting the stated initial conditions,

$$\bar{x}(s)[s^2 + \omega^2] = \dot{x}_0$$

Solving for $\bar{x}(s)$,

$$\bar{x}(s) = \frac{\dot{x}_0}{s^2 + \omega^2} = \frac{\dot{x}_0}{\omega}\left(\frac{\omega}{s^2 + \omega^2}\right)$$

In view of (8.18*b*) we conclude that

$$x(t) = \frac{\dot{x}_0}{\omega} \sin \omega t$$

Example 8.11 An Inhomogeneous First-Order ODE. As the next example, consider the inhomogeneous first-order linear system

$$\dot{x} = ax + e^t; \qquad x(0) = x_0; \qquad a \neq 1$$

Transforming the entire equation, we find

$$s\bar{x}(s) - x(0) = a\bar{x}(s) + \frac{1}{s - 1}$$

$$\bar{x}(s)(s - a) = x_0 + \frac{1}{s - 1}$$

Now solve for $\bar{x}(s)$ to obtain

$$\bar{x}(s) = \underbrace{\frac{x_0}{s - a}}_{\text{Gives } x_0 e^{at}} + \underbrace{\frac{1}{(s - a)(s - 1)}}_{\text{Needs a closer look}} \tag{8.23}$$

The first term of $\bar{x}(s)$ is easy to invert, as indicated. The second term of $\bar{x}(s)$ could be inverted using the table of transforms (transform pair 7). We will, however, work it out directly in order to illustrate a basic method of inversion. The second term of (8.23) can be inverted by first realizing that it is possible to write this term in the following way:

$$\frac{1}{(s-a)(s-1)} = \frac{A}{s-1} + \frac{B}{s-a}$$

where the constants A and B have to be determined. Noting that

$$\frac{A}{s-1} + \frac{B}{s-a} = \frac{A(s-a)+B(s-1)}{(s-a)(s-1)} = \frac{s(A+B)+(-aA-B)}{(s-1)(s-a)}$$

we observe that if we pick A and B so that $A + B = 0$ and $-aA - B = 1$, then we will reproduce the original term appearing in $\bar{x}(s)$. Solving the stated algebraic equations for A and B, we find $A = 1/(1-a)$, $B = -1/(1-a)$, so that $\bar{x}(s)$ may now be written as

$$\bar{x}(s) = x_0\left(\frac{1}{s-a}\right) + \frac{1}{1-a}\left(\frac{1}{s-1}\right) - \frac{1}{1-a}\left(\frac{1}{s-a}\right)$$

$$x(t) = x_0 e^{at} + \frac{e^t}{1-a} - \frac{e^{at}}{1-a}$$

with the inversion leading to the solution as indicated:

$$x(t) = x_0 e^{at} + \frac{1}{1-a}(e^t - e^{at}) \tag{8.24}$$

The technique of inversion illustrated above is referred to as *partial fraction expansion*.

In order to illustrate a feature of the solutions obtained via Laplace transformation, let's consider now the general first-order problem

$$\dot{x} = ax + f(t) \tag{8.25}$$

where $f(t)$ is any input function. Transformation of this equation leads to the following form for $\bar{x}(s)$:

$$\bar{x}(s) = \frac{x_0}{s-a} + \frac{\bar{f}(s)}{s-a} \tag{8.26}$$

Gives "zero-input response" Gives "zero-state response"

The first term of (8.26), when inverted to the time domain, produces what we call the *zero-input response*. This is the response that would occur if there were no input. The second term of (8.26) produces the *zero-state response*. This is the response that would occur, due to the given input, if all initial values (here x_0) were zero; that is, the input acts on a system which is initially at rest.

It is useful to note that the zero-input response is not necessarily the homogeneous solution to (8.26), and the zero-state response is not necessarily the particular solution. This is evident in (8.24), which may also be written as follows:

$$x(t) = \underbrace{\left(x_0 - \frac{1}{1-a}\right)e^{at}}_{\text{Homogeneous solution}} + \underbrace{\frac{1}{1-a}e^{t}}_{\text{Particular solution}} \tag{8.27}$$

This is the result one would obtain if one first determined the homogeneous solution as $x_H(t) = Ce^{at}$, then assumed a particular solution of the form $x_p(t) = Be^t$ [which would lead to $B = 1/(1-a)$], and finally applied the initial condition $x(0) = x_0$ to the entire solution.

Of course (8.24) and (8.27) represent the same function; the terms are just arranged differently. The point is that LTs will produce a solution which *naturally* is expressed as the sum of the zero-input and the zero-state responses. On the other hand, using the method of undetermined coefficients, the solution is naturally expressed as the sum of the homogeneous and the particular solutions.

Each form is a useful way to view the response in different situations. For instance, control system performance is often assessed by subjecting a system initially at rest to a specified input. The response that results is the zero-state response (but not necessarily the particular solution). So the LT solution is well suited to this analysis.

8.2.5 The General *n*th-Order Linear ODE

We know that any time invariant linear system model of nth order can be expressed as a single nth-order ODE of the form

$$\frac{d^n x}{dt^n} + a_1\frac{d^{n-1}x}{dt^{n-1}} + \cdots + a_{n-1}\frac{dx}{dt} + a_n x = u(t) \tag{8.28}$$

where $a_1, \ldots, a_n$ are given constants. In (8.28), $x(t)$ is the output and $u(t)$ is the input. We'll now use the LT to establish a relation, in the s domain, connecting the input and the output for an nth-order system model expressed in the form (8.28). Our goal is to develop some general results which will be applicable to any nth order linear system of this form.

Let's first look at the general form we get when we take the LT of (8.28). The student may verify that the result of transforming (8.28) can be written in the following way:

$$\bar{x}(s)\left(s^n + a_1 s^{n-1} + \cdots + a_{n-1}s + a_n\right) - \left(b_1 s^{n-1} + b_2 s^{n-2} + \cdots + b_{n-1}s + b_n\right) = \bar{u}(s) \tag{8.29}$$

where the constants $b_1, \ldots, b_n$ are functions of the initial conditions

$$b_1 = x(0)$$
$$b_2 = \dot{x}(0) + a_1 x(0)$$
$$b_3 = \ddot{x}(0) + a_1 \dot{x}(0) + a_2 x(0)$$
$$\vdots$$
$$b_n = \frac{d^{n-1}x}{dt^{n-1}}(0) + a_1 \frac{d^{n-2}x}{dt^{n-2}}(0) + \cdots + a_{n-1}x(0)$$

In (8.29) the terms have been grouped into three types: (1) those terms involving only $\bar{x}(s)$, the transform of the output $x(t)$ which is to be found, (2) those terms involving initial conditions, and (3) those terms involving the input. Now to simplify the form of (8.29), we can redefine it as

$$\bar{x}(s)A(s) = I(s) + \bar{u}(s) \tag{8.30}$$

where the polynomial functions of s, $A(s)$, and $I(s)$, are given by

$$A(s) = s^n + a_1 s^{n-1} + a_2 s^{n-2} + \cdots + a_{n-1}s + a_n$$
$$I(s) = b_1 s^{n-1} + b_2 s^{n-2} + \cdots + b_{n-1}s + b_n$$

Note that $I(s)$ involves initial conditions and $A(s)$ contains all of the information about the "system properties" [the constants a_i, along with the various powers of s, which tell us the order of the derivatives appearing in (8.28)]. Also, $I(s)$ will always be a polynomial of degree 1 less than is $A(s)$. From (8.30) the transform $\bar{x}(s)$ may be expressed as

$$\bar{x}(s) = \underbrace{\frac{I(s)}{A(s)}}_{\text{Gives zero-input response}} + \underbrace{\frac{\bar{u}(s)}{A(s)}}_{\text{Gives zero-state response}} \tag{8.31}$$

We see that, if the input is zero, $\bar{u}(s) = 0$, then $\bar{x}(s) = I(s)/A(s)$, so that inversion will give us the zero-input response; also, when all initial conditions are zero, $I(s) = 0$, so that inversion will give us the zero-state response. In general [at least for most inputs $u(t)$ that we will encounter], each of the two terms on the right-hand side of (8.31) will appear as the ratio of two polynomials, and our job is to do the inversion.

Thus, we are led to consider the inversion of the ratio of two polynomials:

$$F(s) = \frac{N(s)}{D(s)} \tag{8.32}$$

We will assume that the degree of the numerator $N(s)$ is at least 1 less than the degree of the denominator $D(s)$. This condition is certainly satisfied for the term in (8.31) which involves the zero-input response, $I(s)/A(s)$. The reason for making this assumption is that it allows us to express $F(s)$ in a partial fraction expansion, as follows: let's assume that the denominator $D(s)$ in (8.32) is a polynomial of degree n and can be factored into the following form:

$$D(s) = (s - s_1)(s - s_2) \cdots (s - s_n)$$

where $s_1, s_2, \ldots, s_n$ are the roots of $D(s) = 0$. (Assume the s_i have been found, perhaps on a computer.) We call the s_i the roots of $D(s)$ and the *poles* of $F(s)$. We'll assume that all of the poles of $F(s)$ are distinct.

In consequence of the conditions assumed [all poles of $F(s)$ distinct, degree of $N(s)$ < degree of $D(s)$], it turns out that we can do a partial fraction expansion of $F(s)$ into the following form:

$$F(s) = \frac{A_1}{s - s_1} + \frac{A_2}{s - s_2} + \cdots + \frac{A_n}{s - s_n} \tag{8.33}$$

where the constants $A_1, A_2, \ldots, A_n$ have to be determined. [The student should verify that if (8.33) were expressed as a single fraction, we would in fact recover (8.32) with the numerator having a degree one less than the denominator.] Observe that, if we can write (8.32) in the form (8.33), it is easy to do the inversion to find $f(t)$, since each term in (8.33), when inverted, will just give us an exponential function; that is, inversion of (8.33) yields

$$f(t) = A_1 e^{s_1 t} + A_2 e^{s_2 t} + \cdots + A_n e^{s_n t} \tag{8.34}$$

So our job now is to devise a way to get the constants A_i. We'll start with A_1. To find A_1, first multiply (8.33) by the factor $(s - s_1)$ to obtain

$$(s - s_1)F(s) = A_1 + A_2 \frac{(s - s_1)}{(s - s_2)} + \cdots + A_n \frac{(s - s_1)}{(s - s_n)}$$

Now evaluate the right-hand side when $s = s_1$; all terms on the right-hand side except the first then drop out. We conclude that

$$A_1 = [(s - s_1)F(s)]\big|_{s_1}$$

A similar exercise reveals that any one of the A_i can be found from

$$A_i = [(s - s_i)F(s)]\big|_{s_i} \tag{8.35}$$

This procedure works only if the poles of $F(s)$ are distinct. Now let's do some examples to illustrate this.

Example 8.12 The Poles of $F(s)$ Are Real. Suppose that $F(s)$ is given by

$$F(s) = \frac{s + 3}{s^2 + 9s + 20}$$

First we find the poles of $F(s)$ from $s^2 + 9s + 20 = 0$, and we find $s_1 = -4$, $s_2 = -5$, so that $F(s)$ may be written as

$$F(s) = \frac{s + 3}{(s + 4)(s + 5)}$$

Next, expand $F(s)$ using partial fractions,

$$F(s) = \frac{A_1}{s+4} + \frac{A_2}{s+5}$$

where, according to (8.35),

$$A_1 = \left[\frac{(s+3)(s+4)}{(s+4)(s+5)}\right]\Bigg|_{s=-4} = -1$$

$$A_2 = \left[\frac{(s+3)(s+5)}{(s+4)(s+5)}\right]\Bigg|_{s=-5} = 2$$

Thus, $$F(s) = \frac{-1}{s+4} + \frac{2}{s+5}$$

and we can invert each term to obtain $f(t)$ as

$$f(t) = -e^{-4t} + 2e^{-5t}$$

Example 8.13 The Poles of $F(s)$ Are Complex. We now suppose that

$$F(s) = \frac{1}{s^2 + 2s + 10}$$

The poles of $F(s)$, found from $s^2 + 2s + 10 = 0$, are $s_{1,2} = -1 \pm 3i$, so the factored form of $F(s)$ is

$$F(s) = \frac{1}{[s-(-1+3i)][s-(-1-3i)]}$$

and the partial fraction expansion is

$$F(s) = \frac{A_1}{s-(-1+3i)} + \frac{A_2}{s-(-1-3i)}$$

where $$A_1 = \left[\frac{1}{s-(-1-3i)}\right]\Bigg|_{s=-1+3i} = \frac{-i}{6}$$

$$A_2 = \left[\frac{1}{s-(-1-3i)}\right]\Bigg|_{s=-1+3i} = \frac{i}{6}$$

Thus, $$F(s) = \frac{-i}{6}\frac{1}{[s-(-1+3i)]} + \frac{i}{6}\frac{1}{[s-(-1-3i)]}$$

and inversion yields

$$f(t) = \frac{-i}{6}e^{(-1+3i)t} + \frac{i}{6}e^{(-1-3i)t}$$

$$f(t) = e^{-t}\left(\frac{1}{6i}\right)(e^{3it} - e^{-3it}) = \frac{1}{3}e^{-t}\sin 3t$$

Observe that this problem could be done directly from the table of transforms (see transform pair 15).

Example 8.14 The Zero-State Response of a Linear Oscillator Subjected to a Step Input. The system model for this example is similar to that of Example 5.6:

$$\ddot{x} + 2\xi\omega\dot{x} + \omega^2 x = \omega^2 u(t)$$

where $x(0) = \dot{x}(0) = 0$, $\xi < 1$, $u(t) = 1$ for $t > 0$, and where an ω^2 has been multiplied into the input in order to make the eventual steady-state response be unity. Laplace transformation of the differential equation leads to

$$\bar{x}(s)\underbrace{\left[s^2 + 2\xi\omega s + \omega^2\right]}_{A(s)\text{ in (8.30)}} - \underbrace{\left[sx_0 + \dot{x}_0 + 2\xi\omega x_0\right]}_{I(s)\text{ in (8.30), zero here}} = \omega^2\bar{u}(s)$$

Thus, noting that $\bar{u}(s) = 1/s$, we have for the zero-state response

$$\bar{x}(s) = \frac{\omega^2}{s(s^2 + 2\xi\omega s + \omega^2)}$$

Note that at this point we could use transform pair 17 to find $x(t)$, but we'll do it the long way in order to illustrate the partial fraction procedure. In factored form $\bar{x}(s)$ is

$$\bar{x}(s) = \frac{\omega^2}{s[s - (-\xi\omega + i\omega_d)][s - (-\xi\omega - i\omega_d)]}$$

where $\omega_d = \omega(1 - \xi^2)^{1/2}$ is the damped natural frequency of the oscillator. Note that one of the poles of $\bar{x}(s)$ is zero, signifying that a constant will appear in $x(t)$. The partial fraction expansion of $\bar{x}(s)$ will be

$$\bar{x}(s) = \frac{A_1}{s} + \frac{A_2}{[s - (-\xi\omega + i\omega_d)]} + \frac{A_3}{[s - (-\xi\omega - i\omega_d)]}$$

and the student may verify that application of (8.35) gives

$$A_1 = 1$$

$$A_2 = \frac{-1}{2\omega_d}(\omega_d - i\xi\omega)$$

$$A_3 = \frac{-1}{2\omega_d}(\omega_d + i\xi\omega)$$

Inversion then yields

$$x(t) = 1 - \frac{1}{2\omega_d}\left[(\omega_d - i\xi\omega)e^{(-\xi\omega + i\omega_d)t} + (\omega_d + i\xi\omega)e^{(-\xi\omega - i\omega_d)t}\right]$$

A little manipulation enables us to rewrite this solution for the step response in terms of the trigonometric functions as

$$x(t) = 1 - e^{-\xi\omega t}\left[\cos\omega_d t + \frac{\xi\omega}{\omega_d}\sin\omega_d t\right] \tag{8.36}$$

The steady-state response is observed to be $x_{ss}(t) = 1$, and the solution (8.36) shows us how this steady-state response is approached when the damping factor $\xi < 1$. The response of the damped oscillator to a step input was shown previously in Fig. 5.9 for several values of the damping factor ξ.

Comments. The student should note that the Laplace transform technique bears similarities to other techniques for solving linear differential equations. In order to illustrate this point, let's consider a second-order autonomous system as analyzed by both state variables and LTs.

Example 8.15 A Two-State System. Consider the two-state system

$$\begin{aligned}\dot{x}_1 &= -4x_1 + 2x_2 \\ \dot{x}_2 &= -3x_1 + x_2;\end{aligned} \qquad \begin{Bmatrix}\dot{x}_1 \\ \dot{x}_2\end{Bmatrix} = \begin{bmatrix}-4 & 2 \\ -3 & 1\end{bmatrix}\begin{Bmatrix}x_1 \\ x_2\end{Bmatrix} \tag{8.37}$$

Following the procedure of Chap. 4, the substitution $\{x\} = \{C\}e^{\lambda t}$ leads to the eigenvalue problem

$$\begin{bmatrix}(-4-\lambda) & 2 \\ -3 & (1-\lambda)\end{bmatrix}\{C\} = \{0\} \tag{8.38}$$

The eigenvalues $\lambda_{1,2}$ are found from the characteristic equation, obtained by setting the determinant of the above coefficient matrix to zero, which leads to

$$\lambda^2 + 3\lambda + 2 = 0 \tag{8.39}$$

This is solved to give $\lambda_{1,2} = -2, -1$. The student may verify that the associated eigenvectors are $\{C\}_1^T = \lfloor 1 \quad 1 \rfloor$, $\{C\}_2^T = \lfloor 1 \quad 3/2 \rfloor$. Thus, the general solution is

$$\begin{aligned}x_1 &= \alpha_1 e^{-2t} + \alpha_2 e^{-t} \\ x_2 &= \alpha_1 e^{-2t} + \tfrac{3}{2}\alpha_2 e^{-t}\end{aligned} \tag{8.40}$$

where α_1 and α_2 are to be found from the initial conditions $x_1(0)$, $x_2(0)$.

Now let's do the problem with LTs. We will pick x_1 as our output variable and recast the system model as a single second-order equation for x_1. To do this, differentiate (8.37*a*) with respect to time, then use (8.37*a* and *b*) to eliminate x_2 in the result, yielding

$$\ddot{x}_1 + 3\dot{x}_1 + 2x_1 = 0 \tag{8.41}$$

Now take the LT of (8.41) to get

$$\underbrace{\bar{x}_1(s)\,[s^2 + 3s + 2]}_{A(s) \text{ in } (8.30)} = \underbrace{[sx_1(0) + 3x_1(0) + \dot{x}_1(0)]}_{I(s) \text{ in } (8.30)}$$

so that

$$\bar{x}_1(s) = \frac{sx_1(0) + 3x_1(0) + \dot{x}_1(0)}{s^2 + 3s + 2} \tag{8.42}$$

[Note that $\dot{x}_1(0)$ is obtained from (8.37*a*) as $\dot{x}_1(0) = -4x_1(0) + 2x_2(0)$]. We now observe a basic feature of LTs: The *poles* of $\bar{x}_1(s)$ [that is, the zeroes of $A(s)$] are just the *eigenvalues* of the state variable form of the system model, since $A(s)$ in (8.42) is just the characteristic polynomial with λ replaced by the Laplace variable s. This always occurs in the zero-input case; thus, the poles of $\bar{x}_1(s)$ give us the same information as the eigenvalues of the state variable form (8.37). Also note that the $A(s)$ appearing in (8.42) for our output variable x_1 would have been the same had we chosen x_2 instead of x_1 as the output variable.

Comments

1. The material presented in this section is not meant to imply that the LT technique is applicable only to single, nth-order ODEs. The method may be used directly on (8.37), for example. If we transform (8.37), we get

$$s\bar{x}_1(s) - x_1(0) = -4\bar{x}_1(s) + 2\bar{x}_2(s)$$
$$s\bar{x}_2(s) - x_2(0) = -3\bar{x}_1(s) + \bar{x}_2(s)$$

These equations can be written in vector-matrix form as

$$\begin{bmatrix} (s+4) & -2 \\ 3 & (s-1) \end{bmatrix} \begin{Bmatrix} \bar{x}_1(s) \\ \bar{x}_2(s) \end{Bmatrix} = \begin{Bmatrix} x_1(0) \\ x_2(0) \end{Bmatrix} \tag{8.43}$$

The coefficient matrix is just the negative of that in (8.38) with s replacing λ. Solving (8.43) for $\bar{x}_1(s)$ and $\bar{x}_2(s)$, we obtain

$$\bar{x}_1(s) = \frac{(s-1)x_1(0) + 2x_2(0)}{s^2 + 3s + 2}; \qquad \bar{x}_2(s) = \frac{(s+4)x_2(0) - 3x_1(0)}{s^2 + 3s + 2}$$

In order to find solutions for x_1 and x_2 via partial fraction expansion, we would first find the poles of $\bar{x}_1(s)$ and $\bar{x}_2(s)$, which is equivalent to finding the eigenvalues using the state variable formulation.

2. It should by now be clear that the poles of the transform $\bar{x}(s)$ will tell us a lot about the behavior of $x(t)$, just as the eigenvalues do in the state variable formulation. For example, consider the general nth-order homogeneous problem (8.28), with the input $u(t) = 0$. Then the transform $\bar{x}(s)$ will have the form

$$\bar{x}(s) = \frac{I(s)}{A(s)}$$

Assume we can find the distinct poles of $\bar{x}(s)$, denoted as $s_1, s_2, \ldots, s_n$, so that $A(s) = (s - s_1) \cdots (s - s_n)$. Then

$$\bar{x}(s) = \frac{I(s)}{(s - s_1) \cdots (s - s_n)}$$

Now $I(s)$ is in general a polynomial in s of degree $n - 1$, so we can do a partial fraction expansion of $\bar{x}(s)$:

$$\bar{x}(s) = \frac{A_1}{s - s_1} + \frac{A_2}{s - s_2} + \cdots + \frac{A_n}{s - s_n} \tag{8.44}$$

where the constants A_i, evaluated from (8.35), will be functions of the initial conditions. The inversion of (8.44) will give the time response as

$$x(t) = A_1 e^{s_1 t} + A_2 e^{s_2 t} + \cdots + A_n e^{s_n t} \tag{8.45}$$

This is of course exactly the same form as we get using the state variable formulation for the autonomous problem; however, the state variable solution automatically gives us the solution for all of the states, whereas the LT method, as applied to the nth-order ODE model, will give us directly only the solution for whichever state we pick as the output variable. In any case, the *poles of $\bar{x}(s)$ (for the zero-input case) govern the stability of the system*; if each pole s_i has a negative real part, then the system is asymptotically stable.

3. A final comment here: Complex poles will occur in complex conjugate pairs and the A's associated with them will also be complex conjugates. Thus, we should view any pair of such terms in (8.44) as providing an exponentially damped (or growing) oscillation in (8.45), just as we did for the state variable formulation of Chap. 4.

8.3 SISO SYSTEMS: BLOCK DIAGRAMS AND TRANSFER FUNCTIONS

In this section we introduce the transfer function and the block diagram. The *transfer function* relates an output to an input in the Laplace domain and is used to characterize mathematically systems and their components. The *block diagram* is a pictorial/mathematical representation of the system in which the various components (system to be controlled, controller, disturbances, error, and so on) and their interactions are displayed. Transfer functions and block diagrams will contain the same information as the corresponding ODE model of the system.

The examples considered in Examples 8.1 through 8.4 involved first- or second-order systems. Here we will develop the basic ideas in the more general context of a linear system of arbitrary (nth) order. The general ODE model for an nth-order, time-invariant linear system may always be put into the following form:

$$\frac{d^n x}{dt^n} + a_1 \frac{d^{n-1} x}{dt^{n-1}} + \cdots + a_{n-1}\dot{x} + a_n x = u(t) \tag{8.46}$$

where $x(t)$ is the variable to be controlled. The input $u(t)$ in (8.46) may actually consist of several contributors; a portion of $u(t)$ may be due to a disturbance, while a portion may define the desired operating condition. Equation (8.46) is the single ODE version of (5.6). Note that the uncontrolled system and the controlled system will each be represented by a different version of (8.46); for example, compare (8.2) and (8.5).

8.3.1 Transfer Functions

The transfer function, denoted $H(s)$, is the ratio of the Laplace transform of the output to that of the input for the case of zero initial conditions (IC):

$$H(s) = \left.\frac{\mathscr{L}[\text{output}]}{\mathscr{L}[\text{input}]}\right|_{\text{zero IC}}$$

The transfer function is identified with the zero-state response. In terms of the basic system model (8.46), in which $x(t)$ is the output and $u(t)$ the input, the transfer function is

$$H(s) = \frac{\bar{x}(s)}{\bar{u}(s)}, \qquad \text{(zero IC)}$$

or
$$\bar{x}(s) = H(s)\bar{u}(s) \tag{8.47}$$
where zero initial conditions are understood. We view the transfer function $H(s)$ as "operating" on the input to produce the output.

The relation (8.47) is just the zero-state response portion of (8.31) of Sec. 8.2. Thus, for the system ODE model given by (8.28) or (8.46), the transfer function $H(s) = 1/A(s)$ and is the reciprocal of the system characteristic polynomial $A(s)$.

We observe that, in the time domain, the input and output are connected by an ordinary differential equation, whereas in the Laplace domain, the input and output are connected by the transfer function. Thus, the statement (8.47) provides the same information as does the system ODE (8.46). The transfer function is just an alternate way to define the system mathematical model.

Example 8.16 The Linear Oscillator. Find the transfer function for the linear oscillator, for which (8.46) has the form

$$\ddot{x} + 2\xi\omega\dot{x} + \omega^2 x = u(t)$$

In physical terms the input $u(t)$ here is the externally applied force per unit mass, and the output $x(t)$ is the resulting displacement. Laplace transforming (8.48) for the case of zero initial conditions, we obtain

$$\bar{x}(s)\left[s^2 + 2\xi\omega s + \omega^2\right] = \bar{u}(s)$$

or
$$\bar{x}(s) = \left(\frac{1}{s^2 + 2\xi\omega s + \omega^2}\right)\bar{u}(s)$$

In view of the definition (8.47) the transfer function $H(s)$ for the linear oscillator is

$$H(s) = \frac{1}{s^2 + 2\xi\omega s + \omega^2} \tag{8.48}$$

The denominator of $H(s)$ is observed to be the characteristic polynomial $A(s)$ of this second-order system.

Example 8.17 Automobile Cruise Control. Here we consider the mathematical model (8.11) of the automobile cruise control (with a proportional controller) for the case of no disturbance ($\alpha = 0$), so that (8.11) reduces to

$$m\dot{v} = -2\beta V_0 v + k_p(v_d - v) \tag{8.49}$$

Here the input is the desired velocity deviation $v_d(t)$ and $v(t)$ is the output. The transfer function is found by first determining the Laplace transform of (8.49) for the case of zero initial conditions:

$$ms\bar{v}(s) = -2\beta V_0\bar{v}(s) + k_p\left[\bar{v}_d(s) - \bar{v}(s)\right]$$

Collecting all terms involving $\bar{v}(s)$ then yields

$$\bar{v}(s)\left(ms + 2\beta V_0 + k_p\right) = k_p\bar{v}_d(s)$$

so that the transfer function of the undisturbed system is seen to be

$$H(s) = \frac{k_p}{ms + \left(2\beta V_0 + k_p\right)} \tag{8.50}$$

The denominator of $H(s)$ is once again observed to be the characteristic polynomial of this first-order system.

Example 8.18. Find the governing differential equation for a system whose transfer function is

$$H(s) = \frac{1}{s^3 + 3s + 10}$$

From (8.47) we have

$$\bar{x}(s) = \left(\frac{1}{s^3 + 3s + 10} \right) \bar{u}(s)$$

or

$$\bar{x}(s)(s^3 + 3s + 10) = \bar{u}(s) \tag{8.51}$$

This implies that the governing differential equation is

$$\dddot{x} + 3\dot{x} + 10x = u(t) \tag{8.52}$$

The student may verify that the Laplace transform of (8.52), with zero initial conditions, will yield (8.51).

Examples 8.16 through 8.18 illustrate that the transfer function $H(s)$ contains the same information as the nth-order ODE model (8.46). The transfer function is simply an alternate way to present this information. An important characteristic of the system transfer function is that its poles determine the stability of the linear system with which it is associated. The denominator of the transfer function, a polynomial in s, is the system characteristic polynomial. Setting this polynomial to zero to find its roots (the poles of the transfer function) is the same as solving the characteristic equation obtained in the state variable formulation. In analyzing control systems, we will see that the uncontrolled system will be defined by a certain transfer function, from which the stability and damping properties of the uncontrolled system will be defined. Likewise, the controlled system will be defined by a different transfer function, with generally different stability and damping properties. In order to design and analyze SISO control systems, we must understand clearly how the controlled system or closed-loop transfer function is related to system stability and performance. This will be discussed in Sec. 8.4.

8.3.2 Block Diagrams

Block diagrams are pictorial/mathematical representations of dynamic systems and are especially useful in the representation of simple systems and their controllers. The basic block is shown in Fig. 8.6.

Each block is characterized by a transfer function which relates, in the Laplace domain, the input to the block and the resulting output. Figure 8.6(b) is equivalent to the equation

$$\bar{x}(s) = H(s)\bar{u}(s) \tag{8.53}$$

The block symbolizes the operation performed on the input in order to obtain the output, in the Laplace domain.

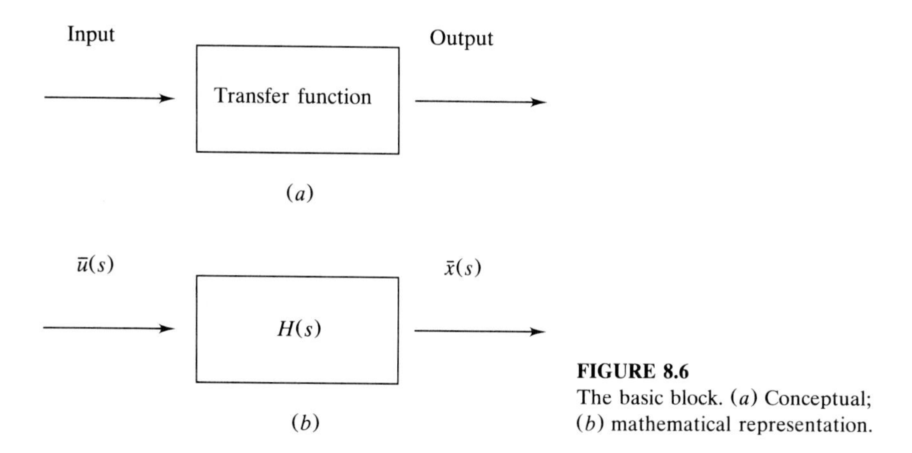

FIGURE 8.6
The basic block. (*a*) Conceptual; (*b*) mathematical representation.

Example 8.19 The Oscillator System of Example 8.16. The oscillator system defined in Example 8.16 is represented by the transfer function $H(s)$ given by (8.48). A block diagram representation would be as shown below:

$$\bar{u}(s) \rightarrow \boxed{\frac{1}{s^2 + 2\xi\omega s + \omega^2}} \rightarrow \bar{x}(s) \tag{B8.1}$$

It commonly occurs that block diagrams contain several blocks, some of which appear in series; the output of one block becomes the input to another block. Such a situation is shown in (B8.2):

$$\bar{u}(s) \rightarrow \boxed{H_1(s)} \rightarrow \bar{y}(s) \rightarrow \boxed{H_2(s)} \rightarrow \bar{x}(s) \tag{B8.2}$$

In (B8.2) the input is $u(t)$, the output is $x(t)$, and $y(t)$ represents an "intermediate output variable," which would often be omitted from the diagram. Diagram (B8.2) is, from the definition of the single block, equivalent to the two equations

$$\bar{x}(s) = H_2(s)\bar{y}(s)$$
$$\bar{y}(s) = H_1(s)\bar{u}(s)$$

so that
$$\bar{x}(s) = H_2(s)H_1(s)\bar{u}(s)$$

We could, therefore, represent the block diagram (B8.2) in the following form:

$$\bar{u}(s) \rightarrow \boxed{H_1(s)H_2(s)} \rightarrow \bar{x}(s) \tag{B8.3}$$

If the elements of (B8.2) are considered to be a system, then the system transfer function $H(s) = H_1(s)H_2(s)$. Diagrams (B8.2) and (B8.3), which represent the same "system," show that the block diagram of a given system is not unique.

In constructing block diagrams, we need to define two elements in addition to the basic block. These are the *pickoff point* and the *summing junction*, or *summer*, shown respectively in Figs. 8.7(a) and (b).

The pickoff point allows us to send a given "signal" [represented by the Laplace transform $a(s)$ in Fig. 8.7(a)] to several places. The summer receives two or more signals and sends out a resultant signal that is the sum or difference of the incoming signals.

The elements of Fig. 8.7 can be used to represent another common situation, the occurrence of blocks in parallel, with feedback, as in diagram ($B8.4$). The block diagram ($B8.4$) can be converted to single block form, and the system transfer function identified, by noting that the mathematical statement equivalent to ($B8.4$) is

$$\bar{x}(s) = G(s)\underbrace{[\bar{u}(s) \pm H(s)\bar{x}(s)]}_{\text{Output of summer}}$$

or

$$\bar{x}(s)[1 \mp G(s)H(s)] = G(s)\bar{u}(s)$$

so that

$$\bar{x}(s) = \underbrace{\left(\frac{G(s)}{1 \mp G(s)H(s)}\right)}_{\text{System transfer function}}\bar{u}(s) \tag{8.54}$$

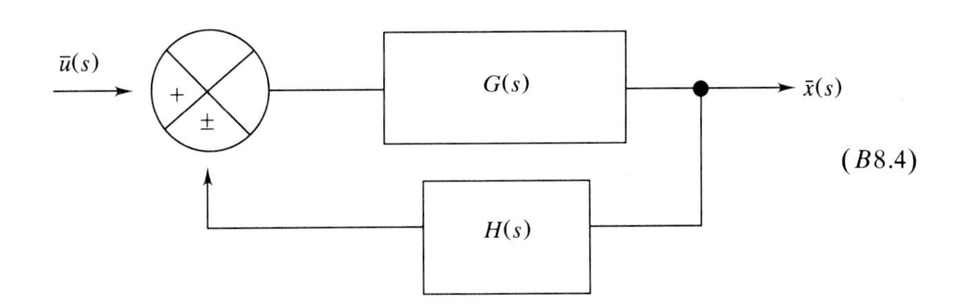

($B8.4$)

The block diagram ($B8.4$) may also be expressed, therefore, as

$$\bar{u}(s) \rightarrow \boxed{\frac{G(s)}{1 \mp G(s)H(s)}} \rightarrow \bar{x}(s) \tag{B8.5}$$

The result ($B8.5$) will prove useful in determining system transfer functions for controlled systems. Some simple examples are now presented to illustrate the construction of system block diagrams.

Example 8.20 An Oscillator System (Fig. 8.8). For the linear system shown, $u(t)$ is the input force and $x(t)$ the displacement of the mass m. A second displacement $y(t)$ is needed to define the motion of the spring/damper connection. We will

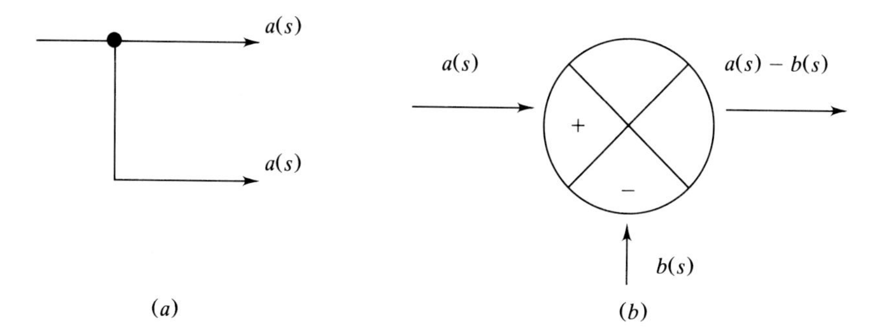

FIGURE 8.7
(a) The pickoff point; (b) the summing junction.

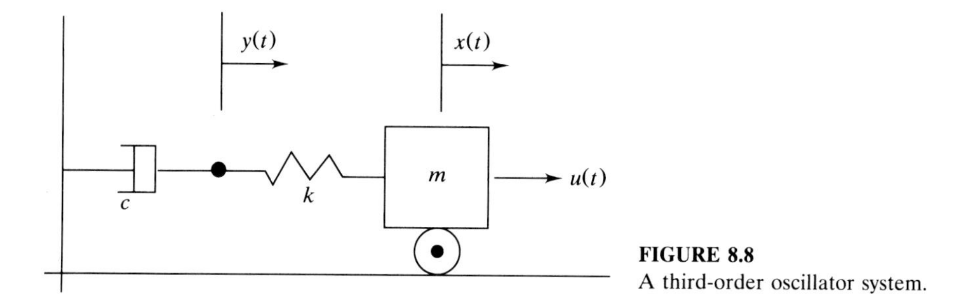

FIGURE 8.8
A third-order oscillator system.

assume $x(t)$ to be the output variable, and our objective will be to develop a block diagram in which only the output $\bar{x}(s)$ and the input $\bar{u}(s)$ appear.

The first step is to determine the equations of motion. The student may verify that they are as follows:

$$
\begin{aligned}
m\ddot{x} + k(x - y) &= u(t) \\
c\dot{y} + k(y - x) &= 0
\end{aligned}
\tag{8.55}
$$

The system is observed to be of third order and the states are x, $\dot{x}$, and y. The output variable $x(t)$ is one of the states.

Next we take the Laplace transform of each of equations (8.55), with all initial conditions zero, the objective being to identify the transfer functions involved. This leads to

$$
\begin{aligned}
\bar{x}(s)[ms^2 + k] &= \bar{u}(s) + k\bar{y}(s) \\
\bar{y}(s)(cs + k) &= k\bar{x}(s),
\end{aligned}
$$

or

$$\bar{x}(s) = \frac{\bar{u}(s)}{ms^2 + k} + \frac{k\bar{y}(s)}{ms^2 + k} \tag{8.56a}$$

$$\bar{y}(s) = \frac{k\bar{x}(s)}{cs + k} \tag{8.56b}$$

Now each of Eqs. (8.56) can be represented in block diagram form. For (8.56b) we have the following simple block diagram connecting x and y:

$$\bar{x}(s) \longrightarrow \boxed{\frac{k}{cs + k}} \longrightarrow \bar{y}(s) \qquad (B8.6)$$

To construct a block diagram of (8.56a), we note that $\bar{x}(s)$ is obtained by operating on the quantity $(\bar{u}(s) + k\bar{y}(s))$ with the transfer function $1/(ms^2 + k)$, so that the block diagram of (8.56a) may be represented as in Diagram (B8.7). As indicated in (B8.7), the quantity coming out of the summing junction is $\bar{u}(s) + k\bar{y}(s)$.

The overall system block diagram can be represented in single input/single output form by combining (B8.6) and (B8.7) so as to eliminate $\bar{y}(s)$, as indicated below (Diagram (B8.8)). The two lower blocks can be combined to yield Diagram (B8.9), which is in the desired form: Only the input and output variables appear, the intermediate variable having been eliminated.

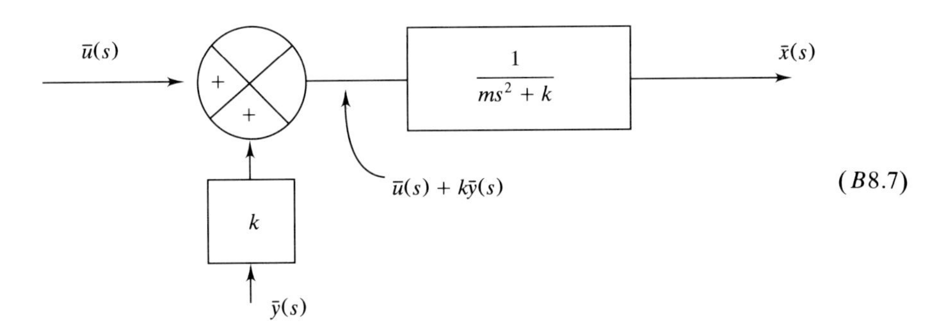

(B8.7)

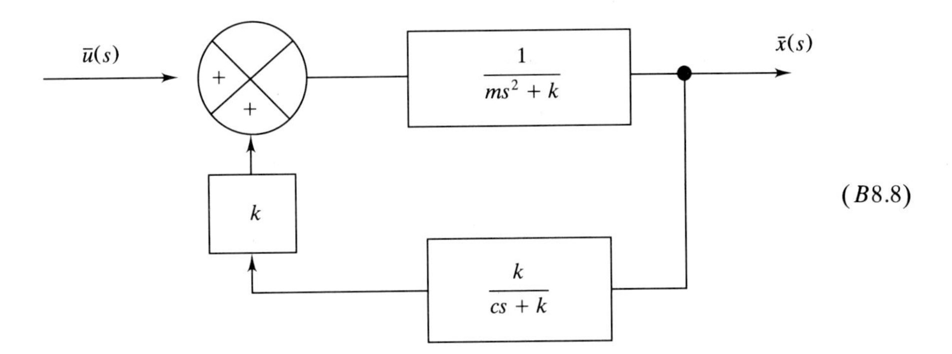

(B8.8)

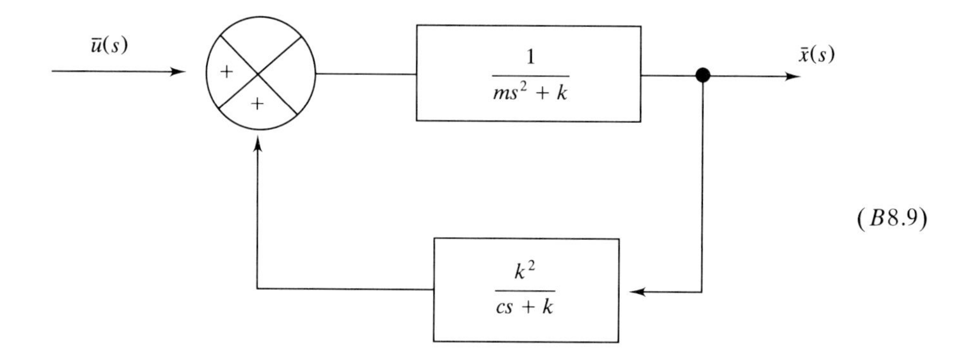

It is also instructive to represent the system block diagram in single block form. This can be done by substituting (8.56*b*) into (8.56*a*) to eliminate $\bar{y}(s)$. The result, which contains only $\bar{x}(s)$ and $\bar{u}(s)$, may be expressed as

$$\bar{x}(s) = \bar{u}(s)\left[\frac{(cs + k)}{(ms^2 + k)(cs + k) - k^2}\right] \tag{8.57}$$

where the overall or closed-loop transfer function $H(s)$ is the coefficient of $\bar{u}(s)$. Thus, the single block form of the block diagram looks like this:

$$\bar{u}(s) \longrightarrow \boxed{\frac{cs + k}{(ms^2 + k)(cs + k) - k^2}} \longrightarrow \bar{x}(s) \tag{B8.10}$$

The stability of this system would be determined by finding the poles of the overall or closed-loop transfer function [i.e., find the zeroes of the characteristic polynomial $(ms^2 + k)(cs + k) - k^2$]. The block form $(B8.10)$ also indicates that, if we were to convert the two differential equations (8.56) comprising the math model to a single third-order ODE in which the output $x(t)$ is the dependent variable, we would obtain the following, denoting the input force as $F(t)$:

$$\dddot{x} + \frac{k}{c}\ddot{x} + \frac{k}{m}\dot{x} = \frac{1}{m}\dot{F}(t) + \frac{k}{mc}F(t) \tag{8.58}$$

According to our basic nth-order ODE model (8.46), the entire right-hand side is the input, $u(t) = \dot{F}/m + kF/mc$. This shows that the mathematical input $u(t)$ appearing in (8.46) may actually be composed of a linear combination of the *physical* input [here $F(t)$] and one or more of its derivatives.

This example also illustrates the utility of Laplace transformation in converting a set of ODEs to the single ODE from (8.46). For instance, the math model (1.1) was previously converted to the single ODE form (1.6) by manipulating the differential equations directly. Use of LTs turns out to be a much more direct way to do this.

Example 8.21 Ship Roll Control (Example 8.1) Revisited. Recall the simple model (8.4) we developed for the ship rolling motion with disturbance torques and with control supplied by roll fins:

$$I\ddot{\theta} + c\dot{\theta} + k\theta = \underset{\substack{\uparrow \\ \text{Disturbance}}}{T_d(t)} + \underset{\substack{\uparrow \\ \text{Control torque}}}{T_c(t)} \tag{8.59}$$

where $k = mgL$. For reference, consider first the block diagram of the *uncontrolled* system ($T_c \equiv 0$) subjected to the disturbance torque $T_d(t)$. The Laplace transform of (8.59) in this case yields

$$\bar{\theta}(s) = \frac{\bar{T}_d(s)}{Is^2 + cs + k}$$

so that the block diagram is

$$\bar{T}_d(s) \longrightarrow \boxed{\frac{1}{Is^2 + cs + k}} \longrightarrow \bar{\theta}(s) \tag{B8.11}$$

To obtain the block diagram of the controlled system, we take Laplace transforms of (8.59) for the zero-state case, leading to

$$\bar{\theta}(s) = \frac{\bar{T}_d(s) + \bar{T}_c(s)}{Is^2 + cs + k}$$

This relation may be represented by the block diagram ($B8.12$).

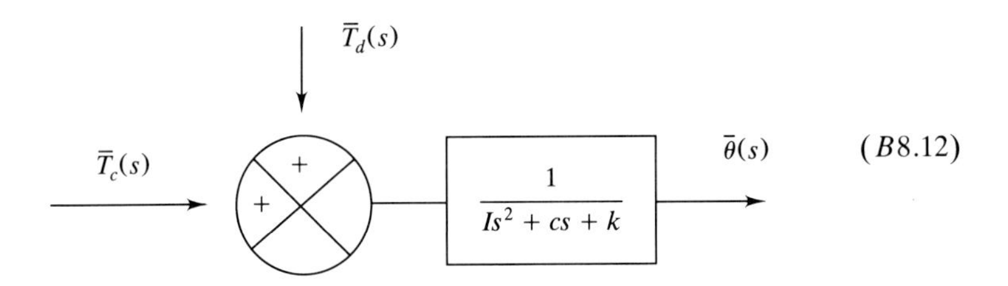

($B8.12$)

In Example 8.1 we defined the control action $T_c(t)$ to be proportional to the instantaneous error $T_c = k_p(\theta_d - \theta)$. A block diagram which represents this control law is then obtained from the relation $\bar{T}_c(s) = k_p[\bar{\theta}_d(s) - \bar{\theta}(s)]$, as shown in ($B8.13$). Diagram ($B8.13$) can be combined with ($B8.12$) to obtain the controlled system block diagram shown in ($B8.14$).

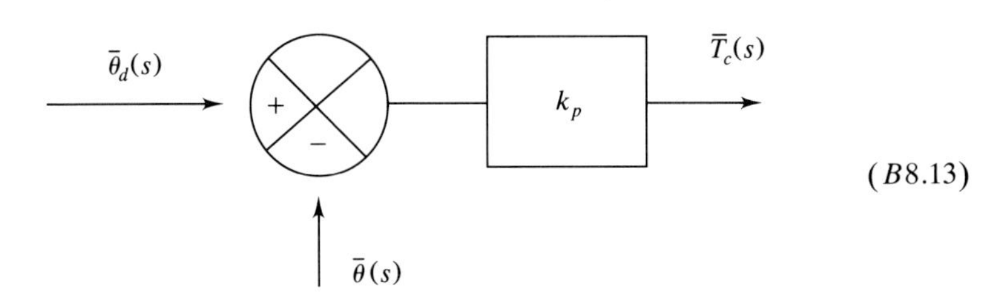

($B8.13$)

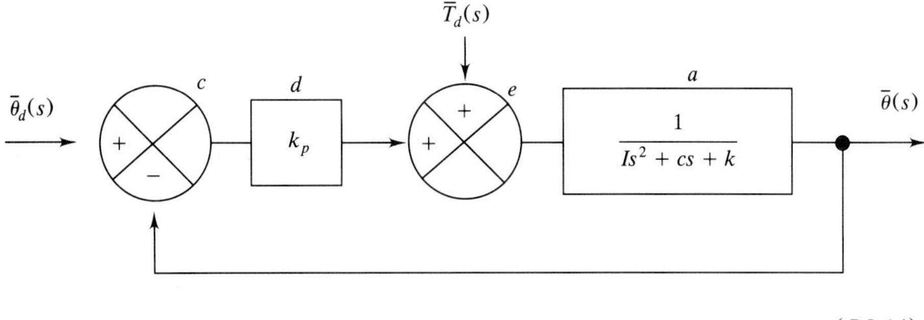

$$(B8.14)$$

Here the input is $\theta_d(t)$, the desired roll orientation (usually zero), and the output is $\theta(t)$, the actual roll orientation. In this diagram we can identify several things which we commonly wish to display in the block diagram: (1) The *uncontrolled* system properties are defined by the transfer function in block a. (2) The *feedback* loop b [involving measurement of $\theta(t)$] is needed in order to determine the error by comparison to the input at c. (3) The transfer function of the *controller* is in block d, indicating the manner in which the error is used to produce actuation to control the system; the output of block d is the control torque T_c. (4) The disturbance appears at the summer e, providing the overall torque acting on the system.

Comparison of ($B8.14$) and ($B8.11$) shows, in block form, the manner in which the addition of the controller has changed the system.

Example 8.22 Automobile Cruise Control (Example 8.3) Revisited. Equations (8.10) and (8.11) define the linear math model of this system; in (8.11) a proportional control strategy was assumed. From (8.10) we obtain the block diagram of the *uncontrolled* system ($f_D = 0$) by transforming (8.10) with zero initial conditions, whence

$$s\bar{v}(s) = \frac{-2\beta V_0}{m}\bar{v}(s) - g\bar{\alpha}(s)$$

or

$$\bar{v}(s) = \left(\frac{-g}{s + \dfrac{2\beta V_0}{m}}\right)\bar{\alpha}(s)$$

Thus, the single block representation of this disturbed, uncontrolled first-order system is

$$\bar{\alpha}(s) \longrightarrow \boxed{\frac{-g}{s + (2\beta V_0/m)}} \longrightarrow \bar{v}(s) \qquad (B8.15)$$

which is analogous to ($B8.11$).

The student may verify that for the controlled system with proportional control, as in (8.11), the block diagram representation is as follows:

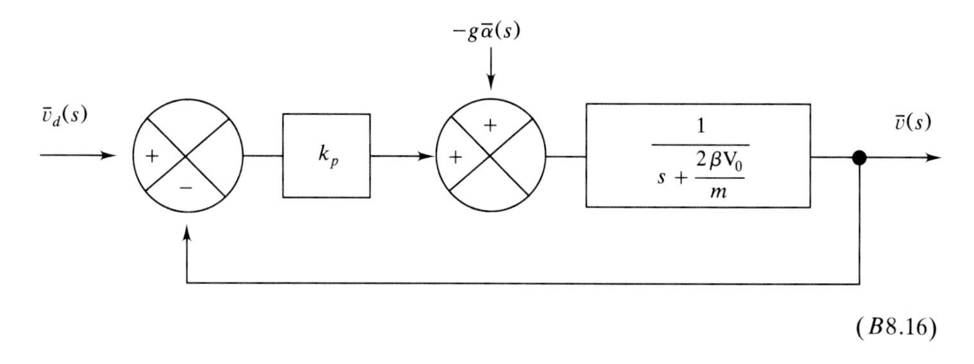

$(B8.16)$

The form of this diagram is identical to $(B8.14)$ of Example 8.21.

The form of the block diagram which we obtained in the two preceding examples is typical of the linear SISO system. We may generalize these examples by considering the following form of the SISO block diagram:

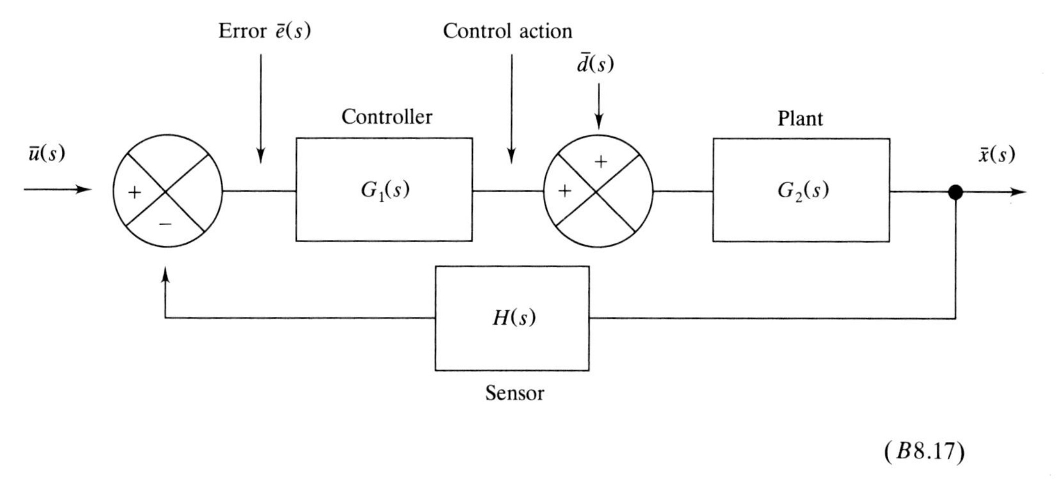

$(B8.17)$

In diagram $(B8.17)$ the "plant" is the uncontrolled system, represented by the transfer function $G_2(s)$. The controller transfer function is $G_1(s)$. The error, $\bar{e}(s) = \bar{u}(s) - H(s)\bar{x}(s)$, is acted upon by the controller to produce the control action (for example, a force or torque), which, along with the disturbance $\bar{d}(s)$, forms the input to the plant. A sensor transfer function $H(s)$ has been included for generality in order to indicate that the input and output need not be the same quantity; for instance, the output of our system might be a position $x(t)$, but we may wish to control the velocity $\dot{x}(t)$, so that $H(s) = s$. We will normally assume here, however, that the output and the variable to be controlled are the same, so that $H(s) = 1$, and the sensor block may be omitted from the diagram. This situation $[H(s) = 1]$ is referred to as *unity feedback*.

The system block diagram ($B8.17$) is a useful representation of the system math model because it connects directly the input and output and clearly separates the various aspects of the SISO controls problem. The diagram shows the model of the system to be controlled (plant), and it indicates the way in which the controller operates. Thus, if we wish to consider several different control strategies for a given plant, only the control block need be altered, and we proceed with the analysis.

Now that we see some of the basic ideas of control systems and have presented the general single input/single output block diagram ($B8.17$), we need to spend some time looking at the following: (1) how the addition of the control system affects the basic properties and stability of the overall system, (2) deciding what it is that the control system should do, and (3) understanding how various control actions will or will not accomplish our objectives.

8.4 BASIC ASPECTS OF PERFORMANCE AND CONTROL

In Sec. 8.1 the three main functions of a controller were noted: (1) to maintain or produce stability and adequate damping, (2) to reject disturbances, and (3) to track closely the input, which we consider to define the desired history of the output variable. In practice, any or all of these functions may be relevant, depending on the particular system and on the design objectives. In this section we will focus on the stability and tracking issues. Thus, we will consider a system defined by ($B8.17$) with the disturbance removed:

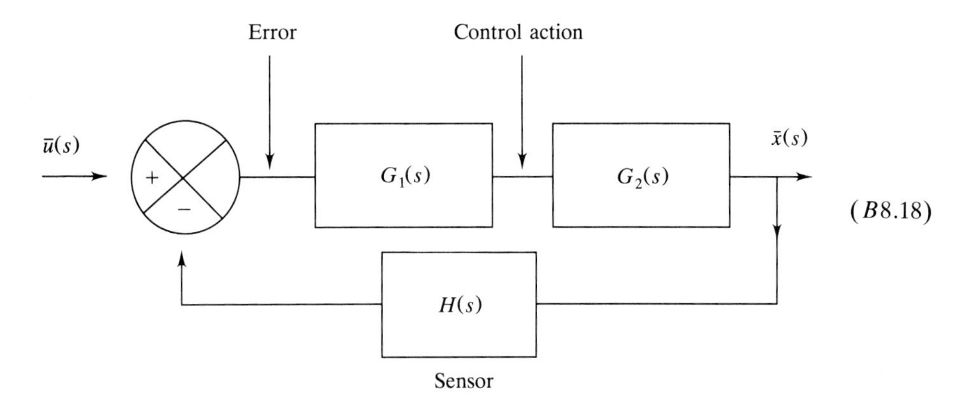

($B8.18$)

The controller and plant, defined by the transfer functions $G_1(s)$ and $G_2(s)$, respectively, may be represented by an overall "feed-forward" or "open loop" transfer function $G(s) = G_1(s)G_2(s)$. Diagram ($B8.18$) is then equivalent to ($B8.19$). Normally we will work with ($B8.18$) in order to delineate both plant and controller characteristics.

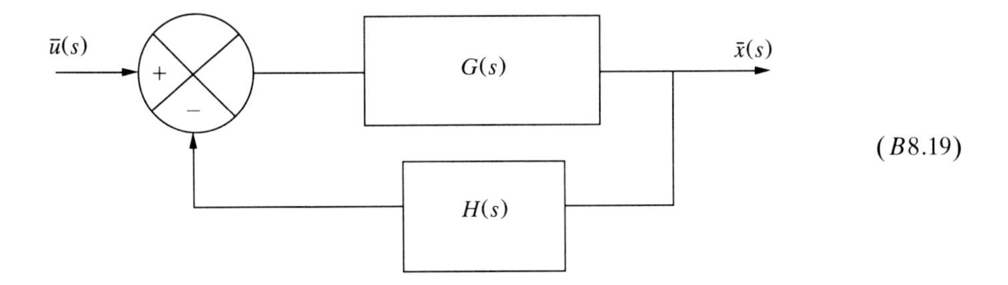

$(B8.19)$

8.4.1 Stability

The stability of the controlled system $(B8.18)$ is governed by the overall or *closed-loop* transfer function, which we'll call $T(s)$, such that $\bar{x}(s) = T(s)\bar{u}(s)$. This closed-loop transfer function can be determined directly from $(B8.18)$ and (8.54), where the $G(s)$ in (8.54) is replaced by the open loop transfer function $G_1(s)G_2(s)$. Thus, for the prototype control system $(B8.18)$ the closed-loop transfer function $T(s)$ is given by

$$T(s) = \frac{G_1(s)G_2(s)}{1 + G_1(s)G_2(s)H(s)} \tag{8.60}$$

The characteristic equation of the controlled system is then

$$1 + G_1(s)G_2(s)H(s) = 0 \tag{8.61}$$

For the types of systems and controllers we will consider, the transfer functions $G_1(s)$, $G_2(s)$ and $H(s)$ may each be written as the ratio of two polynomials. Let us define the transfer functions in (8.61) as follows:

$$G_1(s) = \frac{N_c(s)}{D_c(s)}; \qquad G_2(s) = \frac{N_p(s)}{D_p(s)}; \qquad H(s) = \frac{N_s(s)}{D_s(s)}$$

where the subscripts denote controller, plant, and sensor, respectively. Then the characteristic equation (8.61) may be written as

$$1 + \frac{N_c(s)N_p(s)N_s(s)}{D_c(s)D_p(s)D_s(s)} = 0$$

or

$$D_c(s)D_p(s)D_s(s) + N_c(s)N_p(s)N_s(s) = 0 \tag{8.62}$$

This relation gives the characteristic equation in terms of the numerators and denominators of the relevant transfer functions. The form (8.62) is useful in obtaining the system characteristic equation directly from the system block diagram. For the case of unity feedback $[H(s) = 1]$, $N_s(s) = D_s(s)$, and (8.76) simplifies to

$$D_c(s)D_p(s) + N_c(s)N_p(s) = 0 \tag{8.63}$$

By contrast, note that the characteristic equation of the uncontrolled system (that is, the plant) is simply $D_p(s) = 0$. The asymptotic stability of the controlled system is guaranteed provided that each pole of the transfer function $T(s)$, that is, each zero of (8.63), has a negative real part.

Example 8.23 Attitude Control of a Lunar Lander. The attitude of a space vehicle descending toward the surface of the moon (Fig. 8.9) must be controlled during descent, because, inherently, the system possesses neither rotational stiffness nor damping. In fact the *uncontrolled* vehicle equation of motion is simply $I\ddot{\theta} = 0$, where I is the moment of inertia about an axis out of paper through the center of mass. In order to control the attitude, gas jets apply moments about the center of mass so as to maintain the rotation angle θ sufficiently close to zero. The system block diagram is shown in ($B8.20$) (verify!), with the desired orientation $\theta_d(t)$ nominally zero. The question we pose is whether this system can be stabilized with a proportional controller, $G_c(s) = k_p$. Note first that the uncontrolled system characteristic equation is simply $Is^2 = 0$, indicating that there is a zero pole of order 2 (as in Example 4.25). In physical terms, if the uncontrolled spacecraft possesses an initial rotational velocity $\dot{\theta}_0$, this rotational velocity will never change. The uncontrolled system is unstable.

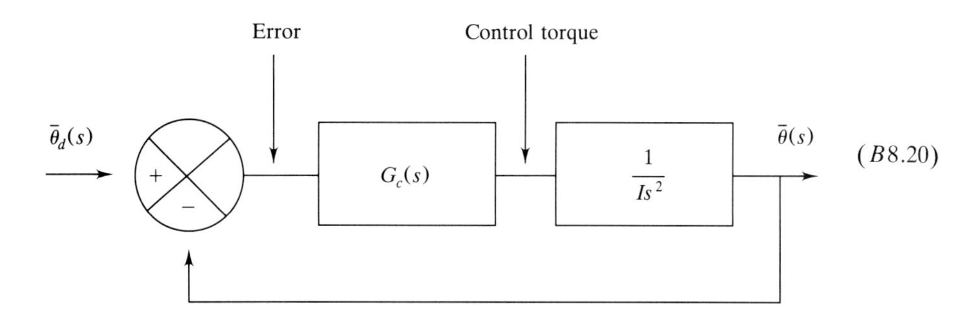

($B8.20$)

The transfer function for the proportional controller is $G_c(s) = k_p$. Thus, with $N_c(s) = k_p$, $D_c(s) = 1$, $N_p(s) = 1$, $D_p(s) = Is^2$, (8.63) provides the controlled system characteristic equation as

$$Is^2 + k_p = 0$$

The poles of the closed-loop transfer function $T(s)$ are then

$$p_{1,2} = \pm i\left(\frac{k_p}{I}\right)^{1/2}$$

The result shows that the controlled system is stable, but not asymptotically stable; the rotation angle $\theta(t)$ will exhibit a sustained oscillation about $\theta = 0$ at a frequency $\omega = \sqrt{k_p/I}$ and an amplitude determined by the initial conditions. As

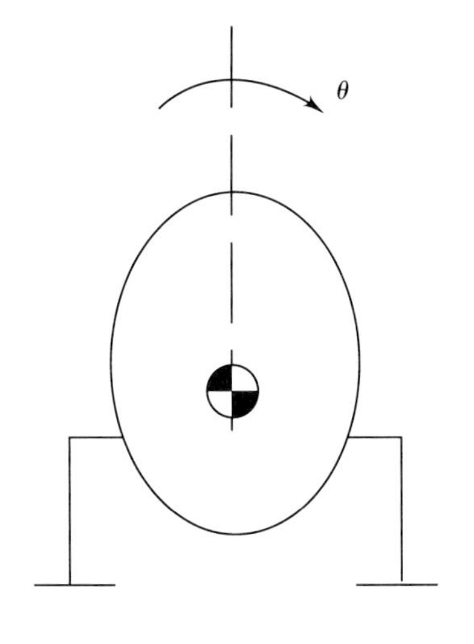

FIGURE 8.9
Schematic of a lunar lander.

noted in Example 8.1, the incorporation of a proportional controller to control a second-order system will alter the system "stiffness," but will have no effect on system damping. Some damping would be needed here in order to render the system asymptotically stable.

8.4.2 Performance Specifications

In this section we consider the tracking problem quantitatively. We assume that the objective of the controller is to cause the system output to follow or track the input sufficiently closely. The input specifies the *desired* history of the output variable. This input may define a predetermined function of time which the system is to follow, as in Example 8.1 (the input or desired output is zero) or Example 8.3 (the input is the desired forward speed V_0). The input may also be some function of time which is unknown ahead of time, as for the radar tracker of Example 8.4. In designing a controller for a system which is to track inputs, it is generally difficult to know a priori exactly what types of inputs the system will be subjected to. For instance, an x-y plotter may have to follow inputs which vary slowly or rapidly and which may have an oscillatory or a nonoscillatory character. What control design engineers often do is to specify adequacy of performance in terms of the response of the system to a few mathematically simple inputs which are designed to provide some measure of the transient and steady-state responses in various circumstances. The idea is that if the system can track certain simple functions closely enough, then it should also do OK for the inputs to be encountered in practice.

Of the simple input functions used in design, perhaps the most common is the *step input*. This input corresponds to a sudden change in the desired operating or equilibrium condition of the system. (For example, the x-y plotter may be recording a temperature or pressure which changes suddenly.) It is usually desirable that the system move quickly to the new position and then attain a steady-state value close to that of the step input. Common performance measures for a step input are shown in Fig. 8.10, in which the input and hypothetical output of a system with at least one pair of complex conjugate poles are shown (the complex poles cause the response of the output variable to be oscillatory).

With reference to Fig. 8.10, the following characteristics provide measures of control system performance:

1. Rise time T_r: The time needed first to reach to within a specified level (say, 90 percent) of the desired steady-state value.
2. Time T_p to first oscillation peak.
3. The percent overshoot, defined as $\text{PO} = (x_{\text{max}} - x_{ss})/x_{ss}$.
4. Steady-state error $e_{ss} \equiv x_d - x_{ss}$.
5. Settling time T_s: The time needed for the response to settle to within a certain percentage (often 2 or 5 percent) of the steady state; for example, if we choose a 2 percent requirement, then for all $t > T_s$, $1.02x_{ss} > x > 0.98x_{ss}$.

The rise time T_r and time T_p to first oscillation peak are measures of the initial quickness of the response to a sudden change in the desired operating condition. The settling time T_s is a measure of the time needed to reach the newly desired operating condition. The steady-state error determines the ability of the controller to move the system accurately to a new operating condition.

Generally we would like to make T_r, T_s, PO, and e_{ss} as small as possible, so that the system responds very quickly, with little or no overshoot, and quickly reaches the desired steady-state conditions. If a system responds in this manner

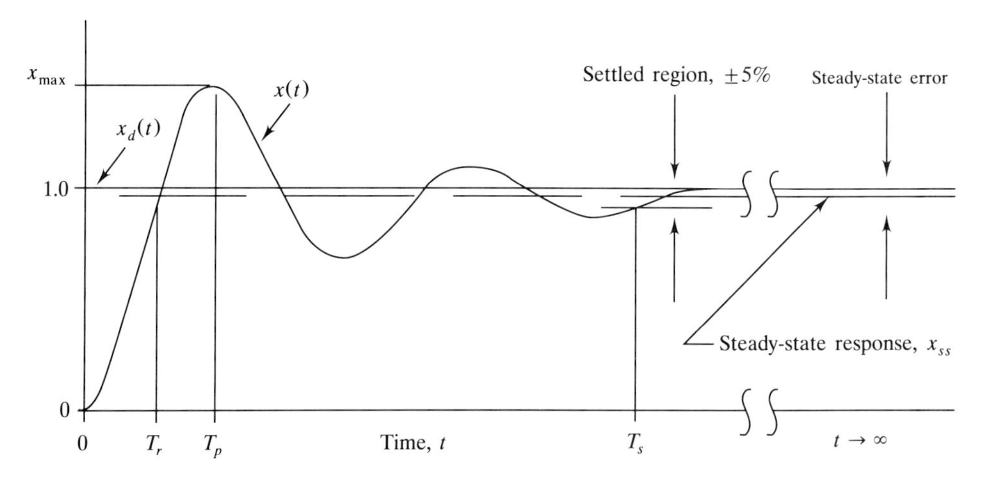

FIGURE 8.10
A unit step input $x_d(t) = 1$ for $t > 0$ and $x_d(t) = 0$ for $t < 0$; the controlled system output $x(t)$, which is supposed to follow the input.

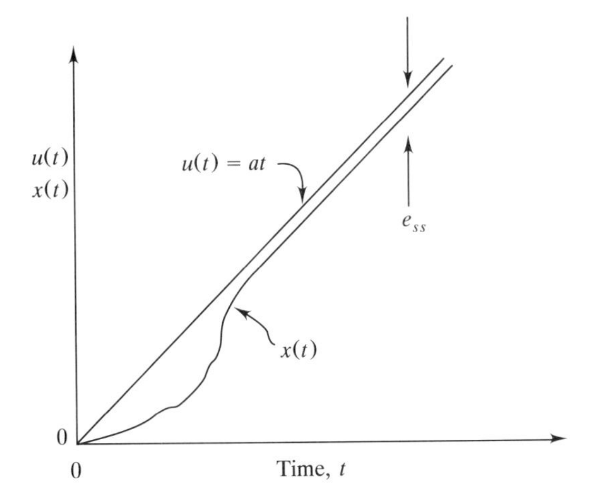

FIGURE 8.11
The ramp input $u(t)$ and a hypothetical response of the output variable $x(t)$.

to a step input, then we can be reasonably sure that it will accurately follow a variety of other inputs to which the system may be subjected in practice.

Reality rears its ugly head, however, because quicker, more accurate response requires greater actuator power. We can nearly always design a controller *on paper* which will satisfy any combination of T_r, T_s, e_{ss}, and PO that we wish to specify. Our design, however, may be completely unrealistic in terms of size and power required to implement the control scheme. What we try to do is the best that we can within established constraints of size, power, and cost.

A second type of input which is commonly used to assess performance is the *ramp* or linear function of time, $u(t) = at$ for $t \geq 0$, as shown in Fig. 8.11, along with a hypothetical zero-state response $x(t)$.

The steady-state error e_{ss} for a ramp input is a measure of the system's ability to track accurately a command signal which is changing, often reasonably gradually, with time.

We will next investigate the responses of first-, second-, and higher-order systems to simple inputs of the type described here. We will also introduce control strategies in addition to proportional control, which is the only type of controller considered so far in the examples. The objective will be to develop an understanding of the ways in which various control strategies affect the response, as well as the associated performance measures, for simple inputs.

8.4.3 The Proportional / Integral / Derivative (PID) Control Family

So far we have presented some examples involving proportional control: The control action is proportional to the instantaneous error, as defined by the controller transfer function $G_c(s) = k_p$. Now we consider two other possibilities which are commonly employed in practice: (1) *derivative* control, in which the control action is proportional to the *rate of change* of the error and (2) *integral* control, for which the control action is proportional to the *integral* of the error. Thus, if the input $u(t)$ defines the desired history of the output variable $x(t)$, the associated control actions $A_c(t)$ would be defined as follows:

$$\text{Proportional control:} \quad A_c(t) = k_p[u(t) - x(t)] = k_p e(t)$$

$$\text{Derivative control:} \quad A_c(t) = k_d[\dot{u}(t) - \dot{x}(t)] = k_d \dot{e}(t)$$

$$\text{Integral control:} \quad A_c(t) = k_I\left[\int_0^t (u(t) - x(t))\, dt\right] = k_I \int_0^t e(t)\, dt$$

In a mechanical system the "action" $A_c(t)$ is often a force or moment produced by an actuation device (hydraulically or electromagnetically, for example) which moves an object. In the Laplace domain the proportional, derivative, and integral control actions would have the following block diagram representations ($B8.21$):

Proportional control

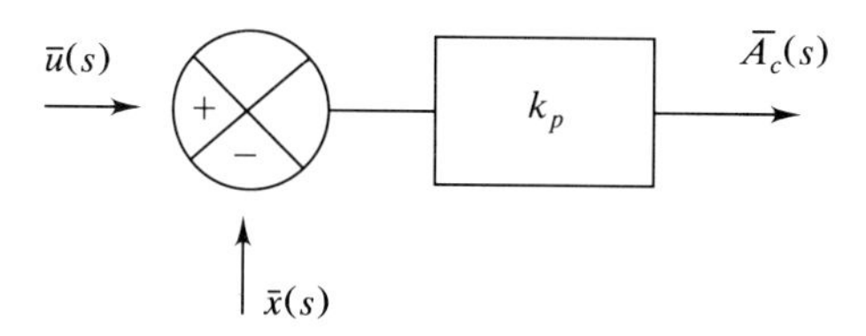

Derivative control

$\bar{u}(s)$ + − $k_d s$ $\bar{A}_c(s)$ $\bar{x}(s)$

$$(B8.21)$$

Integral control

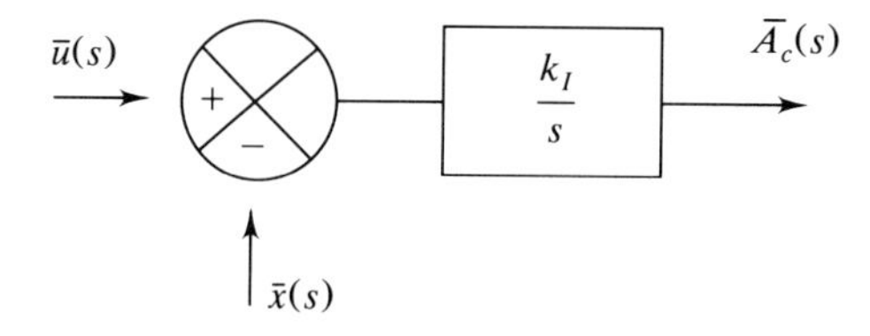

In many situations of practical importance, combinations of two or all three of the above strategies are employed. Thus we might use proportional plus derivative (PD) control, proportional plus integral (PI) control, or proportional plus integral plus derivative (PID) control. These possible control strategies would, respectively, be represented by the following block diagrams ($B8.22$):

PD control

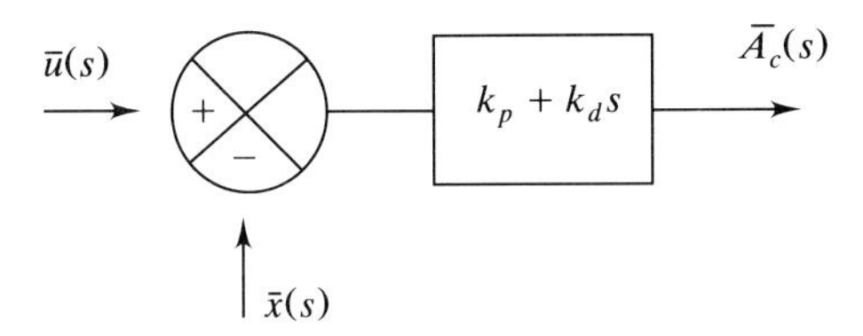

PI control

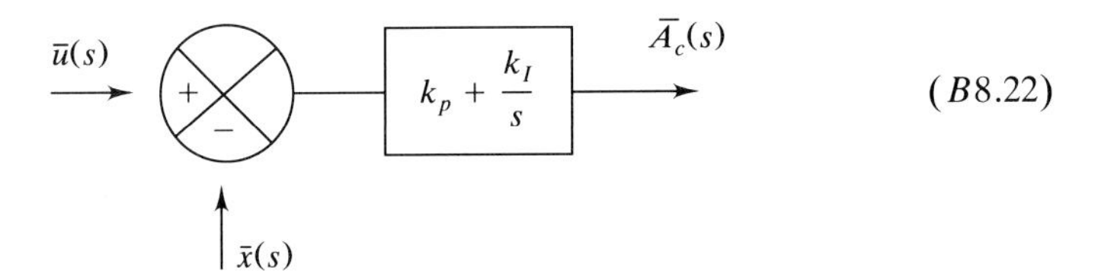

PID control

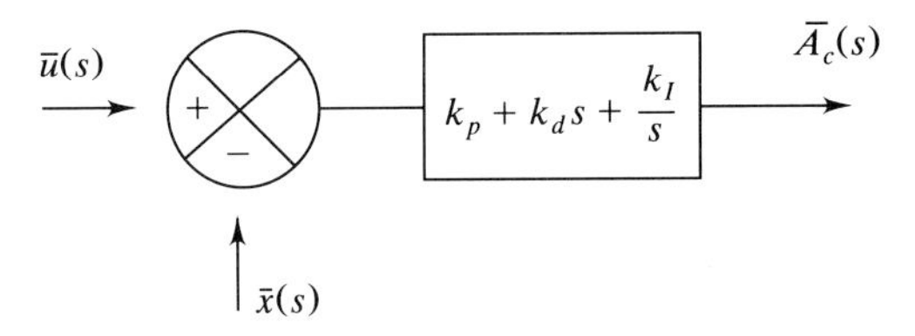

Having stated the mathematical and block diagram representations of the PID control family, let us now analyze these control strategies in order to understand their effect on system response and performance. In this study we will focus mainly on the response to a step input.

8.4.4 PID Control of First-Order Systems

Here we investigate various combinations of the PID control family as applied to the control of a first-order system. The objective will be to establish how each of the three control strategies (P, I, D), as well as combinations of them, may or may not be used to improve the transient and steady-state responses of the system when it is subjected to a step input. We take the basic control system model to be of the following form ($B8.23$):

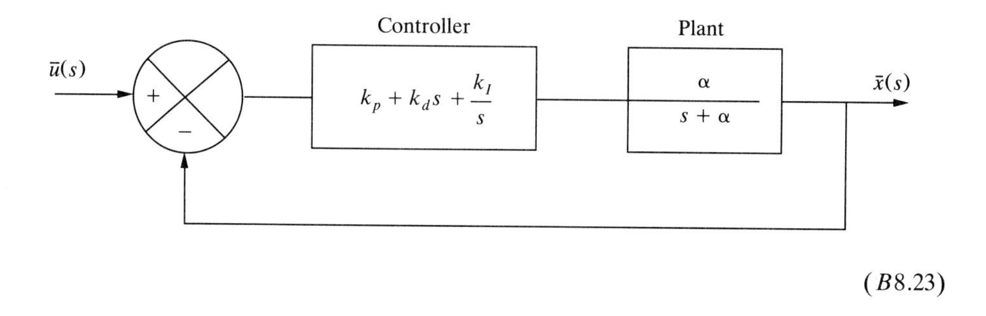

($B8.23$)

Here $\alpha > 0$ and PID control has been assumed for generality. In studying the response to a step input, let us, as a reference, first consider the response of the uncontrolled system to a unit step $\bar{u}(s) = 1/s$:

$$\bar{u}(s) = \frac{1}{s} \longrightarrow \boxed{\frac{\alpha}{s+\alpha}} \longrightarrow \bar{x}(s)$$

or

$$\bar{x}(s) = \frac{1}{s}\left(\frac{\alpha}{s+\alpha}\right)$$

Inversion yields the uncontrolled step response as

$$x(t) = 1 - e^{-\alpha t}$$

So the input and the response are as shown in Fig. 8.12.

In the plant transfer function $G_p(s) = \alpha/(s+\alpha)$ in ($B8.23$), an α has been included in the numerator so that the steady-state response of the uncontrolled system to the unit step will be unity. Using a 2 percent criterion, the settling time T_s of the uncontrolled system may be determined approximately as $T_s \cong 4/\alpha$ (note that $(1 - e^{-4}) = 0.98$). The quantity $1/\alpha$, which has units of time, is referred to as the *system time constant* (Sec. 4.3.3). Thus, for exponential behavior, about four time constants are needed in order that the system response settle to within 2 percent of the final value, and this rule of thumb is commonly used to determine the settling time of first-order systems.

Next, consider the PID control system of ($B8.23$). The student may verify that the closed-loop transfer function $T(s)$, the error transfer function $E(s)$,

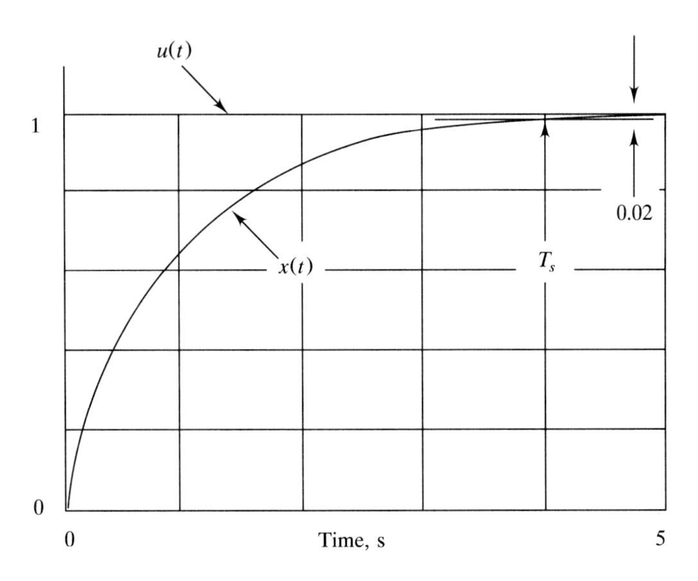

FIGURE 8.12
Response of the uncontrolled version of ($B8.23$) to the unit step input. Results shown for the case $\alpha = 1$. The setting time $T_s = 4$ s.

and the system characteristic polynomial are given by the following:

$$T(s) = \frac{\alpha\left(k_d s^2 + k_p s + k_I\right)}{s^2(1 + \alpha k_d) + s\left(\alpha + \alpha k_p\right) + \alpha k_I}$$

$$E(s) = \frac{s(s + \alpha)}{s^2(1 + \alpha k_d) + s\alpha\left(1 + k_p\right) + \alpha k_I} \tag{8.64}$$

$$\text{Characteristic polynomial:} \quad s^2(1 + \alpha k_d) + s\alpha\left(1 + k_p\right) + \alpha k_I$$

Note that the error transfer function $E(s)$ will allow us to determine directly the steady-state error e_{ss}. The response $x(t)$ and the error $e(t)$ for a step input will be defined in the Laplace domain by

$$\bar{x}(s) = \frac{1}{s}T(s), \qquad \bar{e}(s) = \frac{1}{s}E(s) \tag{8.65}$$

Let's first consider the case of proportional control. Settling $k_d = k_I = 0$ in (8.64) leads to the closed-loop transfer function

$$T(s) = \frac{\alpha k_p}{s + \alpha\left(1 + k_p\right)}$$

The response $x(t)$ to a step input is then defined in the Laplace domain, from (8.65), as

$$\bar{x}(s) = \frac{\alpha k_p}{s[s + \alpha(1 + k_p)]} \tag{8.66}$$

Inversion of (8.66), a good exercise for the student, yields the response $x(t)$ to the step input as

$$x(t) = \frac{k_p}{1 + k_p}\left(1 - e^{-\alpha(1+k_p)t}\right) \tag{8.67}$$

The result shows the system to respond more quickly than does the uncontrolled system. The new system time constant $\tau = 1/\alpha(1 + k_p)$. We also observe from (8.67) that the steady-state value of $x(t)$ is $x_{ss} = k_p/(1 + k_p)$, so that the steady-state error

$$e_{ss} = u(t) - x_{ss} = 1 - \frac{k_p}{1 + k_p} = \frac{1}{1 + k_p}$$

In order to minimize the steady-state error, the control gain k_p should be as large as possible. Thus, the larger the gain k_p, the smaller the steady-state error and the quicker the response. Figure 8.13 shows a step response for the case $\alpha = 1$, $k_p = 10$.

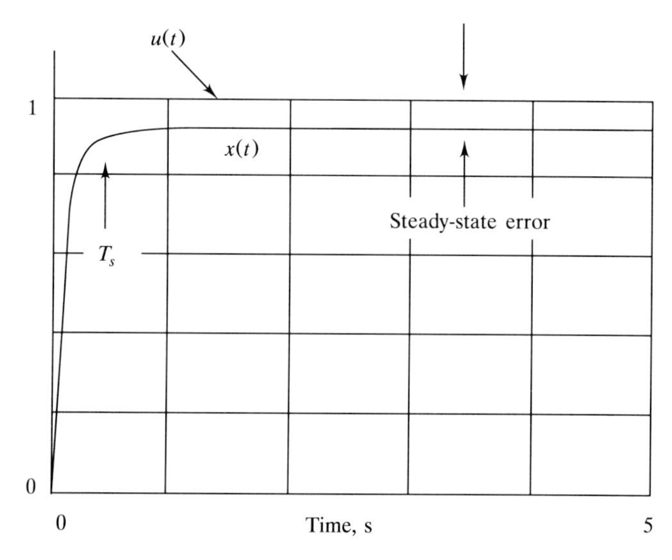

FIGURE 8.13
Step response of the proportionally controlled first-order system for the case $\alpha = 1$, $k_p = 10$. The scale is the same as in Fig. 8.12 to emphasize how much more quickly the system responds. The steady-state error $e_{ss} \approx 9$ percent.

The simple strategy of proportional control works well in the control of first-order systems because the effect of an increase in the control gain k_p is to improve *all* of the relevant performance measures: steady-state error, rise time, and settling time. For this reason the control of first-order systems is conceptually simple. By comparison we'll see that for second- and higher-order systems, improvement in one performance measure may come at the expense of degradation of another, so that design tradeoffs must be carefully considered.

Other first-order system control strategies from the PID family are summarized in Table 8.2. For each type of control strategy, the controller transfer function, the system closed-loop transfer function, the step response, and the steady-state error for a step input are shown. The intent of this table is only to provide some idea as to the effects of various possibilities, as summarized below:

1. *Derivative control:* Observe that derivative-only control actually makes things much worse. The system time constant is increased from $1/\alpha$ (uncontrolled) to $(1 + k_d\alpha)/\alpha$, signifying more sluggish response. The steady-state error is unity and the tracking properties are abysmal. These results indicate a general trend, applicable to many systems of first and higher order: Derivative-only control is generally not advisable.
2. *Integral control:* An important effect of integral control is to increase by 1 the order of the system, so that a first-order plant with integral control will exhibit second-order system response. That the controlled system is of the second order is clearly seen in the closed-loop transfer function, the denominator of which is the characteristic polynomial, which is of degree 2. One advantage of integral control generally is that it improves steady-state behavior. From Table 8.1 we observe that for a step input, the steady-state error is, in fact, zero. In physical terms, as the steady-state condition is approached (and the error becomes small), a proportional controller will provide only a small corrective action. The integral control, however, will continue to provide appreciable control action, because it is based on the *integral* of the error. Even a small error, when integrated over a long time, results in an appreciable effect, eventually driving the error for a step input to zero.

On the other hand, the transient response of a first-order system with integral-only control turns out actually to be degraded. The poles of the closed-loop system are given by

$$p_{1,2} = -\frac{\alpha}{2} \pm \left(\frac{\alpha^2}{4} - k_I\alpha\right)^{1/2}$$

If $0 < k_I < \alpha/4$, the controlled system has two real, negative poles, but one will always be *less negative* than the single pole $-\alpha$ of the uncontrolled system. The transient response may in fact be fairly sluggish if k_I is close to $\alpha/4$, so that one of the poles is close to zero. Furthermore, if $k_I > \alpha/4$, then the poles are

TABLE 8.2
PID Control of first-order systems

Type of control	Controller TF	CLTF T(s)	Step response	Steady-state error
Proportional	k_p	$\dfrac{\alpha k_p}{s + \alpha(1 + k_p)}$	$\dfrac{k_p}{1 + k_p}[1 - e^{-\alpha(1 + k_p)t}]$	$\dfrac{1}{1 + k_p}$
Derivative	$k_d s$	$\dfrac{\alpha k_d s}{s(1 + \alpha k_d) + \alpha}$	$\dfrac{\alpha k_d}{1 + \alpha k_d} e^{-[\alpha/(1 + \alpha k_d)]t}$	1
Integral	k_I/s	$\dfrac{k_I \alpha}{s^2 + \alpha s + k_I \alpha}$	$1 + \dfrac{e^{-(\alpha/2)t}}{\left(1 - \dfrac{\alpha}{4k_I}\right)^{1/2}} \sin\left[\sqrt{\alpha k_I}\left(1 - \dfrac{\alpha}{4k_I}\right)^{1/2} t - \psi\right]$	0
PD	$k_p + k_d s$	$\dfrac{\alpha(k_p + k_d s)}{s(1 + \alpha k_d) + \alpha(1 + k_p)}$	$\dfrac{k_p}{1 + k_p} - \left(\dfrac{k_p - \alpha k_d}{(1 + \alpha k_d)(1 + k_p)}\right) e^{-(\alpha(1 + k_p)t)/(1 + \alpha k_d)}$	$\dfrac{1}{1 + k_p}$
PI	$k_p + k_I/s$	$\dfrac{\alpha(k_p s + k_I)}{s^2 + \alpha s(1 + k_p) + \alpha k_I}$	$1 - e^{-at}\left(\cos \omega t + \dfrac{\alpha - a}{\omega} \sin \omega t\right)$ $a = \dfrac{\alpha(1 + k_p)}{2}$ $\omega^2 = \alpha k_I - a^2$ assumes $\alpha k_I > a^2$	0

Notes: CLTF = System Closed-Loop Transfer Function $T(s)$. Steady-state error is for step input. Step response for integral control assumes poles of T(s) are complex.

complex conjugates, and the transient response will be oscillatory. The time constant in this case is defined by the real part of the pole, which governs the exponential decay to the steady-state condition. This time constant is $\tau = 2/\alpha$, which is twice as large as for the uncontrolled system. (*Remember*: the larger the time constant, the more sluggish the response.)

Based on the preceding discussion, integral control of a first-order system is not advisable unless the transient response performance is unimportant.

8.4.5 PD Control of Second-Order Systems

The preceding section introduced the basic ideas of PID control as applied to the simplest type of linear system, that of first order. We now explore the control of systems of second order. Second-order linear systems are especially important for two reasons: (1) Many simple systems can be modeled, at least approximately, as second-order systems, and (2) as we shall see in Sec. 8.4.6, many systems of higher order behave in essentially the same way as a system of second order. This latter result turns out to be very useful in practice because it means that design principles developed for second-order systems can be applied to many (but not all!) higher-order systems, even if the order is significantly greater than 2.

Let us illustrate some of the basic ideas by considering the PD control of the second-order system shown in ($B8.24$). In ($B8.24$) the plant transfer function is defined in terms of the system parameters ξ and ω (Sec. 5.3.1), which identify any inherent damping and stiffness possessed by the plant. An ω^2 is arbitrarily included in the numerator of the plant transfer function in order that the steady-state response of the plant to a unit step input be unity. Thus, the plant without any control will exhibit response to inputs as described in Sec. 5.3.1. For example, see Fig. 5.9, which shows the response to a step input for

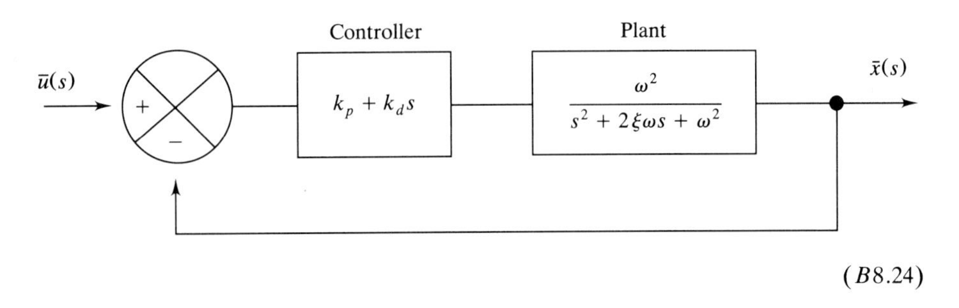

($B8.24$)

various levels of damping factor ξ.

Now we know that the addition of the controller will do two things: (1) alter the basic properties of the system so as to produce (ideally) an improvement in the transient and/or steady-state responses, and (2) specify as input the desired history of the variable to be controlled. Let us first consider the effect of the PD controller on the basic system properties. To do this, we determine the closed-loop transfer function $T(s)$ and then study how the poles of $T(s)$ are altered by the presence of the controller. The reader may verify that the closed-loop transfer function $T(s)$ for the system of (B8.24) is

$$T(s) = \frac{\omega^2(k_p + k_d s)}{s^2 + s(2\xi\omega + \omega^2 k_d) + \omega^2(1 + k_p)} \tag{8.68}$$

In this relation (8.68), it is important to note the presence of the factor $(k_p + k_d s) = k_d(s + k_p/k_d)$ in the *numerator* of the closed-loop transfer function. The transfer function $T(s)$ now has a *zero* z_1, located at $z_1 = -k_p/k_d$ in the complex plane. This zero of $T(s)$ is produced by the derivative control. The effect of this zero of $T(s)$ on the transient response of the system will be considered later in this section.

Let us now define the following quantities:

$$2\xi_c\omega_c = 2\xi\omega + \omega^2 k_d \tag{8.69a}$$

$$\omega_c^2 = \omega^2(1 + k_p) \tag{8.69b}$$

These relations define the damping factor ξ_c and undamped natural frequency ω_c of the *controlled* system in terms of the same quantities ξ and ω of the plant and in terms of the control constants k_p and k_d. Observe that the proportional control k_p affects the overall system stiffness, while the derivative control k_d affects the overall system damping. Thus, both the "quickness" of the system response (ω) and the rate of decay of any oscillations can be controlled by properly choosing the control constants k_p and k_d.

It is useful in control systems analysis to display in the complex plane the poles of the closed-loop transfer function $T(s)$. These poles are just the eigenvalues of the equivalent state variable model of the system, so that such a plot displays the same information as Fig. 4.23 or 4.27 from linear system theory. In the present application of the PD control of the second-order system (8.68), the poles $p_{1,2}$ of the closed-loop transfer function $T(s)$ are given in terms of the parameters ω_c, ξ_c, by

$$p_{1,2} = -\xi_c\omega_c \pm i\omega_c(1 - \xi_c^2)^{1/2} \qquad (\xi_c < 1) \tag{8.70a}$$

$$p_{1,2} = -\xi_c\omega_c \pm \omega_c(\xi_c^2 - 1)^{1/2} \qquad (\xi_c \geq 1) \tag{8.70b}$$

In the complex plane the poles will appear as in Fig. 8.14(a) and (b). The case $\xi_c < 1$, for which the poles are complex and the response oscillatory, is shown in Fig. 8.14(a), while the case $\xi_c > 1$, for which the poles are real and the response exponential in character, is represented in Fig. 8.14(b).

The type of display shown in Fig. 8.14 turns out to be very useful in the design of control systems of order 2 and greater. Proper "pole placement" will produce the desired response characteristics of the closed-loop system.

Example 8.24 PD Control of a Second-Order System. A second-order plant has mass, damping, and stiffness properties that result in an undamped natural

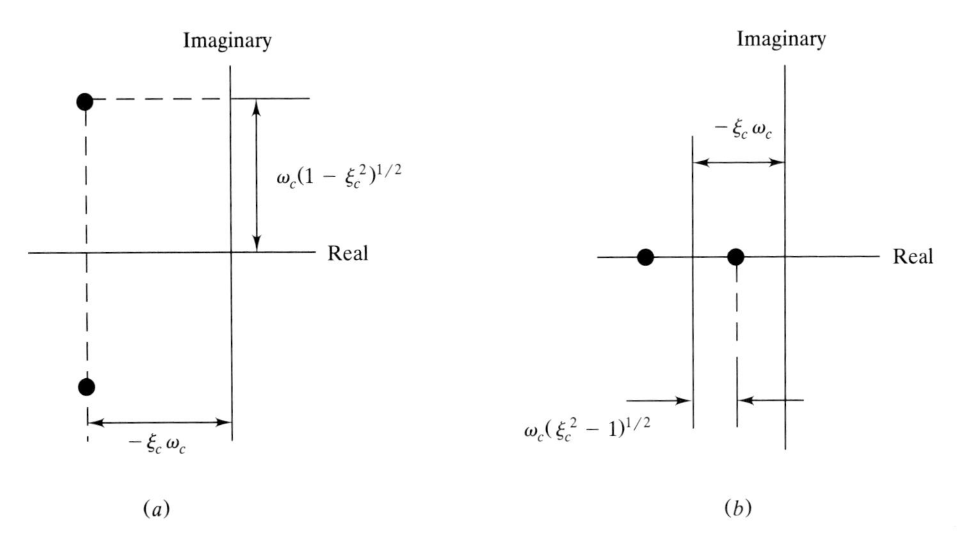

FIGURE 8.14
(*a*) The poles $p_{1,2}$ of the system (*B*8.24) for the case $\xi_c < 1$; (*b*) poles $p_{1,2}$ of (*B*8.24) for the case $\xi_c > 1$.

frequency $\omega = 5$ rad/s and a damping factor $\xi = 0.1$. These properties result in a plant which responds relatively slowly and for which transient behavior does not decay as quickly as desired. In order to render the transient response characteristics acceptable, an analyst determines that the system should exhibit the properties $\omega_c = 25$ rad/s and $\xi_c = 0.5$. Determine the required control constants k_p and k_d for a PD controller, such that the specified values of ξ_c and ω_c are realized.

From (8.69*b*) the required proportional control k_p can be calculated as $k_p = (\omega_c/\omega)^2 - 1 = 24$. Then (8.69*a*) can be used to determine the required derivative control constant k_d as $k_d = (2\xi_c\omega_c - 2\xi\omega)/\omega^2 = 24/25$. In a physical application the size, weight, and power requirements of a controller that will accomplish this substantial change in system properties would have to be assessed.

In order to visualize the effect of the control system on the overall system properties in this example, let us plot the poles in the complex plane for both the plant and the controlled system. For the plant the system parameters $\omega = 5$ rad/s and $\xi = 0.1$ lead from (8.70) to the poles $p_{1,2} = -0.5 \pm 4.975i$. The controlled system parameters $\omega_c = 25$, $\xi_c = 0.5$ lead to the pole locations

$$p_{1,2} = -12.5 \pm 21.65i \qquad \text{(controlled system)}$$

Thus, the addition of the controller has moved the poles in the complex plane as shown in Fig. 8.15. The controller has changed the basic system properties substantially.

Remark. *On the Step Response of the System (B8.24) with PD Control.* As noted in Sec. 8.4.2, the desired performance of a control system is often specified in terms of the response to a step input, so that certain quantities such as rise time, overshoot, settling time, and steady-state error are to be within specified limits in

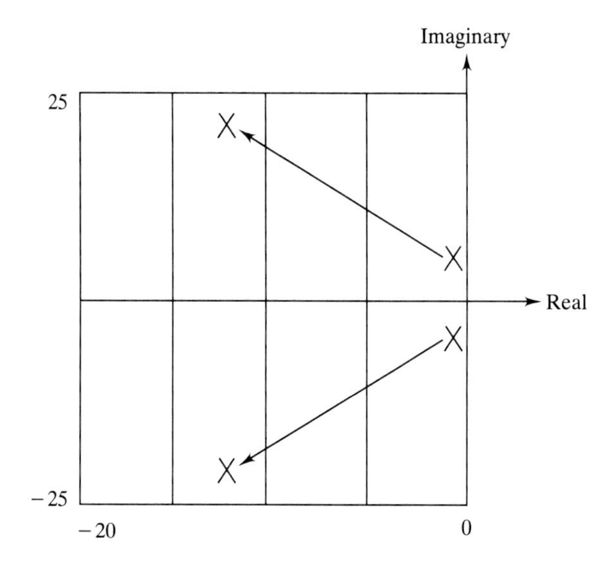

FIGURE 8.15
Pole placement for the plant and for the controlled system of (Example 8.24). Arrows indicate the change in pole locations after addition of the controller.

order that the system performance be deemed acceptable. We have previously studied the step response of second-order systems in Sec. 5.3.1, and Figs. 5.9 and 5.10 are useful in determining the systems properties ξ and ω (or ξ_c, ω_c) needed to achieve specified overshoot and time to first oscillation peak. There is, however, a fundamental difference between the step response studied in Sec. 5.3.1 and the step response of the system ($B8.24$) with PD control. The reason for this difference can be seen in the forms of the mathematical models for the two situations. The "standard" unit step response $x(t)$ was presented in Sec. 5.3.1 as the solution to (5.48)

$$\ddot{x} + 2\xi\omega\dot{x} + \omega^2 x = \omega^2 u(t) \tag{8.71}$$

where the constant g_0/m in (5.48) has been replaced by $\omega^2 u(t)$, so that the steady-state response is unity. The solution to (8.71) for the case of zero initial conditions is

$$x(t) = 1 - \frac{e^{-\xi\omega t}}{(1-\xi^2)^{1/2}} \sin(\omega_d t + \phi) \qquad (\xi < 1) \tag{8.72}$$

where $\omega_d = \omega(1-\xi^2)^{1/2}$ and $\tan\phi = (1-\xi^2)^{1/2}/\xi$. This is the response plotted in Fig. 5.9.

For the system ($B8.24$) with PD control, however, the transfer function (8.68) yields the differential equation model

$$\ddot{x} + 2\xi_c\omega_c\dot{x} + \omega_c^2 x = \omega^2\left[k_p u(t) + k_d\dot{u}(t)\right]$$

which may be rewritten, using (8.69b), as

$$\ddot{x} + 2\xi_c\omega_c\dot{x} + \omega_c^2 x = \omega_c^2\left[\frac{k_p}{1+k_p}u(t) + \frac{k_d}{1+k_p}\dot{u}(t)\right] \tag{8.73}$$

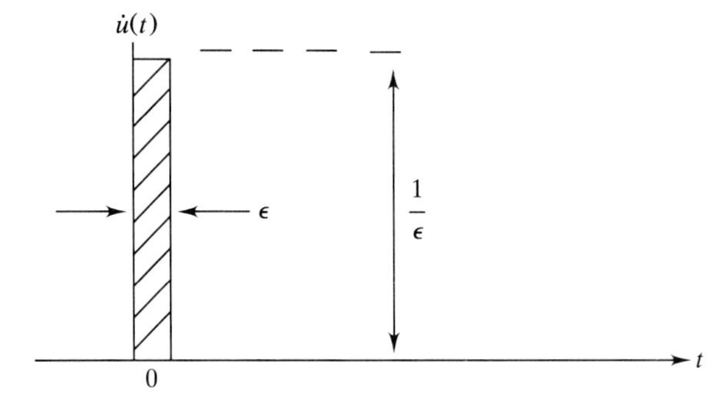

FIGURE 8.16
The unit impulse function is the function shown in the limit $\varepsilon \to 0$.

The primary difference between (8.73) and the "standard" model (8.71) is the addition of the forcing term

$$\omega_c^2\left(\frac{k_d}{1+k_p}\right)\dot{u}(t)$$

which arises from the derivative control. For a step input the function $\dot{u}(t)$ will be zero for all times except $t = 0$, at which it is undefined (it is effectively infinite). This derivative $\dot{u}(t)$ of the unit step function $u(t)$ is termed the *unit impulse function*. Intuitively we can visualize the unit impulse function as in Fig. 8.16 as a function which is very large for a very short period of time, such that the area under the function is unity.

In physical terms, as the input step $u(t)$ changes from zero to unity at $t = 0$, the derivative control senses a very large rate of change in $u(t)$ and provides a large action of very short duration. This additional action due to the derivative control may have a significant effect on the transient response, as we now demonstrate.

Consider first the response to a unit impulse alone, according to the system model

$$\ddot{x} + 2\xi\omega\dot{x} + \omega^2 x = \omega^2\dot{u}(t) \tag{8.74}$$

Laplace transformation of (8.74) for the zero-state case, with $\mathscr{L}[\dot{u}(t)] = s\mathscr{L}[u(t)] = 1$, yields

$$\bar{x}(s) = \frac{\omega^2}{s^2 + 2\xi\omega s + \omega^2} \tag{8.75}$$

Inversion then provides the *unit impulse response* $x(t)$ as

$$x(t) = \frac{\omega}{\left(1-\xi^2\right)^{1/2}}e^{-\xi\omega t}\sin\omega_d t \qquad (\xi < 1) \tag{8.76}$$

Note that this response $x(t) \to 0$ as $t \to \infty$, as the impulse $\dot{u}(t) = 0$ everywhere but at $t = 0$; that is, there is no input for $t > 0$. Physically, the unit impulse merely imparts an initial velocity $\dot{x}(0)$, and the system then moves freely with no outside influences present. The student is invited to show that the impulse response (8.76) for the zero-state case is the same as the zero-input response with an initial velocity $\dot{x}(0) = \omega^2$ and with $x(0) = 0$.

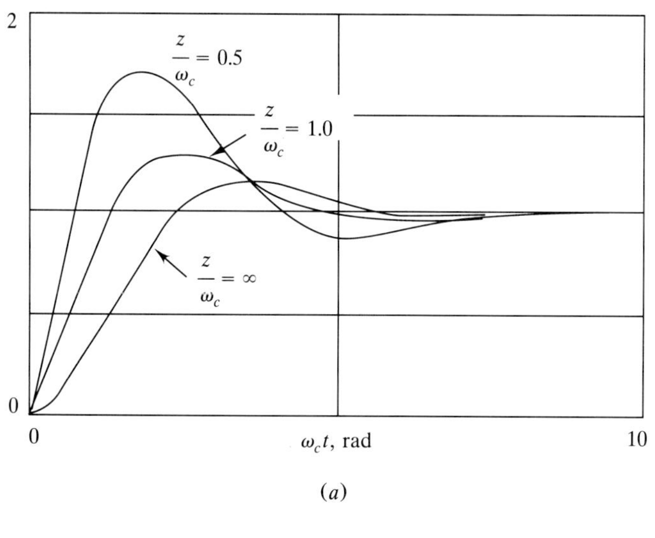

(*a*)

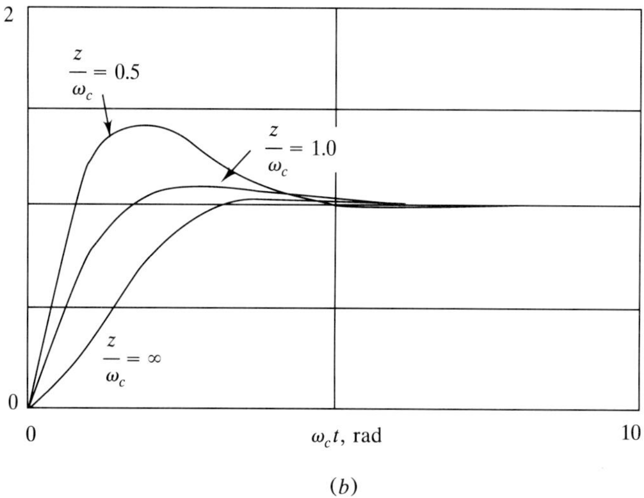

(*b*)

FIGURE 8.17
The effect of a zero ($z_1 = -k_p/k_d$) on the step response of the second-order system with PD control: (*a*) $\xi_c = 0.5$; (*b*) $\xi_c = 0.7$. In each case the vertical axis is

$$\frac{x(t)}{k_p/(1 + k_p)}$$

The parameter $z = -z_1$ so that $z/\omega_c > 0$.

Because the system ($B8.24$) is linear, the step response of this system, defined by (8.73), is obtained by superposition of the step response (8.72) and the impulse response (8.76). We thus obtain the step response of the system ($B8.24$) with PD control as

$$x(t) = \left(\frac{k_p}{1+k_p}\right)\left[1 - \frac{e^{-\xi_c \omega_c t}}{\left(1-\xi_c^2\right)^{1/2}} \sin(\omega_d t + \phi)\right]$$
$$+\left(\frac{k_d}{1+k_p}\right)\left[\frac{\omega_c}{\left(1-\xi_c^2\right)^{1/2}} e^{-\xi_c \omega_c t} \sin \omega_d t\right] \tag{8.77}$$

where $\tan \phi = (1-\xi_c^2)^{1/2}/\xi_c$ and $\omega_d = \omega_c(1-\xi_c^2)^{1/2}$. The result (8.77) may also be expressed as follows:

$$x(t) = \left(\frac{k_p}{1+k_p}\right)\left[1 - \frac{e^{-\xi_c \omega_c t}}{\left(1-\xi_c^2\right)^{1/2}}\left(1 - \frac{2\xi_c \omega_c}{z} + \frac{\omega_c^2}{z^2}\right)^{1/2} \sin(\omega_d t + \psi)\right]$$

where $z = k_p/k_d$ is the negative of the zero $z_1 = -k_p/k_d$ of $T(s)$, (8.68), and where $\tan \psi = (1-\xi_c^2)^{1/2}/(\xi_c - \omega_c/z)$. As the derivative control $k_d \to 0$, $z \to \infty$

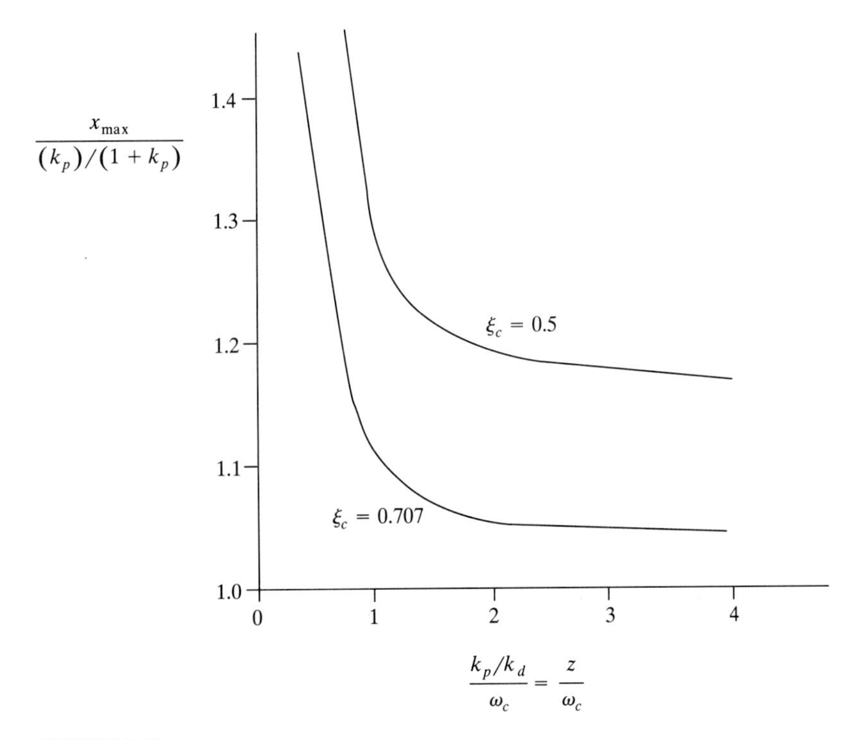

FIGURE 8.18
The effect of a zero ($z_1 = -k_p/k_d$) on the overshoot for two values of controlled system damping factor ξ_c. The parameter $z = -z_1$.

and the response (8.77) reduces to (8.72), that is the "standard" case of proportional only control. From (8.77) we also observe that the steady-state response is $x_{ss}(t) = k_p/(1 + k_p)$, rather than unity, so that there will be a steady-state error equal to $1/(1 + k_p)$.

The effect of the derivative control on the transient response is illustrated in Figs. 8.17 and 8.18. Shown in Fig. 8.17 are the step responses for three cases: (1) PD control with $z = k_p/k_d = 0.5$ and 1.0, which represent a relatively large k_d, and (2) $z \to \infty$, so that we reduce to the case of proportional control. It is clear that when $z = 0.5$ the derivative control causes a significant difference in the transient response relative to the "standard case." The large derivative control action, providing an impulsive velocity $\dot{x}(0)$, causes the system to respond more quickly near $t = 0$. The attendant overshoot is, however, much greater than in the "standard" case, and this possibility is the main point to be noted here.

Figure 8.18 shows the variation of the overshoot with the ratio z/ω_c for two values of controlled system damping: $\xi_c = 0.5$ and $\xi_c = 0.707$. Here it is observed that, provided $z/\omega_c \gtrsim 3$, the overshoot is not much different from what it would be for the "standard" case. That is, the presence of the zero $z_1 = -k_p/k_d$ in the closed-loop transfer function $T(s)$ does not have a significant effect on the transient response performance measures.

DESIGN CONSIDERATIONS FOR PD CONTROL OF SECOND-ORDER SYSTEMS

1. *Derivative-only control:* Use of derivative-only control will result in a step response consisting of the second of the two terms in (8.77). As $t \to \infty$, $x(t) \to 0$, so that the steady-state error is unity. There is no tracking ability, and this illustrates the fundamental problem of derivative-only control: If the error is unchanging, as occurs here as $t \to \infty$, the controller cannot tell whether the error is large or small. Sensing only a zero rate of change of error, the controller provides no corrective action. Thus, derivative-only control is generally ineffective.
2. *PD control:* With PD control the poles of the closed-loop transfer function $T(s)$ can theoretically be located anywhere we desire to place them, as is clear from (8.69) and (8.70). Thus, we can theoretically shape the transient response as we wish by selecting the appropriate pole locations and by taking proper account of the effect on the step response of the zero $z_1 = -k_p/k_d$ of $T(s)$, as noted in the preceding remark.
3. The controlled system damping and frequency parameters ξ_c and ω_c define the locations of the poles of $T(s)$, according to (8.70). These parameters are generally to be selected so that the overshoot and settling time are acceptably small. The overshoot is determined by ξ_c, while the settling time T_s depends on both ξ_c and ω_c; for a 2 percent settling criterion, $T_s \approx 4\tau$, where the time constant τ for a second-order system with $\xi_c < 1$ is $\tau = 1/\xi_c\omega_c$. In order to maintain the overshoot below about 10 percent, we generally select a damping factor ξ_c in the range $0.5 < \xi_c < 0.707$. Then the frequency ω_c can be selected so that the settling time is sufficiently small. Theoretically, of course, we can on paper select ω_c to be very large and ξ_c to be close to

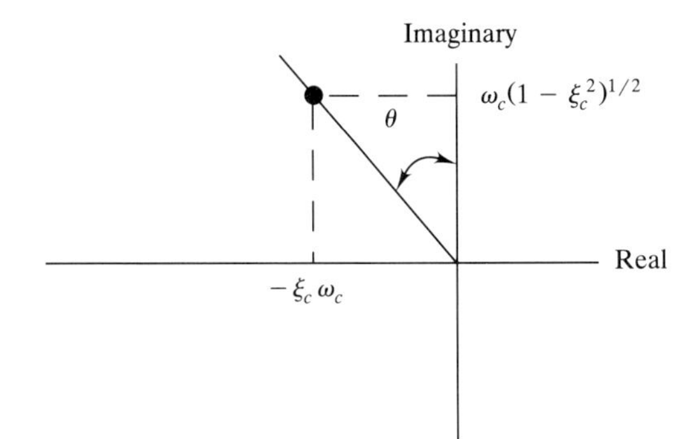

FIGURE 8.19
The angle θ determines the damping factor ξ_c for a pair of conjugate poles according to $\tan\theta = \xi_c/(1 - \xi_c^2)^{1/2}$.

unity, so that the system would track a step input very closely. From a practical standpoint, however, the size and power requirements of the control system increase with both ξ_c and ω_c, so compromise is necessary in order to achieve a design which is realistic in terms of size, power, and economy.

Note that the rule of thumb damping factor $\xi_c = 1/\sqrt{2} = 0.707$ corresponds to the 45° line in the complex plane, as shown in Fig. 8.19, because $\tan\theta = \xi_c/(1 - \xi_c^2)^{1/2}$.

Also note from Fig. 8.19 that, as the angle θ approaches zero, so does the damping factor ξ_c. Poles defined by a small angle θ in the complex plane will, therefore, exhibit relatively large overshoot and relatively slow decay of any oscillations which are induced. Furthermore, for given θ (hence given damping ξ_c), as we move away from the origin, the frequency ω_c increases, causing the system to respond faster. The ability to visualize the relation between the location of poles in the complex plane and the resulting response is very helpful in the design of control systems.

8.4.6 Application of Second-Order Principles to Systems of Higher Order

It is in the nature of things that many linear systems of order higher than 2 behave in essentially the same way as linear systems of second order. If a given higher-order system does exhibit this "second-order tendency," then the ideas used to design controllers for second-order systems can also be used for systems of higher order. The purpose of this section is to show, mainly by example, how the second-order tendency comes to be, how to determine whether it is valid, and how to define a second-order system model that approximates the actual model of higher order. The student should keep in mind that nowadays the use of software packages for computerized analysis of control systems is widespread, and this makes it reasonably straightforward to analyze higher-order control systems via computer. Thus, the discussion in this section is not meant to imply that one should always attempt to simplify down to second order. An understanding of the reasons underlying the second-order tendency, however, is essential in understanding the behavior of dynamic systems. Furthermore, the second-order tendency is exhibited by a wide range of phenomena in engineering and science. It is not limited to control systems.

In the two following examples, we will look at the responses of two third-order systems to a step input. The intent of these examples will be to illustrate how a system of third order may exhibit a response that is approximately the same as that of an "equivalent" system of second order.

Example 8.25 Step Response of a Third-Order System. Consider a linear third-order system characterized by the transfer function

$$T(s) = \frac{20}{(s+10)(s+1)(s+2)} \tag{8.78}$$

The linear system (8.78) does not necessarily arise from a controls problem; $T(s)$ is simply viewed as the transfer function defining some third-order linear system. The three poles of $T(s)$ are real and negative, $p_1 = -1$, $p_2 = -2$, $p_3 = -10$, and the system is asymptotically stable.

If we denote the system response by $x(t)$ and the input by $u(t)$, we then have from the definition of the transfer function

$$\bar{x}(s)[(s+10)(s+1)(s+2)] = 20\bar{u}(s)$$

so that the equivalent third-order differential equation model is

$$\dddot{x} + 13\ddot{x} + 32\dot{x} + 20x = 20u(t) \tag{8.79}$$

Note that the 20 appearing in the numerator of $T(s)$ causes the steady-state response to a unit step input to be unity. In the complex plane the three poles of the system transfer function will lie along the real axis, as shown in Fig. 8.20.

In the preceding commentary we have described the characteristics of the third-order system being considered. Let us now calculate the response to a unit

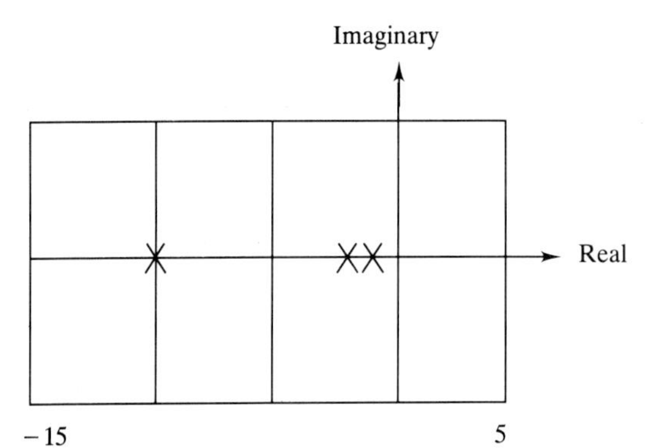

FIGURE 8.20
The poles of the transfer function of Example 8.25 in the complex plane.

step, $\bar{u}(s) = 1/s$. The step response is obtained from

$$\bar{x}(s) = \frac{20}{s(s+10)(s+1)(s+2)}$$

The technique of partial fraction expansion (Sec. 8.2.5) can be used to rewrite $\bar{x}(s)$ in the form

$$\bar{x}(s) = \frac{A_1}{s} + \frac{A_2}{s+1} + \frac{A_3}{s+2} + \frac{A_4}{s+10} \tag{8.80}$$

Calculating $A_1, \ldots, A_4$ from (8.35) yields $A_1 = 1$, $A_2 = -2.222$, $A_3 = 1.25$, $A_4 = -0.028$. Inversion of (8.80) then yields the following step response of this third-order system:

$$x(t) = 1 - 2.222e^{-t} + 1.25e^{-2t} - 0.028e^{-10t} \tag{8.81}$$

In the result (8.81) we observe the following:

1. The "contribution" of each of the three poles is evident in the three exponential terms.
2. The last exponential term, associated with the pole at -10, will die out much more rapidly than the exponential terms associated with the poles at -1 and -2.
3. The coefficient (-0.028) associated with the pole at -10 is much smaller than the coefficients associated with the other two poles.

The net effect of observations 2 and 3 is that, except for a *very brief* time interval near $t = 0$, the pole at -10 contributes very little to the response, which is dominated by the contributions from the poles at -1 and -2. Thus, the system behaves in essentially the same way as a *second-order system* having poles at -1 and -2.

To reinforce this point, let us determine an "equivalent" second-order system transfer function that will produce a step response approximately the same as (8.81). A simple way to do this is to select the second-order system that preserves the following features of the actual system: (1) The steady-state response to a step input should be the same (here unity), and (2) the locations of the "dominant poles," here -1 and -2, should be the same. The transfer function $T(s)$ having these properties is

$$T(s) = \frac{2}{(s+1)(s+2)}$$

This transfer function was obtained by merely deleting the s in the factor $(s + 10)$ in the original transfer function (8.78). The second-order ODE model "equivalent" to (8.79) is then found to be

$$\ddot{x} + 3\dot{x} + 2x = 2u(t) \tag{8.82}$$

and the response to the unit step input is

$$x(t) = 1 - 2e^{-t} + e^{-2t} \tag{8.83}$$

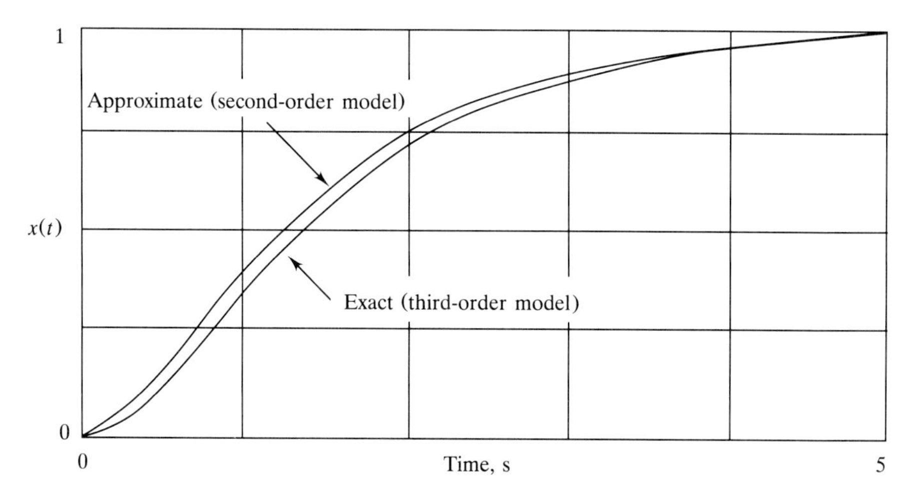

FIGURE 8.21
Step responses of the third-order system (8.79) and of the "equivalent" second-order system (8.82).

The result (8.83) is of course not the same as the actual step response (8.81), as (8.82) is only an approximation to the actual third-order model (8.79), and some information is lost in the order reduction process. A comparison of the actual and approximate solutions is shown in Fig. 8.21. The reduced-order model, while certainly not exact, captures the essential behavior of the actual system.

Example 8.26 Step Response of a Third-Order System Having a Pair of Complex Conjugate Poles. Here we summarize the model reduction for a third-order system having the transfer function

$$T(s) = \frac{20}{(s+10)(s^2+2s+2)} \tag{8.84}$$

The poles of this transfer function are located at $p_1 = -10$, $p_{2,3} = -1 \pm i$. The pair of complex conjugate poles with negative real part signifies that the system response will exhibit an exponentially decaying oscillation. The response $x(t)$ to a unit step input, $\bar{u}(s) = 1/s$, will be determined by inversion of

$$\bar{x}(s) = \frac{20}{s(s+10)[s-(-1+i)][s-(-1-i)]}$$

A partial fraction expansion enables this inversion to be represented as

$$x(t) = A_1 + A_2 e^{-10t} + A_3 e^{(-1+i)t} + A_4 e^{(-1-i)t}$$

Evaluating the constants $A_1, \ldots, A_4$ using the same procedure as in Example 8.25, followed by some trigonometric manipulation to eliminate the complex exponential, eventually leads to the step response as follows:

$$x(t) = 1 - 0.0244e^{-10t} + e^{-t}(-0.9756 \cos t - 1.2196 \sin t) \tag{8.85}$$

Observe that, as in Example 8.25, the contribution associated with the pole $p_1 = -10$ is small to begin with and then dies out very rapidly, so that after a very short time the response is dominated by the contribution associated with the complex poles.

The second-order system that preserves the location of the dominant poles and has the same steady-state response to a step input is defined by the transfer function

$$T(s) = \frac{2}{s^2 + 2s + 2}$$

and the step response of this approximate, reduced-order model is

$$x(t) = 1 + e^{-t}(-\cos t - \sin t)$$

which is a reasonably good approximation to (8.85) in the essential aspects of behavior.

Comments

1. Sometimes higher-order systems do not behave as second-order systems, and it can be very misleading to attempt to define an equivalent second-order model which mimics the actual one. The validity of the model reduction idea depends on the disposition of the poles in the complex plane. For instance, for the third-order system defined by the transfer function

$$T(s) = \frac{6}{(s+3)(s+2)(s+1)}$$

the leftmost pole at -3 is not sufficiently to the left of the other poles that model reduction is valid. For this system we would not be able to analyze the response as if it were similar to that of some second-order system. We would just have to bite the bullet and use the third-order model.

2. Many linear systems are characterized by a pair of complex conjugate poles located relatively close to the imaginary axis, with all other poles well to the left of this complex pair. For such systems we term this pair of complex poles the "dominant poles," which define the "dominant mode" or "least-damped mode" of the response (Sec. 4.3.4). For example, suppose we consider a seventh-order system having the following pole locations (see also Fig. 8.22):

$$p_{1,2} = -0.5 \pm 10i$$

$$p_3 = -20$$

$$p_4 = -30$$

$$p_{5,6} = -50 \pm 20i$$

$$p_7 = -200$$

By analogy with Examples 8.25 and 8.26, the unit step response of this system may be expressed in the general form

$$x(t) = 1 + A_1 e^{-0.5t}\cos(10t - \phi_1) + A_3 e^{-20t} + A_4 e^{-30t}$$
$$+ A_5 e^{-50t}\cos(20t - \phi_2) + A_7 e^{-200t}$$

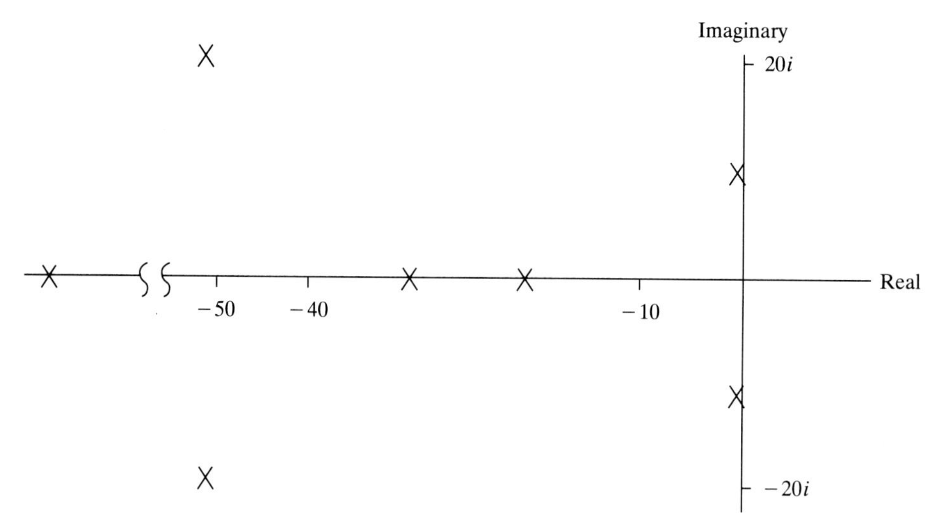

FIGURE 8.22
Pole locations for a hypothetical seventh-order system.

where $A_1, \ldots, A_7$, ϕ_1, and ϕ_2 are constants determined via the partial fraction expansion. The result indicates that, except for an interval of very short duration near $t = 0$, only the dominant mode (defined by $p_{1,2}$) will be observable in the response, and the behavior will be approximately the same as that of a second-order system having the poles $p_{1,2} = -0.5 \pm 10i$.

3. In the examples considered, it has been shown that some higher-order systems behave in approximately the same way as systems of second order. Other possibilities exist, however. For example, consider the fourth-order linear system modeled by the transfer function

$$T(s) = \frac{A}{(s + 200)(s + 100)(s + 50)(s + 1)}$$

For this hypothetical system, the pole $p_1 = -1$ is close to the imaginary axis in comparison to the other three poles, which are well to the left of the imaginary axis. Thus, this system will exhibit behavior which is approximately the same as that of a *first-order system* with a pole at -1. This example illustrates that, generally, a system of order n can be approximately modeled by a system of reduced order $m < n$ if one can identify m dominant poles, with the remaining $n - m$ poles well to the left of these in the complex plane.

4. The idea of model reduction can be visualized in the context of the phase or state space motions studied in Chaps. 4 and 5. Suppose that a given system is of order n and that it is possible to identify $m < n$ dominant poles. What happens is that orbits in the n-dimensional state space, initiated perhaps via a step input, will quickly collapse onto the m-dimensional subspace defined by the dominant modes. What we actually observe, except during a brief time interval near $t = 0$, is a motion in this m-dimensional subspace.

5. *On the use of reduced-order models:* Suppose that our objective is to design/analyze a control system for which we have developed a linear model of order n. If there is any doubt as to whether a reduced-order model can be used to approximate the behavior of the actual, nth-order system, then it is necessary to employ the original nth-order model in the analysis. In such circumstances a rough idea of the system response may be discernible from the least-damped modes, but accurate assessment of overshoot, settling time, and so on must come from the full nth-order model. Even if we are reasonably certain that a reduced-order model will be valid, it is good practice to establish verification via computer simulation of the nth-order system. In any event the principal utility of the reduced-order or second-order tendency is to enable us to understand why the system acts like it does and to establish some possible control strategies, based on an understanding of first- and second-order system response.

8.4.7 Steady-State Error

We close this section on basic aspects of performance and control with a discussion of the effect of the various control strategies of the PID family on the steady-state error in response to a step input. For first-order systems this steady-state error is listed in Table 8.2. Here we consider systems of arbitrary order n, as follows: the system block diagram is assumed to be ($B8.18$) with unity feedback, $H(s) = 1$. The plant transfer function $G_2(s)$ is assumed to be given by $G_2(s) = a_n/(s^n + a_1 s^{n-1} + \cdots + a_{n-1}s + a_n)$. The a_n has been included in the numerator of $G_2(s)$ so that the steady-state response of the plant will be unity. In this scenario it is assumed that none of the n open-loop poles s_i of $G_2(s)$ is zero.

Denoting the controller transfer function as $G_1(s)$, as in ($B8.18$), the closed-loop transfer function $T(s)$ for the system ($B8.18$) is obtained as

$$T(s) = \frac{a_n G_1(s)}{s^n + a_1 s^{n-1} + \cdots + a_{n-1}s + a_n + a_n G_1(s)}$$

The characteristic equation is obtained by setting the denominator of $T(s)$ to zero, and we assume that the poles p_i of the closed-loop transfer function all have negative real parts, so that the controlled system is asymptotically stable. The response $x(t)$ to a step input is then obtained from inversion of $\bar{x}(s) = T(s)/s$, or

$$\bar{x}(s) = \frac{a_n G_1(s)}{s[s^n + \cdots + a_{n-1}s + a_n + a_n G_1(s)]}$$

Now the partial fraction expansion of $\bar{x}(s)$ will be of the form

$$\bar{x}(s) = \frac{A_1}{s} + \sum_{i=2}^{n} \frac{A_i}{s - p_i}$$

Because the poles p_i of $T(s)$ in the summation all have negative real parts, these terms will die out exponentially, and in the steady-state, $x(t) \to A_1$. For

$\bar{x}(s)$ as given above, A_1 is evaluated from (8.35) as $A_1 = s\bar{x}(s)|_{s=0}$, or

$$A_1 = \frac{a_n G_1(0)}{a_n + a_n G_1(0)} = \frac{G_1(0)}{1 + G_1(0)}$$

Now let us evaluate A_1, and the resulting steady-state error $e_{ss} = 1 - A_1$, for the following control strategies from the PID family:

1. Proportional control, $G_1(s) = k_p$: here $A_1 = k_p/(1 + k_p)$ and $e_{ss} = 1/(1 + k_p)$. The steady-state error is made small by utilizing a large control gain.
2. Derivative control, $G_1(s) = k_d s$: here $G_1(0) = 0$, so that the steady-state response $A_1 = 0$, and the steady-state error is unity. This is one reason why derivative-only control is generally not a good idea.
3. Integral control, $G_1(s) = k_I/s$: here $A_1 = (k_I/s)/(1 + k_I/s)|_{s=0}$. Multiplying numerator and denominator by s results in $A_1 = k_I/(s + k_I)|_{s=0}$, so that $A_1 = 1$ and the steady-state error is zero. Generally integral control is an effective way to eliminate steady-state error.
4. Combinations: the reader is invited to show that for PD control, the steady-state error is the same as for proportional control; for PI control, the steady-state error is the same as for integral control (zero); and for PID control, the steady-state error is the same as for integral control (zero). Thus, for combinations from the PID family, the steady-state error is governed by the type of control which would produce the smallest error if employed alone.

Comment. The preceding results assumed the plant to be of order n and to be characterized by open-loop poles s_i which were all nonzero. In cases for which the plant has a pole at zero, so that the denominator of $G_2(s)$ is given by $s^n + \cdots + a_{n-1}s$, the steady-state response to a step input will be given by inversion of

$$\bar{x}(s) = \frac{G_1(s)}{s[s^n + \cdots + a_{n-1}s + G_1(s)]}$$

Thus, for derivative control the steady-state response $A_1 = k_d/(1 + k_d)$ so that $e_{ss} = 1/(1 + k_d)$.* For both proportional and integral control, the steady-state response will be unity, so that there is zero steady-state error for proportional and integral control. The basic result is that, if the plant has a pole at $s = 0$, then proportional or integral control will result in zero steady-state error. (For example, see Example 8.30 in Section 8.5.3.)

*This result assumes that the transfer function $G_2(s)$ has the form $G_2(s) = \alpha/(s^n + \cdots + \alpha s)$.

8.5 THE ROOT LOCUS

The fact that some higher-order systems behave in essentially the same way as systems of second order is the basis of a control systems design method known as the root locus. The *root locus* is a plot, in the complex plane, showing the movement of the poles of the closed-loop transfer function $T(s)$ as a parameter is varied. In control systems analysis this parameter is usually a control gain, and the root locus plot allows us to see clearly how the system poles are placed for various values of this control gain. The idea is to place the dominant poles where we want them by proper selection of the control gain.

In this section we first introduce the root locus plot using the example of a simple second-order system with proportional control. Then we discuss the construction of root locus plots for higher-order systems. Finally, we show how the root locus method is used for control system design.

8.5.1 Some Examples of the Root Locus

Example 8.27 Root Locus of a Second-Order System. Let us consider the following idealized second-order system ($B8.25$) with proportional control: The plant transfer function shows that the uncontrolled system characteristic polynomial is $s^2 + 2s$. Hence we may view the plant as a mechanical system possessing mass and damping but not stiffness. The jet airplane roll control problem (Example 8.2) is of this type; the plant is stable but not asymptotically stable. The objective of this example is to gain an understanding of control system behavior for the complete range of possible values of the proportional control gain k, $0 \leq k < \infty$.

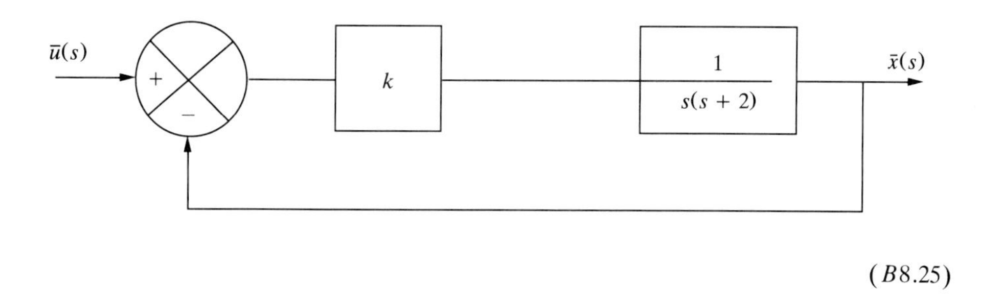

($B8.25$)

For the system ($B8.25$) the closed-loop transfer function $T(s)$ is (verify!)

$$T(s) = \frac{k}{s^2 + 2s + k} \tag{8.86}$$

The poles of this transfer function will be functions of the control constant k; each value of k defines a pair of points (the poles) in the complex plane. If we determine the poles of $T(s)$ for each value of k in the range $0 \leq k < \infty$, we generate a locus of points defining the changes in the poles of $T(s)$ as we change the control constant k. This plot is the root locus.

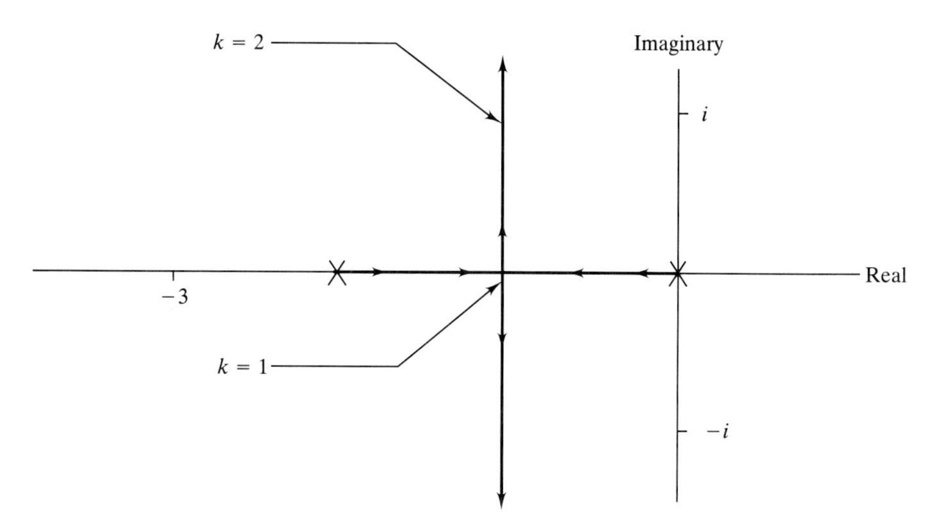

FIGURE 8.23
Root locus for the system ($B8.25$).

From (8.86) the system characteristic equation is $s^2 + 2s + k = 0$. The poles $p_{1,2}$ are then obtained as functions of the gain k as

$$p_{1,2} = -1 \pm (1 - k)^{1/2} \tag{8.87}$$

In order to construct the root locus, we first mark the locations of the poles for the case of zero gain, $k = 0$, so that $p_{1,2} = 0, -2$. These are just the poles of the plant transfer function and are depicted as x's in Fig. 8.23.

Now consider how the pole locations change as the gain k is increased from zero. The pole starting at zero becomes slightly negative, while the pole starting at -2 becomes less negative. For instance, if $k = 0.1$, then $p_{1,2} = -0.051, -1.949$. As k increases toward unity, the poles migrate toward each other as shown, coalescing at -1 on the real axis when $k = 1$. For $k > 1$ the poles are complex, and as k is increased well beyond 1, the poles move along the vertical line indicated.

Standard practice is to identify values of k at certain locations on the root locus plot, as is done in Fig. 8.23. Thus, if we wish to design a proportional controller with a damping factor $\xi_c = 0.7$, then $p_{1,2} = -1 \pm i$, and $k = 2$. Of course, in this simple example equation (8.87) allows us to obtain all the information we need, and a root locus plot is not necessary. The example is intended only to show how the locus is constructed. For systems of higher order, however, formulas analogous to (8.87) are difficult to find, and the root locus is of substantial utility in understanding the effect of the controller on system behavior.

Example 8.28 A Third-Order System. As a second example, we consider a third-order system with proportional control, as defined in ($B8.26$):

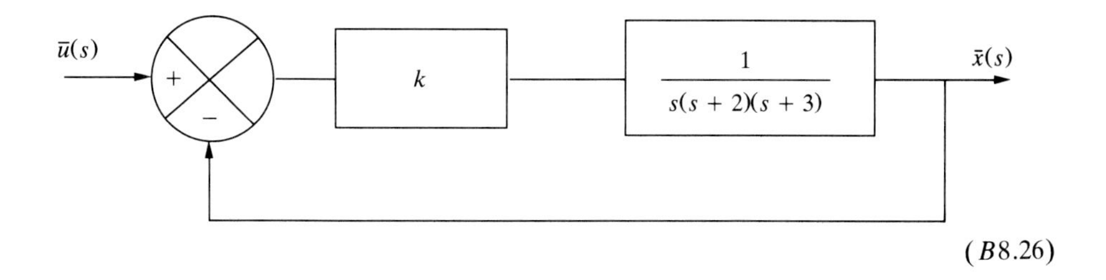

$(B8.26)$

Here the poles of the plant transfer function are 0, -2, and -3. The closed-loop system transfer function $T(s)$ is

$$T(s) = \frac{k}{s(s+2)(s+3)+k}$$

from which the characteristic equation is obtained as

$$s^3 + 5s^2 + 6s + k = 0$$

Note that the plant, if subjected directly to an input, will not exhibit the second-order tendency because the leftmost pole (at -3) is not sufficiently far to the left of the other two.

The root locus for the system ($B8.26$) could be constructed numerically as follows: first denote by x's the poles of the plant, as in Fig. 8.24. Then obtain computer solutions of the characteristic equation, using an algebraic equation solver, for numerous values of the gain k as k is increased from zero. This information allows the plot of Fig. 8.24 to be generated.

Discussion of Fig. 8.24: As k is increased from zero, the poles originating at 0 and -2 move toward each other along the real axis, coalescing at approximately -0.78, for which $k \cong 2.1$. For $k > 2.1$ the poles "break away" from the real axis and become complex. The behavior of these two poles is similar to that of the second-order system of Fig. 8.23, with one important difference. Whereas the complex poles in the second-order system of Fig. 8.23 migrate vertically as k is increased, the poles of the third-order system ($B8.26$) move also toward the imaginary axis, crossing it at $\pm i\sqrt{6}$, for which $k = 30$. For $k > 30$, the poles lie in the right half plane, indicating that the system is unstable for $k > 30$. This type of result commonly obtains for some higher-order systems: Too much control gain may destabilize the system.

As k is increased from zero, the pole originating at -3 migrates along the real axis toward more negative values. (Figure 8.24 is discussed further in Example 8.29.)

Having shown the basic idea of the root locus in the two preceding examples, we next explore the graphical construction of root loci.

8.5.2 Some Rules for Root Locus Construction

In this section we describe some (but not all) of the rules by which root locus plots are constructed. The objective will be to develop an understanding of how

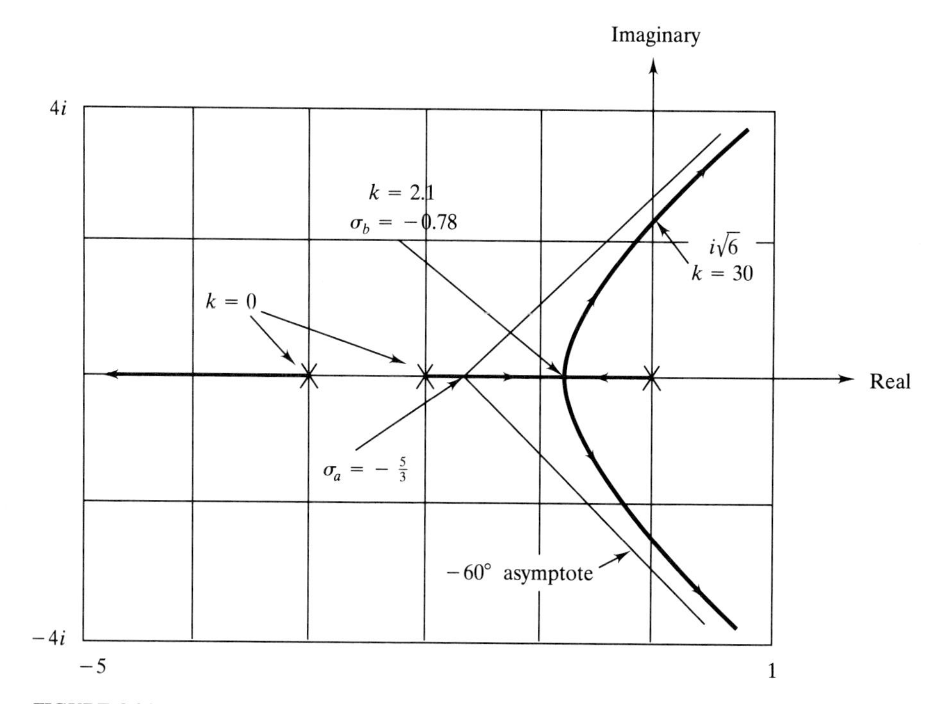

FIGURE 8.24
Root locus for the third-order system ($B8.26$). The real and imaginary scales differ so that the asymptotes are not at 60° angles.

the root locus plot behaves in various circumstances and to develop the capability to determine quickly the qualitative nature of the root locus plot for a given system. Generally, packaged routines that find the solutions to the system characteristic equation and thereby generate the root locus would be employed for detailed analysis. But it is essential to understand why root loci behave as they do, and this can be done approximately without actually solving the characteristic equation.

The general problem considered is the control system ($B8.27$) with unity feedback: The system ($B8.27$) does not accommodate all situations of interest, as the control block contains only a proportional control. The system ($B8.27$) is,

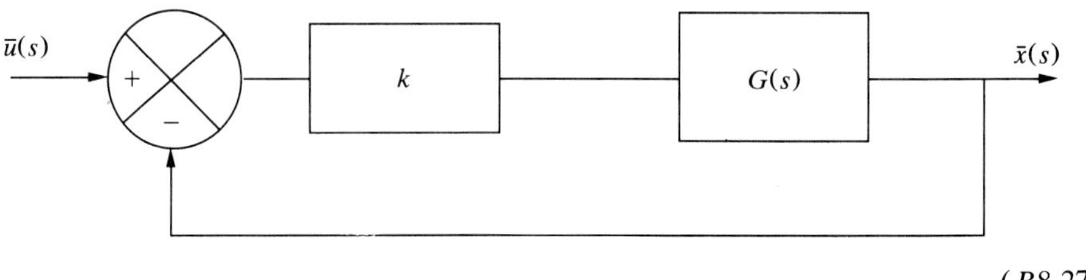

($B8.27$)

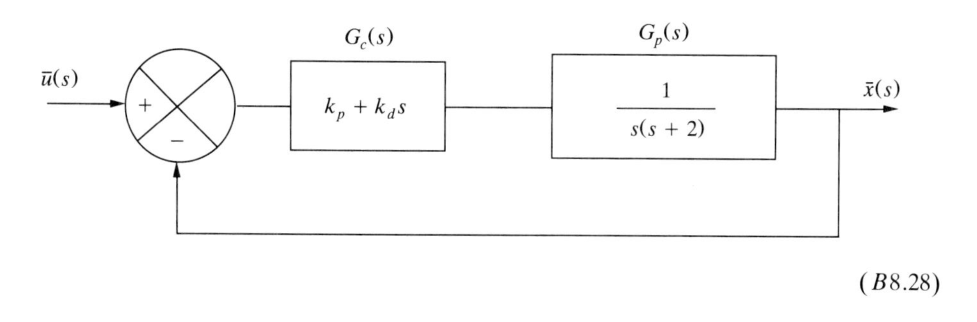

$(B8.28)$

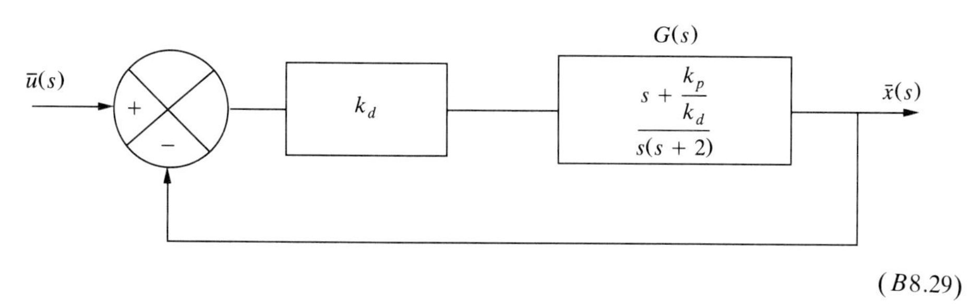

$(B8.29)$

however, more general than it may appear. Specifically, consider the block diagram $(B8.28)$, in which the system of Example 8.27 is assumed to be controlled using PD control and in which the control and plant transfer functions are denoted by $G_c(s)$ and $G_p(s)$, respectively. We can use the representation $(B8.27)$ for more general situations provided that we can write $G_c(s)G_p(s) = kG(s)$, that is, a gain k can be factored out of the product $G_c(s)G_p(s)$. In $(B8.28)$ the control block can be rewritten as $k_d(s + k_p/k_d)$, so that $(B8.28)$ can be converted to $(B8.29)$, which is in the "standard" root locus form $(B8.27)$ but with the distinction between the plant and the controller lost.

Notice that in $(B8.29)$ the transfer function $G(s)$ contains, in addition to the poles $p_{1,2} = 0, -2$, a *zero*, $z_1 = -k_p/k_d$, which is determined by the ratio of the control constants. As we shall see, the zeroes of the transfer function $G(s)$ in $(B8.27)$ may have a significant effect on the root locus and hence on the system response.

In order to see how the root locus is constructed, we must analyze the characteristic equation of the system $(B8.27)$. The closed-loop transfer function $T(s)$ is

$$T(s) = \frac{kG(s)}{1 + kG(s)}$$

so that the characteristic equation is

$$1 + kG(s) = 0 \tag{8.88}$$

This relation can also be written in the form

$$k = \frac{-1}{G(s)} \tag{8.89}$$

Furthermore, we will assume that the transfer function $G(s)$ in $(B8.27)$ comprises a polynomial numerator of degree m and a polynomial denominator of degree $n > m$. If these polynomials are expressed in factored form in terms of the poles s_i and zeroes z_i of $G(s)$, then

$$G(s) = \frac{(s - z_1) \cdots (s - z_m)}{(s - s_1)(s - s_2) \cdots (s - s_n)} \tag{8.90}$$

Here z_i $(i = 1, \ldots, m)$ are the zeroes of the transfer function $G(s)$, and s_i are the poles of $G(s)$, not to be confused with the poles p_i of the closed-loop transfer function $T(s)$. We assume that the poles s_i and the zeroes z_i are known.

The notation of (8.90) allows the characteristic equation (8.89) to be expressed as

$$(s - s_1) \cdots (s - s_n) + k(s - z_1) \cdots (s - z_m) = 0 \tag{8.91}$$

Equations (8.88), (8.89), and (8.91) represent three ways to write the system characteristic equation; each will prove useful in what follows. A point s in the complex plane will lie on the root locus provided that it satisfies the characteristic equation. From (8.89) we observe that in order for a point s in the complex plane to lie on the root locus, $G(s)$ must be real and negative, as the gain k is assumed to be real and positive. Thus, suppose we pick some complex number $s = \sigma + i\omega$ and put it into (8.89). If the result is real and negative, then the point $s = \sigma + i\omega$ is on the root locus, and (8.89) defines the associated value of the gain k.

With the preceding discussion as background, we can now proceed to deveiop some of the rules by which root loci are constructed. The objective will be to learn enough to determine the general character of root loci, mainly for systems of order 3 and higher. Keep in mind that a canned computer package would generally be used to obtain numerically accurate results for analysis and design.

SYMMETRY OF THE ROOT LOCUS. The first thing to notice is that the root locus plot must be *symmetric with respect to the real axis*, because complex poles of any closed-loop transfer function will always occur in conjugate pairs.

ORIGIN ($k = 0$) AND TERMINATION ($k \to \infty$) OF LOCI. A second feature of the root locus can be determined by inspection of the form (8.91) of the characteristic equation. As the gain $k \to 0$, (8.91) approaches

$$(s - s_1)(s - s_2) \cdots (s - s_n) \to 0 \qquad (k \to 0)$$

As $k \to 0$, the solutions for the poles p_i will be $p_i \to s_i$. This shows that as $k \to 0$, the loci will approach the poles s_i of the transfer function $G(s)$ of ($B8.27$). Thus, if we visualize the migration of the poles of $T(s)$ in the complex plane as k is increased from zero, we will see a locus originating ($k = 0$) at each pole of $G(s)$. If there are n poles s_i of $G(s)$, and hence n poles of $T(s)$, there will be n separate loci, one emanating from each pole of $G(s)$.

At the other extreme, as $k \to \infty$, the characteristic equation (8.91) will approach

$$k(s - z_1) \cdots (s - z_m) \to 0 \qquad (k \to \infty)$$

As $k \to \infty$, m of the solutions for the poles p_i will be $p_i \to z_i$. Thus, as $k \to \infty$, *some* of the loci (m of them) will asymptotically approach m of the zeroes z_i of the transfer function $G(s)$. If there are n poles s_i and m zeroes z_i of $G(s)$, m of the n loci will move from a pole ($k = 0$) to a zero ($k \to \infty$). The remaining $n - m$ loci will "move to infinity" (that is, infinitely far from the origin of the complex plane) along directions we next determine.

The movement, as $k \to \infty$, of the $n - m$ loci which do not terminate at a zero of $G(s)$ can be deduced by using (8.91) to express the characteristic equation in the form

$$\left(s^n + a_1 s^{n-1} + \cdots + a_n\right) + k\left(s^m + b_1 s^{m-1} + \cdots + b_m\right) = 0 \quad (8.92)$$

where the coefficients a_i and b_i are determined by the poles s_i and zeroes z_i of $G(s)$. Solutions p_i to (8.92) for which $|p_i| \to \infty$ as $k \to \infty$ may be asymptotically found as follows: If $|p_i| \to \infty$ and $k \to \infty$, then each of the two polynomials in (8.92) will be dominated by the highest power of s appearing in each, so that asymptotically,

$$s^n + ks^m \sim 0 \qquad (\text{as } |s|, k \to \infty)$$

This relation may be rewritten as

$$s^{n-m} \sim -k \qquad (|s|, k \to \infty)$$

Defining the integer $\alpha \equiv n - m$ to be the excess of poles over zeroes of $G(s)$, then

$$s^\alpha \sim -k \qquad (|s|, k \to \infty) \qquad (8.93)$$

Now s is generally complex and may be expressed in the polar from $s = Re^{i\theta}$, where θ is measured counterclockwise from the positive real axis, with the understanding that the magnitude $R \to \infty$. Using complex notation for s, (8.93) becomes $(Re^{i\theta})^\alpha \sim -k$, or

$$e^{i\alpha\theta} \sim \frac{-k}{R^\alpha} \qquad (R, k \to \infty) \qquad (8.94)$$

The right-hand side of (8.94) is a negative real number, as both R and k are real. This indicates that $e^{i\alpha\theta}$ must lie along the negative real axis, which in turn means that the product $\alpha\theta$ must equal $180°, 540°, 900°, \ldots$. The result enables

TABLE 8.3
Angles of asymptotes of loci which "approach infinity"

α	Angle θ of asymptote
1	$180°$
2	$\pm 90°$
3	$\pm 60°, 180°$
4	$\pm 45°, \pm 135°$
5	$\pm 36°, \pm 108°, 180°$
$\vdots$	$\vdots$

us to determine the angles θ of the asymptotes along which the $\alpha = n - m$ loci will move as they approach infinity. These angles are summarized in Table 8.3.

Observe that α, when multiplied into any of the listed asymptotes θ, will produce a ray lying along the negative real axis, as required by the condition (8.94).

The preceding results are exemplified in Examples 8.27 and 8.28. In Example 8.27 $G(s)$ has two poles and no zeroes, so $\alpha = 2$ and the loci approach ∞ along $\pm 90°$ asymptotes (Fig. 8.23). In Example 8.28 $G(s)$ has three poles and no zeroes, so the loci approach ∞ along $\pm 60°$ asymptotes and along the negative real axis ($\theta = 180°$) (Fig. 8.24). In these two examples neither transfer function $G(s)$ has a zero, so we did not encounter any loci that originated at a pole of $G(s)$ and terminated at a zero of $G(s)$. Examples of such cases appear later on in this chapter.

As a final comment note that, for any situation in which the excess of poles over zeroes $\alpha \geq 3$, as k becomes sufficiently large the system will become unstable as a pair of conjugate loci cross the imaginary axis in the complex plane (for example, Fig. 8.24, $k \geq 30$).

INTERSECTION OF ASYMPTOTES WITH THE REAL AXIS. The $\alpha = n - m$ asymptotes discussed above will originate (intersect) at some point on the real axis. Denoting this point as σ_a, we state without proof the following formula which enables σ_a to be determined from the poles s_i and zeroes z_i of the transfer function $G(s)$:

$$\sigma_a = \frac{\left(\sum_{i=1}^{n} s_i - \sum_{i=1}^{m} z_i\right)}{\alpha} \qquad (\alpha \geq 1) \tag{8.95}$$

As an example, consider Example 8.28, the third-order system for which the poles of $G(s)$ are 0, -2, and -3. There are no zeroes in $G(s)$, so $\alpha = 3$. Use of (8.95) then yields $\sigma_a = -\frac{5}{3}$, and this location is noted in Fig. 8.24.

THE REAL AXIS. In constructing root loci, it is important to determine the portions of the real axis on which one or more loci will lie. To see how this is

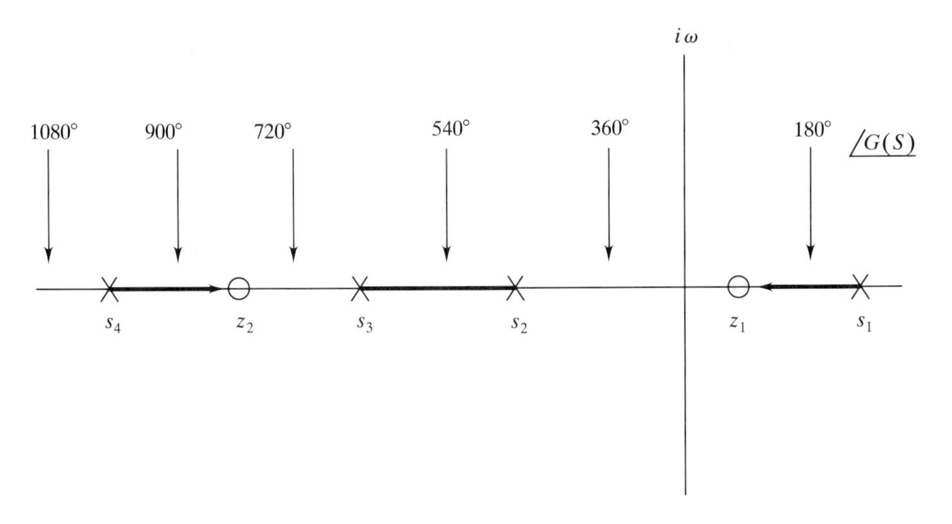

FIGURE 8.25
Portions of the real axis occupied by a hypothetical system for which $G(s)$ has four poles and two zeroes, all on the real axis (scale arbitrary). The angles noted are $\angle G(s)$ on the real axis.

done, consider the hypothetical situation shown in Fig. 8.25, which depicts a fourth-order plant $G(s)$ having four poles (denoted by x's) and two zeroes (denoted by small circles), all on the real axis.

We know that loci will emanate from each of the four poles, and the problem is to determine which two of these four loci will migrate to the two zeroes as $k \to \infty$. The answer can be seen as follows: For a given point s in the complex plane (whether or not the point s is part of the root locus), denote as $\angle G(s)$ the angle, measured counterclockwise from the positive real axis, of the complex number $G(s)$. Then (8.89) indicates that for the point s to be on the root locus, $\angle G(s)$ must equal $180°, 540°, 900°, \ldots$. From the definition (8.90) of $G(s)$, we may express $\angle G(s)$ in terms of the angles of the individual factors in (8.90) as follows:

$$\angle G(s) = \angle s - z_1 + \cdots + \angle s - z_m - \angle s - s_1 - \angle s - s_2 - \cdots - \angle s - s_n \tag{8.96}$$

This follows because each factor may be expressed in polar form, so that the angles of the factors add (zeroes) or subtract (poles) to determine the angle of $G(s)$.

Now if we compute any of the angles appearing in (8.96) for a point s on the real axis, the result will be $0°$ if the point s is to the right of the pole or zero being considered and will be $180°$ if s is to the left of the pole or zero being considered. Shown in Fig. 8.25 are the results for $\angle G(s)$ obtained from (8.96) for various segments of the real axis. The darkened portions, for which $\angle G(s) = 180°, 540°, 900°, \ldots$, will, therefore, lie on the root locus. We conclude that for the system of Fig. 8.25, as k is increased from zero, the locus originating at s_1 will move to the zero at z_1, and the locus originating at s_4 will move to z_2. The loci originating at s_2 and s_3 will move toward each, will coalesce at some point between s_2 and s_3, and will then become complex, eventually asymptoting to $\pm 90°$ lines (as $\alpha = 2$). The general result can be summed up as follows: *The root locus includes all points on the real axis to the left of an odd number of poles and zeroes of $G(s)$.*

SUMMARY OF RULES DEVELOPED SO FAR. Listed below is a summary of the rules developed up to this point for construction of the root locus for a control system represented in the form ($B8.27$):

1. The root locus is symmetric with respect to the real axis.
2. There will be n loci, one originating ($k = 0$) from each of the n poles s_i of $G(s)$.
3. m of the n loci will terminate ($k \to \infty$) at the m zeroes z_i of the transfer function $G(s)$.
4. The remaining $\alpha = n - m$ loci will approach infinity along asymptotes defined according to Table 8.3.
5. The asymptotes emanate from the point σ_a on the real axis defined by (8.95).
6. Portions of the real axis to the left of an odd number of poles and zeroes of $G(s)$ will lie on the root locus.

While these rules allow us to determine much of the structure of the root locus, several aspects have not been addressed. For instance, we would like to be able to determine the location of any "breakaway points," that is, points on the real axis at which two loci coalesce for some value k_b of the gain k, and for which the loci are complex for $k > k_b$. We would also like to be able to determine the locations of any locus crossings of the imaginary axis, as such crossings signify the onset of instability. For systems of higher order the analytical techniques used to obtain this additional information can be fairly involved, and we address these points in ad hoc fashion in the following examples (see the references for more detail).

Example 8.29 The System of Example 8.28 Revisited (Fig. 8.24). Based on the six preceding rules for root locus construction, much of Fig. 8.24 can be understood. A locus emanates from each of the three poles $(0, -2, -3)$ of the transfer function $G(s)$. Because $G(s)$ has three poles and no zeroes, $\alpha = 3$, and the loci terminate at infinity along the asymptotes $\pm 60°$ and $180°$. The three asymptotes emanate at the point $\sigma_a = -\frac{5}{3}$ on the real axis, as given by (8.95). The portion of the real axis between the poles at 0 and -2 and to the left of the pole at -3 will be on the root locus. We also know that the loci originating at 0 and -2 will move toward each other, coalesce at some intermediate point on the real axis, and then become complex, eventually approaching the $\pm 60°$ asymptotes as $k \to \infty$. In order to sketch the root locus reasonably accurately, let us now determine the breakaway point σ_b between 0 and -2 and the location ω_{cr} at which the complex loci cross the imaginary axis.

To determine the breakaway point σ_b and the associated value of k_b of the gain, we first note that the system characteristic equation (8.91) is

$$s^3 + 5s^2 + 6s + k = 0 \tag{8.97}$$

At the breakaway point σ_b two poles have coalesced into a pole of order 2 [i.e., there is a repeated root of (8.97)]. Thus, at the breakaway point the characteristic equation has the form

$$(s - \sigma_b)^2(s - a) = 0$$

where a is the pole to the left of -3. Expanding this relation, we obtain the form

$$s^3 + (-a - 2\sigma_b)s^2 + (\sigma_b^2 + 2a\sigma_b)s - a\sigma_b^2 = 0 \tag{8.98}$$

Equating coefficients of the various powers of s in (8.98) and (8.97), we obtain the following three equations in the three unknowns σ_b, a, and k:

$$-a - 2\sigma_b = 5$$

$$\sigma_b(\sigma_b + 2a) = 6$$

$$-a\sigma_b^2 = k$$

Elimination of a from the first two equations yields a quadratic equation for σ_b:

$$\sigma_b^2 + \frac{10}{3}\sigma_b + 2 = 0$$

The solutions are $\sigma_b = -0.785, -2.55$, but only one of these (-0.785) is on the root locus. Selecting this value $\sigma_b = -0.785$ as the desired one, we then obtain $a = -3.43$ and $k_b = 2.11$. These results establish the breakaway point, as noted in Fig. 8.24.

A similar procedure enables us to determine the points $\pm i\omega_{cr}$ at which the complex loci cross the imaginary axis. For this condition the characteristic equation may be expressed as

$$(s - i\omega_{cr})(s + i\omega_{cr})(s - a) = 0$$

or

$$(s^2 + \omega_{cr}^2)(s - a) = 0 \tag{8.99}$$

Expanding and equating coefficients of like powers of s in (8.99) and (8.97) yields

$$-a = 5$$

$$\omega_{cr}^2 = 6$$

$$-a\omega_{cr}^2 = k_{cr} = 30$$

Thus, we obtain directly the result $\omega_{cr} = \sqrt{6}$, $a = -5$, $k_{cr} = 30$, as indicated in Fig. 8.24.

All of the preceding information was used in construction of the root locus of Fig. 8.24. The technique used to determine σ_b and ω_{cr} is easily applicable to third-order systems, but not for systems of order higher than 3.

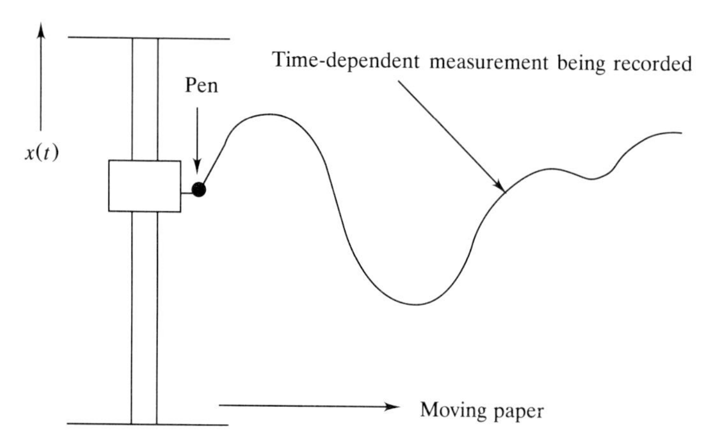

FIGURE 8.26
Schematic of a plotting device. The pen carriage is driven along the support rod by a dc motor (not shown) while the paper advances at a fixed rate.

8.5.3 Controller Design Using the Root Locus

Example 8.30 An Example in Detail Using the Root Locus: A Plotting Device.* Here we analyze a plotting device (Fig. 8.26) designed to record time-dependent measurements such as temperature in a solid, pressure in a gas, position, velocity or acceleration of a moving or vibrating object, angular velocity of a rotating body, and so on. Calling the quantity to be recorded $f(t)$, we assume that $f(t)$ is measured by an instrument such as a thermocouple, pressure transducer, accelerometer, or rate gyro. The output from the instrument is a time-dependent voltage $v(t)$, which when multiplied by a calibration factor β, yields the physical quantity being measured, $f(t) = \beta v(t)$. The objective is to design a plotter for which the displacement $x(t)$ of the pen is always a fixed multiple of the quantity $f(t)$ being measured. This is a good example of the tracking problem: The pen displacement $x(t)$ should faithfully follow the input $f(t)$ to be recorded, no matter how $f(t)$ may fluctuate with time.

In practice we may use a dc motor as the actuator to drive the pen carriage along the support rod. From Example 3.8, Sec. 3.4, we have seen that the dc motor behaves as a third-order system; two orders due to the combined inertia of the motor shaft and pen carriage, and one order due to the electrodynamics of the motor circuitry. Taking a candidate motor with representative values of the parameters, we can represent the block diagram of the third-order plotter system as in ($B8.30$).

*See also Ref. 8.2.

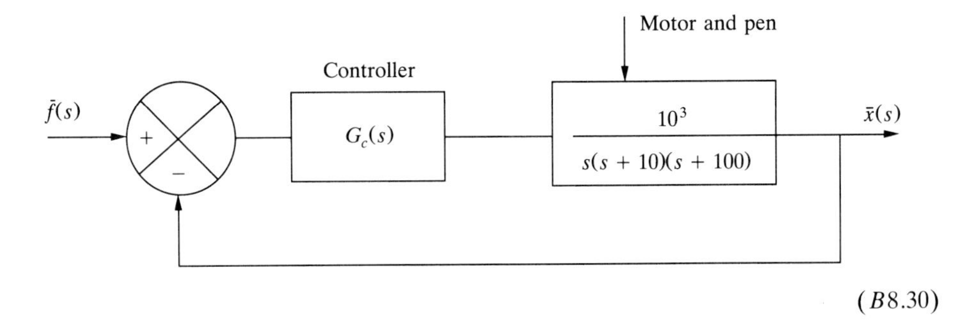

$(B8.30)$

The third-order model of the motor/pen carriage shows the poles of the motor/pen system to be located at $s_1 = 0$, $s_2 = -10$, $s_3 = -100$. The objective is to specify the control action $G_c(s)$ which will ensure that $x(t)$ is always sufficiently close to the actual value $f(t)$ of the quantity to be recorded. In practice the control is implemented using a microprocessor, which receives both the desired position $f(t)$ of the pen and the actual position $x(t)$ of the pen (as measured, for example, using an optical encoder), forms the error, and then, according to the control strategy $G_c(s)$, determines the armature voltage supplied to the dc motor.

Next we state the design goals to be implemented. We will design the system so that, if subjected to a step input, the system responds very quickly with little overshoot and with zero steady-state error. Let us select the following goals for the transient response to the step input: The overshoot should be less than 5 percent, and the settling time should be less than 0.2 s. These goals represent fast, accurate response to the step input, so that we would also expect fast, accurate response for most arbitrary functions of time to be recorded.

We will consider two design strategies to define $G_c(s)$ in $(B8.30)$: (1) proportional control and (2) proportional plus derivative control. In each case the root locus method will be applied, and we will try to select the controller $G_c(s)$ which will locate the closed-loop poles in the complex plane so that the design goals are met.

Recall (Fig. 5.10) that for a *second-order* system subjected to a step input, the 5 percent overshoot condition is satisfied if the poles are complex with a damping factor $\xi_c = 0.707$. Also, using the rule of thumb that the settling time equals four system time constants, the largest system time constant should be ≤ 0.05 s. This in turn implies a pole location at or to the left of -20 in the complex plane. Thus, if the control system possess a pair of complex dominant poles, we should try to locate these poles at $p_{1,2} = -20 \pm 20i$ in order to barely satisfy the transient response design goals. This is illustrated in Fig. 8.27. We have to keep in mind that the preceding design discussion is based on the assumption that the system will behave like a second-order system, and this may or may not actually occur.

We first consider the simpler case of proportional control, $G_c(s) = k$ in $(B8.30)$. The root locus plot for this case is shown in Fig. 8.28.

The plant transfer function has three poles and no zeroes ($\alpha = 3$, Table 8.3). The three loci comprising the root locus originate ($k = 0$) at 0, -10, and -100 and terminate ($k \to \infty$) along ± 60 and 180° asymptotes. The loci originating at 0 and -10 move toward each other and coalesce at the breakaway point $\sigma_b = -4.85$

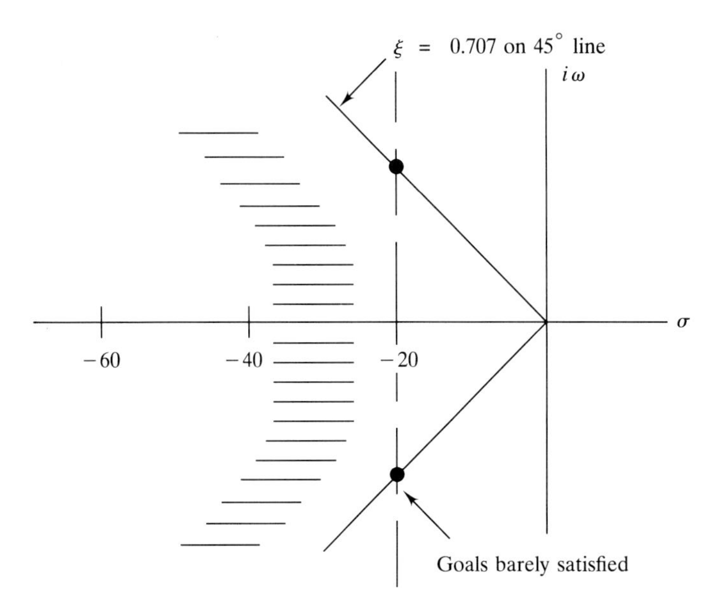

FIGURE 8.27
Assuming a pair of dominant poles to exist, they should be located to the left of −20 in order to satisfy the settling time requirement in Example 8.30. Also, the poles should be on or below the 45° line in order to satisfy the overshoot requirement. The shaded region results.

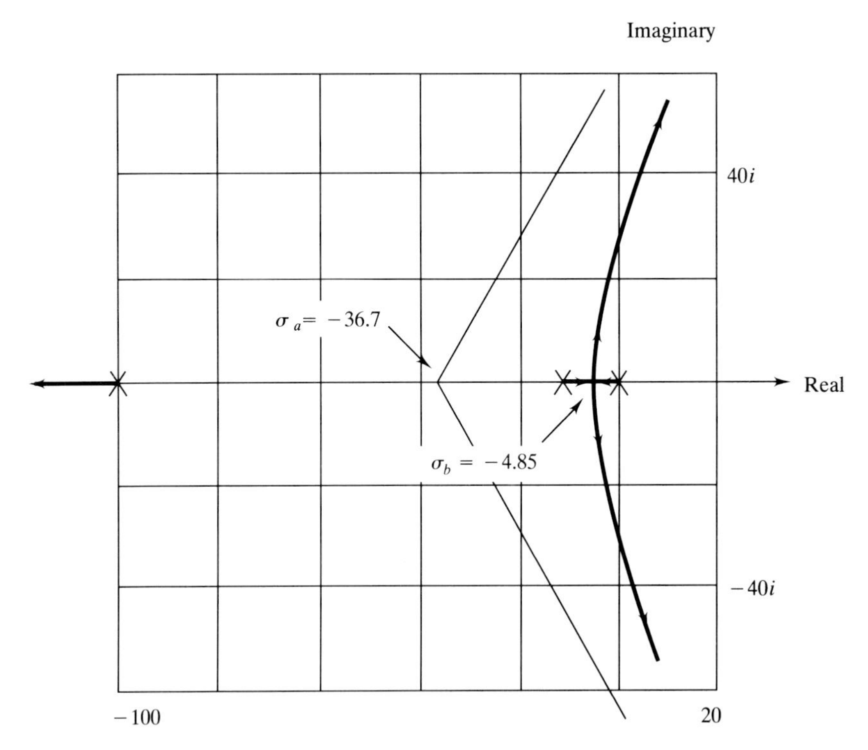

FIGURE 8.28
Root locus plot for (B8.30) with proportional control, $G_c(s) = k$.

($k = 2.37$), calculated using the procedure of Example 8.29. The complex loci which break away from the real axis at σ_b eventually cross the imaginary axis at $\pm\sqrt{1000}\,i$. Meanwhile, the locus originating at -100 moves further to the left along the negative real axis as the gain k is increased. The root locus plot which results is shown in Fig. 8.28.

From Fig. 8.28 we identify the candidate design point, along the 45° line so that $\xi_c = 0.707$, to be located at $p_{\text{des}} \cong -4.85 \pm 4.85i$ ($k = 5$). The following conclusions can be drawn: (1) The second-order tendency principle is valid in this case because the leftmost pole is well to the left of the dominant pair; thus the overshoot and settling time calculated for a second-order system with poles at p_{des} should be close to the overshoot and settling time of the actual third-order system being analyzed. (2) The overshoot design goal will be met because $\xi_c = 0.707$. (3) The settling time requirement will *not* be met; in fact the settling time will be approximately 0.8 s, or a factor of 4 too large (the dominant poles in the complex plane are only about a quarter as far to the left as they should be). Inspection of Fig. 8.28 indicates that, at least for the particular motor/pen properties assumed in ($B8.30$), the system will respond too slowly no matter what the gain (it will also be unstable if the gain is too large!)

Having deduced that a proportional control will not work satisfactorily, we next try a PD controller, $G_c(s) = k_p + k_d s$ in ($B8.30$). The system block diagram is then as follows:

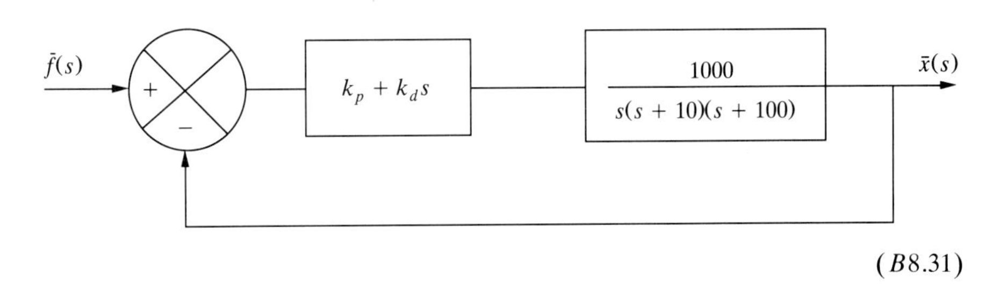

($B8.31$)

This is converted to the standard root locus form ($B8.27$) by writing $k_p + k_d s = k_d(s + k_p/k_d)$, so that we have ($B8.32$) below:

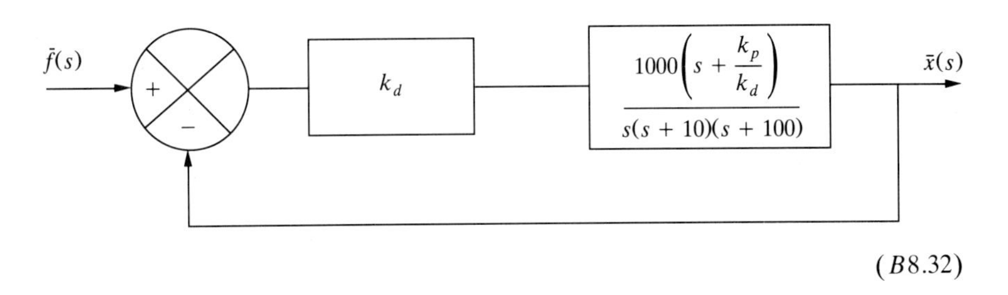

($B8.32$)

The result is now in the form ($B8.27$). Notice that the transfer function $G(s)$ of ($B8.32$) now contains a zero, at $z_1 = -k_p/k_d$, determined by the ratio of the two control gains. In establishing a PD controller design, we will proceed as follows: First, we will (somewhat arbitrarily) fix the location of the zero z_1 at

$z_1 = -20$; later on the effect of other zero locations will be considered. Second, we will generate the root locus using the derivative control gain k_d as the gain to be varied. As k_d is varied, the proportional control gain k_p will also vary, so as to maintain the ratio $z_1 = -k_p/k_d = -20$.

The root locus of the system ($B8.32$) with $z_1 = -20$ is shown in Fig. 8.29. In comparison to Fig. 8.28, proportional control, we see that the addition of the zero z_1 arising from the PD controller has significantly altered the root locus. $G(s)$ now has three poles (0, -10, -100) and one zero (-20), and two of the three loci will follow the $\pm 90°$ asymptotes originating at $\sigma_a = -45$ as $k \to \infty$. Furthermore, the locus originating at the leftmost pole (-100) now moves to the right and terminates at the zero $z_1 = -20$. The poles originating at 0 and -10 move toward each other and coalesce at the breakaway point $\sigma_b \cong -5.7$. These poles then leave the real axis as a pair of complex conjugate poles, curving toward the $\pm 90°$ asymptotes and eventually approaching them.

For design purposes we select the intersection of the complex loci with the 45° lines shown at $p_{\text{des}} \cong -41.3 \pm 41.3i$, for which $k \cong 4.67$. (Note that the complex loci actually cross the 45° line at two other locations, and the leftmost of the three has been selected.) For this design condition the real pole originating at -100 has moved to $p_3 \cong -27.4$. The system time constants are now $\tau_{1,2} =$

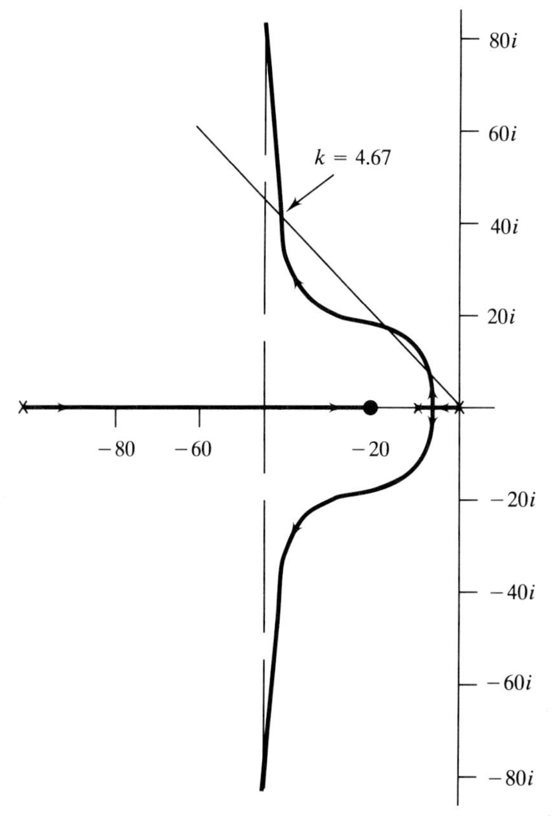

FIGURE 8.29
The root locus for the case of PD control of the system ($B8.32$).

$1/41.3 = 0.024$ s, associated with the complex poles, and $\tau_3\, 1/27.4 = 0.0365$, associated with the real pole. Each is less than the value $\tau = 0.05$ s required to achieve a settling time of 0.2 s. If we assume that the settling time will be governed by the largest time constant, then the "four time constant" rule of thumb yields $T_s \cong 4(0.0365) = 0.146$ s for the settling time, and this satisfies the design goal. It is obvious that the addition of the derivative control has had a significant effect on the transient behavior, and that PD control differs substantially from proportional control for this system. But there is still a problem, as the following comments illustrate.

Comments on Example 8.30

1. For the PD design condition selected ($k \cong 4.67$), the second-order tendency is not realized, as the complex poles are no longer dominant. The real pole is actually to the right of the complex pair, and all of the poles will have a significant influence on the transient response. The system behavior will be fundamentally of third order.
2. In view of comment 1, some care is needed in the interpretation of the results. The original design conditions were based on selection of complex poles with $\xi_c = 0.707$ in order to satisfy the 5 percent overshoot criterion. This criterion (i.e., $\xi_c = 0.707$ will give ≤ 5 percent overshoot) is based on the step response of a *second-order* system with *no zeroes* in the transfer function $G(s)$, that is, it applies for a closed-loop transfer function

$$T(s) = \frac{\omega^2}{s^2 + 2\xi\omega s + \omega^2}$$

   The actual system in Example 8.30, however, exhibits third-order behavior and has a zero in the transfer function. Thus, the analysis here can serve only as a guide in establishing the design. The step response and overshoot should be calculated in order to determine whether the design goals have actually been met.
3. In order to illustrate the point of comment 2, consider the actual step response of the PD system ($B8.32$) being studied. The closed-loop transfer function $T(s)$ is found to be

$$T(s) = \frac{1000k(s + 20)}{s^3 + 110s^2 + 1000(1 + k)s + 20{,}000k}$$

   so that the step response $x(t)$ is defined by inversion of

$$\bar{x}(s) = \frac{1000k(s + 20)}{s\left[s^3 + 110s^2 + 1000(1 + k)s + 20{,}000k\right]}$$

   where $k = 4.67$. For the design condition the poles are $p_{1,2} = -41.3 \pm 41.3i$, $p_3 = -27.4$. Using the partial fraction expansion or a suitable software package, the response $x(t)$ is determined by inversion of $\bar{x}(s)$ to be

$$x(t) = 1 + 0.68e^{-27.4t} - e^{-41.3t}[1.68\cos(41.3t) + 1.23\sin(41.3t)] \quad (8.100)$$

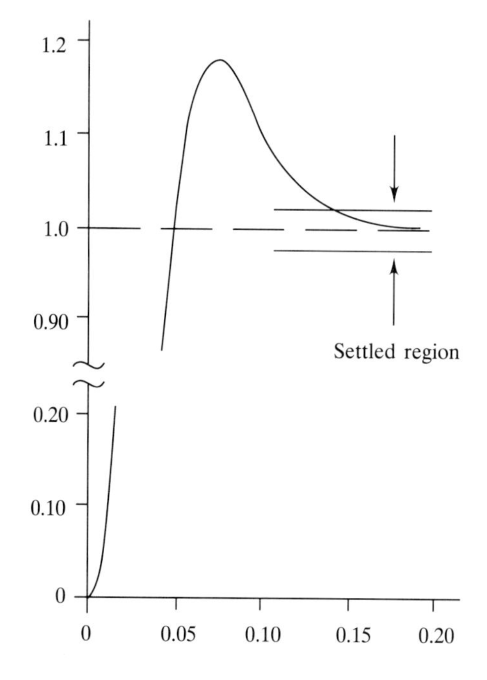

FIGURE 8.30
The step response of the third-order system ($B8.32$) for the design condition with PD control ($k = 4.67$, $p_{1,2} = -41.3 \pm 41.3i$, $p_3 = -27.4$).

A graph of this response appears in Fig. 8.30. (Note the break in the vertical scale.) Equation (8.100) indicates that the steady-state error is zero, as $x(t) \to 1$ as $t \to \infty$.

Figure 8.30 indicates that, while the settling time condition is satisfied pretty much as expected, the overshoot, which we hoped would be a little less than 5 percent because $\xi_c = 0.707$ for the complex poles, is *much* larger than this, invalidating the design. This large overshoot is due to (1) the presence of the zero in $T(s)$, which provides an impulsive input to the system as discussed in Sec. 8.4.5 and as illustrated in Figs. 8.17 and 8.18, and (2) the fact that the system behaves as a third-order rather than a second-order system, so that perceptions and/or predictions based on second-order behavior may not be too meaningful.

4. Based on the preceding comment, which shows the PD design arrived at to be unacceptable, we now propose a simple way to fix the problem. Rather than selecting the control gains k_p and k_d so as to locate the zero of $G(s)$ at $z_1 = -20$, let us select $k_p/k_d = 10$, so that $z_1 = -10$. The open-loop transfer function $G(s)$ in ($B8.32$) is then

$$k_d G(s) = \frac{1000 k_d \cancel{(s + 10)}}{s \cancel{(s + 10)} (s + 100)}$$

As indicated, the selection of the zero at -10 causes cancellation of the factors $(s + 10)$ in numerator and denominator, so that the transfer function reduces to

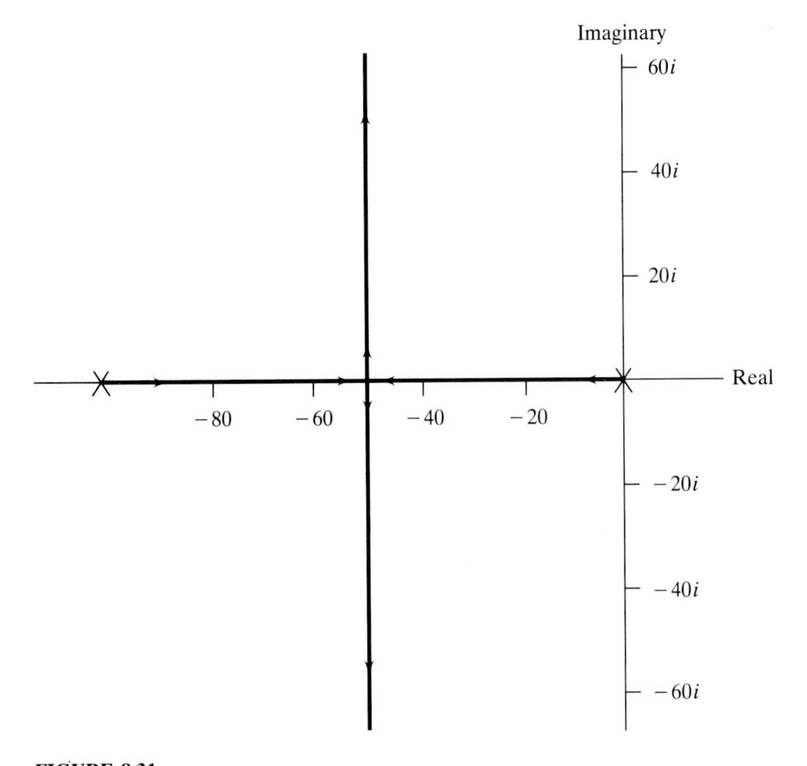

FIGURE 8.31
The root locus of the system ($B8.32$) with pole-zero cancellation ($z_1 = -10$).

one of second order with no zeroes:

$$k_d G(s) = \frac{1000 k_d}{s(s+100)} \tag{8.101}$$

This design strategy is referred to as *pole-zero cancellation*, and it causes a substantial change in the system behavior. One may verify that, with $G(s)$ given by (8.101), the closed-loop transfer function $T(s)$ for the system is

$$T(s) = \frac{1000 k_d}{s^2 + 100s + 1000 k_d} \tag{8.102}$$

Thus, pole-zero cancellation results in a system of second order, and it is straightforward to devise an acceptable design. The poles of the system are located at

$$p_{1,2} = -50 \pm (2500 - 1000 k_d)^{1/2} \tag{8.103}$$

and the root locus is as shown in Fig. 8.31. It has the same form as that of Fig. 8.23. Loci originating at the two poles 0 and -100 move toward each other and coalesce at -50, where breakaway occurs. The design condition, with $\xi_c = 0.707$, can be obtained directly from (8.70). Letting $k_d = 5$, we obtain $p_{1,2} = -50 \pm 50i$, and this system will perform acceptably in response to a step input. The step response may be shown to be

$$x(t) = 1 - e^{-50t}[\cos(50t) + \sin(50t)] \tag{8.104}$$

The overshoot is less that 5 percent, the settling time is about 0.08 s, and there is zero steady-state error.

5. In an actual PD controller design of the system (B8.32), an analyst might begin with the pole-zero cancellation discussed in comment 4. Our initial placement of the zero at $z_1 = -20$ was done mainly to illustrate some pitfalls that may occur if we treat a third-order system using second-order ideas.
6. There are two other aspects of pole-zero cancellation that deserve comment. The first of these is that, in an actual, real-world design, it is probably impossible to *exactly* cancel a pole with a zero. This is because of inaccuracies which nearly always exist in the system plant model, so that the poles s_i of $G(s)$ in the *physical* system will differ from those in the system mathematical model. In addition, the zero location supplied by the controller will actually differ from that which is intended. Generally we would have to assume that it would be impossible to achieve exact pole-zero cancellation. Fortunately, unless we do a poor job of modeling, it is usually possible to come close enough to a cancellation that the system response differs little from what would occur if we had achieved an exact cancellation. In Example 8.30 a *near*-pole-zero cancellation will result in transient response which is very close to what we get with an exact cancellation. This is illustrated in Example 8.31, which follows.

Example 8.31 "Near"-Pole-Zero Cancellation in Example 8.30. In this example we continue the design analysis of the plotter system of Example 8.30. We suppose that our intent is to achieve a pole-zero cancellation as in comment 4, Fig. 8.31, but that because of inaccuracy in the system and controller models, the location of the zero z_1 differs slightly from the pole s_2 location. We represent this inaccuracy by placing the zero at $z_1 = -11$ in order to see the effect on the response. In practice we should be able to do better than this. For this example, then, the open-loop and closed-loop transfer functions $G(s)$ and $T(s)$ will be

$$kG(s) = \frac{1000k(s+11)}{s(s+10)(s+100)} \tag{8.105}$$

$$T(s) = \frac{1000k(s+11)}{s(s+10)(s+100) + 1000k(s+11)} \tag{8.106}$$

and the characteristic equation defining the root locus is

$$s(s+10)(s+100) + 1000k(s+11) = 0 \tag{8.107}$$

Thus, the assumption is that the *actual* system behaves according to (8.105) through (8.107), but that our *model* of the system is (8.101) and (8.102), with exact pole-zero cancellation. If we implement a design based on the model (8.101) and (8.102), the effect of the hypothetical inaccuracies considered here will reveal the significance of the postulated inexact pole-zero cancellation. The root locus plot of the system (8.106) is shown in Fig. 8.32, and it exhibits an interesting character.

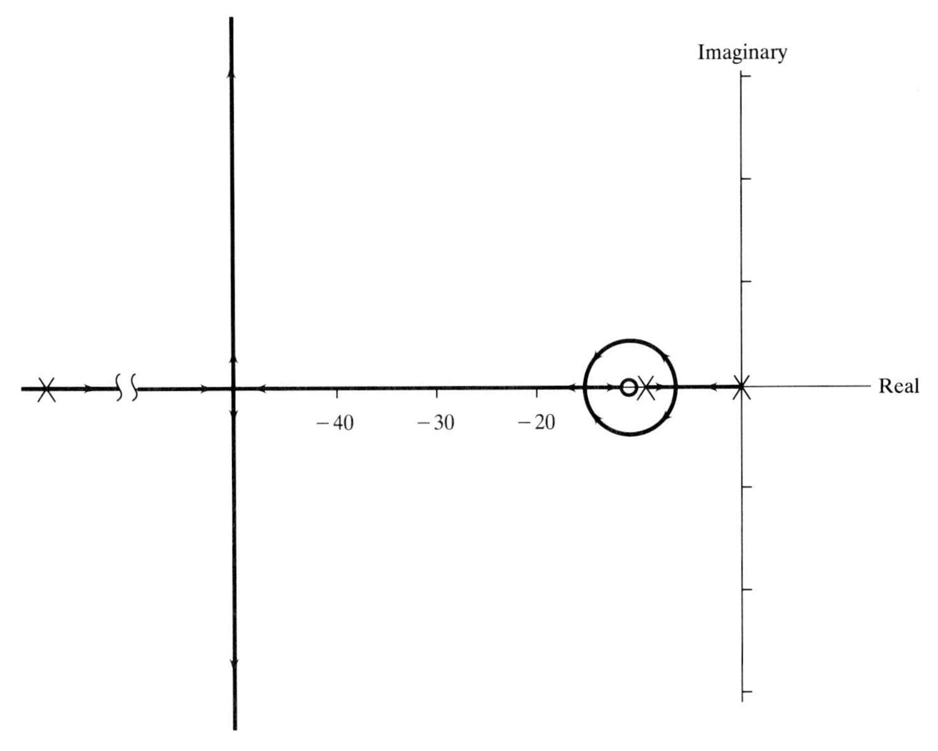

FIGURE 8.32
Root locus plot with inexact pole-zero cancellation, $z_1 = -11$, $s_2 = -10$, for Example 8.31.

The loci originating at $s_1 = 0$ and $s_2 = -10$ move toward each other, coalesce at the breakaway point $\sigma_b \cong -7.7$, then move away from the real axis as a complex pair as k_d is increased. Rather than moving toward the $\pm 90°$ asymptotes originating at $\sigma_a = -49.5$, however, these complex loci circle around the pole at -10 and the zero at -11 and come back down to the real axis at the *break-in* point $\sigma_b \cong -14.5$. From here one locus moves to the right along the real axis, terminating at the zero $z_1 = -11$ as $k_d \to \infty$. The other locus moves to the left toward the locus coming in from the pole $s_3 = -100$. These two loci coalesce at the breakaway point $\sigma_b \cong -49.2$, break away as a complex pair, and then move vertically along (and essentially on) the $\pm 90°$ asymptotes. There are now three locations on the real axis at which complex loci break away or break in.

For design purposes we select the value of control gain $k_d = 5.0$ determined previously for the model system (8.102). For the hypothetical actual system of (8.105) through (8.107), we then obtain the following pole locations: $p_{1,2} = -49.37 \pm 49.37i$, $p_3 = -11.25$. The complex pair is quite close to the design poles of the model system, $p_{\text{des}} = -50 \pm 50i$. There is, however, the third pole $p_3 = -11.25$, which will contribute to the transient response. To see the effect of the postulated inexact pole cancellation, let us compare the step responses of the model system, given by (8.104), and the hypothetical actual system. For the "actual" system a partial fraction expansion or use of a suitable software routine yields the step response as follows:

$$x(t) = 1 + 0.028e^{-11.25t} - e^{-49.37t}[1.028\cos(49.37t) + 1.028\sin(49.37t)] \tag{8.108}$$

This compares closely to the result (8.104) obtained with exact pole-zero cancellation. There is now a small exponential response component, $0.028e^{-11.25t}$, due to the inexact pole-zero cancellation, but the overall transient response differs by only a small amount, and the actual system will exhibit nearly the same response as the model system. This example is a second illustration of a way in which a third-order system may exhibit essentially second-order behavior: the near cancellation of a pole by a zero.

7. There is a second aspect of pole-zero cancellation of which we need to be aware. Suppose that the plant transfer function $G(s)$ in (B8.27) contains a real pole or a complex pair in the right half of the complex plane, so that the plant is unstable. Naively, we may try to stabilize the system by canceling this pole (or pair of poles) with a zero (or zeroes) which is designed into the controller transfer function $G_c(s)$ of (B8.27). Based on the preceding comment, however, we cannot hope to achieve exact pole-zero cancellation. In the actual system response there will always appear small contributions from the modes we are trying to cancel [for example, the $0.028e^{-11.25t}$ in (8.108)]. If these modes are unstable, they will grow with time and, at the least, render system response unacceptable, and, at the worst, cause damage to or destruction of the system. Poles of $G_p(s)$ lying in the right half plane are dealt with by shaping the root locus so that the loci originating at these poles will migrate to the left half plane.

8.5.4 Closing Comments on the Root Locus

1. The purpose of this introductory section on the root locus has been to present some of the basic ideas of the method and to develop an understanding of how the general features of the root locus may be graphed quickly. Not all aspects of the root locus have been considered here, and the references should be consulted for additional information.
2. For higher-order systems the dominant pole idea is useful, but it is not always valid. Due to the availability of modern software packages for control systems analysis, it is fairly easy to generate the root locus plots for higher-order systems and to obtain the step and other types of responses, so that one does not have to rely on the second-order tendency in actual design. The dominant pole/second-order tendency, however, occurs frequently enough that it is a very useful and simple route to an understanding of the dynamic behavior of many systems.
3. The root locus is closely related to the eigenvalue/eigenvector methods of system dynamics analysis presented in Chap. 4, and an intuition for the relation between the closed-loop pole (eigenvalue) locations in the complex plane and the resulting system transient response is a useful thing to develop.

8.6 CONTROLLER DESIGN USING FREQUENCY RESPONSE METHODS

8.6.1 Bode Plot Construction and Examples

In the preceding section we used the root locus as a way to monitor the movement of the poles of the closed-loop transfer function $T(s)$ as a control gain was varied. The basic approach was to locate the dominant poles so that stability is achieved and so that transient response characteristics such as settling time and peak overshoot resulted in acceptable performance. In this section we introduce some of the basic ideas in the analysis and design of control systems via one of several commonly used frequency response methods: the Bode plot. It is noted at the outset that a comprehensive study of all of the various frequency response methods used is not our intent. Rather, we will concentrate on some of the basic ideas involved in using the Bode plot to understand control system behavior and design. More comprehensive treatments of frequency response methods are contained in the references.

The Bode plot was introduced in Sec. 5.4 as a logarithmic plot of the frequency response magnitude, measured in decibels (db). Thus, for an arbitrary linear dynamic system for which the input and output are related by a transfer function $H(s)$, the steady-state frequency response is defined by the complex frequency response function $H(i\Omega)$ according to (5.83), repeated below:

$$H(i\Omega) = |H(i\Omega)|e^{i\phi} \tag{8.109}$$

where $|H(i\Omega)|$ and ϕ are the amplitude and phase of $H(i\Omega)$. In Sec. 5.4 only the magnitude $|H(i\Omega)|$ was used to describe system frequency response; the behavior of the phase ϕ was not considered. In using Bode plots for control system design, the magnitude and phase are equally important, and the behavior of the phase ϕ will be discussed in this section. It is suggested that the reader review Sec. 5.4 as needed before proceeding.

We introduce the Bode plot methods here for a specific version of the SISO controls problem: the unit feedback system defined by ($B8.27$), for which a control gain k and a transfer function $G(s)$ comprise the feed-forward portion of the system block diagram. As applied to control systems analysis, the Bode plot will consist of the magnitude and phase of this *feed-forward* or *open-loop* transfer function $kG(s)$, rather than the magnitude and phase of the closed-loop transfer function $T(s) = kG(s)/[1 + kG(s)]$, which defines the actual frequency response of the system. Thus, in constructing Bode plots for control systems analysis, we will not actually be working with the overall or closed-loop system frequency response $T(i\Omega)$ but with the open-loop frequency response $kG(i\Omega)$. The reason for doing this is that the open-loop frequency response is related to the stability and damping properties of the system, as will be discussed in Sec. 8.6.2. In the remainder of this section we first discuss the behavior of the phase of the complex frequency response function, and then we'll display a few Bode plots for the example systems considered in Sec. 8.5. Then in Secs. 8.6.2 and 8.6.3 we discuss the application of Bode plots to control system analysis and design.

BODE PLOTS OF THE PHASE ϕ. The phase ϕ appearing in (8.109) is determined from the real and imaginary parts of $H(i\Omega)$ as (see Fig. 5.16)

$$\tan\phi = \frac{\text{Im}\left[H(i\Omega)\right]}{\text{Re}\left[H(i\Omega)\right]} \tag{8.110}$$

Summarized below is the behavior of ϕ for various factors which commonly appear in the open-loop transfer function $kG(s)$. Note from (8.109) that a negative phase angle ϕ indicates a phase *lag*, that is, if (8.109) represents a system transfer function, then the output lags the input by the angle $|\phi|$ if ϕ is negative. A positive ϕ indicates a phase *lead*.

Pole of $G(s)$. As in (8.90) denote the poles of $kG(s)$ as $s_1, s_2, \ldots, s_n$. Then factors in $kG(s)$ defining simple real poles are of the form $1/(s - s_i)$, so that

$$kG(i\Omega) = \frac{k}{i\Omega - s_i} = \frac{k(-s_i - i\Omega)}{\left(\Omega^2 + s_i^2\right)}$$

resulting from (8.110) in

$$\tan\phi = \left(\frac{-\Omega}{-s_i}\right) \tag{8.111}$$

If s_i is negative, then ϕ will be between 0 and $-90°$, as follows: As $\Omega \to 0$, $\phi \to 0$ from below; when $\Omega = -s_i$, $\phi = -45°$, and as $\Omega \to \infty$, $\phi \to -90°$. This behavior is shown in Fig. 8.33 for the case $s_1 = -10$. Observe that, approximately, ϕ changes from 0 to $-90°$ linearly over 2 decades of frequency

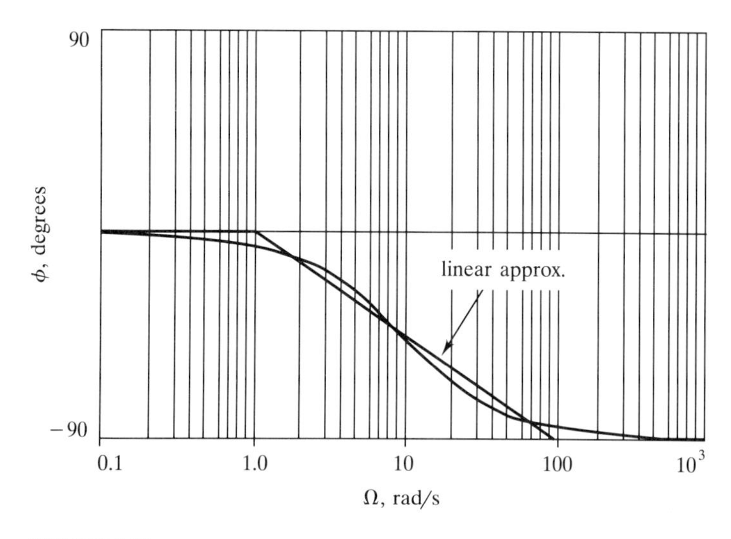

FIGURE 8.33
Bode plot of the phase ϕ for the factor $1/(s + 10)$. The straight-line approximation is also shown.

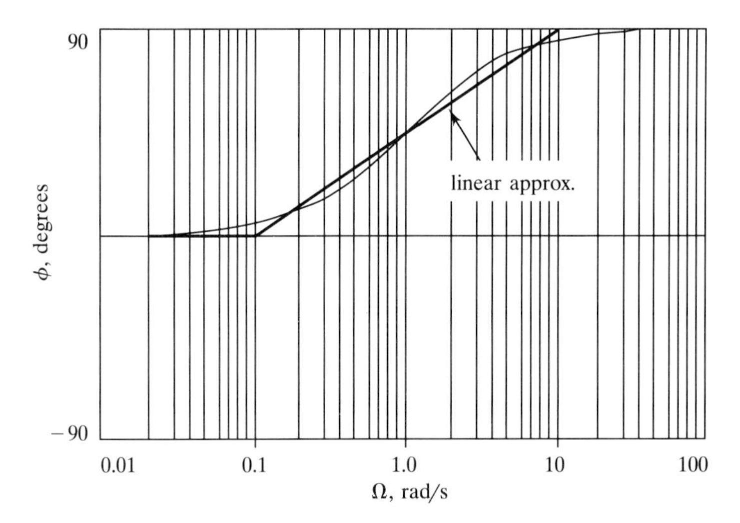

FIGURE 8.34
Bode plot of the phase ϕ for the factor $(s + 1)$. The straight-line approximation is also shown.

centered at $\Omega = 10$. The straight-line approximation is indicated in the figure. When $\Omega = 10$, $\phi = -45°$. Note that for a *positive* real pole, the phase ϕ varies from -180 to $-90°$ as Ω is increased from zero.

A real zero of $kG(s)$. Denoting the zeroes of $kG(s)$ as $z_1, z_2, \ldots, z_m$, as in (8.90), factors of the type $(s - z_1)$ will appear in the numerator of $G(s)$. The phase ϕ is then defined by the real and imaginary parts of $(i\Omega - z_1)$, so that, for the case in which z_1 is real,

$$\tan \phi = \left(\frac{\Omega}{-z_1} \right) \tag{8.112}$$

If z_1 is negative, then ϕ will be between 0 and $+90°$, as follows: As $\Omega \to 0$, $\phi \to 0$ from above, $\phi = 45°$ when $\Omega = -z_1$, and as $\Omega \to \infty$, $\phi \to 90°$. This behavior is illustrated in Fig. 8.34 for the case $z_1 = -1$. The straight line defining the approximate change in ϕ over 2 decades of frequency is also shown.

Based on the preceding discussion we note the following additional results for the phase ϕ:

1. For a pole s_1 at the origin $(s_1 = 0)$, Eq. (8.111) yields $\phi = -90°$ for all Ω, while for a zero at the origin $(z_1 = 0)$, Eq. (8.112) yields $\phi = +90°$ for all Ω.
2. In view of (8.109) the phase angle of a function $G(s)$ consisting of two or more factors involving combinations of poles and zeroes can be obtained by merely adding the phase angles associated with the individual factors. Thus,

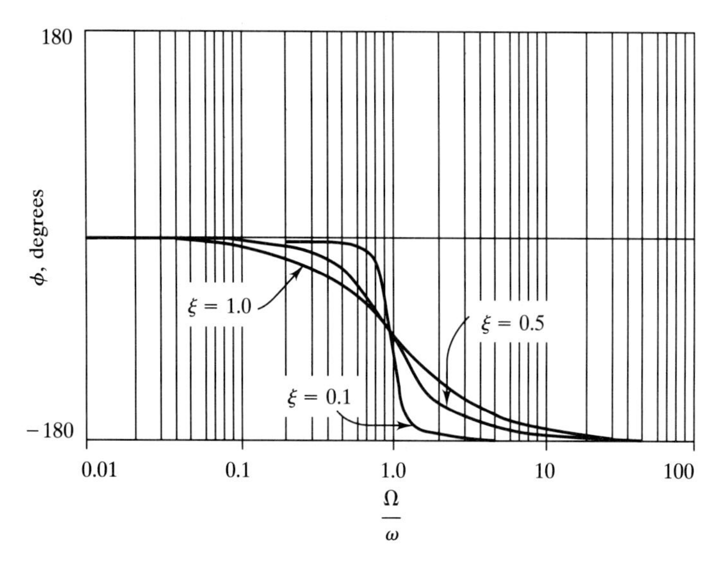

FIGURE 8.35
Bode plot of the phase ϕ for the second-order factor $1/(s^2 + 2\xi\omega s + \omega^2)$ for three values of damping factor ξ: $\xi = 0.1$, 0.5, and 1.0.

the phase ϕ of the transfer function $(s + 1)/(s + 10)$ is the sum of the phase angles appearing in Figs. 8.33 and 8.34.

3. For the second-order factor defining a pair of complex conjugate poles, $1/(s^2 + 2\xi\omega s + \omega^2)$, the phase ϕ is defined by (5.91):

$$\tan\phi = \frac{-2\xi(\Omega/\omega)}{1 - [(\Omega/\omega)]^2} \tag{8.113}$$

Assuming the poles to have a negative real part, the phase ϕ will vary from 0 to $-180°$, as follows: $\phi \to 0$ as $(\Omega/\omega) \to 0$, $\phi = -90°$ when $\Omega = \omega$, and $\phi \to -180°$ as $(\Omega/\omega) \to \infty$. This is shown in Fig. 8.35 for several values of the damping factor ξ.

4. If the open-loop transfer function $G(s)$ has m zeroes and n poles, then as $\Omega \to \infty$, the phase ϕ will approach $-90\,(n - m)°$.

We now consider two example systems already analyzed using the root locus in Sec. 8.5. The objective at this point is to develop an understanding of the Bode plot construction and behavior for simple systems.

Example 8.32 The Second-Order System of Example 8.27. Shown in Fig. 8.36 are the magnitude and phase Bode plots for the system of Example 8.27, for which the

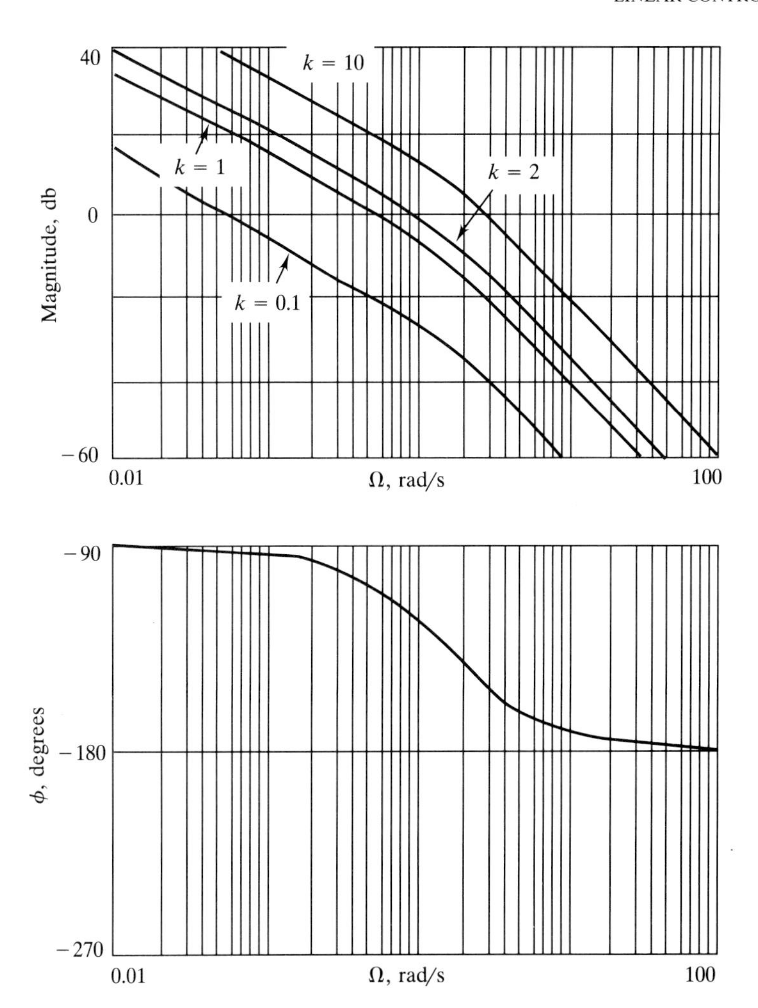

FIGURE 8.36
Bode plots of magnitude and phase of the open-loop transfer function $kG(s) = k/[s(s+2)]$. The phase ϕ is the same for all four cases of gain: $k = 0.1, 1, 2, 10$.

open-loop transfer function, from ($B8.25$), is

$$kG(s) = \frac{k}{s(s+2)} \tag{8.114}$$

The magnitude and phase of the open-loop frequency response function for this system are given by

$$|kG(i\Omega)| = \frac{k}{(\Omega^4 + 4\Omega^2)^{1/2}}$$

$$\tan(\phi) = \frac{-2}{-\Omega}$$

This function has poles at zero and -2. The magnitude plot shows results for four values of the gain k: 0.1, 1.0, 2, and 10. With reference to the system root locus of Fig. 8.23, the case $k = 1$ represents the breakaway condition, for which the poles of the *closed-loop* transfer function $T(s)$ coalesce at $p_{1,2} = -1$. The condition $k = 2$ represents a pair of dominant poles having a damping factor $\xi = 0.707$. The case $k = 10$ represents a damping factor $\xi = 0.32$, while for $k = 0.1$, the poles of $T(s)$ are real and negative, $p_{1,2} = -1.95, -0.05$. At low frequencies ($\Omega \ll 1$), the magnitude Bode plots change at the rate -20 db/decade due to the pole $s_1 = 0$ of $G(s)$. Then, as Ω passes through $\Omega = 2$ (that is, $-s_2$), the change asymptotes to -40 db/decade.

Figure 8.36 illustrates an important characteristic of the Bode plot: For open-loop transfer functions of type $kG(s)$, the gain k has no effect on the phase angle ϕ, which is the same for all four cases of gain shown. The phase $\phi = -90°$ as $\Omega \to 0$, due to the pole of $G(s)$ at the origin. ϕ then decreases from $-90°$ to $-180°$ over the 2 decades centered at $\Omega = 2$, at which $\phi = -135°$.

Figure 8.36 illustrates a useful property of the Bode plot: A change in the gain k merely shifts the entire magnitude plot up or down by a fixed amount. Thus, if we have a Bode plot for one value k_1 of the gain, the magnitude plot for any other gain k_2 is obtained by shifting the plot by an amount $20 \log(k_2/k_1)$. For instance, the plots for $k = 0.1$ and $k = 1.0$ in Fig. 8.36 are separated by 20 db, as are the plots for $k = 1$ and $k = 10$. This useful property makes it straightforward to determine quickly the Bode plots if one wishes to consider a change in the control gain k.

Example 8.33 The System of Example 8.28 (*B*8.26). Here we consider the third-order system of Example 8.28, for which the open-loop transfer function is given by

$$kG(s) = \frac{k}{s(s+2)(s+3)} \tag{8.115}$$

for which the magnitude and phase of the open-loop frequency response are defined by

$$|kG(i\Omega)| = \frac{k}{\left[25\Omega^4 + (6\Omega - \Omega^3)^2\right]^{1/2}}$$

$$\tan \phi = \frac{\Omega^2 - 6}{-5\Omega}$$

The open-loop transfer function has no zeroes and three poles at $s_1 = 0$, $s_2 = -2$, and $s_3 = -3$. The Bode plots are shown in Fig. 8.37 for four values of the gain k: 1, 2.1, 3.6, and 30. With reference to the system root locus plot of Fig. 8.24, the case $k = 30$ corresponds to the case of neutral stability, as the poles of the closed-loop transfer function $T(s)$ move into the right half plane for $k > 30$. The case $k = 2.1$ corresponds to the breakaway, shown in Fig. 8.24 at the point $\sigma_b = -0.78$ on the real axis. The case $k = 3.6$ defines the condition for which a pair of complex poles with damping factor $\xi = 0.707$ occur, with the poles located approximately at $p_{1,2} = -0.7 + 0.7i$. The phase ϕ starts at $-90°$ for $\Omega \to 0$, due to the pole of $G(s)$ at the origin of the complex plane. As $\Omega \to \infty$, $\phi \to -270°$, because the excess of poles over zeroes in (8.115) is 3. The magnitude plot changes at the rate -20 db/decade for $\Omega \ll 1$. Then, as Ω passes through the values 2 and 3 (associated with s_2 and s_3), the change in magnitude asymptotes to -60 db/decade.

Comment. The purpose of the preceding examples has been only to introduce the construction of the Bode plots and to describe the observed variations of magnitude and phase. We have not as yet discussed the relation of the Bode plots to the controls problem. Keep in mind that in controls applications we will use the Bode

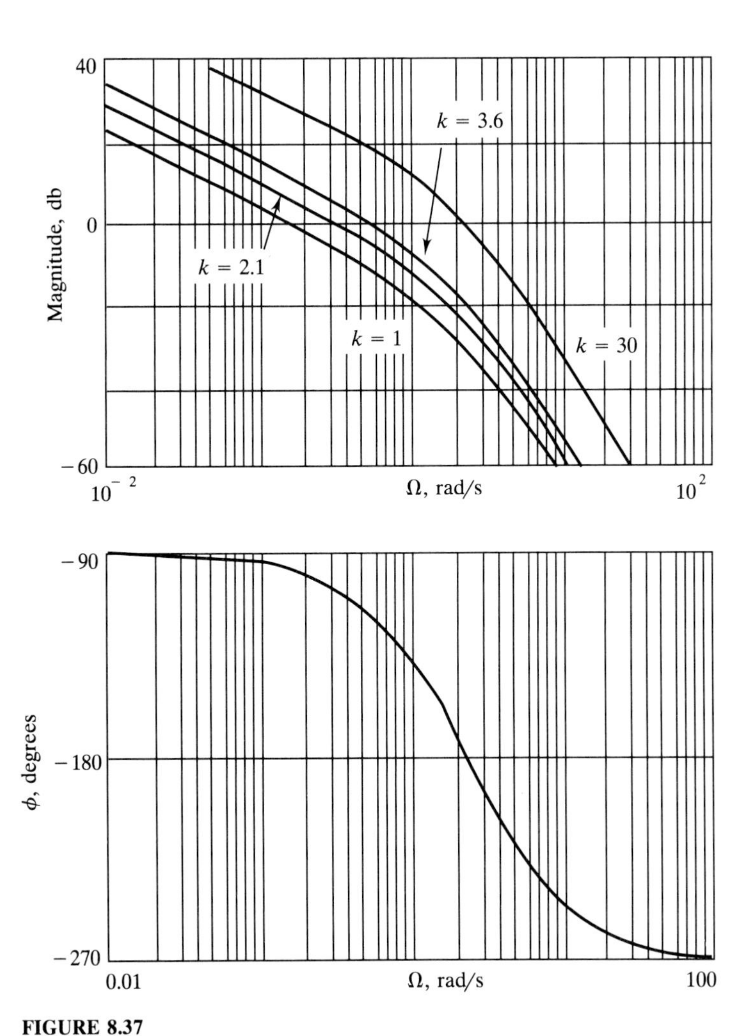

FIGURE 8.37
Bode plots for the third-order system of Example 8.28, for which the open-loop transfer function is $G(s) = k/[s(s+2)(s+3)]$. The phase ϕ is the same for the four gains shown: $k = 1$, $k = 2.1$, $k = 3.6$, $k = 30$.

plots of the open-loop transfer function $kG(s)$, rather than of the closed-loop transfer function $T(s)$. In the following sections we explore the connection between the (open-loop) Bode plots and the (closed-loop) behavior of the controlled system.

8.6.2 Gain and Phase Margins

In order eventually to see how Bode plots are applied to control system design, we explore here the relation between the Bode plots of the open-loop transfer function $kG(s)$ and the stability and damping properties of the closed-loop system. The response which is of interest is the closed-loop response, since this is the actual response of the controlled system. The transfer function $T(s)$ defining this closed-loop response is given, for the unity feedback system ($B8.27$) being considered, as

$$T(s) = \frac{kG(s)}{1 + kG(s)} \tag{8.116}$$

Thus, the characteristic equation for the closed-loop system is $1 + kG(s) = 0$, or $kG(s) = -1$, as was stated in Sec. 8.5 as Eqs. (8.88) and (8.89). From Sec. 8.5 we know that any point s in the complex plane is a pole of the closed-loop system if $kG(s) = -1$. Because the Bode plot is a plot of the magnitude and phase of $kG(i\Omega)$, we conclude that if $kG(i\Omega_c) = -1$ for some frequency Ω_c, then $s = i\Omega_c$ is on the root locus of the closed-loop system. This is an important condition, for it means that a condition of neutral stability exists: The poles of the closed-loop transfer function lie on the imaginary axis for any gain k and frequency Ω_c such that $kG(i\Omega_c) = -1$. For this condition the magnitude $|kG(i\Omega_c)| = 1$ and $\phi = -180°$. This condition, therefore, is represented by 0 db on the magnitude Bode plot and by the phase $\phi = -180°$ on the phase plot. If, for a candidate control system design, a Bode plot shows the magnitude to be 0 db and the phase to be $-180°$ at the same frequency, then the design is deficient, as the system is not asymptotically stable. Such an occurrence is observed in Fig. 8.37 for a gain $k = 30$. The associated root locus plot (Fig. 8.24) shows that when the gain k passes through 30, the poles of the closed-loop transfer function move into the right half plane, rendering the system unstable.

The 0 db, $\phi = -180°$ condition discussed in the preceding paragraph has lead to the definition of the gain and phase "margins" as a way to describe the relative proximity of the system closed-loop poles to the critical condition of neutral stability defined by $kG(i\Omega_c) = -1$. The gain and phase margins are defined as shown in Fig. 8.38. The *gain margin* (GM) is the amount, in decibels, by which the gain k would have to be increased in order for the 0 db condition to occur at the frequency for which $\phi = -180°$. That is, GM defines the increase (or decrease) in gain which would be required to render the system neutrally stable. For example, consider the Bode plots shown in Fig. 8.37 (root locus in Fig. 8.24) for the third-order system of Example 8.33. For a gain $k = 3.6$, for which $p_{1,2} = -0.7 \pm 0.7i$, with $\xi = 0.707$, the gain margin is

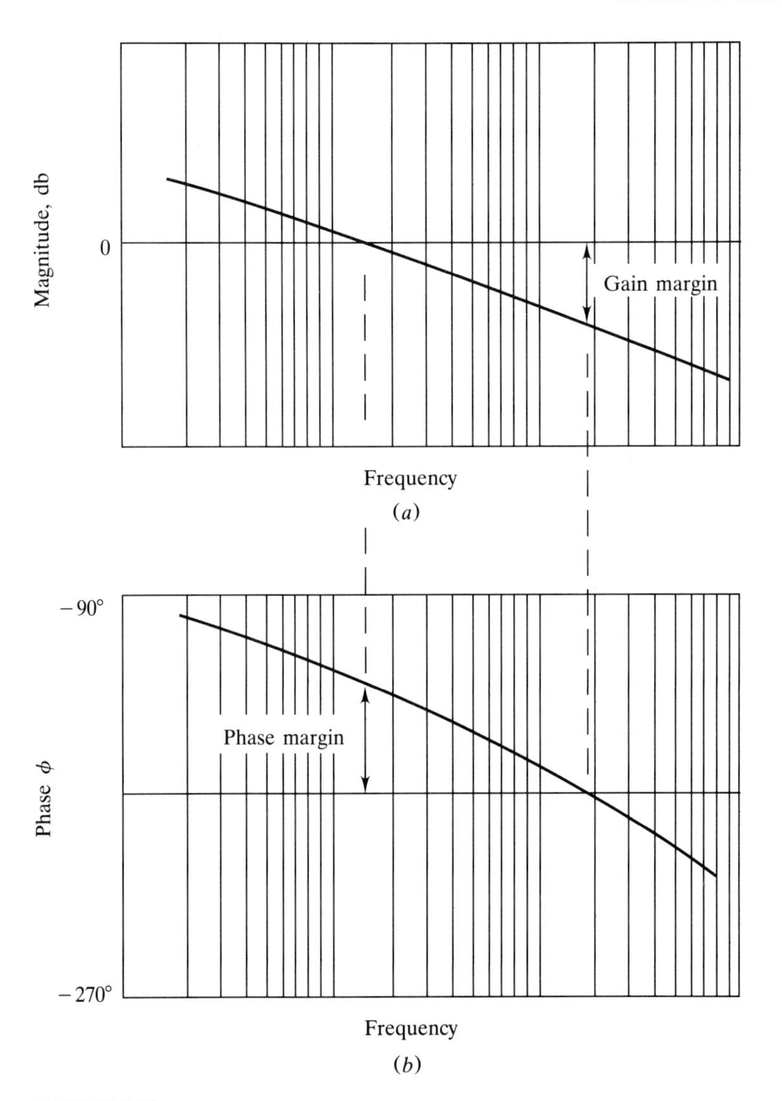

FIGURE 8.38
(*a*) Definition of the gain margin; (*b*) phase margin.

determined from Fig. 8.37 to be GM = 18 db. This is a relatively large value and indicates that a factor of about 8 increase in gain (to a value of $k = 30$) would be needed in order to destabilize the system.

The phase margin ϕ_m is defined as $\phi_m = (180° + \phi_c)$, where ϕ_c is the phase occurring at 0 db magnitude. The phase margin turns out to be a fairly direct measure of the system damping factor ξ in cases for which a pair of dominant conjugate poles can be identified, so that the system behavior is similar to that of a second-order system. For example, consider Fig. 8.37 for the case $k = 3.6$, which corresponds to $\xi = 0.707$. The phase margin is measured from the figure to be approximately $\phi_m = 65°$. This result is typical for systems dominated by a pair of conjugate poles, as is approximately the case here when $k = 3.6$, for which the real pole p_3 of the closed-loop transfer function $T(s)$ is located at $p_3 = -3.6$. In such cases a useful approximation relating the phase margin ϕ_m and the dominant pole damping factor ξ is (Ref. 8.3)

$$\phi_m \approx 100\xi \tag{8.117}$$

This approximate relation allows the dominant pole damping factor to be obtained reasonably accurately.

We next illustrate the relation between the Bode and root locus plots for the plotter control system of Example 8.30.

Example 8.34 The Plotter Control Problem (Example 8.30) Revisited. Consider first the case of proportional control, for which the root locus plot was obtained as Fig. 8.28. The open-loop transfer function for this case is

$$kG(s) = \frac{10^3 k}{s(s + 10)(s + 100)} \tag{8.118}$$

The Bode plot for this third-order system is shown in Fig. 8.39 for four values of the gain k: 1, 2.37, 5, and 110. The condition $k = 2.37$ corresponds to breakaway of the loci from the real axis at $\sigma_b = -4.85$; the condition $k = 5$ is the design condition, for which the dominant complex poles have a damping factor $\xi = 0.707$; the condition $k = 110$ corresponds to the condition of neutral stability, as the closed-loop poles cross the imaginary axis. For this latter condition the gain margin is 0 db, as discussed in the preceding paragraphs. At the design condition $k = 5$, the gain margin is found from Fig. 8.39 to be GM = 27 db and the phase margin $\phi_m = 70°$. Note that as $\Omega \to 0$, the phase $\phi \to -270°$ and the magnitude change asymptotes to -60 db/decade, which is characteristic of a third-order system with no zeroes in $G(s)$.

The preceding example illustrates the manner in which the gain and phase margins are related to stability margin, or relative stability. The term *stability margin*, or *relative stability*, is a qualitative indicator of the relative nearness of a pair of dominant poles to the imaginary axis. Thus, with reference to Fig. 8.19, as the angle $\theta \to 0°$, the GM $\to$ 0 db and $\phi_m \to 0°$. As θ increases, so do the gain and phase margins, with the phase margin being proportional to the damping factor ξ as in (8.117) if there is a pair of dominant poles. A commonly used rule of thumb is that, in order to achieve adequate stability margin, the gain margin should be approximately 10 db and the phase margin above approximately 45°.

An important note regarding the preceding discussion of relative stability is the following: While the gain and phase margins are indicators of relative stability, they say little about *how far* to the left of the imaginary axis the closed-loop poles will lie. This is important information, however, because the

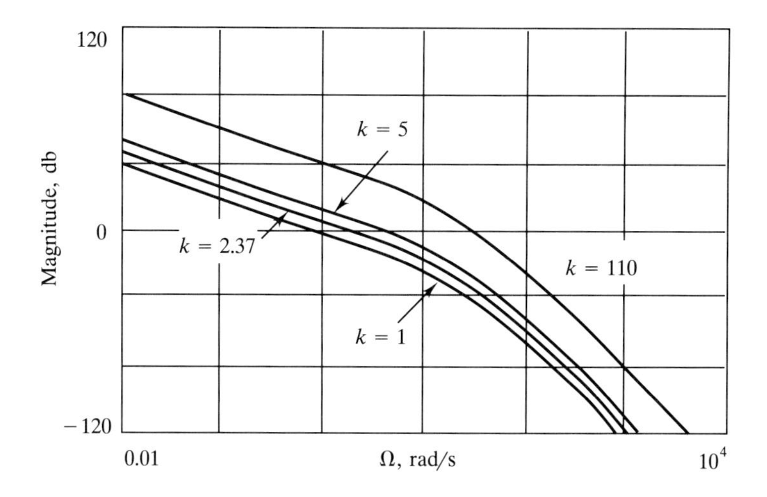

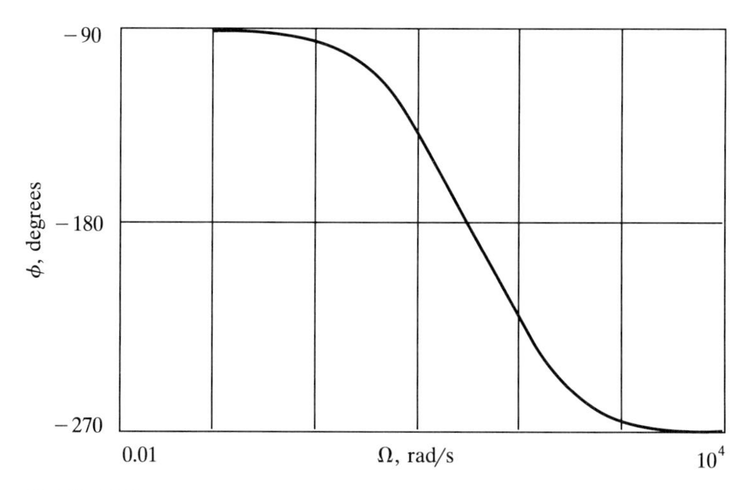

FIGURE 8.39
Bode plots for the plotter system of Example 8.30 with proportional control, for which the open-loop transfer function is $G(s) = 10^3k/[s(s + 10)(s + 100)]$. The phase is the same for the four cases of control gain shown: $k = 1, 2.37, 5, 110$.

system time constants and attendant response times are defined by the magnitudes of the real and imaginary parts of the dominant poles, and the gain and phase margins tell us nothing directly about these magnitudes. The result is that two different systems may each have gain and phase margins of, say, 10 db and 70°, indicating adequate relative stability, but one system may respond 10 times more quickly than the other to a step input. This problem of assessing response time via frequency domain analysis is considered in Sec. 8.6.3.

Example 8.35 The Plotter (Example 8.30) with PD Control. Figure 8.29, from Example 8.30, shows the root locus plot of the third-order plotter system with PD control. The open-loop transfer function for this case is

$$kG(s) = \frac{k(s + 20)}{s(s + 10)(s + 100)} \tag{8.119}$$

The PD controller selected causes the open-loop transfer function $G(s)$ to have a zero at $z_1 = -20$. The associated Bode plot is shown in Fig. 8.40 for four values of the gain k. The design condition, $k = 4.67$, produces a pair of "dominant poles" with damping factor $\xi = 0.707$. Recall from Fig. 8.30, however, that the transient response is also governed by the real pole at $p_3 = -27.4$ ($k = 4.67$). The Bode plot of Fig. 8.40 is characteristic of a third-order system with one zero in $G(s)$ and a pole at the origin. The phase varies from $-90°$ to $-180°$, and at large frequencies Ω the magnitude changes at the rate -40 db/decade. Note that because the phase ϕ never passes through $-180°$, the "gain margin," which is really undefined in this case, is positive for all values of the gain k. This is a consequence of the asymptotic behavior of the root loci as $k \to \infty$. With PD control the closed-loop transfer function $T(s)$ has three poles and one zero, so that the asymptotes along which the complex loci approach infinity are oriented at angles of $+90°$, so that the loci never cross the imaginary axis. Thus, the neutral stability condition $kG(i\Omega) = -1$ is never realized for this system. This is indicated by the phase plot of Fig. 8.40, for which $\phi \to -180°$ asymptotically from above as $\Omega \to \infty$. Similar behavior is observed in Fig. 8.36 (Example 8.27; root locus plot in Fig. 8.23).

For the design condition $k = 4.67$ the phase margin $\phi_m = 60°$. From (8.117) a dominant pole damping factor $\xi = 0.6$ is implied. This is reasonably close to the actual value of $\xi = 0.707$, but the presence of the real pole at $p_3 = -27.4$ renders the result less meaningful than it would be if the complex poles were truly dominant, as they are not here.

Comment. The utility of the Bode plot is illustrated by comparing the design solutions for the two cases of proportional control (Fig. 8.39, $k = 5$) and PD control (Fig. 8.40, $k = 4.67$). For these two cases, defined by the open-loop transfer functions of (8.118) and (8.119), the open-loop transfer functions differ slightly in gain k and in the factor $(s + 20)$ which appears in the numerator for the PD case. The Bode plots for these two cases are shown together in Fig. 8.41. Because multiplicative factors in $G(s)$ appear *additively* in the Bode plots, the Bode plot for the case of PD control may be obtained directly from the proportional control Bode plot by adding in the contribution due to the factor $(s + 20)$, the magnitude and phase of which are

$$|(20 + i\Omega)| = (400 + \Omega^2)^{1/2} = 10\log(400 + \Omega^2) \qquad \text{(db)}$$

$$\tan\phi = \frac{\Omega}{20} \tag{8.120}$$

Thus, the PD control magnitude plot shown in Fig. 8.41 is simply the proportional control magnitude plot with $10\log(400 + \Omega^2)$ added to it. This function causes an

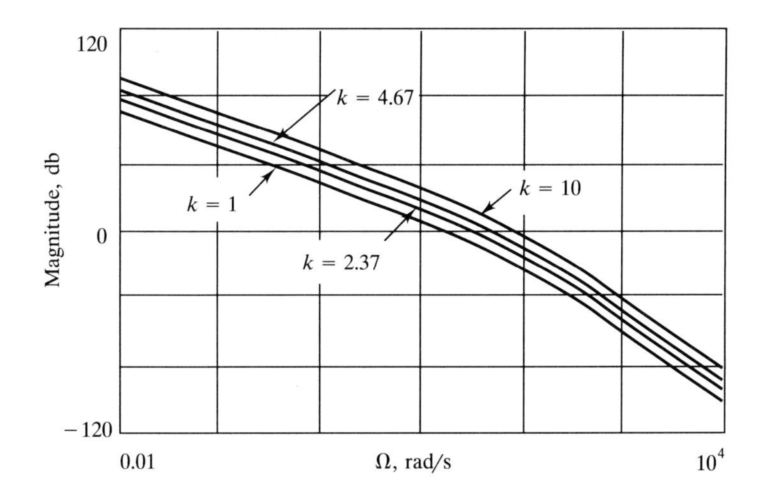

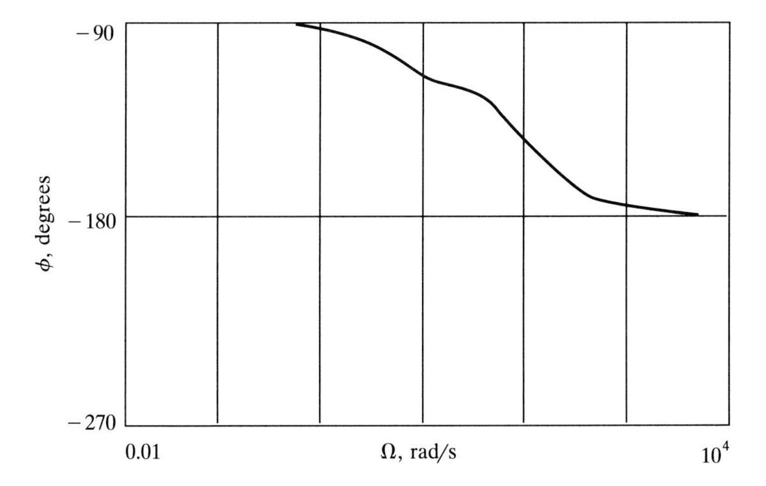

FIGURE 8.40
Bode plots of the plotter system of Example 8.30 for the case of PD control, for which the open-loop transfer function is $G(s) = 10^3k(s + 20)/[s(s + 10)(s + 100)]$. The phase is the same for the four cases of gain shown: $k = 1, 2.37, 4.67, 10$.

upward shift of $10 \log 400 = 26$ db at very low frequencies. Then as Ω passes through 20, the Ω^2 appearing in (8.120) causes the high-frequency rolloff to occur at -40 db/decade with PD control, as opposed to -60 db/decade with proportional control. In a similar manner the phase ϕ_{PD} for the PD control case is equal to the phase ϕ_p of the proportional control case, with the addition of the phase ϕ defined in (8.120). This addition, due to the factor $(s + 20)$ in (8.119), is essentially the same as that shown in Fig. 8.34, except for the difference in zero location (-1 for Fig. 8.34, -20 here).

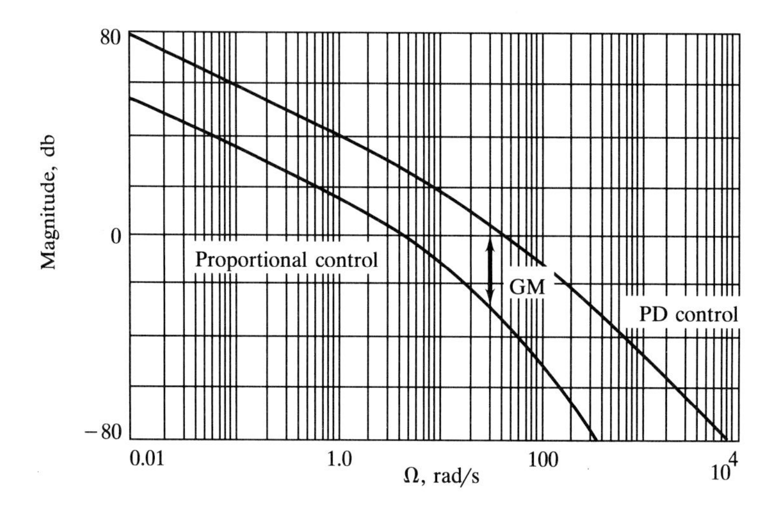

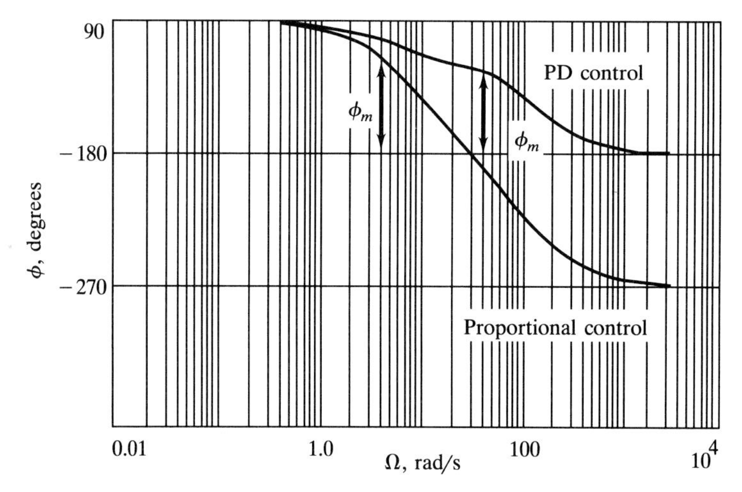

FIGURE 8.41
Bode plots for the plotter system for the two design conditions: proportional control with $k = 5$ and PD control with $k = 4.67$. The phase and gain margins are indicated. The gain margin is undefined for the PD case, as the phase ϕ does not pass through $-180°$, indicating that the system will be stable for all gains k.

Notice that the phase margins for the two cases, indicated in Fig. 8.41, differ only slightly and indicate from (8.117) that the dominant pole damping factor is 0.6 − 0.7. Based on the discussion so far, however, the results do not indicate the large difference in transient response time for the two cases (for the PD control design condition the system responds about 10 times faster than for the proportional control design condition).

We reiterate the utility of the additive property of the Bode plot. Changes in the open-loop transfer function $G(s)$ involving multiplicative factors can be accommodated easily to construct new Bode plots for alternate design conditions. It is important to note that the *closed-loop* transfer function $T(s)$ does not enjoy this property. Thus, in (8.116), if $G(s)$ were multiplied by some factor, such as $(s + 20)$, the closed-loop frequency response $T(i\Omega)$ is fundamentally altered and has to be recalculated.

In summary, for systems of the type (B8.27) considered here, the controller design is based on using the Bode plot to first establish adequate gain and phase margins for the controlled system. The phase margin provides an estimate of the dominant pole damping factor, from which an estimate of peak overshoot can be made if the system is dominated by a pair of complex poles. The next step, to which we now turn, is to establish from the frequency response a quantitative way to determine response time. This is critical because the gain and phase margins alone give no direct indication of the system time constants, which govern the rise and settling times of the system.

8.6.3 Response Time and Controller Design

According to Sec. 8.6.2 adequate stability margin and dominant pole damping can be obtained by realizing appropriate values for the gain and phase margins. In this section we address the determination from the frequency response of important performance measures such as rise time and settling time, which are not directly forthcoming from the Bode plots.

The "response time" characteristics such as rise time, settling time, and time to first oscillation peak were defined in Fig. 8.10. These measures are governed by the disposition of the system closed-loop poles p_i in the complex plane. Here we focus mainly on the settling time T_s. If we can satisfy a specified settling time condition and achieve a desired damping factor (as in Sec. 8.6.2) to control overshoot, then the transient response will generally be acceptable.

The discussion in this section is based on the assumption that the system of interest possesses a pair of conjugate dominate poles that govern the essential behavior of the system. Denote these poles as $p_{1,2} = -\xi\omega \pm i\omega(1 - \xi^2)^{1/2}$. The response time of the system will then be dictated by the real part $-\xi\omega$ of these dominant poles. We assume that a specified damping factor can be achieved, at least approximately, through design of the appropriate phase margin. We then have to arrange the design so that the frequency ω is sufficiently large that the product $\xi\omega$ results in sufficiently fast response. As an approximation, we can use the rule of thumb (Sec. 5.2) that the settling time is approximately equal to four system time constants, where the time constant τ for a second-order system is $\tau = \xi\omega$. Thus,

$$T_s \approx \frac{4}{\xi\omega} \tag{8.121}$$

may be used as an indicator of the quickness of the response.

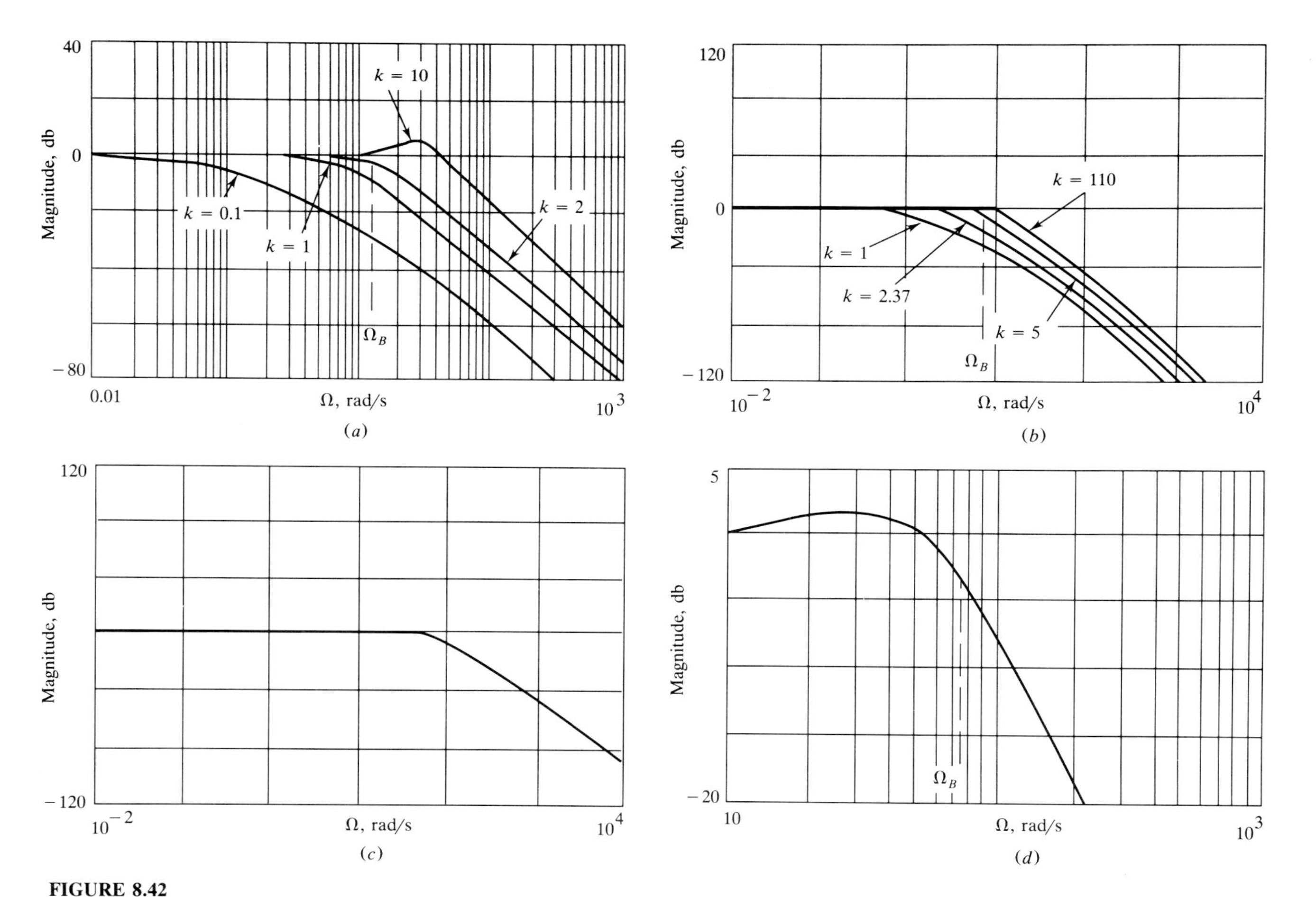

FIGURE 8.42

Plots of the magnitude $|T(i\Omega)|$ of the closed-loop frequency response $T(i\Omega)$ for three systems: (*a*) The second-order system defined by (8.114), (8.122), and (8.123), with the bandwidth noted for the case $k = 2$; (*b*) the plotter system with proportional control, (8.124), with the bandwidth noted for the case $k = 5$; (*c*) the plotter system with PD control, (8.125), (*d*) blowup of (*c*), with the bandwidth noted for the case

Warning. There are two important restrictions which must be kept in mind here:

1. If the closed-loop system behavior is not governed by a pair of dominant poles, then we cannot expect quantitative accuracy from (8.121), which is based on results for a second-order system with complex poles.
2. If the system has zeroes in the closed-loop transfer function, as in Example 8.30 with PD control, these zeroes can have a significant effect on the step response characteristics; for example, see Figs. 8.17 and 8.18 and the related discussion.

SYSTEM BANDWIDTH AND RESPONSE TIME. In the frequency domain the system response time may be estimated by using the system *bandwidth* to estimate the natural frequency ω, as we now describe. The bandwidth Ω_B is defined by the *closed-loop frequency response* $T(i\Omega)$ as follows: Ω_B is the frequency at which the magnitude $|T(i\Omega)|$ is $1/\sqrt{2}$ as large as the constant, asymptotic value $\lim_{\Omega \to 0} |T(i\Omega)|$. This level of decrease in magnitude corresponds to a 3-db drop in $|T(i\Omega)|$. For a second-order system with closed-loop damping factor ξ and natural frequency ω, the closed-loop frequency response was shown in Fig. 5.20. If the damping factor ξ is in the approximate range $\xi = 0.5 - 0.7$, as will often be the case, Fig. 5.20(b) shows that the bandwidth $\Omega_B \approx \omega$. In physical terms this is reasonable, for the frequency rolloff occurs as Ω passes through ω for a second-order system. Thus for a second-order system one may estimate the system natural frequency by determining the bandwidth of the closed-loop frequency response $T(i\Omega)$.

Some examples are shown in Fig. 8.42. Figure 8.42(a) shows the closed-loop frequency response magnitude for the second-order system of Example 8.32, for which (8.114) defines the open-loop transfer function. The closed-loop transfer function is given by (8.116) as

$$T(s) = \frac{k}{s(s+2)+k} \tag{8.122}$$

and the magnitude of the closed-loop frequency response is then found to be

$$|T(i\Omega)| = \frac{k}{\left[(k-\Omega^2)^2 + 4\Omega^2\right]^{1/2}} \tag{8.123}$$

In Fig. 8.42(a) the bandwidth is indicated for the case $k = 2$, for which the closed-loop poles are $p_{1,2} = -1 + i$, so that $\xi = 0.707$ and $\omega = \sqrt{2}$. The bandwidth $\Omega_B = \omega$ for this case. Note in passing that the closed-loop magnitude plots shown in Fig. 8.42(a) illustrate the point that a change in the gain k does not simply shift the plot up or down by a fixed amount, as is the case for the magnitude of the *open-loop* frequency response.

Shown in Fig. 8.42(b) are the closed-loop frequency responses for the plotter system with proportional control; the open-loop frequency response is plotted for the same values of the gain k in Fig. 8.39. For this system the open-loop transfer function is defined by (8.118), so that the closed-loop transfer function and the closed-loop frequency response are given by

$$T(s) = \frac{10^3 k}{s(s+10)(s+100) + 10^3 k}$$
$$|T(i\Omega)| = \frac{10^3 k}{\left[(10^3 k - 110\Omega^2)^2 + \Omega^2(1000 - \Omega^2)^2\right]^{1/2}} \tag{8.124}$$

For the design condition $k = 5$, the closed-loop poles are located at $p_{1,2} = -4.85 + 4.85i$, with associated values $\xi = 0.707$ and $\omega = 6.9$ rad/s. From Fig. 8.42(b) the bandwidth for the design gain is indicated in the figure and for this third-order system is approximately 8 to 9 rad/s. The Bode plot of Fig. 8.39 combined with the closed-loop frequency response of Fig. 8.42(b) allow the response times to be determined fairly accurately, even though this is a third-order system. The root locus plot of Fig. 8.28 shows that for the design condition the complex poles are essentially dominant, so the approximation based on second-order principles is reasonable.

Shown in Fig. 8.42(c) and (d) are magnitude plots of the closed-loop frequency response for the plotter system with PD control for the design gain of $k = 4.67$. For PD control the closed-loop transfer function and the magnitude of the closed-loop frequency response are given by the following:

$$T(s) = \frac{1000k(s+20)}{s^3 + 110s^2 + 1000s(1+k) + 20000k} \tag{8.125a}$$

$$|T(i\Omega)| = \frac{1000k(400 + \Omega^2)^{1/2}}{\left[(20000k - 110\Omega^2)^2 + (1000\Omega(1+k) - \Omega^3)^2\right]^{1/2}} \tag{8.125b}$$

From Fig. 8.42(d) the closed-loop system bandwidth is seen to be approximately 75 rad/s, so that the estimated natural frequency of the closed-loop system would be $\omega = 75$. For the design condition shown ($k = 4.67$), the closed-loop poles are defined by $p_{1,2} = -41.3 + 41.3i$, so that $\xi = 0.707$ and $\omega = 58$. Thus, the natural frequency, estimated from the system bandwidth assuming second-order behavior, is within 25 percent of the actual frequency associated with the pair of complex poles on which the design is based. Recall for this case of PD control that the complex poles are not dominant, as there is a real pole at $p_3 = -27.4$ for the design condition $k = 4.67$. Nevertheless, the use of the closed-loop bandwidth to estimate the "dominant pole" frequency provides a reasonable estimate of the response time.

The preceding discussion has illustrated the use of the closed-loop frequency response to estimate the system dominant pole natural frequency in situations for which the system behavior is essentially second order. Even if the second-order tendency is not realized, the closed-loop bandwidth may provide a reasonable estimate of the response times.

8.6.4 Closing Comments on Bode Design

Here we have given only a brief introduction to some of the basic ideas involved in the construction and use of the Bode plot in control systems design. For a more comprehensive treatment, the references listed at the end of this chapter should be consulted (for example, see Chaps. 6 and 7 of Ref. 8.1). In particular, it should be kept in mind that we have considered only the PID family of controllers in this chapter. There is another commonly used family, the "lag-lead" family, with which all practicing controls engineers are familiar, but which has not been considered here (a nice discussion of lag, lead, and lag-lead controllers is contained in Chap. 7 of Ref. 8.1). The Bode plot method is well suited and is commonly used in lag-lead controller design.

8.7 ADDITIONAL TOPICS IN CONTROLS

8.7.1 Disturbance Rejection

In Examples 8.1, 8.2, and 8.3, the need for a control system was based on the need to reject disturbances that would tend to move the system away from a desired equilibrium condition or operating point. In most of this chapter, however, we have concentrated our efforts on the tracking problem. We now, therefore, discuss briefly the disturbance rejection problem, as defined generally by the system block diagram ($B8.17$), repeated below for the case of unity feedback.

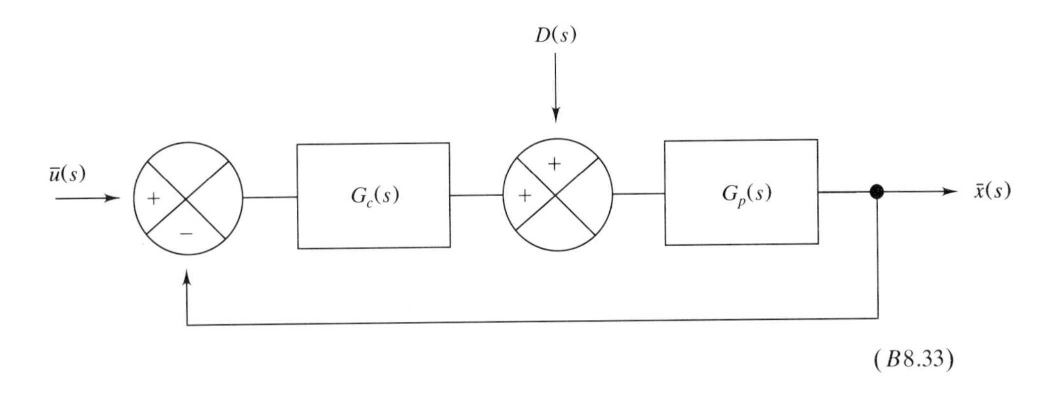

($B8.33$)

This block diagram is equivalent to the algebraic statement

$$\bar{x}(s) = G_p(s)\{G_c(s)[\bar{u}(s) - \bar{x}(s)] + D(s)\}$$

which can be rewritten as

$$\bar{x}(s) = \underbrace{\left[\frac{G_c(s)G_p(s)}{1 + G_c(s)G_p(s)}\right]}_{T(s)}\bar{u}(s) + \underbrace{\left[\frac{G_p(s)}{1 + G_c(s)G_p(s)}\right]}_{T_D(s)}D(s) \quad (8.126)$$

or

$$\bar{x}(s) = T(s)\bar{u}(s) + T_D(s)D(s) \quad (8.127)$$

Here $T(s)$ is the usual system closed-loop transfer function, and $T_D(s)$ is the "disturbance transfer function." If we assume, as usual, that $u(t)$ defines the desired history of the controlled variable which is to be tracked, then the disturbance will superpose some additional response that we would like to keep acceptably small. This "disturbance response" $x_D(t)$ is defined by $\bar{x}_D(s) = T_D(s)D(s)$ and will depend on the properties of the disturbance transfer function $T_D(s)$. Because we generally have no control over the disturbance, design practice is to ensure that the disturbance transfer function has the characteristics needed to keep $x_D(t)$ sufficiently small. This may generally be accomplished by employing a large control gain, so that $|G_c(s)| \gg 1$. If this condition is satisfied, then $|1 + G_c(s)G_p(s)| \approx |G_c(s)G_p(s)|$, and the disturbance transfer function magnitude reduces approximately to $|T_D(s)| \approx 1/|G_c(s)| \ll 1$, ensuring that the disturbance has a small effect on the controlled variable. We now illustrate this with an example.

Example 8.36 Automobile Cruise Control (Examples 8.3 and 8.22) Revisited. The block diagram for this first-order system, including the disturbance due to a nonzero grade α, is (B8.16), repeated below as (B8.34). Recall that $\bar{v}(s)$ represents the deviation from the desired speed V_0, so that $v(t)$ should be sufficiently close to the desired value $v_d(t) = 0$ at all times. The disturbance is $-g\alpha(t)$, where α is the grade along which the car is traveling (Fig. 8.43).

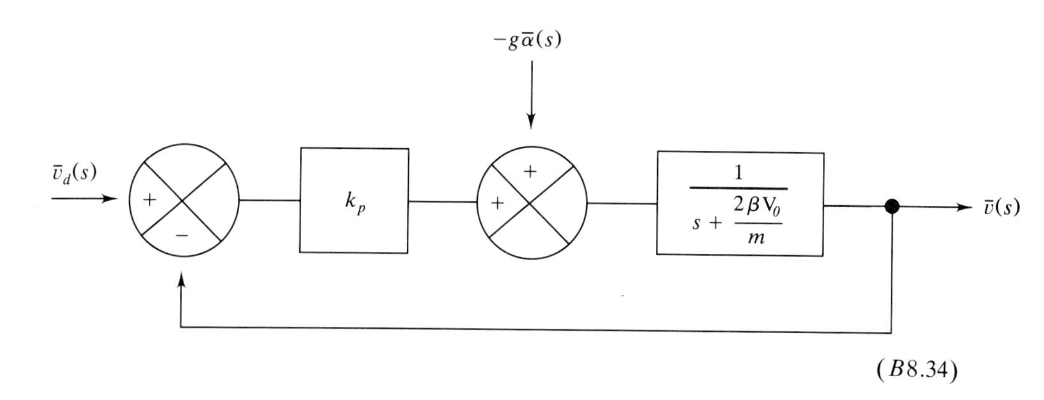

(B8.34)

To study the effect of a disturbance on the system response, consider the situation shown in Fig. 8.43. For the idealized situation shown in Fig. 8.43, the disturbance is represented mathematically by the step function shown in Fig. 8.44.

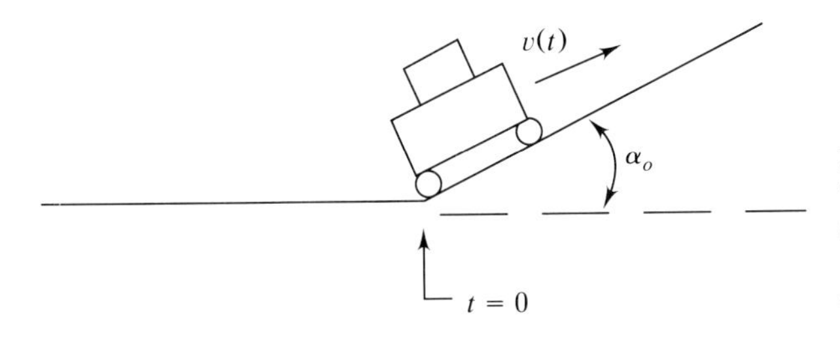

FIGURE 8.43
The car travels along a flat road at the desired speed V_0. At $t = 0$, a constant grade α_0 is suddenly encountered.

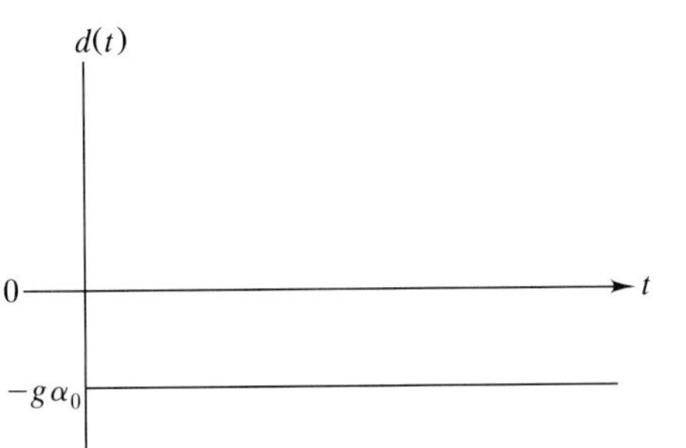

FIGURE 8.44
The disturbance function $d(t)$ for the situation of Fig. 8.43.

The disturbance response $v_D(t)$ is thus the response to a step input, as dictated by the disturbance transfer function $T_D(s)$. According to (8.126) the disturbance response will be determined from the relation

$$\bar{v}_D(s) = \left[\frac{G_p(s)}{1 + G_c(s)G_p(s)}\right] D(s)$$

with $G_p(s) = 1/(s + 2\beta V_0/m)$, $G_c(s) = k_p$, and $D(s) = -g\alpha_0/s$, the above relation yields

$$\bar{v}_D(s) = \left[\frac{1}{s + \left(\dfrac{2\beta V_0}{m} + k_p\right)}\right]\left(\frac{-g\alpha_0}{s}\right) \tag{8.128}$$

The denominator of $T_D(s)$ is the system characteristic polynomial and indicates that this first-order system has a pole $p_1 = -(2\beta V_0/m + k_p)$. Inversion of (8.128) yields the disturbance response $v_D(t)$ as

$$v_D(t) = \frac{-g\alpha_0}{\left(\dfrac{2\beta V_0}{m} + k_p\right)}\left[1 - e^{-(2\beta V_0/m + k_p)t}\right] \tag{8.129}$$

This disturbance response is shown schematically in Fig. 8.45. From (8.129), the steady-state disturbance response $v_{D_{ss}}(t)$ is observed to be

$$v_{D_{ss}}(t) = \frac{-g\alpha_0}{\left(\dfrac{2\beta V_0}{m} + k_p\right)} \tag{8.130}$$

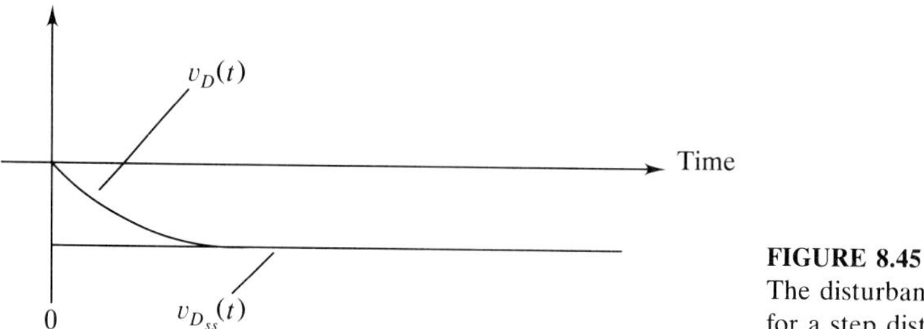

FIGURE 8.45
The disturbance response $v_D(t)$ for a step disturbance in grade.

This steady-state disturbance induced error is proportional to the disturbance magnitude α_0 and to the system time constant $\tau = -1/p_1 = 1/(2\beta V_0/m + k_p)$. The larger the control gain k_p of the proportional controller, the smaller the steady-state disturbance induced error. Note also that $v_{D_{ss}}(t) < 0$, indicating that the car will travel up the grade α_0 at a speed slightly less than the desired speed V_0. Actual automobile cruise controllers employ proportional control with a gain k_p such that an error of a few miles per hour may exist if the grade is fairly steep (say, 6 percent or so).

The preceding discussion and example are intended only to provide some insight into the disturbance rejection problem. For a more detailed analysis of this problem, the references should be consulted.

8.7.2 State Space Analysis of Control Systems: Some Basic Ideas

So far in this chapter only the single input/single output controls problem has been studied, and all of the mathematical analysis has been based on Laplace transformation, with the mathematical models of the form of single differential equations (8.46) of the system order n. In the example models we have studied, the ODE model dependent variable $x(t)$ is the variable to be controlled. We have introduced the PID family of controllers as a means to tailor certain aspects of the system behavior to our desires. In this section we explore briefly an alternative to the classical control systems analysis via Laplace transformation. This alternative is based on the state variable form of the system mathematical model. Analysis and design of control systems using state variable models has become widespread over the past 30 years because state variable models are well suited to analysis via computer. In addition, state variable techniques allow the analyst to handle problems which are too complex to be amenable to classical methods, and state variable methods allow easy consideration of certain control strategies which are difficult to implement using the SISO formulations we have studied. In this section we introduce state variable control methods because of their importance and utility in the two following situations:

1. *Pole placement:* For a system of arbitrary order n, the root locus method has been presented as a way to understand system behavior and to design simple controllers. The essential idea underlying this method is the second-order tendency principle. Implementation of control design via root locus involves selection of a control gain (as well as location of zeroes of the controller transfer function if PD or PID control is to be used) which places the dominant poles in desired locations in the complex plane. The success of this method depends on our ability to vary a *single* control gain such that for some value of this gain the system pole locations will result in acceptable transient and steady-state responses. State variable methods, however, turn out to allow us to use easily more complex control strategies which allow *all* n of the system poles to be placed where we want them. The "pole placement" scheme using state variable models will be summarized shortly for the SISO problem.
2. *MIMO systems:* All of our work so far has dealt with the single input/single output linear control problem. The viewpoint implicit in SISO analysis is that it is sufficient to control a single quantity. That is, if a single system state can be made to behave as desired, then the overall system response will be acceptable. This SISO approach breaks down, however, in many situations. One of these was introduced in Example 8.5 (missile guidance), for which it is necessary to control six independent quantities (three positions and three velocities); the SISO philosophy is simply unworkable. For such MIMO systems, the modern controls approach utilizing state variable methods is used in the analysis and design of the control system.

In the remainder of this section we introduce the state variable methods of control system analysis by considering the pole placement problem for SISO systems. The objectives are (1) to show the mathematical form of the controls problem from the state variable viewpoint and (2) to illustrate the pole placement procedure, whereby all n of the system poles are placed where we want them in the complex plane. We start with the following example.

Example 8.37 State Variable Analysis of a Fourth-Order System. Here we consider the idealized fourth-order system of two masses and two linear springs shown in Fig. 8.46. Newton's law will yield two second-order ODEs for the system displacements v_1 and v_2, so that the model is of fourth order.

The problem is stated as follows: A single control force $u(t)$, produced by an actuator not shown and acting on the mass m_2, is to be designed so that all four system poles are placed in *specified* locations in the complex plane. Let us first develop the mathematical model in state variable form and then determine how the control would be designed.

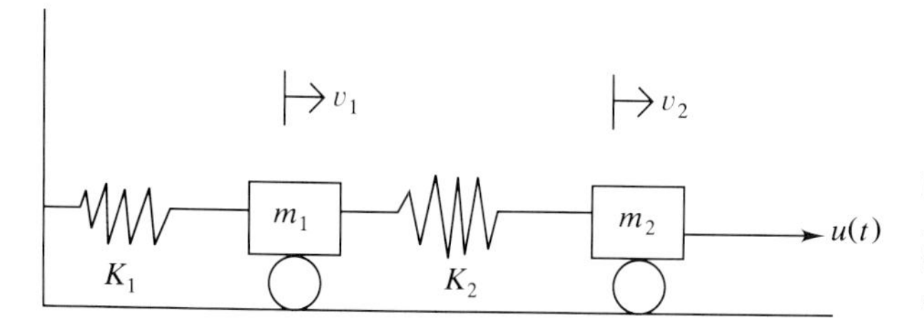

FIGURE 8.46
A fourth-order system consisting of mass and linear spring elements.

Application of Newton's second law or Lagrange's equations yields the following equations of motion for the displacements v_1 and v_2:

$$m_1\ddot{v}_1 = -K_1v_1 - K_2(v_1 - v_2)$$

$$m_2\ddot{v}_2 = -K_2(v_2 - v_1) + u(t)$$

The system states are $v_1, \dot{v}_1, v_2, \dot{v}_2$. Letting $x_1 = v_1$, $x_2 = \dot{v}_1$, $x_3 = v_2$, and $x_4 = \dot{v}_2$ define the four states, the preceding equations of motion may be expressed in the following state variable form:

$$\begin{Bmatrix} \dot{x}_1 \\ \dot{x}_2 \\ \dot{x}_3 \\ \dot{x}_4 \end{Bmatrix} = \begin{bmatrix} 0 & 1 & 0 & 0 \\ -\dfrac{1}{m_1}(K_1 + K_2) & 0 & \dfrac{K_2}{m_1} & 0 \\ 0 & 0 & 0 & 1 \\ \dfrac{K_2}{m_2} & 0 & \dfrac{-K_2}{m_2} & 0 \end{bmatrix} \begin{Bmatrix} x_1 \\ x_2 \\ x_3 \\ x_4 \end{Bmatrix} + \begin{Bmatrix} 0 \\ 0 \\ 0 \\ \dfrac{1}{m_2} \end{Bmatrix} u(t) \tag{8.131}$$

The following general form may be used to represent this math model:

$$\{\dot{x}\} = [A]\{x\} + \{B\}u(t) \tag{8.132}$$

where, for this example, the state vector $\{x\}$ is 4×1, the 4×4 matrix $[A]$ defines the plant, and u is to be specified. Note that the poles of the uncontrolled system or plant, the properties of which are defined by the matrix $[A]$, will lie on the imaginary axis in the complex plane, as shown in Fig. 8.47. This follows because the system has no damping, and the analysis of Sec. 7.3 shows the system eigenvalues (poles) to be purely imaginary.

The plant is stable but not asymptotically stable. Our job is to design a controller that will move the four poles shown in Fig. 8.47 to locations which we

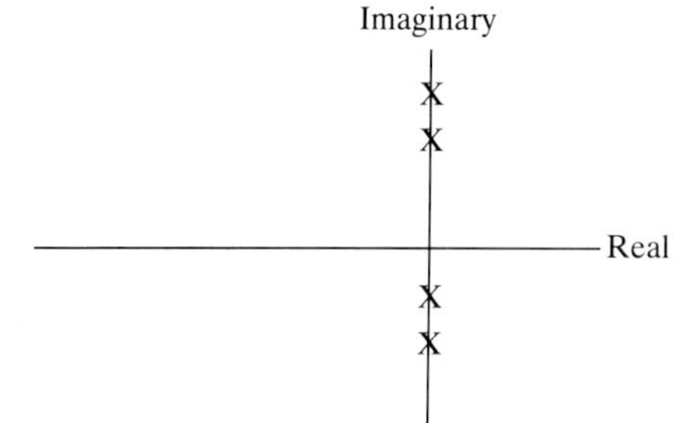

FIGURE 8.47
Schematic of the poles of the plant defined by (8.131) with $u(t) = 0$. These poles define the two natural frequencies for the 2-DOF vibratory system with no damping and no control.

specify. To accomplish this purpose, the pole placement procedure of modern control theory utilizes "full-state feedback," that is, the control $u(t)$ is a linear function of *all of the states*. Thus, in the present example a control u would be assumed in the following form:

$$u = -\lfloor k_1 \quad k_2 \quad k_3 \quad k_4 \rfloor \begin{Bmatrix} x_1 \\ x_2 \\ x_3 \\ x_4 \end{Bmatrix} \tag{8.133}$$

or notationally,

$$u = -\lfloor k \rfloor \{x\} \tag{8.134}$$

The relation (8.134) says that the control force u applied to the mass m_2 is to be a linear function of the four systems states. From a mathematical standpoint, the idea is that if we have four control constants at our disposal, then we should be able to control the placement of the four system poles. This can be seen more clearly if we use (8.134) in (8.132) to write the system model as follows:

$$\{\dot{x}\} = [A]\{x\} - \{B\}\lfloor k \rfloor \{x\}$$

or

$$\{\dot{x}\} = [[A] - \{B\}\lfloor k \rfloor]\{x\} \tag{8.135}$$

The poles of the controlled system are the eigenvalues λ_i of the matrix $[A] - \{B\}\lfloor k \rfloor$. The controlled system characteristic equation is

$$|[A] - \{B\}\lfloor k \rfloor - \lambda[I]| = 0 \tag{8.136}$$

When expanded, (8.136) will yield a polynomial of fourth degree in λ. The four control constants $k_1, \ldots, k_4$ will be present in the coefficients of this characteristic equation. The idea is that, if the four pole locations $\lambda_1, \ldots, \lambda_4$ are *specified* so as to achieve desired performance of the system, then the control gains $k_1, \ldots, k_4$ needed to produce these pole locations can theoretically be determined from the system characteristic equation. We now illustrate this procedure for the system of Fig. 8.46.

In order to simplify the analysis numerically, we assume that all masses and stiffnesses in the system of Fig. 8.46 are equal to unity. Thus, the mathematical model (8.131) becomes

$$\{\dot{x}\} = \underbrace{\begin{bmatrix} 0 & 1 & 0 & 0 \\ -2 & 0 & 1 & 0 \\ 0 & 0 & 0 & 1 \\ 1 & 0 & -1 & 0 \end{bmatrix}}_{[A]} \{x\} - \underset{\substack{\uparrow \\ \{B\}}}{\begin{Bmatrix} 0 \\ 0 \\ 0 \\ 1 \end{Bmatrix}} \lfloor k_1 \quad k_2 \quad k_3 \quad k_4 \rfloor \{x\} \tag{8.137}$$

or, combining the matrices $[A]$ and $-\{B\}\lfloor k \rfloor$,

$$\{\dot{x}\} = \begin{bmatrix} 0 & 1 & 0 & 0 \\ -2 & 0 & 1 & 0 \\ 0 & 0 & 0 & 1 \\ (1 - k_1) & -k_2 & (-1 - k_3) & -k_4 \end{bmatrix} \{x\} \tag{8.138}$$

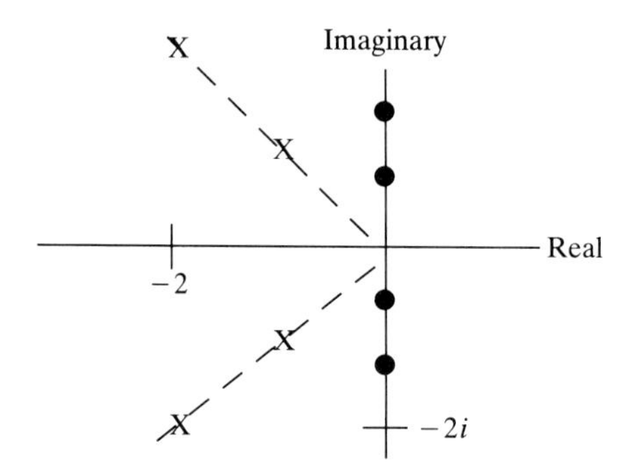

FIGURE 8.48
Desired pole placements (x) for the four-state system of Example 8.37. The pole locations of the plant are denoted by ($\cdot$).

Next, consider the specification of the pole locations. The uncontrolled system or plant, defined by $\{\dot{x}\} = [A]\{x\}$, has poles located on the imaginary axis at $p_{1,2} = \pm 0.618i$, $p_{3,4} = \pm 1.618i$, as indicated schematically in Fig. 8.47. Thus, the uncontrolled system will exhibit oscillatory behavior characterized by the two system natural frequencies $\omega_1 = 0.618$, $\omega_2 = 1.618$ (see Sec. 7.3.1 for an analysis of this same system from the standpoint of free vibration of a multi-degree-of-freedom system). The system is neutrally stable. For purposes of this simple example, we suppose that it is desired to place the poles of the controlled system at the locations $\lambda_{1,2} = -1 \pm i$, $\lambda_{3,4} = -2 \pm 2i$, as shown in Fig. 8.48. So we need to select the control gains k_1, k_2, k_3, and k_4 so that the eigenvalues $\lambda_1, \ldots, \lambda_4$ of the coefficient matrix $[A] - \{B\}\lfloor k \rfloor$ in (8.138) have these specified values. To do this, form the characteristic equation of (8.138):

$$\begin{vmatrix} -\lambda & 1 & 0 & 0 \\ -2 & -\lambda & 1 & 0 \\ 0 & 0 & -\lambda & 1 \\ (1 - k_1) & -k_2 & (-1 - k_3) & (-\lambda - k_4) \end{vmatrix} = 0$$

Expansion of this determinant yields the following fourth-order polynomial in λ:

$$\lambda^4 + k_4\lambda^3 + (3 + k_3)\lambda^2 + (k_2 + 2k_4)\lambda + (1 + k_1 + 2k_3) = 0 \qquad (8.139)$$

The k's must be selected so that the coefficients of the various powers of λ match those in the following factored form of the characteristic equation, obtained directly from the specified pole placement:

$$\underbrace{(\lambda^2 + 2\lambda + 2)}_{\substack{\uparrow \\ \text{Produces specified poles} \\ p_{1,2} = -1 \pm i}} \quad \underbrace{(\lambda^2 + 4\lambda + 8)}_{\substack{\uparrow \\ \text{Produces specified poles} \\ p_{3,4} + -2 \pm 2i}} = 0$$

which when expanded yields

$$\lambda^4 + 6\lambda^3 + 18\lambda^2 + 24\lambda + 16 = 0 \qquad (8.140)$$

Comparison of (8.140) and (8.139) then yields the following four linear algebraic equations for the control gains $k_1, \ldots, k_4$:

$$\begin{aligned} k_4 &= 6 \\ 3 + k_3 &= 18 \\ k_2 + 2k_4 &= 24 \\ 1 + k_1 + 2k_3 &= 16 \end{aligned} \tag{8.141}$$

These equations are solved for the required control gains: $k_4 = 6$, $k_3 = 15$, $k_2 = 12$, and $k_1 = -15$.

This introductory example has been intended only to show the basic idea of pole placement via the state variable formulation. Associated with this problem are several important issues which have not been addressed here. Notable among these is the problem of *controllability*: it may turn out that for a given system, defined by the matrices $[A]$ and $\{B\}$, it is not possible to specify the closed-loop pole locations, even with full state feedback. In mathematical terms it may occur that the equations (8.141) which are solved to determine the control gains are singular (that is, if written in matrix-vector form as a set of simultaneous algebraic equations with the k's the unknowns to be solved for, the coefficient matrix has a zero determinant; see Exercise 8.32). The references should be consulted for a more serious look at the state space analysis of control systems; for example, see Chap. 10 of Ref. 8.1 to get started.

A final comment on pole placement using full state feedback is that, generally, the calculation of the required control gains is not done as in Example 8.37. The procedure of Example 8.37 was used only because it allowed the basic ideas to be illustrated in a straightforward way.

8.8 CLOSING COMMENTS ON CONTROL SYSTEMS

In this chapter the simplest controls problem has been considered: control of the single input/single output linear system. Furthermore, control strategies were limited to those of the PID family, with a concentration on PD controllers. The methods introduced for control systems analysis were the classical techniques of root locus and frequency response. In the control systems field, there are many additional areas with which practicing engineers are typically familiar. These include the following: (1) control strategies other than those of the PID family, (2) the multi-input/multi-output problem, (3) the state variable formulation of the linear controls problem (touched on briefly here in Sec. 8.7.2), (4) control of nonlinear systems, (5) optimal control, and (6) discrete time control, which is important when digital computers are used in the measurement and control of dynamic systems, as is common. Developing an understanding of these areas generally requires years of study and engineering practice.

REFERENCES

There may be shortages of many things in life, but books on control systems is not one of them. The following is a sampling at the beginning to intermediate level.

8.1. Ogata, K.: *Modern Control Engineering*, 2d ed., Prentice-Hall, Englewood Cliffs, N.J. (1990). An excellent and comprehensive reference.
8.2. Dorf, R. C.: *Modern Control Systems*, 5th ed., Addison-Wesley, Reading, Mass. (1989). Contains numerous real-world control system examples and problems.
8.3. Phillips, C. L., and R. D. Harbor: *Feedback Control Systems*, Prentice-Hall, Englewood Cliffs, N.J. (1988).
8.4. Thaler, G. J.: *Automatic Control Systems*, West, St. Paul, Minn., (1989). See Ch. 5 for some good examples of the use of Bode plots in control system design.
8.5. Kuo, B. J.: *Automatic Control Systems*, 5th ed., Prentice-Hall, Englewood Cliffs, N.J. (1987).
8.6. Palm III, W. J., *Modeling, Analysis and Control of Dynamic Systems*, Wiley, New York (1983).
8.7. D'Souza, A. F., *Design of Control Systems*, Prentice-Hall, Englewood Cliffs, N.J. (1988).
8.8. Hale, F. J., *Introduction to Control System Analysis and Design*, 2d ed., Prentice-Hall, Englewood Cliffs, N.J. (1988).
8.9. Shinners, S. M., *Modern Control System Theory and Application*, 2d ed., Addison-Wesley, Reading, MA (1978).
8.10. Raven, F. H., *Automatic Control Engineering*, 3d ed., McGraw-Hill, New York (1978).

EXERCISES

1. The ship of Example 8.1, initially at rest ($\theta = \dot{\theta} = 0$), is subjected to a sudden broadside wind gust which produces a disturbance moment $M_d(t)$ as shown in Fig. *P*8.1 (assume that T_0 is large compared to the roll oscillation period of the ship). The response $\theta(t)$ then behaves as that of a second-order system subjected to a truncated step input, as in Example 5.7. Assuming a proportional control strategy, is the percentage overshoot affected by the value of the control constant k_p? Is the time T_p to the first oscillation peak affected by k_p? Discuss.

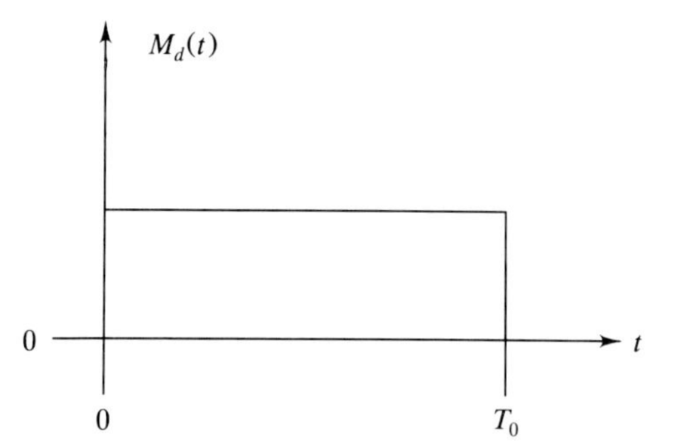

FIGURE *P*8.1

2. Consider the *uncontrolled* jet airplane of Example 8.2 and suppose that a large disturbance torque $M_d(t)$ acts over a small time interval, imparting an initial angular velocity $\dot{\theta}_0$. $M_d(t) = 0$ thereafter. If $\theta_0 = 0$, determine the histories $\dot{\theta}(t)$ and $\theta(t)$ and show that the final value $\theta(t \to \infty)$ may take on any value, depending on $\dot{\theta}_0$, c, and I.

3. For the system of Example 8.2 determine the differential equation which describes the dynamics of the controlled system with a proportional control. If $c = 10$, $I = 100$, and $k_p = 50$, plot in the complex plane the eigenvalues of the controlled and the uncontrolled systems.

4. Verify the solution (8.36) to Example 8.14 by working out the partial fraction expansion, for which the results are quoted above Eq. (8.36).

5. In Example 8.15, redo the example using x_2, rather than x_1, as the output variable, and verify that the poles of $\bar{x}_2(s)$ are the same as the eigenvalues of the state variable formulation.

6. A function $v(t)$ has the Laplace transform

$$\bar{v}(s) = \frac{(s+1)}{(s+5)(s+3)}$$

Construct the partial fraction expansion and determine $v(t)$.

7. For the following second-order system, use Laplace transformation to find the zero-input and the zero-state responses:

$$\ddot{x} + \omega^2 x = \omega^2 u(t)$$

where $u(t)$ is the unit step function.

8. Consider the second-order damped oscillator

$$\ddot{x} + 10\dot{x} + 50x = 0$$

Determine the poles of $\bar{x}(s)$, and plot them in the complex plane.

9. Determine the governing differential equation for a system whose transfer function is

$$H(s) = \frac{s+2}{(s+1)(s^2+2s+10)}$$

where $\bar{x}(s) = H(s)\bar{u}(s)$. Is the system stable?

10. Find the transfer function $H(s)$ for the radar tracking system of Example 8.4, with proportional control, as defined by (8.13). Here $\theta_A(t)$ is the input.

11. A certain linear system has the transfer function

$$H(s) = \frac{10}{(s^2+9)(s^2+4s+3)}$$

Find the poles of $H(s)$ and determine the stability of the system.

12. Verify that ($B8.10$) follows from ($B8.9$).

13. For the three-state oscillator system of Example 8.20, assume that $y(t)$ is the output variable and develop a block diagram which contains only $\bar{y}(s)$ and the input $\bar{u}(s)$.

14. Obtain the single block representation of the block diagram in Fig. $P8.14$.

15. Construct the block diagram representation of a jet airplane (Example 8.2) equipped with a proportional control system.

16. Consider the proportional plus derivative (PD) control of a lunar lander, for which the block diagram is shown in Fig. $P8.16$.

Determine values of the control constants k_p and k_d so that the controlled system exhibits a damped frequency of 0.25 Hz and a damping factor $\xi = 0.50$. (Assume $I = 100 \text{ kg} - \text{m}^2$.)

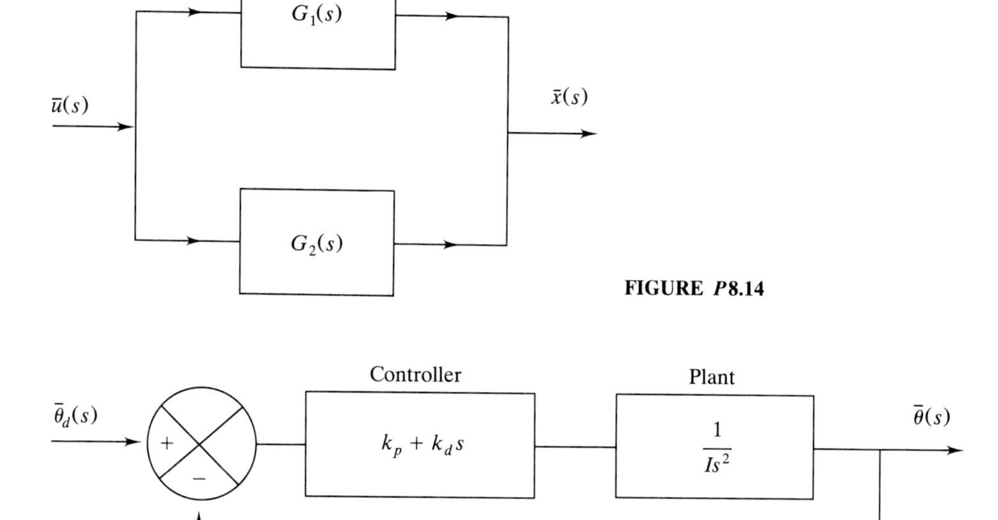

FIGURE *P*8.14

FIGURE *P*8.16

17. Verify the step response given in Table 8.2 for the case of PD control.

18. The first-order system shown in Fig. *P*8.18 is to be controlled such that the steady-state error in response to a step input is zero, there is no overshoot, and the settling time $T_s \leq 0.2$ s.

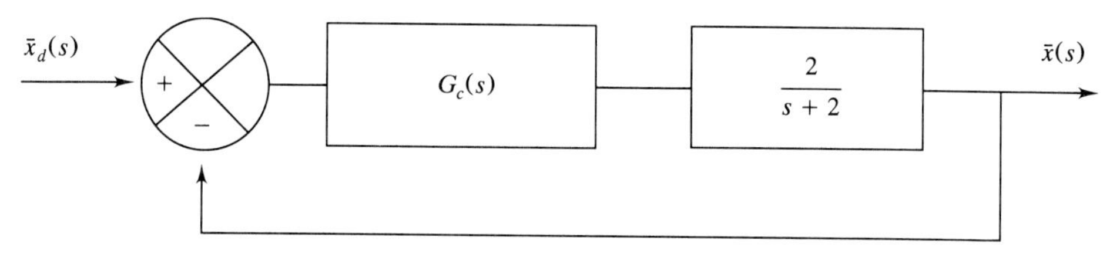

FIGURE *P*8.18

Select a controller from the PID family and determine acceptable value(s) of the control constant(s).

19. Consider the two third-order systems defined by the transfer functions

a. $T(s) = \dfrac{200}{(s+1)(s+2)(s+100)}$

b. $T(s) = \dfrac{8}{(s+1)(s+2)(s+4)}$

Determine the unit step response in each case. Determine the "equivalent" second-order system in each case, assuming the leftmost pole in the complex plane can be ignored. In each case comment on the validity of attempting to view the system as if it behaves like a second-order system.

20. A system is modeled as a mass m restrained by a linear spring of stiffness k and a linear damper having damping constant c. The values of the system constants are $m = 1$ kg, $k = 1$ N/m, and $c = 4$ N-s/m. The system responds very sluggishly to a step input. In order to produce quicker response, a proportional plus derivative controller is added, so that the block diagram of the controlled system is as shown in Fig. $P8.20$. Determine the values of the control constants k_p and k_d which will result in a settling time T_s less than 1 s and a peak overshoot of less than 20 percent.

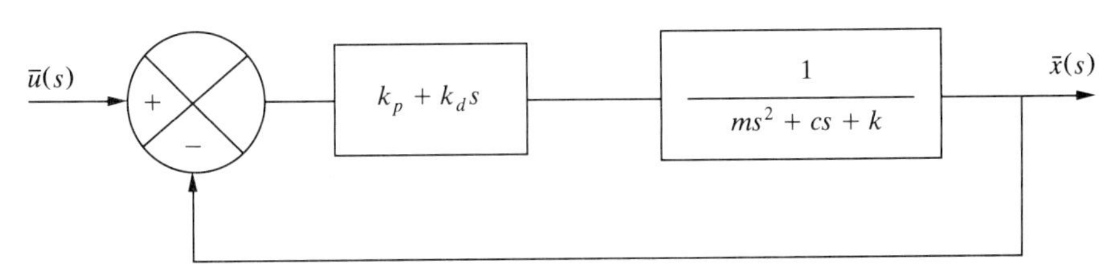

FIGURE ***P*8.20**

21. The rolling motion of a high-performance airplane is to be controlled. The system block diagram, including the proposed controller, is shown in Fig. $P8.21$. Determine the system closed-loop transfer function and the order of the controlled system, and explain how the stability of the controlled system would be determined.

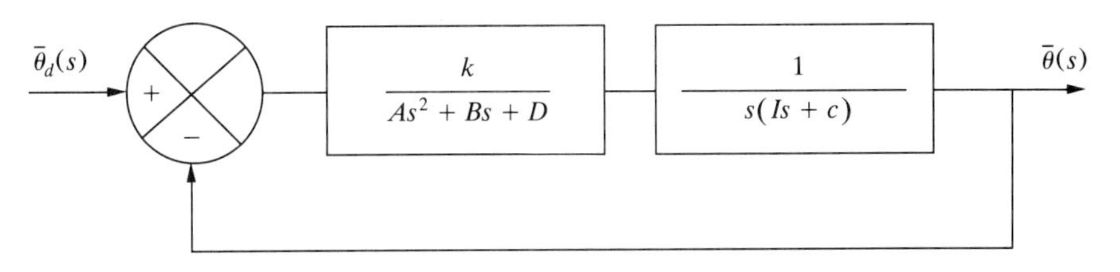

FIGURE ***P*8.21**

22. A mass is restrained by a linear spring and a damper as shown in Fig. $P8.22$. The spring is connected to a rigid base, while the damper is connected to a moving plate which undergoes a specified motion $u(t)$. Determine the transfer function $H(s)$ which relates the input $u(t)$ and the output $x(t)$. Determine the steady-state response to a ramp input, $u(t) = at$, where a is a specified constant.

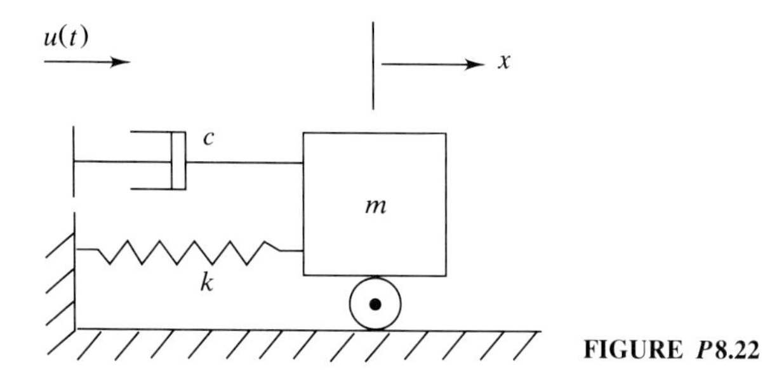

FIGURE P8.22

23. The attitude control of a lunar lander is modeled by the block diagram of Fig. P8.23. $G_c(s)$ is the transfer function which specifies the control action. I is the lander moment of inertia, $\theta(t)$ is the lander orientation angle, which is to be controlled, and $\theta_d(t)$ is the desired orientation angle.
   a. Suppose that $G_c(s) = k_p$ (proportional control). Determine the stability of the controlled system.
   b. Suppose that $G_c(s) = k_d s$ (derivative control). Determine the stability of the controlled system.
   c. Would you recommend either (a) or (b) as a control strategy? Explain.

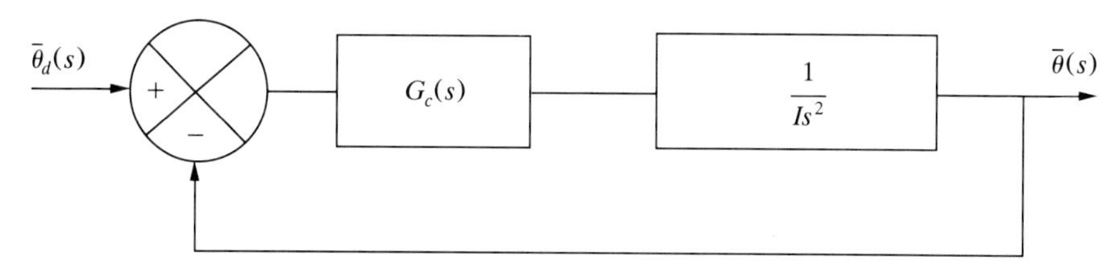

FIGURE P8.23

24. The mathematical model of a third-order linear system is given below:

$$\ddot{x} + 2\xi\omega\dot{x} + \omega^2 x = \alpha y$$

$$\dot{y} + \beta y + \alpha\dot{x} = V(t)$$

The states are x, $\dot{x}$, and y. The input is $V(t)$ and the output is $x(t)$. Construct a block diagram of the system, and find the system transfer function.

25. The block diagram of a single input/single output system is shown in Fig. P8.25. Determine the closed-loop transfer function and determine the range of the control constant k for which the system is stable. Determine the range of k for which the system will exhibit overshoot in response to a step input. Does k have any effect on the settling time which results for a step input? Explain.

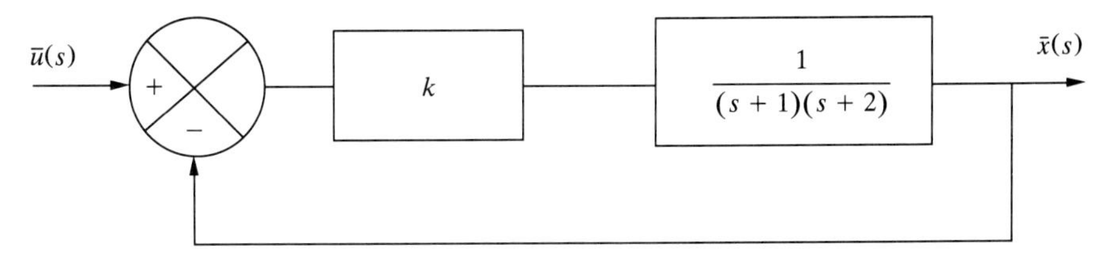

FIGURE ***P*8.25**

26. The open-loop transfer function $G(s)$ for a certain unity feedback system has a zero at -3 and poles at 0, -1, and -23 in the complex plane. Plot the approximate root locus of the system as the gain k is increased from 0 to ∞. Identify the asymptotes and the location at which these asymptotes intersect the real axis. Identify any breakaway points. Will the system be stable for very large gains? Explain.

27. A radar tracking dish is to track an airplane, such that if the airplane moves across the sky at the fixed angular rate $\theta_A = 3°/\text{s}$, the steady-state tracking error is 0.1° or less. In addition, the largest system time constant should be less than or equal to 0.5 s (Fig. *P*8.27). A PD controller is proposed to control the radar dish motion, $G_c(s) = k_p + k_d s$. If $I = 20$, $c = 5$, $k_p = 10$, and $k_d = 1$, all in appropriate units, determine whether the design specifications are realized.

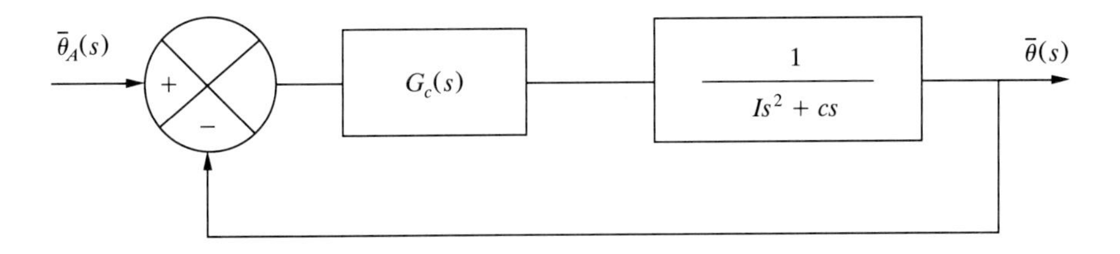

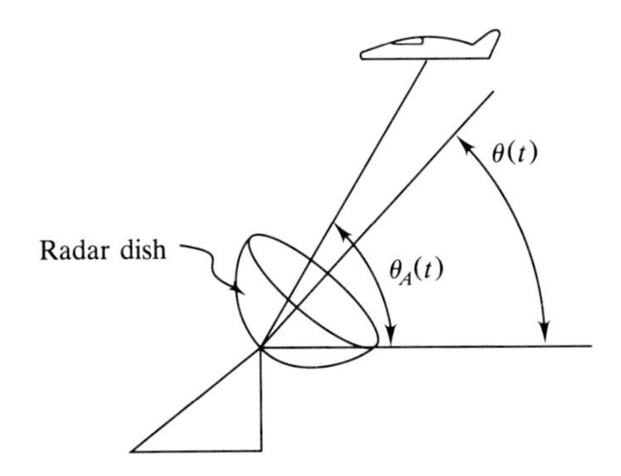

FIGURE ***P*8.27**

28. The open-loop transfer function $G(s)$ for a certain unity feedback system has zeroes $z_1 = -1.5$, $z_2 = -5.5$, and poles $s_1 = 0$, $s_2 = -1$, $s_3 = -5$. Sketch the root locus for this system.

29. Sketch, or determine via computer calculation, the Bode plots for the system of Exercise 8.28 for the case in which the control gain equals unity. Identify the gain and phase margins and determine the approximate gain k required to achieve a controlled system having a pair of dominant poles with damping factor $\xi = 0.707$.

30. The open-loop transfer function for a unity feedback system is given by

$$G(s) = \frac{2.5k(s+4)}{(s^2+2s+10)(s+1)}$$

where k is the control gain. Construct the Bode plots for the magnitude and phase and determine the gain and phase margins for the case $k = 1$. Determine whether the system is stable for this condition, and estimate the percentage overshoot if the system were subjected to a step input. Would you expect the overshoot to increase or decrease for gains $k > 1$? Explain.

31. Verify that (8.66), when inverted, leads to the step response given by (8.67), for the first-order system with proportional control.

32. Consider the 4 state system of Example 8.37, Fig. 8.46, with the following variation. Let a control force $u(t)$ be applied to mass 2, as in Fig. 8.46. Let a second control force, given by $0.618u(t)$, be applied to mass 1, i.e., the control force applied to mass 1 is always to be 0.618 times as large as that applied to mass 2. Work through the analysis of Example 8.37, with the same desired pole placements, and show that the algebraic equations which replace (8.141) have no solution; thus the system is not controllable.

CHAPTER 9

NUMERICAL METHODS FOR DYNAMIC SYSTEMS ANALYSIS

9.1 INTRODUCTION

In the first eight chapters the fundamentals of dynamics systems modeling and analysis, as well as applications in classical dynamics, vibrations, and controls, were covered. In obtaining solutions to the ODE models to describe the free and forced response of these systems, we have considered mostly low-order linear systems for which analytical solutions could be found easily. Analytical solutions have several advantages: (1) The qualitative behavior of the response is directly perceived. (2) The solutions show how each system parameter affects the response; thus the effect of changing a system parameter can be deduced directly. (3) Analytical solutions allow simple system identification techniques to be employed to find approximate values for (at least some of) these parameters, as in Sec. 7.2.1.

Many real engineering systems encountered, however, are of high order and/or exhibit nonlinear behavior. This usually necessitates that at least some of the analysis be implemented on a digital computer. Section 9.2 contains a summary of the main types of numerical analysis required in the study and computer simulation of dynamic systems. The remaining sections of this chapter then address the more important of these items in more detail. The student should always keep in mind that computer analysis/simulation will be successful only if the analyst has a firm grasp of the fundamental ideas of dynamic systems. It is easy to develop a tendency to allow computer simulation to substitute for thought or to become so enmeshed in the details of the computation that one begins to lose sight of the original objectives of the analysis. The computer is merely a very fast calculator and should be treated accordingly. Many sophisticated and novel dynamic systems were developed and analyzed in precomputer days, largely as a result of the creative abilities of the designers/analysts involved.

9.2 SUMMARY OF SOME COMPUTER APPLICATIONS

Here we describe some of the more commonly encountered generic dynamic systems problems for which numerical analysis via digital computation is useful. In the brief descriptions that follow, actual numerical techniques are not discussed. The intent is only to provide insight into the *types* of situations in which computer analysis is typically used.

9.2.1 Finding Roots of Polynomial Equations

In state variable form the free response of a linear time invariant system is governed by the autonomous math model (4.62). An equivalent description for SISO control systems is defined by the system closed-loop transfer function $T(s)$. The time course of the free response and the stability of equilibrium are then defined by the eigenvalues λ_i of the system matrix $[A]$ (Sec. 4.2.1) or by the poles p_i of the closed-loop transfer function $T(s)$. For a given linear system, the poles p_i and eigenvalues λ_i are the same and are found by solving the system characteristic equation, which may be expressed in the following form for an nth-order system [see (4.76)]:

$$\lambda^n + a_1\lambda^{n-1} + a_2\lambda^{n-2} + \cdots + a_{n-1}\lambda + a_n = 0 \tag{9.1}$$

Each coefficient a_i in (9.1) is generally a function of one or more of the system parameters, and the solutions to (9.1) may be real or complex. If $n \geq 3$, the characteristic equation (9.1) is usually solved numerically. A root locus analysis (Sec. 8.5) will necessitate that solutions to (9.1) be obtained many times to generate the root locus plot.

The roots of (9.1) allow one to infer system stability and to see the qualitative nature of the free response. A similar problem arises in determining the natural frequencies ω_i of a linear, multi-degree-of-freedom vibrating structure (Sec. 7.3). Here the characteristic equation is of the form (7.56) for an n-DOF system:

$$\omega^{2n} + a_1\omega^{2n-2} + \cdots + a_{n-1}\omega^2 + a_n = 0 \tag{9.2}$$

The solutions for ω^2, which are real and positive, define the natural frequencies ω_i of the n-DOF vibratory system. This information is always needed in a structural dynamics analysis.

9.2.2 Solving Algebraic Equations

The problem of finding solutions x_i to inhomogeneous linear algebraic equations of the form

$$[A]\{x\} = \{b\} \tag{9.3}$$

with the solution formally written as $\{x\} = [A]^{-1}\{b\}$, may arise in several ways: (1) finding the equilibrium solution for a linear dynamic system, as in Sec. 4.1.2, (2) finding the particular solution for a driven linear system by the method of undetermined coefficients [these coefficients are the entries in the vector $\{x\}$ in (9.3)], and (3) in using certain system identification procedures, as discussed in Sec. 9.5. Numerical solutions to (9.3) might be found using Gauss elimination, Cholesky decomposition, or other methods.

Another problem that sometimes arises is finding solutions to *nonlinear* algebraic equations. An example of this type of problem would be finding the various equilibrium solutions for a nonlinear system. This problem is defined, for an n-state system, by (4.6):

$$f_i(\bar{x}_1, \ldots, \bar{x}_n) = 0 \qquad i = 1, \ldots, n$$

where $\bar{x}_1, \ldots, \bar{x}_n$ are the equilibrium values of the states.

9.2.3 The Algebraic Eigenvalue Problem

We have encountered two forms of this problem. The first defines the free response of a linear n-state system, as in Sec. 4.2.1,

$$[[A] - \lambda[I]]\{C\} = \{0\} \tag{9.4}$$

and the second defines the natural frequencies and mode shapes of an undamped, linear vibratory system (Sec. 7.3.1),

$$[[k] - \omega^2[m]]\{C\} = \{0\} \tag{9.5}$$

In general, we would like to know both the eigenvalues λ_i (or ω_i^2) and the associated eigenvectors $\{C\}_i$. There are numerous ways to solve (9.4) or (9.5) via numerical analysis. In practice, "canned routines" are available on most computer systems, and these can be accessed to provide the complete solution. Thus, in analyzing the eigenvalue problem on the computer, one does not normally find the eigenvalues by actually forming the characteristic equation as described in Sec. 9.2.1. We would typically specify the entries in the matrix $[A]$ (or of $[k]$ and $[m]$) and call the canned routine, which then finds both the eigenvalues and eigenvectors for us.

9.2.4 Numerical Integration of ODEs

Numerous techniques have been developed to find numerical solutions to ordinary differential equations. The most commonly used methods require that the ODE model be put into the state variable form (4.2):

$$\dot{x}_i = f_i(x_1, \ldots, x_n, t) \qquad i = 1, \ldots, n \tag{9.6}$$

The system parameters and initial conditions $x_i(0)$ are specified and the solution generated for each of the n states. The idea behind these techniques is based on the Taylor series: If we know the value of a function $x(t)$ at some time t, and if we also know its first few derivatives $\dot{x}, \ddot{x}, \ldots,$ at time t, then we can use the Taylor series to obtain an approximate value of the function $x(t + \Delta t)$, that is, a short time Δt later.

Numerical integration of the system math model (9.6) can be of substantial aid in the following situations (among others):

1. *If the system (9.6) is nonlinear*: We have discussed in Chap. 4 how to linearize the system model (9.6) to study the motions near an equilibrium solution. Frequently, however, we encounter problems in which the behavior "far from equilibrium" (another way to say that nonlinear effects are important) is important, or for which equilibrium solutions are not relevant. An example of this latter case is the calculation of the trajectory of a projectile or space vehicle reentering the atmosphere. Here, although equilibrium solutions may exist mathematically, they are not relevant in physical terms. The difficulty encountered in analyzing nonlinear systems is that, even for "simple" nonlinearities, it is usually impossible to obtain exact analytical solutions.
2. *If inputs or system characteristics are not available in analytical form*: An example of this type would be determination of the response of a structure (such as a building or offshore platform) to an excitation having the characteristics of Fig. 5.25. In a real application excitation "data" such as in Fig. 5.25 would exist as a series of recorded values versus time which would be stored on a tape, disk, or in a data file. Even though the system may be linear, we cannot obtain an analytical solution to the problem because the input is not described by a mathematical expression. Numerical integration is commonly used in this type of situation.

Another situation of this type that is commonly encountered is in the description of the constitutive behavior of system components. For example, the flow of hydraulic fluid through a control valve orifice, as in Example 3.3, was idealized as being proportional to the product $x_v\sqrt{\Delta p}$, where x_v is the control valve opening and Δp is the pressure difference across the valve. In actuality, however, nonideal effects such as viscosity may render this constitutive model inaccurate for some operating conditions, and tests would be conducted to measure the actual flows for various valve openings and pressure differences. These data would be stored in a data file or table and retrieved as needed by doing a "table lookup," which involves interpolation between values in the table as the calculations are being performed.

Numerical integration of ODEs is one of the most, if not the most, important computer tools of the dynamics analyst and nowadays is assumed to be within the capabilities of any practicing dynamic systems analyst.

9.2.5 System Identification

System identification involves the formulation of constitutive models to describe all cause-and-effect relationships of relevance and to then identify, in specific cases, the values of the coefficients appearing in those models. We can categorize this process in two ways:

1. *Separable components*: Many systems contain components that can be analyzed, tested, and modeled on their own. Examples include shock absorbers, coil springs, servovalves, and electromagnetic devices. The behavior of such components can be studied and quantified in isolation from the system of which the components are a part.
2. *Nonseparable effects*: Some system characteristics cannot be identified until the entire system is put together and operated. A good example in this category is system damping in structural systems. For instance, the damping properties of buildings, structural supports, and airplane wings, can be determined accurately only for the actual system as a whole because there are several mechanisms responsible. They cannot be modeled in isolation, and their overall effect is dependent on the configuration of the system as a whole. For single degree of freedom vibratory systems (Sec. 7.2.1), a method for identification of the system damping factor ξ via free vibration testing was described, and this represents a simple example of the system identification process.

In Sec. 9.5 we shall address some examples of system identification for the latter of the two preceding cases. The objective will be to present ways to select system constants so that the system ODE model closely matches the results of actual system behavior observed during tests. Computerized methods are indispensable in this process if the system is at all complicated.

9.2.6 Finite Elements

There are numerous applications of computerized analysis in addition to those described so far. We will close this chapter with an introduction to one of these techniques, the finite element method. This is a method for obtaining approximate solutions to system math models that consist of *partial* differential equations. Partial differential equations arise in cases for which the unknown variables depend not only on time but also on location within the system. Our purpose in Sec. 9.6 will be to present only some of the basic ideas underlying the finite element method, and we will limit the analysis to a simple structural problem: the vibrating beam.

9.2.7 Implementation

In implementing computer solutions to the types of problems described here, there are essentially three levels at which the analyst may work, as follows.

First, one may decide (or have) to write a computer program from scratch to implement all of the numerical calculations needed. This requires an understanding of the techniques of numerical analysis and computer coding.

At the second level the analyst utilizes *canned routines* available on the system being used to do most of the difficult computations. For instance, if you have access to a mainframe computer or to a workstation, there will generally be available numerous canned routines that can be called to do the bulk of the specialized calculations needed. Routines that find the solutions to simultaneous linear or nonlinear algebraic equations, or find the roots of polynomial equations, or solve the algebraic eigenvalue problems (9.4) or (9.5), or do numerical integration of ODEs, or calculate the Fourier transform of a time series, are common. If you are using a personal computer, software packages such as MATHCAD or MATHEMATICA will contain some of these options, as well as others. Thus, in utilizing canned routines, the major programming effort is often expended in specifying input information and in storage and manipulation of output data files to obtain printout and plots of the results.

At the third level are *general-purpose programs*, defined here as any software package that is run without having to do any programming. Examples of general-purpose programs include finite element packages such as ANSYS and IDEAS, control system design packages such as LAS, and computer-aided design packages such as CADKEY.

A cautionary note. In the mechanically oriented engineering profession, computerization is (easily) the most significant change to have occurred in the past 40 years. Modern computer analysis/design tools such as general-purpose programs allow engineers to accomplish things that would have been impossible 40 years ago. Computerized analysis, however, is a two-edged sword. It is easy to develop a tendency to use the computer as a substitute for understanding of the underlying mechanics of the systems being modeled, designed, or analyzed. In order to use the computer successfully, one must be able to (1) check the results to make sure they are correct and (2) understand the significance of those results. Neither of these can be accomplished without an understanding of the fundamental aspects of the underlying engineering science.

9.3 THE ALGEBRAIC EIGENVALUE PROBLEM

We have encountered the algebraic eigenvalue problem in analyzing two types of linear dynamic systems: (1) the state variable form (4.62), which describes undriven motions in the vicinity of equilibrium,

$$\{\dot{x}\} = [A]\{x\} \Rightarrow [[A] - \lambda[I]]\{C\} = \mathbf{0} \tag{9.7}$$

where $\{x\} = \sum_{i=1}^{n} \alpha_i \{C\}_i e^{\lambda_i t}$, and (2) the undamped free vibration problem (7.52),

$$[m]\{\ddot{x}\} + [k]\{x\} = \mathbf{0} \Rightarrow \big[[k] - \omega^2[m]\big]\{C\} = \mathbf{0} \tag{9.8}$$

where $\{x\} = \sum_{i=1}^{n} \alpha_i \{C\}_i \cos(\omega_i t - \phi_i)$. In either case the eigenvalues and eigenvectors of (9.7) or (9.8) define the solution to the problem.

The algebraic eigenvalue problem is also encountered in several applications that we have not discussed, most notably in finding the principal stresses (and their axes) in a loaded elastic solid and in finding the principal moments of inertia (and their axes) of a rotating rigid body. In this section we briefly outline a way to solve (9.7) or (9.8) by iteration. We then discuss a procedure that is commonly used to study the stability of near-equilibrium motions for systems which are mathematically "complex." This numerical stability analysis involves the direct numerical determination of the system Jacobian matrix $[A]$.

9.3.1 Solution of the Algebraic Eigenvalue Problem

Here we describe a conceptually simple, iterative way to find the eigenvalues and eigenvectors of (9.7) and (9.8) for the case in which the eigenvalues are *real*. Starting with the form (9.7), the procedure is implemented by writing (9.7) as (9.7) as

$$[A]\{C\} = \lambda\{C\} \tag{9.9}$$

We then come up with an initial guess, or "trial vector," for the eigenvector $\{C\}$; denote this guess as $\{\hat{C}\}^{(0)}$, where the hat indicates an approximation or guess and the zero superscript indicates that this is the initial approximation. Following our usual convention that the top entry of an eigenvector is to be unity, $\{\hat{C}\}^{(0)}$ has the form

$$\{\hat{C}\}^{(0)} = \begin{Bmatrix} 1 \\ \vdots \\ \vdots \end{Bmatrix}$$

The next step is to plug this trial vector into the left-hand side of (9.9). The result on the right-hand side of (9.9) generates a new trial vector $\{\hat{C}\}^{(1)}$ according to

$$[A]\{\hat{C}\}^{(0)} = \lambda^{(1)}\{\hat{C}\}^{(1)}$$

where the iterative approximation $\lambda^{(1)}$ to the eigenvalue λ is the scale factor needed to make the top entry of $\{\hat{C}\}^{(1)}$ be unity. This procedure is continued iteratively, so that after n iterations the result may be written as

$$[A]\{\hat{C}\}^{(n-1)} = \lambda^{(n)}\{\hat{C}\}^{(n)} \qquad (n \geq 1) \tag{9.10}$$

The iterative procedure is carried out until convergence is achieved, that is $\{\hat{C}\}^{(n-1)}$ is sufficiently close to $\{\hat{C}\}^{(n)}$.

It turns out that this iterative procedure will converge to the eigenpair $\lambda, \{C\}$ associated with the eigenvalue of *largest* magnitude. Thus, the initial guess $\{\hat{C}\}^{(0)}$ is not too important, but faster convergence will be achieved if $\{\hat{C}\}^{(0)}$

is "close to" the actual eigenvector. This procedure is illustrated below with a simple example.

Example 9.1 The System of Example 4.19 Revisited. The problem is to determine the eigenvalue of largest magnitude, and the associated eigenvector, for the matrix $[A]$ of Example 4.19:

$$[A] = \begin{bmatrix} 5 & 3 \\ -6 & -4 \end{bmatrix}$$

For reference we state the exact results obtained for Example 4.19:

$$\lambda_1 = -1, \qquad \{C\}_1 = \begin{Bmatrix} 1 \\ -2 \end{Bmatrix}$$

$$\lambda_2 = 2, \qquad \{C\}_2 = \begin{Bmatrix} 1 \\ -1 \end{Bmatrix}$$

The iterative procedure will converge to $\lambda_2, \{C\}_2$. We start the procedure by choosing the trial vector $\{\hat{C}\}^{(0)} = \begin{Bmatrix} 1 \\ 1 \end{Bmatrix}$, a common starting point [but note that $\{\hat{C}\}^{(0)}$ is not very close to $\{C\}_2$]. The first iteration, (9.10), yields

$$\begin{bmatrix} 5 & 3 \\ -6 & -4 \end{bmatrix} \begin{Bmatrix} 1 \\ 1 \end{Bmatrix} = \begin{Bmatrix} 8 \\ -10 \end{Bmatrix} = 8 \begin{Bmatrix} 1 \\ -1.25 \end{Bmatrix}$$

The ensuing iterations are summarized below:

$$n = 2: \quad \begin{bmatrix} 5 & 3 \\ -6 & -4 \end{bmatrix} \begin{Bmatrix} 1 \\ -1.25 \end{Bmatrix} = \begin{Bmatrix} 1.25 \\ -1.0 \end{Bmatrix} = 1.25 \begin{Bmatrix} 1 \\ -0.8 \end{Bmatrix}$$

$$n = 3: \quad \begin{bmatrix} 5 & 3 \\ -6 & -4 \end{bmatrix} \begin{Bmatrix} 1 \\ -0.8 \end{Bmatrix} = \begin{Bmatrix} 2.6 \\ 2.8 \end{Bmatrix} = 2.6 \begin{Bmatrix} 1 \\ -1.077 \end{Bmatrix}$$

$$n = 4: \quad \begin{bmatrix} 5 & 3 \\ -6 & -4 \end{bmatrix} \begin{Bmatrix} 1 \\ -1.077 \end{Bmatrix} = \begin{Bmatrix} 1.769 \\ -1.692 \end{Bmatrix} = 1.769 \begin{Bmatrix} 1 \\ -0.9565 \end{Bmatrix}$$

$$\vdots$$

$$n = 7: \quad \begin{bmatrix} 5 & 3 \\ -6 & -4 \end{bmatrix} \begin{Bmatrix} 1 \\ -0.9895 \end{Bmatrix} = \begin{Bmatrix} 2.0306 \\ -2.042 \end{Bmatrix} = \underset{\substack{\uparrow \\ \lambda^{(7)}}}{2.0306} \underset{\substack{\uparrow \\ \{\hat{C}\}^{(7)}}}{\begin{Bmatrix} 1 \\ -1.0056 \end{Bmatrix}}$$

We observe that the procedure does converge to the eigenpair $\lambda = 2, \{C\}_2 = \begin{Bmatrix} 1 \\ -1 \end{Bmatrix}$.

A similar procedure may be employed to determine the eigenpair for the eigenvalue having the *smallest* magnitude. This is done by premultiplying (9.9) by $[A]^{-1}$ and then dividing by λ (assuming it not to be zero) to obtain

$$[A]^{-1}\{C\} = \frac{1}{\lambda}\{C\} \tag{9.11}$$

with the associated iterative formula

$$[A]^{-1}\{\hat{C}\}^{(n-1)} = \frac{1}{\lambda^{(n)}}\{\hat{C}\}^{(n)} \tag{9.12}$$

Comparison of (9.11) and (9.9) reveals that the eigenvalues of $[A]^{-1}$ are just the reciprocals of the eigenvalues of $[A]$. Thus, the iterative sequence (9.12) finds the smallest (in magnitude) eigenvalue of $[A]$ by finding the largest eigenvalue of $[A]^{-1}$.

Similar ideas are often applied to the linear, multi-DOF free vibration problem (9.8). First, write (9.8) as follows:

$$\omega^2[m]\{C\} = [k]\{C\} \tag{9.13}$$

Next premultiply (9.13) by $[k]^{-1}$ and divide by ω^2 to obtain

$$[k]^{-1}[m]\{C\} = \frac{1}{\omega^2}\{C\}$$

or

$$[D]\{C\} = \frac{1}{\omega^2}\{C\} \tag{9.14}$$

Where the "dynamic matrix" $[D] \equiv [k]^{-1}[m]$ contains all the information pertaining to the system mass and stiffness properties. The form (9.14) is just (9.12) with $[D]$ replacing $[A]^{-1}$ and ω^2 replacing λ. Thus, if we express (9.14) in the iterative form

$$[D]\{\hat{C}\}^{(n-1)} = \frac{1}{\omega^{2(n)}}\{\hat{C}\}^{(n)} \tag{9.15}$$

the iterations will converge to the mode having the *lowest* natural frequency ω_1. This procedure can be quite useful, as the mode shapes in the free vibration problem have an obvious physical meaning, and we can often formulate a reasonable guess $\{\hat{C}\}^{(0)}$ from physical considerations. Furthermore, knowledge of the lowest natural frequency of a vibratory system is very useful information in terms of assessing the responses likely to occur.

Example 9.2 The Structure of Example 7.3 Revisited. The problem is to determine the lowest natural frequency and mode shape for the three-story building structure of Example 7.3, shown in Fig. 9.1, for the case of equal masses and stiffnesses, with numerical values noted on the figure.

The mass and stiffness matrices for this system are (verify!)

$$[m] = m\begin{bmatrix} 1 & 0 & 0 \\ 0 & 1 & 0 \\ 0 & 0 & 1 \end{bmatrix}; \qquad [k] = k\begin{bmatrix} 2 & -1 & 0 \\ -1 & 2 & -1 \\ 0 & -1 & 1 \end{bmatrix}$$

The inverse of the stiffness matrix turns out to be

$$[k]^{-1} = \frac{1}{k}\begin{bmatrix} 1 & 1 & 1 \\ 1 & 2 & 2 \\ 1 & 2 & 3 \end{bmatrix}$$

and the dynamic matrix $[D] = [k]^{-1}[m]$ is then

$$[D] = \frac{m}{k}\begin{bmatrix} 1 & 1 & 1 \\ 1 & 2 & 2 \\ 1 & 2 & 3 \end{bmatrix}$$

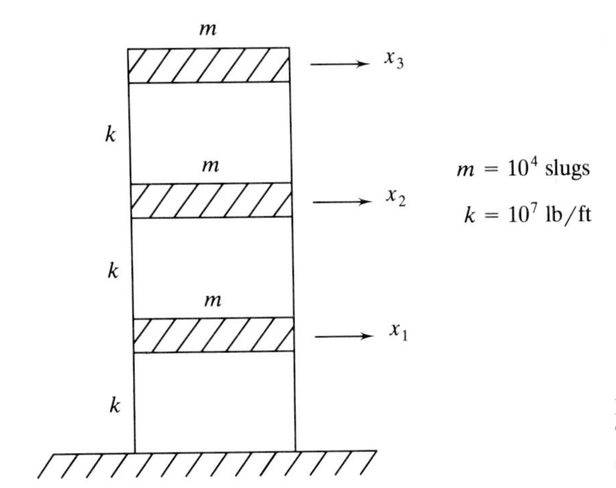

FIGURE 9.1
Three-story building structure of Example 9.2.

The iterative sequence (9.15) is then defined by

$$\begin{bmatrix} 1 & 1 & 1 \\ 1 & 2 & 2 \\ 1 & 2 & 3 \end{bmatrix} \{\hat{C}\}^{(n-1)} = \frac{10^3}{\omega^{2(n)}} \{\hat{C}\}^{(n)}$$

where $k/m = 10^3/\text{s}^2$. We select a trial vector $\{\hat{C}\}^{(0)}$ based on the expectation that, for the lowest-frequency mode, all three masses move together, with mass 2 moving with greater amplitude than mass 1, and with mass 3 moving with greater amplitude than mass 2. A reasonable starting point is

$$\{\hat{C}\}^{(0)} = \begin{Bmatrix} 1 \\ 2 \\ 3 \end{Bmatrix}$$

which leads to the first iterative result

$$\begin{bmatrix} 1 & 1 & 1 \\ 1 & 2 & 2 \\ 1 & 2 & 3 \end{bmatrix} \begin{Bmatrix} 1 \\ 2 \\ 3 \end{Bmatrix} = \begin{Bmatrix} 6 \\ 11 \\ 14 \end{Bmatrix} = 6 \begin{Bmatrix} 1 \\ 1.833 \\ 2.333 \end{Bmatrix}$$

This leads to the first estimate $\omega_1^{(1)}$ of the lowest natural frequency as

$$\frac{10^3}{\omega_1^{2(1)}} = 6 \Rightarrow \omega_1^{(1)} = 12.91 \text{ rad/s}$$

Results for the second and third iterations are given below:

$$n = 2: \quad \begin{bmatrix} 1 & 1 & 1 \\ 1 & 2 & 2 \\ 1 & 2 & 3 \end{bmatrix} \begin{Bmatrix} 1 \\ 1.833 \\ 2.333 \end{Bmatrix} = \begin{Bmatrix} 5.166 \\ 9.333 \\ 11.666 \end{Bmatrix} = 5.166 \begin{Bmatrix} 1 \\ 1.8065 \\ 2.2581 \end{Bmatrix}$$

$$\omega_1^{(2)} = 13.91 \text{ rad/s}$$

$$n = 3: \quad \begin{bmatrix} 1 & 1 & 1 \\ 1 & 2 & 2 \\ 1 & 2 & 3 \end{bmatrix} \begin{Bmatrix} 1 \\ 1.8065 \\ 2.2581 \end{Bmatrix} = \begin{Bmatrix} 5.0646 \\ 9.1291 \\ 11.3873 \end{Bmatrix} = 5.0646 \begin{Bmatrix} 1 \\ 1.8026 \\ 2.2484 \end{Bmatrix}$$

$$\omega_1^{(3)} = 14.05 \text{ rad/s}$$

After only three iterations we have achieved close convergence to the exact first-mode solution $\omega_1 = 14.08$ rad/s; $\{C\}_1^T = \lfloor 1 \quad 1.8018 \quad 2.2472 \rfloor$.

The preceding procedure can be extended to calculate the higher modes of vibration. This requires a bit of work, however, and we only discuss the general idea here. For a readable account of the details, see Ref. 7.2.

In order to determine the higher modes, we first need to recall [as in (7.59)] that, at any stage of the iteration process, the iterate $\{\hat{C}\}^{(n)}$ may be represented mathematically as a linear combination of the N actual mode shapes, which span N space. Thus, we can write

$$\{\hat{C}\}^{(n)} = \gamma_1^{(n)}\{C\}_1 + \gamma_2^{(n)}\{C\}_2 + \cdots + \gamma_N^{(n)}\{C\}_N \tag{9.16}$$

In the preceding procedure, as the iteration number increases, $\gamma_1^{(n)} \to 1$, and the rest of the γ's $\to 0$: The procedure converges to the lowest mode. Actually, the iterative procedure will converge to the lowest mode *present* in the iterates $\{\hat{C}\}^{(n)}$. Normally this will be $\{C\}_1$ because our chances of selecting a $\{\hat{C}\}^{(0)}$ which contains no contribution from $\{C\}_1$ are extremely remote. If, however, we can "subtract out" the contribution of $\{C\}_1$ to $\{\hat{C}\}^{(n)}$ at each stage of iteration, then $\gamma_1^{(n)} \equiv 0$ and the procedure will converge to the second mode. Similar ideas can be applied for the higher modes, and the interested reader should consult Clough and Penzien (Ref. 7.2) for details.

In this Sec. 9.3.1 we have presented the rudiments of one iterative technique for finding eigenvalues and eigenvectors numerically. Several other techniques exist, as noted in the references. Modern numerical eigenvalue/eigenvector methods can be fairly involved to code, so that canned routines are used extensively to solve such problems.

9.3.2 Computer Stability Analysis

The discussion in this section does not relate to how the eigenvalue problem might be solved. Rather, it deals with a numerical procedure for constructing the system Jacobian matrix $[A]$ for situations in which this is difficult to do analytically. For an n-state system undergoing free motion, the generally nonlinear mathematical model describing these motions has the form (4.5), repeated here:

$$\begin{aligned} \dot{x}_1 &= f_1(x_1, \ldots, x_n) \\ \dot{x}_2 &= f_2(x_1, \ldots, x_n) \\ &\vdots \\ \dot{x}_n &= f_n(x_1, \ldots, x_n) \end{aligned} \tag{9.17}$$

The linearized motions in the vicinity of an equilibrium solution $(\bar{x}_1, \bar{x}_2, \ldots, \bar{x}_n)$ are then defined by the vector-matrix form (4.62)

$$\{\dot{x}\} = [A]\{x\} \tag{9.18}$$

where the x_i in (9.18) now represent small deviations from their equilibrium values $\bar{x}_i$. The system or Jacobian matrix $[A]$ is defined by the following matrix of partial derivatives:

$$[A] = \begin{bmatrix} \frac{\partial f_1}{\partial x_1} & \frac{\partial f_1}{\partial x_2} & \cdots & \frac{\partial f_1}{\partial x_n} \\ \frac{\partial f_2}{\partial x_1} & \frac{\partial f_2}{\partial x_2} & \cdots & \frac{\partial f_2}{\partial x_n} \\ \vdots & \vdots & & \vdots \\ \frac{\partial f_n}{\partial x_1} & \frac{\partial f_n}{\partial x_2} & \cdots & \frac{\partial f_n}{\partial x_n} \end{bmatrix} \tag{9.19}$$

Each of the partial derivatives in (9.19) is to be evaluated at equilibrium. In Chap. 4 (Examples 4.15 through 4.17) we considered examples for which the calculation of the Jacobian matrix entries $(\partial f_i/\partial x_j)$ was straightforward. It sometimes occurs, however, that the system order n is sufficiently large and/or the individual ODEs in (9.17) are sufficiently complex that analytical calculation of all the entries $(\partial f_i/\partial x_j)$ is impractical. For such system models the terms appearing in the Jacobian matrix (9.19) may be calculated numerically by the following procedure:

1. First determine the equilibrium solution to be investigated; this may involve finding solutions to the nonlinear algebraic equations $f_i(x_1, \ldots, x_n) = 0$. Denote the equilibrium values as $\bar{x}_1, \bar{x}_2, \ldots, \bar{x}_n$.
2. Add a small increment Δx_1 to the first state, so that $x_1 = \bar{x}_1 + \Delta x_1$, and hold the remaining states x_i at their equilibrium values $\bar{x}_i$. This moves the system state slightly from equilibrium in the x_1 direction in phase space.
3. Evaluate, on the computer, all n of the functions $f_i(\bar{x}_1 + \Delta x_1, \bar{x}_2, \ldots, \bar{x}_n)$ and define the resulting changes in the f_i as Δf_i; that is

$$\Delta f_i = f_i(\bar{x}_i + \Delta x_1, \bar{x}_2, \ldots, \bar{x}_n) - f_i(\bar{x}_1, \bar{x}_2, \ldots, \bar{x}_n)$$

4. Next form the n ratios $\Delta f_i/\Delta x_1$. These ratios will be approximations to the derivatives $(\partial f_i/\partial x_1)$ that compromise the *first column* of the Jacobian matrix $[A]$.
5. To generate the second column of $[A]$, evaluate the functions $f_i(\bar{x}_1, \bar{x}_2 + \Delta x_2, \bar{x}_3, \ldots, \bar{x}_n)$, which will define the partial derivatives $(\partial f_i/\partial x_2)$. Continue this procedure until all of the columns of $[A]$ are filled out.

The result is a numerically determined Jacobian matrix for the specific set of parameter values being considered. The eigenvalues and eigenvectors for the associated eigenvalue problem (9.7) can then be found by calling a canned routine, and the stability of equilibrium can be determined. The procedure can be repeated as necessary for different values of the system parameters to check their effect on stability.

Comment: For the procedure to work accurately, the imposed state changes Δx_j must be sufficiently small that the Δf_i are determined by the first-order terms in the Taylor series expansion (4.43). This can be checked easily by computing the Δf_i for state changes Δx_j and $2\Delta x_j$. If for all i $\Delta f_i(2\Delta x_j) \cong 2\Delta f_i(\Delta x_j)$, then Δx_j is small enough to be in the linear range.

Example 9.3 A Two-State Nonlinear System. The following two-state nonlinear system is to be analyzed:

$$\begin{aligned} \dot{x}_1 &= x_1 - x_1^3 - \frac{4}{3}x_1x_2 \\ \dot{x}_2 &= -x_2 + 2x_1x_2 \end{aligned} \tag{9.20}$$

Note: For this system the Jacobian matrix $[A]$ is easily calculated analytically, and it is not necessary to determine $[A]$ using the numerical procedure of this section. This example was chosen as a simple one with which to illustrate the calculation procedure. The analytically determined Jacobian matrix is given by

$$[A] = \begin{bmatrix} \left(1 - 3x_1^2 - \frac{4}{3}x_2\right) & \frac{-4}{3}x_1 \\ 2x_2 & (-1 + 2x_1) \end{bmatrix}_{\text{equil}} \tag{9.21}$$

where the entries in $[A]$ are to be evaluated at the equilibrium solution being analyzed.

The first step in the analysis is to find the equilibrium solutions by finding all of the solutions to the two nonlinear algebraic equations

$$\begin{aligned} x_1 - x_1^3 - \frac{4}{3}x_1x_2 &= 0 \\ -x_2 + 2x_1x_2 &= 0 \end{aligned}$$

Writing the above as

$$\begin{aligned} x_1\left(1 - x_1^2 - \frac{4}{3}x_2\right) &= 0 \\ x_2(-1 + 2x_1) &= 0 \end{aligned}$$

it is straightforward to show that there are actually *four* equilibrium solutions, given by

$$\begin{aligned} &\bar{x}_1 = \bar{x}_2 = 0 \\ &\bar{x}_1 = \pm 1, \qquad \bar{x}_2 = 0 \\ &\bar{x}_1 = \frac{1}{2}, \qquad \bar{x}_2 = \frac{9}{16} \end{aligned}$$

These are shown as points in the phase plane in Fig. 9.2.

For a complete description of the near-equilibrium motions, we should determine the four associated Jacobian matrices $[A]$ and their eigenvalues and eigenvectors.

To illustrate the numerical procedure leading to (9.19), we will evaluate the Jacobian matrix $[A]$ for the equilibrium solution $\bar{x}_1 = 1$, $\bar{x}_2 = 0$. To generate the

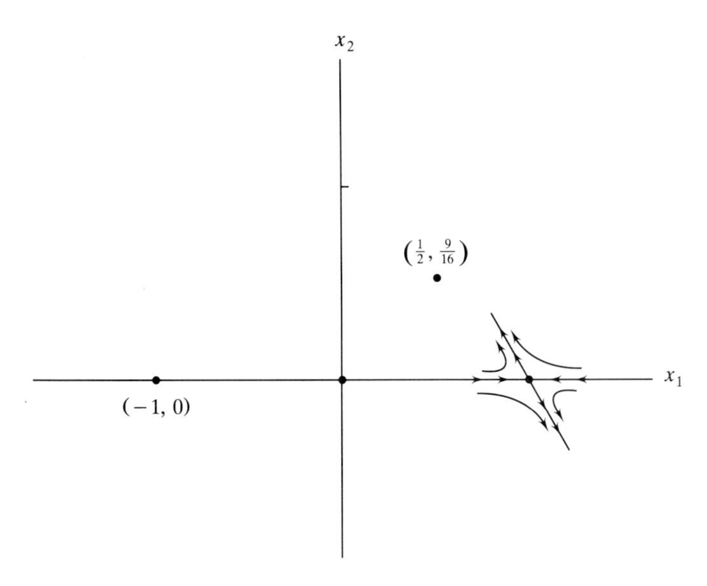

FIGURE 9.2
Location of equilibrium solutions for the system (9.20). The linearized motions near $\bar{x}_1 = 1$, $\bar{x}_2 = 0$ are also indicated.

first column of $[A]$, we let $\Delta x_1 = 0.01$ and $\Delta x_2 = 0$, so that we plug $x_1 = \bar{x}_1 + \Delta x_1 = 1.01$, $x_2 = \bar{x}_2 = 0$ into the right-hand sides of (9.20). Numerical evaluation then yields $\Delta f_1 = -0.0202$, $\Delta f_2 = 0$, so that the partial derivatives in the first column of $[A]$ are given (approximately) by $(\partial f_1/\partial x_1) \cong (\Delta f_1/\Delta x_1) = -2.02$; $(\partial f_2/\partial x_1) \approx (\Delta f_2/\Delta x_1) = 0$. We next let $\Delta x_1 = 0$, $\Delta x_2 = 0.01$ to generate the second column of $[A]$, with the results $\Delta f_1 = -0.01333$, $\Delta f_2 = 0.01$, leading to $(\partial f_1/\partial x_2) \approx -1.333$, $(\partial f_2/\partial x_2) \approx 1.0$. The numerically determined Jacobian matrix for the equilibrium solution $(1, 0)$ is then given by

$$[A]_{(1,0)} \cong \begin{bmatrix} -2.02 & -1.333 \\ 0 & 1.0 \end{bmatrix}$$

For comparison the exact matrix, obtained using the equilibrium values $\bar{x}_1 = 1$, $\bar{x}_2 = 0$ in (9.21), is

$$[A]_{(1,0)} = \begin{bmatrix} -2.0 & -1.333 \\ 0 & 1.0 \end{bmatrix} \tag{9.22}$$

Use of smaller values for Δx_i (for example, 10^{-4}) will yield more accurate results.

In the same way the Jacobian matrices for the other three equilibrium solutions may be determined, and the stability and nature of the phase portraits in the neighborhood of each equilibrium point may be assessed. For instance, the eigenvalues and eigenvectors obtained from (9.22) are $\lambda_1 = 1$, $\{C\}_1^T = \lfloor 1 \quad -\frac{9}{4} \rfloor$, $\lambda_2 = -2$, $\{C\}_2^T = \lfloor 1 \quad 0 \rfloor$. This equilibrium is unstable, with orbits as indicated in Fig. 9.2.

9.4 NUMERICAL INTEGRATION OF ODEs

9.4.1 Introduction

Obtaining solutions to ODE models via numerical integration is one of the most common computer tasks performed by dynamic systems analysts. Over the years many numerical integration schemes have been developed. In this section we present only the basic idea behind the techniques and then concentrate on the fourth-order Runge-Kutta (RK4) method, the foot soldier of numerical integration schemes. RK4 is stable, accurate, and easy to program. A typical subroutine (canned program) library will contain several numerical integrators which can be called. The fourth-order Runge-Kutta procedure, however, is so easy to implement that it is commonly programmed by the analyst.

The basic idea of numerical integration is first illustrated for the simplest situation possible, the first-order autonomous ODE:

$$\dot{x} = f(x); \qquad x(t_0) = x_0 \tag{9.23}$$

Given an initial value $x_0 = x(t_0)$ at some time t_0, we wish to generate the solution $x(t)$ for times $t > t_0$. In numerical integration schemes the procedure is to define a time increment, or *integration step size*, $\Delta t = h$ and then to obtain estimates of the solution at times $t = t_0 + nh$ (where $n = 1, 2, \dots$); that is, we find $x(t_0 + \Delta t), x(t_0 + 2\Delta t), \dots$. Thus, we do not obtain an analytical solution. Rather, we obtain approximations at discrete times $t > t_0$. The time step $\Delta t = h$ is generally made sufficiently small that necessary detail of the solution is obtained, and plotting routines can then construct a continuous curve $x(t)$ based on the discrete values calculated. But we emphasize that solutions will be found only at discrete times.

In (9.23) note that, geometrically, $f(x)$ is the *slope* of the function $x(t)$. Thus, if we know $x(t)$, we can compute its slope $\dot{x}$ from (9.23). Initially ($t = t_0$), then, we know both x and its slope $\dot{x} = f(x)$. The basic idea of numerical integration can be seen by writing (9.23) in the differential form

$$dx = f(x)\,dt$$

If we integrate this equation from $t = t_0$ to $t = t_0 + h$, that is, over one integration step, we obtain

$$\int_{x=x(t_0)}^{x=x(t_0+h)} dx = \int_{t=t_0}^{t=t_0+h} f(x)\,dt$$

or

$$x(t_0 + h) - x(t_0) = \int_{t_0}^{t_0+h} f(x)\,dt \tag{9.24}$$

Now define $x(t_0 + h) - x(t_0) \equiv \Delta x(h)$, and rewrite the right-hand side of (9.24) as follows:

$$\Delta x(h) = \left[\frac{1}{h}\int_{t_0}^{t_0+h} f(x)\,dt\right]h \equiv f_{\text{av}}h \tag{9.25}$$

In (9.25) the quantity in brackets is the *average slope* f_{av} of the function $f(x)$

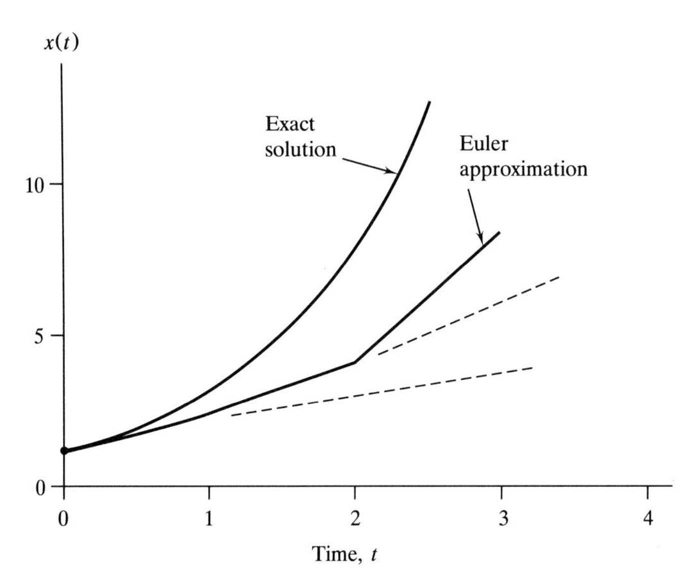

FIGURE 9.3
The exact solution $x(t) = e^t$ of (9.26) and the Euler approximation with a time step $h = 1$ s.

over the integration interval $t_0 \to t_0 + h$. Equation (9.25) tells us what we know intuitively: The *change* $\Delta x(h)$ in x, over some time interval h, can be obtained by multiplying the *average* slope in that time interval by the length h of the interval. Essentially, numerical integration schemes employ various strategies to estimate the average slope in the current integration interval. Note that in (9.25) the integral on the right-hand side cannot generally be found analytically, because f is a function of x rather than of the integration variable t.

The simplest numerical integration scheme is the Euler method, which is now described. To fix ideas, we consider a specific version of (9.23) for which the exact solution is known:

$$\dot{x} = x \Rightarrow x(t) = x_0 e^t \tag{9.26}$$

where we have assumed $t_0 = 0$. We will also assume that $x_0 = 1$, so that $x(t) = e^t$ is the exact solution, as shown in Fig. 9.3.

In the Euler scheme the average slope in the interval $(0, h)$ is approximated by the slope at the *beginning* of the interval, that is $f_{\text{av}} \approx f(0)$. The approximation for $x(h)$ will then be given by

$$x(h) = x(0) + f(0)h \tag{9.27}$$

Geometrically, we calculate the slope at the beginning of the interval and extend it to the end of the interval. The relation (9.27) is recognized as simply the leading terms of a Taylor series expansion for $x(h)$,

$$x(h) = x(0) + \left(\frac{dx}{dt}\right)_0 h + \frac{1}{2}\left(\frac{d^2x}{dt^2}\right)_0 h^2 + \frac{1}{6}\left(\frac{d^3x}{dt^3}\right)_0 h^3 + \cdots$$

where $(dx/dt)_0 = f(0)$. For the exponential function e^t, we know the form of the Taylor series to be as follows:

$$x(t) = 1 + t + \frac{t^2}{2} + \frac{t^3}{6} + \cdots + \frac{t^n}{n!} + \cdots$$

Thus, the Euler method applied to this particular problem approximates the actual function $x(t)$ with the truncated Taylor series $x(t) = 1 + t$. Clearly this will not be very accurate if $t = h$ is "large." For instance, suppose we select an integration step size $h = 1$ s. Then the derivative $f(0) = 1$ and from (9.27) we obtain x (1 s) = 2, whereas the exact value is $x(1) = e$. There is a large error. In order to continue the approximation, we would take the calculated value at $t + h, x(h) = 2$, and determine the slope $f(h)$ based on this calculated value $x(h)$. As shown in Fig. 9.3, the calculated slope $f(h) = f(2) = 2$. We then extend this slope from $t = h$ to $t = 2h$ to obtain the approximation $x(2h) = x(h) + f(h)h = 4$. A third step in this integration sequence leads to a value $x(3h) = x(3) = 8$, as shown in Fig. 9.3. Generally, the result after n integration steps may be written as

$$x_n = x_{n-1} + f_{n-1}h \tag{9.28}$$

where $x_n = x(nh), x_{n-1} = x((n-1)h)$, and $f_{n-1} = f(x_{n-1})$. This procedure is very easy to implement on the computer, but it is not very accurate unless the integration step size is very small.

The Euler scheme is easily extended to the case of an arbitrary number n of state equations of the form (9.6). We consider as an example the case $n = 2$, so that (9.6) reads

$$\dot{x}_1 = f_1(x_1, x_2, t)$$
$$\dot{x}_2 = f_2(x_1, x_2, t)$$

Note that the possibility that the state derivatives f_1 and f_2 are explicitly time dependent has been included. This presents no problem in numerical integration.

We assume that initial values $x_1(0)$ and $x_2(0)$ are specified. This information serves to define both state derivatives f_1 and f_2 at $t = 0$. Thus, we can determine approximately the solution at $t = h$, that is, after one integration step,

$$x_1(h) = x_1(0) + hf_1[x_1(0), x_2(0), 0]$$
$$x_2(h) = x_2(0) + hf_2[x_1(0), x_2(0), 0]$$

The solution is then extended stepwise, so that at any time $t_n = nh$, we have

$$x_1(t_n) = x_1(t_{n-1}) + hf_1[x_1(t_{n-1}), x_2(t_{n-1}), t_{n-1}]$$
$$x_2(t_n) = x_2(t_{n-1}) + hf_2[x_1(t_{n-1}), x_2(t_{n-1}), t_{n-1}]$$

The extension of the procedure to a system with any number of states is straightforward.

In the Euler scheme, as with any numerical integration scheme, there are two sources of error. First, by approximating the average slope f_{av} by the slope at the beginning of the integration interval, some error results for x_n, even if the value x_{n-1} is correct (as it is here only at $t = 0$); we are ignoring higher-order terms in the Taylor series. This type of error we refer to as *local error*. The second type of error is *global*, or *accumulated*, *error*. Errors in x_n will accumulate because of errors that have built up during all preceding integration steps. The local error of the Euler method will be proportional to h^2, the order of the neglected terms in the Taylor series. The accumulated error, however, will be proportional to h. In summary, the Euler method uses the incorrect slope (producing local error), and this slope is calculated using an incorrect value of x, due to accumulated error.*

In order to remedy the accuracy problem of the Euler method, we have to use a sufficiently small step size h, which usually turns out to be very small, making the computation relatively costly. For this reason, the Euler method is seldom used in dynamic systems simulation, and we now consider more accurate techniques.

The Runge-Kutta family of integration schemes is now described. The basic idea is illustrated by the *second-order scheme* (RK2), as follows: Suppose that at some stage t_n in the integration process, we have the solution $x(t_n)$ for a single-state system. We wish to extend the solution to $t_{n+1} = t_n + h$. First, calculate the derivative $f[x(t_n)]$ at the beginning of the integration interval. Then extend this derivative to obtain a value $x(t_{n+1}) = x(t_n) + f(t_n)h$, just as in the Euler method. Now, however, calculate an approximation to the derivative at the *end* of the integration interval using the Euler value of x; that is, determine $f(x_n + hf_n)$. Then take the *average* of the two derivatives to obtain an improved estimate of the average slope in the integration interval. Using this average slope to extend the solution to t_{n+1} results in the second-order Runge-Kutta solution:

$$x_{n+1} = x_n + \frac{1}{2}h\Big[\underbrace{f(x_n)}_{\substack{\text{Slope at beginning}\\ \text{of interval, based}\\ \text{on current value } x_n}} + \underbrace{f(x_n + hf_n)}_{\substack{\text{Approx. slope at end of}\\ \text{interval, based on Euler}\\ \text{extrapolation } x_n + hf_n}}\Big] \tag{9.29}$$

It turns out that the approximation (9.29) is equivalent to computing $x(t_n + h)$ from the second-order Taylor series

$$x(t_n + h) \cong x(t_n) + hf(t_n) + \frac{h^2}{2}f'(t_n)$$

*The discussion of this paragraph assumes infinite precision in all numerical calculations. Actually, computer roundoff is yet another source of error, and it may be significant if accurate results are required over many (for example, 10^4) integration steps. If in any doubt, use double precision.

TABLE 9.1
Illustration of second-order Runge-Kutta

n	t	Exact solution	f_n	$f(x_n + hf_n)$	$\langle f \rangle$	x_n	(Euler, 0.25) x_n
1	0.0	1.0	1.0	1.5	1.25	1.0	1.0
2	0.5	1.649	1.625	2.438	2.0313	1.625	1.5625
3	1.0	2.718	2.641	3.961	3.301	2.641	2.441
4	1.5	4.482	4.292	6.437	5.365	4.292	3.814
5	2.0	7.389	6.924	10.461	8.718	6.974	5.659
6	2.5	12.182	11.333	17.00	14.166	11.333	8.839
7	3.0	20.086				18.416	13.811

The accuracy is improved in comparison to the Euler method. The local error is of the order h^3, while the accumulated error is proportional to h^2. The generation of the numerical solution is illustrated in Table 9.1, which shows results for the earlier example $\dot{x} = x$, $x(0) = 1$, using an integration step size $h = 0.5$ s. In this table the approximate average slope in the integration interval is denoted by $\langle f \rangle$.

Shown in the last column is the Euler solution for a time step $h = 0.25$ s. We note that, for a given h, RK2 requires essentially twice as much computation as the Euler method. Thus, in comparing the accuracy of the two methods, we should compare RK2 results with step size h to Euler results with step size $h/2$. Table 9.1 illustrates the superiority of RK2. Nevertheless, the accuracy of the RK2 results in Table 9.1 is poor, and a much smaller step size h would be needed for an accurate numerical solution.

9.4.2 The Fourth-Order Runge-Kutta Method

As the name is meant to imply, the fourth-order Runge-Kutta (RK4) is a procedure for obtaining accuracy which is the same as that of the fourth-order Taylor series; all terms through h^4 are retained. The idea of RK4 is similar to RK2: We calculate the derivative $\dot{x} = f$ in four different ways and then average them to obtain an estimate of the average slope in the integration interval.

The four derivative evaluations are summarized below for the case of a single-state equation $\dot{x} = f(x, t)$. We assume we have a value x_n at $t_n = nh$, and we wish to extend the solution to find x_{n+1} at t_{n+1}.

1. Let $r_1 = hf(x_n, t_n)$; r_1 is the *change* in x we would get using the Euler method, i.e., based on the slope at the beginning of the integration interval.
2. Let $r_2 = hf[x_n + (r_1/2), t_n + (h/2)]$; r_2 is the change in x based on an approximation to the slope at the *middle* of the integration interval, that is, at $t_n + (h/2)$.
3. Let $r_3 = hf[x_n + (r_2/2), t_n + (h/2)]$; r_3 is the change in x based on an updated approximation to the slope at the middle of the integration interval.

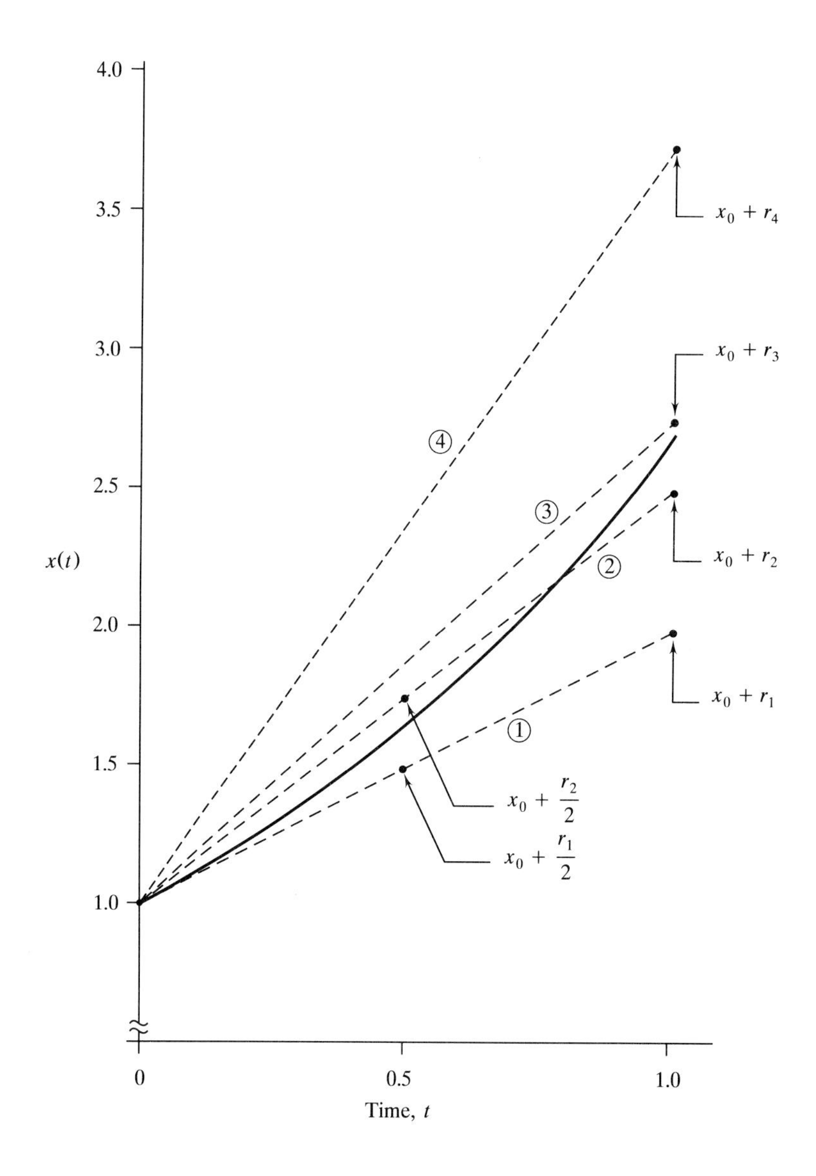

FIGURE 9.4
Graphical depiction of RK4 steps: 1 = Extension based on the slope $f(x_0) = f(1) = 1$; 2 = Extension based on slope $f(x_0 + \frac{r_1}{2}) = f(1.5) = 1.5$; 3 = Extension based on slope $f(x_0 + \frac{r_2}{2}) = f(1.75) = 1.75$; and 4 = Extension based on slope $f(x_0 + r_3) = f(2.75) = 2.75$. The calculated value from (9.30) is $x_1 = 2.7083$.

4. Let $r_4 = hf(x_n + r_3, t_n + h)$; r_4 is the change in x based on the estimated slope at the *end* of the integration interval.

The preceding definitions of r_1, r_2, r_3, and r_4 represent the changes in x, over one integration step, resulting from four ways of calculating the slope f. The RK4 approximation for x_{n+1} is then obtained as the following weighted average:

$$x_{n+1} = x_n + \frac{1}{6}(r_1 + 2r_2 + 2r_3 + r_4) \tag{9.30}$$

Notice that r_2 and r_3, which are based on the estimated slopes at the middle of the integration step, are weighted more heavily than r_1 and r_4, which are based on slopes at the ends of the interval. The various steps outlined above are shown in Fig. 9.4, which assumes $h = 1.0$, $\dot{x} = x$, and $x(0) = 1$.

The value x_1 which results is $x_1 = 2.7083$, reasonably close to the exact value $x_1 = e = 2.7183$, even though $h = 1.0$ would be considered a large time step. Note also that the results are much better than are obtained using RK2 with $h = 0.5$ or Euler with $h = 0.25$ (Table 9.1).

Example 9.4 Coding RK4 to Study a Nonlinear Oscillator. In Sec. 10.2 we consider the free oscillation of a single DOF oscillator for which the damping force is proportional to the square of the velocity, as may occur due to fluid dynamic effects. The system model is defined by (10.3), repeated below:

$$\ddot{x} + c\dot{x}|\dot{x}| + x = 0 \qquad (c > 0) \tag{9.31}$$

The response $x(t)$ versus t and phase portraits x versus $\dot{x}$ are shown in Figs. 10.3 and 10.4 for the case $c = 0.16$ and initial conditions $x(0) = 10$, $\dot{x}(0) = 0$. The nonlinearity makes it difficult to obtain an analytical solution, and the results shown in Figs. 10.3 and 10.4 were obtained using the numerical integration procedure RK4 described here. To implement RK4 for this system, we first convert the model (9.31) to state variable form. Letting $x_1 = x$ and $x_2 = \dot{x}$, the state equations are

$$\begin{aligned} \dot{x}_1 &= x_2 \\ \dot{x}_2 &= -cx_2|x_2| - x_1 \end{aligned} \tag{9.32}$$

A computer code, in Fortran, which implements the fourth-order Runge-Kutta procedure, is described below (input and output information does not appear). The following definitions are adopted:

`I = 1 or 2`	will identify the state x_1 or x_2.
`J = 1, 2, 3 or 4`	will identify which of the four RK steps is being done.
`X(I), I = 1 or 2,`	will define the values of the two states at the *end* of the integration interval.
`Y(I), I = 1 or 2,`	will define the values of the two states at the beginning of the integration interval.

`F(I), I=1 or 2,`	defines the two-state derivatives according to (9.32).
`R(I, J)`	identifies r_1, r_2, r_3, or r_4 (J) for each of the two states (I).
`H`	is the integration time step.

A typical RK4 Fortran code, with annotations, is given below. The code assumes single-precision arithmetic.

```
t = 0.0 (initialize time)
X(1) = 10.0 (initial conditions)
X(2) = 0.0 (initial conditions)

10    CONTINUE
       DO 20 I = 1,2  } Reinitialization for next
       Y(I) = X(I)    } integration step.
20    CONTINUE
      DO 200 J = 1, 4 → RK4 loop starts.
       F(1) = X(2)                          } State equations.
       F(2) = -C*X(2)*ABS(X(2))- X(1)       }
      If (J-1) 110, 100, 110 → goes to 100 first time through.
100   DO 120 I = 1, 2
      R(I, 1) = F(I)*H → Determines r1 for each state.
120   X(I) = Y(I) + F(I)*H / 2.0 → calculates states at middle of step in
      preparation for midslope calculation.
      GO TO 200 → recalculates F(I) based on midstep values of X(I).
110   IF (J-3) 130, 140, 150
130   DO 160 I = 1, 2
      R(I, 2) = F(I)*H → finds r2 for each state.
160   X(I) = Y(I) + F(I)*H / 2.0 → recalculates states at midstep in
      preparation for another midslope calculation.
      GO TO 200
140   DO 170 I = 1, 2
      R(I, 3) = F(I)*H → finds r3 for each state.
170   X(I) = Y(I) + F(I)*H → Calculate states at end of step in preparation
      for end slope calculations.
      GO TO 200
150   DO 180 I = 1, 2
      R(I, 4) = F(I)*H
180   X(I) = Y(I)+(R(I, 1)+2.*(R(I, 2)+R(I, 3))+R(I, 4))/6.0
      [Values of x1 and x2 at end of step as in (9.30)].
200   CONTINUE
      T = T + H → update time.
      (Print or save output if desired.)
      GO TO 10 → goes back to start and reinitializes for next integration step.
```

The results shown in Figs. 10.3 and 10.4, as well as many others in this book, were generated using this code.

Trick: In the state equations (9.32) of this example, the time t does not appear explicitly. If the time *were* to appear explicitly in any of the state equations, coding would have to be added to ensure usage of the proper time in the four evaluations of the derivatives $F(I)$. Specifically, if t is the current time, R(I, 1) is based on t, R(I, 2) and R(I, 3) are based on $t + h/2$, and R(I, 4) is based on $t + h$. An easy way to take care of this is to define time t as an extra "state," via the equation $\dot{t} = 1$; that is, $F(3) = 1$, which would be added to the code following the relations for $F(1)$ and $F(2)$. All loops involving the state counter I then change to $DO - I = 1, 3$. This procedure will automatically keep track of the correct time and is commonly used by analysts.

Comments

1. Doing numerical integration requires that an appropriate integration time step h be selected. A simple procedure that can often be used to establish an acceptable h follows. Pick a candidate step size h_0 and define the sequence of step sizes $h_1 = h_0/2$, $h_2 = h_0/4, \ldots, h_n = h_0/2^n$. Start by comparing results obtained using the step size h_0 to those obtained using h_1. If there are noticeable differences, then h_0 is too large, so compare the two cases h_1 and h_2. This procedure is continued until results for step sizes h_n and h_{n+1} are essentially identical, i.e., we reach the point where a reduction in step size causes essentially no improvement in the results, so we use h_n. For practical purposes we obtain the exact solution at the integration times $t_n = nh$. Many canned ODE integration routines will have the option of determining the appropriate time step internally, based on a user-supplied error criterion.
2. There are many numerical integration schemes in common usage; RK4 is just one possibility. For alternatives consult the references at the end of this chapter and your system software library, if available.

Concluding remarks

1. In order to obtain ODE solutions via numerical integration, one has to specify values of the system constants and the initial conditions. If one wishes to study the response for different initial conditions and/or different system constants, another simulation has to be done. Performing design-oriented numerical integration studies often requires that results be obtained for a large number of individual cases, with an attendant large number of computer simulations. A thoughtful evaluation of the results requires a thorough understanding of basic system dynamics.
2. In this book two ways to solve ODE models have been discussed: exact analytical solution techniques for linear systems and numerical integration methods, as discussed in this section. The reader may have inferred that the only way to analyze nonlinear systems is by numerical integration of the governing ODEs. This is not necessarily true. For many nonlinear problems in dynamics and other areas of engineering and science, there exist a variety of methods that can be used to obtain *approximate* analytical solutions to the governing ODE model. Such methods can provide considerable insight into the behavior of nonlinear systems. One of the simpler of these techniques is

introduced in Chap. 10 to study nonlinear oscillation phenomena. Several references are also listed in Chap. 10.

9.5 SYSTEM IDENTIFICATION

In this section are presented a few of the techniques used to do system identification. The discussion is limited to those situations in which the identification is made through experimentation on the system as a whole, as opposed to identifying the properties of individual components that can be isolated from the overall system and tested on their own (refer to the discussion in Sec. 9.2.5).

9.5.1 Solution-Based Techniques

In this class of problems the relevant ODE model is solved analytically, and the system constants are determined so that the analytical solution and the experimentally determined response agree to within some acceptable error. We have already been exposed to this procedure in Sec. 7.2.1 for the single degree of freedom, damped oscillator: The solution (7.10) leads to the relation (7.16), which is used to interpret the experimental amplitude data to estimate the system damping factor ξ, as in Figs. 7.4 and 7.6.

Analytical procedures can also be used for the harmonically excited oscillator. An example of this type of system identification is considered at the end of Sec. 7.2 (see Figs. 7.16 and 7.17 and the associated discussion). By exciting the system harmonically and then measuring the frequency response, the system natural frequency ω and damping factor ξ can be estimated.

We next consider an example for a system that differs from the oscillator systems discussed in the preceding paragraphs.

Example 9.5 Determination of Subsonic Drag Coefficients C_D. In the late 1970s planetary probes were launched to the planet Venus for the purpose of measuring the properties of the Venusian atmosphere.* Properties to be measured included atmospheric temperature, pressure and density, chemical composition, and optical properties. The probes were blunt bodies with instruments mounted on a bulbous afterbody, as shown schematically in Fig. 9.5.

The blunt body design provided high aerodynamic drag, quickly decelerating the probe from its atmospheric entry velocity of about 38,000 ft/s to its terminal velocity** below the speed of sound. Thus, the probe was to descend slowly,

---

*In an early example of Soviet-American scientific cooperation, the Russians measured the dirt, and we agreed to do the atmosphere.

**The terminal velocity occurs when the aerodynamic drag force and gravitational force are in balance, so that there is no change in velocity.

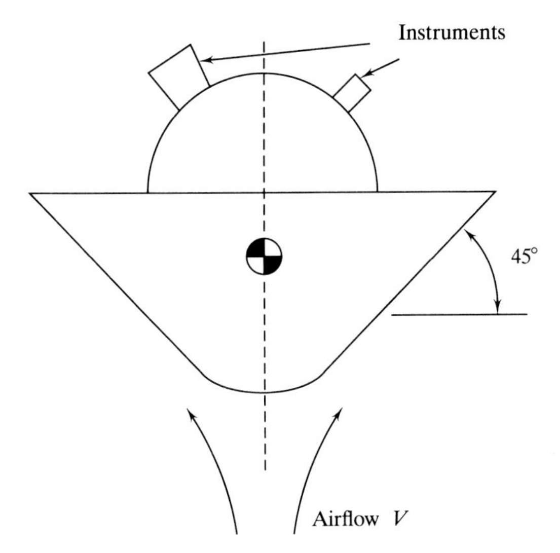

FIGURE 9.5
Schematic of Venusian probe in vertical descent.

allowing sufficient time for extensive measurements to be made; the actual descent took about a half-hour, as the atmosphere of Venus is *very* dense. In order to estimate the descent trajectory ahead of time, an accurate knowledge of the subsonic drag coefficient C_D was needed. This subsonic drag coefficient was difficult to calculate, in part due to the complex (from a fluid flow standpoint) shape of the instrumentation mounted on the aft end of the probe. The only reliable way to determine C_D at subsonic speeds was to devise a test to measure it experimentally.

The test facility used for this purpose was the NASA Langley Spin Tunnel, which is shown schematically in Fig. 9.6.

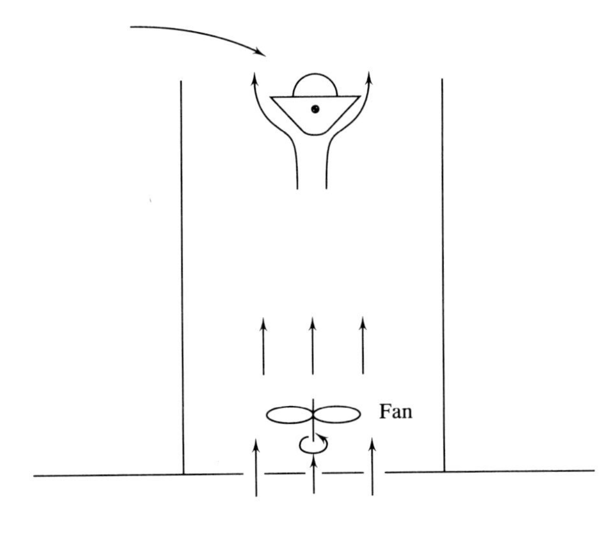

FIGURE 9.6
Schematic of NASA Langley Spin Tunnel.

This facility is a large cylindrical structure with a fan at the bottom. The fan forces air upward through the tunnel at an average velocity V which is controllable. A scale model of the probe is launched into the top of the tunnel and the air velocity adjusted so that the model remains essentially stationary. The measured air velocity V is then used to determine the subsonic drag coefficient as follows: From Example 2.7, the mathematical model for vertical descent near the surface of the earth applies (2.63). For the experimental conditions described, there is no vertical acceleration; the drag and gravity forces balance, so that

$$mg = \tfrac{1}{2}\rho C_D A V^2 \tag{9.33}$$

Equation (9.33) may be solved for the drag coefficient C_D as

$$C_D = \frac{2mg}{\rho A V^2}$$

where ρ is the tunnel air density, A is the cross-sectional area of the model, and mg is the model weight. The test revealed a value of subsonic drag coefficient $C_D \cong 0.75$. Note that, while the mathematics involved in this system identification problem were quite simple, the resulting information obtained was critical to the success of the mission. Further, in a variety of aerospace applications involving flight of vehicles through an atmosphere, system identification is nearly always critical to the design process. This is mainly due to the difficulty in calculating a priori the aerodynamic forces and moments that act on the vehicle. For instance, for the Pioneer Venus probes discussed here, numerous aerodynamic coefficients in addition to the drag had to be determined experimentally. These other coefficients are related mainly to the rigid body rotational dynamics of the probes.

9.5.2 ODE-Based Techniques

The "solution-based" techniques discussed in the preceding subsection require that the ODE model of the system be solvable analytically. In some situations, however, the ODEs may not be solvable in analytical form. This may occur if the model is nonlinear or if there are time-varying effects in the model. For such systems it is common to implement an iterative procedure, solving the ODEs by numerical integration, as follows: First we make an initial guess for the values of the system parameters to be found; then the governing ODEs are integrated numerically and the numerical solutions compared to the results of a test. The system parameters are then adjusted, and the numerical integration/comparison to test results repeated until we obtain good agreement between the ODE model and test results. One such method, based on a least squares approach, is described below.

Let us assume that we have the math model for an n-state system in the form (9.6),

$$\dot{x}_i = f_i(x_1, \ldots, x_n, t; c_1, c_2) \qquad i = 1, \ldots, n \tag{9.34}$$

In (9.34) c_1 and c_2 are two system constants or parameters which we wish to identify. Generally, there may be an arbitrary number of these constants. To keep the mathematics relatively simple, we will work out the least squares procedure for the special case of only two system constants. Once this result is obtained, we will see how to extend it to the case of an arbitrary number of system parameters to be identified. The objective, then, is to make some measurements of one or more of the states x_i and to find values of c_1 and c_2 which will cause the solutions of (9.34) to match closely the results of the experiment.

To fix ideas, let us suppose that during some time interval $[0, T]$ in which the system is in motion, we measure, at known times, the history of *one* of the states, say, x_m. We denote these measured values of x_m as $v_1, v_2, \ldots, v_N$; there is a total of N measured values made at the times $t_1, t_2, \ldots, t_N$; these times $t_1, \ldots, t_N$ at which the measurements are made are assumed to be known. The number N of data points should be sufficiently large.

Next, suppose that we make an initial guess for the values c_1 and c_2 of the system constants to be identified. Call these guessed values $\tilde{c}_1$ and $\tilde{c}_2$. If we put these guessed values into (9.34) and generate solutions for the states $x_1(t), \ldots, x_m(t), \ldots, x_n(t)$, typically via numerical integration of (9.34), then we can obtain *calculated* values of the state x_m at the measurement times $t_1, \ldots, t_N$. We will call these calculated values $y_1, y_2, \ldots, y_N$. To summarize: $v_1, \ldots, v_N$ are the N *measured values* of the state x_m, and $y_1, \ldots, y_N$ are the N *calculated values* of x_m, according to the model (9.34). We would like to find system parameters c_1 and c_2 which will render the y's and v's as "close as possible" to each other.

We now have to quantify "as close as possible." There are several ways to do this, and we adopt the least squares approach. To implement this approach, we first define an error E as follows:

$$E \equiv \sum_{k=1}^{N} (y_k - v_k)^2 \tag{9.35}$$

The error E is the sum, over all measurements $k = 1, \ldots, N$, of the squares of the differences between the calculated y_k and measured v_k values, so that $E \geq 0$. Observe that the values y_k calculated from the ODE model (9.34) will be a function of the system constants c_1 and c_2: If c_1 and c_2 are changed, then the solution to (9.34) will change. Thus, the error E will also be a function of c_1 and c_2, $E = E(c_1, c_2)$. The least squares method dictates that the best set of system parameters (c_1, c_2) is the one that minimizes E. Thus, we require that c_1 and c_2 be chosen so that

$$\begin{aligned} \frac{\partial E}{\partial c_1} &= 0 \\ \frac{\partial E}{\partial c_2} &= 0 \end{aligned} \tag{9.36}$$

Our initial guess $(\tilde{c}_1, \tilde{c}_2)$ will generally not render E minimum, so we need to develop a systematic, iterative way to improve c_1 and c_2 to drive the error toward its minimum. This process is shown schematically in Fig. 9.7.

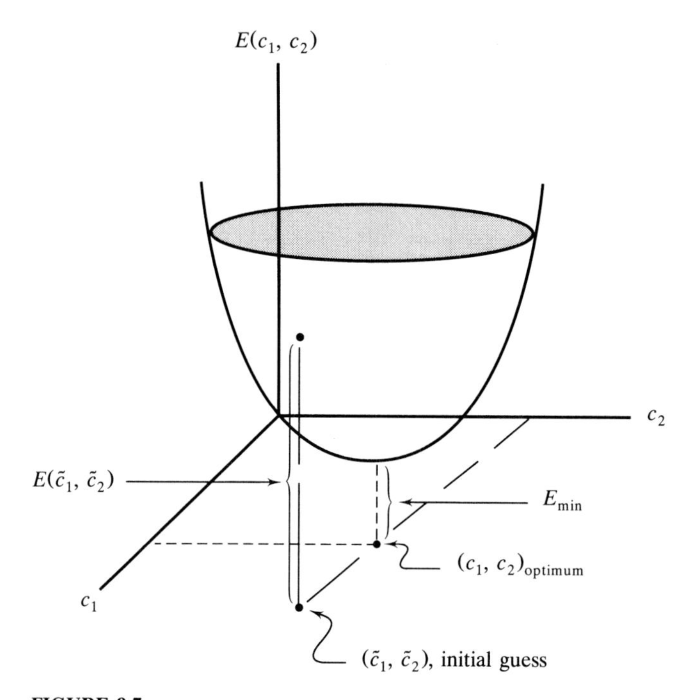

FIGURE 9.7
The initial guess $(\tilde{c}_1, \tilde{c}_2)$ for the system parameters produces an error $E(\tilde{c}_1, \tilde{c}_2)$ which may be large. We have to move from $(\tilde{c}_1, \tilde{c}_2)$ to $(c_1, c_2)_{opt}$ in order to minimize E. The conditions $(\partial E/\partial c_1) = (\partial E/\partial c_2) = 0$ dictate that the plane tangent to E will be horizontal at $(c_1, c_2)_{opt}$.

Given the initial guess $(\tilde{c}_1, \tilde{c}_2)$, we need a systematic way to determine increments $\Delta c_1, \Delta c_2$ such that $\tilde{c}_1 + \Delta c_1, \tilde{c}_2 + \Delta c_2$ will be an improved estimate of the system parameters. This can be done by combining (9.36) and (9.35) and noting that the y_k's (but not the v_k's) are functions of c_1 and c_2:

$$\begin{aligned} \frac{\partial E}{\partial c_1} &= \sum_{k=1}^{N} 2(y_k - v_k)\frac{\partial y_k}{\partial c_1} = 0 \\ \frac{\partial E}{\partial c_2} &= \sum_{k=1}^{N} 2(y_k - v_k)\frac{\partial y_k}{\partial c_2} = 0 \end{aligned} \tag{9.37}$$

Next let us use the first-order Taylor series to approximate the change in y_k that would result from small changes Δc_1 and Δc_2 in the system constants c_1 and c_2:

$$y_k = \tilde{y}_k + \frac{\partial y_k}{\partial c_1}\Delta c_1 + \frac{\partial y_k}{\partial c_2}\Delta c_2 \tag{9.38}$$

where $\tilde{y}_k = y_k(\tilde{c}_1, \tilde{c}_2)$ is the value of y_k resulting for the guessed values $\tilde{c}_1, \tilde{c}_2$ of the system constants. It is important to note that the partial derivatives $\partial y_k/\partial c_1$

and $\partial y_k/\partial c_2$ appearing in (9.38) can be calculated numerically, as follows: First, determine the $\tilde{y}_k = y_k(\tilde{c}_1, \tilde{c}_2)$ by integrating (9.34) numerically using the guessed values of the system constants. Then impose a small change Δc_1 in c_1 and calculate $y_k(\tilde{c}_1 + \Delta c_1, \tilde{c}_2)$. Determine the resulting changes Δy_k in the y_k's, $\Delta y_k = y_k(\tilde{c}_1 + \Delta c_1, \tilde{c}_2) - y_k(\tilde{c}_1, \tilde{c}_2)$. The partial derivatives with respect to c_1 are then approximated by $\partial y_k/\partial c_1 \approx \Delta y_k/\Delta c_1$. A similar procedure generates the partial derivatives with respect to c_2. This procedure for numerical calculation of partial derivatives is essentially the same as that used in Sec. 9.3.2 to determine, via computer, the system Jacobian matrix.

Substitution of (9.38) into (9.37) leads to the two equations

$$\sum_{k=1}^{N} \left(\tilde{y}_k + \frac{\partial y_k}{\partial c_1} \Delta c_1 + \frac{\partial y_k}{\partial c_2} \Delta c_2 - v_k \right) \frac{\partial y_k}{\partial c_1} = 0$$

$$\sum_{k=1}^{N} \left(\tilde{y}_k + \frac{\partial y_k}{\partial c_1} \Delta c_1 + \frac{\partial y_k}{\partial c_2} \Delta c_2 - v_k \right) \frac{\partial y_k}{\partial c_2} = 0$$

or

$$\begin{aligned} \sum_{k=1}^{N} \left[\left(\frac{\partial y_k}{\partial c_1} \right)^2 \Delta c_1 + \frac{\partial y_k}{\partial c_1} \frac{\partial y_k}{\partial c_2} \Delta c_2 \right] &= - \sum_{k=1}^{N} (\tilde{y}_k - v_k) \frac{\partial y_k}{\partial c_1} \\ \sum_{k=1}^{N} \left[\frac{\partial y_k}{\partial c_1} \frac{\partial y_k}{\partial c_2} \Delta c_1 + \left(\frac{\partial y_k}{\partial c_2} \right)^2 \Delta c_2 \right] &= - \sum_{k=1}^{N} (\tilde{y}_k - v_k) \frac{\partial y_k}{\partial c_2} \end{aligned} \tag{9.39}$$

The right-hand sides of (9.39) are recognized from (9.37) as merely half the negative partial derivatives of the error E, evaluated at the point $(\tilde{c}_1, \tilde{c}_2)$ in the "parameter space," here the c_1, c_2 plane. As Fig. 9.7 shows, these partial derivatives will be nonzero unless E happens to be at its minimum value. Also note that the parameter increments Δc_1 and Δc_2 can be moved outside the summations in (9.39), so that (9.39) can be written as follows:

$$\begin{aligned} \left[\sum_{k=1}^{N} \left(\frac{\partial y_k}{\partial c_1} \right)^2 \right] \Delta c_1 + \left[\sum_{k=1}^{N} \frac{\partial y_k}{\partial c_1} \frac{\partial y_k}{\partial c_2} \right] \Delta c_2 &= -\frac{1}{2} \frac{\partial \tilde{E}}{\partial c_1} \\ \left[\sum_{k=1}^{N} \frac{\partial y_k}{\partial c_1} \frac{\partial y_k}{\partial c_2} \right] \Delta c_1 + \left[\sum_{k=1}^{N} \left(\frac{\partial y_k}{\partial c_2} \right)^2 \right] \Delta c_2 &= -\frac{1}{2} \frac{\partial \tilde{E}}{\partial c_1} \end{aligned} \tag{9.40}$$

Note that $\partial \tilde{E}/\partial c_1$ and $\partial \tilde{E}/\partial c_2$ can be computed numerically in the same way that the partial derivatives $\partial y_k/\partial c_i$ are determined.

Equations (9.40) comprise a pair of simultaneous linear algebraic equations of the form

$$[A]\{\Delta c\} = \{r\} \tag{9.41}$$

where the elements a_{ij} of $[A]$ and r_i of $\{r\}$ are defined by

$$a_{ij} = \sum_{k=1}^{N} \frac{\partial y_k}{\partial c_i} \frac{\partial y_k}{\partial c_j}$$
$$r_i = -\frac{1}{2} \frac{\partial \tilde{E}}{\partial c_i} \tag{9.42}$$

If we solve these simultaneous equations for Δc_1 and Δc_2, we obtain a correction to the original estimates $\tilde{c}_1$ and $\tilde{c}_2$, so that an improved estimate is obtained as

$$c_1 = \tilde{c}_1 + \Delta c_1$$
$$c_2 = \tilde{c}_2 + \Delta c_2 \tag{9.43}$$

The result (9.43), while (we hope!) an improvement on the initial guess, will not generally render E minimum. This is because the Taylor expansion (9.38) is only an approximation. The correction given by (9.43) will generally move us from $(\tilde{c}_1, \tilde{c}_2)$ *toward* $(c_1, c_2)_{\text{opt}}$, but we will not land exactly on $(c_1, c_2)_{\text{opt}}$: We have to do more iterations. The analysis is redone using the values defined by (9.43) as the initial guess, and another correction is obtained. This process is repeated until convergence is achieved. As we get very close to $(c_1, c_2)_{\text{opt}}$, the right-hand side of (9.41) will approach zero.

We next consider a very simple example using this technique and then make some concluding remarks.

Example 9.6 Application to a Simple First-Order System. Consider a single-state system governed by the first-order linear model

$$\dot{x} = \alpha x \tag{9.44}$$

Suppose that we want to determine a value of the single system constant α from two measurements made for the case in which the initial condition $x(0) = 1$ is known exactly. We make measurements of x at $t_1 = 0.5$ s and at $t_2 = 1.0$ s and find $v_1 = 0.60653$, $v_2 = 0.36788$ (these "data" are exact results for the case $\alpha = -1$, which is the answer to the problem). The situation is shown in Fig. 9.8.

Clearly, this problem is easily solvable using the analytical solution to (9.44) to determine α, as described in Sec. 9.5.1; no analyst would use the least squares ODE method outlined here. We consider this example in order to illustrate the steps of the analysis for a very simple case.

Because there is only one system constant α to be found, (9.41) reduces to the single scalar equation

$$\left[\sum_{k=1}^{N} \left(\frac{\partial y_k}{\partial \alpha}\right)^2\right] \Delta\alpha = -\frac{1}{2} \frac{\partial \tilde{E}}{\partial \alpha} \tag{9.45}$$

which defines the correction $\Delta\alpha$ to be made in the system parameter. Furthermore,

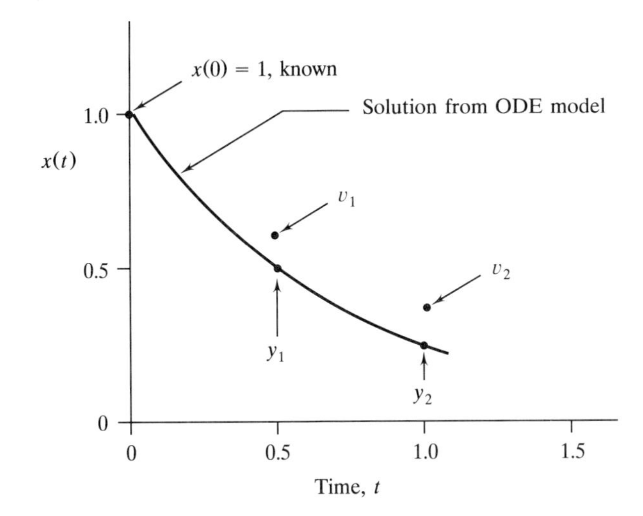

FIGURE 9.8
The known initial value $x(0)$ and the two measured values of the state x: $v_1 = x(\frac{1}{2})$, $v_2 = x(1.0)$. Also shown is the solution from the ODE model (9.44) for the initial guess $\tilde{\alpha} = -1.4$.

there are only $N = 2$ two data points, so (9.45) may be expanded to read

$$\left[\left(\frac{\partial y_1}{\partial \alpha}\right)^2 + \left(\frac{\partial y_2}{\partial \alpha}\right)^2\right]\Delta\alpha = -\frac{1}{2}\frac{\partial \tilde{E}}{\partial \alpha} \tag{9.46}$$

Let us begin the analysis by guessing a value $\tilde{\alpha} = -1.4$. Then the solution to the ODE model (9.44), based on the guessed value of α, is $x(t) = x(0)e^{\tilde{\alpha}t} = e^{-1.4t}$. This produces calculated values $y_1 = e^{-1.4(0.5)} = 0.49659$, $y_2 = e^{-1.4} = 0.24660$. This solution $x(t)$, based on the initial guess $\tilde{\alpha}$, is shown in Fig. 9.8. The error $E(\tilde{\alpha})$ is then found from (9.35) to be $\tilde{E} = (y_1 - v_1)^2 + (y_2 - v_2)^2 = 0.02680$. Note that the numerical value of this error, small here, has little meaning by itself. The next step in the analysis is to calculate the two partial derivatives $\partial y_1/\partial\alpha, \partial y_2/\partial\alpha$ in (9.46). We impose a small change of 0.01 in α, so that $\alpha = -1.39$, which implies $y_1 = 0.49907$, $y_2 = 0.24908$. This provides numerically obtained values for the partial derivatives as $\partial y_1/\partial\alpha \approx (0.49907 - 0.49659)/0.01 = 0.2480$, $\partial y_2/\partial\alpha \cong (0.24908 - 0.24660)/0.01 = 0.2480$. The error partial derivative on the right-hand side of (9.46) is likewise found approximately from $[E(\tilde{\alpha} + 0.01) - E(\tilde{\alpha})]/0.01 = (0.02566 - 0.02680)/0.01 = -0.114$. Thus, the first correction $\Delta\alpha$ is obtained from (9.46); $(0.123)\Delta\alpha = +0.114/2$, which implies that $\Delta\alpha \cong +0.46$. This yields an updated system constant as $\alpha \cong -1.4 + 0.46 = -0.94$, which is much closer to the actual value $\alpha = -1.0$.

A second calculation with the newly obtained iterate $\tilde{\alpha} = -0.94$ then yields the following results: $y_1 = 0.625$, $y_2 = 0.39063$, $\tilde{E} = 0.00118$, $\partial y_1/\partial\alpha \approx 0.314$, $\partial y_2/\partial\alpha \approx 0.393$, $\partial\tilde{E}/\partial\alpha \approx 0.032$, and $\Delta\alpha = -0.063$, so that the updated parameter value is $\alpha = -1.003$, very close to the actual result. The error $E(\alpha)$ for this problem is shown in Fig. 9.9. Note the drastic decrease in E (percentage wise) from the initial error $E(\alpha = 1.4) = 0.0268$ to $E(-0.94) = 0.00086$ for the first updated value. Further, note that the solution for the parameter, $\alpha = -1.00$, will result in zero error.

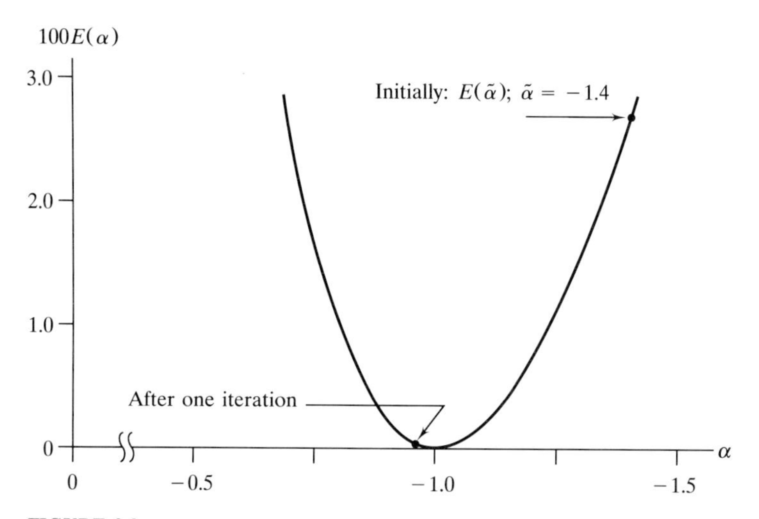

FIGURE 9.9
The error (9.35) as a function of the parameter α for Example 9.6.

Comments

1. The result in (9.41) and (9.42), which was derived for the case of two system parameters, turns out to be applicable generally, regardless of how many system constants are to be identified. All partial derivatives are calculated numerically, and the basic procedures are the same as for the two-parameter case.
2. In Example 9.6 it was assumed that the initial condition was known exactly. Sometimes this will not be the case. In general, those initial conditions that are not well known can be viewed as system parameters and identified in the same way as the system constants c_i.
3. If both the measurements and the model are exact, the least squares technique will lead to the exact values for the system constants, with a resulting error $E = 0$. The integration of (9.34) produces simulated results that are in exact agreement with the experiments.
4. No mention has been made of measurement error. For instance, in Example 9.6 the "measurements" v_1 and v_2 were assumed to be exact. Real measurements, however, will be contaminated with noise and instrument inaccuracy and will be only approximations to reality. The least squares method works well in the presence of typically encountered noise, however, provided that sufficient amounts of data are available. The noise may cause the identified values for the system constants to differ from their true values (see also Exercise 9.5).
5. A second source of error is that the math model (9.34) itself may only represent approximately the actual process being studied. This always occurs to some degree. For example, suppose the data of Example 9.6 were actually generated by a process for which the dynamics unfold according to a "true" model of the form $\dot{x} = \alpha_1 x + \alpha_2 x^2$, with α_2 small, so that the behavior is close to that of the associated linear system $\dot{x} = \alpha x$. Not knowing any better, we may attempt to do

system identification with the linear model $\dot{x} = \alpha x$. We will then get an answer for α (it will differ from the true α_1 in the true model), but we may be unaware that we have mismodeled the process. The importance of having confidence in the mathematical model, or at least to be aware of its limitations, should be evident.

6. The least squares–ODE method of system identification is only one of numerous computer techniques in use for identification purposes. Many of these methods are sophisticated. A lot of research has been done on the problem of system identification in the presence of both measurement noise and "process noise" (i.e., using a slightly incorrect mathematical model). For further information, consult References 9.7 through 9.9.

9.6 INTRODUCTION TO FINITE ELEMENTS IN STRUCTURAL DYNAMICS

In Sec. 7.3 we studied the vibratory characteristics of linear mechanical systems having $n \geq 2$ degrees of freedom. The mathematical model for such systems consists of n second-order ODEs [Eqs. (7.51) for free vibrations and (7.67) for forced vibration]. It was stated that the most important physical properties of these systems are their natural frequencies, which are determined from the eigenvalue-eigenvector analysis of the free vibration problem (7.51). These natural frequencies determine how the system will respond to external stimuli.

In practice, the systems for which vibratory characteristics are important in design can be much more complex than those considered in Sec. 7.3. This is especially true of systems which are deformable, i.e., do not move as rigid bodies. Examples in everyday life include the swaying of tree branches in a wind, vibration of automobile radio antennas, the vibration (audible) resulting when you rub a moistened finger around the top of a wine glass, the vibration of a stretched rubber band or stringed instrument, and vibration of wall panels that transmit sound. Practical examples include the vibration of airplane wings, rail vehicles, automobile panels, bridges, large flexible space structures, and structures that support machinery or house instrumentation.

A key property of these systems is that the mass and stiffness are *distributed continuously* throughout them, rather than being lumped as discrete mass and stiffness elements, as in Chap. 7. Nowadays the most common way to analyze the structural dynamics of systems possessing continuously distributed mass and stiffness is the *finite element method* (FEM). The basic idea of this technique is outlined in the remainder of this section through the analysis of a simple structural dynamics problem: the free vibration of a uniform beam. For more general applications of FEM in structural dynamics and other areas of engineering science, the references should be consulted.

9.6.1 Free Vibration of a Uniform Beam

In this subsection we consider some of the basic aspects of the free vibration of a uniform beam. This problem is easy to formulate and solve exactly, and we do

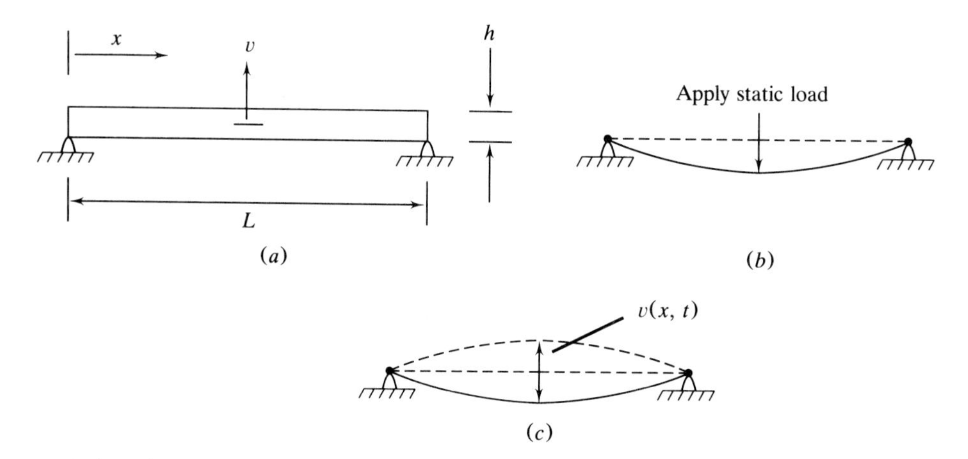

FIGURE 9.10
(a) The basic setup for the simply supported beam. (b) We induce a static displacement. (c) Release the beam from the static displacement in (b) and it vibrates.

so here in order to provide basic understanding and to have exact reference results to which the finite element results can be compared. The finite element analysis of this problem starts in Sec. 9.6.2.

The beam system to be considered is shown in Fig. 9.10 and consists of a simply supported beam having constant properties along its length. The length of the beam is L, and its cross section is rectangular with height h and depth b.

The material properties of relevance are: E = Young's modulus (in psi or N/m^2), I = area moment of inertia = $bh^3/12$ (in m^4 or cm^4) and ρ = beam mass per unit length (in slugs/ft or kg/m). The "simply supported" conditions at each end indicate that the vibratory displacement must always be zero at these boundaries but that the ends are free to rotate.

Suppose we initiate a vibratory motion in the following way: The beam is initially at rest in the static equilibrium configuration of Fig. 9.10(a). We push down on the beam at some location in Fig. 9.10(b), establishing a static displacement pattern away from the original equilibrium. We then suddenly remove the static load. The beam will vibrate as indicated schematically in Fig. 9.10(c).

Notice that the vibratory displacement will depend on both location x along the beam and on time t: $v = v(x, t)$; at a given location x, the displacement will vary with time, and, at a given time t, the displacement will vary with x. This functional dependence of the displacement differs significantly from that of all dynamic systems modeled so far in this book. In everything we have done so far, the unknowns to be found were functions only of time.

The mathematical model for the vibrating beam system can be obtained by first recalling the differential equation that governs the deflection of a *statically* loaded beam, as studied in your strength of materials class:

$$\frac{d^2}{dx^2}\left(EI\frac{d^2v}{dx^2}\right) = p(x) \tag{9.47}$$

where $p(x)$ is the applied static load per unit length and the displacement $v = v(x)$ is now a function of x only. Equation (9.47) is nothing more than Newton's second law ($F = ma$), in the vertical direction, per unit length of beam. There is no acceleration, so that the net vertical force on every infinitesimal segment of the beam must vanish. This is seen more easily if we write (9.47) as

$$p(x) - \frac{d^2}{dx^2}\left(EI\frac{d^2v}{dx^2}\right) = 0 \tag{9.48}$$

In this equation, $p(x)$ is the applied static force per unit length, and $(-d^2/dx^2)[EI(dv/dx^2)]$ is the net shear force acting on the segment, so that (9.48) merely says $F = 0$, per unit length.

Now consider the dynamic case, for which the inertia of vibratory motion must be included in the model. This is done by inserting the mass times vertical acceleration, per unit length of beam, on the right-hand side of (9.48), yielding

$$p(x,t) - \frac{\partial^2}{\partial x^2}\left(EI\frac{\partial^2 v}{\partial x^2}\right) = \rho\frac{\partial^2 v}{\partial t^2} \tag{9.49}$$

where partial derivatives are now used because v is a function of both x and t. Furthermore, the possibility of a time-dependent external force is indicated by writing $p(x, t)$. Equation (9.49) is often written in the form

$$\rho\frac{\partial^2 v}{\partial t^2} + \frac{\partial^2}{\partial x^2}\left(EI\frac{\partial^2 v}{\partial x^2}\right) = p(x,t) \tag{9.50}$$

This relation is analogous to that of the simple, undamped oscillator, in that it shows the interaction of inertia, stiffness, and external forcing. For the case of *free vibration*, the external load $p(x, t) = 0$, and (9.50) reduces to

$$\rho\frac{\partial^2 v}{\partial t^2} + \frac{\partial^2}{\partial x^2}\left(EI\frac{\partial^2 v}{\partial x^2}\right) = 0 \tag{9.51}$$

Equation (9.51) or (9.50) is our first example of a *partial* differential equation, which occurs if the unknown to be found depends on more than one independent variable. All of our previous work has involved dynamic systems for which the math models naturally arise as sets of ordinary differential equations. But the student should not get the idea that this is the only possibility. Many important dynamics problems are formulated naturally in terms of partial differential equations.

Shortly we will set up a finite element solution for the beam system of Fig. 9.10. Before doing this, however, let us obtain the exact solution to (9.51) for the simply supported case in order to eventually interpret and evaluate the accuracy of the finite element results.

We have seen that for ODE models it is necessary to specify *initial conditions* in order to obtain solutions for specific cases. For the beam vibration problem, we must also specify *boundary* conditions, which will state the manner in which the beam is supported (or not supported) at the ends. For the simply supported beam of Fig. 9.10, the boundary conditions may be stated in words as (1) the displacement v must vanish at each end, for all time, and (2) the bending moment $M = EI(\partial^2 v/\partial x^2)$ must vanish at each end, for all times. Mathematically, we have

$$
\begin{aligned}
x = 0: \qquad & v(0,t) = 0 = \frac{\partial^2 v}{\partial x^2}(0,t) \\
x = L: \qquad & v(L,t) = 0 = \frac{\partial^2 v}{\partial x^2}(L,t)
\end{aligned}
\tag{9.52}
$$

There are four boundary conditions because the governing PDE is of fourth order in the spatial variable x.

Let us attempt to find an exact solution to (9.51) and (9.52) by assuming that it is possible for the beam to execute motions which are simple harmonic in time; that is,

$$v(x,t) = \phi(x)\cos(\omega t - \phi) \tag{9.53}$$

where the natural frequency ω is at this stage unknown, as is the function $\phi(x)$. The assumed solution is similar to (7.53), which was assumed for the multi-DOF systems of Chap. 7. Here, however, the harmonic term $\cos(\omega t - \phi)$ is multiplied by the function $\phi(x)$, as we must find the solution at each of the (infinite set of) points between $x = 0$ and $x = L$. Effectively we may consider the present system to possess an infinite number of degrees of freedom.

From (9.53),

$$
\begin{aligned}
\frac{\partial^2 v}{\partial t^2} &= -\omega^2\phi(x)\cos(\omega t - \phi) \\
\frac{\partial^4 v}{\partial x^4} &= \phi'''' \cos(\omega t - \phi)
\end{aligned}
$$

where a prime denotes an ordinary derivative with respect to x, $(\ )' \equiv d(\ )/dx$. Plugging these relations into the governing PDE (9.51) and canceling the $\cos(\omega t - \phi)$ terms in the result, we obtain an ODE for the function $\phi(x)$

$$EI\phi'''' - \rho\omega^2\phi = 0$$

or

$$\phi'''' - \left(\frac{\rho\omega^2}{EI}\right)\phi = 0 \tag{9.54}$$

Our assumption of harmonic motion has removed the time dependence from the problem. In order to solve (9.54) for $\phi(x)$, we have to convert the original boundary conditions (9.52) so that they apply for $\phi(x)$. The student should convince herself that in order for (9.52) to be satisfied at all times, it is sufficient that ϕ satisfy the following:

$$\begin{aligned} \phi(0) &= \phi(L) = 0 \\ \phi''(0) &= \phi''(L) = 0 \end{aligned} \tag{9.55}$$

Equation (9.54) with the conditions of (9.55) constitutes a *boundary value problem*. Because (9.54) is a linear ODE of fourth order, there will be four linearly independent solutions, which turn out to be (verify!) $\cosh \lambda x$, $\sinh \lambda x, \cos \lambda x, \sin \lambda x$, where $\lambda^4 = (\rho\omega^2/EI)$. The general solution may be expressed as a linear combination of the four solutions noted, so that

$$\phi(x) = c_1 \cosh \lambda x + c_2 \sinh \lambda x + c_3 \sin \lambda x + c_4 \cos \lambda x \tag{9.56}$$

Application of the boundary conditions (9.55) then leads to the following set of algebraic equations for the constants c_1, c_2, c_3, c_4:

$$\begin{aligned} &\left.\begin{aligned} \phi(0) &= 0 = c_1 + c_4 \\ \phi''(0) &= 0 = \lambda^2 c_1 - \lambda^2 c_4 \end{aligned}\right\} \Rightarrow c_1, c_4 \equiv 0 \\ &\phi(L) = c_2 \sinh \lambda L + c_3 \sin \lambda L = 0 \\ &\phi''(L) = \lambda^2 c_2 \sinh \lambda L - \lambda^2 c_3 \sin \lambda L = 0 \end{aligned} \tag{9.57}$$

The last two conditions of (9.57) are satisfied if $c_2 = 0$ (because $\sinh \lambda L$ is nonzero) and if *either* $c_3 = 0$ *or* $\sin \lambda L = 0$. The case $c_3 = 0$ corresponds to the trivial solution $v = 0$ for which no vibratory motion occurs. To obtain nontrivial solutions, we have to select λ so that $\sin \lambda L = 0$. This means that

$$\lambda L = \pi, 2\pi, 3\pi, \ldots \tag{9.58}$$

Observe that mathematically there are an infinite number of solutions, one for each possible λ. To keep track of them, denote the nth solution as λ_n, so that $\lambda_n = (n\pi/L)\,(n \geq 1)$. Each solution λ_n defines an associated natural frequency of vibration ω_n according to the relation $\lambda_n^4 = \rho\omega_n^2/EI$, which implies

$$\omega_n = \lambda_n^2 \sqrt{\frac{EI}{\rho}} = n^2\pi^2 \sqrt{\frac{EI}{\rho L^4}} \tag{9.59}$$

Each frequency ω_n has associated with it a function $\phi_n(x)$ given by

$$\phi_n(x) = \sin(\lambda_n x) = \sin\left(\frac{n\pi x}{L}\right) \tag{9.60}$$

Each of the solutions defined by (9.60) defines a particular mode of vibration, just as in the multi-DOF analysis of Sec. 7.3. The only difference is that the mode shapes or eigenvectors of Sec. 7.3 are now replaced by the *continuous* mode shapes or *eigenfunctions* $\sin(n\pi x/L)$. These continuous mode shapes

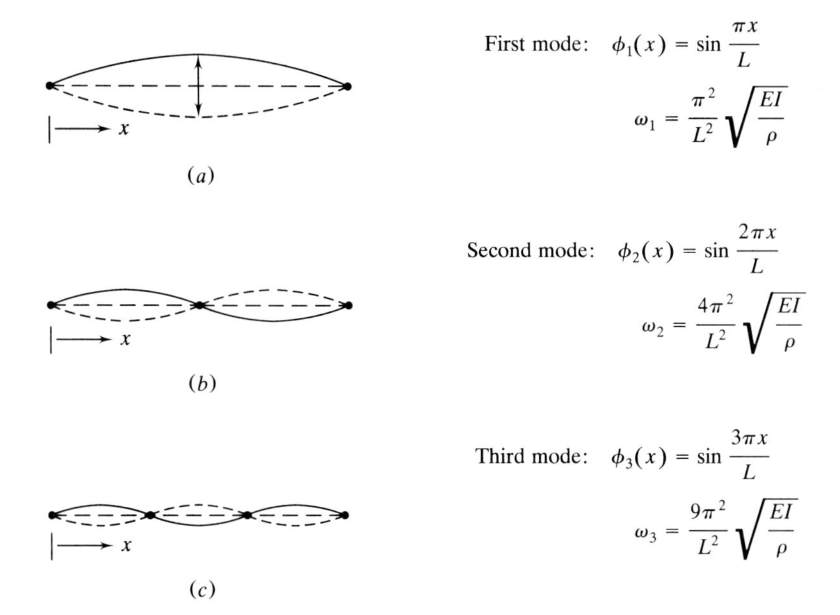

FIGURE 9.11
Schematic of the first three modes of the simply supported beam.

define the shape that the beam assumes during vibration in a particular mode. The lowest several modes are illustrated in Fig. 9.11.

The most general free vibration solution consists of a superposition of the individual solutions (9.53):

$$v(x,t) = \sum_{n=1}^{\infty} c_n \sin\left(\frac{n\pi x}{L}\right)\cos(\omega_n t - \phi_n) \tag{9.61}$$

This solution is analogous to (7.58) for the multi-DOF case. In general, that is, for arbitrary initial conditions, a given free vibrational motion will contain all modes, although often only the lowest few contribute significantly.

Remark. The natural frequencies and mode shapes of a uniform beam will change if we change the boundary conditions. For instance, for a cantilever beam (built rigidly into a wall at $x = 0$ and free at $x = L$), the boundary conditions are: (1) Displacement and rotation are prevented at $x = 0$, so that $v(0, t) = v'(0, t) = 0$ and (2) the end $x = L$ is shear and moment free, so that $v''(L, t) = v'''(L, t) = 0$. The analysis of the free vibration of a cantilever proceeds exactly as done for the simply supported beam. All steps through (9.56) are the same. But the new boundary conditions, used to determine the c's and λ in (9.56), will cause the mode shapes $\phi_n(x)$ and natural frequencies ω_n to differ significantly from the simply supported case. The importance of boundary conditions in problems of this type must be noted.

Using the results of this section as a reference, we next investigate the application of the finite element method to the solution of the simply supported beam problem. Keep in mind that the intent is only to illustrate the basic ideas of the method, rather than to analyze realistic structures commonly encountered in practice.

9.6.2 Discretization and Approximation Using Finite Elements

Finite elements is a technique for finding approximate solutions to boundary value problems. Here we will outline the basic steps for the specific problem under consideration, the uniform beam. We begin by identifying two points, or *nodes*, on the beam, one at $x = 0$ and one at $x = L$, as shown in Fig. 9.12. Denote the vertical displacements at these nodes as v_1 and v_2 as shown. Also, denote the *rotations* at these nodes as θ_1 and θ_2. For small displacements, to which the present analysis is restricted, the rotation θ is just the slope v' of the displacement, as is indicated in Fig. 9.13. From the geometry, $\tan \theta = dv/dx = v'$, and for small θ, $\tan \theta \approx \theta$, so that $\theta \approx v'$.

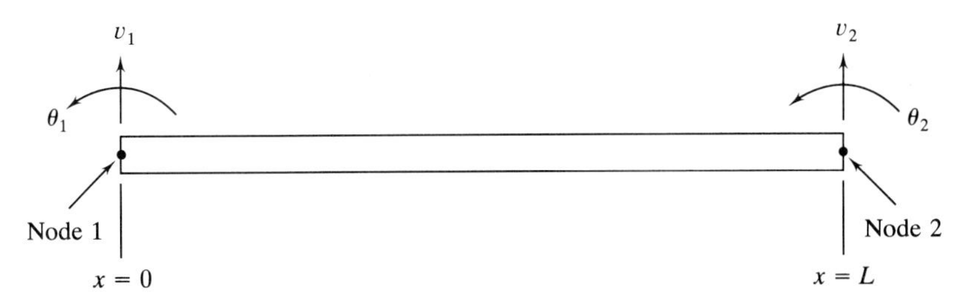

FIGURE 9.12
Definition of nodal displacements and rotations for a beam.

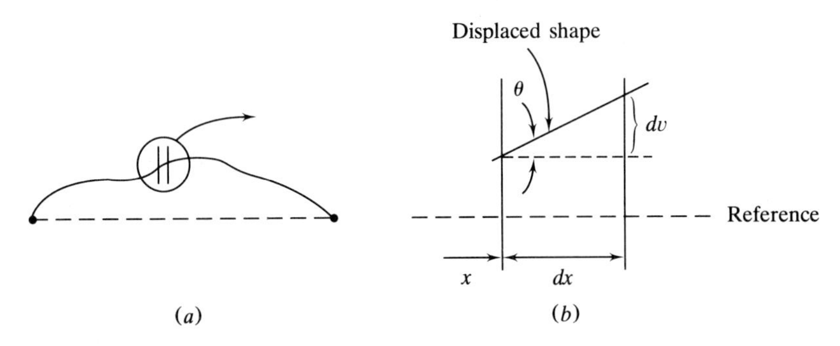

FIGURE 9.13
The relation between rotation θ and slope v'.

It is important to note that as the vibrational motion evolves, the *nodal values* $v_1, \theta_1, v_2, \theta_2$ will be functions of time only, as each is identified with a specific location on the beam.

Suppose that at some instant t we are given the four discrete values $v_1(t), \theta_1(t), v_2(t), \theta_2(t)$, so that we know exactly how the beam is displaced and rotated at its ends. According to FEM, what we then do is devise a way to determine, at least approximately, how the beam is displaced *everywhere* between $x = 0$ and $x = L$. To see how we might do this, let us visualize qualitatively what the deflected shape of the beam would be if we considered the following nodal values: $v_1 = 1$, $\theta_1 = v_2 = \theta_2 = 0$; The end $x = 0$ is displaced one unit, but all other nodal values are zero. Then the displacement would have the general shape shown in Fig. 9.14.

Mathematically we can represent the displacement pattern shown in Fig. 9.14 as

$$v(x,t) = N_1(x)v_1(t) \tag{9.62}$$

Here $N_1(x)$ defines the displaced shape everywhere in the beam, per unit of displacement v_1, with the other three nodal displacements zero. This relation also provides an approximation to the slope $\theta(x,t) = v'(x,t)$ everywhere in the beam:

$$\theta(x,t) = v'(x,t) = N_1'(x)v_1(t) \tag{9.63}$$

Note that in (9.63) only N_1 is differentiated, because v_1 is a function of time only. The statements (9.62) and (9.63) indicate that $N_1(x)$ must satisfy the following four conditions:

$$\begin{aligned} N_1(0) &= 1 \qquad (v_1 = 1) \\ N_1'(0) &= 0 \qquad (\theta_1 = 0) \\ N_1(L) &= 0 \qquad (v_2 = 0) \\ N_1'(L) &= 0 \qquad (\theta_2 = 0) \end{aligned} \tag{9.64}$$

The first and third conditions in (9.64) come from (9.62), evaluated at $x = 0$ and $x = L$, while the second and fourth conditions in (9.64) come from (9.63), evaluated at $x = 0$ and $x = L$.

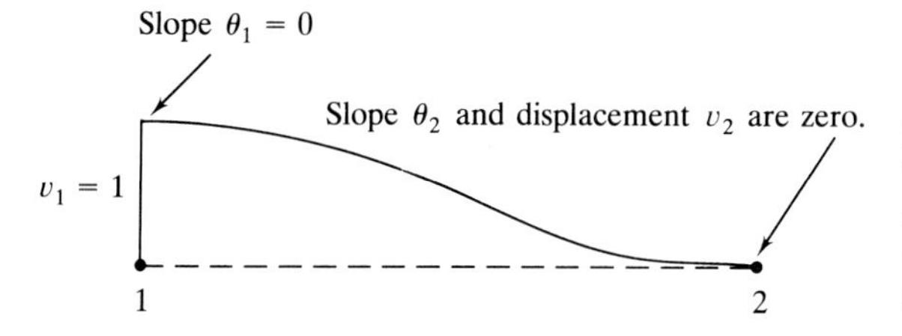

FIGURE 9.14
The displaced shape if the slope and displacement vanish at $x = L$, the slope is zero at $x = 0$, and $v_1 = 1$.

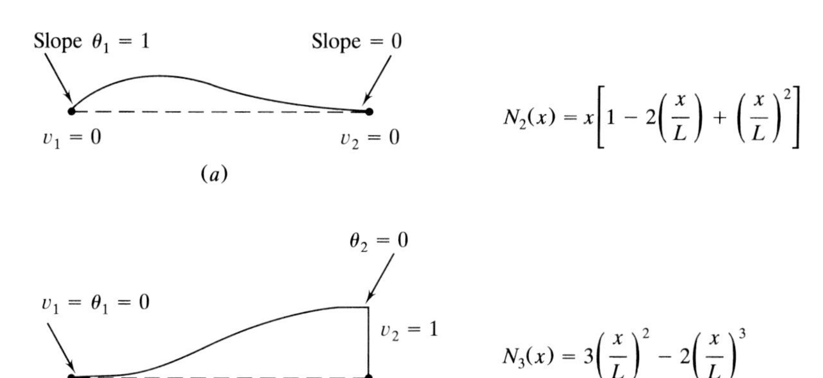

$$N_2(x) = x\left[1 - 2\left(\frac{x}{L}\right) + \left(\frac{x}{L}\right)^2\right]$$

$$N_3(x) = 3\left(\frac{x}{L}\right)^2 - 2\left(\frac{x}{L}\right)^3$$

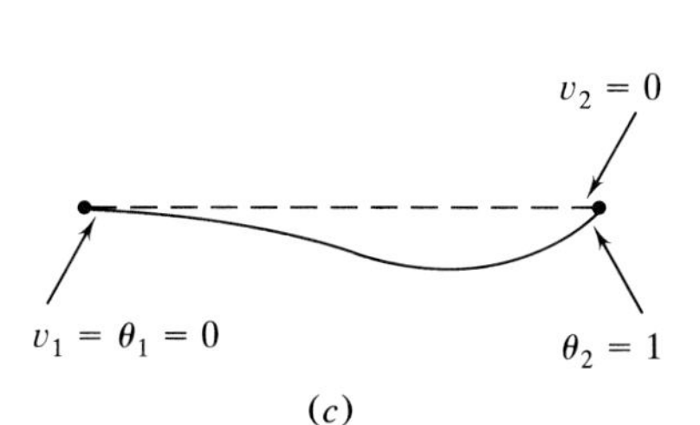

$$N_4(x) = x\left[\left(\frac{x}{L}\right)^2 - \left(\frac{x}{L}\right)\right]$$

FIGURE 9.15
(*a*) The cubic function $N_2(x)$ is shown, obtained from the conditions $\theta_1 = 1$, $v_1 = v_2 = \theta_2 = 0$. (*b*) The function $N_3(x)$ is shown, obtained from the conditions $v_1 = \theta_1 = \theta_2 = 0$, $v_2 = 1$. (*c*) The function $N_4(x)$, obtained from the conditions $v_1 = \theta_1 = v_2 = 0$, $\theta_2 = 1$.

A simple function of x which satisfies the conditions (9.64) is the cubic polynomial*

$$N_1(x) = 1 - 3\left(\frac{x}{L}\right)^2 + 2\left(\frac{x}{L}\right)^3 \tag{9.65}$$

The result of (9.62) and (9.65) is that the "shape function" $N_1(x)$ allows us to approximate the displaced shape, everywhere in the beam, due to a displacement v_1 at node 1. It is important to note, however, that this approximation was developed independently of the physical system being analyzed; it is strictly a mathematical construction.

A similar analysis can be done to define the contribution of the other three nodal values $(v_2, \theta_1, \theta_2)$ to the overall displacement pattern. Figure 9.15 shows the general shape that the beam assumes for the following cases:

---

*The result (9.65) is obtained by assuming $N_1(x)$ to be a general cubic polynomial of the form $N_1(x) = \alpha_1 + \alpha_2 x + \alpha_3 x^2 + \alpha_4 x^3$, and then selecting the α_i so that the stated conditions (9.64) are satisfied. We use a cubic function because a cubic function is uniquely determined by *four* conditions.

(a) $\theta_1 = 1$ and $v_1 = v_2 = \theta_2 = 0$; (b) $v_2 = 1$ and $\theta_1 = \theta_2 = v_1 = 0$; (c) $\theta_2 = 1$, $v_1 = v_2 = \theta_1 = 0$.

During an actual vibratory motion, we generally expect that all four of the nodal displacements may be nonzero. At any instant the displaced shape of the beam $v(x,t)$ is thus defined as a superposition of the four shape functions $N_i(x)$, as follows:

$$v(x,t) = N_1(x)v_1(t) + N_2(x)\theta_1(t) + N_3(x)v_2(t) + N_4(x)\theta_2(t) \quad (9.66)$$

If we know the histories of the four nodal displacements $v_1, \theta_1, v_2, \theta_2$, then (9.66) tells us the displacement everywhere in the beam. Note that (9.66) is an approximation. For given v_i and θ_i, $v(x,t)$ in (9.66) will be by construction a cubic polynomial. The cubic approximation (9.66) was defined independently of any physical considerations or governing mathematical model, and the accuracy of (9.66) will depend on whether the actual displacement pattern during vibration can be represented reasonably well by a cubic function.

Example 9.7 Use of (9.66) to Find Approximately $v(x,t)$. Suppose that, at a given instant t_0, the following four nodal values occur: $v_1 = 0.05L$, $v_2 = 0.01L$, $\theta_1 = 0.10$, $\theta_2 = -0.05$. Then, according to (9.66), the continuous displacement $v(x,t_0)$ at this time t_0 will be given approximately by

$$v(x,t_0) = 0.05LN_1(x) + 0.10N_2(x) + 0.01LN_3(x) - 0.05N_4(x)$$

Using the preceding relations for the shape functions $N_i(x)$, as defined by (9.65) and in Fig. 9.15, we obtain the following cubic function for $v(x,t_0)$:

$$v(x,t_0) = 0.05L + 0.10x - 0.27\left(\frac{x^2}{L}\right) + 0.13\left(\frac{x^3}{L^2}\right)$$

The resulting displacement pattern is shown in Fig. 9.16.

Note: We must be aware of the units of terms in (9.66). The nodal displacements v_1 and v_2 have units of length, and the associated shape functions $N_1(x)$ and $N_3(x)$ are dimensionless, because the products v_1N_1 and v_2N_3 must have dimensions of length. On the other hand, θ_1 and θ_2 are dimensionless (actually, understood to be measured in radians), so the associated shape functions $N_2(x)$ and $N_4(x)$ must have dimensions of length. In Example 9.7 v_1 and v_2 were stated as fractions of beam length L, while θ_1 and θ_2 were specified in radians.

So far we have introduced the use of four nodal values, and their associated shape functions, to approximate the displaced shape of the beam during vibration. The resulting formulation (9.66) approximates the displacement as a cubic function. Whether this representation will yield an accurate displacement, however, will depend on whether a cubic function can adequately approximate the actual continuous mode shapes $\phi_n(x)$. For instance, consider the actual mode shapes $\phi_n(x)$ for the simply supported beam, as shown in Fig. 9.11. While a cubic function will describe the first two mode shapes reasonably well, it will not adequately represent the third mode nor any mode

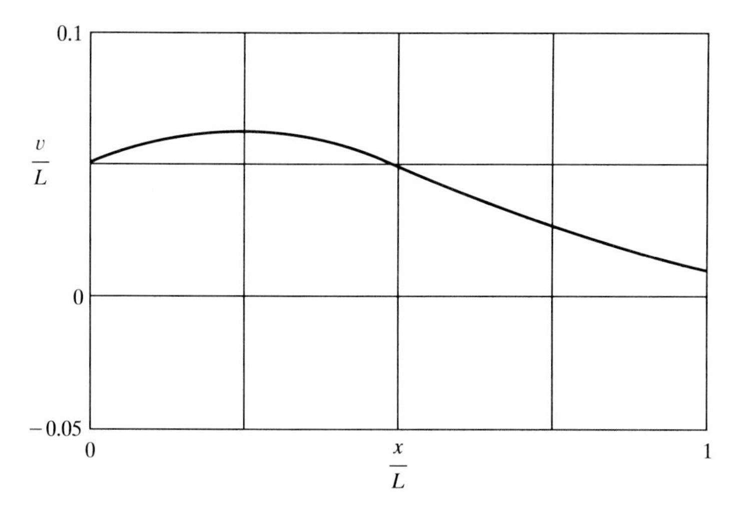

FIGURE 9.16
Displacement pattern for Example 9.7; (v/L) versus (x/L).

higher than the third. What we need is a way to improve and control the accuracy of the analysis.

A more detailed and accurate representation of the continuous displacement pattern is obtained by defining an arbitrary number of nodes, each having an associated displacement v and rotation θ. The general scheme is shown in Fig. 9.17. In this figure there is a total of N nodes, numbered consecutively.

There are $2N$ nodal degrees of freedom: the N displacements $V_1, V_2, \ldots, V_N$ and the N rotations $\Theta_1, \ldots, \Theta_N$. Any two adjacent nodes define an *element*, that is, a (possibly small) segment of the beam. The elements in Fig. 9.17 are numbered from 1 to E, where $E = N - 1$. Each element is of specified length L_e; the elements may or may not be of equal length, depending on the wishes of the analyst.

For a configuration such as shown in Fig. 9.17, the approximation of the displaced shape during vibratory motion is made *element by element*. Specifically, *within* a given element, the same cubic shape functions $N_1, \ldots, N_4$ defined

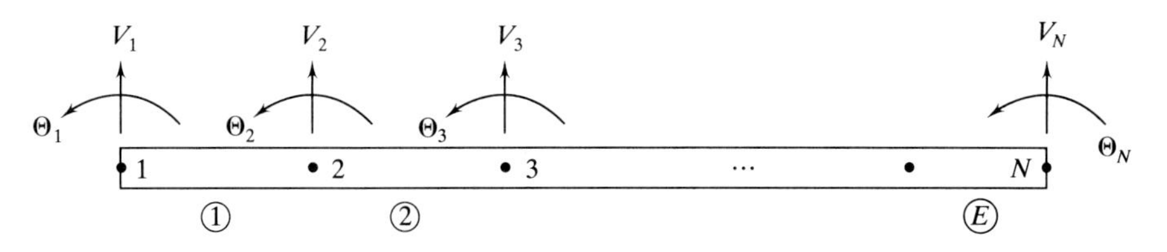

FIGURE 9.17
Identification of N nodes; the beam is now split into $E = N - 1$ "elements." There are a total of $2N$ nodal degrees of freedom. This is a "global" representation.

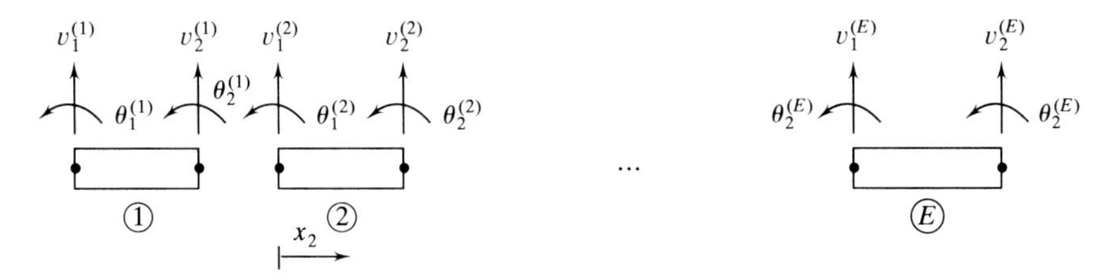

FIGURE 9.18
The "local" representation, in which the elements are isolated from each other. The superscripts on the nodal displacements and rotations define the element number.

previously are used to construct the approximation, which applies within that element. In order to define this construction unambiguously, the individual elements may be visualized as being isolated from each other, as shown in Fig. 9.18.* If we consider, for example, the displaced shape $v^{(2)}(x,t)$ within element 2, this displaced shape will be defined by the four nodal displacements associated with element 2, according to

$$v^{(2)}(x,t) = v_1^{(2)}N_1(x_2) + \theta_1^{(2)}N_2(x_2) + v_2^{(2)}N_3(x_2) + \theta_2^{(2)}N_4(x_2) \quad (9.67)$$

where the shape functions are given by

$$\begin{aligned} N_1(x_2) &= 1 - 3\left(\frac{x_2}{L_2}\right)^2 + 2\left(\frac{x_2}{L_2}\right)^3 \\ N_2(x_2) &= x_2\left[1 - \frac{2x_2}{L_2} + \left(\frac{x_2}{L_2}\right)^2\right] \\ N_3(x_2) &= 3\left(\frac{x_2}{L_2}\right)^2 - 2\left(\frac{x_2}{L_2}\right)^3 \\ N_4(x_2) &= x_2\left[\left(\frac{x_2}{L_2}\right)^2 - \frac{x_2}{L_2}\right] \end{aligned} \quad (9.68)$$

Here L_2 is the specified length of element 2, and x_2 is a "local" coordinate which is zero at node 1 of element 2 (global node 2) and is L_2 at node 2 of element 2 (global node 3). In terms of the actual (global) x coordinate, $x_2 = x - L_1$.

The preceding description describes the *finite element* approximation for the (cubic) displacement pattern within a given element. By piecing together the displacement patterns for all of the elements, we obtain the continuous, global

*Generally, capital V's and Θ's are used here to denote *global* displacements and rotations, while lowercase v's and θ's are used to denote displacements and rotations at the local or element level.

displacement pattern for the whole beam. The basic idea is that, if we use enough elements, we should be able to get a good approximation within any element with a cubic function.

Comments

1. The global displacement pattern we have constructed consists of a piecewise cubic approximation. Within any element the beam displacement $v^{(e)}(x,t)$ is given by the cubic function (9.67) and (9.68). In the different elements, however, the displacements will be represented by different cubic functions, since the four nodal displacements will differ from element to element. Thus, we do not obtain a single equation which can be used to describe the global displacement pattern; we obtain E different cubic approximations, one for each element.
2. From Fig. 9.18, observe that each "left-hand" pair of local displacements for the eth element, $v_1^{(e)}, \theta_1^{(e)}$, is identified with (i.e., is the same as) the corresponding "right-hand" pair $v_2^{(e-1)}, \theta_2^{(e-1)}$ of element $(e-1)$. For example, $v_1^{(2)} \equiv v_2^{(1)}$ and $\theta_1^{(2)} \equiv \theta_2^{(1)}$. This guarantees that the global solution, when pieced together, will be continuous (since $v_1^{(2)} = v_2^{(1)}$, and so on) and will have a continuous slope (since $\theta_1^{(2)} = \theta_2^{(1)}$, and so on). On physical grounds, displacement and slope continuity are necessary; if, for example, the global slope were discontinuous at some point, the implication is that the beam would be "broken" at that point, and this is not allowed. Note, however, that generally the second and third derivatives v'' and v''' will be discontinuous at the nodes, according to the approximate solution, so that the approximate moment and shear will also be discontinuous. This is OK. Such discontinuities may actually occur.
3. Each set of four element nodal displacements is identified with four of the global displacements. For instance, for element 2 of Fig. 9.18, we have

$$v_1^{(2)} \equiv V_2$$

$$\theta_1^{(2)} \equiv \Theta_2$$

$$v_2^{(2)} \equiv V_3$$

$$\theta_2^{(2)} \equiv \Theta_3$$

We have to keep track of this local/global correspondence.

In the following section we use the approximation developed here to analyze beam vibration using a single finite element (Fig. 9.12). Then in Example 9.10 the analysis is extended to the case of an arbitrary number of elements.

9.6.3 Developing the System Math Model Using a Single Element

The preceding section outlined the way in which the approximation for the continuous time-dependent displacement $v(x,t)$ is formulated in terms of the nodal displacements $v_i(t)$ and $\theta_i(t)$ and the shape functions $N_i(x)$. In devising

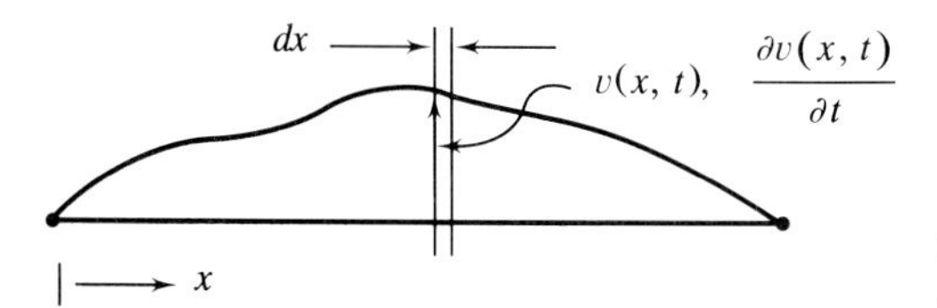

FIGURE 9.19
An infinitesimal section of beam.

this finite element approximation, no account was taken of any underlying physical principles or governing mathematical model [here (9.51)]. This must now be done. The objective in this section will be to develop the differential equations that govern the time dependence of the nodal displacements $v_i(t)$ and $\theta_i(t)$. Because these nodal displacements are functions of time only, we will obtain ODEs in the resulting math model. The method used here to obtain these governing ODEs is via Lagrange's equations (Sec. 6.3), which are stated below for the case of free vibration considered here (no generalized forces):

$$\frac{d}{dt}\left(\frac{\partial T}{\partial \dot{q}_i}\right) - \frac{\partial T}{\partial q_i} + \frac{\partial V}{\partial q_i} = 0 \qquad i = 1, \ldots, n \tag{9.69}$$

where the q_i are the generalized coordinates. Note that Lagrange's equations (9.69) cannot be applied directly to any system for which the displacement is a continuous function of both location and time. In order to use (9.69), the system kinetic and potential energies must be expressed in terms of *discrete coordinates*, which are functions of time only. In the present application these discrete coordinates will be the nodal displacements $v_i(t)$ and rotations $\theta_i(t)$.

In order to apply Lagrange's equations, we now formulate the general relations for the kinetic and potential energies of the beam. The kinetic energy T is found by considering an infinitesimal section of length dx, as shown in Fig. 9.19.

The incremental kinetic energy dT of the mass contained in the segment dx shown is $dT = \frac{1}{2}(\rho\, dx)(\partial v/\partial t)^2$, where $\rho\, dx$ is the mass in the segment and $\partial v/\partial t$ is the velocity of the segment. The kinetic energy T of the entire beam is obtained by integration over the beam length L, so that

$$T = \frac{1}{2}\int_0^L \rho\left(\frac{\partial v}{\partial t}\right)^2 dx \tag{9.70}$$

The elastic potential energy V stored due to bending arises as a result of beam *curvature*; in any length of beam that is straight, there is no elastic energy stored. The curvature for small vibrational displacements is defined by $(\partial^2 v/\partial x^2)$, and incremental elastic energy dV stored in a segment dx is proportional to the square of the curvature and is given by $dV = \frac{1}{2}EI(\partial^2 v/\partial x^2)^2$, where EI is the flexural stiffness of the beam. The potential energy V for the beam as a whole is obtained by integration as

$$V = \frac{1}{2}\int_0^L EI\left(\frac{\partial^2 v}{\partial x^2}\right)^2 dx \tag{9.71}$$

To calculate T and V at any instant t, we would need to know the velocity and curvature at all points on the beam. Here we develop the finite element approximation to T and V for the simple case of a single element, as in Fig. 9.12. For a single element the displacement $v(x,t)$ is approximated by (9.66). For notational simplicity we now write (9.66) in vector form as

$$v(x,t) = \lfloor N_1 \quad N_2 \quad N_3 \quad N_4 \rfloor \begin{Bmatrix} v_1 \\ \theta_1 \\ v_2 \\ \theta_2 \end{Bmatrix}$$

or notationally,

$$v(x,t) = \lfloor N \rfloor \{v\} \tag{9.72}$$

where the $(1x4)$ row vector $\lfloor N \rfloor$ consists of the four shape functions, and the $(4x1)$ column vector $\{v\}$ contains the nodal displacements and rotations. Because $v(x,t)$ defined by (9.72) is a scalar, (9.72) could also be written as

$$v(x,t) = \{v\}^T \lfloor N \rfloor^T = \lfloor v_1 \quad \theta_1 \quad v_2 \quad \theta_2 \rfloor \begin{Bmatrix} N_1 \\ N_2 \\ N_3 \\ N_4 \end{Bmatrix} \tag{9.73}$$

Let us now utilize the finite element approximation (9.72) or (9.73) to reexpress the kinetic energy T in terms of the nodal velocities. With $v(x,t)$ given by (9.72) and (9.73), the velocity $(\partial v/\partial t)$ is

$$\frac{\partial v}{\partial t} = \lfloor N \rfloor \{\dot{v}\} = \{\dot{v}\}^T \lfloor N \rfloor^T$$

The notation $\{\dot{v}\}$ implies ordinary (not partial) derivatives, as $\{v\}$ is a function of time only. The velocity squared may then be written as

$$\left(\frac{\partial v}{\partial t}\right)^2 = (\lfloor N \rfloor \{\dot{v}\})^2 = \{\dot{v}\}^T \lfloor N \rfloor^T \lfloor N \rfloor \{\dot{v}\} \tag{9.74}$$

Aside: The form (9.74) is useful and common in finite element work and can be easily verified by the following example: Let a scalar c be defined by

$c = \lfloor N_1 \quad N_2 \rfloor \begin{Bmatrix} u_1 \\ u_2 \end{Bmatrix}$, so that $c = N_1 u_1 + N_2 u_2$. Then $c^2 = (N_1 u_1 + N_2 u_2)^2 = N_1^2 u_1^2 + 2N_1 N_2 u_1 u_2 + N_2^2 u_2^2$. The vector form (9.74) would be stated as

$$c^2 = \lfloor u_1 \quad u_2 \rfloor \begin{Bmatrix} N_1 \\ N_2 \end{Bmatrix} \lfloor N_1 \quad N_2 \rfloor \begin{Bmatrix} u_1 \\ u_2 \end{Bmatrix} = \lfloor u_1 \quad u_2 \rfloor \begin{bmatrix} N_1^2 & N_1 N_2 \\ N_1 N_2 & N_2^2 \end{bmatrix} \begin{Bmatrix} u_1 \\ u_2 \end{Bmatrix}$$

$$= \lfloor u_1 \quad u_2 \rfloor \begin{Bmatrix} N_1^2 u_1 + N_1 N_2 u_2 \\ N_1 N_2 u_1 + N_2^2 u_2 \end{Bmatrix}$$

$$= u_1\left(N_1^2 u_1 + N_1 N_2 u_2\right) + u_2\left(N_1 N_2 u_1 + N_2^2 u_2\right)$$

which is the correct result.

Substitution of (9.74) into (9.70) leads to the following form for the kinetic energy T:

$$T = \tfrac{1}{2} \int_0^L \rho \{\dot{v}\}^T \lfloor N \rfloor^T \lfloor N \rfloor \{\dot{v}\}\, dx$$

Because $\{\dot{v}\}$ is a function of time only, it can be moved outside the integral, so that

$$T = \frac{1}{2} \{\dot{v}\}^T \Bigg[\underbrace{\int_0^L \rho \underset{4\times 1}{\lfloor N \rfloor^T} \underset{1\times 4}{\lfloor N \rfloor}\, dx}_{4 \times 4 \text{ matrix}} \Bigg] \{\dot{v}\}$$

As indicated, the integral is a 4×4 matrix, which we designate as $[m]$, the "mass matrix," so that

$$T = \tfrac{1}{2} \{\dot{v}\}^T [m] \{\dot{v}\} \tag{9.75}$$

The entry m_{ij} in the ith row and jth column of the 4×4 mass matrix $[m]$ will be given by the integral

$$m_{ij} = \int_0^L \rho N_i N_j \, dx \tag{9.76}$$

These integrals can be calculated because the shape functions N_i have been specified in Sec. 9.6.2.

Example 9.8 Calculate the Mass Matrix Entry m_{12} for a Uniform Beam. This entry is defined as

$$m_{12} = \int_0^L \rho N_1(x) N_2(x)\, dx$$

Noting that the relevant shape functions $N_1(x)$ and $N_2(x)$ are

$$N_1(x) = 1 - 3\left(\frac{x}{L}\right)^2 + 2\left(\frac{x}{L}\right)^3$$

$$N_2(x) = x\left(1 - \frac{2x}{L} + \frac{x^2}{L^2}\right)$$

we have for constant ρ

$$m_{12} = \rho\int_0^L \left[1 - 3\left(\frac{x}{L}\right)^2 + 2\left(\frac{x}{L}\right)^3\right]\left[1 - \frac{2x}{L} + \left(\frac{x}{L}\right)^2\right] x\,dx$$

The student may verify that carrying out the integration yields the result

$$m_{12} = \frac{11\rho L^2}{210}$$

If the remaining entries in the 4×4 mass matrix are calculated as in Example 9.8, for a beam of uniform mass per unit length ρ, the following result is obtained for the mass matrix:

$$[m] = \frac{\rho L}{420}\begin{bmatrix} 156 & 22L & 54 & -13L \\ 22L & 4L^2 & 13L & -3L^2 \\ 54 & 13L & 156 & -22L \\ -13L & -3L^2 & -22L & 4L^2 \end{bmatrix} \tag{9.77}$$

The result (9.75) and (9.76) shows that the kinetic energy is now expressed [approximately, since (9.72) is an approximation] as *a* quadratic function of the four nodal velocities $\dot{v}_1$, $\dot{\theta}_1$, $\dot{v}_2$, and $\dot{\theta}_2$. This represents a substantial mathematical simplification in comparison to (9.70), which requires that we know $\partial v/\partial t$ everywhere along the beam. Our use of the shape functions in (9.72) has allowed us to "integrate out" the x dependence of the kinetic energy, leaving a function of only the nodal velocities. This removal of the spatial dependence is characteristic of all finite element applications.

A similar result is obtained for the elastic potential energy V defined by (9.71). With $v(x,t) \cong \lfloor N \rfloor\{v\}$, then the curvature is

$$\frac{\partial^2 v}{\partial x^2} = \lfloor N_1'' \quad N_2'' \quad N_3'' \quad N_4'' \rfloor \begin{Bmatrix} v_1 \\ \theta_1 \\ v_2 \\ \theta_2 \end{Bmatrix}$$

so that, notationally,

$$\left(\frac{\partial^2 v}{\partial x^2}\right)^2 = \{v\}^T \lfloor N'' \rfloor^T \lfloor N'' \rfloor \{v\}$$

Using this result in (9.71) and moving the terms $\{v\}^T$ and $\{v\}$ outside the integral as before, there results for the potential energy V

$$V = \frac{1}{2}\{v\}^T \Bigg[\underbrace{\int_0^L EI \lfloor N'' \rfloor^T \lfloor N'' \rfloor \, dx}_{4 \times 4 \text{ matrix}} \Bigg] \{v\}$$

We now identify the 4×4 matrix indicated as the stiffness matrix $[k]$. The potential energy thus becomes a quadratic function of the nodal displacements

$$V = \tfrac{1}{2}\{v\}^T [k] \{v\} \tag{9.78}$$

An entry k_{ij} in the ith row and jth column of the stiffness matrix $[k]$ will be defined by

$$k_{ij} = \int_0^L EI \frac{d^2 N_i}{dx^2} \frac{d^2 N_j}{dx^2} \, dx \tag{9.79}$$

For example, to calculate k_{12} for a beam of uniform stiffness EI, we note that

$$\frac{d^2 N_1}{dx^2} = \left(\frac{-6}{L^2} + \frac{12x}{L^3} \right) \quad \text{and} \quad \frac{d^2 N_2}{dx^2} = \left(\frac{-4}{L} + \frac{6x}{L^2} \right)$$

so that

$$k_{12} = EI \int_0^L \left(\frac{12x}{L^3} - \frac{6}{L^2} \right) \left(\frac{6x}{L^2} - \frac{4}{L} \right) dx = \frac{6EI}{L^3}$$

For a uniform beam, calculation of all of the entries in $[k]$ results in the following:

$$[k] = \frac{2EI}{L^3} \begin{bmatrix} 6 & 3L & -6 & 3L \\ 3L & 2L^2 & -3L & L^2 \\ -6 & -3L & 6 & -3L \\ 3L & L^2 & -3L & 2L^2 \end{bmatrix} \tag{9.80}$$

Equations (9.75) and (9.78) show that the functional dependence of T and of V are now as follows: $T = T(\dot{v}_1, \dot{\theta}_1, \dot{v}_2, \dot{\theta}_2)$, and $V = V(v_1, \theta_1, v_2, \theta_2)$. Thus, the nodal displacements v_1, θ_1, v_2, and θ_2 qualify as generalized coordinates, and we may use Lagrange's equations to derive the equations of motion governing the histories of these four nodal displacements. These four equations of motion will appear as follows:

$$\begin{gathered} \frac{d}{dt}\left(\frac{\partial T}{\partial \dot{v}_1} \right) - \frac{\partial T}{\partial v_1} + \frac{\partial V}{\partial v_1} = 0 \\ \vdots \\ \frac{d}{dt}\left(\frac{\partial T}{\partial \dot{\theta}_2} \right) - \frac{\partial T}{\partial \theta_2} + \frac{\partial V}{\partial \theta_2} = 0 \end{gathered} \tag{9.81}$$

where it is noted in (9.81) that T does not depend explicitly on the nodal displacements (but only on the nodal velocities). If we were to expand the quadratic forms (9.75) and (9.78), perform the operations dictated by Lagrange's equations (9.81), and express the four resulting ODEs in vector-matrix form, we would get the following system ODE model for undamped free vibration:

$$[m]\{\ddot{v}\} + [k]\{v\} = \mathbf{0} \tag{9.82}$$

This form is the same as that obtained in the study of the undamped free vibration of multi-DOF discrete systems considered in Sec. 7.3 [see (7.52)]. Finite elements, therefore, converts the problem of vibration of a *continuous system* to the simpler problem of vibration of a *discrete*, *multi-DOF system*. This feature of FE analysis is a primary reason for the importance of the material covered in Sec. 7.3. The finite element method is *the* basic technique of structural dynamics analysis, so the material of Sec. 7.3 is of much wider applicability than might be evident upon first exposure to the subject.

Aside: That (9.82) results from the energies (9.75) and (9.78) when we apply Lagrange's equations (9.81) can be seen clearly if we consider a system with only 2 degrees of freedom.

Suppose that the potential energy V has the form (9.78)

$$V = \frac{1}{2}\begin{Bmatrix} q_1 \\ q_2 \end{Bmatrix}^T \begin{bmatrix} k_{11} & k_{12} \\ k_{21} & k_{22} \end{bmatrix} \begin{Bmatrix} q_1 \\ q_2 \end{Bmatrix}$$

where q_1 and q_2 are the generalized coordinates. Expansion of the above yields

$$V + \tfrac{1}{2}\left[k_{11}q_1^2 + (k_{12} + k_{21})q_1 q_2 + k_{22}q_2^2\right]$$

Then the stiffness terms in the equations of motion are given by

$$\frac{\partial V}{\partial q_1} = k_{11}q_1 + \frac{1}{2}(k_{12} + k_{21})q_2$$

$$\frac{\partial V}{\partial q_2} = \frac{1}{2}(k_{12} + k_{21})q_1 + k_{22}q_2$$

Because the stiffness matrix $[k]$ is symmetric ($k_{12} = k_{21}$), the above may be written as

$$\frac{\partial V}{\partial q_1} = k_{11}q_1 + k_{12}q_2$$

$$\frac{\partial V}{\partial q_2} = k_{21}q_1 + k_{22}q_2$$

or

$$\begin{Bmatrix} \dfrac{\partial V}{\partial q_1} \\ \dfrac{\partial V}{dq_2} \end{Bmatrix} = [k]\{q\}$$

A similar result obtains for the inertia terms.

In the preceding development nothing has been said about the boundary conditions, which will certainly have an effect on the free vibration solution. Let us now consider two examples, with different boundary conditions, of the single-element analysis of the vibrating beam.

Example 9.9 Single-Element Analysis of the Undamped, Freely Vibrating, Simply Supported Beam (Fig. 9.10). The analysis begins by writing out in detail (9.82), using (9.77) and (9.80) for the uniform beam:

$$\frac{\rho L}{420}\begin{bmatrix} 156 & 22L & 54 & -13L \\ 22L & 4L^2 & 13L & -3L^2 \\ 54 & 13L & 156 & -22L \\ -13L & -3L^2 & -22L & 4L^2 \end{bmatrix}\begin{Bmatrix} \ddot{v}_1 \\ \ddot{\theta}_1 \\ \ddot{v}_2 \\ \ddot{\theta}_2 \end{Bmatrix}$$

$$+\frac{2EI}{L^3}\begin{bmatrix} 6 & 3L & -6 & 3L \\ 3L & 2L^2 & -3L & L^2 \\ -6 & -3L & 6 & -3L \\ 3L & L^2 & -3L & 2L^2 \end{bmatrix}\begin{Bmatrix} v_1 \\ \theta_1 \\ v_2 \\ \theta_2 \end{Bmatrix} = \mathbf{0} \qquad (9.83)$$

This single-element result is obtained with no consideration of the boundary conditions, which must now be invoked. For the simply supported beam, $v_1 = v_2 = 0$, so that there are actually only two nontrivial nodal degrees of freedom, θ_1 and θ_2. This means that in the Lagrangian context, we actually have only two generalized coordinates, θ_1 and θ_2. We must, therefore, extract from (9.82) the two equations of motion associated with these coordinates, in this case the second and fourth equations of (9.83). Invoking the boundary conditions thus leads to a pair of ODEs for the nontrivial degrees of freedom, as follows:

$$\frac{\rho L}{420}\begin{bmatrix} 4L^2 & -3L^2 \\ -3L^2 & 4L^2 \end{bmatrix}\begin{Bmatrix} \ddot{\theta}_1 \\ \ddot{\theta}_2 \end{Bmatrix} + \frac{2EI}{L^3}\begin{bmatrix} 2L^2 & L^2 \\ L^2 & 2L^2 \end{bmatrix}\begin{Bmatrix} \theta_1 \\ \theta_2 \end{Bmatrix} = \mathbf{0} \qquad (9.84)$$

We now proceed exactly as in Sec. 7.3 to determine the two natural frequencies and mode shapes which define the solutions to (9.84). Multiplying (9.84) by $L^3/2EI$ and canceling L^2, which is common to all terms, we obtain

$$\left(\frac{\rho L^4}{840EI}\right)\begin{bmatrix} 4 & -3 \\ -3 & 4 \end{bmatrix}\begin{Bmatrix} \ddot{\theta}_1 \\ \ddot{\theta}_2 \end{Bmatrix} + \begin{bmatrix} 2 & 1 \\ 1 & 2 \end{bmatrix}\begin{Bmatrix} \theta_1 \\ \theta_2 \end{Bmatrix} = \mathbf{0}$$

Assuming the harmonic motion solution (7.53) converts the above ODEs to the algebraic eigenvalue problem (7.54), which here reads

$$\left[\begin{bmatrix} 2 & 1 \\ 1 & 2 \end{bmatrix} - \lambda\begin{bmatrix} 4 & -3 \\ -3 & 4 \end{bmatrix}\right]\begin{Bmatrix} c_1 \\ c_2 \end{Bmatrix} = \mathbf{0} \qquad (9.85)$$

where λ is defined by

$$\lambda = \frac{\rho L^4 \omega^2}{840EI} \qquad (9.86)$$

and is proportional to the square of the natural frequency ω. We next solve for the two eigenvalues λ_i and then find the associated natural frequencies ω_i from (9.86):

$$\omega_i = \left(\frac{840EI\lambda_i}{\rho L^4}\right)^{1/2} \tag{9.87}$$

Rewriting (9.85) as

$$\begin{bmatrix} (2-4\lambda) & (1+3\lambda) \\ (1+3\lambda) & (2-4\lambda) \end{bmatrix} \begin{Bmatrix} c_1 \\ c_2 \end{Bmatrix} = 0 \tag{9.88}$$

the characteristic equation, obtained by requiring the determinant of the coefficient matrix in (9.88) to vanish, is

$$(2-4\lambda)^2 - (1+3\lambda)^2 = 0$$

and this yields the eigenvalues $\lambda_1 = \frac{1}{7}$, $\lambda_2 = 3$, with associated natural frequencies ω_i given below [the exact natural frequencies given by (9.59) are also shown for comparison]:

Finite element	**Exact**	**Error, %**
$\omega_1 = 10.95\left(\frac{EI}{\rho L^4}\right)^{1/2}$	$\omega_1 = 9.87\left(\frac{EI}{\rho L^4}\right)^{1/2}$	11
$\omega_2 = 50.2\left(\frac{EI}{\rho L^4}\right)^{1/2}$	$\omega_2 = 39.48\left(\frac{EI}{\rho L^4}\right)^{1/2}$	27

Observe that the finite element results are in error by about 11 percent (ω_1) and 27 percent (ω_2), indicating that use of a single element does not produce very accurate results for this problem.

To see why these inaccuracies have occurred, we need to investigate the mode shapes associated with the two solutions. For mode 1, use of $\lambda_1 = \frac{1}{7}$ in the first of (9.88) leads to

$$\left(2 - \tfrac{4}{7}\right)c_1 + \left(1 + \tfrac{3}{7}\right)c_2 = 0$$

or $c_2 = -c_1$, so the first mode shape is $\begin{Bmatrix} 1 \\ -1 \end{Bmatrix}$. This discrete mode shape is next used to determine, via (9.66), the finite element approximation for the continuous mode shape $\phi_1(x)$. Letting $\theta_1 = 1$ and $\theta_2 = -1$ in (9.66), with $v_1 = v_2 = 0$, leads to

$$\phi_1(x)_{FE} \sim N_2(x) - N_4(x) = x\left(1 - \frac{x}{L}\right)$$

The first mode shape is proportional to $N_2(x) - N_4(x)$ (see Fig. 9.15 to visualize this sum). The exact first mode shape $\phi_1(x)$, on the other hand, is given by (9.60) as

$$\phi_1(x) = \sin\left(\frac{\pi x}{L}\right)$$

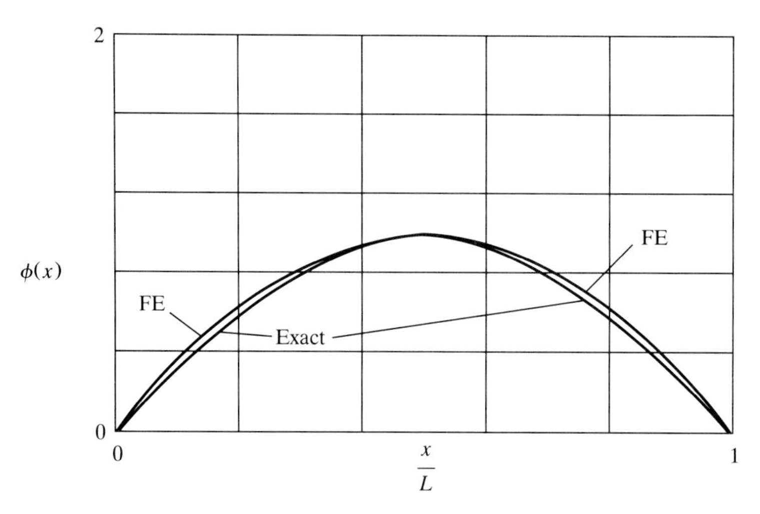

FIGURE 9.20
Finite element approximation and exact first mode shapes for the simply supported beam. (Magnitude of displacement is arbitrary to within a multiplicative constant.)

In order to compare the exact and finite element mode shapes, we take

$$\phi_1(x)_{\text{FE}} = 4\frac{x}{L}\left(1 - \frac{x}{L}\right)$$

so that $\phi_1(x)_{\text{FE}} = 1$ at $x = L/2$; This is the same value as the exact mode shape at $L/2$. A comparison of the exact and FE continuous mode shapes is shown in Fig. 9.20.

In using a single-element FE model, we are attempting to approximate the actual mode shape $\sin(\pi x/L)$ with the *quadratic function* $4(x/L)[1 - (x/L)]$. The agreement shown in Fig. 9.20 is not too bad, considering the crudeness of the approximation. The error in the mode shape does, however, result in an 11 percent error in the associated natural frequency ω_1. In essence, the accuracy of the results will depend on how well the actual mode shapes can be represented by linear combinations of the shape functions.

We can see a similar situation with mode 2, for which the exact mode shape is $\phi_2(x) = \sin(2\pi x/L)$. Using the second eigenvalue $\lambda_2 = 3$ in (9.88) yields the second mode eigenvector as $\{C\}_2 = \begin{Bmatrix} 1 \\ 1 \end{Bmatrix}$. The finite element approximation, $\phi_2(x)_{\text{FE}}$ is then proportional to $N_2(x) + N_4(x)$. Requiring $\phi_2(x)_{\text{FE}}$ to be equal to the exact mode shape $\phi(x)$ at $x = L/4$ (so that we can compare the two), we find

$$\phi_2(x)_{\text{FE}} = \frac{32x}{3L}\left(1 - \frac{3x}{L} + \frac{2x^2}{L^2}\right)$$

A comparison of the exact and approximate (FE) mode shapes for the second mode is shown in Fig. 9.21.

Here we observe a substantial discrepancy between the exact and approximate (FE) mode shapes. The reason is that it is inaccurate to approximate

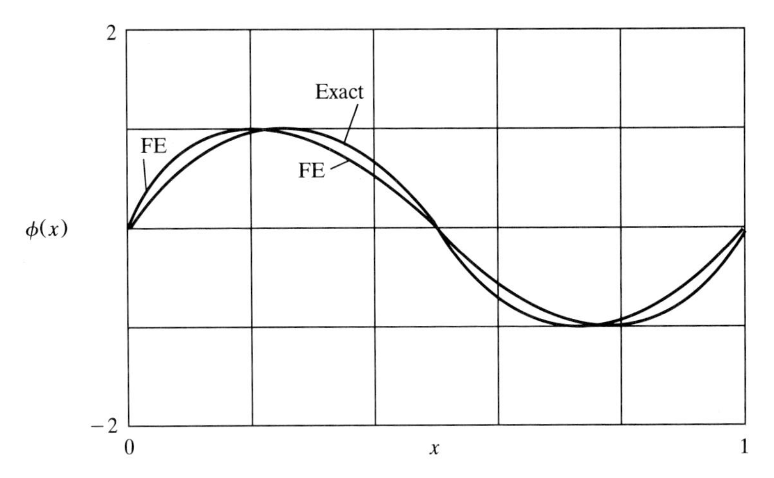

FIGURE 9.21
Finite element approximation and exact second mode shapes for the simply supported beam. (Magnitude of displacement is arbitrary to within a multiplicative constant.)

the function $\sin(2\pi x/L)$, from $x/L = 0$ to $x/L = 1$, with a cubic function. The attendant accuracy in calculated natural frequency deteriorates.

Comment. In the preceding example there were only 2 nontrivial degrees of freedom, so we obtained approximations to the first two modes of the system. No information was obtained for the third and higher modes. As a general rule, if there are n nontrivial degrees of freedom in the finite element formulation, we obtain solutions for the first n modes of motion. As a rule of thumb, it is often assumed that approximately the first $n/2$ modes will be determined accurately using finite elements.

Example 9.10 Two-Element Analysis of a Uniform Beam Cantilevered at Both Ends. As a second example, we consider the vibration of a uniform beam cantilevered at both ends, as shown in Fig. 9.22(a). For these boundary conditions the displacement v and the slope $v' = \theta$ must vanish at $x = 0$ and at $x = L$. This means that a single-element analysis would be inappropriate because all 4 nodal degrees of freedom would be identically zero. We choose, therefore, to model the beam as two equally spaced finite elements, as shown in Fig. 9.22(b). Here the global degrees of freedom V_2 and Θ_2, representing the displacement and slope at the midpoint of the beam, are the "nonzero degrees of freedom." Thus, we expect to end up with a model of the type (9.82), consisting of two second-order ODEs, with V_2 and Θ_2 the unknowns.

In formulating the approximate math model (9.82) for the setup of Fig. 9.22(b), we must now take into account that two elements are being used. Since the global degrees of freedom V_2 and Θ_2 are common to both elements, each element will contribute to the eventual global model (9.82). To see how the contributions from the two elements are to be combined, consider the kinetic energy T of the entire beam. For a single-element model, (9.75) defines the beam

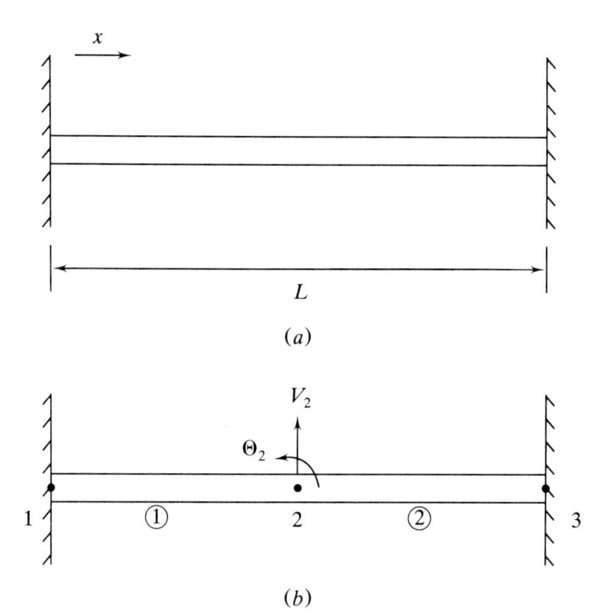

FIGURE 9.22
(*a*) A uniform beam cantilevered at each end. (*b*) Two-element model of the uniform beam cantilevered at each end.

kinetic energy. For the two-element model used here, the total kinetic energy will be the sum of the kinetic energies of the individual elements, $T = T^{(1)} + T^{(2)}$. Use of (9.75) to write this out yields

$$T = \frac{1}{2}\begin{Bmatrix} \dot{v}_1^{(1)} \\ \dot{\theta}_1^{(1)} \\ \dot{v}_2^{(1)} \\ \dot{\theta}_2^{(1)} \end{Bmatrix}^T [m]^{(1)} \begin{Bmatrix} \dot{v}_1^{(1)} \\ \dot{\theta}_1^{(1)} \\ \dot{v}_2^{(1)} \\ \dot{\theta}_2^{(1)} \end{Bmatrix} + \frac{1}{2}\begin{Bmatrix} \dot{v}_1^{(2)} \\ \dot{\theta}_1^{(2)} \\ \dot{v}_2^{(2)} \\ \dot{\theta}_2^{(2)} \end{Bmatrix}^T [m]^{(2)} \begin{Bmatrix} \dot{v}_1^{(2)} \\ \dot{\theta}_1^{(2)} \\ \dot{v}_2^{(2)} \\ \dot{\theta}_2^{(2)} \end{Bmatrix} \tag{9.89}$$

where the superscripts denote element 1 or element 2 and where $[m]^{(1)}$ and $[m]^{(2)}$ are the 4×4 mass matrices defined by (9.77), with the L's in (9.77) replaced by $(L/2)$'s. Next, note that each of the local nodal displacements is equal to one of the global nodal displacements, as defined below:

$$v_1^{(1)} = V_1$$

$$\theta_1^{(1)} = \Theta_1$$

$$v_2^{(1)} = V_2 = v_1^{(2)}$$

$$\theta_2^{(1)} = \Theta_2 = \theta_1^{(2)}$$

$$V_3 = v_2^{(2)}$$

$$\Theta_3 = \theta_2^{(2)}$$

The preceding equalities allow the kinetic energy (9.89) to be rewritten in terms of the global coordinates as follows:

$$T = \frac{1}{2}\begin{Bmatrix} \dot{V}_1 \\ \dot{\Theta}_1 \\ \dot{V}_2 \\ \dot{\Theta}_2 \end{Bmatrix}^T [m]^{(1)} \begin{Bmatrix} \dot{V}_1 \\ \dot{\Theta}_1 \\ \dot{V}_2 \\ \dot{\Theta}_2 \end{Bmatrix} + \frac{1}{2}\begin{Bmatrix} \dot{V}_2 \\ \dot{\Theta}_2 \\ \dot{V}_3 \\ \dot{\Theta}_3 \end{Bmatrix} [m]^{(2)} \begin{Bmatrix} \dot{V}_2 \\ \dot{\Theta}_2 \\ \dot{V}_3 \\ \dot{\Theta}_3 \end{Bmatrix} \tag{9.90}$$

The next step is to combine the two terms in (9.90) by using the full six-entry global displacement vector and by adding in the element 4×4 matrices in the appropriate locations, so that the following form reproduces (9.90):

$$T = \frac{1}{2}\begin{Bmatrix} \dot{V}_1 \\ \dot{\Theta}_1 \\ \dot{V}_2 \\ \dot{\Theta}_2 \\ \dot{V}_3 \\ \dot{\Theta}_3 \end{Bmatrix}^T \begin{bmatrix} & [m]^{(1)} & & 0 & 0 \\ & & & 0 & 0 \\ & & & & \\ 0 & 0 & & [m]^{(2)} & \\ 0 & 0 & & & \end{bmatrix} \begin{Bmatrix} \dot{V}_1 \\ \dot{\Theta}_1 \\ \dot{V}_2 \\ \dot{\Theta}_2 \\ \dot{V}_3 \\ \dot{\Theta}_3 \end{Bmatrix} \tag{9.91}$$

The 6×6 matrix appearing in (9.91) is the global mass matrix $[m]^G$. Notice that it is obtained by starting with the 4×4 matrix for element 1 in the upper-left-hand corner, with the 4×4 mass matrix for element 2 added into the lower-right corner of the global matrix. The two element matrices overlap in the central 2×2 submatrix of $[m]^G$. In this portion of the global matrix the element matrices are merely added to each other. The reader may verify that the 6×6 global matrix that results is as follows:

$$[m]^G = \frac{\rho L_e}{420}\begin{bmatrix} 156 & 22L_e & 54 & -13L_e & 0 & 0 \\ 22L_e & 4L_e^2 & 13L_e & -3L_e^2 & 0 & 0 \\ 54 & 13L_e & 312 & 0 & 54 & -13L_e \\ -13L_e & -3L_e^2 & 0 & 8L_e^2 & 13L_e & -3L_e^2 \\ 0 & 0 & 54 & 13L_e & 156 & -22L_e \\ 0 & 0 & -13L_e & -3L_e^2 & -22L_e & 4L_e^2 \end{bmatrix} \tag{9.92}$$

where $L_e = L/2$ is the length of each element. A similar procedure may be used to obtain the global stiffness matrix for the configuration of Fig. 9.22(*b*) (see Exercise 9.7). The matrix defined by (9.92), along with the associated 6×6 stiffness matrix, can be used as the starting point for any analysis of a uniform beam modeled as two equally spaced elements. We next apply the boundary conditions for the cantilever-cantilever system of this example. Noting that only V_2 and Θ_2 will be nonzero, we extract from the global mass and stiffness matrices the central 2×2 submatrices, involving rows and columns 3 and 4, to obtain the ODE model that applies for the given boundary conditions, as follows:

$$\frac{\rho L_e}{420}\begin{bmatrix} 312 & 0 \\ 0 & 8L_e^2 \end{bmatrix}\begin{Bmatrix} \ddot{V}_2 \\ \ddot{\Theta}_2 \end{Bmatrix} + \frac{2EI}{L_e^3}\begin{bmatrix} 12 & 0 \\ 0 & 4L_e^2 \end{bmatrix}\begin{Bmatrix} V_2 \\ \Theta_2 \end{Bmatrix} = \begin{Bmatrix} 0 \\ 0 \end{Bmatrix} \tag{9.93}$$

The zero off-diagonal entries in the matrices of (9.93) indicate that the two ODEs are *uncoupled*. This means that the two ODEs in (9.93) may be solved separately. The ODE defining V_2 is as follows, after clearing the coefficient of V_2:

$$\ddot{V}_2 + \left(\frac{420EI}{13\rho L_e^4}\right)V_2 = 0 \tag{9.94}$$

The ODE defining the midspan rotation Θ_2 is also obtained as follows:

$$\ddot{\Theta}_2 + \left(\frac{420EI}{\rho L_e^4}\right)\Theta_2 = 0 \tag{9.95}$$

The coefficients of the static terms in (9.94) and (9.95) are the squares of the two system natural frequencies which are obtained from the finite element analysis. Substituting $L_e = L/2$ in these coefficients, we obtain the following results:

$$\omega_1 = 22.7\left(\frac{EI}{\rho L^4}\right)^{1/2} \qquad \omega_2 = 82\left(\frac{EI}{\rho L^4}\right)^{1/2}$$

For comparison, the exact results for the two lowest natural frequencies of this system are stated below:

$$\omega_{1,\,\text{exact}} = 22.37\left(\frac{EI}{\rho L^4}\right)^{1/2} \qquad \omega_{2,\,\text{exact}} = 61.67\left(\frac{EI}{\rho L^4}\right)^{1/2}$$

The low-frequency mode of vibration, obtained from (9.94), consists of a motion not involving rotation at the beam midspan. The mode shape for this motion is shown in Fig. 9.23. The finite element displacement pattern in this mode consists of the shape function $N_3(x_1)$ in element 1 and $N_1(x_2)$ in element 2. The finite element result is quite accurate for this low-frequency mode. The second mode, however, is not modeled very accurately. This mode, which arises from (9.95), involves a midspan rotation only, with no midspan displacement. The finite element mode shape is shown in Fig. 9.24 and consists of $N_4(x_1)$ in element 1 and $N_2(x_2)$ in element 2.

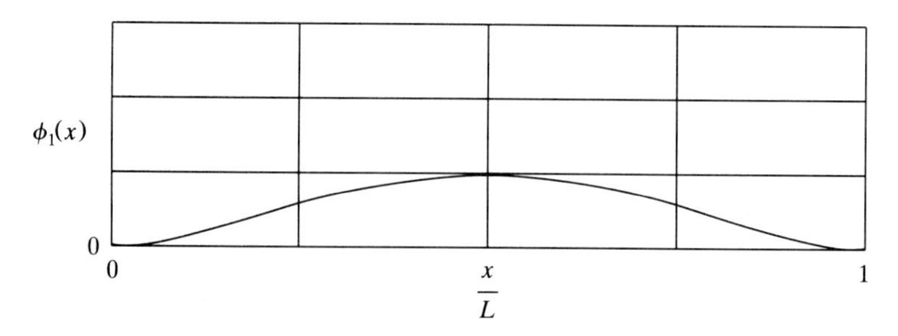

FIGURE 9.23
The lowest-frequency mode shape from a two-element analysis of the system of Fig. 9.22(*b*). The displaced shape is arbitrary to within a multiplicative constant.

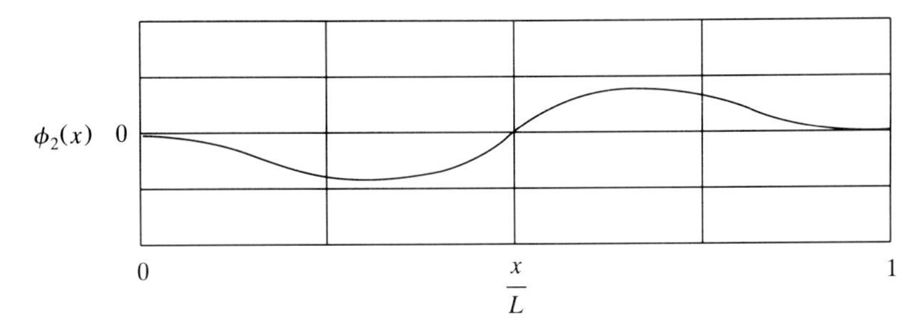

FIGURE 9.24
The second mode shape from a two-element analysis of the system of Fig. 9.22(b). The displaced shape is arbitrary to within a multiplicative constant.

Closing comments

1. The preceding examples utilized very crude finite element approximations in which the beam being analyzed was modeled as only one or two elements. Realistic structural dynamics finite element modeling is typically performed with a general-purpose finite element code. The complexity of real structural systems often necessitates the use of hundreds or even thousands of degrees of freedom in the finite element model. The examples considered here are not representative of real-world structural dynamics and have been presented in order to demonstrate the basic ideas of the finite element method. It should be kept in mind, however, that a clear understanding of the basic problem (9.82) for the simple examples considered here will go a long way toward enabling one to make sense of finite element results obtained for much more complex systems.
2. Example 9.10 illustrates a nice property of the finite element method: The global mass and stiffness matrices can be obtained by merely adding in the 4×4 element matrices into the correct locations in the global matrices. For example, suppose we had modeled the system of Fig. 9.22(a) using 10, rather than 2, equally spaced elements. Then the setup would have 11 nodes, and the global mass and stiffness matrices would be, before application of boundary conditions, 22×22. These global matrices are formed exactly as in (9.91): $[m]^{(1)}$ goes into the upper-left-hand corner of $[m]^G$; then $[m]^{(2)}$ is placed by moving over two columns and down two rows, as in (9.91); $[m]^{(3)}$ is then placed by moving over an additional two columns and down an additional two rows, so that $[m]^{(3)}$ occupies a 4×4 region consisting of rows and columns 5 through 8 in $[m]^G$, and so on. The additive property of the element matrices, which is a result of (9.89), is very convenient from a computational standpoint.
3. In the practical application of the finite element method, there are numerous additional aspects of the analysis that have not been covered here, and the references should be consulted for more detailed material.

REFERENCES

General references on numerical analysis:

9.1. James, M. L., G. M. Smith, and J. C. Wolford: *Applied Numerical Methods for Digital Computation*, 3d ed., Harper & Row, New York (1985).

9.2. Maron, M. J.: *Numerical Analysis, A Practical Approach*, Macmillan, New York (1982).

9.3. Press, W. H., Flannery, B. P., Teukolsky, S. A., and W. T. Vetterling: *Numerical Recipes; The Art of Scientific Computing*, Cambridge University Press, Cambridge (1986). Heavily used nowadays by people who do numerical work.

For finite elements in structural dynamics, see Refs. 7.2, 7.3, 7.5, and 7.6, as well as the following texts on the finite element method.

9.4. Weaver, W., Jr., and P. R. Johnston: *Structural Dynamics by Finite Elements*, Prentice-Hall, Englewood Cliffs, N.J. (1987).

9.5. Bickford, W. B.: *A First Course in the Finite Element Method*, Irwin, Boston (1990).

9.6. Huebner, K. H., and E. A. Thornton: *The Finite Element Method for Engineers*, 2d ed., Wiley, New York (1982).

Material on system identification is contained in the following references:

9.7. Inman, D. J.: *Vibration, with Control, Measurement and Stability*, Prentice-Hall, Englewood Cliffs, N.J. (1989).

9.8. Graupe, D.: *Identification of Systems*, Krieger, Melbourne, Fla. (1976).

9.9. Ewins, D. J.: *Modal Testing: Theory and Practice*, Wiley, New York (1984).

EXERCISES

1. A matrix

$$[A] = \begin{bmatrix} 1 & 3 \\ -2 & -4 \end{bmatrix}$$

defines the algebraic eigenvalue problem (9.9) for a given system [see Fig. 4.25(b) for the exact solution]. Use (9.10) to estimate the eigenvalue λ having the largest magnitude. Use an initial guess $\{\hat{C}\}^{(0)} = \begin{Bmatrix} 1 \\ 1 \end{Bmatrix}$ to start the iteration.

2. Consider the 2 degree of freedom vibratory system (7.33). Use the iterative sequence (9.15), with an initial guess $\{\hat{C}\}^{(0)} = \begin{Bmatrix} 1 \\ 1 \end{Bmatrix}$ for the first mode shape, to determine the lowest frequency and mode shape of the system.

3. Use the Euler method to determine the first few values of x and $\dot{x}$ for the linear oscillator model $\ddot{x} + (2\pi)^2 x = 0$, with initial conditions $x(0) = 1.0$, $\dot{x}(0) = 0$. The exact solution for these initial conditions is $x(t) = \cos(2\pi t)$, so that the oscillation has unit amplitude and a period of 1 s. Start with a time step $\Delta t = 0.02$ s and see how many integration steps are required for the numerical solution to exhibit noticeable error.

4. Same as Exercise 3, but use the second-order Runge-Kutta method (9.29) to generate the numerical solution. Use available software packages, or code your own, to do the computation.

5. Consider the system identification problem of Example 9.6. Assume that, due to measurement noise, the measured values v_1 and v_2 are $v_1 = 0.615$ and $v_2 = 0.36$. Do the system identification as in Example 9.6 to determine the system constant α in (9.44), and compare the result to that obtained in Example 9.6.

6. Determine the four second-order ODEs of the form (9.82) which describe the vibratory motion of a uniform cantilever beam (Fig. *P*9.6), modeled using *two* finite elements of equal length.

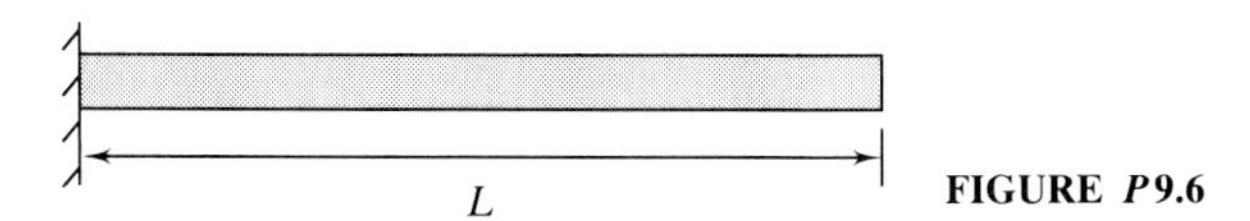

FIGURE *P*9.6

7. Determine the 6×6 global stiffness matrix for a beam of length L which is modeled as two elements of equal length $L/2$. The mass matrix for this case is given by (9.92).

8. Use the finite element method with a single element to estimate the lowest frequency and vibration mode shape for a beam cantilevered at the end $x = 0$ and supported at $x = L$ in such a way that translation $v(L)$ is allowed but rotation $\theta(L)$ is constrained to be zero (Fig. *P*9.8).

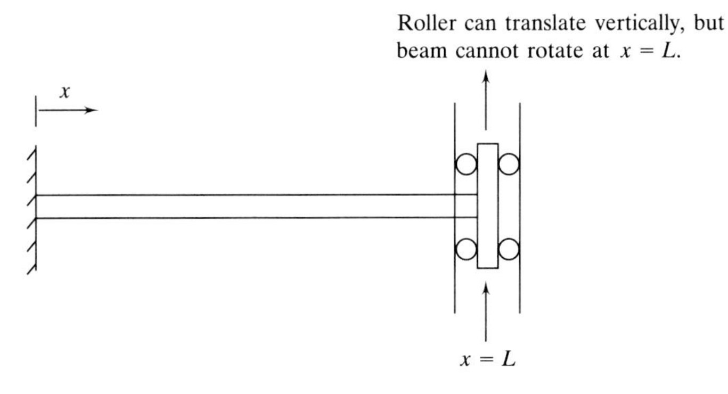

FIGURE *P*9.8

9. Use a single-element model to estimate the two lowest natural frequencies and mode shapes of a beam cantilevered at $x = 0$ and free at $x = L$ (Fig. *P*9.6).

10. Same as Exercise 9, but the system now has a linear spring of stiffness k attached to the end $x = L$ (Fig. *P*9.10). First determine how the spring alters the stiffness matrix $[k]$ in (9.82).

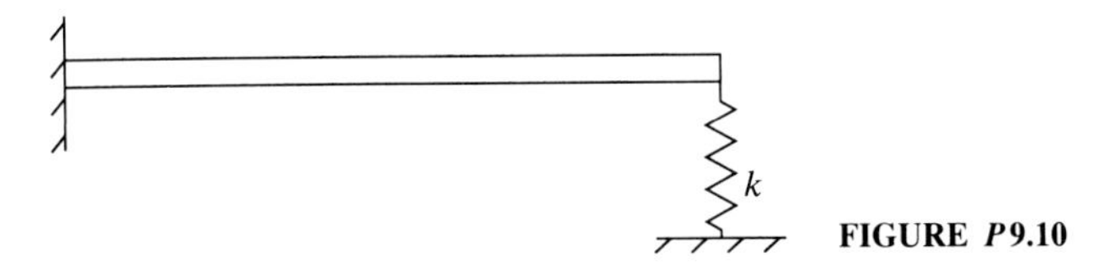

FIGURE *P*9.10

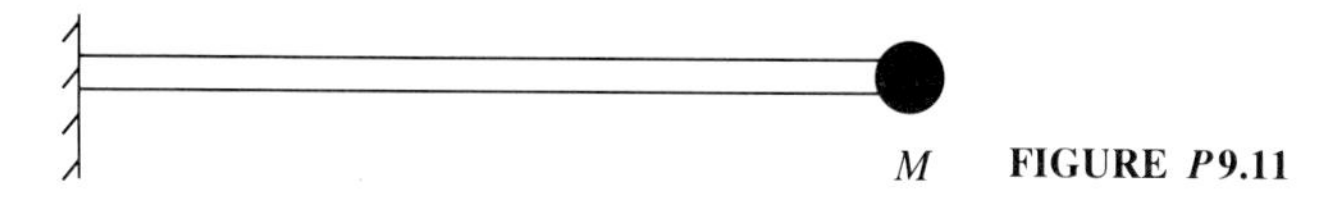

FIGURE ***P*****9.11**

11. Same as Exercise 9, but the system now has a mass M attached to the end $x = L$ (Fig. *P*9.11). First determine how the mass M alters the mass matrix in (9.82).

12. A nonuniform beam of length L has flexural stiffness EI and mass per unit length ρ which vary with position x as follows: $\rho = \rho_0(1 + x/L)$, $EI = (EI)_0(1 + x/L)^3$, where ρ_0 and $(EI)_0$ are values at $x = 0$. If the beam is modeled as a single finite element, determine the entry k_{11} in the element stiffness matrix $[k]$.

13. A vibrating beam of length $L = 300$ cm has the following values measured for the displacements and rotations at either end, at some time t: $v_1 = 2$ cm, $\theta_1 = -5°$, $v_2 = 0$, $\theta_2 = 10°$. Use the cubic finite element approximation (9.66), assuming the beam is modeled as a single finite element, to determine the approximate displaced shape at time t. Plot the result.

14. Consider the two-state nonlinear system of Example 9.3 (Eqs. (9.20), Fig. 9.2). Use the numerical procedure of Example 9.3 to generate the jacobian matrix which describes motions in the vicinity of the equilibrium solution $\bar{x}_1 = \frac{1}{2}$, $\bar{x}_2 = \frac{9}{16}$.

CHAPTER 10

INTRODUCTION TO NONLINEAR DYNAMICS

10.1 INTRODUCTION

In Chaps. 2 and 3 we encountered many example systems for which the governing differential equations were nonlinear. In mechanical systems two types of nonlinearities are common: (1) nonlinearities due to the geometry of motion, as in Example 2.1 (pendulum), Example 2.10 (planar manipulator), and Example 4.6 (rolling ball in a double well); and (2) nonlinear constitutive behavior, i.e., a nonlinear dependence of force or moment on position or velocity, as in Example 2.7 (aerodynamic drag and gravitational forces for the vertical descent problem). In HPE systems nonlinearities often arise in constitutive relations. For instance, in Bernoulli's equation the relation between pressure p (and/or altitude z) and velocity V is nonlinear, and this results in nonlinearities in Example 3.1 (water tank), Example 3.2 (V^2 damper), and Example 3.3 (hydraulic servovalve). Likewise, the nonlinear relation connecting pressure and density for an isentropic thermodynamic process results in nonlinearities in Example 3.4 (airspring) and Example 3.5 (air cushion landing system).

Nonlinear systems can exhibit interesting and surprising responses that have no counterpart in the linear world. The objective of this chapter is to introduce some of these responses and to show why they occur. The discussion here will be limited to a specific system: the single degree of freedom nonlinear oscillator. The idealized physical system to be considered is thus similar to that of Fig. 2.23, but we will now assume that the stiffness and damping forces may be nonlinear functions of the position x and/or of the velocity $\dot{x}$. The intent will be only to introduce some of the basic phenomena and the ideas used to analyze and understand them. The reader is cautioned that this chapter only scratches the surface of the subject. Important nonlinear phenomena occur in most branches of engineering, science, and biology, and the nonlinear oscillator is just one illustrative paradigm.

It was pointed out in Chaps. 4 and 5 that nonlinear effects generally have to be taken into account if the system phase space motions are sufficiently far from an equilibrium solution. This type of situation typically arises in mechanical systems in three ways: (1) The initial conditions cause a motion to be *initiated* far from equilibrium, (2) an external excitation (input) *drives* the system far enough from a stable equilibrium solution that nonlinear effects become important, and (3) a motion, with or without an input, is initiated near an *unstable* equilibrium solution and moves far from equilibrium due to the instability. In this latter case usually the nonlinearities will limit the motion, preventing the system from moving arbitrarily far from equilibrium, as would be predicted by the linear model. So far in this book we have not studied the *effect* of nonlinearities on the system response. We have merely said that such effects exist and may have to be taken into account. The purpose of this chapter is to show how some simple nonlinearities will affect the response of the nonlinear oscillator.

Central to the discussion of nonlinear dynamics is the idea of an attractor, discussed previously in Sec. 4.3.9. An *attractor* is a set of points in the phase space, such as a single point or a closed curve, to which all nearby orbits will asymptote as $t \to \infty$, that is, in the steady state. Thus, any orbit initiated in the "basin of attraction" of the attractor will asymptotically approach the attractor. By their nature attractors are defined by asymptotically stable steady-state solutions to the governing ODE model of the system under consideration. Let us recall the types of attractors we have encountered so far. First, consider the autonomous linear system (4.62), $\{\dot{x}\} = [A]\{x\}$, for which the origin of the phase space is the equilibrium solution. If this equilibrium solution is asymptotically stable, then the equilibrium solution is an attractor, and the entire phase space is the basin of attraction. For the autonomous linear system (4.62), an asymptotically stable equilibrium point is the only type of attractor that can exist.

A second type of attractor we have seen comes from the harmonically driven oscillator (5.67)

$$\ddot{x} + 2\xi\omega\dot{x} + \omega^2 x = \frac{P}{m}\cos\Omega t \tag{10.1}$$

The steady-state solution, which is realized for all initial conditions, is given by (5.69) and (5.78):

$$x_{ss}(t) = \frac{(P/k)\cos(\Omega t - \phi)}{\left[\left(1 - \dfrac{\Omega^2}{\omega^2}\right)^2 + \left(2\xi\dfrac{\Omega}{\omega}\right)^2\right]^{1/2}} \tag{10.2}$$

This steady-state solution defines a *periodic attractor*, and the entire phase space is the basin of attraction. Periodic attractors are always closed curves in the phase space, but not all closed curves are attractors, as we'll see in Sec. 10.3. A periodic solution is an attractor only if the periodic solution is asymptotically stable. Another type of periodic attractor, the limit cycle, is discussed in Sec. 10.2. This type of periodic attractor commonly arises in systems possessing competing stability mechanisms: a linear, destabilizing mechanism that controls motions near equilibrium, and a nonlinear damping mechanism that becomes important far from equilibrium and that limits the motion. (Some analysts refer to any periodic attractor as a *limit cycle*; here, however, the descriptor *limit cycle* will refer only to a response occurring in an autonomous system with nonlinear damping.)

An interesting nonlinear phenomenon that cannot occur in a linear system is the possibility that several different attractors may coexist, each with its own basin of attraction. Thus, the attractor that a given phase space orbit goes to will depend on where the orbit is initiated. Furthermore, the basins of attraction of multiple attractors are separated by "basin boundaries," which generally are organized by unstable solutions or "repellers." The organization of phase space flows for nonlinear systems can be quite interesting. From a practical standpoint attractors are of obvious significance because they are what we actually observe during steady-state operation of the system.

In the following sections we consider three basic phenomena associated with the nonlinear motions of the single degree of freedom oscillator:

1. Limit cycles (Sec. 10.2), which arise in autonomous systems having the following traits: (a) The eigenvalues defining motions near equilibrium are complex with a positive real part, so that equilibrium is unstable, and orbits initiated near equilibrium will tend to spiral outward in the phase plane; and (b) the system possesses nonlinear damping, which comes into play far from equilibrium and limits the motion, resulting in a periodic attractor.
2. Periodic attractors in harmonically driven oscillators having linear damping and *nonlinear stiffness* (Secs. 10.3 through 10.5).
3. *Chaotic motions* (Secs. 10.6 and 10.7) in damped, harmonically driven nonlinear oscillators. Chaotic, or "strange", attractors define motions which appear to have characteristics of randomness but which are actually highly organized. The existence of such attractors in mechanical systems was only discovered in the 1970s.

10.2 NONLINEAR DAMPING AND LIMIT CYCLES

10.2.1 Introduction: Nonlinear Damping

We start by introducing the effect of nonlinear damping with an example.

Example 10.1 The V^2 Damper. Consider a mass m restrained by a linear spring and by the V^2 damper of Example 3.2. The equation of motion, taking $m = k = 1$

for simplicity, is

$$\ddot{x} + c\dot{x}|\dot{x}| + x = 0 \qquad c > 0 \tag{10.3}$$

The nonlinear damping force F_d for this system is $F_d = -c\dot{x}|\dot{x}|$, and the work W_{12} done by this force during a motion of the system between any two times t_1 and t_2 is (2.36)

$$W_{12} = -\int_{t_1}^{t_2} c\dot{x}^2|\dot{x}|\,dt \tag{10.4}$$

Because $\dot{x}^2|\dot{x}| \geq 0$ for all $\dot{x}$, the work W_{12} will always be negative (the force always opposes the motion), indicating that energy is being dissipated, as we expect.

In order to have a reference with which to compare the response of (10.3), let us also consider the mass-spring system with a *linear viscous damper*, for which the mathematical model is (2.51), with $m = k = 1$:

$$\ddot{x} + c\dot{x} + x = 0 \qquad (c > 0) \tag{10.5}$$

If we work out the solution to (10.5) using the state variable form

$$\begin{Bmatrix} \dot{x}_1 \\ \dot{x}_2 \end{Bmatrix} = \begin{bmatrix} 0 & 1 \\ -1 & -c \end{bmatrix} \begin{Bmatrix} x_1 \\ x_2 \end{Bmatrix}$$

we obtain the eigenvalues and eigenvectors as follows:

$$\lambda_{1,2} = -\frac{c}{2} \pm i\left(1 - \frac{c^2}{4}\right)^{1/2} \equiv \sigma + i\omega$$

$$\sigma = -\frac{c}{2}$$

$$\omega = \left(1 - \frac{c^2}{4}\right)^{1/2}$$

$$\{C\}_1 = \{C\}_2^* = \begin{Bmatrix} 1 \\ \sigma + i\omega \end{Bmatrix}$$

According to (4.109) the general solution for the first state $x_1 = x$ will be

$$x(t) = Ae^{-(c/2)t}\cos(\omega t + \theta) \tag{10.6}$$

This shows that $x(t)$ consists of an exponentially damped oscillation with frequency

$$\omega = \left(1 - \frac{c^2}{4}\right)^{1/2}$$

For the particular case in which the damping constant $c = 0.16$, the solution (10.6) is shown in Fig. 10.1 for initial conditions such that $A = 1$, $\theta = 0$.

Figure 10.1 exhibits the exponential oscillation decay characteristic of the damped linear oscillator. The damping factor $\xi = c/2 = 0.08$, and this implies from (7.16) that successive peaks of oscillation will diminish by a factor of approximately 0.6; that is, $(A_n/A_{n-1}) \approx 0.6$, where A_n is the displacement x at the nth oscillation peak.

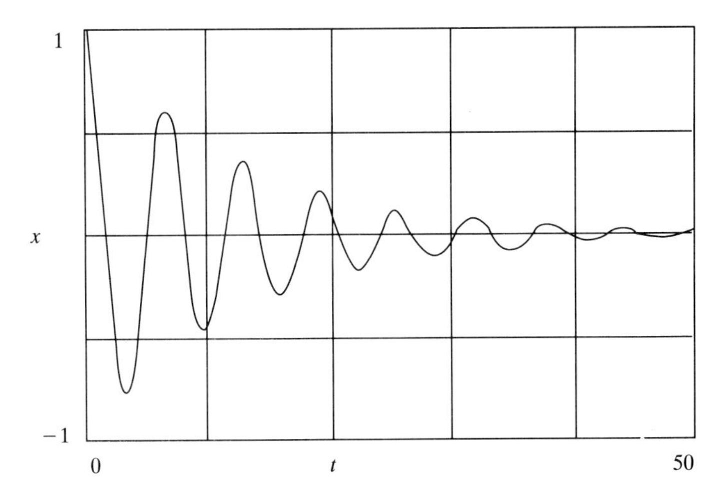

FIGURE 10.1
Solution $x(t)$ for the damped linear oscillator (10.5) with $c = 0.16$ and $x(0) = 1$, $\dot{x}(0) = 0$.

It is important to note that the solution $x(t)$ for *any* initial amplitude A (with $\theta = 0$) will look exactly the same as Fig. 10.1 except for an expansion or contraction of scale of the x axis. We see the familiar asymptotic approach of $x(t)$ to zero, characteristic of a complex eigenvalue with a negative real part. In the phase plane, orbits will spiral into the equilibrium solution at the origin, succeeding spirals shrinking exponentially (Fig. 10.2).

Now let us investigate the solutions to the nonlinear model (10.3), for the same value of $c = 0.16$. We'll see first whether we can guess what to expect based on physical grounds. To start, note that if we remove the damping c from each of (10.3) and (10.5), then the two systems reduce to the same linear systems, $\ddot{x} + x = 0$, which has a pair of imaginary eigenvalues $\lambda_{1,2} = \pm i$. Thus, the

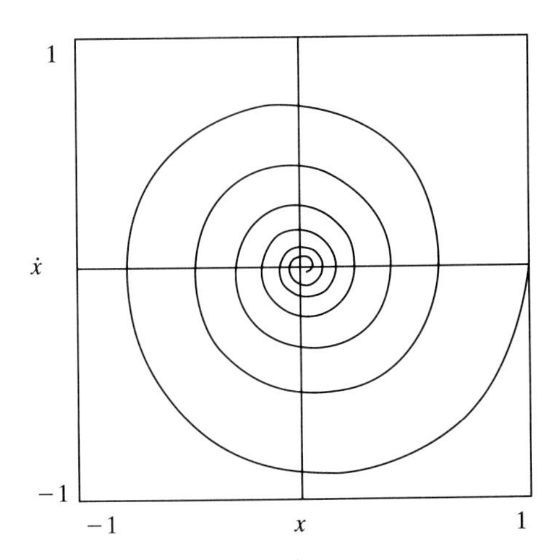

FIGURE 10.2
Phase plane orbits of the damped linear oscillator (10.5) with $c = 0.16$ and $x(0) = 1$, $\dot{x}(0) = 0$.

"associated undamped systems" will each exhibit periodic behavior characterized by closed orbits in phase space and constant amplitudes if we plot x versus time. We know that the addition of linear damping $c\dot{x}$ will cause the oscillations to decay exponentially. Because the nonlinear damping in (10.3) also dissipates energy continuously, we would expect (10.3) also to exhibit orbits in the phase plane which spiral in to the origin. However, the rate of energy dissipation for the linear damper is proportional to $\dot{x}^2$, while for the nonlinear damper it is proportional to $\dot{x}^2|\dot{x}|$. This means that, although we expect to see decaying oscillations for (10.3), the decay will not be exponential; the only way to get exponentially decaying oscillations is with linear damping. We can see further what to expect by considering two special cases:

1. Suppose we compare orbits of (10.3) and (10.5) initiated at the same point in the phase plane *far from* equilibrium, so that $(x_0^2 + \dot{x}_0^2)^{1/2} \gg 1$. Then as the orbits of the two systems spiral in toward the origin, we would expect orbits of the nonlinear system (10.3) to decay faster initially, because the rate of dissipation is $c\dot{x}^2|\dot{x}|$ for the nonlinear system and is $c\dot{x}^2$ for the linear system, and $\dot{x}$ is on average much larger than unity. Thus, the oscillations of the nonlinear system (10.3) will decay faster than those of the linear system (10.5) if the two systems are far from equilibrium.
2. Suppose we now compare orbits of (10.3) and (10.5) initiated *very close* to equilibrium, so that $(x_0^2 + \dot{x}_0^2)^{1/2} \ll 1$. Then we know the linear system still exhibits exponential decay to the equilibrium solution $(0, 0)$. For the nonlinear system, however, the rate of energy dissipation $c\dot{x}^2|\dot{x}|$ is now much smaller than for the linear system, because $\dot{x}$ is on average much less than unity. Thus, for orbits initiated close to equilibrium, oscillations of the nonlinear system will

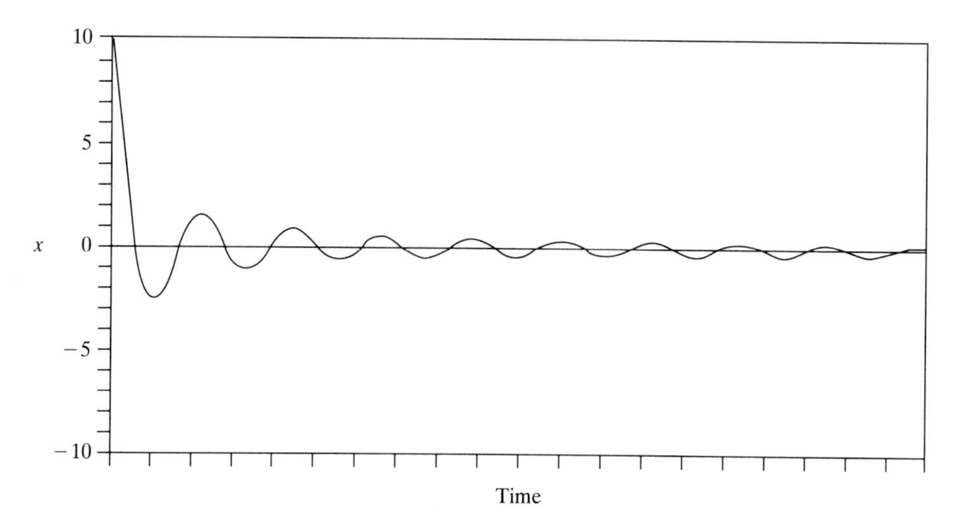

FIGURE 10.3
Solution $x(t)$ for the oscillator (10.3) with nonlinear damping, with initial conditions $x(0) = 10$, $\dot{x}(0) = 0$ and with $c = 0.16$.

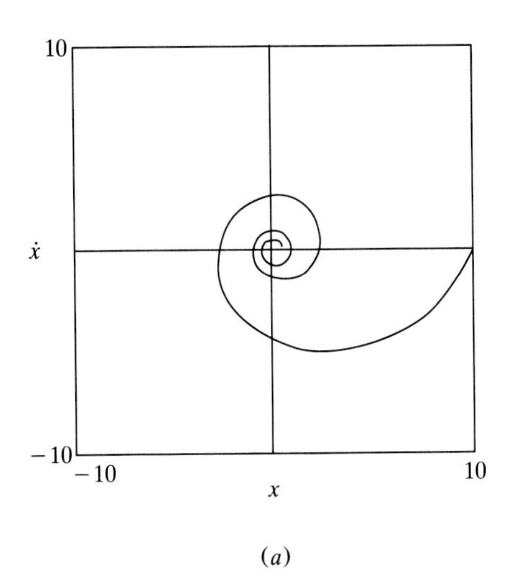

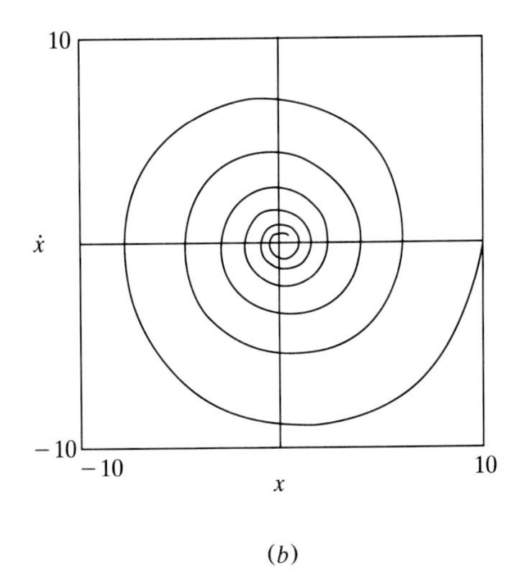

(*a*) (*b*)

FIGURE 10.4
(*a*) Phase plane orbits of the oscillator (10.3) with nonlinear damping. (*b*) For the linear oscillator (10.5). Same conditions as in Fig. 10.3.

decay more slowly than those of the linear system. This is the opposite of what happens far from equilibrium. In fact, if we linearize the nonlinear model (10.3) to study motions near the equilibrium solution $(0, 0)$, we find the linearized model to be $\ddot{x} + x = 0$, which says the oscillations do not decay at all. Once again (as in Sec. 4.3.7) we encounter a nonlinear system for which the near-equilibrium linearized model has eigenvalues with a zero real part. The orbits of the nonlinear system, nevertheless, do asymptotically approach equilibrium, which is asymptotically stable, but the approach to equilibrium is "slower than exponential."

Figure 10.3 shows a graph of the solution $x(t)$ to the nonlinear model (10.3) for initial conditions $x_0 = 10$, $\dot{x}_0 = 0$. Note the rapid oscillation decay initially and the very slow decay as the solution approaches equilibrium. Figure 10.4 shows the phase portraits for the linear and nonlinear systems for the same far from equilibrium initial conditions $x_0 = 10$, $\dot{x}_0 = 0$. Figure 10.5 shows the solution $x(t)$ for the nonlinear oscillator (10.3) for initial conditions $x_0 = 0.1$, $\dot{x}_0 = 0$, that is, close to equilibrium. While the oscillations decay, they do so much more slowly than for the linear oscillator.

Two comments regarding this example are worthwhile:

1. The rate of oscillation decay for the linear and nonlinear systems (10.3) and (10.5) differs markedly, depending on whether the systems are operating far from or close to equilibrium. Thus, we observe fairly substantial *quantitative* differences in the responses. *Qualitatively*, however, the responses of the two systems are similar, in the sense that both exhibit a single, asymptotically stable equilibrium at the origin of the phase space, and both exhibit orbits

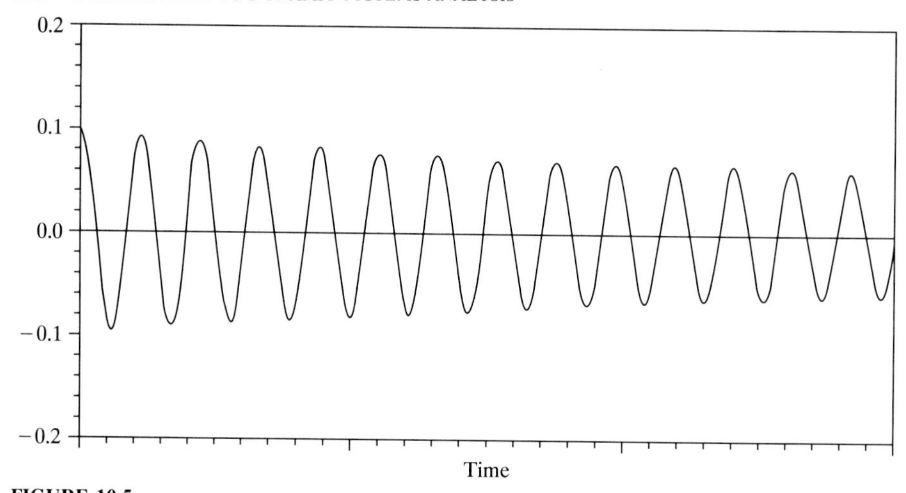

FIGURE 10.5
Solutions $x(t)$ for the oscillator (10.3) with nonlinear damping, for near-equilibrium initial conditions, $x(0) = 0.1$, $\dot{x}(0) = 0$, $c = 0.16$.

that spiral into this equilibrium from all regions of the phase space. Thus, the presence of the nonlinearity has not caused any fundamentally new phenomena to appear.

2. In any autonomous *linear system* with distinct eigenvalues which exhibits decaying oscillations, the oscillation decay will always be *exponential*. This is an obvious consequence of the general linear solution (4.80). This fact enables one to determine, in specific cases, whether any energy dissipation mechanisms present can be modeled by the linear viscous damper. One sets the system into motion, perhaps by "hitting" it, measures the decay of the oscillations, and checks whether the decay is exponential. If the decay is clearly nonexponential, one knows that a nonlinear damping mechanism is at work.

10.2.2 Limit Cycles in Systems with Nonlinear Damping

In the preceding subsection we studied a simple oscillator with a nonlinear damper, for which the qualitative behavior was similar to that of the linear oscillator. We'll now take a look at a much different situation, for which a system with nonlinear damping exhibits a phenomenon that cannot occur in an autonomous linear system: the limit cycle. We will again begin the study with a specific example.

Example 10.2 Nonlinear Damping and Limit Cycles. The system to be considered is a mass restrained by a linear spring, as in Example 2.3, but now we assume a different form for the system damping. The mathematical model is (with $m = k = 1$)

$$\ddot{x} - c\dot{x}\left[1 - (x^2 + \dot{x}^2)\right] + x = 0 \qquad c > 0 \tag{10.7}$$

Observe that for this system there are actually two damping forces present. One damping force $F_{D_1} = c\dot{x}$ is linear. The second damping force $F_{D_2} = -c\dot{x}(x^2 + \dot{x}^2)$ is nonlinear. Let us again see whether we can determine intuitively how the system is likely to behave. First, if we investigate the motion near the single equilibrium solution $(0, 0)$, we obtain the linearized model

$$\ddot{x} - c\dot{x} + x = 0 \tag{10.8}$$

The eigenvalues $\lambda_{1,2}$ for the linearized system are $\lambda_{1,2} = (c/2) \pm i(1 - c^2/4)^{1/2}$, indicating that the phase plane orbits initiated near equilibrium will spiral outward; the equilibrium solution $(0, 0)$ is unstable. In physical terms the force $F_{D_1} = c\dot{x}$ is proportional to the velocity but is exerted in the *same direction* in which the mass is moving. The work W_1 done by this force in any time interval (t_1, t_2) is

$$W_1 = \int_{t_1}^{t_2} c\dot{x}^2\, dt \tag{10.9}$$

Because $W_1 > 0$, the force F_{D_1} is "adding energy" to the system, rather than dissipating it. We speak of "undamping," or "negative damping." So, based on the linear theory, orbits initiated near equilibrium will spiral outward, and the system tends to move far from equilibrium in the phase plane.

Now let us see what effect we would expect the nonlinearity to have. First, the work W_2 done by the nonlinear damping force F_{D_2} in a time interval (t_1, t_2) is

$$W_2 = \int_{t_1}^{t_2} -c\dot{x}^2(x^2 + \dot{x}^2)\, dt \tag{10.10}$$

We see that $W_2 < 0$, as the integrand is always negative, so the nonlinearity is continually dissipating energy. Furthermore, near equilibrium, x and $\dot{x}$ are small compared to unity, so $|W_2| \ll |W_1|$; the rate of energy dissipation by the nonlinear force F_{D_2} is much smaller than the rate of energy addition by the linear force F_{D_1}, and the system moves in accordance with the linear model (10.8). Far from equilibrium, however [say, $(x^2 + \dot{x}^2) \gg 1$], the rate of dissipation due to the nonlinear force F_{D_2} is much larger than the rate of energy addition by the linear force F_{D_1}. Thus, orbits initiated far from equilibrium will tend to spiral *inward* in phase space. We see that there are two competing mechanisms present: (1) the destabilizing effect of F_{D_1}, which is dominant near equilibrium, and (2) the stabilizing effect of the nonlinear force F_{D_2}, which is dominant very far from equilibrium. What we might expect to happen, then, is that for some intermediate region of the phase space, neither too close nor too far from equilibrium, the two competing mechanisms will balance on average, resulting in a closed orbit in the phase plane. As the system moves around this closed orbit, the amount of energy added by F_{D_1} is balanced by the energy dissipated by F_{D_2}, and the oscillation neither grows nor decays. This is exactly what happens in (10.7); the resulting closed orbit is called a *limit cycle*, and such behavior cannot occur in an autonomous linear system.

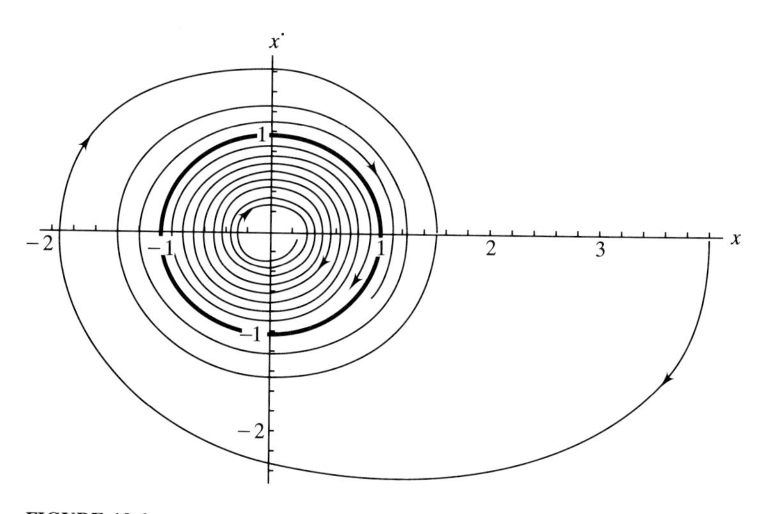

FIGURE 10.6
Phase plane orbits of the system (10.7), showing the asymptotic approach to the limit cycle, which is the unit circle ($c > 0$).

For the system (10.7) we can actually find the limit cycle analytically by noting that a solution to (10.7) is $x(t) = \cos t$, which, if plugged into (10.7), yields*

$$-\cos t + c \sin t\left[1 - (\cos^2 t + \sin^2 t)\right] + \cos t \equiv 0$$

Thus, the limit cycle is a circle of unit radius in the phase plane. Shown in Fig. 10.6 are two orbits of (10.7), one initiated near equilibrium, and one initiated far from equilibrium; these orbits were obtained through numerical integration of (10.7), as discussed in Chap. 9. Both orbits asymptotically approach the limit cycle.

In fact, we can see that an orbit initiated *anywhere* in the phase plane, except at the origin (0, 0), will eventually reach the limit cycle. Thus, the limit cycle is an *attractor*, and the domain of attraction is the entire phase plane, less the origin. We also note that any *area* of the phase plane will eventually shrink onto the limit cycle.

We have uncovered a completely new phenomenon that has no counterpart in the autonomous linear world. The limit cycle is very important from a practical standpoint because, being an attractor, it represents the observed long-term state of the system (10.7), just as the asymptotically stable equilibrium is the long-term realization for many other systems.

There are several things to note about Example 10.2. First we see that, in order for a limit cycle to exist, two competing physical mechanisms must be present. The first of these is a destabilizing effect which is dominant near

*Actually, $x(t) = \sin t$, $x(t) = \cos(t + \phi)$, $x(t) = \sin(t + \phi)$, with ϕ arbitrary, will all work too, as the student may verify.

equilibrium, "feeding energy into the system" and causing it to move away from equilibrium. The second mechanism is a stabilizing effect, which dissipates energy and comes into play as the system moves an appreciable distance from equilibrium.

A second comment relates to the *stability* of the limit cycle. We have seen that, in the case of equilibrium solutions, we will not actually observe in nature an equilibrium solution unless it is asymptotically stable (i.e., unless it is an attractor). In the same way, it is possible that we can find periodic solutions to equations similar to (10.7), but such periodic solutions will not be observed unless they are asymptotically stable: Any orbit initiated near the limit cycle must asymptotically approach it. For example, suppose we alter the sign of the damping term in (10.7), so that it reads

$$\ddot{x} + c\dot{x}\left[1 - (x^2 + \dot{x}^2)\right] + x = 0 \qquad (c > 0) \tag{10.11}$$

The linear term $c\dot{x}$ in (10.11) is now dissipative, while the nonlinear term $-c\dot{x}(x^2 + \dot{x}^2)$ adds energy to the system. The origin of the phase space is now asymptotically stable; orbits initiated near equilibrium will asymptotically approach it. The system (10.11) still admits the solution $x(t) = \cos t$, just as (10.7) does, but now the periodic solution is unstable. Orbits initiated anywhere inside the limit cycle will spiral into the origin, and orbits initiated outside the limit cycle will spiral off to infinity (at least, according to the model). We have an "unstable limit cycle". Figure 10.7 shows the behavior in the phase plane.

Limit cycles may also exist in systems of order higher than 2. If a limit cycle is asymptotically stable, then it is an attractor.

The stability of limit cycles may be tested by linearization in a manner similar to that used to test the stability of equilibrium. For example, suppose we

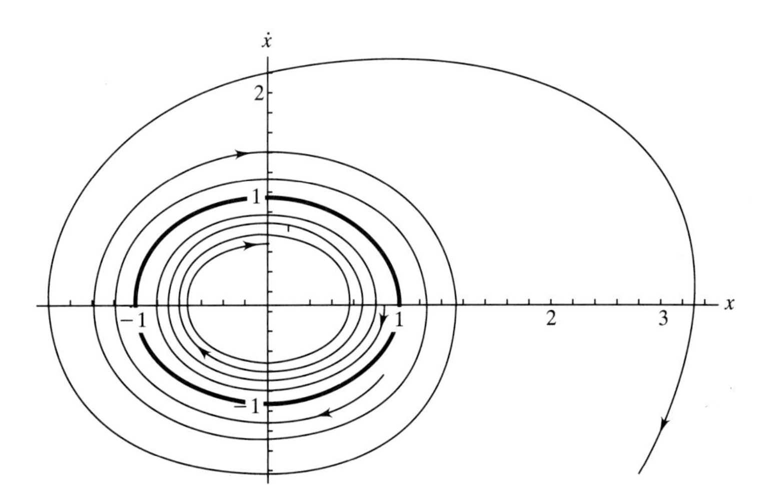

FIGURE 10.7
Phase plane orbits of the system (10.7) for the case $c < 0$. In this case the limit cycle is unstable.

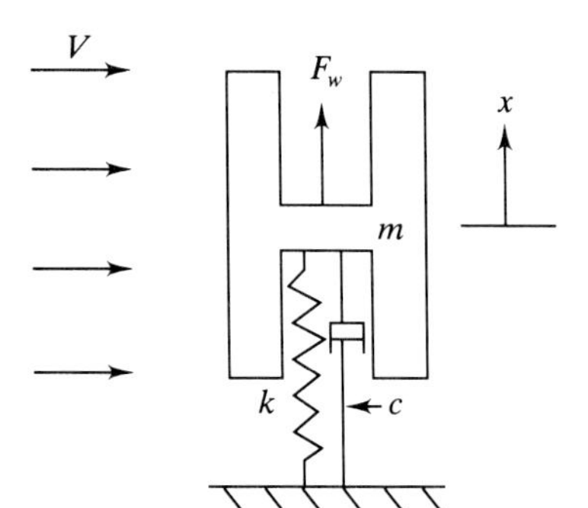

FIGURE 10.8
A damped linear oscillator excited by a crosswind of speed V.

investigate the nature of "near limit cycle" orbits of (10.7). In a manner analogous to (4.28), we put

$$x(t) = \cos t + \xi(t) \tag{10.12}$$

where $\cos t$ is the limit cycle solution whose stability is to be tested, analogous to the equilibrium solution $\bar{x}$ in (4.28), and $\xi(t)$ represents an arbitrarily small, time-dependent deviation from the limit cycle. We plug (10.12) into (10.7), and the resulting equation for the deviation $\xi(t)$ is

$$\ddot{\xi} + c(\dot{\xi} - \sin t)(\xi^2 + \dot{\xi}^2 + 2\xi\cos t - 2\dot{\xi}\sin t) + \xi = 0$$

Because ξ is supposed to be small, we next delete all nonlinear terms, analogous to dropping higher terms in the Taylor series (4.30), to obtain

$$\ddot{\xi} + 2c\sin t(\dot{\xi}\sin t - \xi\cos t) + \xi = 0 \tag{10.13}$$

The result (10.13) is a linear second-order ODE governing the evolution of the small deviation $\xi(t)$. For asymptotic stability of the limit cycle, we have to show that $\xi(t) \to 0$ as $t \to \infty$. Note, however, that although (10.13) is linear, it contains periodic coefficients in the middle term, and this renders the process of solving (10.13) considerably more involved than anything we wish to tackle here. The basic ideas, however, are similar to what we do for an ordinary equilibrium solution.

The preceding, somewhat contrived,* example served to illustrate the basic mechanisms by which limit cycles occur. Now let us turn to an example of the type actually occurring in engineering.

Example 10.3 Galloping Vibration. We now consider the vertical motion of a mass restrained by a linear spring and a linear damper, as in Example 2.3 (Fig. 10.8). The mass m is assumed to have a symmetrical shape typical of structural components such as box beams, I beams, or H beams. The system is subjected to a horizontal wind of uniform speed V. This wind will exert a horizontal force on m,

*We usually do not find nice, neat solutions like $x(t) = \cos t$ for the limit cycle.

but we'll assume the system to be very stiff in the horizontal direction, so that it moves only vertically, in the plane of the paper. Denoting the wind force in the vertical direction as F_W, the equation of motion of the system is

$$m\ddot{x} + c\dot{x} + kx = F_W \qquad (c > 0) \tag{10.14}$$

Let us explore the functional dependence of F_W. First, if the mass m is displaced a distance x but is *stationary* ($\dot{x} = 0$), then $F_W = 0$, and we reason that F_W will not depend on the position x. If the mass m is in motion, however, we would expect a vertical force to be generated, so that $F_W = F_W(\dot{x})$. The force F_W should not depend explicitly on time, so we have a vertical wind force which will depend only on velocity $\dot{x}$. If we expand this force F_W in a Taylor series, the first few terms are

$$F_W(\dot{x}) = a_1\dot{x} + a_2\dot{x}^2 + a_3\dot{x}^3 + \cdots \tag{10.15}$$

We can reason that $a_2 = 0$ because, for a given velocity magnitude $|\dot{x}|$, the force magnitude $|F_W|$ should be the same whether the mass is moving up or down. This can occur only if F_W is an odd function of $\dot{x}$, that is, if $-F_W(-\dot{x}) = F_W(\dot{x})$. Thus, retaining only terms through the cubic, (10.15) reduces to

$$F_W(\dot{x}) \cong a_1\dot{x} + a_3\dot{x}^3 \tag{10.16}$$

Now intuition may imply that $a_1 < 0$; this would say physically that the vertical wind force, for small $\dot{x}$, is in the opposite direction as the motion of m. If this were true, then the wind provides a damping force which augments the natural damping c of the system. Tests reveal, however, that for masses of certain shape, $a_1 > 0$, so that the vertical wind force is in the same direction as the motion and actually feeds energy into the system when $|\dot{x}|$ is small. This possibility may appear reasonable when we consider that the airflow around a moving structure of reasonably complex shape is very complicated, and that the net wind force is a consequence of integrating, over the entire body surface, the fluid pressures that are generated. In any event, let us pursue the dynamic behavior of (10.14) assuming that the shape of the structure of interest results in $a_1 > 0$, so that, for small velocities $\dot{x}$, a portion of the energy of the flowing fluid is converted to the mechanical energy of motion of the mass m. We usually find, in addition, that for relatively large velocities, for which the nonlinear term $a_3\dot{x}^3$ is dominant, the vertical wind force opposes the direction of motion ($a_3 < 0$), so that the overall wind force $F_W(\dot{x})$ might vary with $\dot{x}$ as shown in Fig. 10.9.

We now combine (10.14) and (10.16) to obtain the system mathematical model in the form

$$m\ddot{x} + (c - a_1)\dot{x} - a_3\dot{x}^3 + x = 0 \qquad \begin{pmatrix} a_1 > 0 \\ a_3 < 0 \end{pmatrix} \tag{10.17}$$

Finally, we note that the coefficients a_1 and a_3 are *proportional to the wind speed* V, as well as the size of the mass m.

Now we are in a position to see how a limit cycle may occur for this system. Suppose the system is at equilibrium, $x = \dot{x} = 0$, and we turn on a wind at *low speed* V. Because a_1 is proportional to V, at low wind speeds the destabilizing effect due to a_1 is weaker than the stabilizing effect due to the inherent system damping c ($c > a_1$); the coefficient $(c - a_1)$ of $\dot{x}$ in (10.17) is positive. So if the

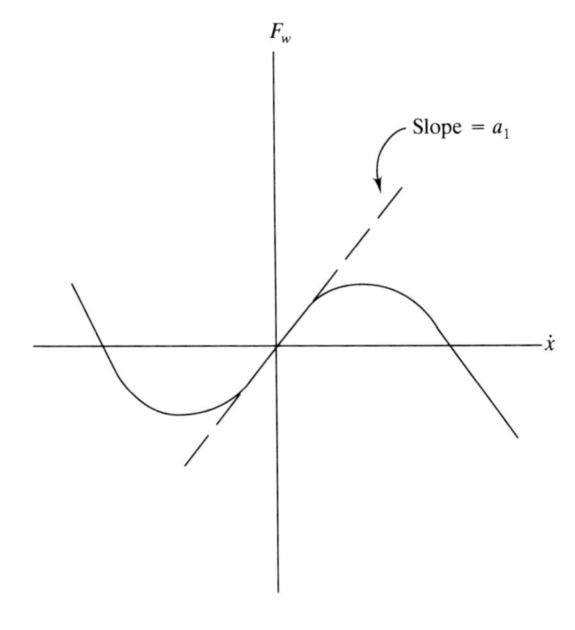

FIGURE 10.9
Typical vertical wind force variation with velocity for a case in which the wind force is destabilizing at small velocity ($a_1 > 0$) and stabilizing at large velocity ($a_3 < 0$).

system is disturbed from equilibrium, decaying oscillations back to equilibrium will result.

Next consider what happens as we increase the wind speed V. At some critical value V_c, the inherent system damping c is exactly offset by the destabilizing effect of a_1; that is, $c = a_1$. In this situation the overall linear damping has disappeared. If V is increased beyond V_c, $c < a_1$; the equilibrium at the origin becomes unstable. The system will exhibit growing oscillations that will eventually be limited by the stabilizing effect of the nonlinear damping $a_3\dot{x}^3$; a limit cycle results. The limit cycle motions can be quite violent. The well-known Tacoma Narrows Bridge failure was due, at least in part, to this effect.

The physical effect just described is called *galloping vibration*. This phenomenon is also exhibited by ice-coated transmission lines in high winds. Transmission lines typically consist of 1- to 1.5-in aluminum cable, stretched in spans of 1000 ft or so. When coated by ice, the aerodynamic properties can be altered drastically, and very large amplitude (10 to 20 ft) limit cycle motions can occur. An excellent treatment of the general flow induced vibrations problem is contained in Ref. 10.1.

It is clear that the destabilizing effect that leads to limit cycle behavior requires a source of energy. In mechanical systems this energy source is most often due to some sustained *unidirectional effect* exerted on the system. For instance, in the galloping vibration problem, the unidirectional energy source is the steady wind flow. A portion of the wind energy is converted to energy of motion of the system, so that a sustained oscillation results. Other examples of limit cycles in mechanical systems with unidirectional energy sources are provided in Table 10.1.

TABLE 10.1
Some sources of limit cycles in mechanical systems

System / phenomena	Vibrating component / motion	Unidirectional energy source
1. Chatter in machining, cutting operations	Machine tool	Steady rotation of workpiece (object which is to be cut)
2. Automobile shimmy	Wheel/axle assembly	Energy of forward motion
3. Rail vehicle axle vibrations	Axle/truck assembly	Energy of forward motion
4. Air cushion landing system (ACLS, Example 3.5)	Vehicle vertical motion	Steady air inflow provided by fan

Each of the above systems is excited by a unidirectional energy source. Each unidirectional energy source is characterized by a certain steady speed or flow rate. As in the case of galloping vibration, limit cycles in the above systems will occur only when the speed or flow rate exceeds a certain critical level, such that the inherent system damping is overcome by the destabilization due to the unidirectional energy source. Furthermore, it does not necessarily occur that the equilibrium solution is unstable for all speeds or flows above critical. Usually there is a certain range (or ranges) of speeds above critical in which limit cycles occur.

As another example, consider the air cushion landing system (Example 3.5), for which the unidirectional energy source is the steady air inflow q_{in} provided by a fan. It is observed that for a certain range of flows q_{in}, a violent vertical vibration of the vehicle occurs. This is illustrated in Figs. 10.10 and 10.11. Figure 10.10 shows the complex eigenvalues of the least-damped mode obtained from a linear analysis describing near-equilibrium behavior of a slightly more complicated version of the mathematical model of Example 3.5.

We observe that, for inflows in the range 758 lb/min $< q_{in} <$ 890 lb/min, the least-damped mode eigenvalue has a positive real part, signifying an unstable equilibrium. For inflows q_{in} in this range, the system will exhibit growing oscillations, with nonlinear damping eventually coming into play to limit these oscillations. Figure 10.11 shows graphs of vertical displacement y and cushion pressure p_c for an inflow $q_{in} = 840$ lb/min, for which the least-damped mode eigenvalue is approximately $\lambda_{1,2} \cong 2.5 \pm 17.4i$.

One observes that for times t less than about 1.25 s, the oscillation growth rate and frequency are consistent with the least-damped mode eigenvalue; the oscillations grow approximately as $e^{2.5t}$, and the frequency is about 3 cycles/s. By about $t = 2$ s, the limit cycle oscillation has been reached.

Comments

1. Based on the discussion of this section, we should be aware of the conditions for which we should be on the lookout for limit cycles. The first is the presence of a unidirectional energy source. Such a source must add energy to the system

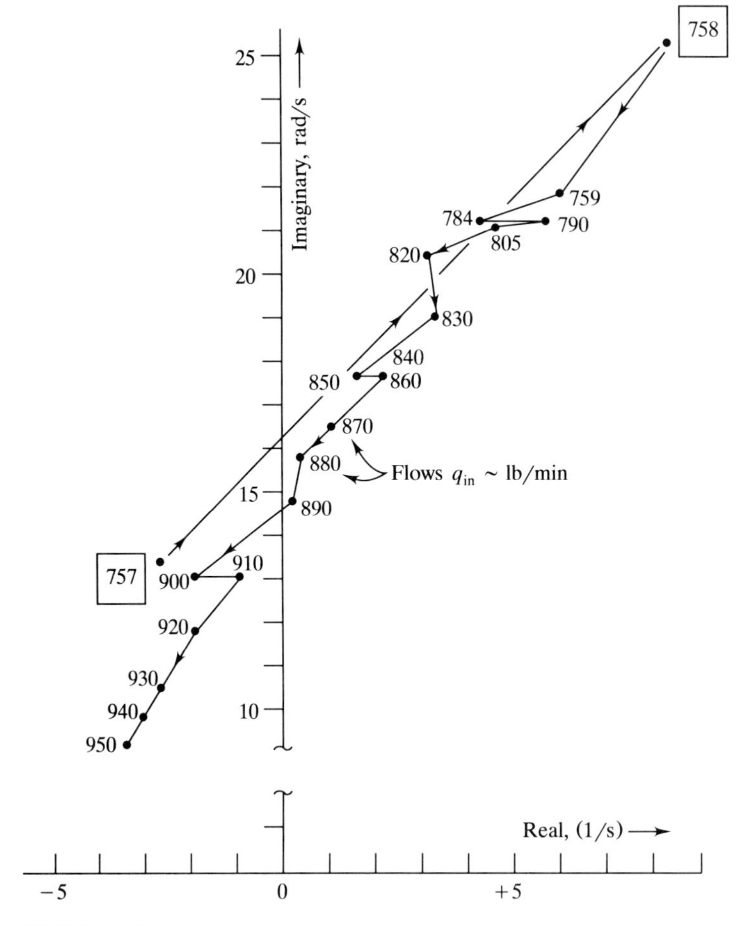

FIGURE 10.10
Eigenvalues of the least-damped mode, as a function of air inflow q_{in}, for an ACLS linearized model.

for motions near equilibrium and dissipate energy far from equilibrium; hence it must be nonlinear. Note that it is often difficult to tell exactly how the energy transfer occurs; furthermore, the presence of a unidirectional energy source does not necessarily mean that a limit cycle *has* to occur, just that, if a limit cycle *does* occur, a unidirectional energy source is usually the culprit.

The second (related) condition that usually signifies the onset of a limit cycle is when a complex eigenvalue crosses the imaginary axis as a system parameter is varied. Thus, suppose we use the linear model (4.62) to analyze the near-equilibrium motions of an autonomous nonlinear system, and suppose we determine the eigenvalues for different values of the parameter which characterizes the strength of the unidirectional energy source, such as wind speed V in Example 10.3 or air inflow q_{in} in Example 3.5. We may find that, at

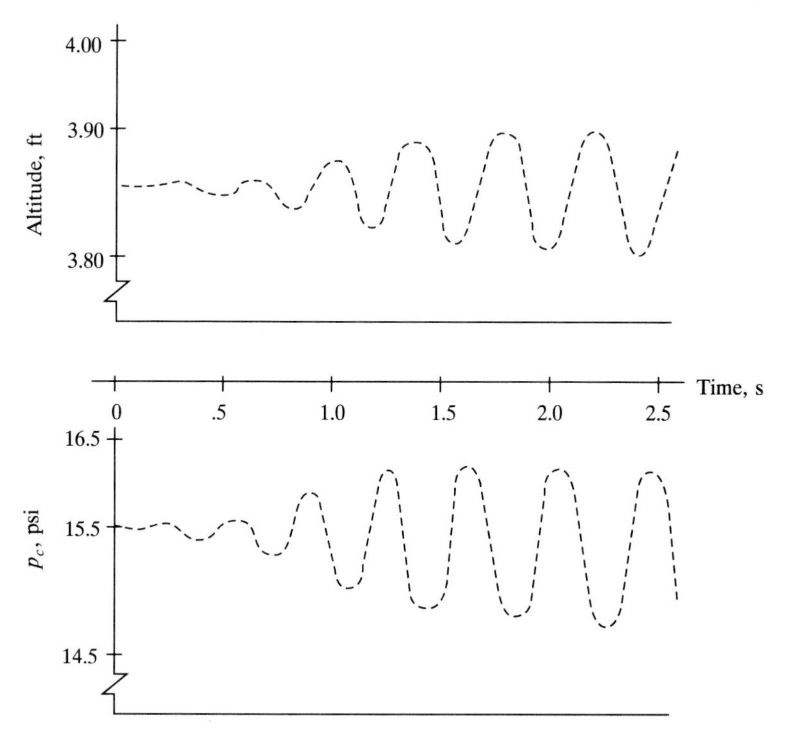

FIGURE 10.11
Histories of vertical displacement y and cushion pressure p_c, as obtained via numerical integration of a nonlinear ACLS model, for an inflow q_{in} for which equilibrium is unstable.

a certain critical value of this parameter, the eigenvalue associated with the least-damped mode crosses the imaginary axis, so that its real part σ becomes positive. What usually happens then is that orbits in phase space spiral outward and are attracted to a limit cycle.*

2. One thing we have to be clear on is that the unidirectional energy sources that produce limit cycles are *not* considered by us to be inputs. This is because these energy sources, while they may appear physically not to be a "part of the system," nevertheless produce forces that are generated by (i.e., are a function of) the system motion, rather than being explicit functions of time. The limit cycle phenomenon that we have studied here, then, is associated with *autonomous systems*.

*Note, however, that if we encounter an eigenvalue with a positive real part, the result does not tell us *what happens*. It only tells us with certainty what *doesn't happen*, i.e., equilibrium is not reached by the system. What actually occurs as the system moves far from equilibrium is, of course, not forthcoming from the linear analysis.

10.3 THE UNDAMPED OSCILLATOR WITH NONLINEAR STIFFNESS

10.3.1 Introduction

In this and the following sections, we shall analyze some of the nonlinear responses exhibited by the single degree of freedom oscillator possessing linear damping and nonlinear stiffness. Because ideal stiffness effects are conservative, the stiffness force F_s may be represented as the derivative of a potential energy function $V_0(x)$, as in (2.40), $F_s = -dV_0/dx$. The oscillator equation of motion may then be expressed as

$$m\ddot{x} + c\dot{x} + \frac{dV_0(x)}{dx} = F(t)$$

For the linear oscillator, $V_0(x) = \frac{1}{2}kx^2$ and the above reduces to (2.52). Division by the mass m leads to the form

$$\ddot{x} + 2\xi\dot{x} + \frac{dV(x)}{dx} = f(t) \tag{10.18}$$

where $V(x) = V_0(x)/m$, $f(t) = F(t)/m$, and $\xi = c/2m$. The potential energy $V(x)$ has been used to characterize stiffness effects because $V(x)$ is itself a useful indicator of the type of unforced response to be expected. Specifically, the graph of $V(x)$ versus x will reveal the locations of equilibrium solutions $(dV/dx = 0)$ and their stability. Oscillatory systems possessing nonlinear stiffness have been discussed earlier: See Examples 2.1 and 2.2 (the pendulum with fixed and moving supports), Example 3.4 (the airspring), and Example 4.6 (the rolling ball).

In the following sections, we'll consider responses for two mathematically simple versions of (10.18) $V(x) = \frac{1}{2}kx^2 + \frac{1}{4}\alpha x^4$, and $V(x) = \frac{1}{4}\alpha x^4$. These potential energies lead to the two mathematical models

$$\ddot{x} + 2\xi\omega\dot{x} + \omega^2 x + \alpha x^3 = p\cos\Omega t \tag{10.19}$$

and

$$\ddot{x} + 2\xi\dot{x} + \alpha x^3 = p\cos\Omega t \tag{10.20}$$

where harmonic excitation has also been assumed. Equations (10.19) and (10.20) differ in the following sense: If motions generated by (10.19) happen to be near equilibrium, the nonlinear stiffness term αx^3 does not have much effect in comparison to the linear stiffness term $\omega^2 x$ and the motion is governed essentially by the linear model (10.1). Equation (10.20), on the other hand, possesses no linear stiffness term, so its behavior will be fundamentally nonlinear both near and far from equilibrium.

In Secs. 10.3.2 and 10.3.3 we consider the response of the *undamped* version of (10.19), and introduce an approximate method of analysis, the method of harmonic balance. In Sec. 10.4 we analyze the harmonically driven model (10.19) with damping effects included. In Sec. 10.5 some additional aspects of the forced problem are presented. Then in Secs. 10.6 and 10.7 we discuss chaotic motion in the model (10.20). Chaos is a type of nonlinear response discovered relatively recently and currently of great interest in dynamics research.

10.3.2 Undamped Free and Forced Response

Here we study the free and forced vibration of the undamped versions of (10.19) and (10.20). The neglect of damping allows some very simple, approximate analytical solutions to be obtained, and these approximations reveal much of the important nonlinear behavior of the systems considered. Generally, we'll concentrate on (10.19) and then deduce from that analysis how (10.20) will behave.

UNDAMPED FREE RESPONSE. We begin with an analysis of (10.19) for the case of no damping ($\xi = 0$) and no forcing ($p = 0$). Furthermore, we will assume the linear natural frequency $\omega = \sqrt{k/m}$ to be unity (no generality is lost in doing this), so that (10.19) becomes

$$\ddot{x} + x + \alpha x^3 = 0 \tag{10.21}$$

The constant α will be assumed for now to be positive. The system (10.21) is conservative, so the sum of the kinetic and potential energies is conserved:

$$\tfrac{1}{2}\dot{x}^2 + \tfrac{1}{2}x^2 + \tfrac{1}{4}\alpha x^4 = E \tag{10.22}$$

where the energy $E = \frac{1}{2}\dot{x}_0^2 + \frac{1}{2}x_0^2 + \frac{1}{4}\alpha x_0^4$ is determined by the initial conditions $\dot{x}(0) = \dot{x}_0$ and $x(0) = x_0$.

For reference, the *linear oscillator*, characterized by $V(x) = \frac{1}{2}x^2$, would result in the relations

$$\ddot{x} + x = 0 \tag{10.23}$$

$$\tfrac{1}{2}\dot{x}^2 + \tfrac{1}{2}x^2 = E \tag{10.24}$$

The solution to (10.23) is simple harmonic, $x(t) = a\cos(t - \phi)$, where a and ϕ are determined from the initial conditions. Furthermore, the phase portraits can be deduced directly from (10.24), which is, in the $x, \dot{x}$ plane, the equation of a circle of radius $\sqrt{2E}$. Typical phase portraits for the linear oscillator are shown in Fig. 10.12. The phase plane orbits for the linear oscillator are closed curves, indicating periodic response, as we know. Note, however, that the orbits shown in Fig. 10.12 are *not* attractors. An orbit initiated a small distance d from any orbit E_n shown will not collapse asymptotically onto the E_n orbit. Rather, it will remain forever a distance d from that orbit.

From the standpoint of comparing the linear and nonlinear responses, there are two characteristics of the linear response which are noteworthy: (1) All orbits, no matter how far from equilibrium, have the same period T, as seen directly from the linear solution for $x(t)$. The frequency $\omega = 2\pi/T$ is independent of the radius of the orbit. (2) The linear solution consists of a single harmonic term.

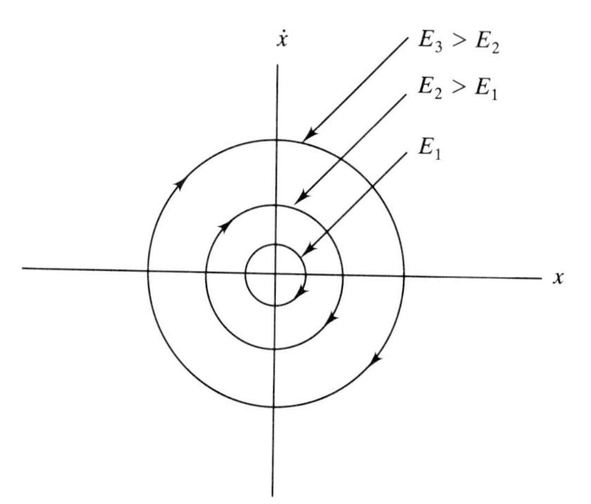

FIGURE 10.12
Phase portraits for the linear oscillator (10.23) and (10.24) for different energies E.

Now let us investigate the behavior of the nonlinear oscillator (10.21) and (10.22), which possesses a cubic nonlinearity αx^3. This nonlinearity represents the effect of a "nonlinear spring," an idealized device which produces a force $F_s = -x - \alpha x^3$. The variation of this nonlinear spring force with displacement x is shown in Fig. 10.13 for the case $\alpha > 0$ under consideration. Notice that, for a given magnitude $|x|$, the magnitude $|F_s|$ of the spring force is *larger* than it would be for the associated linear model ($\alpha = 0$). For this reason the spring is

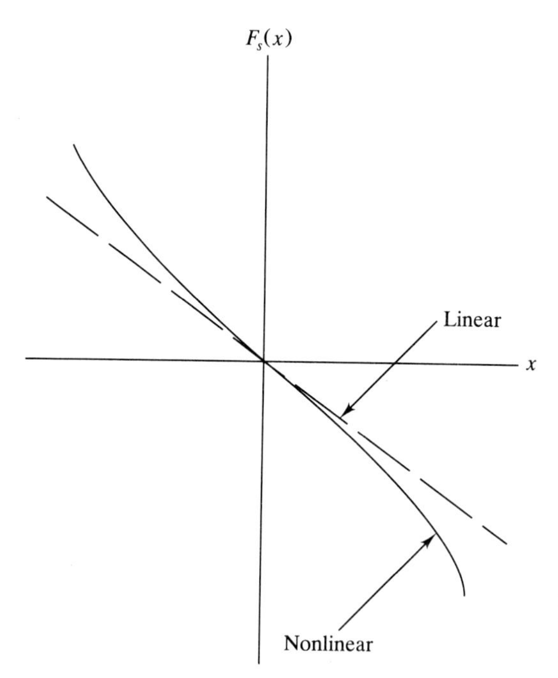

FIGURE 10.13
Nonlinear spring force $F_s = -x - \alpha x^3$, ($\alpha > 0$).

called a *stiffening spring*; the larger the amplitude of oscillation, the larger the average stiffness of the spring.

While the force model $F_s = -x - \alpha x^3$ may appear to be merely a mathematical concoction devised to illustrate nonlinear effects, it is actually of fairly general utility. A force model $F_s = -x - \alpha x^3$ is often used to represent stiffening effects that are symmetric, that is, for which $F_s(x) = -F_s(-x)$; the magnitude F_s is the same for both positive and negative values of x. For any symmetric force, only odd powers of x will appear in the Taylor series representation of F_s. Thus, the force model $F_s = -x - \alpha x^3$ is viewed as a Taylor expansion of any symmetric nonlinear spring, where only the first nonlinear term (the cubic) of the series is retained.

Phase plane orbits for the nonlinear system (10.21) can be determined from the associated energy conservation relation (10.22), from which $\dot{x}$ can be solved for as a function of x:

$$\dot{x} = \pm\left(2E - x^2 - \tfrac{1}{2}\alpha x^4\right)^{1/2} \tag{10.25}$$

where $2E \geq x^2 + \frac{1}{2}\alpha x^4$. By substituting into (10.25) all values of x for which $2E \geq x^2 + \frac{1}{2}\alpha x^4$, the phase orbits are generated. For given energy E the phase orbit is a closed curve in the phase plane. The family of such closed orbits obtained for various values of E will, however, not have a circular shape as in the linear case, as shown in Fig. 10.14 for the case $\alpha = 1$. As E is increased, so that the orbits are initiated further from equilibrium, the noncircularity becomes more pronounced. Note, however, that orbits initiated *near* equilibrium (small $|x|$) have essentially circular shape, conforming to the linear model (10.23) and (10.24).

The closed orbits of the nonlinear system (10.21) and (10.22) indicate that the free oscillations are periodic. The stiffening behavior of the nonlinear spring leads us to expect that the oscillation period $T = 2\pi/\omega$ will depend on the amplitude a of the oscillation.* Orbits characterized by larger amplitudes a will, over the course of an oscillation period, be subjected to an average stiffness which is greater the larger the amplitude. Thus, we should expect the period T (frequency $\omega = 2\pi/T$) to decrease (increase) with a.

Let us next seek a simple, approximate analytical solution to (10.21). The objective will be to determine how the oscillation period T (or frequency ω) depends on the amplitude a. The initial conditions will be taken as $x(0) = a$, $\dot{x}(0) = 0$; these will suffice to specify a given orbit in the phase plane. If the system were linear, the solution for the stated initial conditions would be simply $x(t) = a \cos t$. The fundamental difficulty in attacking the nonlinear problem is

*A given orbit in phase space is defined by specifying the energy E or the amplitude a, which is merely the value of x occurring at $\dot{x} = 0$, as shown in Fig. 10.14. In terms of a, $E = \frac{1}{2}a^2 + \frac{1}{4}a^4$. Often it is more convenient to use a rather than E to designate a particular orbit.

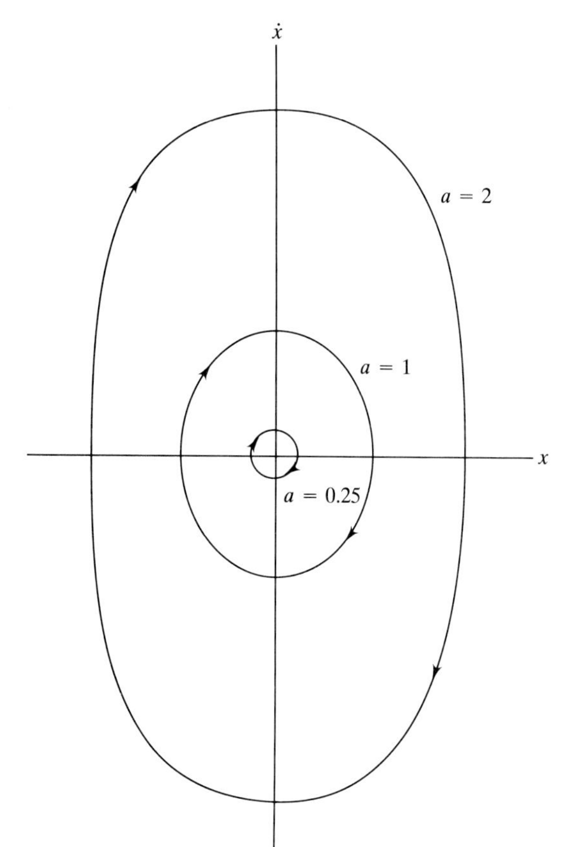

FIGURE 10.14
Phase plane orbits for the nonlinear oscillator (10.21) and (10.25) with $\alpha = 1$. Results are shown for three cases: (*a*) $a = 0.25$ ($E = 0.03223$); (*b*) $a = 1.0$ ($E = 0.75$); and (*c*) $a = 2.0$, ($E = 6.0$).

that the oscillation frequency $\omega = 2\pi/T$ is unknown and depends on the amplitude a, according to preceding intuitive arguments.

To start the analysis of (10.21), let us see what would happen if we were to assume a harmonic solution to (10.21) of the form

$$x(t) = a \cos \omega t \tag{10.26}$$

where a is known from the initial conditions and the frequency ω is the unknown to be found. Substitution of (10.26) into (10.21) would lead to

$$-a\omega^2 \cos \omega t + a \cos \omega t + \alpha(a \cos \omega t)^3 = 0$$

The trigonometric identity $\cos^3 \theta = \frac{3}{4} \cos \theta + \frac{1}{4} \cos 3\theta$ allows us to expand the $\cos^3 \omega t$ term, so that

$$-a\omega^2 \cos \omega t + a \cos \omega t + \tfrac{3}{4}\alpha a^3 \cos \omega t + \tfrac{1}{4}\alpha a^3 \cos 3\omega t = 0$$

or

$$\cos \omega t\left[-a\omega^2 + a + \tfrac{3}{4}\alpha a^3\right] + \cos 3\omega t\left(\tfrac{1}{4}\alpha a^3\right) = 0 \tag{10.27}$$

The result (10.27) is of the form

$$C_1 \cos \omega t + C_3 \cos 3\omega t = 0$$

Thus, if (10.26) is to be the exact solution to (10.21), we must require a and ω to be such that $C_1 = C_3 = 0$, because $\cos \omega t$ and $\cos 3\omega t$ are linearly independent functions. For any nonzero a, we cannot satisfy these conditions $C_1 = C_3 = 0$, which tells us that (10.26) cannot be the exact solution to (10.21).

In (10.27) we see an important property of nonlinearities: They produce higher harmonics or overtones if a harmonic function is substituted into them. The cubic nonlinearity here has produced an overtone having frequency 3ω. This suggests that we add a term $a_3 \cos 3\omega t$ to the "solution" (10.26) to try to account for this overtone, so that we would try

$$x(t) = a_1 \cos \omega t + a_3 \cos 3\omega t \tag{10.28}$$

The problem is that, if we substitute (10.28) into (10.21), even more overtones will be produced by the nonlinearity. Specifically, use of the trigonometric identities stated in the appendix to this chapter shows that the quantity $(a_1 \cos \omega t + a_3 \cos 3\omega t)^3 = a_1^3 \cos^3 \omega t + a_3^3 \cos^3 3\omega t + 3a_1^2 a_3 \cos^2 \omega t \cos 3\omega t + 3a_1 a_3^2 \cos \omega t \cos^2 3\omega t$, will contain harmonics having frequencies ω, 3ω, 5ω, 7ω, and 9ω. The result of this is that, if we want to extend (10.28) to find an exact solution to (10.6), we will have to use an infinite series of the form

$$x(t) = \sum_{n \text{ odd}} a_n \cos n\omega t \tag{10.29}$$

This is just the Fourier series representation (Sec. 7.4) used to represent analytically a symmetric periodic function. That we would need (10.29) to represent formally the exact solution to (10.21) should not be too surprising because we know that the Fourier series can be used to represent any periodic function, and we know the solution to (10.21) to be periodic.

Let us now find an *approximate* solution by assuming that, at least for some a, the periodic response of (10.21) will be dominated by the fundamental harmonic in (10.29), that is, $a_3, a_5, \ldots,$ will be small compared to the fundamental a_1. This may be justified on the grounds that, as $a \to 0$, $\omega \to 1$, and the response approaches that of the associated linear oscillator, $x(t) = a \cos t$. We reason that as a is increased from near-zero values, there is likely to be some range of values $0 < a < a_0$ for which nonlinear effects will be "small but noticeable" and that these effects can be described, at least approximately, by using the single-term approximation (10.26). Use of the approximation (10.26) in (10.21) leads to (10.27). We now require that (10.27) be satisfied only approximately as follows: For given amplitude a (known from the specified initial conditions), require the frequency ω to be such that the coefficient of the fundamental harmonic $\cos \omega t$ in (10.27) be zero, ignoring the effect of the third harmonic generated by the nonlinearity. This results in the requirement that

$$a - a\omega^2 + \left(\tfrac{3}{4}\right)\alpha a^3 = 0$$

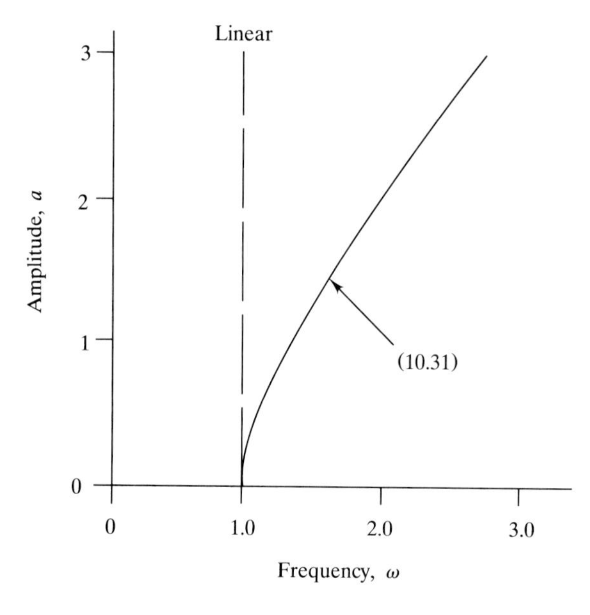

FIGURE 10.15
The approximate frequency-amplitude dependence given by (10.31) for the oscillator (10.21) with $\alpha = 1$.

or

$$a\left(1 - \omega^2 + \tfrac{3}{4}\alpha a^2\right) = 0$$

The case $a = 0$ is not of interest, so that we define the frequency ω to satisfy

$$1 - \omega^2 + \tfrac{3}{4}\alpha a^2 = 0$$

this relation can be solved for ω^2:

$$\omega^2 \cong 1 + \tfrac{3}{4}\alpha a^2 \tag{10.30}$$

and this is the basic result of the analysis. It tells us, approximately, how the oscillation frequency ω depends on the amplitude a of the oscillation,

$$\omega \cong \left(1 + \tfrac{3}{4}\alpha a^2\right)^{1/2} \tag{10.31}$$

The result says that if $\alpha > 0$, the frequency ω increases with amplitude as defined by (10.31). This is consistent with the stiffening effect noted previously on qualitative grounds. The variation $\omega(a)$ given by (10.31) is shown in Fig. 10.15. Note that as $a \to 0$, $\omega \to 1$ in accordance with the linear theory.

Comments

1. The approximate method of solution we have used is known as the *harmonic balance* (HB) *method*. This technique is commonly used to obtain approximate solutions for nonlinear responses that are periodic, as follows: (1) A truncated version of the Fourier series is assumed as the approximate solution (we have used a single harmonic term here but more can be used if one is willing to do the algebra). (2) This approximate solution is plugged into the governing ODE model, and trigonometric identities are used to rewrite powers of the $\sin \omega t$ and $\cos \omega t$ terms that result. (3) Terms involving like harmonics are grouped together, and the coefficients of all harmonics that appear in the original approximate solution are made to vanish. (4) The influence of other harmonics is ignored (thus rendering the solution approximate only).

2. We should expect the HB solution (10.26) used here to provide a good approximate solution only if the actual solution turns out to be dominated by the fundamental harmonic. We cannot always tell ahead of time whether this will occur, and HB should, therefore, be used with caution.

We can get some idea how well the approximate solution (10.26) and (10.30) works for the system (10.21) by comparing the actual and approximate phase portraits. Noting that, according to the approximate solution (10.26), $x(t) = a\cos\omega t$, then $\dot{x}(t) = -a\omega^2 \sin\omega t$, where ω is defined by (10.31). Thus, the approximate solution satisfies

$$x^2 + \frac{\dot{x}^2}{\omega^2} = a^2$$

This result indicates that the phase portraits generated by the one-term HB solution will be ellipses with semimajor and minor axes a (along x) and $a\omega$ (along $\dot{x}$). The approximate and exact phase portraits are compared in Fig. 10.16 for the case $\alpha = 1$, for various amplitudes a. The agreement is remarkably

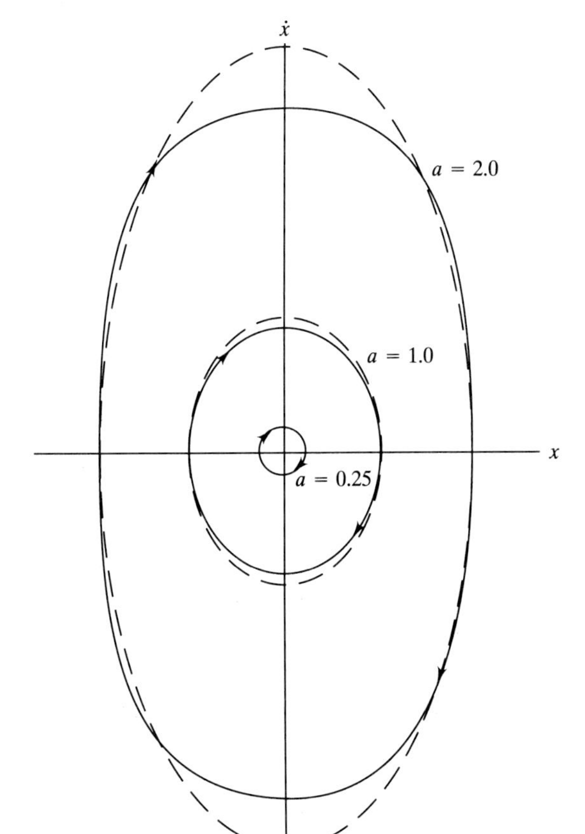

FIGURE 10.16
Exact (———) and approximate (– – – –) phase portraits of the system (10.21) with $\alpha = 1$. The three cases shown are the same as in Fig. 10.14. For $a = 0.25$, the exact and approximate phase portraits are indistinguishable on this scale.

good, even very far from equilibrium. It turns out that *for this system* with $\alpha > 0$, the frequency-amplitude relation (10.31) is within about 2 percent of the exact frequency for all α and a, no matter how far from equilibrium.

Let us summarize the results of this section. The simple cubic nonlinearity in (10.21) causes the free oscillation frequency (hence also the period of the oscillation) to be a function of the oscillation amplitude a, as given approximately by the HB result (10.31) and as shown in Fig. 10.15. The neophyte may not view this frequency-amplitude dependence to be of much significance. We will see in the sequel, however, that frequency-amplitude dependence due to nonlinear stiffness is responsible for a variety of important and interesting nonlinear phenomena.

10.3.3 Undamped Harmonically Forced Oscillation

The objective of this section will be to begin to examine the steady-state response of the undamped nonlinear oscillator subjected to a harmonic excitation, so that the ODE model considered is

$$\ddot{x} + x + \alpha x^3 = p \cos \Omega t \tag{10.32}$$

Damping has been ignored here only because the analysis is algebraically simple without it; much of the interesting behavior is illustrated by the undamped problem. In this section we will develop an approximate solution for the undamped nonlinear frequency response, that is, the relation between the steady-state vibration amplitude a and the driving frequency Ω, as in Fig. 5.15. A more detailed investigation is contained in Sec. 10.4 where the damped problem is analyzed. The present section is essentially an introduction to Sec. 10.4.

We begin by citing previously obtained results for the *linear* problem associated with (10.32):

$$\ddot{x} + x = p \cos \Omega t \tag{10.33}$$

Some review of Sec. 5.3.2 would prove beneficial. For the driven linear oscillator (10.33), the steady-state frequency response is given by (5.76*a*) with $m = \omega = 1$ and with $\cos \phi = 1$:

$$a = \frac{p}{1 - \Omega^2}$$

Note that a may be positive or negative, depending on whether $\Omega < 1$ or $\Omega > 1$. A graph of a versus Ω would be as shown in Fig. 5.15 except that, at $\Omega = 1$ (resonance), the undamped response would theoretically asymptote to infinity as $\Omega \to 1$ due to the absence of damping.

We can see how the cubic nonlinearity in (10.32) will modify the frequency response by using HB to obtain an approximate solution to (10.32). We will assume the steady-state response to be at the driving frequency Ω, so that a one-term approximate solution will be of the form

$$x(t) = a \cos \Omega t \tag{10.34}$$

Notice here that the frequency Ω of the response is *known*, as we assume the system to respond at the driving frequency. The steady-state vibration amplitude a is the unknown to be found. This is the opposite of what occurred for the free oscillation problem of Sec. 10.3.2. There the amplitude a was given as an initial condition, and the resulting natural frequency ω was to be determined.

Substitution of (10.34) into (10.32) leads to

$$-a\Omega^2 \cos \Omega t + a \cos \Omega t + \alpha a^3\left(\tfrac{3}{4} \cos \Omega t + \tfrac{1}{4} \cos 3\Omega t\right) = p \cos \Omega t$$

or

$$\cos \Omega t\left(a - a\Omega^2 + \tfrac{3}{4}\alpha a^3 - p\right) + \tfrac{1}{4}\alpha a^3 \cos 3\Omega t = 0 \tag{10.35}$$

According to the HB method, the approximate solution is obtained by requiring the coefficient of $\cos \Omega t$ to vanish, while ignoring the higher harmonic in (10.35). Thus, we require that

$$a - a\Omega^2 + \tfrac{3}{4}\alpha a^3 - p = 0 \tag{10.36}$$

This is a cubic algebraic equation that defines the nonlinear frequency response $a = a(\Omega, \mathrm{p})$. This result (10.36) may be written in the implicit form

$$\Omega^2 = 1 + \frac{3}{4}\alpha a^2 - \frac{p}{a} \tag{10.37}$$

and this is the central result of the analysis.

In interpreting (10.37), we make the following comments:

1. Both positive and negative solutions for the amplitude a are possible. If $a > 0$, then the response, given approximately by (10.34), is *in phase* with the excitation. If $a < 0$, then the response is *out of phase* (by 180°) with the excitation.
2. Both of the possibilities in 1 above can be taken into account by defining a to be positive and amending (10.37) to read

$$\Omega^2 = 1 + \frac{3}{4}\alpha a^2 \pm \frac{p}{a} \tag{10.38}$$

   where the $+(-)$ sign yields the out-of-phase (in-phase) solution (note that whether $a > 0$ or $a < 0$ has no effect on the quadratic term $\tfrac{3}{4}\alpha a^2$). Equation (10.38) says that a given amplitude a can be produced by either of two driving frequencies.
3. The frequency response $a = a(\Omega)$ can be constructed easily from (10.38) as follows: Pick an amplitude a, now defined to be > 0, and determine from (10.38) the two driving frequencies that will produce it; the lower of these two driving frequencies is defined by $\Omega^2 = 1 + \tfrac{3}{4}\alpha a^2 - p/a$, while the higher of the two driving frequencies is defined by $\Omega^2 = 1 + \tfrac{3}{4}\alpha a^2 + p/a$. This process is illustrated in Fig. 10.17. Do this for enough a's to generate the complete frequency response curve.

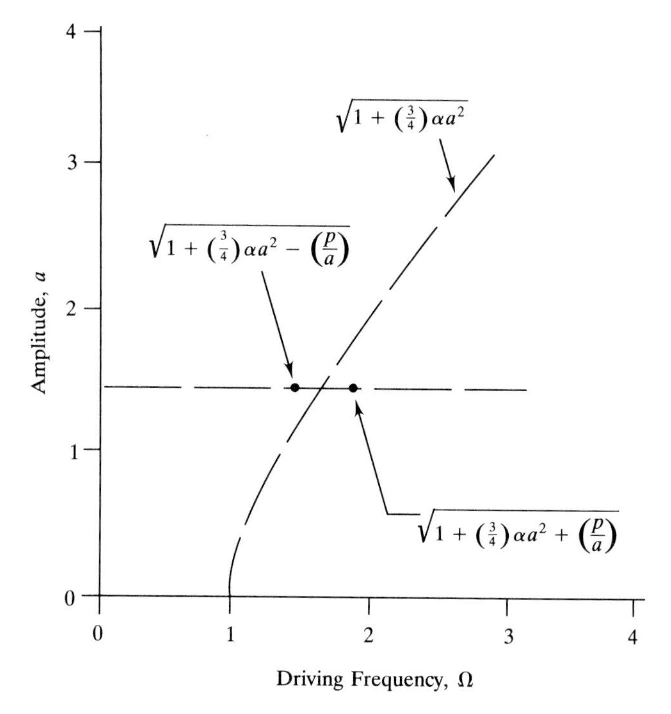

FIGURE 10.17
Construction of the approximate frequency response curve given by (10.38).

The resulting frequency response obtained is shown in Fig. 10.18 for the case $\alpha = p = 1$ in (10.32). Of special note in Fig. 10.18 is that at larger amplitudes a, both branches of the solution follow closely the function $(1 + \frac{3}{4}\alpha a^2)^{1/2}$. This function defines the natural frequency ω of free oscillation, as determined approximately in Sec. 10.3.2; see (10.31). The curve $\omega(a) = (1 + \frac{3}{4}\alpha a^2)^{1/2}$ shown in Fig. 10.18 is often called the *backbone curve* for the system. At large amplitudes the forced response follows this curve because, as seen in (10.38), if a is large, then p/a is small, and the two driving frequencies are both close to $\omega(a)$. For reference in Fig. 10.18 the linear frequency response $(\alpha = 0)$ is also plotted, and we observe the nonlinear response to resemble the linear response but to be "bent over" along the backbone curve.

We close this section with several remarks regarding the nonlinear behavior shown in Fig. 10.18:

1. The importance of the dependence of *free* oscillation frequency ω on amplitude a is now evident: At larger amplitudes a, the forced response closely follows this free oscillation backbone curve. Thus, the bent-over nature of the nonlinear frequency response is a direct result of how the free

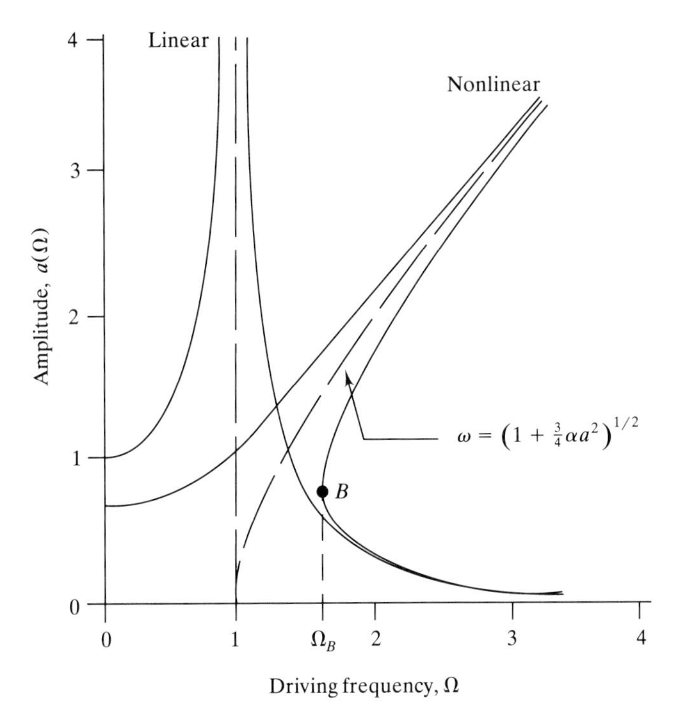

FIGURE 10.18
The (undamped) nonlinear frequency response of the oscillator (10.32) for the case $\alpha = p = 1$. The linear response ($\alpha = 0$) is also shown for reference. The nonlinear solution is approximate, based on the HB result (10.38).

oscillation frequency ω varies with amplitude a for the particular system under study.

2. The response in the neighborhood of the backbone curve constitutes what is usually known as a *nonlinear resonance*, by analagy with the associated linear phenomenon.
3. Notice that the nonlinear solutions characterized by small amplitudes a do not differ much from the linear response. These small amplitude solutions represent conditions away from the main nonlinear resonance. This linear/nonlinear agreement makes sense because if a is sufficiently small, the influence of the nonlinearity should not be too important, and the behavior should be essentially linear.
4. A very important difference between the linear and nonlinear responses is the possibility of multiple steady-state solutions for the nonlinear case. For $\Omega > \Omega_B$ in Fig. 10.18, we observe *three* solutions at any driving frequency $\Omega > \Omega_B$. This type of behavior cannot occur in a linear system and is explored in a fair amount of detail in the following section in which the same system (10.32) is analyzed but with damping included.

10.4 THE DAMPED HARMONICALLY FORCED OSCILLATOR

In the preceding section, the basic features of the nonlinear steady-state response of (10.32) were shown in Fig. 10.18 for the undamped case. We now explore the effect of small damping, with particular attention to the range of frequencies Ω for which multiple steady-state solutions exist. The mathematical model to be considered is (10.19) with the linear natural frequency $\omega = 1$:

$$\ddot{x} + 2\xi\dot{x} + x + \alpha x^3 = p \cos \Omega t \tag{10.39}$$

In qualitative terms the influence of a small amount of damping can be guessed by analagy with the driven linear oscillator ($\alpha = 0$). We have seen in Sec. 5.3.2 (Fig. 5.15) and in Sec. 7.2.3 (see the comments following Example 7.1) that, for a driven linear oscillator, the main effect of a small amount of damping is to limit the response amplitude at and very near resonance. The damping, however, has little effect on the response away from resonance. In a similar fashion the damping ξ in the nonlinear model (10.39) will have the effect of preventing the response amplitude a from becoming arbitrarily large as we move along the backbone curve; there will be some maximum amplitude reached.

The nonlinear oscillator (10.39) will exhibit both transient and steady-state responses. In Sec. 10.3.3 no mention was made of the transient versus the steady-state response. For the driven undamped oscillator, the "transient" will in fact not die out (thus, it is not really a transient), and the only way to realize the "steady-state" responses found in Sec. 10.3.3 is to arrange for the initial conditions to lie exactly on the steady-state solution found in the phase space. For the damped model (10.39), both transient and steady-state responses occur, but the damping causes the transients to die out. This is what also happens for the linear oscillator. In the linear case, however, it is possible to separate the transient and steady-state contributions to the response, as in (7.22). The two contributions then sum to form the overall response. For the nonlinear system, however, we cannot mathematically separate the transient and steady-state responses as two contributions which combine to form the overall response, as the principle of superposition does not apply for nonlinear systems. Our interest here will be in the steady-state response only.

10.4.1 Harmonic Balance Solution

To obtain an approximate steady-state solution to (10.39), we will use the harmonic balance method. Now, however, we must account for the fact that, in contrast to the undamped case, the damping may cause the response phase lag to be anywhere between 0 and 180°. Thus, we assume a simple harmonic steady-state solution with phase lag, $x(t) = a \cos(\Omega t - \phi)$. To simplify the algebra, let us include the phase ϕ in the excitation term, so that the mathematical model (10.39) and steady-state solution become

$$\ddot{x} + 2\xi\dot{x} + x + \alpha x^3 = p \cos(\Omega t + \phi) \tag{10.40a}$$

$$x(t) = a \cos \Omega t \qquad \text{(steady state)} \tag{10.40b}$$

Aside: The way in which the phase ϕ is handled in (10.40) is the same as was done in Sec. 5.3.2 in the study of the steady-state response of the linear oscillator [see (5.70)]. In fact, the analysis which follows here is essentially the same as the "method of undetermined coefficients" analysis given by Eqs. (5.70) through (5.78). Here, however, we have a nonlinearity included, and we neglect higher harmonics generated by this nonlinearity.

The approximation (10.40*b*) is expected to be a good one if it turns out that the steady-state response is dominated by the fundamental harmonic. Proceeding with the HB solution, we plug (10.40*b*) into (10.40*a*), resulting in

$$-a\Omega^2 \cos \Omega t - 2\xi\Omega a \sin \Omega t + a \cos \Omega t + \alpha a^3\left(\tfrac{3}{4}\cos \Omega t + \tfrac{1}{4}\cos 3\Omega t\right) = p\cos(\Omega t + \phi)$$

Noting that $\cos(\theta + \psi) = \cos\theta\cos\psi - \sin\theta\sin\psi$, the above may be written as

$$\cos \Omega t\left(-a\Omega^2 + a + \tfrac{3}{4}\alpha a^3 - p\cos\phi\right) + \sin\Omega t(-2\xi\Omega a + p\sin\phi) + \tfrac{1}{4}\alpha a^3 \cos 3\Omega t = 0 \tag{10.41}$$

Note the similarity to (5.73). According to the HB method, the approximate solution is required to balance the fundamental harmonics in (10.41). This requires that the unknowns a and ϕ be chosen so that

$$-a\Omega^2 + a + \tfrac{3}{4}\alpha a^3 - p\cos\phi = 0$$

$$-2\xi\Omega a + p\sin\phi = 0$$

Division by a and transposition lead to

$$1 - \Omega^2 + \frac{3}{4}\alpha a^2 = \frac{p}{a}\cos\phi$$

$$2\xi\Omega = \frac{p}{a}\sin\phi$$

The steady-state response amplitude a is of primary interest, so we eliminate the phase ϕ by squaring and adding the preceding equations to obtain

$$\left(1 - \Omega^2 + \frac{3}{4}\alpha a^2\right)^2 + (2\xi\Omega)^2 = \left(\frac{p}{a}\right)^2 \tag{10.42}$$

This result is a nonlinear algebraic equation that relates the parameters Ω, α, ξ, and p to the steady-state response amplitude a. Note that (10.42) reduces to the undamped result (10.38) if $\xi = 0$. In order to interpret (10.42), it is convenient to express (10.42) in the following form, which is analagous to (10.38) for the undamped oscillator. Rewriting (10.42), we have

$$\Omega^2 = 1 + \frac{3}{4}\alpha a^2 \pm \left[\left(\frac{p}{a}\right)^2 - (2\xi\Omega)^2\right]^{1/2} \tag{10.43}$$

This shows, as does (10.38), that the nonlinear frequency response follows the backbone curve $(1 + \frac{3}{4}\alpha a^2)^{1/2}$, according to the reasoning behind Fig. 10.17. We observe, however, that the damping ξ will cause the two branches of the solution to coalesce at the point along the backbone curve at which $p/a = 2\xi\Omega$, at which the radical in (10.43) becomes zero. (For any point further along the backbone curve, this radical will be negative, indicating that no solution exists, as Ω must be real.) Thus, as we would expect, the presence of damping limits the response as it does in the linear case.

We also observe that it is inconvenient to use (10.43) to actually generate a plot of the nonlinear frequency response because Ω appears on both sides of the equation. If we expand (10.42), we obtain a quadratic algebraic equation for Ω^2, which can be solved using the quadratic formula. The student is invited to show that, upon expansion of the terms in (10.42), the quadratic formula yields the following for Ω^2:

$$\Omega^2 = \underbrace{1 + \frac{3}{4}\alpha a^2 - 2\xi^2}_{\text{Backbone curve}} \pm \left[\left(\frac{p}{a} \right)^2 - 4\xi^2 \left(1 + \frac{3}{4}\alpha a^2 - \xi^2 \right) \right]^{1/2} \tag{10.44}$$

This result reduces to (10.38) if $\xi = 0$. The backbone curve is seen to have a small correction due to the damping. If this damping is light (say, $\xi < 0.05$), however, the backbone curve is virtually the same as for the undamped case. Equation (10.44) is used to construct a plot of the nonlinear frequency response as follows: For given p, α, and ξ, select an amplitude a and determine the two associated values of Ω, one on either side of the backbone curve (just as in Fig. 10.17). Do this for enough values of a to generate the frequency response curve.

Figure 10.19 shows the resulting frequency response curve for the case $p = 1.0$, $\alpha = 1.0$, and $\xi = 0.05$. The parameters p and a are the same as those used to generate Fig. 10.18 for the undamped case, so a comparison of Figs. 10.18 and 10.19 shows directly the effect of the damping. The following qualitative features of this nonlinear frequency response are noteworthy:

1. Away from the point A the response is essentially the same as for the undamped case (Fig. 10.18). The effect of damping, provided that it is light, is mainly to limit the response at some point along the backbone curve.
2. There are two points, denoted A and B in Fig. 10.19 with associated driving frequencies Ω_A and Ω_B, at which the tangent to the response curve is vertical. For driving frequencies Ω in the range $\Omega_B < \Omega < \Omega_A$, there are observed to be *three* solutions for the amplitude a, implying three different steady-state responses for a given set of parameters Ω, ξ, and p.
3. In the analysis of this section no mention has been made of the *stability* of the periodic solutions we have found. The fact that we can find a periodic function $x(t)$ which will solve (approximately, in our case) (10.40a) does not guarantee that this solution will actually be observed in the steady state. In order to be realized, a periodic solution must be asymptotically stable.

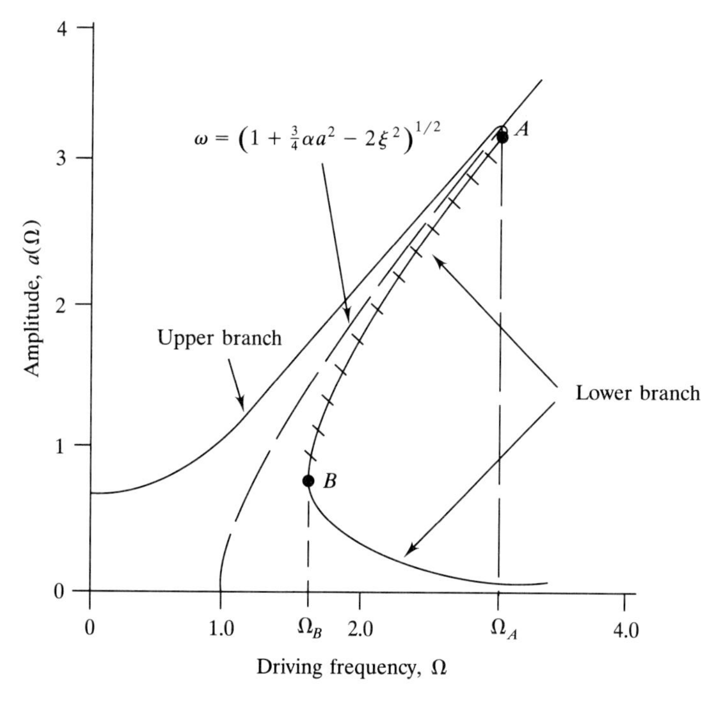

FIGURE 10.19
The nonlinear frequency response of the damped oscillator (10.39) with $p = \alpha = 1$, $\xi = 0.05$. The solution is approximate, based on the HB result (10.44). The hatched region between points B and A is unstable.

Aside: In Chap. 4 the concept of stability was introduced for *equilibrium* solutions to the governing ODE model. But the same ideas apply as well to other types of steady-state solutions, such as the periodic ones we are finding here.

When we assume the harmonic balance solution (10.40*b*), there is nothing in the HB analysis we have used to indicate whether the approximate solutions we find will be stable or unstable. Unfortunately, we cannot determine the stability of the periodic solutions shown in Fig. 10.19 without a fair amount of additional effort. Here only the results will be stated: It turns out that solutions on the portion of the response curve between points A and B on the lower branch are unstable, while those on the remainder of the response curve are asymptotically stable (at A and at B the solutions are stable but not asymptotically stable). The unstable solution branch is denoted by the hatched line in Fig. 10.19. The stability result means that for $\Omega_A < \Omega < \Omega_B$, there are two periodic attractors and one periodic repeller. Which of the two attractors will be observed will depend on the initial conditions. For some initial conditions, orbits will move asymptotically to one attractor, while for other initial conditions, orbits will move asymptotically to the other attractor. This aspect of the response is discussed further later on. This type of situation (multiple attractors) never arises in a linear system.

4. The stable "small amplitude" solutions (say, $a < 0.5$) in Fig. 10.19 do not differ very much from what the linear theory ($\alpha = 0$) would predict. This is because the phase plane orbits associated with these solutions remain close enough to equilibrium that nonlinear effects are not too important (but see Sec. 10.5.1).

In order to illustrate the nature of the nonlinear response in more detail, we now present some numerical integration results which show some of the steady-state solutions and their phase portraits. We first consider the case $p = 1.0$, $\alpha = 1.0$, $\xi = 0.05$ (as in Fig. 10.19) and $\Omega = 2.0$. For these conditions, according to Fig. 10.19, there are two attractors and one repeller. Figures 10.20(a) and (b) show the responses for two different sets of initial conditions: (a) $x(0) = \dot{x}(0) = 0$, so that the system is initially at rest, and (b) $x(0) = 3$, $\dot{x}(0) = 0$, so that the system is initially far from equilibrium.

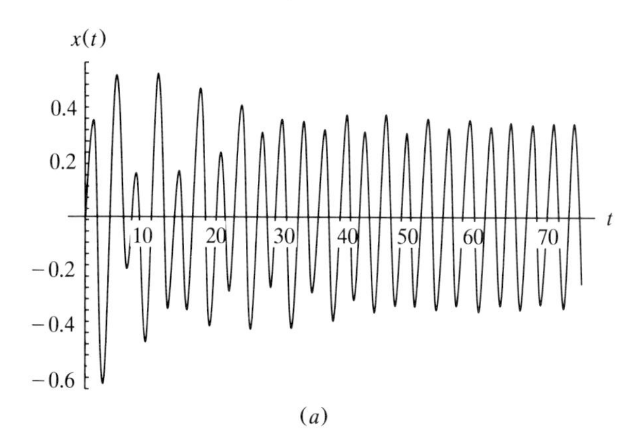

(a)

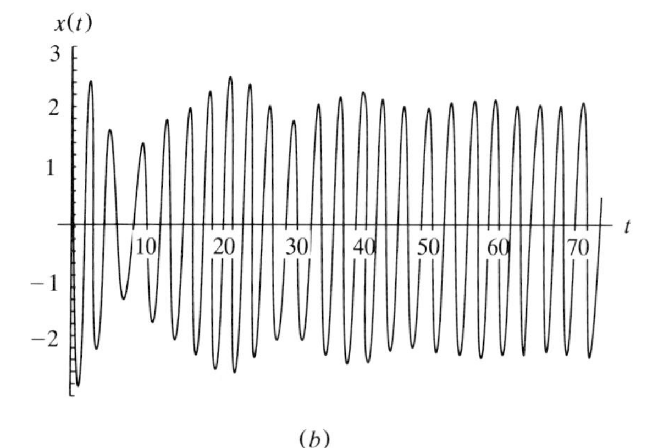

(b)

FIGURE 10.20
Numerical solutions to (10.40) for the case $p = \alpha = 1$, $\xi = 0.05$, $\Omega = 2$. In (a) the initial conditions $x(0) = 0$, $\dot{x}(0) = 0$ lead to the small amplitude attractor. In (b) the initial conditions $x(0) = 3$, $\dot{x}(0) = 0$ lead to the large amplitude attractor. Both responses have the same period as the excitation.

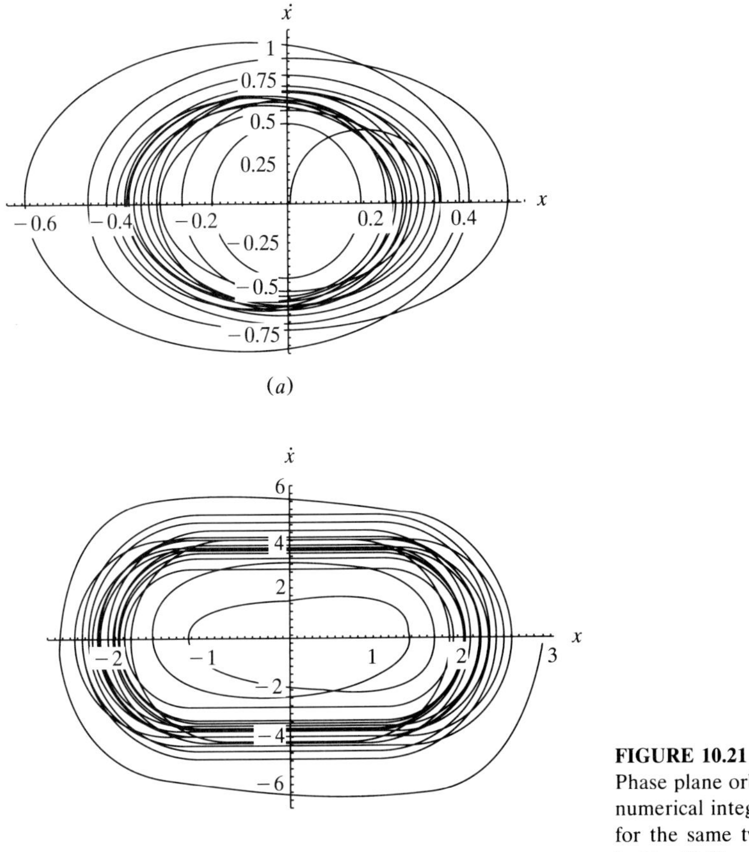

FIGURE 10.21
Phase plane orbits, obtained via numerical integration of (10.40) for the same two conditions as in Fig. 10.20.

In Figs. 10.20(a) and (b) the transient and steady-state regimes are easily identified. The simple harmonic balance theory, leading to (10.44), is also verified, as two different steady-state solutions are obtained for the same system parameters. It is also clear that these solutions are dominated by the fundamental harmonic, and the numerical values for the oscillation amplitudes are close to the HB results shown in Fig. 10.19. The phase portraits for the two solutions shown in Fig. 10.20 are shown in Fig. 10.21.

10.4.2 The Jump Phenomenon

The existence of multiple attractors results in a very interesting phenomenon that occurs if we do the following numerical experiment on (10.39) (see Fig. 10.22). Suppose we pick a driving frequency Ω_S and solve (10.39) numerically, eventually attaining the steady-state solution denoted by the point S in Fig. 10.22. We then *slowly* increase the driving frequency Ω at some small,

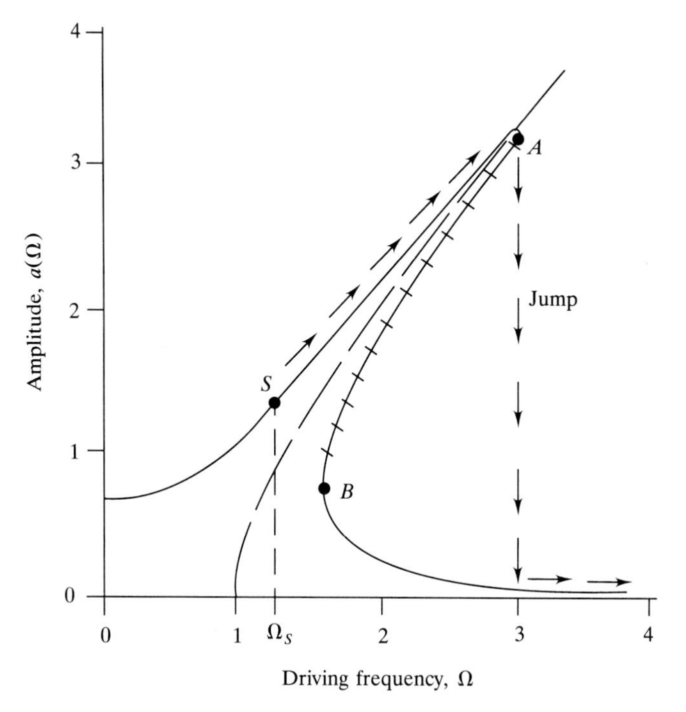

FIGURE 10.22
Schematic of the jump phenomenon for the oscillator (10.40) as the driving frequency Ω is slowly increased from the starting value Ω_S (see text for discussion).

constant rate. As the frequency increases, the oscillation amplitude will slowly grow as the solution essentially moves along the upper branch of the response curve, as shown. As the point A is reached, with Ω passing through Ω_A, the upper solution *disappears*. For $\Omega > \Omega_A$ there is only one attractor. The response will "jump" down to the lower branch of the response curve (after some transient behavior) and will remain there as Ω is increased further. This behavior is known as the *jump phenomenon* and can be very dramatic in actual mechanical systems. The time history for one such case is shown in Fig. 10.23.

The jump phenomenon also occurs if the driving frequency Ω is *decreased* from a value $\Omega > \Omega_A$, as shown schematically in Fig. 10.24. In this case, however, the jump occurs at B from the lower branch of the response curve to the upper branch.

All the nonlinear phenomena described in this section, including the existence of multiple solutions and the jump phenomenon, are caused primarily by the nonlinear dependence of the *free* oscillation frequency ω on the amplitude a (Sec. 10.3). The frequency response $a(\Omega)$ discussed in this section follows closely the free oscillation dependence $\omega(a)$, the so-called backbone curve. In many forced nonlinear oscillator systems the interesting, nonlinear

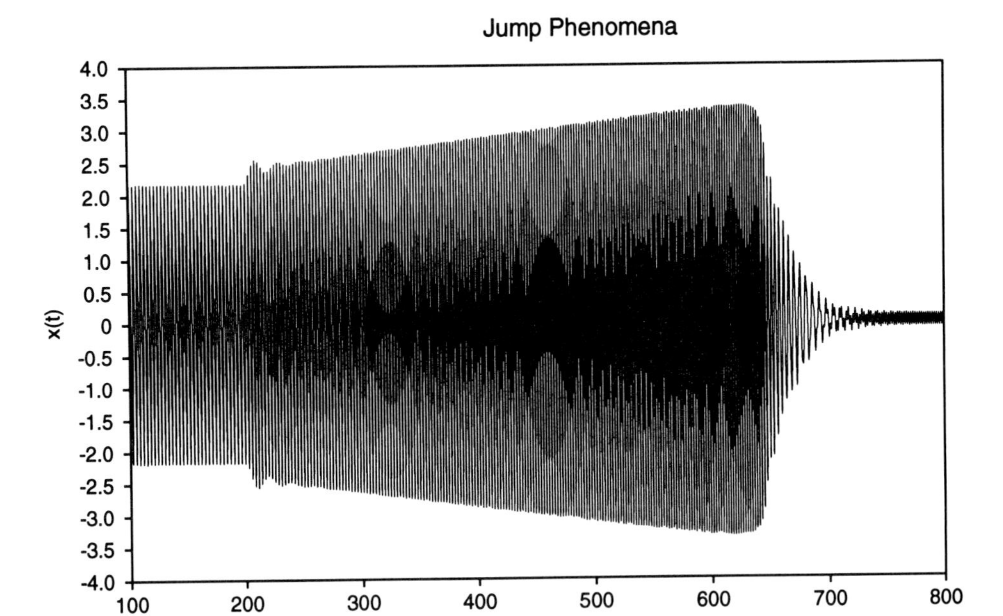

FIGURE 10.23
Numerical integration of (10.39), illustrating the jump phenomenon as the driving frequency Ω is increased. For $t < 200$ s, $\Omega = 2$, leading to the steady oscillation of amplitude slightly greater than 2. At $t = 200$ s, Ω is slowly increased, at the rate $\dot{\Omega} = 0.002$ rad/s^2. At about 640 s, for which $\Omega = 2.88$ rad/s, the large amplitude solution disappears and the jump to the small amplitude attractor takes place. Steady-state conditions are reached shortly after 700 s.

aspects of behavior are a direct consequence of the particular free oscillation frequency-amplitude dependence for the system under study.*

10.5 OTHER TYPES OF PERIODIC RESPONSES

In Secs. 10.3.2 and 10.4 we studied the harmonically forced response associated with the "primary nonlinear resonance." This primary resonance occurs when

*Nonlinear free oscillation frequency-amplitude dependence arises from nonlinear stiffness characteristics, which in turn are defined by the system potential energy. Thus, the cited sentence may also be viewed as: The interesting nonlinear behavior is a consequence of the particular potential energy $V(x)$ which characterizes conservative forces in the system.

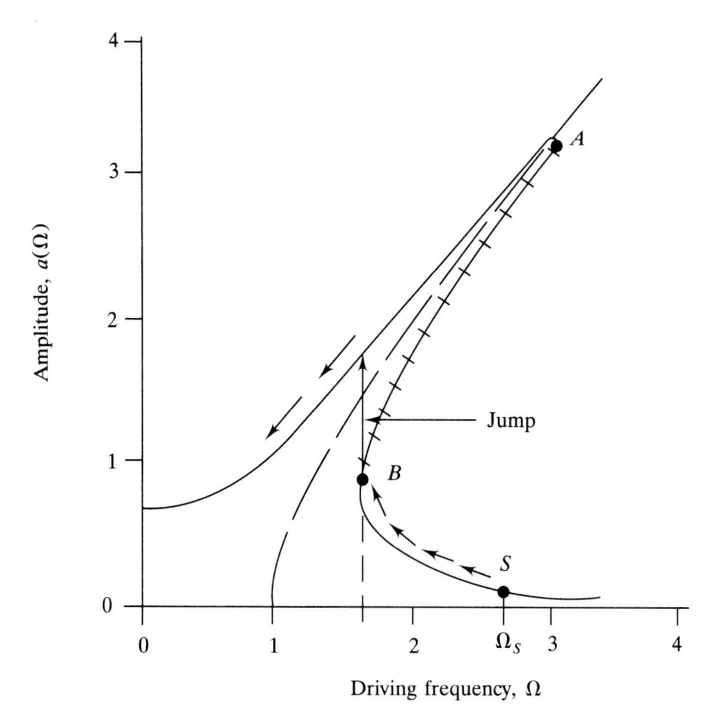

FIGURE 10.24
Schematic of the jump phenomenon for the oscillator (10.40) as the driving frequency Ω is slowly decreased from a starting value $\Omega_S > \Omega_B$.

the driving frequency Ω is close to the backbone curve, which defines the free oscillation frequency. The primary steady-state response has the same period as the excitation and is dominated by the fundamental harmonic. From a qualitative standpoint the important features of the primary nonlinear resonance are those indicated in Figs. 10.19, 10.22, and 10.24: the existence of multiple periodic attractors and the jump phenomenon. It turns out, however, that this primary nonlinear resonance is not the only important effect of the nonlinearity; other types of nonlinear resonances are possible. (Here the word *resonance* implies that the driving frequency Ω is tuned so as to produce some type of "large" amplitude response.) In this section we investigate a few of these other nonlinear resonances that may occur when the driving frequency Ω is *not* near the primary resonance condition. Two situations will be of special interest: (1) the excitation of a steady-state response having the same period as the excitation but having a large Fourier component of frequency 3Ω, and (2) the excitation of a steady-state response having a period which is *three times* the period of the excitation. The existence of such responses will be uncovered using harmonic balance, as in the case of the primary resonance. We will also consider the undamped case in order to keep the algebra relatively simple.

10.5.1 The Superharmonic Resonance of Order 3

The description "superharmonic resonance of order 3" denotes a periodic steady-state response having, in addition to the fundamental harmonic of frequency Ω, a significant Fourier component of frequency 3Ω. Let us now see how such a response may arise in the undamped oscillator (10.32):

$$\ddot{x} + x + \alpha x^3 = p \cos \Omega t \tag{10.45}$$

If an order 3 superharmonic resonance exists, the steady-state response will contain significant harmonics having frequencies Ω and 3Ω, and this implies that, at least to start, an approximate HB approximation is*

$$x(t) = a_1 \cos \Omega t + a_3 \cos 3\Omega t \tag{10.46}$$

Our objective is to see whether excitation conditions (p, Ω) may exist such that the superharmonic amplitude a_3 is significant. If a_3 turns out to be negligible, then we merely recover the fundamental harmonic approximation (10.34) used in Secs. 10.3.2 and 10.4 to study the primary resonance.

Following the HB procedure, we substitute (10.46) into (10.45), use trigonometric identities, and balance the terms involving $\cos \Omega t$ and $\cos 3\Omega t$ (because there is no damping, no sine terms will appear) to obtain

$$\begin{aligned} &\cos \Omega t\left[a_1(1 - \Omega^2) + \tfrac{3}{4}\alpha a_1^3 + \tfrac{3}{2}\alpha a_1 a_3^2 + \tfrac{3}{4}\alpha a_1^2 a_3 - p\right] \\ &\quad + \cos 3\Omega t\left[a_3(1 - 9\Omega^2) + \tfrac{3}{4}\alpha a_3^3 + \tfrac{3}{2}\alpha a_1^2 a_3 + \tfrac{1}{4}\alpha a_1^3\right] + \mathrm{HH} = 0 \end{aligned} \tag{10.47}$$

where HH stands for *higher harmonics*, having frequencies of 5Ω, 7Ω, and 9Ω. Ignoring the effect of the higher harmonics, (10.47) is satisfied approximately by requiring the coefficients of $\cos \Omega t$ and of $\cos 3\Omega t$ to vanish. This leads to a pair of nonlinear algebraic equations which are to be solved for the harmonic amplitudes a_1 and a_3:

$$a_1(1 - \Omega^2) + \tfrac{3}{4}\alpha a_1^3 + \tfrac{3}{2}\alpha a_1 a_3^2 + \tfrac{3}{4}\alpha a_1^2 a_3 = p \tag{10.48}$$

$$a_3(1 - 9\Omega^2) + \tfrac{3}{4}\alpha a_3^3 + \tfrac{3}{2}\alpha a_1^2 a_3 + \tfrac{1}{4}\alpha a_1^3 = 0 \tag{10.49}$$

Notice that if $a_3 = 0$, then (10.48) reduces to (10.35) and (10.36); that is, we recover the results of Sec. 10.3.2 for the primary resonance. In order to establish conditions for which a_3 may be significant compared to the fundamental a_1, let us rewrite (10.49) in the following way:

$$a_3\left[(1 - 9\Omega^2) + \tfrac{3}{4}\alpha a_3^2 + \tfrac{3}{2}\alpha a_1^2\right] = -\tfrac{1}{4}\alpha a_1^3$$

*Harmonics of other frequencies will also be present, but we assume them to be small, so that (10.46) provides a reasonable approximation.

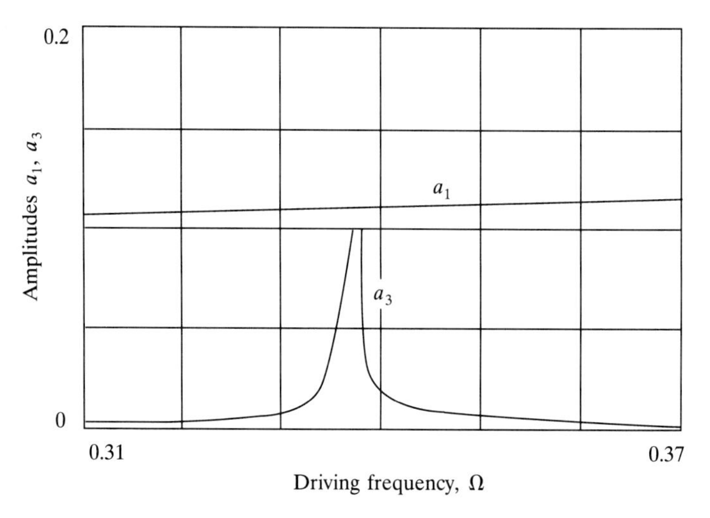

FIGURE 10.25
The superharmonic frequency response of the undamped oscillator (10.45) for excitation frequencies Ω that are approximately one-third the system linear natural frequency.

or
$$a_3 = \frac{-\frac{1}{4}\alpha a_1^3}{\left(1 - 9\Omega^2 + \frac{3}{4}\alpha a_3^2 + \frac{3}{2}\alpha a_1^2\right)} \tag{10.50}$$

In order to interpret (10.48) and (10.50), let us first suppose that the amplitude a_1 of the fundamental harmonic turns out to be small compared to unity. This will occur if the excitation amplitude p is small and if the driving frequency Ω is away from the primary resonance. Next, observe that in (10.50), a_3 will be approximately proportional to a_1^3. Thus, if a_1 is small compared to unity, a_3 will be much smaller yet, *unless* the denominator in (10.50) is also very small. This, in turn, can only occur if the driving frequency Ω is such that

$$1 - 9\Omega^2 + \tfrac{3}{4}\alpha a_3^2 + \tfrac{3}{2}\alpha a_1^2 \cong 0$$

or
$$\Omega \cong \tfrac{1}{3}\left(1 + \tfrac{3}{2}\alpha a_1^2 + \tfrac{3}{4}\alpha a_3^2\right)^{1/2} \tag{10.51}$$

If a_1 and a_3 are small, the condition (10.51) is approximately $\Omega = \frac{1}{3}$. Thus, if $\Omega \cong \frac{1}{3}$, there is a possibility that a_3 will be significant compared to a_1, even though each is small compared to unity. This turns out to be the case, as shown in Fig. 10.25, in which the fundamental harmonic a_1 and the order 3 superharmonic amplitude a_3 are shown for the case $\alpha = 1.0$, $p = 0.1$. The results shown in Fig. 10.25 represent the approximate solution to (10.48) and (10.49), obtained using the linear solution $a_1 = p/(1 - \Omega^2)$ from (10.48) and then solving (10.49) to determine a_3. The results should be viewed as qualitative only.

We make the following observations regarding Fig. 10.25:

1. The fundamental harmonic a_1 is nearly constant over the excitation frequency range shown, $a_1 \cong 0.11$. Furthermore, because p is small and Ω is away from the primary resonance, the fundamental harmonic amplitude a_1 is defined essentially by the linear theory, $a_1 = p(1 - \Omega^2)$ (see comment 3 following Fig. 10.18).

2. A significant (compared to a_1) superharmonic a_3 is seen to occur only in a very small frequency range centered at $\Omega = 0.336$. Away from this exciting frequency range, $a_3 \cong 0$.
3. In general, (10.49) and (10.50) indicate that the superharmonic a_3 response is qualitatively similar to what happens for a_1 near the primary resonance. This can be seen if we rewrite (10.49) so that it is analogous to (10.38):

$$\Omega^2 = \underbrace{\left(\frac{1}{9}\right)\left[1 + \frac{3}{2}\alpha a_1^2 + \frac{3}{4}\alpha a_3^2\right.}_{\text{Backbone curve}} \left.\pm \frac{\frac{1}{4}\alpha a_1^3}{a_3}\right] \tag{10.52}$$

There is a backbone curve, as indicated, which emanates from the frequency axis at $\Omega = \frac{1}{3}(1 + \frac{3}{2}\alpha a_1^2)^{1/2}$ and then bends to the right as a_3 increases, as shown schematically in Fig. 10.26. Note that in (10.52) the last term is given a $\pm$ sign to account for the fact that a_3 may be positive (in phase with the excitation) or negative (out of phase with the excitation). Note also that in Fig. 10.25 the response curve does not bend to the right very much because a_3 is not large enough to have much of an effect.

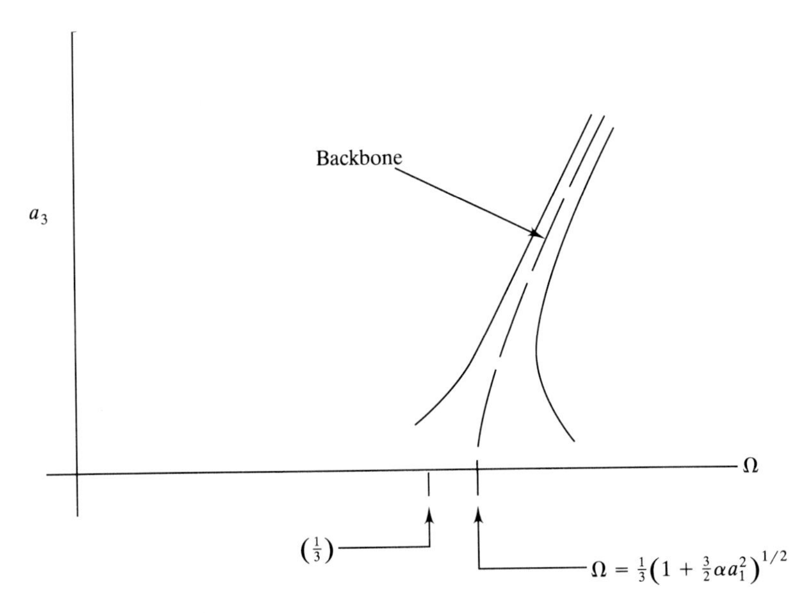

FIGURE 10.26
Schematic of the order 3 superharmonic frequency response for the case of p, a_1, and a_3 small compared to unity.

4. The "excitation" of a_3 is provided by the fundamental harmonic a_1, as can be seen in (10.50); if a_1 were zero, then a_3 would also be zero.

COMMENTS / SUMMARY: THE SUPERHARMONIC RESONANCE.

1. The system (10.39) considered here has a single static, hardening cubic nonlinearity. For such systems the superharmonic resonance of order 3 arises when the exciting frequency Ω is near one-third the linear natural frequency [which is unity in (10.39) and ω in (10.19)]. If the system possesses no linear stiffness, as in (10.20), then order 3 superharmonic resonances can also occur, but the tuning condition (Fig. 10.26) becomes $\Omega = \frac{1}{3}(\frac{3}{2}\alpha a_1^2)^{1/2}$. If there is very light damping ξ and/or strong excitation p, the superharmonic response may be multivalued and may exhibit jump phenomena in a way similar to that of the primary resonance.

2. A superharmonic resonance of order 3 is not the only possibility. For the system (10.19) or (10.20) it is not too difficult to excite superharmonic resonances of odd orders $3, 5, 7, 9, \ldots$. For instance, the order 5 superharmonic resonance may occur in (10.39) if the driving frequency Ω is near one-fifth the linear natural frequency.

3. There are two generic situations that can occur in this type of problem:

a. *The "small" or "weak" excitation case*: For this case, as exemplified by Fig. 10.25, the fundamental harmonic a_1 is determined essentially from the linear theory. This fundamental harmonic excites the third harmonic a_3. The chain of cause and effect here is in one direction only: The excitation p excites a_1, which in turn excites a_3. Mathematically, we can solve (10.48) for a_1, ignoring a_3, and then solve (10.49) for a_3; the a_1 equation is essentially uncoupled from the a_3 equation.

b. *The "strong" excitation case, which we have not considered here*: In this case there is a full nonlinear interaction between the harmonics a_1 and a_3. By this we mean that p excites a_1, which excites a_3 (as in the weak excitation case), but a_3 then has an influence on a_1. Mathematically (10.48) and (10.49) are fully coupled and have to be solved simultaneously as a pair of nonlinear algebraic equations.

10.5.2 The Subharmonic Resonance of Order $\frac{1}{3}$

An important characteristic of the superharmonic response of the preceding section is that the steady-state response has the *same period* as the harmonic excitation. We merely get a large overtone of frequency 3Ω if the exciting frequency Ω is correctly tuned. Now we investigate the possibility of steady-state responses which have periods T that *differ* from the excitation period. Specifically, we will look for periodic solutions having a period *three times* that of the excitation. If such steady-state responses occur, they will contain a Fourier component having frequency $\Omega/3$, as well as integer multiples of $\Omega/3$. We then speak of a "subharmonic resonance of order $\frac{1}{3}$."

The order 3 *superharmonic resonance* occurs when the driving frequency Ω is near $\frac{1}{3}$ the natural frequency. We shall see that the *subharmonic resonance* of order $\frac{1}{3}$ occurs when the driving frequency is near *three times* the natural frequency. In order to investigate the possibility of a subharmonic resonance of order $\frac{1}{3}$ in the system (10.19) or (10.20), let us employ HB with an assumed steady-state solution of the form

$$x(t) = a_1 \cos \Omega t + a_{1/3} \cos\left(\frac{\Omega t}{3}\right) \tag{10.53}$$

Equation (10.53) is similar to (10.46), which we assumed to study the superharmonic resonance of order 3, except (10.53) says we are now seeking a steady-state response which may have a large Fourier component of frequency $\Omega/3$ in addition to the fundamental. If we substitute (10.53) into the undamped model (10.45), use trigonometric identities, and then require the coefficients of $\cos \Omega t$ and of $\cos(\Omega t/3)$ to vanish in accord with HB, we obtain the following results, which are analagous to (10.48) and (10.49):

$$a_1(1 - \Omega^2) + \frac{3}{4}\alpha a_1^3 + \frac{3}{2}\alpha a_1 a_{1/3}^2 + \frac{1}{4}\alpha a_{1/3}^3 = p \tag{10.54}$$

$$a_{1/3}\left(1 - \frac{\Omega^2}{9}\right) + \frac{3}{4}\alpha a_{1/3}^3 + \frac{3}{2}\alpha a_{1/3} a_1^2 + \frac{3}{4}\alpha a_1 a_{1/3}^2 = 0 \tag{10.55}$$

These nonlinear algebraic equations define (approximately) the amplitudes a_1 and $a_{1/3}$ of the steady-state response. Examining (10.55), we observe that each term in (10.55) contains an $a_{1/3}$, so that (10.55) may be written as

$$a_{1/3}\left[\left(1 - \frac{\Omega^2}{9}\right) + \frac{3}{4}\alpha a_{1/3}^2 + \frac{3}{2}\alpha a_1^2 + \frac{3}{4}\alpha a_1 a_{1/3}\right] = 0 \tag{10.56}$$

In order to satisfy (10.56) there are two possibilities: (1) $a_{1/3} = 0$ and (2) the bracketed term multiplying $a_{1/3}$ is zero. The first case, $a_{1/3} = 0$, simply indicates that it is possible to find steady-state solutions which *do not have* a Fourier component of frequency $\Omega/3$. This case is the same as the primary resonance studied in Sec. 10.4, in which the periodic solution has the same period as the excitation and is dominated by the fundamental harmonic.

The second possibility is that

$$\frac{3}{4}\alpha a_{1/3}^2 + \frac{3}{4}\alpha a_1 a_{1/3} + \frac{3}{2}\alpha a_1^2 + \left(1 - \frac{\Omega^2}{9}\right) = 0$$

This equation is quadratic in $a_{1/3}$ and may be solved for $a_{1/3}$ as a function of a_1 and Ω. The result is

$$a_{1/3} = -\frac{a_1}{2} \pm \left[\left(\frac{4}{27\alpha}\right)(\Omega^2 - 9) - \frac{7}{4}a_1^2\right]^{1/2} \tag{10.57}$$

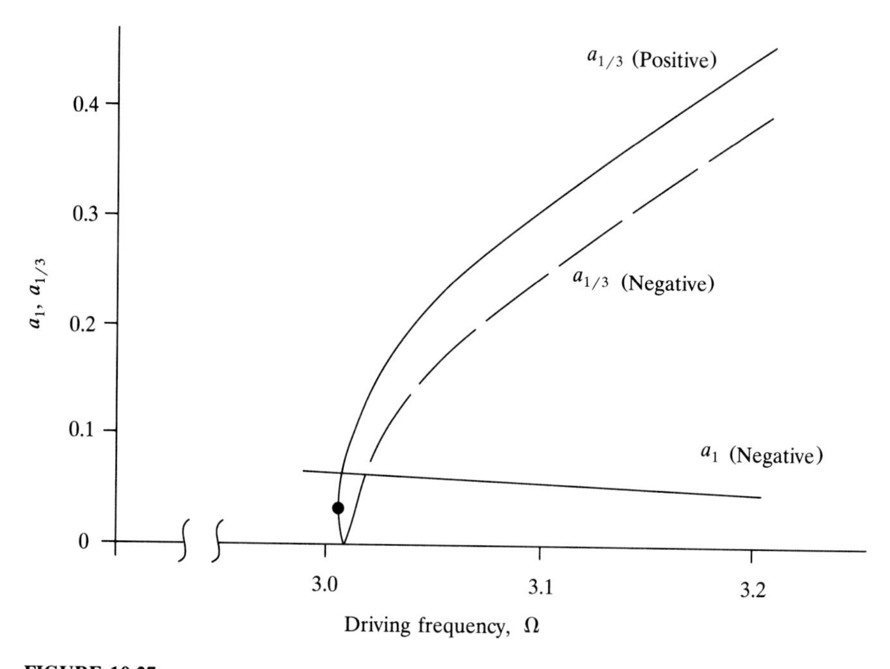

FIGURE 10.27
The fundamental harmonic a_1 and the subharmonic $a_{1/3}$, versus driving frequency, for $p = 0.5$, $\alpha = 1$ in (10.45). See text for discussion.

Now because $a_{1/3}$ must be real, the radical in (10.57) must be positive in order for the subharmonic solution to exist. This means that a one-third subharmonic response can occur only if

$$\Omega^2 > 9 + \frac{189\alpha a_1^2}{16} \tag{10.58}$$

If we consider the case of "weak" excitation (p small), so that a_1 is small compared to unity, then the condition (10.58) shows that the one-third subharmonic solutions will appear for driving frequencies Ω slightly greater than 3. Furthermore, if (10.58) is satisfied, there will be two subharmonic solutions possible, as given by (10.57). Only one of these solutions, however, turns out to be stable. Some results for the undamped case considered are shown in Fig. 10.27 for the case $p = 0.5$, $\alpha = 1.0$.

Discussion of Fig. 10.27: The results shown were generated by solving (10.54) and (10.57) as follows: Because p is "small" and Ω is away from the primary resonance, a_1 is also small and can be determined quite accurately from the linear theory, $a_1 = p/(1 - \Omega^2)$. Notice that a_1 will be negative, as $\Omega > 1$, so that the fundamental harmonic of the response is out of phase with the excitation. In the region of interest $|a_1| \cong 0.06$. As Ω is increased beyond 3, the condition (10.58) is satisfied at $\Omega_c \cong 3.0076$, for which the radical in (10.57) vanishes, and we get a single subharmonic solution $a_{1/3} = -a_1/2$. As Ω is increased beyond Ω_c, there are two solutions for $a_{1/3}$ in accordance with (10.57). Both solutions for $a_{1/3}$ are positive (in phase with the excitation) until just before $\Omega \cong 3.009$, after which one solution is negative and one is positive, as noted in the figure. As Ω is increased further, the subharmonic amplitude becomes *much larger* than the fundamental. Because there is no damping in the analysis, none of the solutions shown in Fig. 10.27 is an attractor. If a small amount of damping were added, the solid (upper) branch of $a_{1/3}$ becomes asymptotically stable, while the broken (lower) branch is unstable. It is especially noteworthy that subharmonic solutions can be excited which are much larger than the fundamental. Also noteworthy is that the qualitative nature of the subharmonic response is much different from that of either the primary resonance or the superharmonic resonance of order 3. This is due to the nature of the Eqs. (10.49) and (10.56). For the superharmonic resonance, (10.49) is a cubic equation, indicating that, generally, one or three real solutions will exist for a_3. On the other hand, (10.56) shows that there will be either none or two nonzero solutions for $a_{1/3}$. This fundamental difference between the subharmonic and superharmonic resonances is typical.

Finally, we note that the presence of some damping will cause the two solution branches in Fig. 10.27 to merge, so that the subharmonic response will disappear at some critical value of driving frequency, just as the solutions suddenly appeared at $\Omega_c \cong 3.0076$.

CONCLUDING REMARKS ON NONLINEAR RESONANCES. We have considered only two of the many types of secondary nonlinear resonances which are possible: the sub- and superharmonic resonances of orders $\frac{1}{3}$ and 3. Furthermore, we have studied only the oscillators (10.19) and (10.20) of Duffing type with linear damping and a cubic, hardening nonlinearity. Many other types of systems and nonlinearities are of relevance in engineering. The intent here has been only to provide a flavor of some of the steady-state responses that can occur. The interested reader should consult the references for more detailed information. The subject of nonlinear resonances is important and interesting, because of the possibility of multiple attractors, jumps, and large amplitude responses that occur at driving frequencies Ω well removed from the ordinary resonance condition one expects to see in a linear oscillator.

In the material presented so far, the harmonic balance method has been used to study approximately the nonlinear resonance phenomenon. HB, combined with numerical integration of the equations of motion, is a powerful line of attack in the analysis of such phenomena. There is, however, another class of approximate, analytical techniques called *perturbation methods*, which differ significantly from HB in the way the approximate solution is constructed. These methods, described in the references, have the advantage that both the steady-state periodic solutions and their stability are determined in a single analysis. HB, by comparison, requires a separate stability analysis, which we have not done in this chapter. The disadvantage of the perturbation methods is that they are limited to the "weakly" nonlinear case. A working knowledge of both perturbation and HB methods is recommended for anyone doing work in nonlinear oscillations. Comprehensive treatment of the nonlinear oscillation phenomena discussed here, as well as other nonlinear phenomena, are contained in Refs. 10.2 and 10.3.

10.6 CHAOS IN NONLINEAR OSCILLATORS

10.6.1 Introduction

In the preceding sections several important nonlinear oscillator phenomena were introduced: the primary, sub-, and superharmonic resonances in the harmonically excited Duffing oscillator with a hardening static nonlinearity. A key aspect of these steady-state responses is that they are *periodic*. The period is *equal* to the excitation period in the case of the primary or superharmonic resonance. The period is an *integer multiple* of the excitation period in the case of the subharmonic resonance; for example, the steady-state response period is three times the period of the excitation for the subharmonic resonance of order $\frac{1}{3}$.

The single degree of freedom nonlinear oscillators (10.19) and (10.20) which we have been studying are characterized by linear damping and by harmonic excitation. Prior to the mid-1970s, the prevailing view among dynamicists was that such harmonically excited, linearly damped, nonlinear single degree of freedom oscillators would *always* exhibit steady-state responses which were periodic. The following quote from Ref. 10.3, which was published in 1979, summarizes this view: "Systems having one degree of freedom . . . always possess steady-state periodic motions when acted upon by periodic excitations in the presence of positive damping."

From the late 1970s to the present, however, researchers began to find steady-state responses that were not only aperiodic but which also had a number of other unusual characteristics which distinguished them from the more mundane periodic responses. These new types of steady-state responses were called "chaotic,"* and the study of them has caused a resurgence in the (to then) fairly moribund subject of dynamics. A new field of science, *chaotic dynamics*, was born and in a decade has reached a fairly mature level. The objective of this section is to present a few of the basic ideas of chaotic dynamics.

---

*The term "chaotic" is perhaps unfortunate, in that it implies a motion which is very disorganized and without pattern. Chaotic motions, however, do possess a high degree of organization, as we shall see.

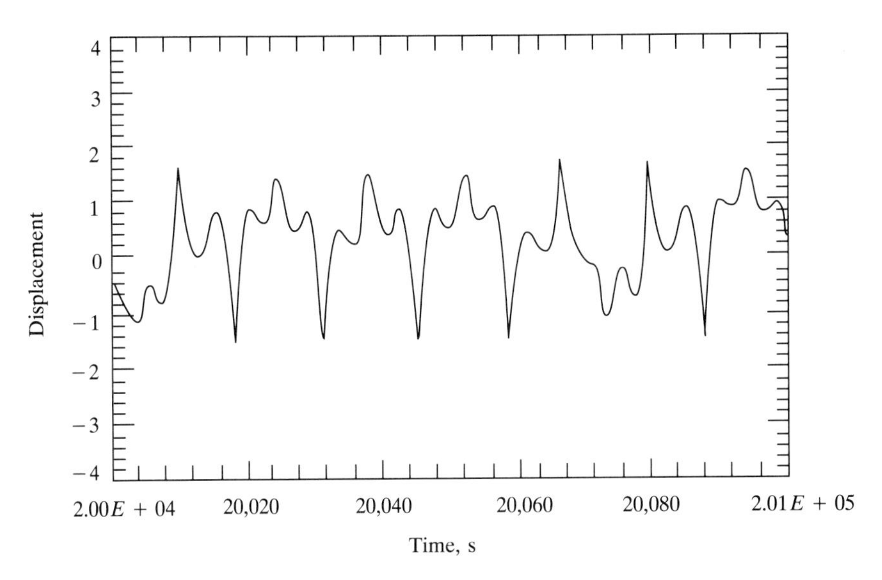

FIGURE 10.28
A hundred seconds of the chaotic response of (10.59) with $p = 1$, $\Omega = 0.44964$, $\xi = 0.02248$. Results obtained via numerical integration. (Reprinted from Ref. 10.9 with permission of the *Journal of Sound and Vibration*.)

For this purpose we will consider the response of (10.20), the Duffing oscillator with no linear stiffness;

$$\ddot{x} + 2\xi\dot{x} + x^3 = p \cos \Omega t \tag{10.59}$$

For some combinations of system parameters (ξ, p, Ω), (10.59) will exhibit *periodic* steady-state responses associated with the nonlinear resonances studied in Secs. 10.3 through 10.5. But for other combinations of ξ, p, and Ω, we will see steady-state responses which are *not periodic* and which exhibit several "unusual" characteristics.

It should be borne in mind that our results will all be obtained for a single system, (10.59). The characteristics we'll see, however, will turn out to be fairly general and the basic ideas applicable to a variety of systems, many of which differ significantly from (10.59).

Let us start by looking at some of the characteristics of a typical chaotic steady-state response of (10.59). Our objective at this point is simply to see what a typical chaotic response looks like. A steady-state chaotic solution is shown in Fig. 10.28 for the parameter values (Ref. 10.13) $p = 1$, $\Omega = 0.44964$, $\xi = 0.02248$ in (10.59). This solution was obtained using the fourth-order Runge-Kutta numerical integration scheme. The steady-state response shown is aperiodic, although it appears by inspection that certain frequencies contribute in an important way. This is verified in Fig. 10.29, which shows the Fourier transform

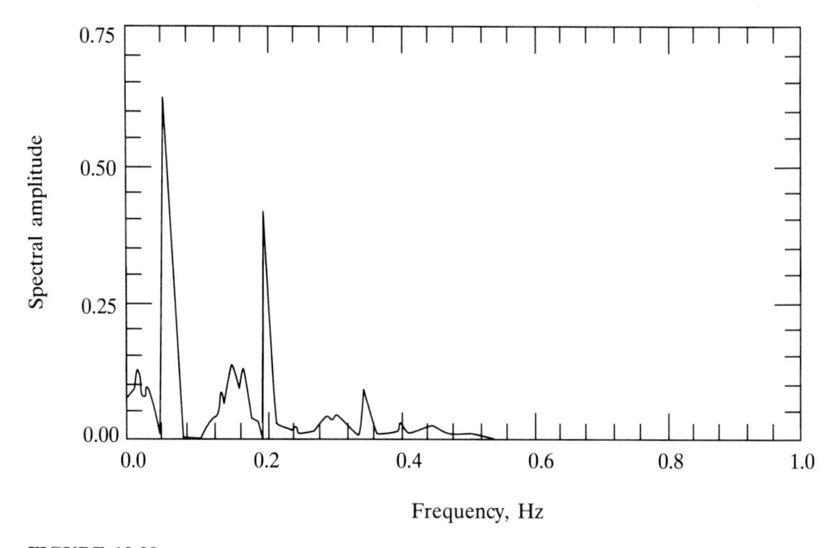

FIGURE 10.29
The frequency spectrum of the response shown in Fig. 10.28. (Reprinted from Ref. 10.9 with permission of the *Journal of Sound and Vibration*.)

of the steady-state response of Fig. 10.28. Large discrete peaks in the spectrum are observed at Ω and at 3Ω, and a smaller peak occurs at 5Ω. There is also some "broadband" or continuous frequency content, centered near $\Omega/3$ and near $5\Omega/3$. The steady-state response, then, is aperiodic but contains a significant periodic content.

The phase portrait of this response is shown in Fig. 10.30. The orbit shown encompasses a time interval of only about three periods of the excitation. (The long time phase portrait is quite convoluted and difficult to understand. The structure of the phase portrait consists of two major "loops," never repeated. If this orbit were plotted over a very long period of time, it would essentially "fill up" a certain portion of the phase plane. Note, however, that the dynamic system (10.59) being studied is actually three dimensional (time is the third dimension), so the phase portrait is a projection of a 3-D orbit onto the plane x, $\dot{x}$. This projected phase portrait reveals that the steady-state motion shown is a lot different from those associated with the periodic attractors considered in Secs. 10.3 through 10.5, each of which is a closed curve in the phase plane. The long time phase portrait, however, would be so jumbled that it would be difficult to see whether there were any "underlying structure" in the steady-state response.

Summarizing Figs. 10.28 through 10.30, we can see the following characteristics of this steady-state response: (1) It is *aperiodic*, never repeating itself, even though the excitation *is* periodic. (2) There is some broadband frequency content, indicating to the casual observer a certain degree of randomness. (3)

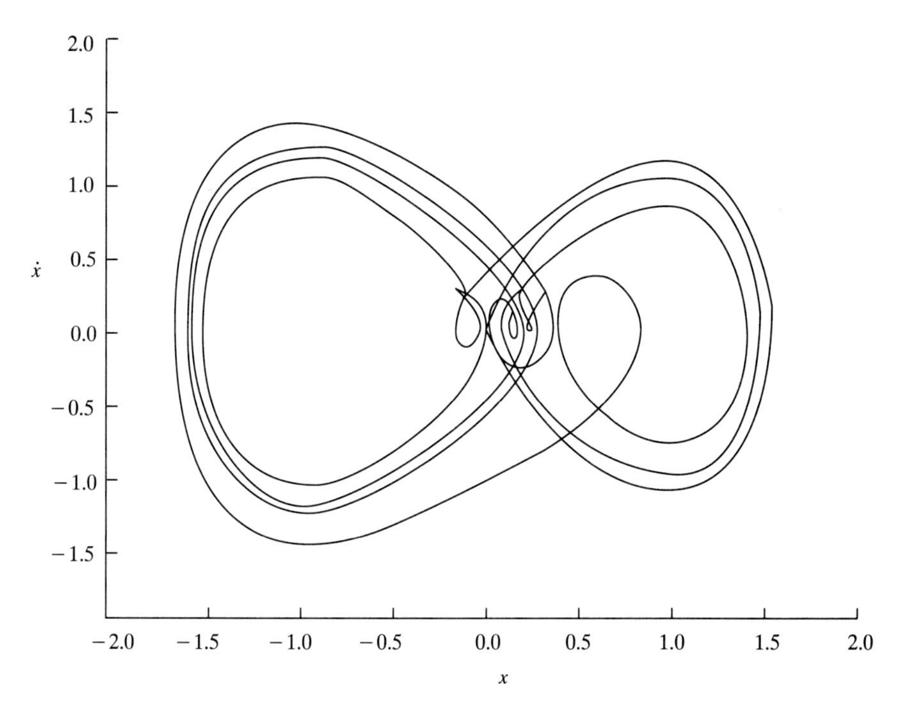

FIGURE 10.30
Several "cycles" of the phase portrait of the response shown in Fig. 10.28. (Courtesy of John Massenburg.)

The long time phase portrait, as well as the time series and frequency spectrum, reveal little about any underlying structure that may exist.

It turns out that the steady-state motion shown in Figs. 10.28 through 10.30 does possess certain interesting properties which are not evident in these figures. In order to uncover these properties, we now introduce a new way to visualize the response, as follows: First, let the system motion evolve over a long enough time that any transients have died out. (The results shown in Figs. 10.28 through 10.30 are for $t \geq 20{,}000$ s, which is certainly adequate for the attainment of steady-state conditions.) Then, pick arbitrarily a starting time t_0, record the point $[(x(t_0), \dot{x}(t_0)]$, and plot this point in the phase plane. Then wait *until the excitation has gone through one full cycle*, so that $t = t_1 = t_0 + 2\pi/\Omega$, and record and plot $x(t_1), \dot{x}(t_1)$. Continue this process of recording and plotting x and $\dot{x}$ at time intervals $\Delta t = 2\pi/\Omega$, that is, at the times $t_n = t_0 + 2\pi n/\Omega$. Do this until a large number of points, say, several thousand, have been plotted. Our plot will then consist of a set of discrete points, rather than a continuous curve, in the phase plane. This process is illustrated in Fig. 10.31. This construction is known as a "Poincaré map," after the eminent French mathematician Henri Poincaré, who explored nonlinear dynamics extensively in the

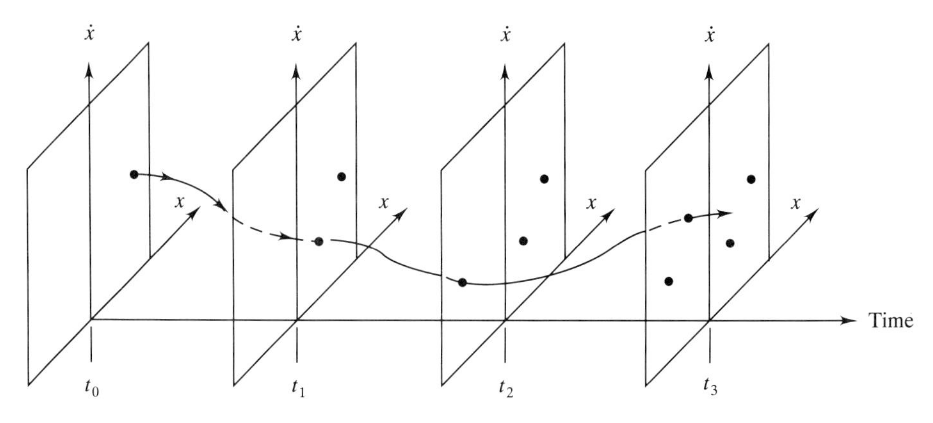

FIGURE 10.31
Construction of the Poincaré map for the driven oscillator (10.59). A hypothetical orbit in $(x, \dot{x}, t)$ space is shown. Each time the driving force completes one period, the point $(x, \dot{x})$ is recorded. At t_3, four such points have been accumulated.

nineteenth century and paved the way for many of the concepts developed more recently.

Now, note the following properties of our Poincaré map:

1. Suppose the steady-state response turned out to be periodic, with the *same period* as the excitation, as, for example, the primary and superharmonic resonances discussed previously. Then the same point $(x, \dot{x})$ would keep repeating itself ad infinitum, and the Poincaré map would consist of a single point in the phase plane, endlessly repeated.
2. Suppose the steady-state response turned out to be periodic, but with a period three times that of the excitation, as, for example, for the order $\frac{1}{3}$ subharmonic resonance of Sec. 10.5. Then the Poincaré map would consist of three points, endlessly repeated in the same order, as it now takes exactly three cycles of the excitation to complete one cycle of the steady-state response.
3. Suppose the steady-state response is *aperiodic*. Then the individual points on the Poincaré map will all be different. No point is ever repeated; if it were, we would have a periodic orbit.
4. Suppose the steady-state response were purely random. Then the Poincaré map would appear as a "cloud" of points with no discernible structure, and the density of points would vary smoothly according to some type of probability distribution.

Figure 10.32 shows the Poincaré map constructed for the response of Figs. 10.28 through 10.30. What is striking about this figure is the obvious, well-defined structure. The Poincaré map consists of certain well-defined "leaves"

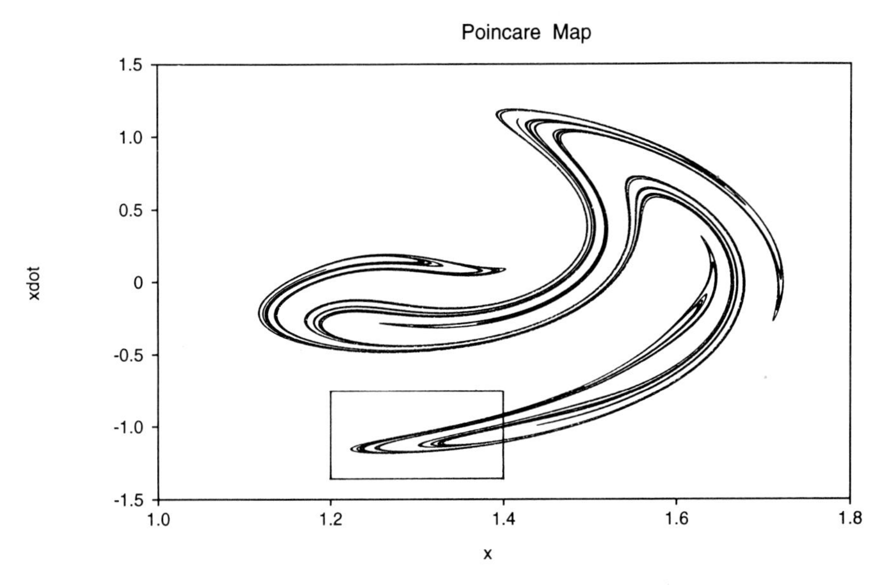

FIGURE 10.32
The Poincaré map for the response of Fig. 10.28. Note the well-defined structure of the chaotic attractor; (After Refs. 10.12 and 10.13.)

which are folded and which are visited by the system point $(x, \dot{x})$. Certain regions of the phase space are inaccessible. The set of points shown in Fig. 10.32 constitutes a *chaotic attractor*, and it is evident that there is significant organization or structure exhibited by this attractor. As we shall see, the *detailed* structure of Fig. 10.32 is even more interesting than it appears here.

COMMENTS ON THE POINCARÉ MAP. Nowadays Poincaré maps are commonly used to study the structure of chaotic attractors. In mathematical terms the Poincaré map is an example of a *discrete mapping*. In the present context this means the following: Let the successively recorded values of x be denoted by $x_0, x_1, x_2, \ldots, x_n$ and the associated successively recorded values of $\dot{x}$ be denoted by $y_0, y_1, y_2, \ldots, y_n$. Then there exist two functions g_1 and g_2 which take a given Poincaré point (x_n, y_n) and generate the next point (x_{n+1}, y_{n+1}) according to

$$\begin{aligned} x_{n+1} &= g_1(x_n, y_n) \\ y_{n+1} &= g_2(x_n, y_n) \end{aligned} \tag{10.60}$$

Equation (10.60) is an example of a two-dimensional *discrete dynamical system*. The word *discrete* implies that the outcome is a set of points, rather than the continuous phase space orbit that we have usually used to describe the motion of a dynamical system.

There are two important characteristics of discrete mappings that we should be aware of:

1. Because the system (10.59) being studied is nonautonomous, the phase space is really three dimensional; time t, or the phase Ωt of the excitation, may be considered the third state variable. The Poincaré map (10.60) is, on the other hand, strictly two dimensional; the explicit time dependence has been removed. Thus, the Poincaré map reduces by 1 the dimension of the space in which the motion is viewed. Here we have a three-dimensional, continuous (in time) dynamical system which is converted to a two-dimensional, discrete dynamical system by the action of the Poincaré map. Unfortunately, the original differential equation (10.59) does not tell us what the functions g_1 and g_2 in (10.60) actually are. All we can say is that they exist. The reduction of the phase space dimension by 1 is a very useful property from the standpoint of interpreting the dynamic behavior.
2. There is a direct correspondence between attractors that exist in the continuous system (10.59) and those that exist in the associated discrete dynamical system (10.60). For example, consider the periodic attractors, having the same period as the excitation, that we studied in Secs. 10.4 and 10.5 (primary and superharmonic resonances). In the phase plane these attractors will consist of closed curves. An orbit initiated near one of these attractors will approach the attractor asymptotically. A similar situation exists for the Poincaré map. In the Poincaré plane the periodic attractors of Sec. 10.4 will be represented by *equilibrium* or *fixed points* (points that repeat endlessly). In the context of (10.60) such an equilibrium or fixed point satisfies $x_{n+1} = x_n$ and $y_{n+1} = y_n$, which conditions define a fixed point of the discrete dynamical system (10.60). Thus, in the Poincaré plane a discrete orbit initiated at some point (x_0, y_0) near but not at the fixed point $(\bar{x}, \bar{y})$ representing the attractor will, under the action of the map (10.60), converge asymptotically to $(\bar{x}, \bar{y})$. This process is illustrated in Fig. 10.33: 10.33(a) shows the phase portrait of the periodic attractor associated with the primary resonance of (10.59). The attractor is identified, as are two orbits, initiated inside and outside the attractor, which converge to the attractor. The associated Poincaré map is shown in Fig. 10.33(b). The periodic attractor of the ODE system (10.59) becomes an asymptotically stable equilibrium of the discrete dynamical system (10.60), and the successive iterates of the Poincaré map move to this equilibrium, as shown. Notice that, although the "orbits" of Fig. 10.33(b) consist of a succession of discrete points, the general structure is reminiscent of the near-equilibrium orbits we studied in Chap. 4 for autonomous systems. The main point here is that a *periodic attractor* of the ODE system (10.59) is equivalent to an *asymptotically stable fixed point* of the Poincaré map. Thus, the dynamic behavior of the 2-D (discrete) Poincaré map is fully equivalent to the dynamic behavior of the 3-D ODE system (10.59), and we can study the dynamics using either representation.

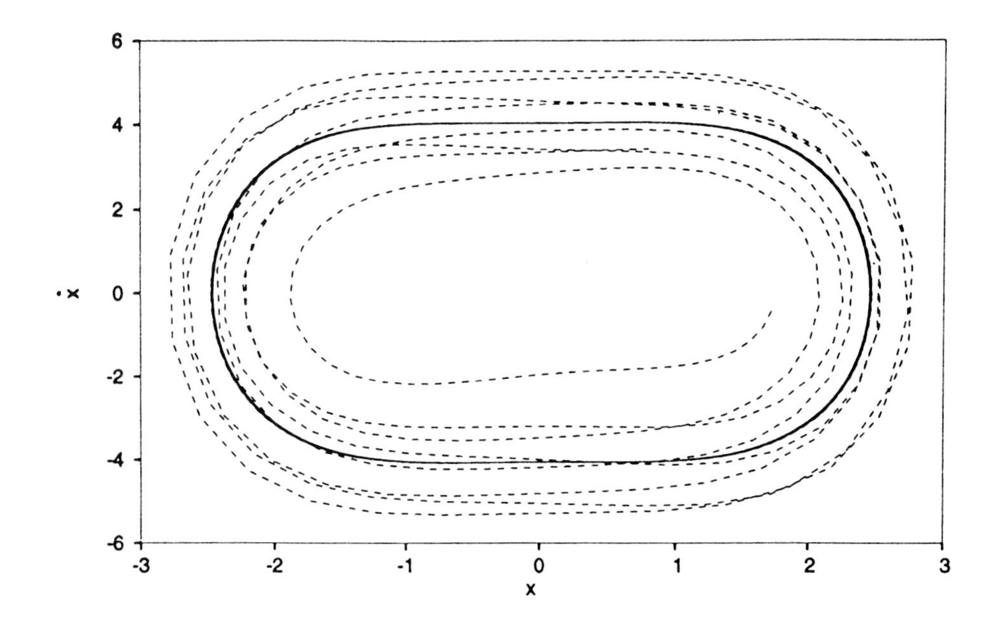

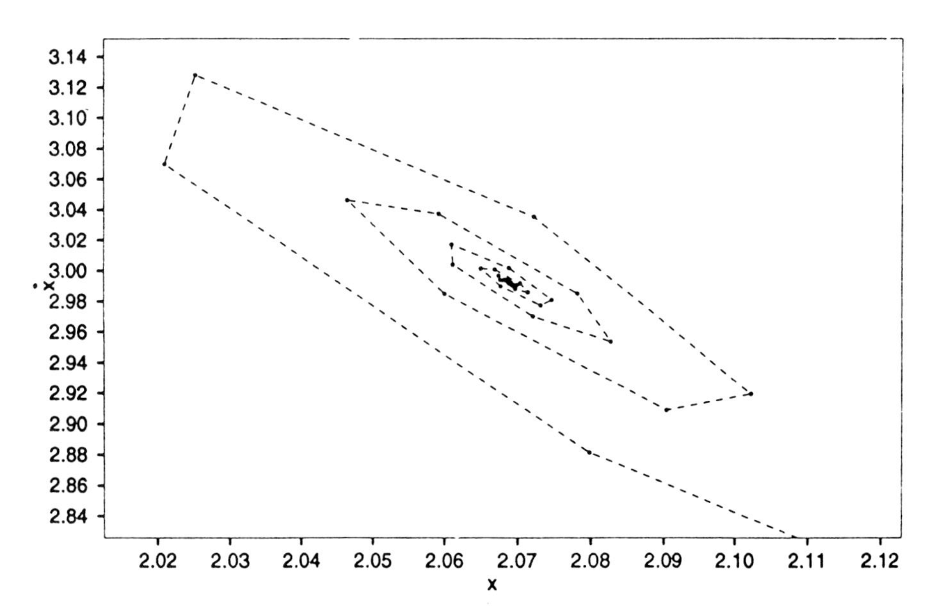

FIGURE 10.33
The equivalence of a stable periodic orbit in phase space and a stable fixed point of the Poincaré map: (a) The periodic attractor (dark solid line) associated with the primary resonance of (10.59) for parameter values $\xi = 0.05$, $p = 1.0$, and $\Omega = 2.0$. The dashed lines show transient behavior occurring when the initial conditions are not on the attractor: (b) The Poincaré map in the vicinity of the attracting point $x = 2.069$, $\dot{x} = 2.992$; the points defining the Poincaré map are connected by a dashed line in order to show the progression of points; in the steady state the stable fixed point of the Poincaré map is endlessly repeated.

In this introductory section the intent has been only to address the question, "What does a typical chaotic attractor look like?" and to introduce the Poincaré map as a useful way to describe such responses. We now turn to the characteristics of chaotic attractors from a more technical viewpoint.

10.6.2 Characteristics of Chaotic Response: Sensitivity to Initial Conditions

So far we have seen that chaotic response is aperiodic and may appear to have an element of randomness, for example, if we look at a spectrum or a phase portrait. But there is also an underlying structure, as is seen clearly in the Poincaré map. In this section we investigate an important characteristic of chaotic response/attractors: the sensitive dependence on initial conditions.

Perhaps the most important characteristic of chaotic attractors is the *sensitive dependence on initial conditions*. By this we mean the following: Suppose that we generate a chaotic motion and then pick some point on the chaotic attractor. This point is defined by the three quantities $t_1, x_1 = x(t_1)$, and $\dot{x}_1 = \dot{x}(t_1)$. Here t_1 is arbitrary, as long as it is large enough that the motion has asymptoted to the chaotic attractor. Next, we consider a second orbit with slightly different starting conditions, $x_2 = x(t_1) + \Delta x$ and $\dot{x}_2 = \dot{x}(t_1) + \Delta \dot{x}$, where Δx and $\Delta \dot{x}$ are very small, so that the two orbits which evolve for $t > t_1$ start out very close to each other. The phrase "sensitive dependence on initial conditions" means that, for *almost all* infinitesimal pairs $(\Delta x, \Delta \dot{x})$, the two orbits will diverge from each other exponentially, *on average*, over a long time. Thus, the two orbits, in detail, eventually bear little resemblance to each other, even though each defines the same chaotic attractor.

For example, suppose that we pick two points very close to each other on Fig. 10.32 and then watch how successive iterates evolve; this is equivalent to comparing the two time continuous orbits after each period of the excitation. An example is shown in Fig. 10.34. The two sets of initial conditions used to generate these results were as follows: Reference orbit: $x = 1.712909$, $\dot{x} = 0.22488$; nearby orbit: $x = 1.712850$, $\dot{x} = 0.22488$. The initial distance d_0 between these points in the phase space is $d_0 = 0.000059$. Observe that the orbits stay close to each other for a short time but then rapidly diverge, so that after about 10 cycles, they bear little relation to each other. This sensitive dependence on initial conditions does not occur for any of the periodic nonlinear phenomena discussed in the preceding sections. The practical importance of this property of chaotic dynamics is that, for a real chaotic motion, long-term exact prediction of behavior is impossible, because we can never establish the initial conditions exactly. We can identify the general structure of the chaotic attractor (as in Fig. 10.32), but over long times we cannot tell exactly where the orbits will go. This type of "unpredictability" is usually ascribed to dynamic systems which are acted upon by random stimuli, such as inputs which inher-

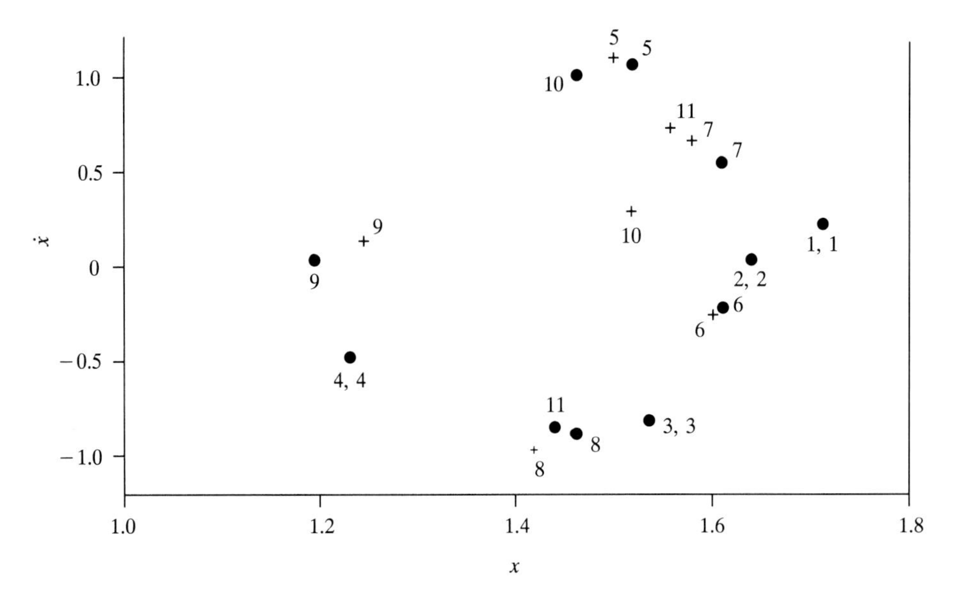

FIGURE 10.34
Comparison of 10 iterates of the Poincaré maps for two points starting very close to each other on the attractor shown in Fig. 10.32. The reference orbit, denoted by $(\cdot)$, is initiated at $x = 1.712909$, $\dot{x} = 0.22488$. The neighboring orbit, denoted by $(+)$, is initiated at $x = 1.712850$, $\dot{x} = 0.22488$. The point labeled 1 is the starting point, with successive iterates numbered. On this scale the first two iterates (points 2, 3) are indistinguishable, but thereafter the two orbits quickly diverge.

ently possess random or noisy components. But we should note that the system being considered here is *completely deterministic*; there is no noise anywhere in the model. (In simulations a small amount of "noise" is introduced due to computer roundoff error, but it can be made insignificant with a sufficiently small integration time step.) To reiterate, the unpredictability of the details of chaotic motions observed in the real world is due to a combination of the sensitive dependence on initial conditions and our inability to measure initial conditions with infinite precision. If we *were* able to measure an initial state of some system with infinite precision, and if our mathematical model of the phenomenon were exact, then we *could* determine exactly where a given orbit would go, as we can do on the computer.

Let us now explore in more detail the divergence of nearby chaotic orbits. Some aspects of this sensitive dependence on initial conditions can be understood in the context of linear theory, a specific example of which is now discussed. Consider the following two-state linear system having real eigenvalues:

$$\begin{aligned} \dot{x}_1 &= -1.2x_1 \\ \dot{x}_2 &= x_2 \end{aligned} \tag{10.61}$$

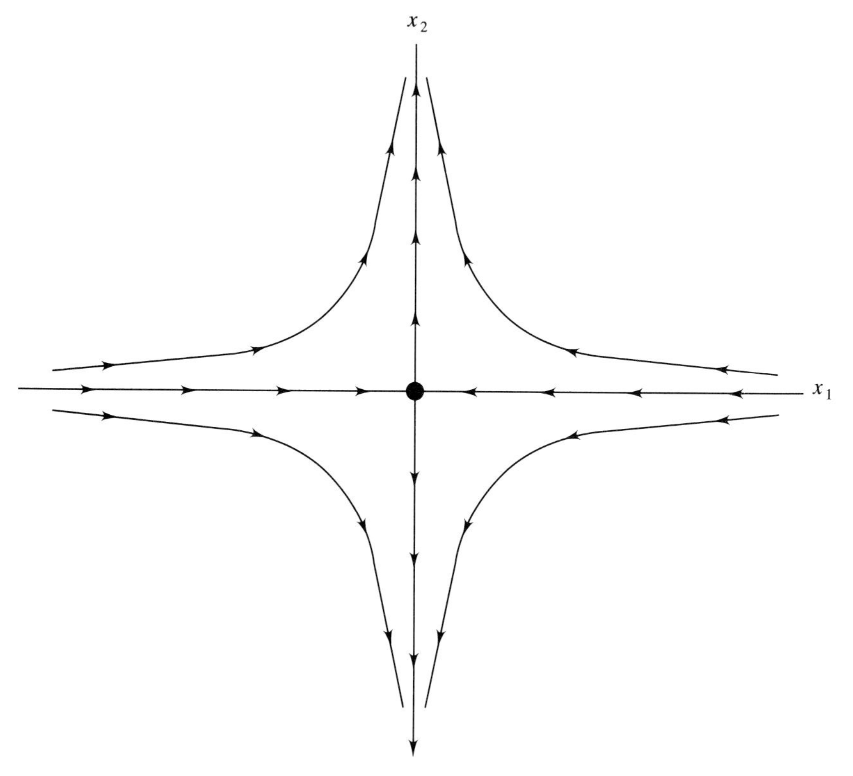

FIGURE 10.35
Phase plane orbits of the two-state linear system (10.61).

The eigenvalues are $\lambda_1 = -1.2$, $\lambda_2 = 1.0$ and the associated eigenvectors lie along the x_1 and the x_2 axes, respectively; the system is in "diagonal form," that is, the two-state equations are uncoupled, done here to make the analysis simple. Typical phase plane orbits are shown in Fig. 10.35. Orbits initiated on the x_1 axis will move asymptotically to the origin, while those initiated on the x_2 axis will move along it to infinity. Any orbit not initiated on the x_1 axis will eventually move to infinity along $\pm x_2$. The equilibrium solution $(0, 0)$ is unstable.

Let us now consider what happens to orbits initiated in the neighborhood of the unstable equilibrium. To do this, consider a circle of unit radius surrounding the origin as shown in Fig. 10.36. Take all points on this circle as initial conditions, and monitor the deformation of the circle as time elapses. The two points starting on the x_1 axis will move to the origin, while the two points starting on the x_2 axis will move away from the origin. The circle deforms into an ellipse, which becomes thinner and thinner as time elapses. Figures 10.36(b) through (d) show the sequence of such ellipses at 0.5-s intervals.

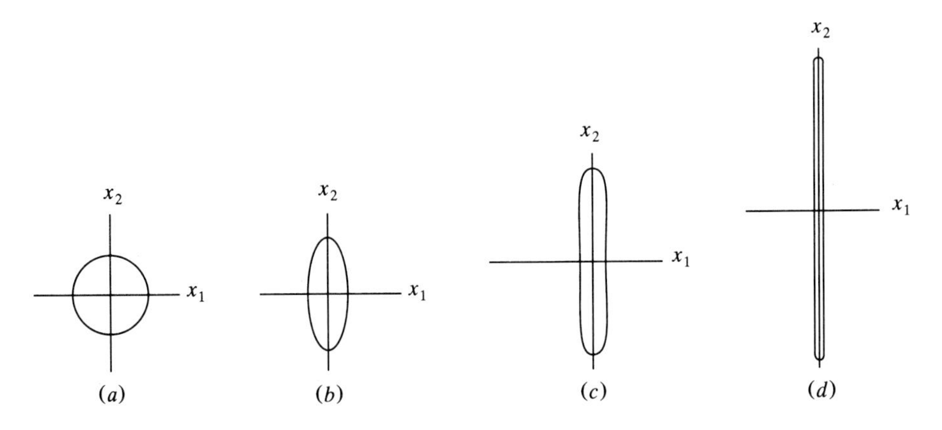

FIGURE 10.36
For the linear system (10.61), the evolution of a circular neighborhood of the unstable equilibrium at the origin. There is shrinkage in one direction and expansion in the other.

Comments on Fig. 10.36

1. As time elapses, the area of the ellipse diminishes. This can be seen by noting that the semimajor axis $a = e^t$, while the semiminor axis $b = e^{-1.2t}$. Thus, the area $A(t) = \pi ab = \pi e^{-0.2t}$ (the initial area is π). Notice that the coefficient of time t in the exponent of $A(t)$ is the *sum of the eigenvalues*. This turns out to be an important general result, as follows: For a general n-state autonomous system, whether linear or nonlinear, the mathematical model has the form (4.5):

$$\begin{aligned} \dot{x}_1 &= f_1(x_1, x_2, \ldots, x_n) \\ \dot{x}_2 &= f_2(x_1, x_2, \ldots, x_n) \\ &\vdots \\ \dot{x}_n &= f_n(x_1, x_2, \ldots, x_n) \end{aligned} \tag{10.62}$$

The functions $f_1, \ldots, f_n$ form a vector field $\mathbf{f}$, as discussed in Sec. 4.1. If we consider an n-dimensional volume V in the phase space, then the relative rate at which this volume grows or shrinks is defined by the divergence of the vector field $\mathbf{f}$. This is analogous to the growth or shrinkage of the volume occupied by a compressible fluid in three space. For a fluid the volume change is governed by the divergence of the velocity vector $\mathbf{v}$. In the case of the dynamical system (10.62), the state derivatives $\dot{x}_i$ may be viewed as "phase space velocities" of an "n-dimensional fluid flow." Mathematically, the statement "the relative volume growth rate equals the divergence of the vector field $\mathbf{f}$" is stated as

$$\frac{\dot{V}(t)}{V(t)} = div(\mathbf{f}) = \frac{\partial f_1}{\partial x_1} + \frac{\partial f_2}{\partial x_2} + \cdots + \frac{\partial f_n}{\partial x_n} \tag{10.63}$$

where $V(t)$ is the phase space volume at time t evolved from an initial n-dimensional phase space volume V_0.

For a linear system, (10.62) becomes $\{\dot{x}\} = [A]\{x\}$, and the divergence of **f** is the trace of the Jacobian matrix $[A]$. The trace of $[A]$, in turn, must equal the sum of the eigenvalues. The trace of $[A]$ is a constant and a basic property of the linear system. Thus, for the linear system (4.62), the phase space volume growth relation becomes

$$\frac{\dot{V}}{V} = \sum \lambda_i \tag{10.64}$$

and this can be solved, with the initial condition $V(0) = V_0$, to yield

$$V(t) = V_0 e^{(\Sigma \lambda_i)t} \tag{10.65}$$

This result shows that the area of the ellipses of Fig. 10.36 will asymptotically approach zero. The initial neighborhood circle is turned asymptotically into a *line* which is essentially the x_2 axis.

2. An important feature of the neighborhood evolution shown in Fig. 10.36 is that *all* points on the initial neighborhood circle, except the two points *on* the x_1 axis, will eventually approach infinity along the $\pm x_2$ axis.
3. The linear system (10.61) exhibits the same sensitivity to (nearly all) initial conditions as is characteristic of the chaotic motion, in the sense that two points in phase space which are initially close to each other, say, a small distance ε apart on the x_2 axis, will diverge from each other exponentially.

Warning! We must be careful in implying too much similarity between the linear system discussed here and the nonlinear systems which exhibit chaos. There are some *very* important differences, which are summarized below.

For a linear system to exhibit sensitivity to initial conditions, there need only be a positive eigenvalue, or a complex pair with positive real part. This implies that the equilibrium solution is unstable and that motions initiated near equilibrium become unbounded. So we have no attractor and an unbounded motion. On the other hand, chaos is a motion which *is* bounded in phase space, and the motion defines a chaotic attractor. So we have an attractor and then by definition a bounded motion. Chaotic attractors are characterized by "local instability" (nearby trajectories diverge) and "global stability" (the motion is bounded and defines an attractor). The other attractors we have considered, equilibrium solutions and periodic orbits, do not exhibit sensitivity to initial conditions.

The purpose in considering our two-state example (10.61) was to show how small volumes in phase space evolve when they are stretched in some directions and compressed in others and to illustrate the idea of sensitivity to initial conditions. We chose the linear system because the way in which phase space volumes deform is easy to determine. Next, we illustrate the sensitivity to initial conditions and deformation of phase space volumes for the nonlinear system (10.59).

The system model (10.59), if expressed in state variable form with the phase $\Omega t \equiv x_3$ as the third state, becomes

$$\begin{aligned} \dot{x}_1 &= x_2 \\ \dot{x}_2 &= -2\xi x_2 - x_1^3 + p\cos x_3 \\ \dot{x}_3 &= \Omega \end{aligned} \tag{10.66}$$

The divergence of the vector field of (10.66) is div(**f**) $= -2\xi$, which is a constant. According to (10.63), then, any volume in the phase space will shrink exponentially as $V(t) = V_0 e^{-2\xi t}$. As there is neither shrinkage nor expansion in the x_3 direction, we need only consider phase space volumes in the 2-D $(x, \dot{x})$ phase plane.

In order to observe the spreading of nearby trajectories associated with a chaotic motion, let us consider Fig. 10.37, which shows a portion of a chaotic orbit of (10.59) and (10.66) in the phase plane. Call this chaotic orbit $[x_c(t), \dot{x}_c(t)]$. At some time t_0 we pick a reference point on the chaotic orbit, as indicated in Fig. 10.37. We surround this reference point with a number of nearby points which lie on a circle about the reference point. This circle defines a "neighborhood" of the chaotic reference point. We want to watch this

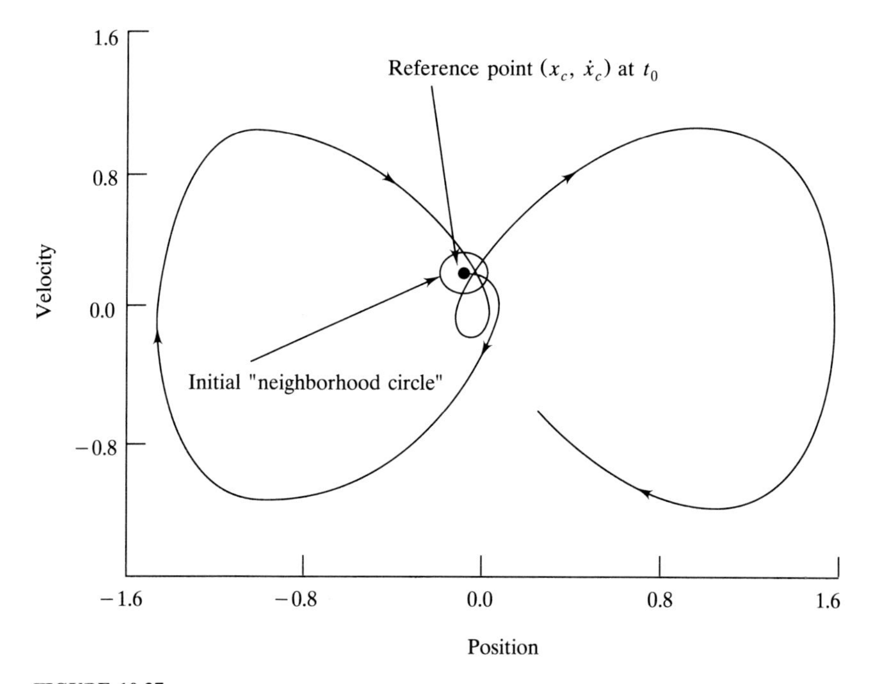

FIGURE 10.37
A representative point $(x_c, \dot{x}_c)$ on the chaotic orbit of Fig. 10.28. This point is surrounded by a circle of nearby points. The evolution of this "neighborhood circle" is shown in Fig. 10.38. (Courtesy of John Massenburg.)

neighborhood evolve for $t > t_0$. Due to the sensitive dependence on initial conditions, we expect that points starting in the neighborhood of $[x_c(t_0), \dot{x}_c(t_0)]$ will not, over long times, remain close to the chaotic reference orbit $[x_c(t), \dot{x}_c(t)]$.

Important note: The radius of the initial neighborhood circle should be *infinitesimal*; we want to look at the evolution of a neighborhood in which all points are infinitesimally close to the chaotic reference orbit. The neighborhood circle shown in Fig. 10.37 is finite only to make this visualization easier. Thus, the neighborhood circle shown in Fig. 10.37, as well as the ellipses which evolve from it in Fig. 10.38, should be mentally scaled down by the reader.

The evolution of the initial neighborhood circle of Fig. 10.37 is shown in Fig. 10.38. As the neighborhood circle evolves for $t > t_0$, it deforms into an ellipse with easily identifiable major and minor axes. The ellipse also rotates, so that the directions of the major and minor axes are continually changing. If we were to follow the evolving neighborhood ellipse for a long time, the major axis will, on average, continue to expand exponentially, while the minor axis will, on average, continue to shrink exponentially. The ellipse will asymptote to a line along the expanding direction. The sensitivity to initial conditions is now clear:

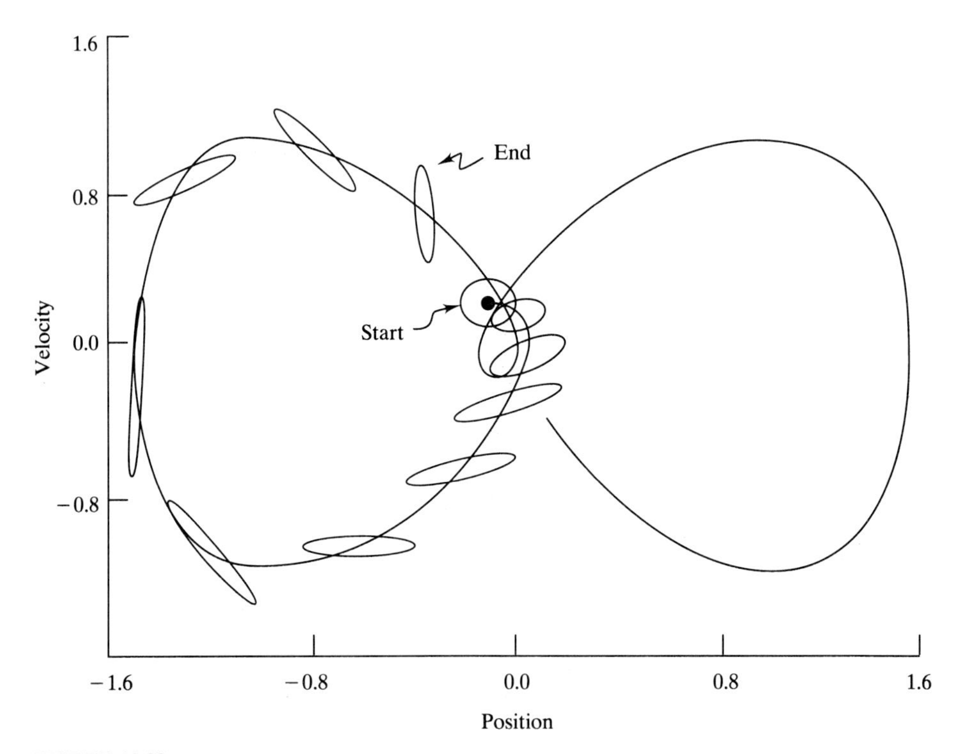

FIGURE 10.38
The deformation of the "neighborhood circle" of Fig. 10.37 into ellipses which on average contract along one axis and expand along the other. Results are shown in 0.5-s intervals, with the ellipse marked "end" occurring at 5 s after the "start." (Courtesy of John Massenburg.)

All points on the initial neighborhood circle, *except* the two points that define the contracting direction, will move, on average, exponentially away from the reference orbit, even though they will be "on the attractor."

> ***Comment:*** Note the use in the preceding paragraph of the words "on average" to describe the expansion and contraction of the major and minor axes of the evolving neighborhood ellipse. This is done because the expanding and contracting axes do not expand and contract at a uniform exponential rate, as occurs in the linear system (10.61), for example. In fact, it is observed in Fig. 10.38 that the major axis may actually *shrink* during certain time intervals. But *on average* over long times it will expand exponentially.
>
> The purpose of this Sec. 10.6.2 has been to introduce, mainly by example, the very important property of sensitivity to initial conditions, which all chaotic motions exhibit. In Sec. 10.7 we will study this phenomenon quantitatively through the introduction of Lyapunov exponents, which are the average exponential rates of expansion (contraction) of the axes of the evolving neighborhood ellipse (or ellipsoid in the case of more than two dimensions).

10.7 LYAPUNOV EXPONENTS AND FRACTAL DIMENSION

10.7.1 Lyapunov Exponents

In Sec. 10.6.2 we saw the following: (1) For the autonomous linear system $\{\dot{x}\} = [A]\{x\}$, the relative rate of growth or shrinkage of volumes in the phase space is the same everywhere in the phase space and is defined by the sum of the eigenvalues according to (10.64) and (10.65). (2) For a nonlinear system exhibiting chaotic behavior, such as (10.66), small phase space volumes defined initially to surround some point on the chaotic attractor will, on average, behave in an analagous fashion, in that there also exist parameters that describe the growth and shrinkage of such phase space volumes for a nonlinear system. These parameters are the *Lyapunov exponents*, and we provide here a brief description of them. Consult Refs. 10.4 and 10.5 for a more complete discussion.

As in Sec. 10.6.2, define the three system states of (10.59) as in (10.66), so that $x_1 = x$, $x_2 = \dot{x}$, and $x_3 = \Omega t$, with the understanding that the excitation phase is defined modulo 2π. To define the Lyapunov exponents, let us suppose that we have a chaotic orbit generated for the system (10.66) in the three-dimensional phase space. Call this chaotic orbit $x_c(t)$, where $x_c(t) = (x_{1_c}, x_{2_c}, x_{3_c})$ stands for the vector composed of the three system states. This chaotic orbit, the properties of which are to be described, will be referred to as the *reference orbit*. We next select arbitrarily a point $x_{c_0}(t_0)$ on the chaotic attractor. We wish to explore the nature of orbits started arbitrarily close to this point. Thus we define a small state vector $w_0(t_0)$ which originates at $x_{c_0}(t_0)$ and which points in an arbitrary direction in the phase space, as shown in Fig. 10.39. Denote the initial length of the vector $w_0(t_0)$ as d_0. As the orbits $x_c(t)$ and $w(t)$

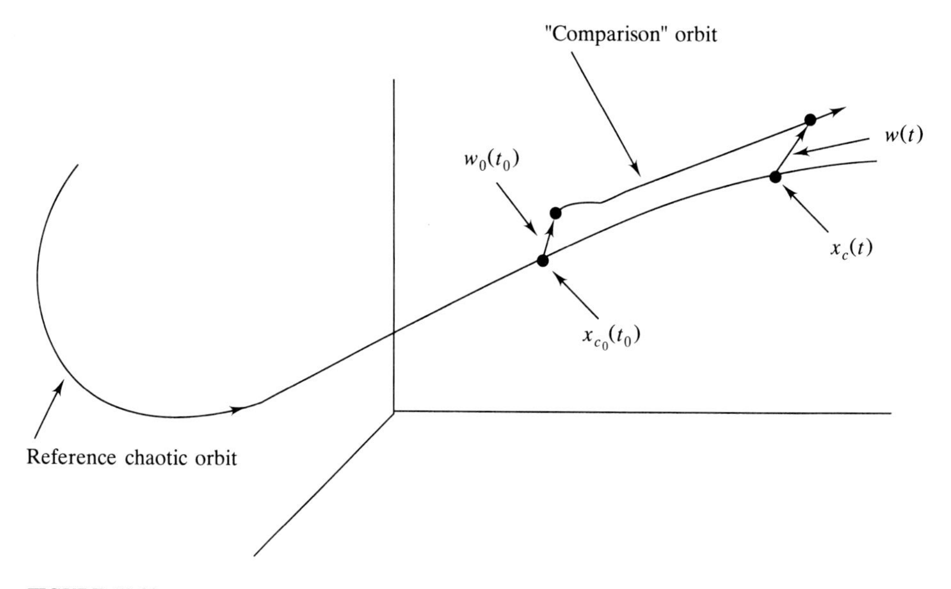

FIGURE 10.39
The reference chaotic orbit and a nearby "comparison" orbit, which evolves from a starting point arbitrarily close to a point on the chaotic reference orbit.

evolve from the initial conditions $x_{c_0}(t_0)$ and $w_0(t_0)$, our interest is in the manner in which $d(t)$ changes, on average, over long times. The function $d(t)$ will define the *separation* of the two nearby points as the orbits evolve.

As applied to the three-state system (10.66) being studied here, the following turns out to occur, stated here without proof: At the point $x_{c_0}(t_0)$ there exists in the phase space a set of three unit vectors e_1, e_2, and e_3, which span three space, and which we can consider to originate at $x_{c_0}(t_0)$ and then to translate along the reference trajectory $x_c(t)$ for $t > t_0$. For the system (10.66), e_1 and e_2 lie in the $x, \dot{x}$ plane, and e_3 is in the direction of time. The directions defined by these unit vectors have the following property: If $w_0(t_0)$ is directed exactly along one of the unit vectors e_i, then $w(t)$ will exhibit a mean exponential rate of convergence or divergence which is characteristic of the particular direction e_i along which the vector $w_0(t_0)$ lies. In a general n-state system there are n parameters, each associated with a particular direction from $x_{c_0}(t_0)$, with each measuring a mean exponential rate of growth or shrinkage of $w(t)$. These parameters are the Lyapunov exponents, denoted here by σ_i and defined formally as follows:

$$\sigma_i(x_{c_0}) = \lim_{\substack{t \to \infty \\ d_0 \to 0}} \ln\left[\frac{d_i(x_{c_0}, t)}{d_i(x_{c_0}, 0)}\right] \tag{10.67}$$

where σ_i is the Lyapunov exponent associated with the direction e_i in phase space. Note that the "Lyapunov directions" e_i will differ at different points in the phase space. Furthermore, as the chaotic reference orbit $x_c(t)$ evolves from $x_{c_0}(t_0)$, the Lyapunov directions will continually change. Finding the Lyapunov directions at an arbitrary point in phase space is not something one does easily. For our purposes, however, it is sufficient to know that such directions, along with the associated Lyapunov exponents σ_i, exist and can be found numerically.

Comments on (10.67)

1. The result (10.67) is based on the long time average of the magnitude $d(t)$ of the n-dimensional state vector $w(t)$. If for a given chaotic attractor the relative rate of growth of d were actually constant (unusual), then (10.67) would simplify to

$$d_i(t) = d_i(0)e^{\sigma_i t} \tag{10.68}$$

which is analogous to the growth or decay of a solution along the direction of an eigenvector associated with a real eigenvalue of a linear system.

2. The limit $d_0 \to 0$ in (10.67) means that the Lyapunov exponent is to be defined such that the magnitude $d(t)$ is always infinitesimally small, as Lyapunov exponents measure local behavior. This means that the actual numerical calculation of Lyapunov exponents is to be based on the model (10.66), linearized about the chaotic reference orbit. Thus, for the system (10.66), let $x_c(t)$ be the chaotic reference orbit being studied, and denote the nearby comparison orbit as $y(t) = x_c(t) + w(t)$, where the magnitude of $w(t)$ is to be infinitesimal. Substitution of the definition of $y(t)$ into (10.66), followed by deletion of nonlinear terms in w in the result, show that $w(t)$ evolves according to

$$\ddot{w} + 2\xi\dot{w} + 3x_c^2(t)w = 0 \tag{10.69}$$

This is a linear ODE for $w(t)$, in which there is a time-dependent coefficient, which is proportional to the square of the displacement $x_c(t)$ of the chaotic orbit.

3. The growth or shrinkage of volumes in the phase space is governed by the Lyapunov exponents in a way similar to (10.64) and (10.65) for a linear system. For an n-state nonlinear system characterized by the Lyapunov exponents $\sigma_1 > \sigma_2 > \cdots > \sigma_n$, the phase space volume V satisfies

$$\sum_{i=1}^{n} \sigma_i = \lim_{\substack{t\to\infty \\ V_0 \to 0}} \ln\left[\frac{V(x_0, t)}{V(x_0, 0)}\right] \tag{10.70}$$

where $V(x_0, 0)$ is the infinitesimal initial phase space volume surrounding the point x_0 in phase space, and which has evolved into $V(x_0, t)$ at some time t.

4. Loosely speaking, we can think of the Lyapunov exponents as being analogous to the real parts of the eigenvalues of a linear system.

From the preceding points we now discuss the nature of the Lyapunov exponents associated with the chaotic attractor of the system (10.66). First we note that the system is dissipative; in fact, the divergence of the vector field f

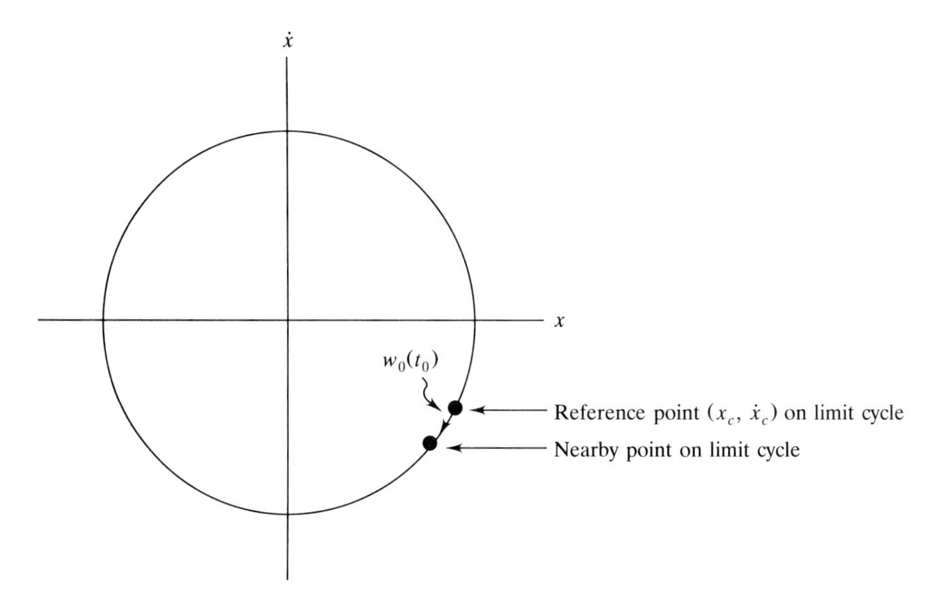

FIGURE 10.40
Reference and nearby points on the limit cycle of the system (10.11). The initial distance between the two points should be infinitesimal and is shown here as finite for visualization purposes.

defined by (10.66) is given by $\nabla f = -2\xi$, indicating that phase space volumes will actually shrink exponentially for this particular system, so that $V(t) = V(0)e^{-2\xi t}$ for any initial volume $V(0)$ considered. According to (10.70) we must have that the sum of the three Lyapunov exponents $\sigma_1 + \sigma_2 + \sigma_3 = -2\xi$ for the system (10.66). In general, for any dissipative system the sum of the Lyapunov exponents must be negative. If this were not so, then we would not have an attractor. The second property required for a chaotic attractor is that there will be a positive exponent, $\sigma_1 > 0$. This is required in order to have the property of sensitivity to initial conditions. The third property is that one of the exponents must be zero. This can be seen by selecting the "test point" $w_0(t_0)$ to be *on* the reference chaotic orbit (take the values of x and $\dot{x}$ on the reference orbit at a time slightly different from t_0). The zero exponent is perhaps more clearly illustrated by considering the limit cycle of Fig. 10.6. Consider the reference point x_0 shown and the test point w_0 near x_0 *on* the limit cycle, as shown in Fig. 10.40. As the orbits started at these two points evolve, the distance $d(t)$ between them will generally change with time, but on average over long times, $d(t)$ will approach a fixed value. For example, after any number m of periods T of oscillation, $t = mT$, the two points have returned to where they started. The associated Lyapunov exponent is zero. To summarize, any chaotic attractor defined in a 3-D phase space will have $\sigma_1 > 0$, $\sigma_2 = 0$, and $\sigma_3 < 0$, with $\Sigma\sigma_i < 0$. An important conclusion can be drawn from this result: The requirements for chaos in a dissipative system that one Lyapunov exponent be

TABLE 10.2
Attractors and signs of Lyapunov exponents for attractors in 1-, 2-, and 3-D phase spaces

n	Attractor	Lyapunov exponents
1	Fixed point	$(-)$
2	Fixed point	$(-, -)$
	Periodic attractor	$(0, -)$
3	Fixed point	$(-, -, -)$
	Periodic attractor	$(0, -, -)$
	Chaotic attractor	$(+, 0, -)$
	Torus	$(0, 0, -)$

positive, one Lyapunov exponent be zero, and the sum of the exponents be negative mean that chaotic attractors can exist only in phase spaces of dimension 3 or greater.

Lyapunov exponents for the types of attractors we have studied (fixed points, periodic attractors, and chaotic attractors) are summarized, according to the signs of the Lyapunov exponents, in Table 10.2. Results are shown for phase spaces of dimensions 1 through 3.

In single-state systems the asymptotically stable fixed point is the only type of attractor that can occur. In two-state systems we may have a fixed point or a periodic attractor, whereas in three-state systems we may find a fixed point, a periodic attractor, a chaotic attractor, or a "torus," which is an extension of the periodic attractor (Ref. 10.4).

10.7.2 Fractal Structure and Fractal Dimension

A fascinating property of chaotic attractors is related to their structure as the attractor is viewed on a finer and finer scale. Consider the Poincaré map of Fig. 10.32 for the system (10.66). In Fig. 10.41(a) the small boxed region of Fig. 10.32 is blown up, and we observe certain "bands" with certain spacing between them. If a small section of Fig. 10.41(a) is blown up in turn, we see the same structure repeated [there are fewer points in Fig. 10.41(b) because we are by now looking at a small region of the phase space; points land in the area shown infrequently, and a long integration time is required in order to obtain a reasonably well-defined Poincaré map]. The same self-similar structure would be repeated at successively smaller scales. This is referred to as *fractal structure*, and we speak of a geometrical object having this structure as a fractal. The fractal structure of chaotic attractors indicates that there is actually a high degree of organization underlying the apparent "randomness."

Fractal structure is observable, at least approximately, in nature and in physical processes. Examples include the branching of fir trees, the dendritic structure of some crystals, and the branching of the arterial system in animals.

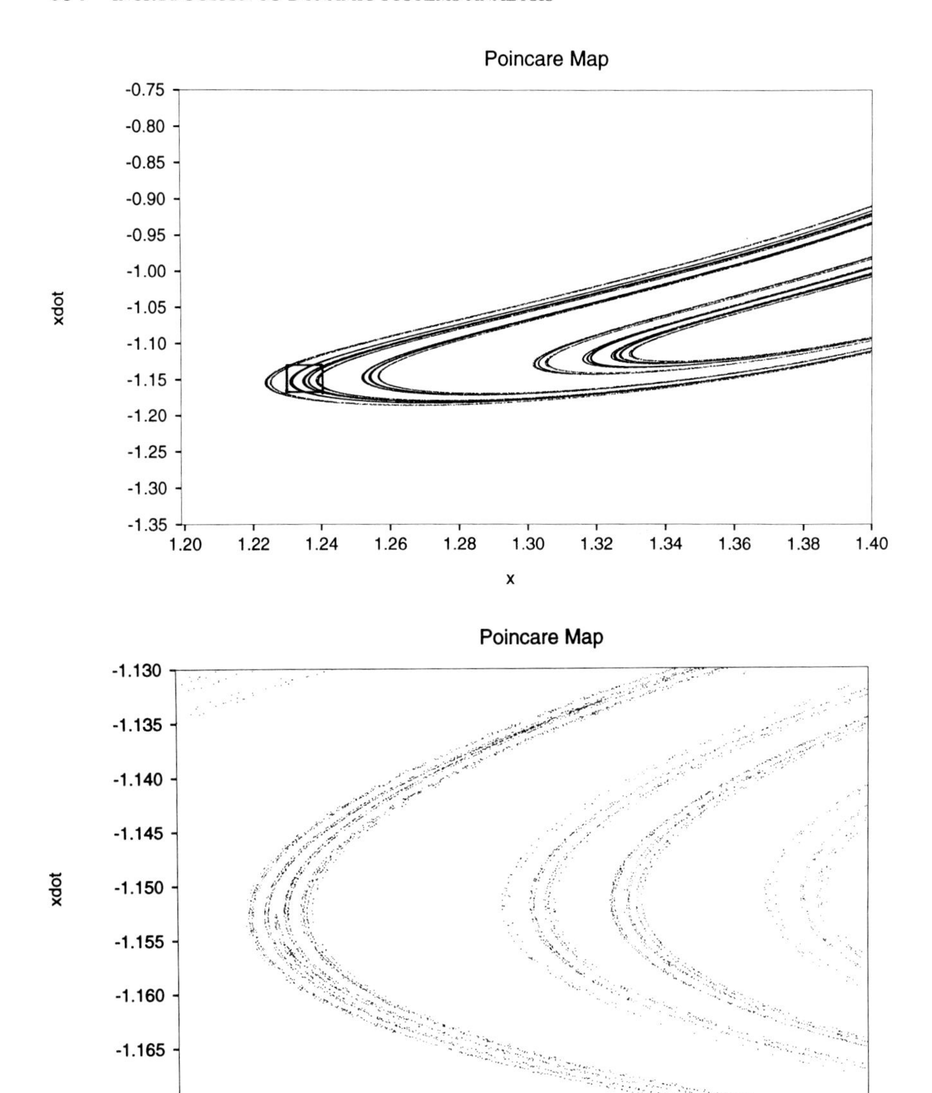

FIGURE 10.41
Poincaré map of the system (10.59). (*a*) Blowup of the boxed region of Fig. 10.32; (*b*) blowup of the boxed region of (*a*).

Many interesting geometrical objects exhibiting this fractal structure may be found in Ref. 10.6.

A characteristic of chaotic attractors is that if we consider any point on a Poincaré map, it will be approached arbitrarily closely, but never exactly repeated, at some later time. Likewise, in the phase space any point will be approached arbitrarily closely at some later time. Points will not be repeated exactly, however, for that would make the attractor periodic.

One way to quantify the fractal structure of a chaotic attractor is the so-called fractal dimension d_F, which is a measure of the "density" of the attractor in phase space. A single point has a dimension zero; a continuous line has a dimension of unity; an area or smooth two-dimensional surface, such as the surface of a sphere, has a dimension of 2; a volume has a dimension of 3; and so on. A characteristic of fractal structures is that they possess noninteger dimension. Loosely speaking, a fractal object with a dimension of, say, 1.37, takes up more space than a line but less space than an area.

The calculation of the fractal dimension of specific geometrical objects is defined in the context of the following example.

Example 10.4 The Koch Curve (Ref. 10.7). Here we illustrate the creation of a fractal object which is known as the *Koch curve*, described in Ref. 10.7. We start with the equilateral triangle with sides of unit length, depicted in Fig. 10.42(a). Call this object F_0 and denote its arc length as $L_0 = 3$. We generate an object having fractal structure by defining a sequence of iterates as follows: First remove the middle third of each side of F_0 and replace it by the two sides of the equilateral triangle, one-third as large, as shown in Fig. 10.42 (b). Call this object F_1 and denote its length by $L_1 = 3 - 3(1/3) + 6(1/3) = 4$. Next, remove the middle third of each of the six triangles of F_1 and replace each with the two sides of the triangles shown in Fig. 10.42(c). This object F_2 has length $L_2 = L_1 - 12/9 + 24/9 = L_1 + 4/3$. Object F_3 is then generated by removing the middle third of each of the triangles of F_2 and replacing each "removed segment" by the triangles shown in Fig. 10.42(d). The length L_3 is then $L_3 = L_2 - 36/27 + 72/27 = L_2 + 4/3$. The next object F_4 in the sequence is shown in Fig. 10.42(e) and has length $L_4 = L_3 - 108/81 + 216/81 = L_3 + 4/3$. The fractal object is generated as the limit F_n, $n \to \infty$. Observe that for $n \geq 2$ each object F_n in the sequence has length 4/3 greater than the preceding object F_{n-1}, that is, $L_n = L_{n-1} + 4/3$. Thus in the limit $n \to \infty$ the length of the object is infinite. We have a line of infinite length which encloses a finite area.

In the preceding example the object generated has the so-called fractal structure, that is, a self-similarity in geometry as the object is viewed on a finer and finer scale. Note that the existence of true fractal objects in the natural world would be unexpected, as the fractal structure is unlikely to be maintained as one passes to the molecular level.

The fractal object of Example 10.4 has in addition the following property enjoyed by fractals: If we attempt to measure the object as a series of *line segments*, then the combined length of the line segments is infinite. If we attempt to measure the object as an *area*, then we clearly get zero. Thus,

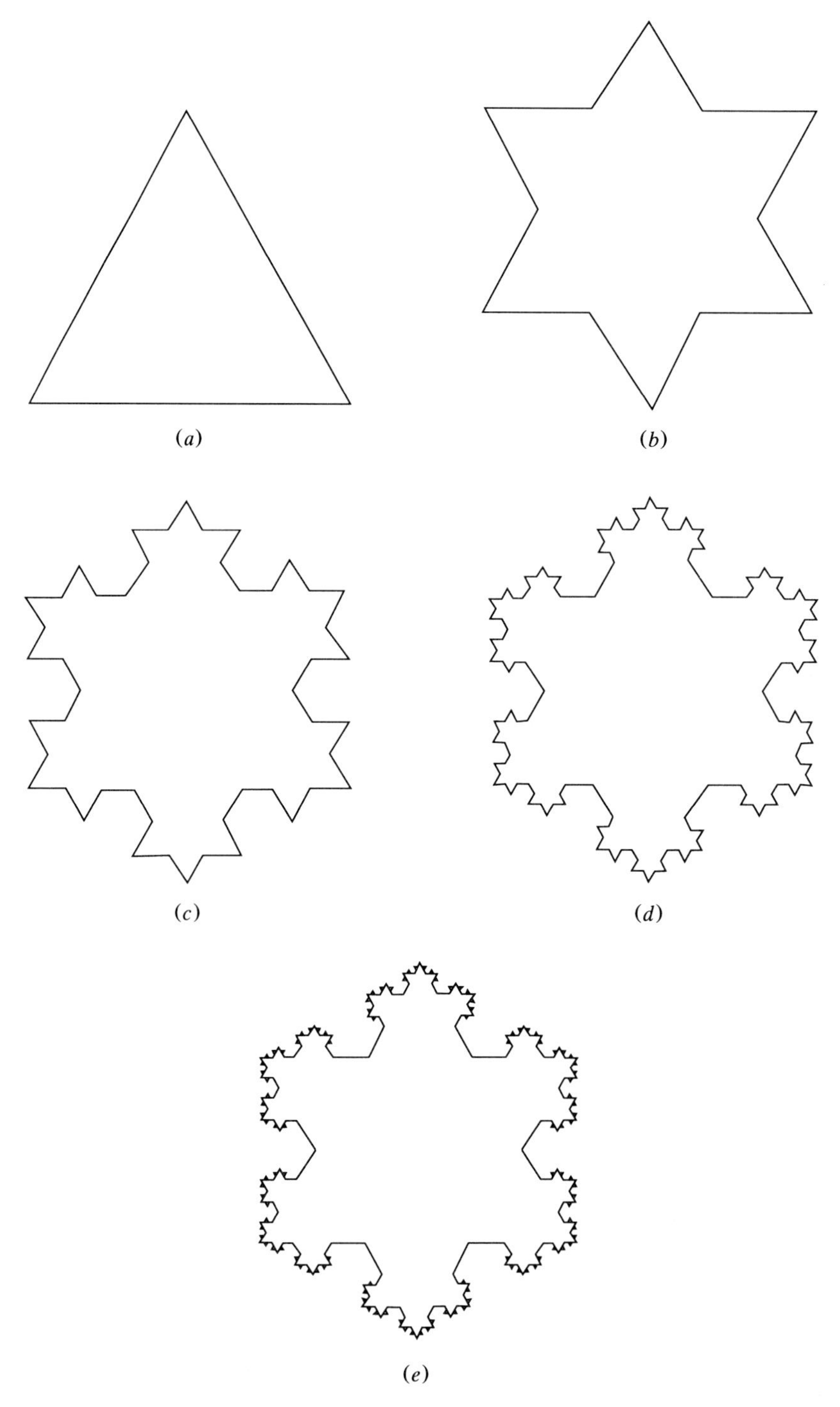

FIGURE 10.42
Generation of the fractal object of Example 10.4. See text for discussion.

loosely, we conclude that the object has a dimension greater than 1 but less than 2, and it is necessary to define the noninteger or fractal dimension d_F of the object. For the fractal object of Example 10.4, the fractal dimension is defined as follows: For each iterate F_n of Example 10.4, determine the minimum number M of squares, each of side ε, which is needed to "cover" the object being measured. Then in the limit $n \to \infty$, $\varepsilon \to 0$, the number M increases asymptotically as

$$M(\varepsilon) \to K\varepsilon^{-d_F} \tag{10.71}$$

where K is a constant. The fractal dimension d_F is then determined in the limit $\varepsilon \to 0$ as

$$d_F = \frac{\ln M(\varepsilon)}{\ln(1/\varepsilon)} \tag{10.72}$$

For the Koch curve of Example 10.4 $M(\varepsilon)$ is proportional to 4^n as n approaches infinity, while the box size ε for given n is $(1/3)^n$. Thus, the fractal dimension of the fractal of Example 10.4 is $d_F = \ln 4/\ln 3 = 1.262$.

Next consider the fractal object defined by the Poincaré map shown in Fig. 10.32 for the system (10.66). This object is composed of an infinite set of points which turn out to have a fractal dimension $d_F = 1.54$. This means that the fractal dimension d_F of the original chaotic attractor in the three-dimensional phase space is $d_F = 2.54$, because the construction of the Poincaré map has reduced the dimension by unity.

There is a close connection between the fractal dimension of a chaotic attractor and the Lyapunov exponents that characterize the attractor. Kaplan and Yorke (Ref. 10.8) have conjectured that the attractor fractal dimension d_F can be calculated from the Lyapunov exponents as follows:

$$d_F = m + \frac{\sum_{i=1}^{m} \sigma_i}{-\sigma_{m+1}} \tag{10.73}$$

where the Lyapunov exponents are ordered such that $\sigma_1 > \sigma_2 > \cdots > \sigma_n$ and where m is the largest integer having the property that

$$\sigma_1 + \sigma_2 + \cdots + \sigma_m > 0 \tag{10.74}$$

In the case of a chaotic attractor in a three-dimensional phase space, (10.73) reduces to

$$d_F = 2 - \frac{\sigma_1}{\sigma_3} \tag{10.75}$$

(Note that $\sigma_3 < 0$, so that $d_F > 2$.) The Kaplan-Yorke conjecture (10.73) turns out to provide an accurate approximation in most cases, as shown, for example, in Ref. 10.4.

10.7.3 Closing Comments on Chaotic Motion

1. Chaotic attractors are common in nature and have been found in nearly all fields of engineering and science. From a practical standpoint, if a dynamic phenomenon in nature normally manifests itself as a chaotic motion, then long-term prediction of the detailed behavior of the system states is essentially impossible, because of the sensitivity to initial conditions and our inability to determine precisely an initial state of the system. The weather may fit into this category.
2. Let us summarize the essential features of, and system properties required for, a chaotic motion to occur. First, the system must be nonlinear and of order 3 or higher. We have also assumed the system to be dissipative; that is, all volumes in phase space will, on average over long times, shrink. Second, chaos may occur for some values of the system parameters but not for others; in fact, the study of the nature of a given system's attractors, and how the nature of these attractors change as the system parameters are changed, is a central aspect of the study of nonlinear systems, but we have not considered this aspect here. Chaotic attractors are aperiodic and generally possess a frequency spectrum that is continuously distributed. This property would classically be thought to imply randomness; but note that chaos occurs in systems which are completely deterministic, i.e., there is no uncertainty in any of the system parameters nor in the excitation. Nevertheless, there is a high degree of underlying organization, manifested as the fractal structure. Chaotic orbits in phase space have fractal dimension greater than 2. Students interested in further study of this fascinating field may want to start with the references listed at the end of this chapter.
3. The intent of this brief introduction has been to provide a flavor of the characteristics of chaotic motion. A number of important aspects of chaotic behavior have not been discussed here, and the reader is cautioned that the treatment here is by no means complete.

REFERENCES

10.1. Blevins, R. D.: *Flow Induced Vibrations*, van Nostrand Reinhold, New York (1977).
10.2. Jordan, D. W., and P. Smith: *Nonlinear Ordinary Differential Equations*, Oxford University Press, Oxford (1977).
10.3. Nayfeh, A. H., and D. T. Mook: *Nonlinear Oscillations*, Wiley, New York (1979).
10.4. Lichtenberg, A. J., and M. A. Lieberman: *Regular and Stochastic Motion*, Springer (1983).
10.5. Wolf, A., et al.: Determination of Lyapunov Exponents from an Experimental Time Series, *Physica 16D*, p. 285 (1985).
10.6. Mandelbrot, B.: *The Fractal Geometry of Nature*, Freeman, San Francisco (1982).
10.7. Schuster, H. G.: *Deterministic Chaos, An Introduction*, Physik-Verlag Weinheim, Germany (1984).
10.8. Kaplan, J., and J. Yorke: *Springer Lecture Notes in Mathematics 730*, p. 204 (1979).
10.9. Burton, T. D., and Anderson, M.: "On Asymptotic Behavior in Cascaded Chaotically Excited Non-Linear Oscillators," *Journal of Sound and Vibration*, **133**(2): 353–358 (1989).

Other references not cited in text

10.10. Moon, F. C.: *Chaotic Vibrations*, Wiley, New York (1987).

10.11. Moon, F. C., and P. J. Holmes: "A Magnetoelastic Strange Attractor," *Journal of Sound and Vibration*, **65**(2), 275–296 (1979). This classic paper offers one of the first experimental investigations of the chaotic motion in a mechanical system.

10.12. Ueda, Y.: "Randomly Transitional Phenomena in a System Governed by Duffing's Equation," *Journal of Statistical Physics*, **20**, 181–196 (1979). This reference contains a detailed investigation of chaos in the system (10.59).

10.13. Rahman, Z. H. M.: *Higher Order Perturbation Methods in Nonlinear Oscillations*, Ph.D. thesis, Department of Mechanical Engineering, Washington State University (1986).

APPENDIX

Some useful trigonometric identities (Ref: *Standard Mathematical Tables*, 14th ed., The Chemical Rubber Co., Cleveland (1965)).

$$\sin(x \pm y) = \sin x \cos y \pm \cos x \sin y.$$
$$\cos(x \pm y) = \cos x \cos y \mp \sin x \sin y.$$

$$\sin x \cos y = \tfrac{1}{2}[\sin(x+y) + \sin(x-y)]$$
$$\cos x \sin y = \tfrac{1}{2}[\sin(x+y) - \sin(x-y)]$$
$$\cos x \cos y = \tfrac{1}{2}[\cos(x+y) + \cos(x-y)]$$
$$\sin x \sin y = \tfrac{1}{2}[\cos(x-y) - \cos(x+y)]$$

$$\sin 2x = 2\sin x \cos x.$$
$$\cos 2x = \cos^2 x - \sin^2 x = 2\cos^2 x - 1 = 1 - 2\sin^2 x.$$
$$\sin 3x = 3\sin x - 4\sin^3 x.$$
$$\cos 3x = 4\cos^3 x - 3\cos x.$$
$$\sin 4x = 8\cos^3 x \sin x - 4\cos x \sin x.$$
$$\cos 4x = 8\cos^4 x - 8\cos^2 x + 1.$$
$$\sin 5x = 5\sin x - 20\sin^3 x + 16\sin^5 x.$$
$$\cos 5x = 16\cos^5 x - 20\cos^3 x + 5\cos x.$$

EXERCISES

1. A commonly studied oscillator with nonlinear damping is the *van der Pol* oscillator:

$$\ddot{x} - \varepsilon\dot{x}(1 - x^2) + x = 0 \qquad (\varepsilon > 0)$$

For small ε, use the method of harmonic balance to find an approximation to the limit cycle amplitude a, by assuming a solution of the form $x(t) = a\cos(\omega t)$. Determine a and ω. (*Hint*: You should find $\omega = 1$ using this approximation.)

2. Use a numerical integration package (or code your own) to determine the periodic attractors in the phase plane for the van der Pol oscillator of Exercise 1 for the cases $\varepsilon = 0.1$, 1.0, and 10. Note the increasing noncircularity of the attractors as ε is increased. Does the result for $\varepsilon = 1$ justify use of the simple harmonic balance approximation recommended in Exercise 1? Explain.

3. Obtain the harmonic balance solution for the forced response of the undamped oscillator

$$\ddot{x} + x + \alpha x^5 = p \cos \Omega t$$

using the HB solution (10.34) and proceeding as in Sec. 10.3.3.

4. Same as Exercise 3 but include linear damping in the model:

$$\ddot{x} + 2\xi\dot{x} + x + \alpha x^5 = p \cos \Omega t$$

5. Use HB to obtain the approximate frequency response (primary nonlinear resonance) of the damped oscillator (10.20).

6. Use HB to determine the approximate natural frequency-amplitude dependence $\omega(a)$ for the undamped nonlinear oscillator $\ddot{x} + \alpha x^3 = 0$, $(\alpha > 0)$. Plot the exact and approximate orbits in the phase plane for the cases $a = 0.5$, 1.0, and 3.0, and use the results to assess the validity of the harmonic balance solution.

7. Determine the fractal dimension of the *Cantor set*, defined as the limit of the following iterative sequence: Start with the line of unit length F_0 and remove the middle third to obtain the object F_1. Next, remove the middle third of each of the two line segments of F_1, obtaining F_2 as the result. Continue removing the middle thirds in this fashion. A Cantor set is the object F_n in the limit $n \to \infty$.

8. A simple two-dimensional nonlinear discrete map which produces a chaotic attractor is the *Henon map*:

$$x_{n+1} = y_n + 1 - \alpha x_n^2$$

$$y_{n+1} = \beta x_n$$

For the case $\beta = 0.3$, $\alpha = 1.4$, use the computer to plot several thousand iterates to define the general structure of the chaotic attractor.

9. Use a numerical integration routine to obtain the Poincaré map of the oscillator (10.59) for parameter values $p = 1.0$, $\xi = 0.01684$, $\Omega = 0.33$ (Ref. 10.13). You may have to play with the initial conditions to land on the chaotic attractor.

10. Same as Exercise 9 but use parameter values $p = 1.0$, $\Omega = 0.229$, $\xi = 0.015$ (Ref. 10.13). The result is the "dancing newt" attractor.

11. Consider Sec. 10.4.1, which analyzes the primary nonlinear resonance of the damped oscillator. Show that (10.44) follows from (10.43).

INDEX